CARPENTRY FUNDAMENTALS

CONTEMPORARY CONSTRUCTION SERIES

CARPENTRY FUNDAMENTALS
  BY GLENN BAKER AND REX MILLER
ELECTRICAL WIRING FUNDAMENTALS
  BY VOLT INFORMATION SCIENCES, INC.
HEATING, VENTILATING, AND AIR CONDITIONING FUNDAMENTALS
  BY RAYMOND HAVRELLA
PLUMBING FUNDAMENTALS
  BY JAMES L. THIESSE
READING CONSTRUCTION DRAWINGS
  BY PAUL I. WALLACH AND DONALD E. HEPLER

CARPENTRY FUNDAMENTALS

CONTEMPORARY CONSTRUCTION SERIES

GLENN BAKER
Texas A&M
University

REX MILLER
State University
College at Buffalo

**GREGG DIVISION
McGRAW-HILL BOOK COMPANY**

New York / Atlanta / Dallas / St. Louis / San Francisco / Auckland
Bogotá / Guatemala / Hamburg / Johannesburg / Lisbon / London
Madrid / Mexico / Montreal / New Delhi / Panama / Paris / San Juan
São Paulo / Singapore / Sydney / Tokyo / Toronto

SPONSORING EDITOR: CARY BAKER
EDITING SUPERVISOR: KATHARINE GLYNN
DESIGN SUPERVISOR: NANCY AXELROD
PRODUCTION SUPERVISOR: S. STEVEN CANARIS
ART SUPERVISOR: GEORGE T. RESCH

COVER DESIGNER: AMPERSAND STUDIO
TECHNICAL ILLUSTRATORS: LAURA HARTMAN,
              VANTAGE ART, INC.

Library of Congress Cataloging in Publication Data

Baker, Glenn E
    Carpentry fundamentals.

   (Contemporary construction series)
   Includes index.
   1. Carpentry. I. Miller, Rex, (date) joint
author. II. Title. III. Series.
TH5606.B24   694   80-20056
ISBN 0-07-003361-7

CARPENTRY FUNDAMENTALS

Copyright © 1981 by McGraw-Hill, Inc. All rights reserved. Printed in the United States of America. No part of this publication may be reproduced, stored in a retrieval system, or transmitted, in any form or by any means, electronic, mechanical, photocopying, recording, or otherwise, without the prior written permission of the publisher.

234567890 VHVH 898765432

ISBN 0-07-003361-7

· CONTENTS ·

PREFACE ix

CHAPTER 1
STARTING THE JOB 1
Safety   Using Carpenter Tools   Following Correct Sequences   Working Successfully in Carpentry   Words Carpenters Use   Study Questions

CHAPTER 2
PREPARING THE SITE 25
Basic Sequence   Locating the Building on the Site   The Builder's Level   Preparing the Site   Providing Access during Construction   Words Carpenters Use   Study Questions

CHAPTER 3
LAYING FOOTINGS AND FOUNDATIONS 40
Sequence for Making Footings and Foundations   Laying Out the Footings   Excavating the Footings   Building the Forms for the Footings   Working with Concrete   Building the Foundation Forms   Concrete Block Walls   Plywood Foundations   Drainage and Waterproofing   Words Carpenters Use   Study Questions

CHAPTER 4
POURING CONCRETE SLABS, FLOORS, AND SURFACES 59
Concrete Slabs   Concrete Floors   Sidewalks and Driveways   Special Finishes and Surfaces   Words Carpenters Use   Study Questions

CHAPTER 5
BUILDING FLOOR FRAMES 76
Sequence   Sill Placement   Joists   Subfloors   Special Joists   Energy Factors   Words Carpenters Use   Study Questions

CHAPTER 6
FRAMING WALLS 97
Sequence   Wall Layout   Cutting Studs to Length   Wall Assembly   Corner Braces   Erect the Walls   Interior Walls   Sheathing   Factors in Wall Construction   Words Carpenters Use   Study Questions

CHAPTER 7
BUILDING ROOF FRAMES 119
Sequence   Erecting Truss Roofs   The Framing Square   Roof Framing   Rafters   Brace Measure   Erecting the Roof with Rafters   Special Rafters   Ceiling Joists   Openings   Decking   Constructing Special Shapes   Roof Load Factors   Laying Out a Stair   Aluminum Soffit   Words Carpenters Use   Study Questions

CHAPTER 8
COVERING ROOFS 166
Basic Sequence   How to Estimate Quantities of Roofing   Roofing Tools   Applying an Asphalt Roof   Putting Down Shingles   Strip Shingles   Steep-Slope and Mansard Roofs   Interlocking Shingles   Roll Roofing   Wood Shingles   Applying the Shingle Roof   Words Carpenters Use   Study Questions

CHAPTER 9
INSTALLING WINDOWS AND DOORS 209
Basic Sequence   Preparing the Rough Opening for a Window   Installation of a Wood Window   Prehung Doors   Installing an Exterior Door   Installing Folding Doors   Door and Window Trim   Installing Locks   Installing a Sliding Door   Energy Factors   Words Carpenters Use   Study Questions

CHAPTER 10
FINISHING THE EXTERIOR WALLS 256
Types of Siding   Sequence for Siding   Prepare for the Job   Erecting Scaffolds   Finishing Roof Edges   Installing the Siding   Panel Siding   Shingle and Shake Siding   Preparation for Other Wall Finishes   Aluminum Siding   Solid Vinyl Siding   Words Carpenters Use   Study Questions

CHAPTER 11
INSULATING FOR THERMAL EFFICIENCY 290
Types of Insulation   How Much Is Enough?   Where to Insulate   Installing Insulation   Vapor Barriers and Moisture Control   Thermal Ceilings   Storm Windows   Storm Doors   Sealants   Winterizing a Home   Words Carpenters Use   Study Questions

CHAPTER 12
PREPARING INTERIOR WALLS AND CEILINGS 314
Sequence   Putting Insulation in Walls   Installing a Moisture Barrier   Putting Up Gypsum Board   Preparing a Wall for Tubs and Showers   Paneling Walls   Preparing a Wall for Plaster   Finishing Masonry Walls   Installing Ceiling Tile   Words Carpenters Use   Study Questions

CHAPTER 13
FINISHING THE INTERIOR 341
Sequence for Finishing the Interior   Interior Doors and Window Frames   Window Trim   Cabinets and Millwork   Applying Finish Materials   Floor Preparation and Finish   Finishing Floors   Installing Carpet   Resilient Flooring   Laying Ceramic Tile   Words Carpenters Use   Study Questions

CHAPTER 14
SPECIAL CONSTRUCTION METHODS 382
Stairs   Fireplace Frames   Post-and-Beam Construction   Words Carpenters Use   Study Questions

CHAPTER 15
MAINTENANCE AND REMODELING 408
Planning the Job   Minor Repairs and Remodeling   Converting Existing Building Spaces to Other Uses   Adding Space to Existing Buildings   Creating New Structures   Words Carpenters Use   Study Questions

CHAPTER 16
THE CARPENTER AND THE CONSTRUCTION INDUSTRY 461
Opportunities in Carpentry   Role of the Carpenter on the Job   Broadening Horizons in Carpentry   Building Codes and Zoning Provisions   Trends and Effects   How Do I Measure Up?   Words Carpenters Use   Study Questions

CHAPTER 17
CONSTRUCTION FOR SOLAR HEATING 483
Tech House   Need for Solar-Heated Housing   Building Modifications   Other Methods of Solar Heat Use   Words Carpenters Use   Study Questions

GLOSSARY 495

INDEX 510

PREFACE

Carpentry Fundamentals is written for everyone who wants or needs to know more about carpentry. Whether remodeling an existing home or building a new one, the rewards from a job well done are many-fold.

This text can be used by students in vocational courses, technical colleges, and apprenticeship programs. The home do-it-yourselfer will find answers to many questions that pop up in the course of getting a job done whether over a weekend or over a year's time.

In order to prepare this text, the authors examined many courses of study in schools located all over the country. An effort was made to take into consideration the geographic differences and the special environmental factors relevant to a particular area.

Notice how the text is organized. The first chapter, "Starting the Job," presents the information needed to get construction underway. The next chapter covers preparing the site. Then the footings and foundation are described. Once the footings are in, the pouring of slabs or floors is discussed for those who do not want a complete basement with poured walls. Floor frames continue the sequence. Wall framing, roof frames, openings in the roof and elsewhere, are discussed before the covering of roofs is described. Once the roof is in place, the next step is the installation of windows and doors. When the windows and doors are in place the exterior siding is applied. Next, it is important that the heating and cooling of the living quarters be given consideration. Once the insulation is in place, the interior walls and ceilings are covered in detail before presenting interior finishing methods.

Special construction methods, maintenance and remodeling, as well as careers in carpentry are then described. The building of solar houses and the design of solar heating are covered to keep the student and do-it-yourselfer up-to-date with the latest developments in energy conservation.

No book can be completed without the aid of many people. The acknowledgments which follow on page x mention some of those who contributed to making this text the most current in design and technology techniques available to the carpenter. We trust you will enjoy studying the book as much as we did writing it.

Glenn Baker
Rex Miller

ACKNOWLEDGMENTS

The authors would like to thank the following manufacturers for their generous efforts. They furnished photographs, drawings, and technical assistance. Without the donations of time and effort on the part of many people this book would not have been possible. We hope this acknowledgment of some of the contributions will let you know that the field you are working in or about to enter is one of the best. (The name given below in parentheses indicates the abbreviated form used in the credit lines for the photographs appearing in this text.)

Abitibi Corporation; ALCOA; American Plywood Association; Town of Amherst, New York (Town of Amherst, N.Y.); Andersen Corporation (Andersen); Armstrong Cork Company (Armstrong Cork); C. Arnold and Sons; Beaver-Advance Company (Beaver-Advance); The Bilco Company (Bilco); Bird and Son, Inc. (Bird and Son); Black and Decker Manufacturing Company (Black & Decker); Boise-Cascade; Butler Manufacturing Company (Butler); Certain-Teed Corporation (Certain-Teed); California Redwood Association (California Redwood Assoc.), Conwed Corporation (Conwed); Corl Corporation; Dewalt, Div. of American Machine and Foundry Company (DeWalt); Dow Chemical Company (Dow Chemical); Dow-Corning Company (Dow-Corning); Duo-Fast Corporation (Duo-Fast); EMCO; Formica Corporation (Formica); Fox & Jacobs Corporation (Fox and Jacobs); General Products Corporation (General Products); Georgia-Pacific Corporation (Georgia-Pacific); Goldblatt Tools, Inc. (Goldblatt Tools); Gold Bond Building Products (Gold Bond); Grossman Lumber Company (Grossman Lumber); Gypsum Association (Gypsum); Hilti Fastening Systems (Hilti); IRL Daffin Company (IRL Daffin); Kelly-Stewart Company, Inc. (Kelly-Stewart); Kenny Manufacturing Company (Kenny); Kirch Company (Kirch); Kohler Company (Kohler); Majestic; Manco Tape, Inc. (Manco Tape); Martin Industries (Martin); Masonite Corporation (Masonite); Milwaukee Electric Tool Company (Milwaukee Electric Tool); National Aeronautical & Space Administration (NASA); National Homes; National Lock Hardware Company; National Oak Flooring Manufacturer's Association (National Oak Flooring Manufacturers); National Wood Manufacturers Association (National Wood Manufacturers); New York State Electric and Gas Corporation (NYSE & G); Owens-Corning Fiberglas Corporation (Owens-Corning); Patent Scaffolding Company (Patent Scaffolding); Pella Windows and Doors, Inc. (Pella); Permograin Products; Plaskolite, Inc. (Plaskolite); Portland Cement Association (Portland Cement); Potlatch Corporation (Potlatch); Proctor Products, Inc. (Proctor Products); Red Cedar Shingles and Handsplit Shake Bureau (RCS & HSB); Reynolds Metals Products; Richmond Screw Anchor Company (Richmond Screw Anchor); Riviera Kitchens, Div. of Evans Products; Rockwell International, Power Tool Division (Rockwell); Sears, Roebuck and Company (Sears, Roebuck); Shakertown Corporation; Simplex Industries (Simplex); Stanley Tools Company (Stanley Tools); State University College at Buffalo; TECO Products and Testing Corporation, Washington, D.C. 20015; U.S. Gypsum Company (U.S. Gypsum); U.S. Department of Energy (U.S. Dept. of Energy); U.S. Bureau of Labor Statistics; U.S. Forest Service, Forest Products Laboratory (Forest Products Lab); Texas A & M University (Texas A & M); Universal Fastenings Corporation; Universal Form Clamp Company (Universal Form Clamp); Valu, Inc. (Valu); Weiser Lock, Division of Norris Industries (Weiser Lock); Weslock; Western Wood Products Association (Western Wood Products); Weyerhauser Company (Weyerhauser); David White Instruments, Div. of Realist, Inc. (David White).

1 STARTING THE JOB

Carpentry can be exciting. Carpentry has all kinds of jobs which offer a challenge. You will have to work with hand tools and power tools. You will have to work with all types of building materials. You can become very skilled at your job. This will make it easier to do a good job. You get a chance to be proud of what you do. You can stand back and look at the building you just helped erect and feel great about a job well done.

One of the exciting things about being a carpenter is watching a building come up. You actually see it grow from the ground up. Many people work with you to make it possible to complete the structure. Being part of a *team* can be rewarding, too. Teamwork is what life is all about. We take part in team activities all the time. This job is no different. You can learn to work with others to make beautiful things.

Carpentry involves cutting, shaping, and fastening materials together. You may be involved in making a new building or repairing or remodeling an old one. Either way you are making something or doing something you can be proud of later.

This book will help you do a good job in carpentry. It will aid you in making the right decisions in building a house. It introduces the basic construction techniques.

You have to do something over and over again to gain skill, and you will find this is true with reading. In some cases you won't get the idea the first time. Go over it again until you understand what you are reading. Then go out and practice what you just read. This way you can see for yourself how the instructions actually work. You *cannot* learn to be a carpenter by reading a book. You have to read, reread, and then *do*. This "do" part is the most important part. You have to take the hammer or saw in hand and actually do the work. There is nothing like good, honest sweat from a hard day's work. At the end of the day you can say "I did that" and be proud that you did.

This chapter should help you build these skills:
• Select personal protective gear
• Work safely as a carpenter
• Measure building materials
• Lay out building parts
• Cut building materials
• Fasten materials
• Shape and smooth materials
• Identify basic hand tools
• Recognize common power tools

• SAFETY •

Figure 1-1 shows a carpenter using one of the latest means of driving nails. The nail driver is not a hammer. It is a compressed-air-driven nail driver.

Figure 1-1. This carpenter is using an air-driven nail driver to nail these framing members. *(Duo-Fast)*

It drives the nails into the wood with a single stroke. The black cartridge that appears to run up near the carpenter's leg is a part of the nailer. It holds the nails and feeds them as needed.

As for safety, look at the carpenter's shoes. They have rubber soles for gripping the wood. This will prevent a slip through the joists and a serious fall. The steel toes in the shoes prevent damage to a foot from falling materials. The soles of the shoes are very thick. This will prevent picking up a nail in the foot. The hard hat protects the carpenter's head from falling lumber, shingles, or other building materials. The carpenter's safety glasses cannot be seen in Figure 1-1, but they are required equipment for the safe worker.

Other Safety Measures

To protect the eyes, it is best to wear safety glasses. Make sure your safety glasses are of tempered glass. They will not shatter and cause eye damage. In some instances you should wear goggles. This prevents splinters and other flying objects from entering the eye from under or around the safety glasses. Ordinary glasses aren't always the best, even if they

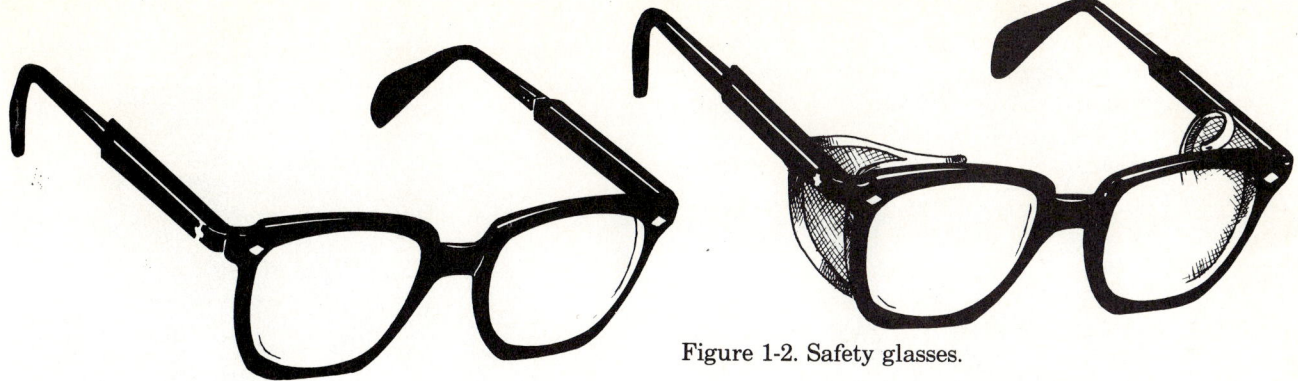

Figure 1-2. Safety glasses.

are tempered glass. Just become aware of the possibilities of eye damage whenever you start a new job or procedure. See Figure 1-2 for a couple of types of safety glasses.

Sneakers are used only by roofers. Sneakers, sandals, and dress shoes do not provide enough protection for the carpenter on the job. Only safety shoes should be worn on the job.

GLOVES • Some types of carpentry work require the sensitivity of the bare fingers. Other types do not require the hands or fingers to be exposed. In cold or even cool weather gloves may be in order. Gloves are often needed to protect your hands from splinters and rough materials. It's only common sense to use gloves when handling rough materials.

Probably the best gloves for carpenter work are a lightweight type. A suede finish to the leather improves the gripping ability of the gloves. Cloth gloves tend to catch on rough building materials. They may be preferred, however, if you work with short nails or other small objects.

BODY PROTECTION • Before you go to work on any job, make sure your entire body is properly protected. The hard hat comes in a couple of styles. See Figure 1-3. Under some conditions the face shield is better protection. See Figure 1-4.

Is your body covered with heavy work clothing? This is the first question to ask before going onto the job site. Has as much of your body as practical been covered with clothing? Has your head been properly protected? Are your eyes covered with approved safety glasses or face shield? Are your shoes sturdy, with safety toes and steel soles to protect against nails? Are gloves available when you need them?

General Safety Rules

Some safety procedures should be followed at all times. This applies to carpentry work especially:

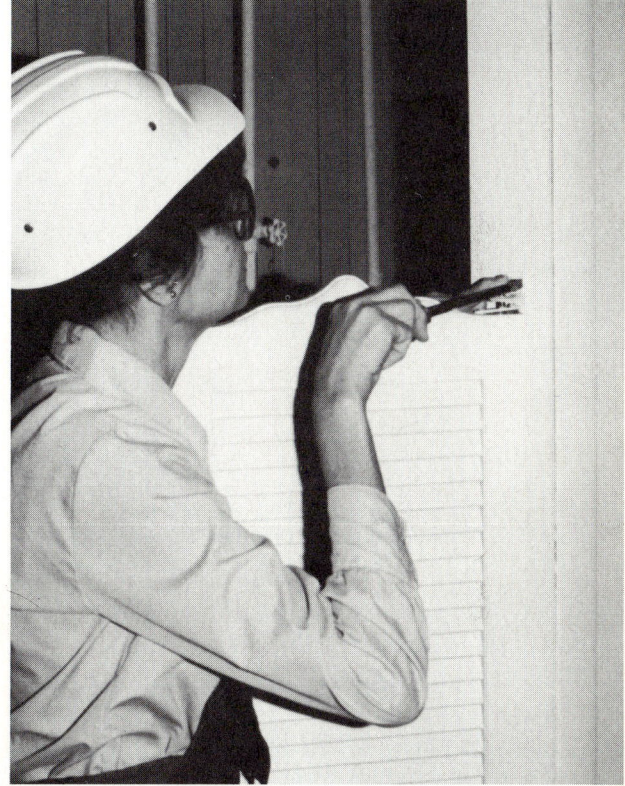

Figure 1-3. A carpenter wears a hard hat while doing all types of work.

Figure 1-4. Face shield.

Starting the Job 3

Figure 1-5. A made-on-the-job ladder.

Figure 1-6. (A) Cluttered work site. (B) A work area can be kept clean if a large dumpster is kept nearby for trash and debris.

Pay close attention to what is being done.
Move carefully when walking or climbing.
(Take a look at Figure 1-5. This type of made-on-the-job ladder can cause trouble.)
Use the leg muscles when lifting.
Move long objects carefully. The end of a carelessly handled 2×4 can damage hundreds of dollars worth of glass doors and windows.
Keep the workplace neat and tidy. Figure 1-6 shows a cluttered working area. It would be hard to walk along here without tripping. If a dumpster is used for trash and debris, as in Figure 1-6B, many accidents can be prevented.
Sharpen or replace dull tools.
Disconnect power tools before adjusting them.
Keep power tool guards in place.
Avoid interrupting another person who is using a power tool.
Remove hazards as soon as they are noticed.

Safety on the Job

A safe working site makes it easier to get the job done. Lost time due to accidents puts a building schedule behind. This can cost many thousands of dollars and lead to late delivery of the building. If the job is properly organized and safety is taken into consideration, the smooth flow of work is quickly noticed. No one wants to get hurt. Pain is no fun. Safety is just common sense. If you know how to do something safely, it will not take any longer than if you did it in an unsafe manner. Besides, why would you deliberately do something that is dangerous? All safety requires is a few precautions on the job. Safety becomes a habit once you get the proper attitude established in your thinking. Some of these important habits to acquire are:

Know exactly what is to be done before you start a job.
Use a tool only when it can be used safely.
Wear all safety clothing recommended for the job.
Provide a safe place to stand to do the work. Set ladders securely. Provide strong scaffolding. Avoid wet, slippery areas.
Keep the working area as neat as practical.

Remove or correct safety hazards as soon as they are noticed. Bend protruding nails over. Remove loose boards.

Remember where other workers are an what they are doing.

Keep fingers and hands away from cutting edges at all times.

Stay alert!

Safety Hazards

Carpenters work in unfinished surroundings. While a house is being built, there are many unsafe places around the building site. You have to stand on or climb ladders. They can be unsafe. You may not have a good footing while standing on a ladder. You may not be climbing a ladder in the proper way. Holding onto the rungs of the ladder is very unsafe. You should hold onto the outside rails of the ladder when climbing.

There are holes which can cause you to trip. They may be located in the front yard where the water or sewage lines come into the building. There may be holes for any number of reasons. These holes can cause you all kinds of problems—especially if you fall into them or turn your ankle.

The house in Figure 1-7 is almost completed. However, if you look closely you can see that some wood has been left on the garage roof. This wood can slide down and hit a person working below. The front porch has not been poured. This means stepping out of the front door can be a rather long step. Other debris around the yard can be a source of trouble. Long slivers of flashing can cause trouble if you step on them and they rake your leg. You have to watch your every step around a construction site.

OUTDOOR WORK • Much of the time carpentry is performed outdoors. This means you will be exposed to the weather. Dress accordingly. Wet weather increases the accident rate. Mud can make a secure place to stand when working outside hard to find. Mud can cause you to slip if you don't clean it off your shoes. Be very careful when it is muddy and you are climbing on a roof or a ladder.

TOOLS • Any tool that can cut wood can cut flesh. You have to keep in mind that tools are an aid to the carpenter. However, they can be a source of injury. A chisel can cut your hand as easily as the wood. In fact, it can do a quicker job on your hand than on the wood it was intended for. Saws can cut wood and bones. Be careful with all types of saws, both hand and electric. Hammers can do a beautiful job on your fingers if you miss the nail head. This hurts very much when the weather is somewhat cooler than usual. Broken bones can be easily avoided if you keep your eye on the nail while you're

Figure 1-7. Even when a house is almost finished, there can still be hazards. Wood left on a roof could slide off and hurt someone, and without the front porch it is a long step down.

hammering. Besides that, you will get the job done more quickly. And, after all, that's why you are there—to get the job done and do it right the first time. Tools can help you do the job right. They can also cause you injury. The choice is up to you.

In order to work safely with tools you should know what they can do and how they do it. The next few pages are designed to help you use tools properly.

• USING CARPENTER TOOLS •

A carpenter is lost without tools. This means you have to have some way of containing them. A tool box is very important. If you have a place to put everything, then you can find the right tool when it is needed. A tool box should have all the tools mentioned here. In fact, you will probably add more as you become more experienced. Tools have been designed for every task. All it takes is a few minutes with a hardware manufacturer's catalog to find just about everything you'll ever need. If you can't find what you need, the manufacturers are interested in making it.

Measuring Tools

FOLDING RULE • When using the folding rule, place it flat on the work. The 0 end of the rule should be exactly even with the end of the space or board to be measured. The correct distance is indicated by the reading on the rule. Figure 1-8 shows workers using a carpenter's folding rule.

A very accurate reading may be obtained by turning the edge of the rule toward the work. In this position, the marked graduations of the face of the rule touch the surface of the board. With a sharp pencil, mark the exact distance desired. Start the

Starting the Job **5**

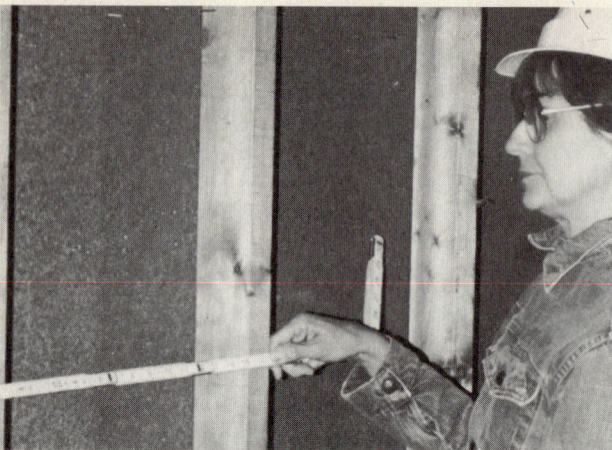

Figure 1-8. Carpenters use a folding rule.

Figure 1-9. Tape measure. *(Stanley Tools)*

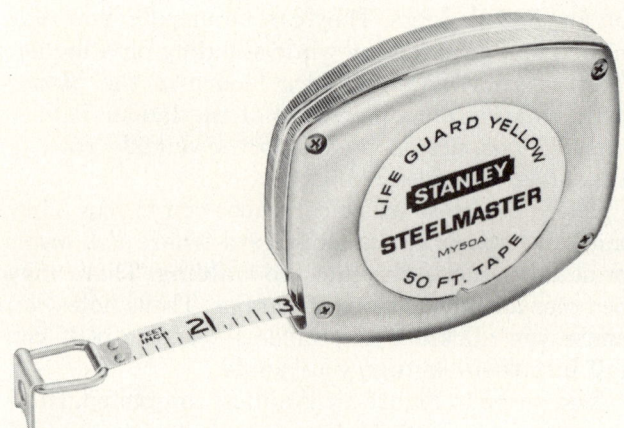

Figure 1-10. A longer tape measure. *(Stanley Tools)*

mark with the point of the pencil in contact with the mark on the rule. Move the pencil directly away from the rule while making the mark.

One problem with the folding rule is that it breaks easily if it is twisted. This happens most commonly when it is being folded or unfolded. The user may not be aware of the twisting action at the time. You should keep the joints oiled lightly. This makes the rule operate more easily.

POCKET TAPE • Beginners may find the pocket tape (Figure 1-9) the most useful measuring tool for all types of work. It extends smoothly to full length. It returns quickly to its compact case when the return button is pressed. Steel tapes are available in a variety of lengths. For most carpentry a rule 6, 8, 10, or 12 feet long is used.

Longer tapes are available. They come in 20-, 50-, and 100-foot lengths. See Figure 1-10. This tape can be extended to 50 feet to measure lot size and the location of a house on a lot. It has many uses around a building site. A crank handle can be used to wind it up once you are finished with it. The hook on the end of the tape makes it easy for one person to use it. Just hook the tape over the end of a board or nail and extend it to your desired length.

Saws

Carpenters use a number of different saws. These saws are designed for specific types of work. Many are misused. They will still do the job, but they would do a better job if used properly. Hand saws take quite a bit of abuse on a construction site. It is best to buy a good quality saw and keep it lightly oiled.

STANDARD SKEW-BACK SAW • This saw has a wooden handle. It has a 22-inch length. A 10-point saw (with 10 teeth per inch) is suggested for crosscutting. Crosscutting means cutting wood *across* the grain. The 26-inch-length, 5½-point saw is suggested

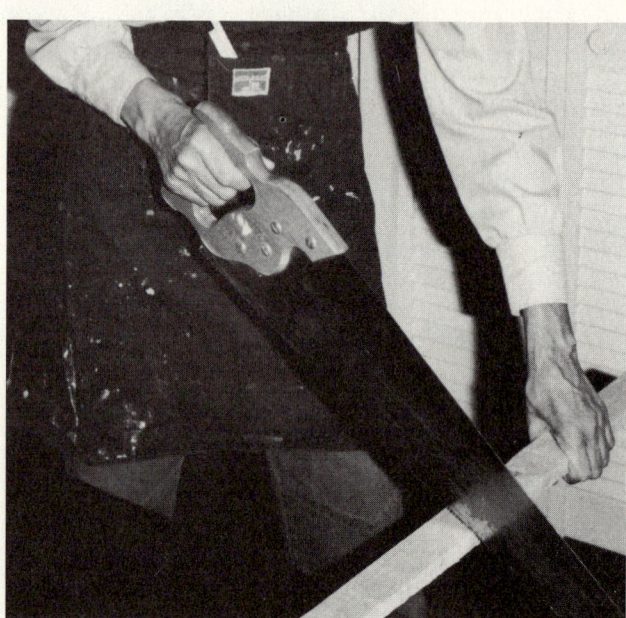

Figure 1-11. Using a hand saw.

for ripping. To rip means to cut *with* the wood grain. Figure 1-11 shows a carpenter using a hand saw.

This saw is used in places where the electric saw cannot be used. Keeping it sharp makes a difference in the quality of the cut and the ease with which it can be used.

6 Carpentry Fundamentals

BACKSAW • The backsaw gets its name from the piece of heavy metal that makes up the top edge of the cutting part of the saw. See Figure 1-12. It has a fine tooth configuration. This means it can be used to cut cross-grain and leave a smoother finished piece of work. This type of saw is used by finish carpenters who want to cut trim or molding.

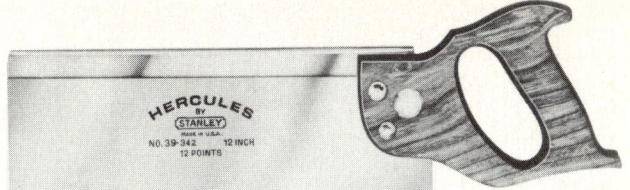

Figure 1-12. Back saw. *(Stanley Tools)*

MITRE BOX • As you can see from Figure 1-13, the mitre box has a backsaw mounted in it. This box can be adjusted using the lever under the saw handle (see arrow). You can adjust it for the cut you wish. It can cut from 90° to 45°. It is used for finish cuts on moldings and trim materials. The angle of the cut is determined by the location of the saw in reference to the bed of the box. Release the clamp on the bottom of the saw support to adjust the saw to any degree desired. The wood is held with one hand against the fence of the box and the bed. Then the saw is used by the other hand. As you can see from the setup, the cutting should take place when the saw is pushed forward. The backward movement of the saw should be made with the pressure on the saw released slightly. If you try to cut on the backward movement, you will just pull the wood away from the fence and damage the quality of the cut.

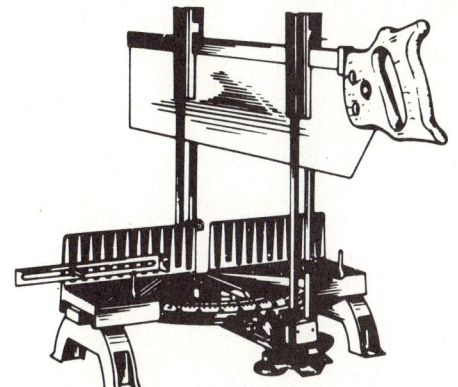

Figure 1-13. Mitre box. *(Stanley Tools)*

COPING SAW • Another type of saw the carpenter can occasionally make use of is the coping saw (Figure 1-14). This one can cut small thicknesses of wood at any curve or angle desired. It can be used to make sure a piece of paneling fits properly or a piece of molding fits another piece in the corner. The blade is placed in the frame with the teeth pointing toward the handle. This means it cuts only on the downward stroke. Make sure you properly support the piece of wood being cut. A number of blades can be obtained for this type of saw. The number of teeth in the blade determines the smoothness of the cut.

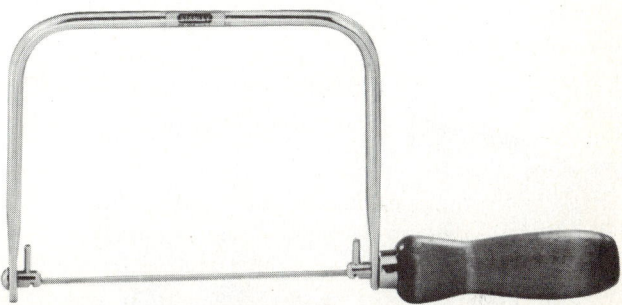

Figure 1-14. Coping saw. *(Stanley Tools)*

Hammers and Other Small Tools

There are a number of different types of hammers. The one the carpenter uses is the *claw* hammer. See Figure 1-15. It has claws which can extract nails from wood if they have been put in the wrong place or have bent while being driven. Hammers can be bought in 20-ounce, 24-ounce, 28-ounce, and 32-ounce weights for carpentry work. The usual carpenter choice is the 20-ounce. You have to work with a number of different weights to find out which will work best for you. Keep in mind that the hammer should be of tempered steel. If the end of the hammer has a tendency to splinter or chip off when it hits a nail, the pieces can hit you in the eye or elsewhere, causing serious damage. It is best to wear safety glasses whenever you use a hammer.

Figure 1-15. Using a claw hammer.

Starting the Job 7

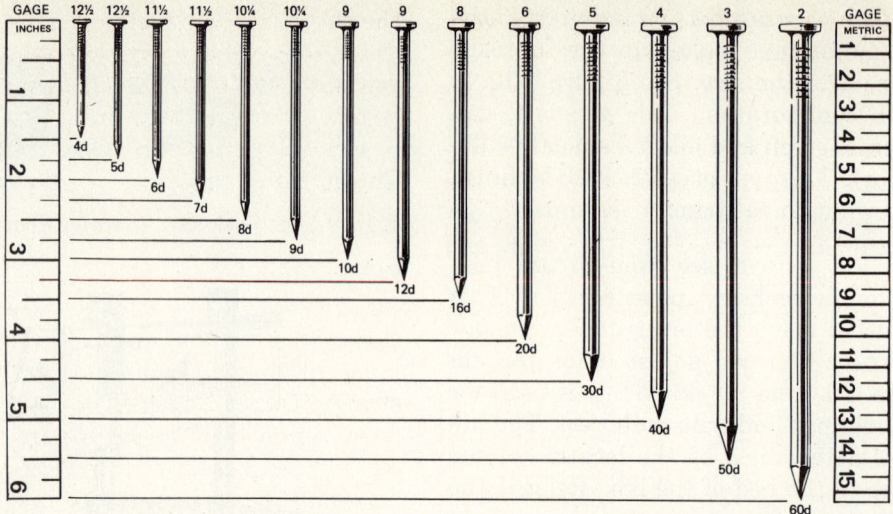

Figure 1-16. Nails. *(Forests Products Laboratory)*

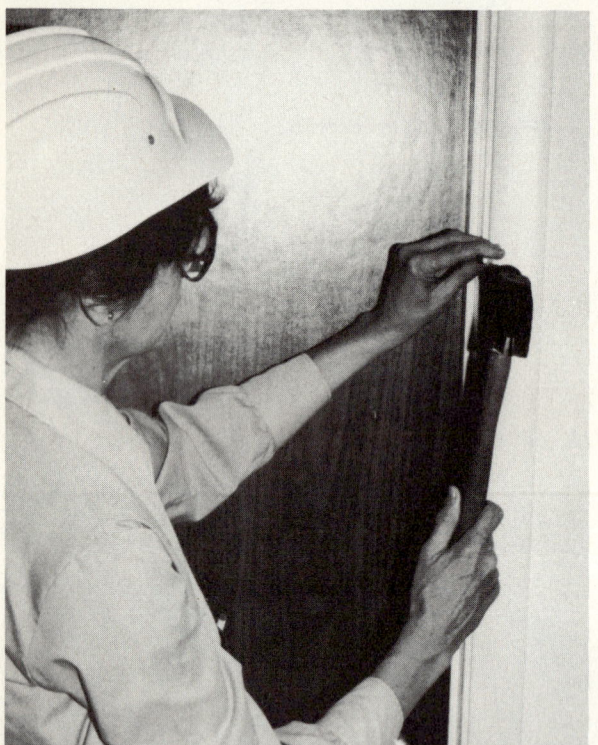

Figure 1-17. Using a nail set.

Nails are driven by hammers. Figure 1-16 shows the gage, inch, and penny relationships for the common box nail. The *d* after the number means *penny*. This is a measuring unit inherited from the English in the colonial days. There is little or no relationship between penny and inches. If you want to be able to talk about it intelligently, you'll have to learn both inches and penny. The gage is nothing more than the American Wire Gage number for the wire that the nails were made from originally. Finish nails have the same measuring unit (penny) but do not have the large, flat heads.

NAIL SET • Finish nails are driven below the surface of the wood by a nail set. The nail set is placed on the head of the nail. The large end of the nail set is struck by the hammer. This causes the nail to go below the surface of the wood. Then the hole left by the countersunk nail is filled with wood filler and finished off with a smooth coat of varnish or paint. Figure 1-17 shows the nail set and its use.

The carpenter would be lost without a hammer. See Figure 1-18. Here the carpenter is placing sheathing on rafters to form a roof base. The hammer is used to drive the boards into place, since they have to overlap slightly. Then the nails are driven by the hammer also.

In some cases a hammer will not do the job. The job may require something known as a hatchet. See Figure 1-19. This device can be used to pry and to drive. It can pry boards loose when they are improperly installed. It can sharpen posts to be driven at the site. The hatchet can sharpen the ends of stakes for staking out the site. It can also withdraw nails. This type of tool can also be used to drive stubborn sections of a wall into place when they are erected for the first time. The tool has many uses. It comes in handy for practically everything.

SCRATCH AWL • An awl is a handy tool for a carpenter. It can be used to mark wood with a scratch mark. It can be used to produce pilot holes for screws. Once it is in your tool box, you can think

8 Carpentry Fundamentals

of a hundred uses for it. Since it does have a very sharp point, it is best to treat it with respect. See Figure 1-20.

WRECKING BAR • This device (Figure 1-21) has a couple of names, depending on which part of the country you are in at the time. It is called a wrecking bar in some parts and a crowbar in others. One end has a chisel-sharp flat surface to get under boards and pry them loose. The other end is so hooked that the slot in the end can pull nails with the leverage of the long handle. This specially treated steel bar can be very helpful in prying away old and unwanted boards. It can be used to help give leverage when you are putting a wall in place and making it plumb. This tool, too, has many uses for the carpenter with ingenuity.

SCREWDRIVERS • The screwdriver is an important tool for the carpenter. It can be used for many things other than turning screws. There are two types of screwdrivers. The standard type has a straight slot-fitting blade at its end. This type is the most common of screwdrivers. The phillips-head screwdriver has a cross or X on the end. This fits a screw head of the same design. Figure 1-22 shows a carpenter using a screwdriver to put in a lock, and also shows the two types of screwdrivers.

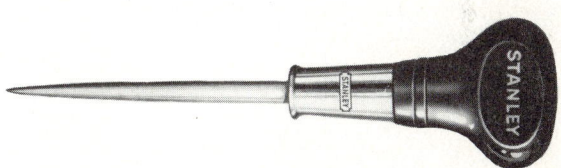

Figure 1-18. Putting on roof sheathing. The carpenter is using a hammer to drive the boards into place.

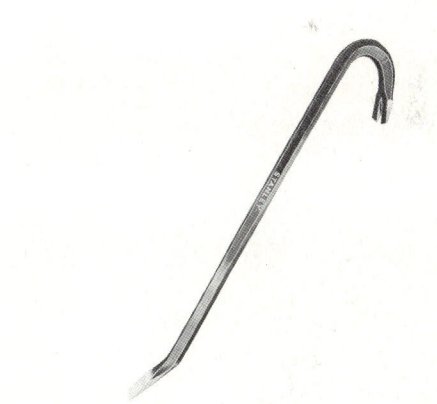

Figure 1-20. Scratch awl. *(Stanley Tools)*

PRY BAR

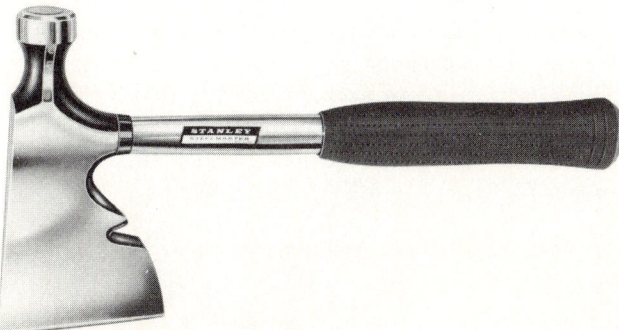

Figure 1-19. Hatchet. *(Stanley Tools)*

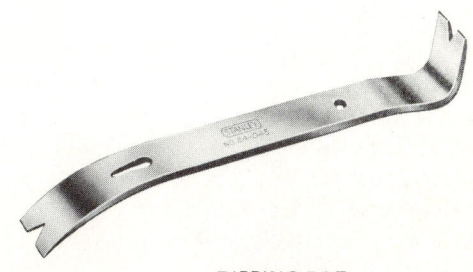

RIPPING BAR

Figure 1-21. Wrecking bars. *(Stanley Tools)*

Starting the Job **9**

Figure 1-22. Using a screwdriver to install a lock.

Squares

In order to make corners meet and standard sizes of materials fit properly, you must have things square. That calls for a number of squares to check that the two walls or two pieces come together at a perpendicular.

TRY SQUARE • The *try square* can be used to mark small pieces for cutting. If one edge is straight and the handle part of the square (Figure 1-23) is placed against this straight edge, then the blade can be used to mark the wood perpendicular to the edge. This comes in handy when you are cutting 2×4s and want them to be square.

FRAMING SQUARE • The framing square is a very important tool for the carpenter. It allows you to make square cuts in dimensional lumber. This tool can be used to lay out rafters and roof framing. See Figure 1-24. It is also used to lay out stair steps.

Later in this book you will see a step-by-step procedure for using the framing square. The tools are described as they are called for in actual use.

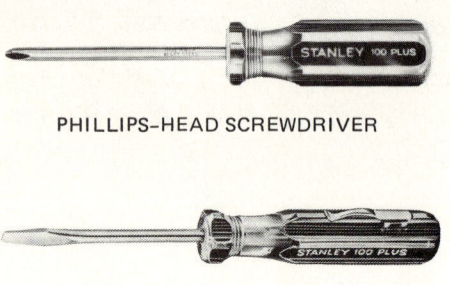

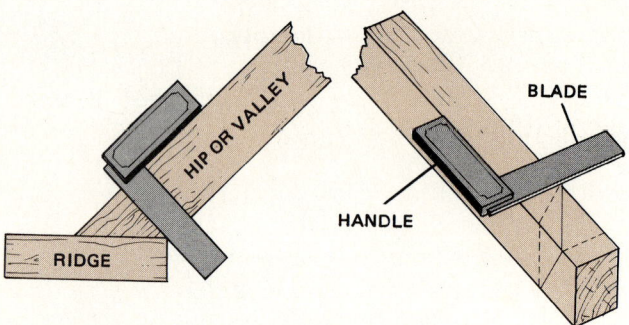

Figure 1-23. Use of a try square. *(Stanley Tools)*

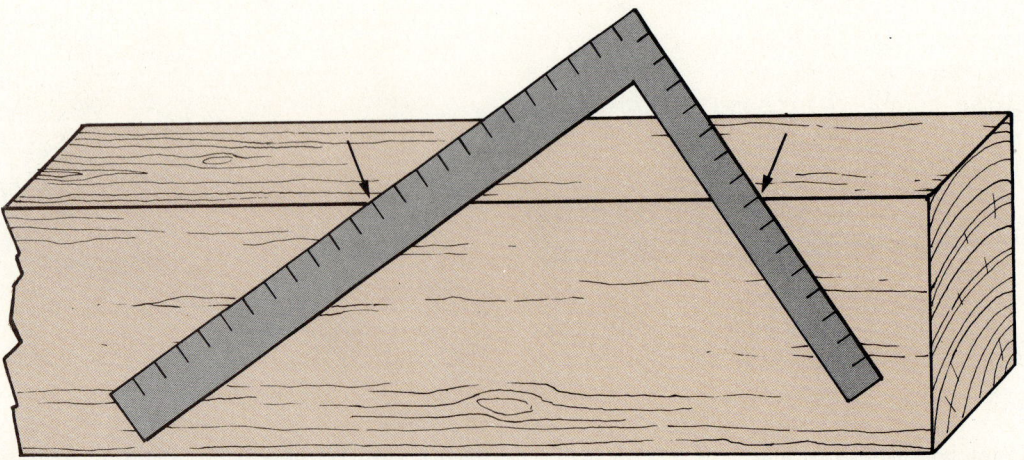

Figure 1-24. Framing square. *(Stanley Tools)*

10   Carpentry Fundamentals

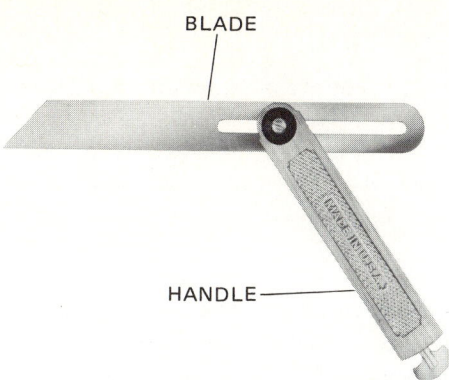

Figure 1-25. Bevel. *(Stanley Tools)*

Figure 1-26. Rafter overhang cut to a given angle.

BEVEL • A bevel can be adjusted to any angle. It helps to make many cuts at the same number of degrees. See Figure 1-25. Note how the blade can be adjusted. Now take a look at Figure 1-26. Here you can see the overhang of rafters. If you want the ends to be parallel with the side of the house, you can use the bevel to mark them before they are cut off. Simply adjust the bevel so the handle is on top of the rafter and the blade fits against the soleplate below. Tighten the screw and move the bevel down the rafter to where you want the cut. Mark the angle along the blade of the bevel. Cut along the mark, and you have what you see in Figure 1-26. It is a good device for transferring angles from one place to another.

CHISEL • Occasionally you may need a wood chisel. It is sharpened on one end. When the other end is struck with a hammer, the cutting end will do its job. That is, of course, if you have kept it sharpened. See Figure 1-27.

The chisel is commonly used in fitting or hanging doors. It is used to remove the area where the door hinge fits. Note how it is used to score the area (Figure 1-27); it is then used at an angle to remove the ridges. A great deal of the work with the chisel is done by using the palm of the hand as the force behind the cutting edge. A hammer can be used. In fact, chisels have a metal tip on the handle so the force of the hammer blows will not chip the handle. Other applications are up to you, the carpenter. You'll find many uses for the chisel in making things fit.

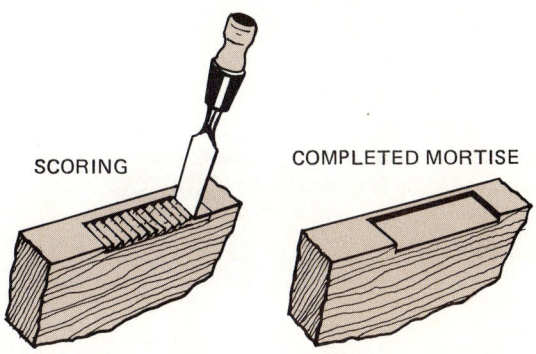

Figure 1-27. Using a wood chisel to complete a mortise.

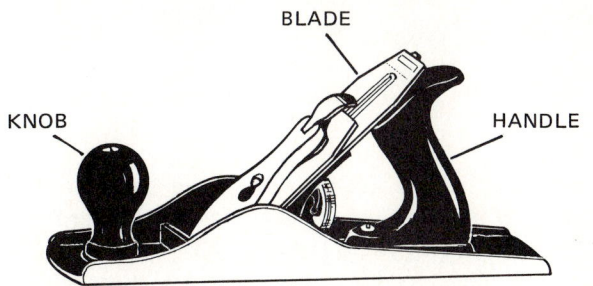

Figure 1-28. Smooth plane. *(Stanley Tools)*

PLANE • Planes (Figure 1-28) are designed to remove small shavings of wood along a surface. One hand holds the knob in front and the other the handle in back. The blade is adjusted so that only a small sliver of wood is removed each time the plane is passed over the wood. It can be used to make sure that doors and windows fit properly. It can be used for any number of wood smoothing operations.

DIVIDERS AND COMPASS • Occasionally a carpenter must draw a circle. This is done with a compass. The compass shown in Figure 1-29A can be converted to a divider by removing the pencil and inserting a straight steel pin. The compass has a

Starting the Job   **11**

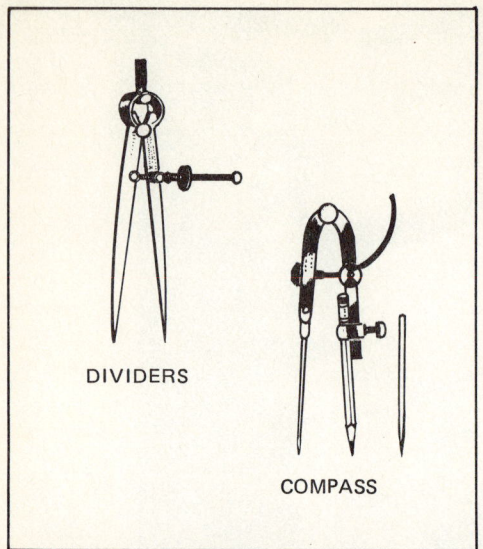

Figure 1-29A. Dividers and compass.

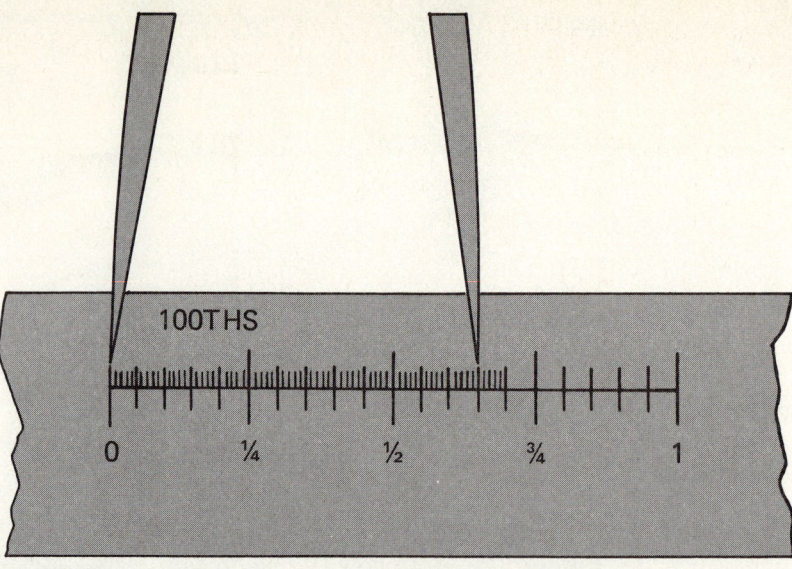

Figure 1-29B. Dividers being used to transfer hundredths of an inch.

Figure 1-30A. Using a level to make sure a window is placed properly before nailing. *(Andersen)*

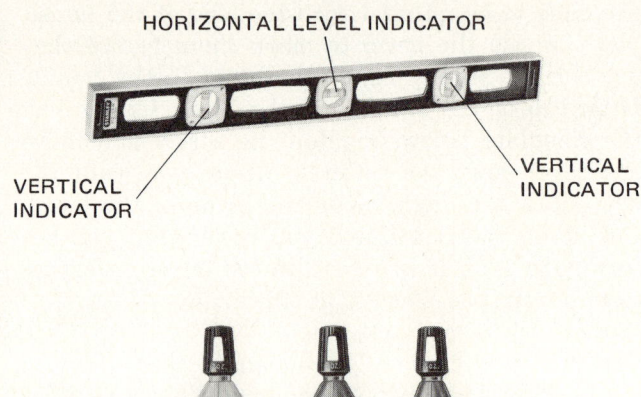

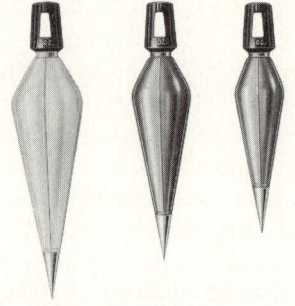

Figure 1-30B. Level and plumb bobs. *(Stanley Tools)*

sharp point which fits into the wood surface. The pencil part is used to mark the circle circumference. It is adjustable to various radii.

The dividers in Figure 1-29A have two points made of hardened metal. They are adjustable. It is possible to use them to transfer a given measurement from the framing square or measuring device to another location. See Figure 1-29B.

LEVEL • In order to have things look as they should, a level is necessary. There are a number of sizes and shapes available. This one shown in Figure 1-30 is the most common type used by carpenters. The bubbles in the glass tubes tell you if the level is obtained. In Figure 1-30A the carpenter is using the level to make sure the window is in properly before nailing it into place permanently.

If the vertical and horizontal bubbles are lined up between the lines, then the window is plumb. Being plumb means that the window is vertical. A plumb bob is a small, pointed weight. It is attached to a string and dropped from a height. If the bob is just above the ground, it will indicate the vertical direction by its string. Keeping windows, doors, and

12 Carpentry Fundamentals

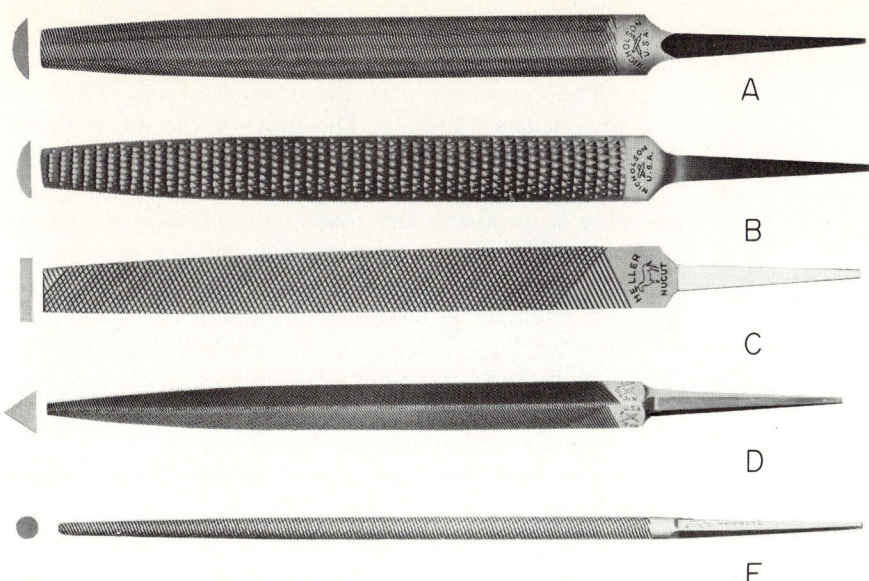

Figure 1-31. Wood and cabinet files. (A) half-round; (B) rasp; (C) flat; (D) triangular; and (E) round. *(Courtesy Millers Falls Division, a division of Ingersol-Rand Co.)*

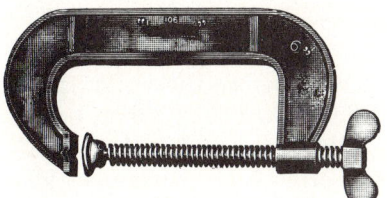

Figure 1-32. C-clamp.

Figure 1-33. Cold chisel. *(Stanley Tools)*

frames square and level makes a difference in fitting. It is much easier to fit prehung doors into a frame that is square. When it comes to placing panels of 4×8-foot plywood sheathing on a roof or on walls, squareness can make a difference as to fit. Besides, a square fit and a plumb door and window look better than those that are a little off. Figure 1-30B shows a level and two plumb bobs.

FILES • A carpenter finds use for a number of types of files. The files have different surfaces for doing different jobs. Tapping out a hole to get something to fit may be just the job for a file. Some files are used for sharpening saws and touching up tool cutting edges. Figure 1-31 shows different types of files. Other files may also be useful. You can acquire them later as you develop a need for them.

CLAMPS • C clamps are used for many holding jobs. They come in handy when placing kitchen cabinets into order. They hold the cabinets together as units till screws can be inserted and properly seated. This type of clamp can be used for an extra hand every now and then when two hands aren't enough to hold a combination of pieces till you can nail them. See Figure 1-32.

COLD CHISEL • It is always good to have a cold chisel around. It has a very sharp cutting edge. The edge can cut metal. This means it is very much needed when you can't remove a nail. Its head may have broken off and the nail must be removed. The chisel can cut the nail and permit the separation of the wood pieces. See Figure 1-33.

If a chisel of this type starts to "mushroom" at the head, you should remove the splintered ends with a grinder. Hammering on the end can produce a mushrooming effect. These pieces should be taken off since they can easily fly off when hit with a hammer. That is another reason for using eye protection when using tools.

CAULKING GUN • In times of energy crisis, the caulking gun gets plenty of use. It is used to fill in around windows and doors and everywhere there may be an air leak. Figure 1-34 shows a caulking gun being used. There are many types of caulk being made today. Another chapter will cover the details

Starting the Job 13

Figure 1-34. Using a caulking gun.

Power Tools

The carpenter uses many power tools to aid in getting the job done. The quicker the job is done, the more valuable the work of the carpenter becomes. This is called productivity. The more you are able to produce, the more valuable you are. This means the contractor can make money on the job. This means you can have a job the next time there is a need for a good carpenter. Power tools make your work go faster. They also help you to do a job without getting fatigued. Many tools have been designed with you in mind. They are portable. They operate from an extension cord.

The extension cord should be the proper size to take the current needed for the tool being used. See Table 1-1. Note how the distance between the outlet and the tool using the power is critical. If the distance is great, then the wire must be larger in size to handle the current without too much loss. The higher the number of the wire, the smaller the diameter of the wire. The larger the size of the wire (diameter), the more current it can handle without dropping the voltage.

Some carpenters run an extension cord from the house next door for power before the building site is furnished power. If the cord is too long or has the wrong size wire, it drops the voltage below 115. This means the saws or other tools using electricity will draw more current and therefore drop the voltage more. Every time the voltage is dropped, the device tries to obtain more current. This becomes a self-defeating phenomenon. You wind up with a saw that has little cutting power. You may have a drill that won't drill into a piece of wood without

of the caulking compounds and their uses.

This gun is easily operated. Insert the cartridge and cut its tip to the shape you want. Puncture the thin plastic film inside. A bit of pressure will cause the caulk to come out the end. The long rod protruding from the end of the gun is turned over. This is so the serrated edge will engage the hand trigger. Remove the pressure from the cartridge when you are finished. Do this by rotating the rod so that the serrations are not engaged by the trigger of the gun.

TABLE 1-1. SIZE OF EXTENSION CORDS FOR PORTABLE TOOLS

Cord Length, Feet	Full-Load Rating of the Tool in Amperes at 115 Volts					
	0 to 2.0	2.10 to 3.4	3.5 to 5.0	5.1 to 7.0	7.1 to 12.0	12.1 to 16.0
	Wire Size (AWG)					
25	18	18	18	16	14	14
50	18	18	18	16	14	12
75	18	18	16	14	12	10
100	18	16	14	12	10	8
200	16	14	12	10	8	6
300	14	12	10	8	6	4
400	12	10	8	6	4	4
500	12	10	8	6	4	2
600	10	8	6	4	2	2
800	10	8	6	4	2	1
1000	8	6	4	2	1	0

If the voltage is lower than 115 volts at the outlet, have the voltage increased or use a much larger cable than listed.

stalling. Of course the damage done to the electric motor is in some cases irreparable. You may have to buy a new saw or drill. Double-check Table 1-1 for the proper wire size in your extension cord.

PORTABLE SAW • This is the most often used and *abused* of carpenter's equipment. The electric portable saw, such as the one shown in Figure 1-35, is used to cut all 2×4s and other dimensional lumber. It is used to cut off rafters. This saw is used to cut sheathing for roofs. It is used for almost every sawing job required in carpentry.

This saw has a guard over the blade. The guard should always be left intact. Do not remove the saw guard. If not held properly against the wood being cut, the saw can kick back and into your leg.

You should wear safety glasses when using this saw. The sawdust is thrown in a number of directions, and one of these is straight up toward your eyes. If you are watching a line where you are cutting, you definitely should have on glasses.

TABLE SAW • If the house has been enclosed, it is possible to bring in a table saw to handle the larger cutting jobs. See Figure 1-36. You can do ripping a little more safely with this type of saw. It has a rip fence. If a push stick is used to push the wood through and past the blade, it is safe to operate. Do not remove the safety guard. This saw can be used for both crosscut and rip. The blade is lowered or raised to the thickness of the wood. It should protrude about ¼ to ½ inch above the wood being

Figure 1-35. Portable hand saw.

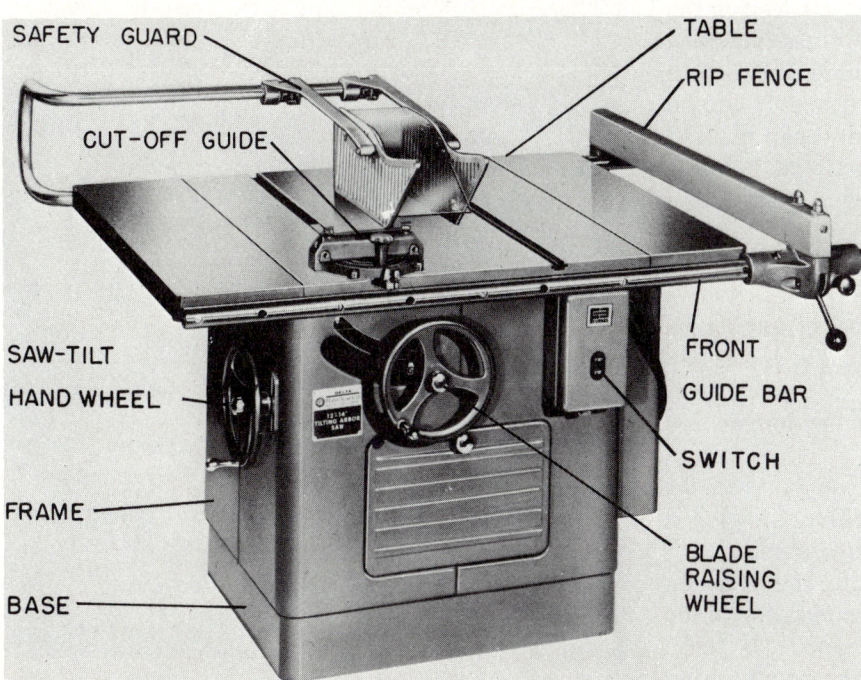

Figure 1-36. Table saw. *(Courtesy of Power Tool Division, Rockwell International)*

Starting the Job 15

cut. This saw usually requires a 1-horsepower motor. This means it will draw about 6.5 amperes to run and over 35 to start. It is best not to run the saw on an extension cord. It should be wired directly to the power source with circuit breakers installed in the line.

RADIAL ARM SAW • This type of saw is brought in only if the house can be locked up at night. The saw is expensive and too heavy to be moved every day. It should have its own circuit. The saw will draw a lot of current when it hits a knot while cutting wood. See Figure 1-37.

In this model the moving saw blade is pulled toward the operator. In the process of being pulled toward you, the blade rotates so that it forces the wood being cut against the bench stop. Just make sure your left hand is in the proper place when you pull the blade back with your right hand. It takes a lot of care to operate a saw of this type. The saw works well for cutting large-dimensional lumber. It will crosscut or rip. This saw will also do miter cuts at almost any angle. Once you become familiar with it, the saw can be used to bevel crosscut, bevel miter, bevel rip, and even cut circles. However, it does take practice to develop some degree of skill with this saw.

Figure 1-37. Radial arm saw. *(DeWalt)*

Figure 1-38. Router. *(Stanley Tools)*

ROUTER • The router has a high-speed type of motor. It will slow down when overloaded. It takes the beginner some time to adjust to *feeding* the router properly. If you feed it too fast, it will stall or burn the edge you're routing. If you feed it too slowly, it may not cut the way you wish. You will have to practice with this tool for some time before you're ready to use it to make furniture. It can be used for routing holes where needed. It can be used to take the edges off laminated plastic on countertops. Use the correct bit, though. This type of tool can be used to the extent of the carpenter's imagination. See Figure 1-38.

SAW BLADES • There are a number of saw blades available for the portable, table, or radial saw. They may be standard steel types or they may be carbide-tipped. Carbide-tipped blades tend to last longer. See Figure 1-39.

Combination blades (those that can be used for both crosscut and rip) with a carbide tip give a smooth finish. They come in 6½-inch diameter with 24 teeth. The arbor hole for mounting the blade on the saw is ¾ to ⅝ inch. A safety combination blade is also made in 10-inch-diameter size with 10 teeth and the same arbor hole sizes as the combination carbide-tipped blade.

The planer blade is used to crosscut, rip, or miter hard or soft woods. It is 6½ or 10 inches in diameter

Figure 1-39. Saw blades. (A) planer blade. (B) framing rip blade. (C) carbide tipped. (D) metal cutting blade. *(Black & Decker)*

16 Carpentry Fundamentals

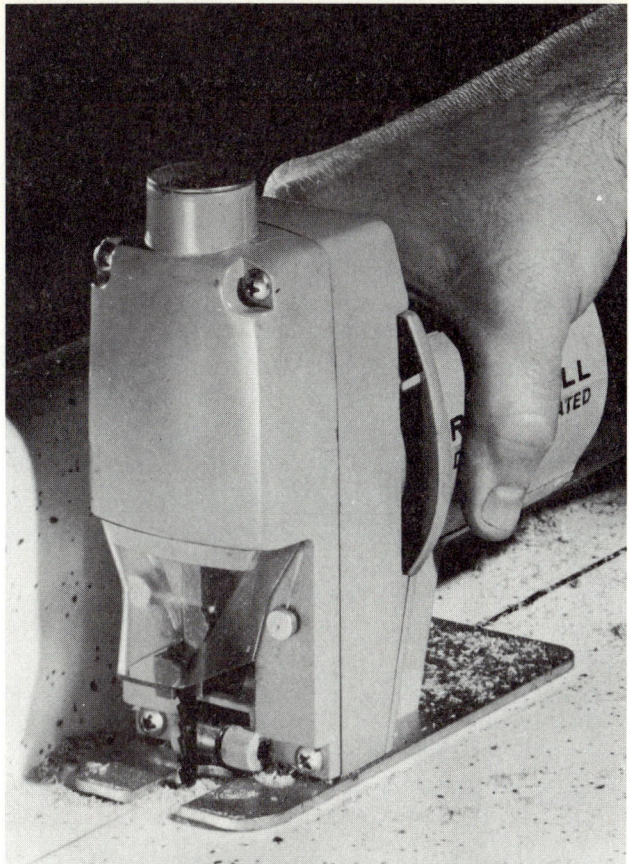

Figure 1-40. Sabre saw.

Figure 1-41. Hand-held portable drill. *(Milwaukee)*

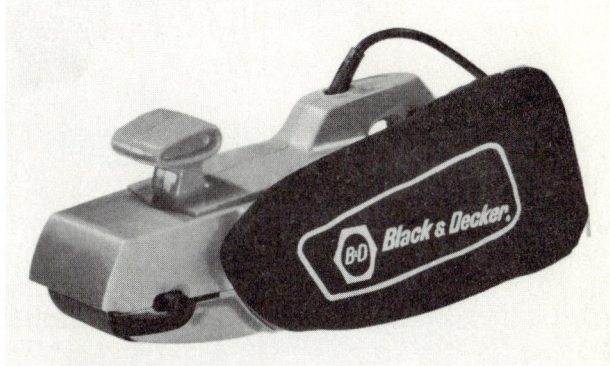

Figure 1-42. Belt sander. *(Black & Decker)*

with 50 teeth. It too can fit anything from ¾- to ⅝-inch arbors.

If you want a smooth cut on plywood without the splinters that plywood can generate, you had better use a carbide-tipped plywood blade. It is equipped with 60 teeth and can be used to cut plywood, Formica, or laminated countertop plastic. It can also be used for straight cutoff work in hard or soft woods. Note the shape of the saw teeth to get some idea as to how each is designed for a specific job. You can identify these after using them for some time. Until you can, mark them with a grease pencil or marking pen when you take them off.

SABER SAW • The saber saw has a blade that can be used to cut circles in wood. See Figure 1-40. It can be used to cut around any circle or curve. If you are making an inside cut, it is best to drill a starter hole first. Then, insert the blade into the hole and follow your mark. The saber saw is especially useful in cutting out holes for heat ducts in flooring. Another use for this type of saw is cutting holes in roof sheathing for pipes and other protrusions. The saw blade is mounted so that it cuts on the upward stroke. With a fence attached, the saw can also do ripping.

DRILL • The portable drill is useful to the carpenter in many ways. There are times when holes must be drilled in sole plates so that anchor bolts can be properly installed. This calls for a drill to do the job in a hurry. Rather than drill by hand, the carpenter finds it easier and much faster to do the job with an electric drill. The drill shown in Figure 1-41 is capable of drilling through all dimensional lumber on the job. With a carbide tip it is capable of drilling into concrete whenever the carpenter needs to anchor a partition into a slab.

The uses for an electric drill are limited only by the imagination of the user. It comes in handy when mounting countertops to cabinets in the kitchen, for example.

SANDERS • The belt sander shown in Figure 1-42 and the orbital sanders shown in Figure 1-43A and B can do almost any required sanding job. The carpenter needs the sander occasionally. It helps align parts properly, especially those that don't fit

Starting the Job **17**

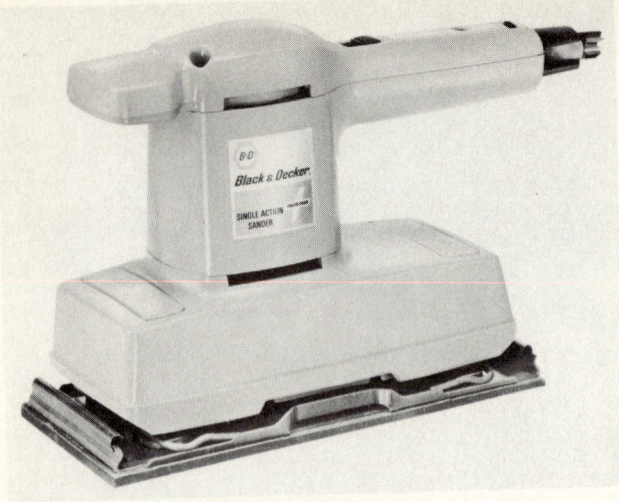

(A)

(B)

Figure 1-43. Orbital sanders. (A) dual action and (B) single action. *(Black & Decker)*

Figure 1-44. Air-powered nailer. *(Duo-Fast)*

pressed air. The staples and nails are especially designed to be driven by the machine. See Tables 1-2 and 1-3 for the variety of fasteners used with this type of machine. The stapler or nailer can also be used to install siding or trim around a window.

The tool's low air pressure requirements (60 to 90 pounds per square inch) allow it to be moved from place to place. Nails for this machine (Figure 1-44) are from 6d to 16d. It is magazine-fed for rapid use. Just pull the trigger.

• FOLLOWING CORRECT SEQUENCES •

One of the important things a carpenter must do is follow a sequence. Once you start a job, the sequence has to be followed properly to arrive at a completed house in the minimum of time.

Preparing the Site

Preparing the site may be expensive. There must be a road or street. In most cases the local ordinances require a sewer. In most locations the storm sewer and the sanitary sewer must be in place before building starts. If a sanitary sewer is not available you should plan for a septic tank for sewage disposal.

by just a small amount. The sander can be used to finish off windows, doors, counters, cabinets, and floors. A larger model of the belt sander is used to sand floors before they are sealed and varnished. The orbital or vibrating sanders are used primarily to put a very fine finish on a piece of wood. Sandpaper is attached to the bottom of the sander. The sander is held by hand over the area to be sanded. The operator has to remove the sanding dust occasionally to see how well the job is progressing.

NAILERS • One of the greatest tools the carpenter has acquired recently is the nailer. See Figure 1-44. It can drive nails or staples into wood better than a hammer. The nailer is operated by com-

18 Carpentry Fundamentals

TABLE 1-2 FINE WIRE STAPLES FOR A PNEUMATIC STAPLE DRIVER

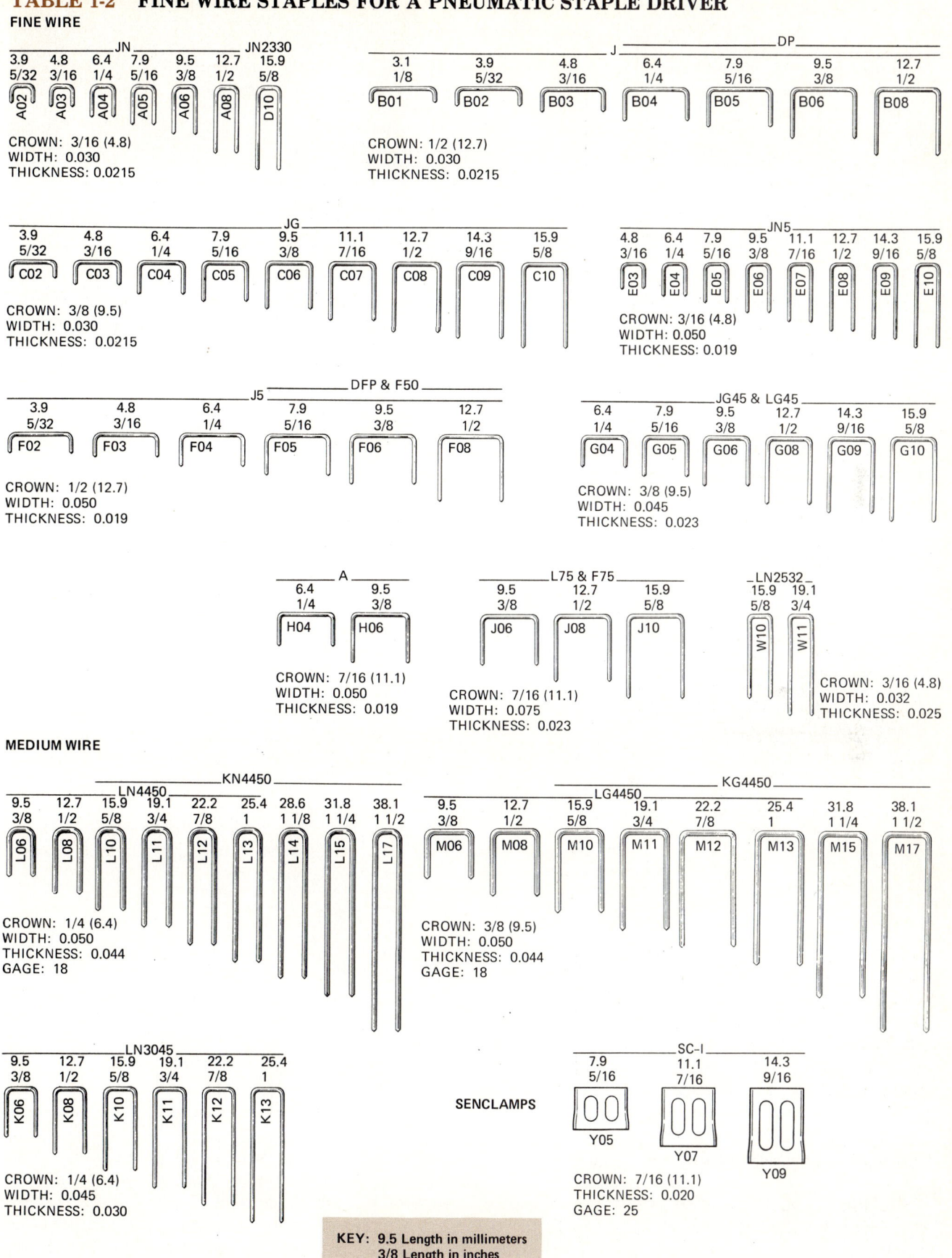

KEY: 9.5 Length in millimeters
3/8 Length in inches

Starting the Job 19

TABLE 1-3   7-DIGIT NAIL ORDERING SYSTEM

1st Digit: Diameter, Inches	2d Digit: Head	3d and 4th Digits: Length, Inches	5th Digit: Point	6th Digit: Wire Chem. and Finish	7th Digit: Finish
A 0.0475	A Brad	08 ½	A Diam.-reg.	A Std. carbon-galv.	A Plain
D 0.072	C Flat	11 ¾	E Chisel	E Std. carb. "Weatherex" galv.	B Sencote
E 0.0915	E Flat/ring shank	13 1			C Painted
G 0.113		15 1¼		G Stainless steel std. tensile	D Painted and sencote
H 0.120	F Flat/screw shank	17 1½			
J 0.105		19 1¾		H Hardened high-carbon bright basic	
K 0.131	Y Slight-headed pin	20 1⅞			
U 0.080		21 2		P Std. carbon bright basic	
	Z Headless pin	22 2⅛			
		23 2¼			
		24 2⅜			
		25 2½			
		26 2¾			
		27 3			
		28 3¼			
		29 3½			

EXAMPLE: 10¼ ga. (K), flat head (C),
KC25AAA—2½" (25), regular point (A), std. carb. galvanized (A), plain, or uncoated (A) Senco-Nail

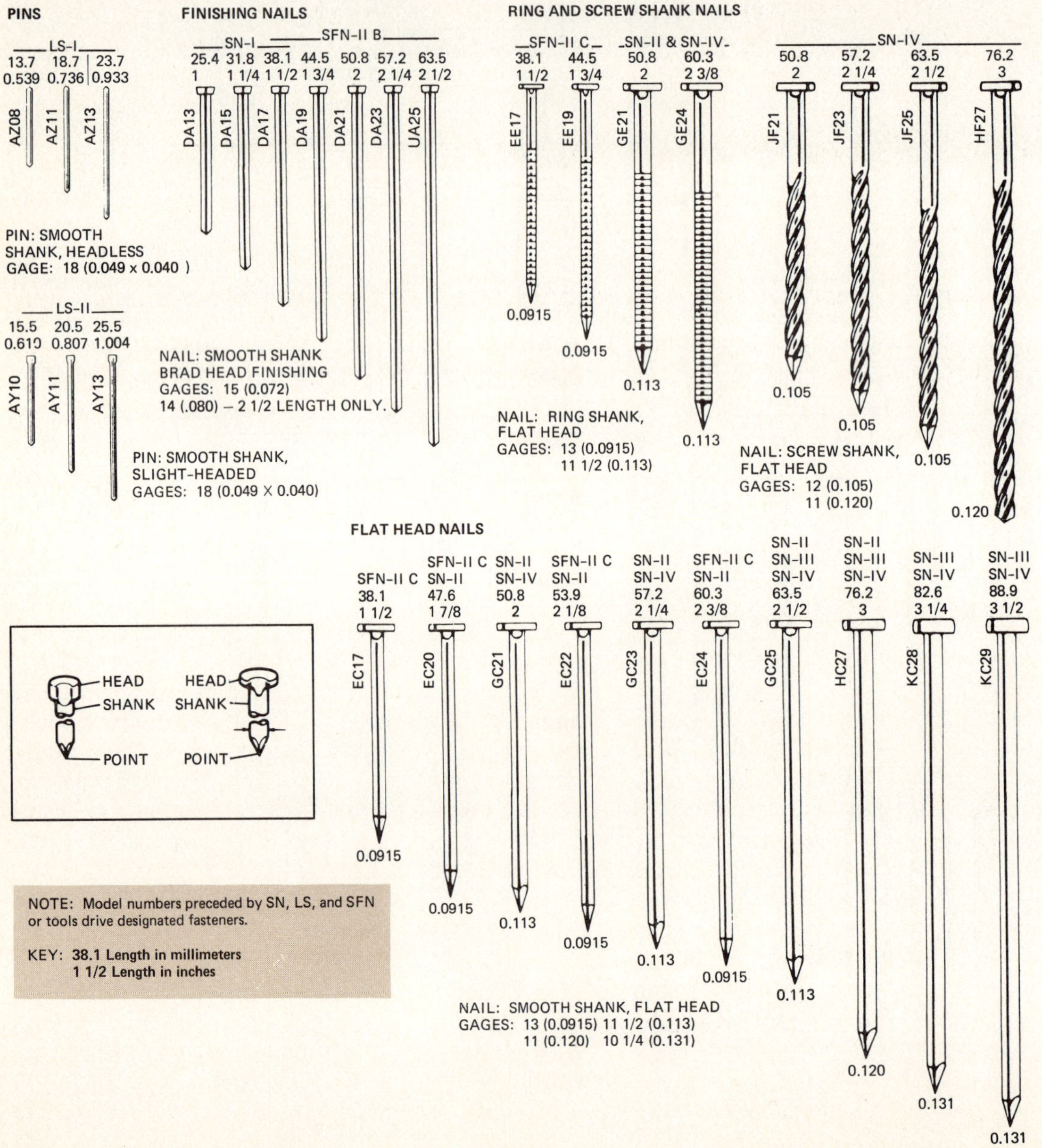

PINS

LS-I
- AZ08 13.7 / 0.539
- AZ11 18.7 / 0.736
- AZ13 23.7 / 0.933

PIN: SMOOTH SHANK, HEADLESS
GAGE: 18 (0.049 × 0.040)

LS-II
- AY10 15.5 / 0.610
- AY11 20.5 / 0.807
- AY13 25.5 / 1.004

PIN: SMOOTH SHANK, SLIGHT-HEADED
GAGES: 18 (0.049 × 0.040)

FINISHING NAILS

SN-I / SFN-II B
- DA13 25.4 / 1
- DA15 31.8 / 1 1/4
- DA17 38.1 / 1 1/2
- DA19 44.5 / 1 3/4
- DA21 50.8 / 2
- DA23 57.2 / 2 1/4
- UA25 63.5 / 2 1/2

NAIL: SMOOTH SHANK BRAD HEAD FINISHING
GAGES: 15 (0.072)
14 (.080) – 2 1/2 LENGTH ONLY.

RING AND SCREW SHANK NAILS

SFN-II C
- EE17 38.1 / 1 1/2   0.0915
- EE19 44.5 / 1 3/4   0.0915

SN-II & SN-IV
- GE21 50.8 / 2   0.113
- GE24 60.3 / 2 3/8   0.113

NAIL: RING SHANK, FLAT HEAD
GAGES: 13 (0.0915)
11 1/2 (0.113)

SN-IV
- JF21 50.8 / 2   0.105
- JF23 57.2 / 2 1/4   0.105
- JF25 63.5 / 2 1/2   0.105
- HF27 76.2 / 3   0.120

NAIL: SCREW SHANK, FLAT HEAD
GAGES: 12 (0.105)
11 (0.120)

FLAT HEAD NAILS

- EC17 SFN-II C   38.1 / 1 1/2   0.0915
- EC20 SFN-II C / SN-II   47.6 / 1 7/8   0.0915
- GC21 SN-II / SN-IV   50.8 / 2   0.113
- EC22 SFN-II C / SN-II   53.9 / 2 1/8   0.0915
- GC23 SN-II / SN-IV   57.2 / 2 1/4   0.113
- EC24 SFN-II C / SN-II   60.3 / 2 3/8   0.0915
- GC25 SN-II / SN-III / SN-IV   63.5 / 2 1/2   0.113
- HC27 SN-II / SN-III / SN-IV   76.2 / 3   0.120
- KC28 SN-III / SN-IV   82.6 / 3 1/4   0.131
- KC29 SN-III / SN-IV   88.9 / 3 1/2   0.131

NAIL: SMOOTH SHANK, FLAT HEAD
GAGES: 13 (0.0915) 11 1/2 (0.113)
11 (0.120) 10 1/4 (0.131)

Diagram labels: HEAD, SHANK, POINT

NOTE: Model numbers preceded by SN, LS, and SFN or tools drive designated fasteners.

KEY: 38.1 Length in millimeters
1 1/2 Length in inches

20 Carpentry Fundamentals

Figure 1-45. Street being extended for a new subdivision.

Figure 1-48. Dirt from the basement excavation is piled high around a building site.

Figure 1-46. The beginning of a street.

Figure 1-49. Hole for a basement.

Figure 1-47. Locating a building site and removing the curb for the driveway.

Figure 1-50. The columns and foundation walls will help support the floor parts.

Figure 1-45 shows a sewer project in progress. This shows a street being extended. The storm sewer lines are visible, as is the digger. Trees had to be removed first by a bulldozer. Once the sewer lines are in, the roadbed or street must be properly prepared. Figure 1-46 shows the building of a street. Proper drainage is very important. Once the street is in and the curbs poured, it is time to locate the house.

Figure 1-47 shows how the curb has been broken and the telephone terminal box installed in the weeds. Note the stake with a small piece of cloth on it. This marks the location of the site.

As you can see in Figure 1-48, the curb has been removed. A gravel bed has been put down for the driveway.

The sewer manhole sticks up in the driveway. The basement has been dug. Dirt piles around it show how deep the basement really is. However, a closer look shows that the hole isn't too deep. That means the dirt will be pushed back against the basement wall to form a higher level for the house. This will provide drainage away from the house when finished. See Figure 1-49 for a look at the basement hole.

The Basement

In Figure 1-50 the columns and the foundation wall have been put up. The basement is prepared in this case with courses of block with brick on the

Starting the Job 21

Figure 1-51. Carpenters are laying plywood subflooring with tongue-and-groove joints. This is stronger. *(American Plywood Association)*

Figure 1-52. Wall frames are erected after the floor frame is built.

outside. This basement appears to be more of a crawl space under the first floor than a full stand-up basement.

Once the basement is finished and the floor joists have been placed, the flooring is next.

The Floor

Once the basement or foundation has been laid for the building, the next step is to place the floor over the joists. Note in Figure 1-51 that the grooved flooring is laid in large sheets. This makes the job go faster and reinforces the floor.

Wall Frames

Once the floor is in place and the basement entrance hole has been cut, the floor can be used to support the wall frame. The 2×4s or 2×6s for the framing can be placed on the flooring and nailed. Once together, they are pushed into the upright position as in Figure 1-52. For a two-story house, the second floor is placed on the first-story wall supports. Then the second-floor walls are nailed together and raised into position.

Sheathing

Once the sheathing is on and the walls are upright, it is time to concentrate on the roof. See Figure 1-53. The rafters are cut and placed into position and nailed firmly. See Figure 1-54. They are reinforced by the proper horizontal bracing. This makes sure they are properly designed for any snow load or other loads that they may experience.

Figure 1-53. Beginning construction of the roof structure. *(Georgia-Pacific)*

Figure 1-54. Framing and supports for rafters.

22 Carpentry Fundamentals

Figure 1-55. Fiberboard sheathing over the wall frame.

Roofing

The roofing is applied after the siding is on and the rafters are erected. The roofing is completed by applying the proper underlayment and then the shingles. If asphalt shingles are used, the procedure is slightly different from that for wooden shingles. Shingles and roofing are covered in another chapter in this book. Figure 1-55 shows the sheathing in place and ready for the roofing.

Siding

After the roofing, the finishing job will have to be undertaken. The windows and doors are in place. Finish touches are next. The plumbing and drywall may already be in. Then the siding has to be installed. In some cases, of course, it may be brick. This calls for bricklayers to finish up the exterior. Otherwise the carpenter places siding over the walls. Figure 1-56 shows the beginning of the siding at the top left of the picture.

Figure 1-57 shows how the siding has been held in place with a stapler. The indentations in the wood show a definite pattern. The siding is nailed to the nail base underneath after a coating of tar paper (felt paper in some parts of the country) is applied to the nail base or sheathing.

Finishing

Exterior finishing requires a bit of caulking with a caulking gun. Caulk is applied to the siding that butts the windows and doors.

Finishing the interior can be done at a more leisurely pace once the exterior is enclosed. The plumbing and electrical work has to be done before the drywall or plaster is applied. Once the wallboard has been finished, the trim can be placed around the edges of the walls, floors, windows, and doors. The flooring can be applied after the finishing of the walls and ceiling. The kitchen cabinets must be installed before the kitchen flooring. There is a definite sequence to all these operations.

As you can imagine, it would be impossible to

Figure 1-56. Siding applied on the left side of the building.

Figure 1-57. Siding applied to a building. Note the pattern of the staples.

place roofing on a roof that wasn't there. It takes planning and following a sequence to make sure the roof is there when the roofing crew comes around to nail the shingles in place. The water must be there before you can flush the toilets. The electricity must be hooked up before you can turn on a light. These are reasonable things. All you have to do is sit down and plan the whole operation before starting. Planning is the key to sequencing. Sequencing makes it possible for everyone to be able to do a job at the time assigned to do it.

• WORKING SUCCESSFULLY IN CARPENTRY •

Being Reliable

Being reliable means that people can rely upon you to do what you say you will do. This does not come overnight. It takes a long time to build people's

Starting the Job 23

confidence in you. You have to work for a long time to have people know that you can be relied upon to do what you say when you say. From the previous paragraphs you can see the importance of getting things done when they should be done. The sequence is important. You are the only one who can make the sequencing work. Carpentry is a team trade. Everyone involved in every aspect of building a house must be reliable. You can be reliable by showing up when you say you will. Do the job you say you will.

Being Careful

It is important for your own health that you be careful. By being careful and following common safety rules you can have much more time to get the job done. The expense of days absent cannot be measured directly. A day off the job when your presence is needed to finish a job can be very expensive to a contractor.

You may not care for the pain and the problems created by carelessness. No one does. The only advice to keep in mind when working with tools and equipment is to know your tools. Know what a tool will do safely. Don't do the unsafe things that happen to pop into your mind. You won't get the job done quicker.

Following Instructions

There are plans made for each building. They exist, in most cases, on paper. These drawings or plans use symbols to get the message across to the carpenter. By following instructions you get the job done the way it was planned. Following instructions also leaves less chance of your being sued or legally tied up in court.

Building Accurately

Building accurately is most important. If you do the job right the first time, you save time and money. Accurate measurements and skillful placement of each part is necessary for a job well done. The safety of a building relies upon the accuracy of each procedure used.

• WORDS CARPENTERS USE •

folding rule	claw hammer	try square	level	saber saw
handsaw	penny (d)	framing square	plumb bob	orbital sander
skew-back saw	nail set	bevel	cold chisel	nailers
crosscut saw	awl	chisel	caulk	sheathing
backsaw	wrecking bar	plane	table saw	roofing
mitre box	carpenter's	divider	radial-arm saw	siding
coping saw	square	compass	router	

• STUDY QUESTIONS •

1-1. What is carpentry involved with in terms of processes?

1-2. Can you learn carpentry from a book?

1-3. Why should a carpenter wear safety-toed shoes?

1-4. Why should a carpenter use safety glasses on the job?

1-5. Why is a hard hat suggested when a carpenter is working on site?

1-6. Why are gloves used on a carpentry job?

1-7. List at least six safety procedures that should be followed at all times.

1-8. List at least five important habits to acquire for safe working on the job.

1-9. How should you climb a ladder?

1-10. Why would a carpenter need a folding rule?

1-11. What is the difference between a rip and a crosscut saw?

1-12. What is a miter box?

1-13. What is a backsaw?

1-14. How is a coping saw used?

1-15. What is a claw hammer used for?

1-16. Why would a carpenter need a nail set?

1-17. What does a carpenter use a wrecking bar for?

1-18. What is a try square?

1-19. What's the difference between a try square and a bevel?

1-20. Where does the carpenter use a caulking gun?

1-21. Why is the length of an extension cord important?

1-22. What's the difference between a table saw and a radial arm saw?

1-23. What are some uses for a router?

1-24. Why is following a sequence important?

1-25. List the sequence needed to build a house.

24 Carpentry Fundamentals

2 PREPARING THE SITE

• DEVELOPING SKILLS •

In this chapter you will learn how to develop these skills:

- Lay out buildings
- Use the builder's level
- Prepare for the start of construction

Each has its importance. The beginning of construction should be preceded by planning. The layout of the building requires that some work be first done on paper. It is much cheaper to make a mistake on paper than on site.

The builder's level will show you how to make sure the building is level. No new leaning tower of Pisa is needed today. One is enough. It would be very hard to sell one today. That means you should choose the site so that the soil will support the weight of the building.

There is a basic sequence to follow in building. It should be followed for the benefit of those who are supposed to operate as part of the team.

• BASIC SEQUENCE •

The basic sequence involves the following operations:

1. Cruise the site and plan the job
2. Locate the boundaries
3. Locate the building area or areas
4. Define the site work that is needed
5. Clear any unwanted trees
6. Lay out the building
7. Establish the exact elevations
8. Excavate the basement or foundation
9. Provide for access during construction
10. Start the delivery of materials to the site
11. Have a crew arrive to start with the footings

• LOCATING THE BUILDING ON THE SITE •

The proper location of a building is very important. It would be embarrassing and costly to move a building once it is built. That means a lot of things have to be checked first.

Property Boundaries

First, a clear deed to the land should be established. This can be done in the county courthouse. Check the records or have someone who is paid for this type of work do it. An abstract of the history of the ownership of the land is usually provided. In Iowa, for instance, the abstract traces ownership back to the Louisiana Purchase of 1803. In New York State the history of the land is traced by owners from the days of the Holland Land Company. Alabama can provide records back to the time the Creek or other Indians owned the land. Each state has its own history and its own procedure for establishing absolute ownership of land. It is best to have proof of this ownership before starting any construction project.

Surveyors should be called in to establish the limits of the property. A plot plan is drawn by the surveyors. This can be used to locate the property. Figure 2-1 is a plot plan showing the location of a house on a lot.

Sidewalks, utilities easements, and other things have to be taken into consideration. The location of the house may be specified by local ordinance. This type of ordinance will usually specify what clearance the house must have on each side. It will probably set the limits of setback from the street. You may also want to plan around a tree or trees. Since trees increase the value of the property, it is important to save as many as possible. Figure 2-2 shows a sketch of some of these considerations.

An *easement* is the right of the utilities to use

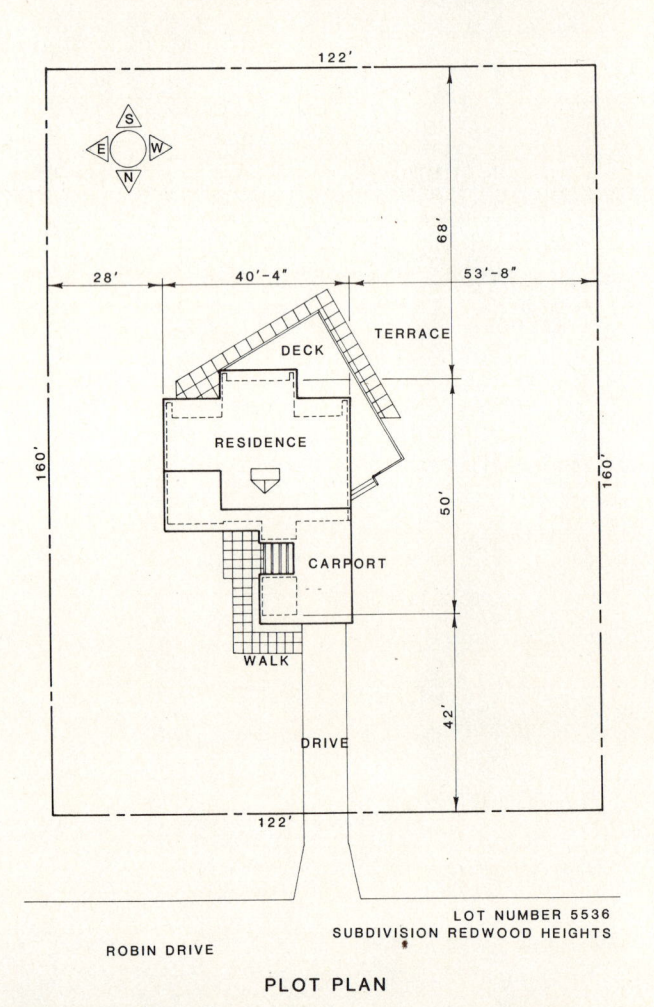

Figure 2-1. Plot plan.

the space to furnish electric power, phone service, and gas to your location and to others nearby. This means you have given them permission to string wires or bury lines to provide their services. Keep in mind also the rights of the city or township to supply water and sewers. These may also cut across the property.

Laying Out the Foundation

Layout of the foundation is the critical beginning in house construction. It is a simple but extremely important process. It requires careful work. Make sure the foundation is square and level. You will find all later jobs, from rough carpentry through finish construction and installation of cabinetry, are made much easier.

1. Make sure your proposed house location on the lot complies with local regulations.
2. Set the house location, based on required setbacks and other factors, such as the natural drainage pattern of the lot. Level or at least rough-clear the site. See Figure 2-3.
3. Lay out the foundation lines. Figure 2-4 shows the simplest method for locating these. Locate each outside corner of the house and drive small stakes into the ground. Drive tacks into the tops of the stakes. This is to indicate the outside line of the foundation wall. This is not the footings limit but the outside wall limit. Next check the squareness of the house by measuring the diagonals, corner to corner, to see that they are equal. If the structure is rectangular, all diagonal measurements will be equal. You can check squareness of any corner by measuring 6 feet down one side, then 8 feet down the other side. The diagonal line between these two end points should measure exactly

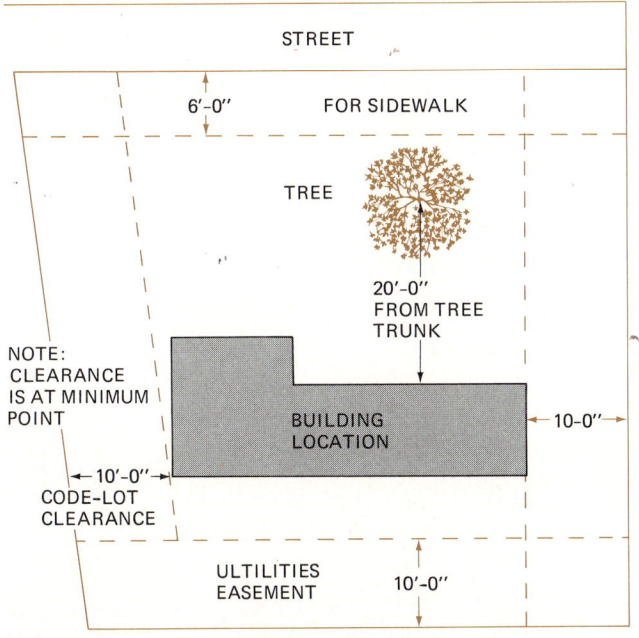

Figure 2-2. Site location must be chosen carefully.

Figure 2-3. Rough-cleared lot. Only weeds need to be taken out, as the basement is dug.

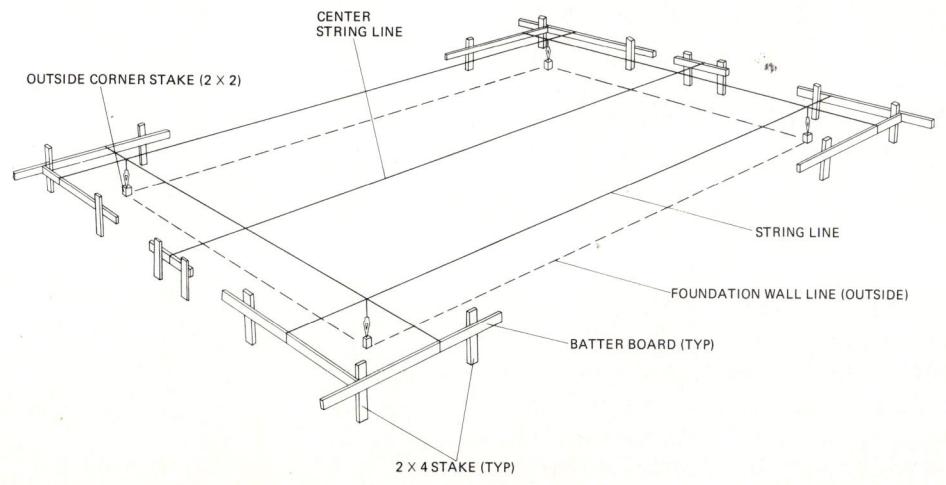

Figure 2-4. Staking out a basement. *(American Plywood Association)*

Preparing the Site 27

10 feet. If it doesn't, the corner isn't truly square. See Figure 2-5.

4. After the corners are located and squared, drive three 2×4 stakes at each corner as shown in Figure 2-4. Locate these stakes 3 feet and 4 feet outside the actual foundation line. Then nail 1×6 batter boards horizontally so that their top edges are all level and at the same grade. Levelness will be checked later. Hold a string line across the tops of opposite batter boards at two corners. Using a plumb bob, adjust the line so that it is exactly over the tacks in the two corner stakes. Cut saw kerfs ¼ inch deep where the line touches the batter boards so that the string lines may be easily replaced if they are broken or disturbed. Figure 2-6 shows how carpenters in some parts of the country use a nail instead of the saw kerf to hold the thread or string. Figure 2-7 shows how the details of the location of the stake are worked out. This one is a 3–4–5 triangle, or 9 feet and 12 feet on the sides and 15 feet on the diagonal. If you use 6, 8, and 10 feet you get a 3–4–5 triangle also. This means 6 ÷ 2 = 3, 8 ÷ 2 = 4, and 10 ÷ 2 = 5. In the other

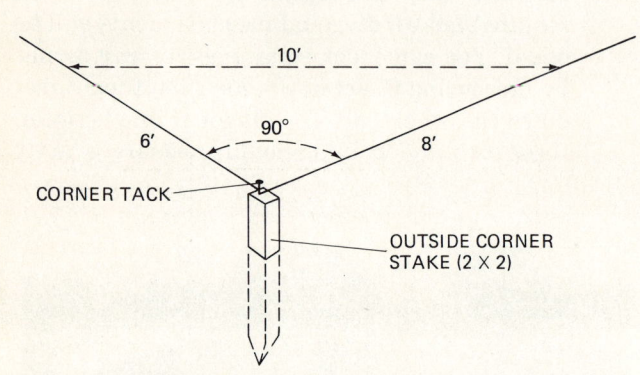

Figure 2-5. Squaring the corner and marking the point. *(American Plywood Association)*

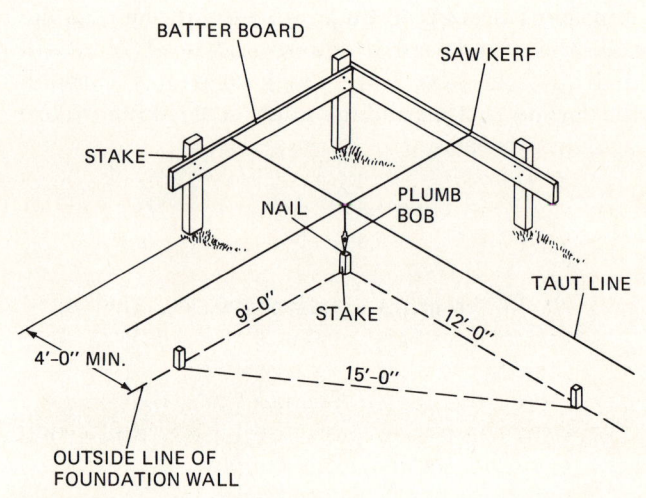

Figure 2-7. Staking and laying out the house. *(Forest Products Lab)*

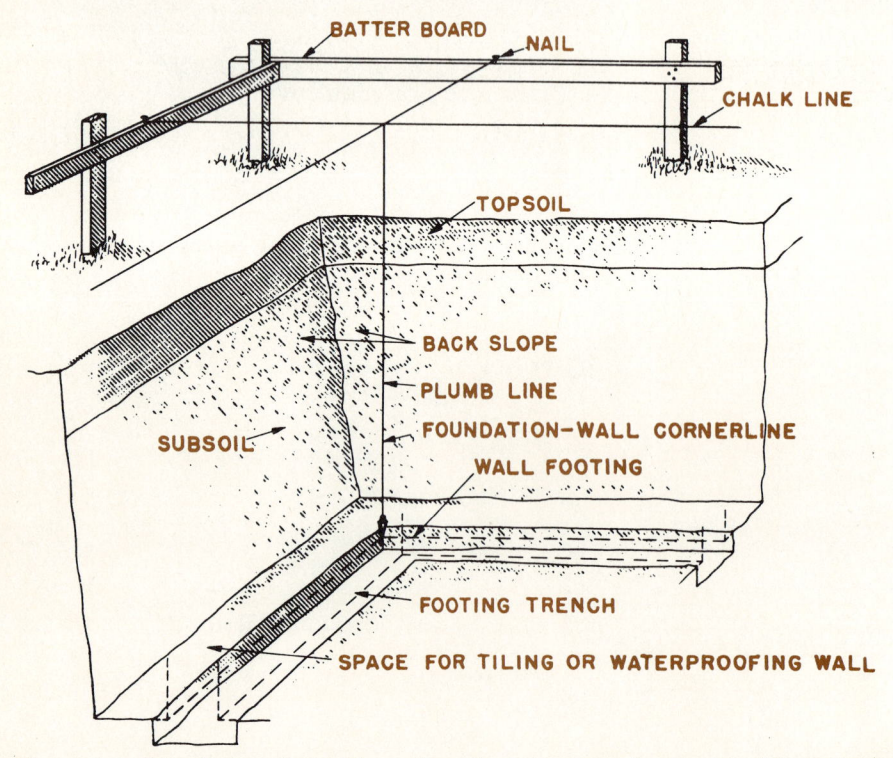

Figure 2-6. Note the location of the nail on the batter board. (Reprinted from *Technical Woodworking*, 2d ed., by C. Groneman and E. Glazener. Copyright © 1976, 1966 with permission of Webster/McGraw-Hill.)

example, 9 feet ÷ 3 = 3, 12 feet ÷ 3 = 4, and 15 ÷ 3 = 5. So you have a 3–4–5 triangle in either measurement. Other combinations can be used but these are the most common. Cut all saw kerfs the same depth. This is because the string line not only defines the outside edges of the foundation but also will provide a reference line. This ensures uniform depth of footing excavation. When you have made similar cuts in all eight batter boards and strung the four lines in position, the outside foundation lines are accurately established.

5. Next, establish the lengthwise girder location. This is usually on the centerline of the house. Double check your house plans for the exact position. This is because occasionally the girder will be slightly off the centerline to support an interior bearing wall. To find the line, measure the correct distance from the corners. Then install batter boards and locate the string line as before.

6. Check the foundation for levelness. Remember that the top of the foundation must be level around the entire perimeter of the house. The most accurate and simplest way to check this is to use a surveyor's level. This tool will be explained later in this chapter. The next best approach is to ensure that batter boards, and thus the string lines, are all absolutely level. You can accomplish this with a 10- to 14-foot-long piece of straight lumber. See Figure 2-8. Judge the straightness of the piece of lumber by sighting along the surface. Use this straightedge in conjunction with a carpenter's level. Then drive temporary stakes around the house perimeter. The distance between them should not exceed the length of the straightedge. Then place one end of the straightedge on a batter board. Check for exact levelness. See Figure 2-8. Drive another stake to the same height. Each time a stake is driven, the straightedge and level should be reversed end for end. This should ensure close accuracy in establishing the height of each stake with reference to the batter board. The final check on overall levelness comes when you level the last stake with the batter board where you began. If the straightedge is level here, then you have a level foundation baseline. During foundation excavation, the corner stakes and temporary leveling stakes will be removed. This stresses the importance of the level batter boards and string line. The corners and foundation levelness must be located using the string line.

• THE BUILDER'S LEVEL •

Practically all optical sighting and measuring instruments can be termed *surveying instruments*. Surveying, in its simplest form, simply means accurate measuring. Accurate measurements have been a construction requirement ever since humans started building things.

How Does It Work?

Even during the days of pyramid building, humans recognized the fact that the most accurate distant measurements were obtained with a perfectly straight line of sight. The basic principle of opera-

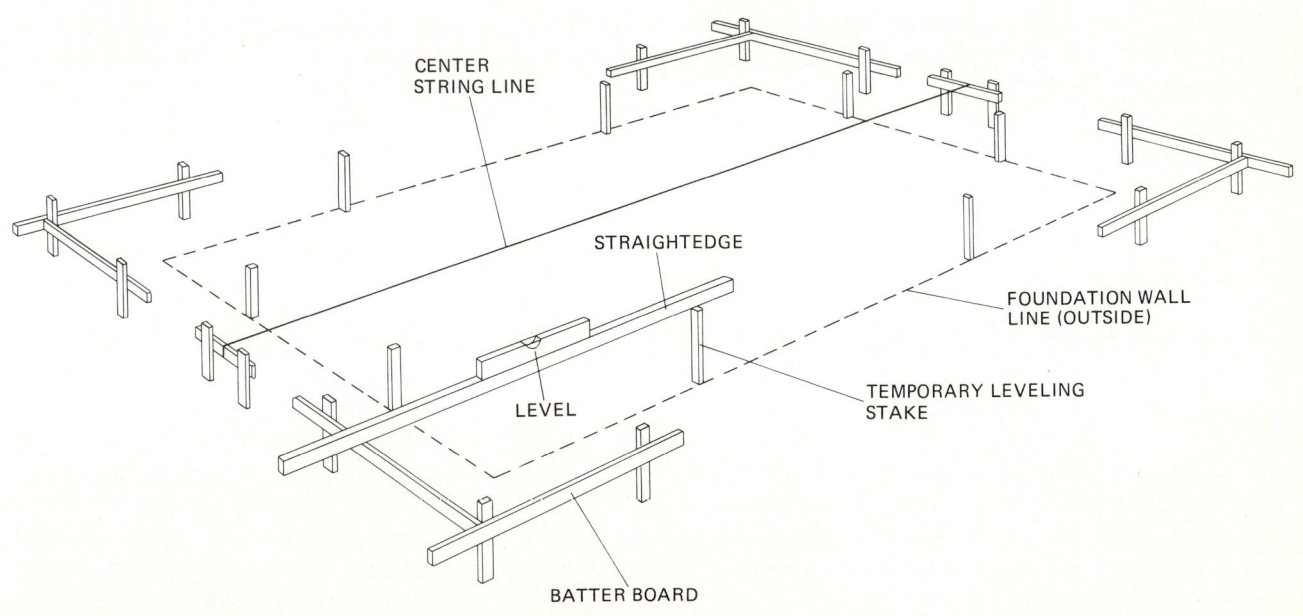

Figure 2-8. Leveling the batter boards. *(American Plywood Association)*

Preparing the Site 29

tion for today's modern instruments is still the same. A line of sight is a perfectly straight line. The line does not dip, sag, or curve. It is a line without weight and is continuous.

Any point along a level line of sight is exactly level with any other point along that line. The instrument itself is merely the device used to obtain this perfectly level line of sight for measurements.

THREE MAIN PARTS OF ANY OPTICAL BUILDER'S LEVEL •

1. The Telescope (Figure 2-9). The telescope is a precision-made optical sighting device. It has a set of carefully ground and polished lenses. They produce a clear, sharp magnified image. The magnification of a telescope is described as its power. An 18-power telescope will make a distant object appear 18 times closer than when viewed with the naked eye. Cross hairs in the telescope permit the object sighted on to be centered exactly in the field of view.
2. The Leveling Vial (Figure 2-10). Also called the "bubble," the leveling vial works just like the familiar carpenter's level. However, it is much more sensitive and accurate in this instrument. Four leveling screws on the instrument base permit the user to center (level) the vial bubble perfectly and thus establish a level line of sight through the telescope. A vital first step in instrument use is leveling. Instrument vials are available in various degrees of sensitivity. In general, the more sensitive the vial, the more precise the results that may be obtained.
3. The Circle (Figure 2-11). The perfectly flat plate upon which the telescope rests is called the circle. It is marked in degrees and can be rotated in any horizontal direction. With the use of an index pointer, any horizontal angle can be measured quickly. Most instruments have a vernier scale. An additional scale is subdivided. It divides degrees into minutes. There are 60 minutes in each degree. There are 360° in a circle.

The Builder's Level

Figure 2-12 shows a builder's level on site. Leveling the instrument is the most important operation in preparing the instrument for use.

LEVELING THE INSTRUMENT • First, secure the instrument to its tripod and proceed to level it as follows. Figure 2-13A shows the type of tripod used to support the instrument. The target pole is shown in Figure 2-13B.

Place the telescope directly over one pair of opposite leveling screws. (See Figure 2-14.) Turn the

Figure 2-9. The telescope on an optical level. *(David White Instruments)*

Figure 2-10. The leveling vial on an optical level. *(David White Instruments)*

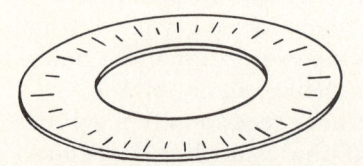

Figure 2-11. The circle on an optical level. *(David White Instruments)*

Figure 2-12. Using the optical (or builder's level) on the job. *(David White Instruments)*

screws directly under the scope in opposite directions at the same time (see step 5 in Figure 2-14) until the level-vial bubble is centered. The telescope is then given a quarter (90°) turn. Place it directly over the other pair of leveling screws (step 2 of Figure 2-14). The leveling operation is then repeated. Then recheck the other positions (steps 3 and 4 of Figure 2-14) and make adjustments. This may not be necessary. Adjust if necessary. When leveling is completed, it should be possible to turn the telescope in a complete circle without any changes in the position of the bubble. See Figure 2-15.

With the instrument leveled, you know that, since the line of sight is perfectly straight, any point on the line of sight will be exactly level with any other point. The drawing in Figure 2-16 shows how exactly you can check the difference in height (elevation) between two points. If the rod reading at B is 3 feet and the reading at C is 4 feet, you know that point B is 1 foot higher than point C. Use the same principle to check if a row of windows is straight or if a foundation is level. Or you can check how much a driveway slopes.

30 Carpentry Fundamentals

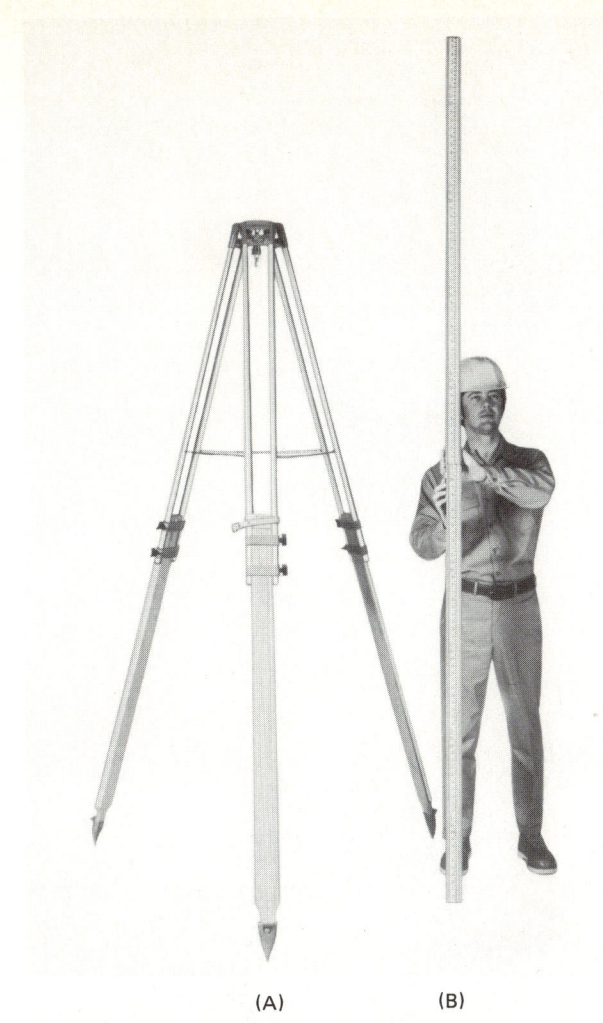

Figure 2-13. (A) The tripod for an optical level. (*David White*) (B) The rod holder for use with an optical level. (*David White*)

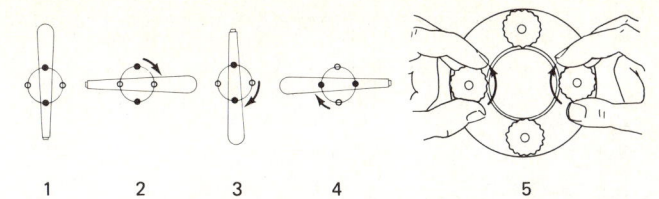

Figure 2-14. Adjusting the screws on the level-transit will level it. Note how it is leveled with two screws, then moved 90° and leveled again. (*David White Instruments*)

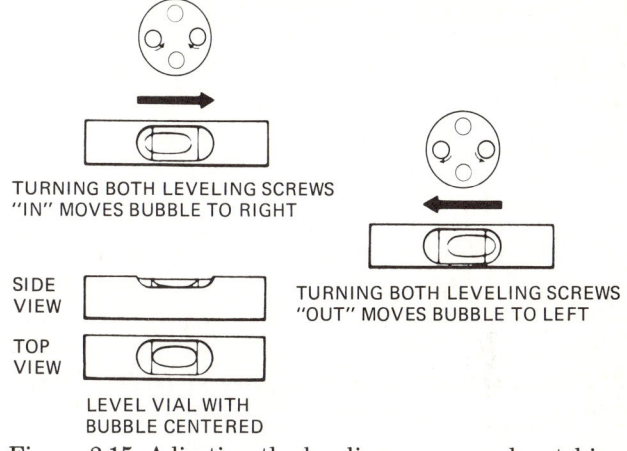

Figure 2-15. Adjusting the leveling screws and watching the bubbles for level. (*David White Instruments*)

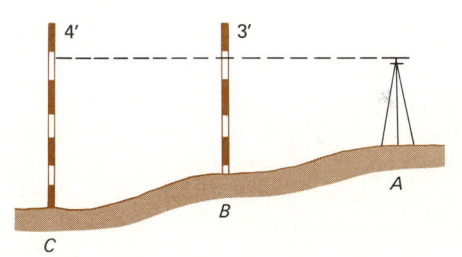

Figure 2-16. Finding the elevation with a level. (*David White Instruments*)

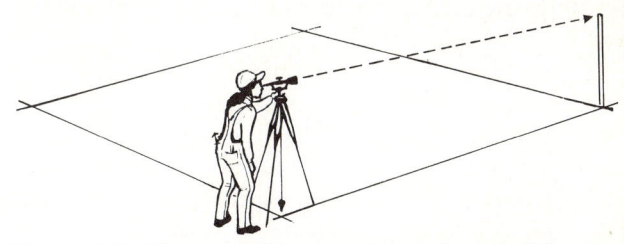

Figure 2-17. How to stake out a house on a building lot using a builder's level. (*David White Instruments*)

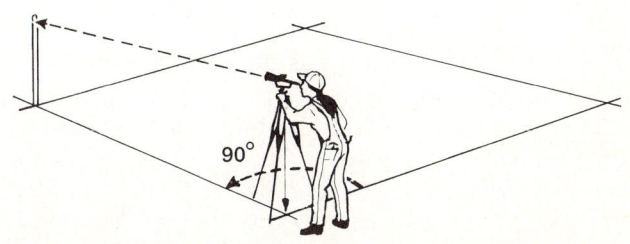

Figure 2-18. Squaring the other corner in laying out a building on a lot. (*David White Instruments*)

STAKING OUT A HOUSE • Start at a previously chosen corner to stake out the house. Sight along line *AB* of Figure 2-17 to establish the front of the house. Measure the desired distance to *B* and mark it with a stake.

Swing the telescope 90° by the circle scale. Mark the desired distance to *D*. This gives you the first corner. All the others are squared off in the same manner. You're sure all foundation corners are square, and all it took was a few minutes setup time. See Figure 2-18.

This method eliminates the use of the old-fashioned string line–tape–plumb bob methods.

The Level-Transit

There are two types of levels used for building sites. The level and the level-transit are the two instruments used. The level has the telescope in a fixed horizontal position but can move sideways 360° to measure horizontal angles. It is usually all that

Preparing the Site **31**

is needed at a building site for a house. See Figure 2-19.

A combination instrument is called a level-transit. The telescope can move in two directions. It can move up and down 45° as well as from side to side 360°. See Figure 2-20. It can measure vertical as well as horizontal angles.

A lock lever or levers permit the telescope to be securely locked in a true level position for use as a level. A full transit instrument, in addition to the features just mentioned, has a telescope that can rotate 360° vertically.

The level-transit is shown in operation in Figure 2-21.

Using the Level and Level-Transit

READING THE CIRCLE AND VERNIER • The 360° circle is divided in quadrants (0 to 90°). The circle is marked by degrees and numbered every 10°. See Figure 2-22.

To obtain degree readings it is only necessary to read the exact degree at the intersection of the zero index mark on the vernier and the degree mark on the circle (or on the vertical arc of the level-transit).

For more precise readings, the vernier scale is used. See Figure 2-23. The vernier lets you subdivide each whole degree on the circle into fractions, or minutes. There are 60 minutes in a degree. If the vernier zero does not line up exactly with a degree mark on the circle, note the last degree mark passed and, reading up the vernier scale, locate a vernier mark that coincides with a circle mark. This will indicate your reading in degrees and minutes.

HANGING THE PLUMB BOB • To hang the plumb bob, attach a cord to the plumb bob hook on the tripod. Knot the cord as shown in Figure 2-24.

If you are setting up over a point, attach the plumb bob. Move the tripod and instrument over the approximate point. Be sure the tripod is set up firmly again. Shift the instrument on the tripod head until the plumb bob is directly over the point. Then set the instrument leveling screws again to level the instrument.

POWER • The power of a telescope is rated in terms of magnification. It may be 24X or 37X. The 24X means the telescope is presenting a view 24 times as close as you could see it with the naked eye. Some instruments are equipped with a feature that lets you zoom in from 24X to 37X. It increases the effective reading range of the instrument more than 42 percent. It also permits greater flexibility in matching range, image, and light conditions. Use low power for brighter images in dim light. Since

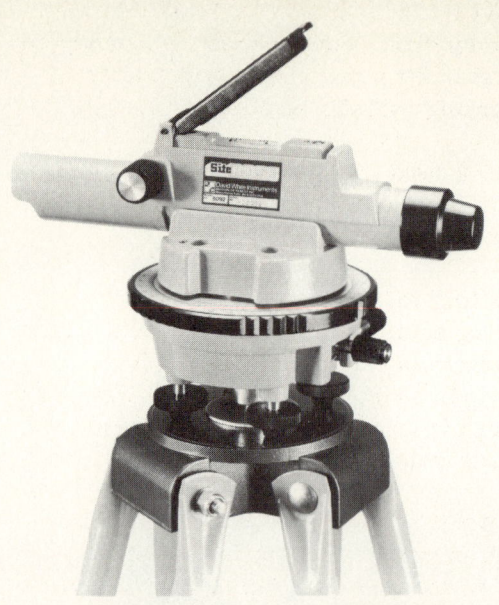

Figure 2-19. Builder's level. *(David White Instruments)*

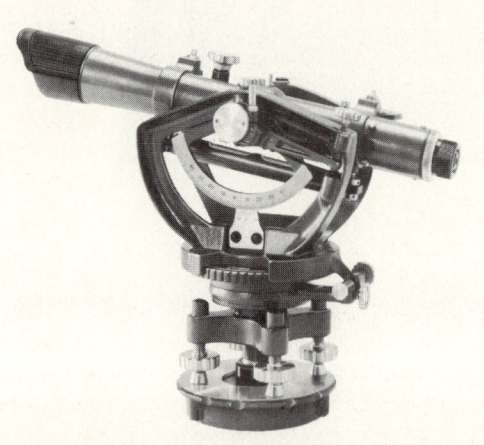

Figure 2-20. Level-transit. *(David White Instruments)*

Figure 2-21. Using the level-transit on the site. *(David White Instruments)*

32 Carpentry Fundamentals

it gives a wider field of view, it is also handy in locating targets. Low power also provides better visibility for sighting through heat waves. See Figure 2-25.

High power is used for sighting under bright light conditions. It is used for long-range sighting and for more precise rod readings.

ROD • Leveling rods are a necessary part of the transit leveling equipment. Rods are direct-reading with large graduations. All rods are equipped with a tough, permanent polyester film scale that will not shrink or expand. This is important when you consider the graduations can be 1/100th of a foot. Figure 2-26 shows a leveling rod with the target attached at 4′5¼″. The target (Figure 2-27) can be moved by releasing a small clamp in the back. Figure 2-28 shows a tape and the graduations. They are ⅛ inch wide and ⅛ inch apart. The tape is marked

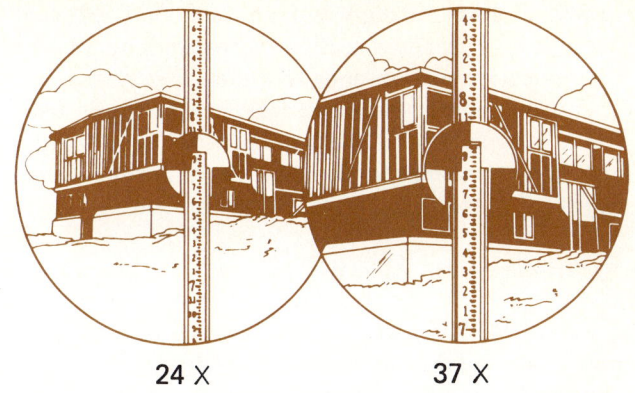

Figure 2-25. Variable instrument power is available. *(David White Instruments)*

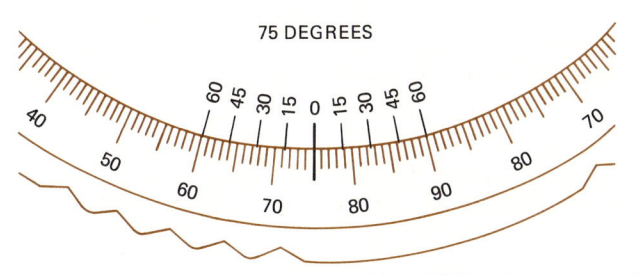

Figure 2-22. Reading the circle. *(David White Instruments)*

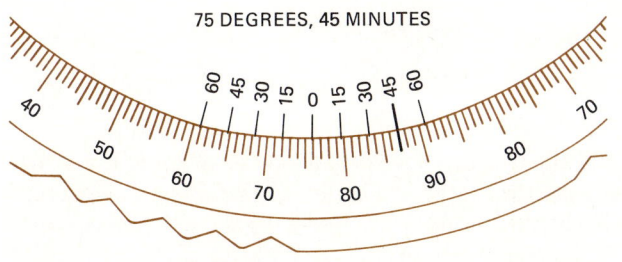

Figure 2-23. Reading the circle and vernier. *(David White Instruments)*

Figure 2-24. To hang the plumb bob, attach a cord to the plumb bob hook on the tripod and knot the cord as shown here. *(David White Instruments)*

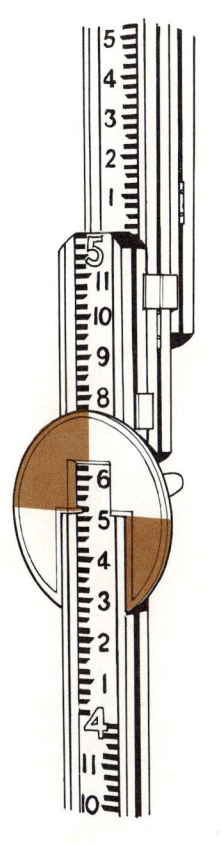

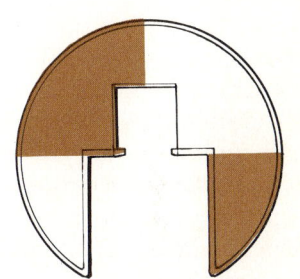

Figure 2-27. Target that fits wood rods. *(David White Instruments)*

Figure 2-26. Leveling rod made of wood. *(David White Instruments)*

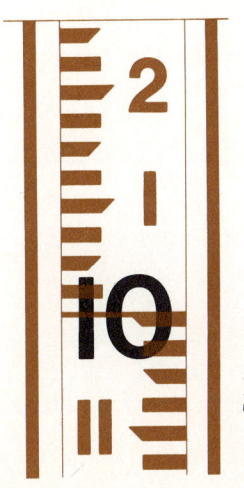

Figure 2-28. Tape face on the rod. This one is marked in feet, inches, and eighths. *(David White Instruments)*

Preparing the Site 33

in feet, inches, and eighths of an inch. Feet are numbered in red. A three-section rod extends to 12 feet. A two-section rod extends to 8 feet 2 inches.

The rod holder is directed by hand signals from the surveyor behind the transit. The hand signals are easy to understand, since they motion in the direction of desired movement of the rod.

Establishing Elevations

Not all lots are flat. That means there is some kind of slope to be considered when digging the basement or locating the house. The level can help establish what these elevation changes are. From the grade line you establish how much soil will have to be removed for a basement. The grade line will also determine the location of the floor.

The bench mark is the place to start. A bench mark is established by surveyors when they open a section to development. This point is a reference to which the lot you are using is tied. The lot is so many feet in a certain direction from a given bench mark.

The bench mark may appear as a mark or point on the foundation of a nearby building. Sometimes it is the nearby sidewalk, street, or curb that is used as the level reference point.

The grade line is established by the person who designed the building. This line must be accurately established. Many measurements are made from this line. It determines the amount of earth removed from the basement or for the foundation footings.

Using the Leveling Rod

Use a leveling rod and set it at any point you want to check the elevation. Sight through the level or transit-level to the leveling rod. Take a reading by using the cross hair in the telescope. Move the rod to another point that is to be established. Now raise or lower the rod until the reading is the same as for the first point. This means the bottom of the rod is at the same elevation as the original point.

One person will hold the rod level. Another will move the target up or down till the cross hair in the telescope comes in alignment with that on the target. The difference between the two readings tells you what the elevation is.

Figure 2-29 shows how the difference in elevation between two points that are not visible from a single point is determined.

If point Z cannot be seen from point W, then you have to set the transit up again at two other points, such as X and Y. Take the readings at each location; then you will be able to determine how much of the soil has to be removed for a basement.

• PREPARING THE SITE •

Clearing

One of the first things to do in preparing the site for construction is to clear the area where the building will be located. Look over the site. Determine if there are trees in the immediate area of the house. If so, mark the trees to be removed. This can be done with a spray can of paint. Put an X on those to be removed or a line around them. In some cases, people have marked those that must go with a piece of cloth tied to a limb.

Also make sure those that are staying are not damaged when the heavy equipment is brought onto the site. Scarring trees can cause them to die later. Covering them more than 12 inches will probably kill them also. You have to cut off a part of the tree top. This helps it survive the covering of the roots.

Don't dig the sewer trench or the water lines through the root system of the trees to be saved. This can cause the tops of the trees to die later, and in some cases will kill them altogether.

Make a rough drawing of the location of the house and the trees to be saved. Make sure the persons operating the bulldozer and digger are made aware of the effort to save trees.

CUTTING TREES • Keep in mind that removing trees can also be profitable. You can cut the trees into small logs for use in fireplaces. This has become

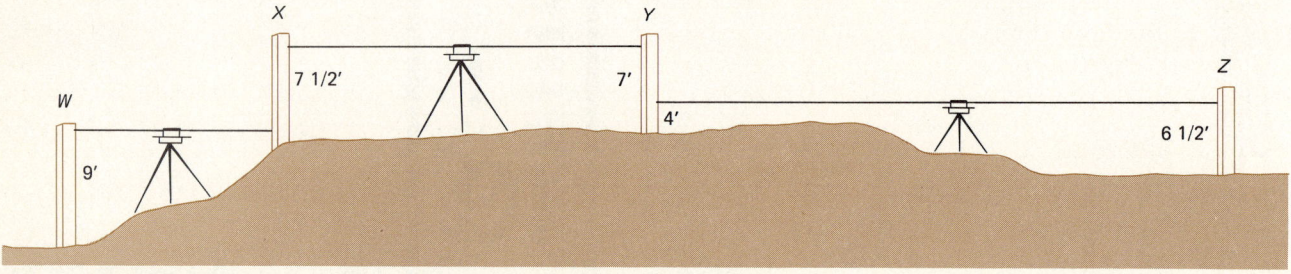

Figure 2-29. Getting the elevation when the two points cannot be viewed from a single point.

an interest of many energy conservationists. The brush and undergrowth can be removed with a bulldozer or other type of equipment. Do not burn the brush or the limbs without checking with local authorities. There is always someone who is interested in hauling off the accumulation of wood.

STUMP REMOVAL • In some cases a tree stump is left and must be removed. There are a number of ways of doing this. One is to use a winch and pull it up by hooking the winch to some type of power takeoff on a truck, tractor, or heavy equipment. You could dig it out, but this can take time and too much effort in most cases.

The use of explosives to remove the stump is not permitted in some locations. Better check with the local police before setting off the blast.

The best way is to use the bulldozer to uproot the entire stump or tree. It all depends upon the size of the tree and the size of the equipment available for the job. Anyway, be sure the lot is cleared so the digging of the basement or footings can take place.

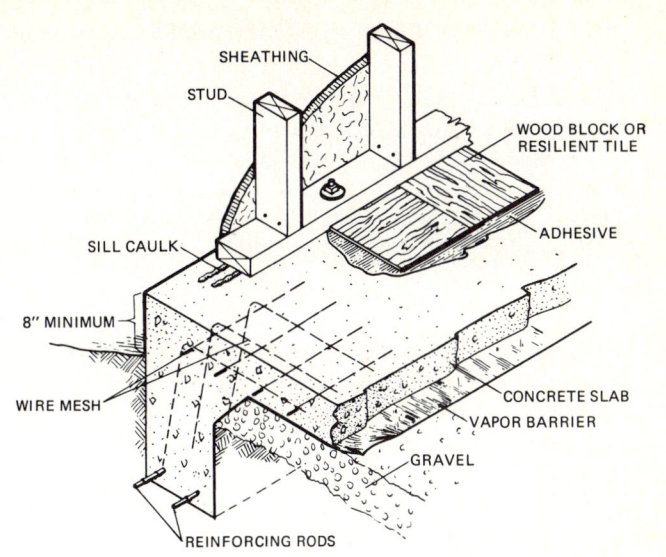

Figure 2-30. Single-slab foundation. *(Forest Products Lab)*

Excavating

A house built on a slab does not require any extensive excavating. Slabs are built in two basic ways. They are poured in one piece or two pieces. They are used on level ground and in warm climates. In cold climates the frost line (the depth to which the ground is frozen during the winter) penetrates deeper. This means another type of slab has to be used in colder areas. This is the two-piece slab. It is also best where much water has to be removed or drained. Take a look at Figures 2-30 and 2-31 for these slabs.

Slab footings must rest beneath the frost line. This gives stability in the soil. The amount of reinforcement needed for a slab varies. The condition of the soil and the weights to be carried determine the reinforcement. Larger slabs and those on less stable soil need more reinforcement.

The top of the slab must be 8 inches above ground. This allows moisture under the slab to drain away from the building. It also gives you a good chance to spot termites building their tunnels from the earth to the floor of the house. The slab should always rest slightly above the existing grade. This is to provide for runoff water during a rainstorm.

BASEMENTS • A basement is the area usually located underground. It provides most homes with a lot of storage. In some it is a place to do the laundry. It also serves as a place to locate the heating and cooling units. If a basement is desired, it must be dug before the house is started. The footings must

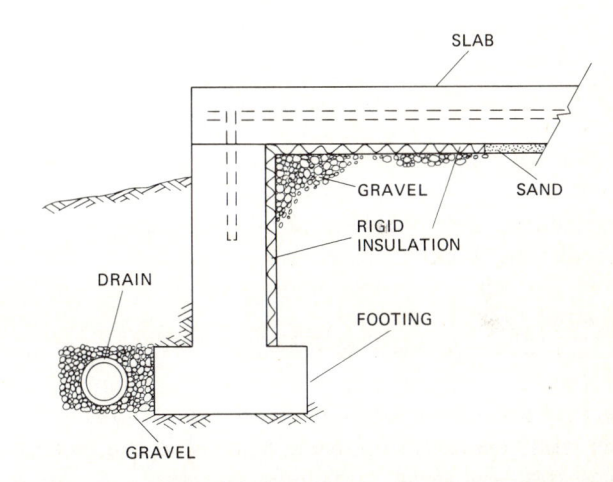

Figure 2-31. Two-slab foundation. *(Forest Products Lab)*

be properly poured and seasoned. Seasoning should be done before poured concrete or concrete block is used for the wall. Some areas now use treated-wood walls for a basement.

Figure 2-32 shows a basement dug for use in colder climates. Trenches from the street to the basement must be provided for the plumbing and water. Utilities may be buried also. If they are, the electric, phone, and gas lines must also be located in trenches or buried after the house is finished. It is a good idea to notify the utility companies so that they can schedule the installation of their services when you are ready for them.

Some shovel work may have to be done to dig the basement trenches for the sewer pipes. This is

Figure 2-32. Excavation for a basement.

done after the basement has been leveled by machines. As you can see from Figure 2-32, the basement may also need shovel work after the digger has left. Note the cave-ins and dirt slides evident in the basement excavation in Figure 2-32.

The basement has to be filled later. Gravel is used to form a base for the poured concrete floor.

The high spots in the basement must be removed by shovel. Proper-size gravel should be spread after the sewer trench is filled. You may have to tamp the gravel to make sure it is properly level and settled. Do this before the concrete mixer is called for the floor job.

The footings have to be poured first. They are boxed in and poured before anything is done in the way of the basement floor. In some instances drain tile must be installed inside or outside the footing. The tile is allowed to drain into a sump. In other locations no drainage is necessary because of soil conditions.

• PROVIDING ACCESS DURING CONSTRUCTION •

The first thing to be established is who is to be on the premises. Check with your insurance company about liability insurance. This is in case someone is hurt on the location. Also decide who should be kept out. You also have to decide how access control is to be set up. It may be done with a fence or by an alert watchman or dog. These things do have to be considered before the construction gets underway. If equipment is left at the site, who is responsible? Who will pay for vandalism? Who will repair damage caused by wind, hail, rain, lightning, or tornado?

Materials Storage

Where will materials for the job be stored? In Figure 2-33 you can see how plywood is stored. What happens if someone decides to haul off some of the plywood? Who is responsible? What control do you have over the stored materials after dark?

Figure 2-34 shows plywood bundles broken open. This makes it easy for single sheets to disappear. With the current price of plywood, it becomes important to plan some type of storage facility on the site.

Some of the shingles in Figure 2-35 may be hard to find if the wind gets to the broken bundle. What's to stop children playing on the site during off hours? They can also take the shingles and spread them over the landscape.

Storage of bricks can be a problem. See Figure 2-36. They are expensive and can easily be removed by someone with a small truck. It is very important to have some type of on-site storage. It is also very important to make sure that materials are not delivered before they are needed. Some type of materials inventory has to be maintained. This may be worth a person's time. The location of the site is a major factor in the disappearance of materials. Location has a lot to do with the liability coverage needed from insurance companies.

TEMPORARY BUILDINGS • Some building sites have temporary structures to use as storage. In some cases the plans for the building are also stored in the tool shed. Covered storage is used in some locations where rain and snow can cause a delay by wetting the lumber or sand or cement. If you are using drywall, you will need to keep it dry. In most instances it is not delivered to the site until the house is enclosed.

Some construction shacks are made on the scene. In other cases the construction shack may be delivered to the site on a truck. It is picked up and moved away once the building can be locked.

Mobile homes have been used as offices for supervisors. This usually is the case when a number of houses are made by one contractor and all are located in one row or subdivision.

Figure 2-33. Storing plywood on the site.

Figure 2-35. Broken bundle of shingles.

Figure 2-36. Storing bricks on site.

Figure 2-34. Broken bundle of plywood sheathing.

Preparing the Site 37

A garage can be used as the headquarters for the construction. The garage is enclosed. It is closed off by doors, so it can be locked at night. Since the garage is easy to close off, it becomes the logical place to take care of paper work. It also becomes a place to store materials that should not get wet.

When building a smaller house, the carpenter takes everything home at the end of the working day. The carpenter's car or truck becomes the working office away from home. Materials are scheduled for delivery only when actually needed. In larger projects some local office is needed, so the garage, tool shed, mobile home, or construction shack is used.

STORING CONSTRUCTION MATERIALS • Storing construction materials can be a problem. It requires a great deal of effort to make sure the materials are on the job when needed. If delivery schedules are delayed, work has to stop. This puts people out of work.

If materials are stored on the job, make sure they are neatly arranged. This prevents accidents such as tripping over scattered materials. Sand should be delivered and placed out of the way. Keep it out of the normal traffic flow from the street to the building.

Everything should be kept in some order. This means you know where things are when you need them. Then you don't have to plow through a mound of supplies just to find a box of nails. Everything should be laid out according to its intended order of use.

Lumber should be kept flat. This prevents warpage, cupping, and twisting. Plywood should be protected from rain and snow if it is interior grade. In any case it should not be allowed to become soaked. Keep it flat and covered.

Humidity control is important inside a house. This is especially true when you're working with drywall. It should be allowed to dry by keeping the windows open. Too much humidity can cause the wood to react and twist or warp.

Temporary Utilities

You will need electricity to operate the power tools. Power can be obtained by using a long extension cord from the house nearby. Or, you may have to arrange for the power company to extend a line to the side and put in a meter on a pole nearby.

Water is needed to mix mortar. The local line will have to be tapped, or you may have to dig the well before you start construction. It all depends upon where the building site is located. If the house is being built near another, you may want to arrange with the neighbors to supply water with a hose to their outside faucet. Make sure you arrange to pay for the service.

Waste Disposal

Every building site has waste. It may be human waste or paper and building-material wrappings. Human waste can be controlled by renting a Porta John or a Johnny-on-the-Spot. This can prevent the house from smelling like a urinal when you enter. The sump basin should not be used as a urinal. It does lend an odor to the place. Besides, it is unsanitary.

Waste paper can be burned in some localities. In others burning is strictly forbidden. You should check before you arrange to have a large bonfire for getting rid of the trash, cut lumber ends, paper, and loose shingles. There are companies that provide a trash-collecting service for construction areas. They leave the place *broom clean*. It leaves a better impression of the contractor when a building is delivered in order, without trash and wood pieces lying around. If you go to the trouble of building a fine home for someone, the least you can do is deliver it in a clean condition. After all, this is going to be a home.

Arranging Delivery Routes

Damage to the construction site by delivery trucks can cause problems later. You should arrange a driveway by putting in gravel at the planned location of the drive. Get permission and remove the curb at the entrance to the driveway. Make sure deliveries are made by this route. Pile the materials so that they are arranged in an orderly manner and can be reached when needed.

Concrete has to be delivered to the site for the basement, foundation, and garage floor. Be sure to allow room for the ready-mix truck to get to these locations. Lumber is usually strapped together. Make allowance for bundles of lumber to be dumped near the location where they will be needed.

Make sure the nearby plants or trees are protected. This may require a fence or stakes. Some method should be devised to keep the trucks, diggers, and earth movers from destroying natural vegetation.

Access to the building site is important. If this is the first house in the subdivision, or if it is located off the road, you have to provide for delivery of materials. You may have to put in a temporary road. This should be a road that can be traveled in wet weather without the delivery trucks becoming bogged down or stuck.

38 Carpentry Fundamentals

As you can see, it takes much planning to accomplish a building program that will come off smoothly. The more planning you do ahead of time, the less time will be spent trying to obtain the correct permissions and deliveries.

The key to a successful building program is planning. Make a checklist of the items that need attention beforehand. Use this checklist to keep yourself current with the delivery of materials and permissions.

It is assumed here that the proper financial arrangements have been made before construction begins.

• WORDS CARPENTERS USE •

foundation	perimeter	builder's level	excavation	site
footings	vernier	surveying	stump	utilities
diagonals	level	level-transit	access	
batter boards	plumb bob	elevation	grade	

• STUDY QUESTIONS •

2-1. What is a builder's level?

2-2. What is the difference between a builder's level and a level-transit?

2-3. Why is it important to have a clear title to a piece of land before building?

2-4. Who establishes the correct property limits?

2-5. What is an easement?

2-6. Why should a house be laid out square and level?

2-7. How do you set up the builder's level?

2-8. What is an elevation?

2-9. What is a vernier?

2-10. What is a plumb bob?

2-11. How is a plumb bob used by a carpenter?

2-12. How is the power of a telescope on a builder's level rated?

2-13. What is a leveling rod?

2-14. What is a target on a leveling rod?

2-15. What methods are used to remove stumps from a building site?

2-16. What does excavate mean?

2-17. Who has access to a house under normal conditions?

2-18. Why is some shovel work needed after the digger has excavated the basement?

2-19. What's the purpose of temporary buildings at a house construction site?

2-20. Why would you want to arrange delivery routes to a building site?

3
LAYING FOOTINGS AND FOUNDATIONS

People often think that footings and foundations are the same. Actually, the footing is the lowest part of the building and carries the weight. The foundation is the wall between the footing and the rest of the building. In this chapter you will learn how to:

- Design footings and foundations
- Locate corners and lines for forms
- Check the level of footing and foundation excavation
- Make the forms for footings
- Make the forms for foundations
- Reinforce the forms as required
- Mix or select concrete for usage
- Pour the concrete into the forms
- Finish concrete in the forms
- Embed anchor systems in forms
- Waterproof foundation walls if needed
- Make necessary drainage systems

• **INTRODUCTION** •

Footings bear the weight of the building. They spread the weight evenly over a wide surface. Figure 3-1 shows the three parts of a footing system. These parts are the bearing surface, the footing, and the foundation. The bearing surface must be located beneath the frost line on firm and solid ground. The frost line is the deepest level to which the ground will freeze in the wintertime. Moisture in the ground above the frost line will freeze and thaw. When it does, the ground moves and shifts. The movement will break or damage the footing or foundation. The location and construction of the footing is very important. Think of the weight of all the lumber, concrete, stone, and furniture that must be supported by this layer. All of these must be supported without sinking or moving.

Footings may be made in several ways. There are flat footings, stepped footings, pillared footings, and pile footings. The flat footing, as in Figure 3-2, is the easiest and simplest footing to make because it is all on one level. The stepped footing is used on sides of hills as in Figure 3-3. The stepped footing is like a series of short flat footings at different levels, much like a flight of steps. By making this type of footing, no special digging (excavation) is needed. The third footing type is the pillared footing. See Figure 3-4. The pillared footing is used in many locations where the soil is evenly packed and little settling occurs. It consists of a series of pads, or feet. Columns are then built on the pads and the building rests upon the columns. Buildings with either flat or stepped footings also usually have pillared footings in the center. This is because buildings are too wide to support the full weight without support in the middle areas.

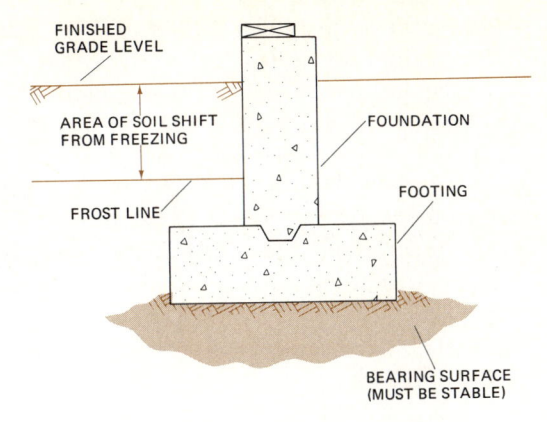

Figure 3-1. Parts of footing and foundation.

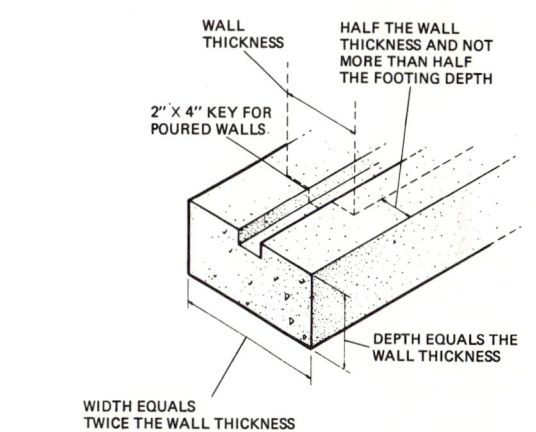

Figure 3-2. Regular flat footing. *(Forest Products Lab)*

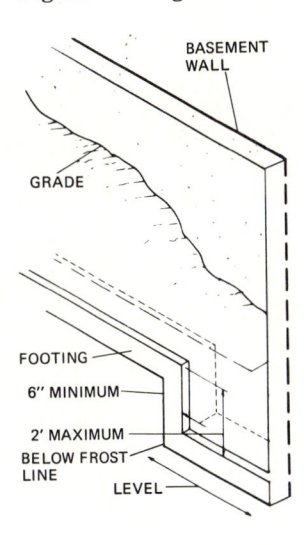

Figure 3-3. A stepped footing is used on hills or slopes. *(Forest Products Lab)*

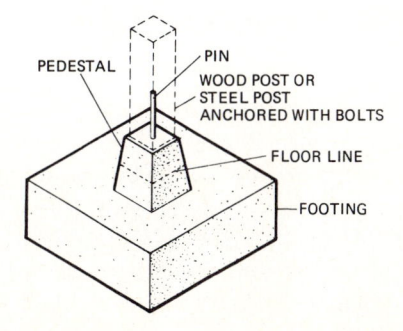

Figure 3-4. The pillar or post footing may be square or round. *(Forest Products Lab)*

Laying Footings and Foundations 41

Pile footings are the fourth type and are used where soil is loose, unstable, or very wet. As in Figure 3-5, long columns are put into the ground. These are long enough to reach solid soil. The columns may be made of treated wood or concrete. The wooden piles are driven into the ground by pounding. The concrete piles are made by drilling a hole and filling it with concrete. Pads or caps are then put over the tops of the piles.

• SEQUENCE FOR MAKING FOOTINGS AND FOUNDATIONS •

No matter what type of footing and foundation is used, a certain sequence should be followed. The sequence can change slightly according to the method involved. For example, the footing and foundation can be poured in one solid concrete piece. However, many footings are made separately and the concrete foundations are built on top of the footings. The sequence is slightly different but similar in both cases. The basic sequence is:

Find the amount of site preparation needed
Lay out footing and foundation shape
Excavate to proper depth
Level the footing corners
Build the footing forms
Reinforce the forms as needed
Estimate concrete needs
Pour the concrete footing
Build the foundation forms
Reinforce the forms as needed
Pour the concrete into forms
Finish the concrete and embed anchors
Remove the forms
Waterproof and drain as required

• LAYING OUT THE FOOTINGS •

Footings are the bottom of the building and must hold up the weight of the building. Two factors are involved in finding the correct shape and size. The first is the strength or solidness of the soil. The second factor is the width and depth of the footing for the weight of the building in that type of soil.

Soil Strength

Soil strength refers to how dense and solid the soil is packed. It also refers to how stable or unmoving the soil is. Some soils are very hard only when dry. Others keep the same strength whether they are wet or dry. In any condition, the soil must be dense and strong enough to support the weight of the building. When soil is soft, the footing is made wider to spread the weight over more surface. In this way, each surface unit holds up less weight. Figure 3-6 shows how much weight various soil types will support. Standard footings should not be poured on loose soil.

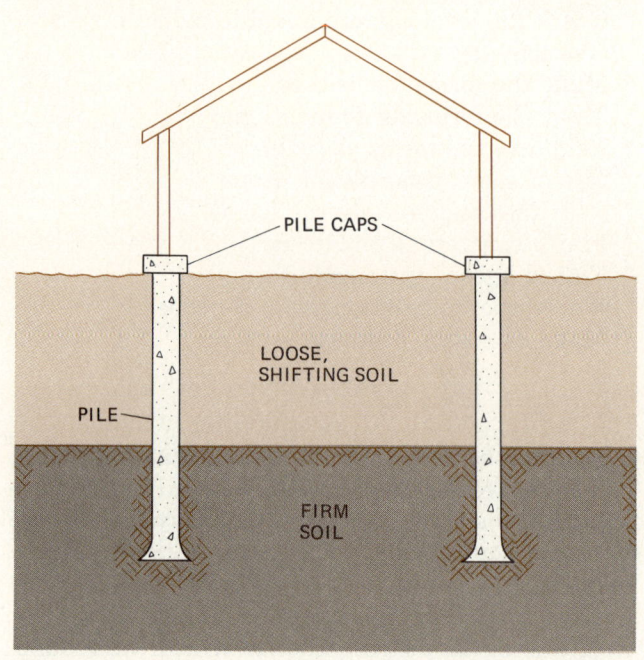

Figure 3-5. Pile footings reach through water or shifting soils.

BEARING CAPACITY OF TYPICAL SOILS

Type of Soil	Bearing Capacity (pounds per ft^2)
Soft Clay loose dirt, etc.	2 000
Loose Sand hard clay, etc.	4 000
Hard Sand or Gravel	6 000
Partially Cemented Sand or Gravel soft stone, etc.	20 000

Figure 3-6. Bearing capacity of typical soils.

Footing Width

The second factor is the width of the footing. As mentioned, the footing should be wider for soft soil. Figure 3-7 shows typical sizes for footings. As a rule, footings are about two times as wide as they are thick. The average footing is about 8 inches thick, and the footing is about the same thickness as the foundation wall.

Locating Footing Depth

Footings are laid out several inches below the frost line. For buildings with basements, place the top of the footing 12 inches below the frost line. For buildings that do not have basements, 4 to 6 inches below the frost line could be deep enough. Local building codes may give exact details.

Footings Under Columns

The footings and foundations that most people see support only the outside walls. But today most houses are wide, and support is needed in the center of a wide building. This support is from footings, pillars, or columns built in the center. Pillars or columns must have a footing just as the outside walls do. For houses with basements, the footings and pillars become part of the basement floor and walls. See Figure 3-8. Many houses do not have basements. Instead, they have a space between the ground and the floor called a *crawl space*. This crawl space provides access to the pipes and utilities. Pillars or columns built on footings are used for supports in the crawl spaces. The footings may be any shape, square, rectangular, or round. Figure 3-9 shows a site prepared in this manner.

TYPICAL FOOTING SIZE

FLOORS	BASE-MENT	ALL WOOD FRAME		WOOD FRAME WITH MASONRY VENEER	
		I	*P*	*I*	*P*
1	None	6″	3″	6″	4″
1	Yes	6″	2″	6″	3″
2	None	6″	3″	6″	4″
2	Yes	6″	4″	8″	5″

NOTE: For soil with 2000 pounds per square foot (PSF) load capacity.

Figure 3-7. Typical footing size.

Figure 3-8. Footings in a basement later become a part of the basement floor.

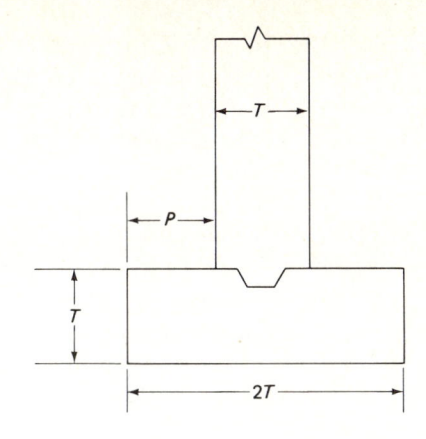

Figure 3-9. Footings and piers must be located in crawl spaces.

Laying Footings and Foundations 43

Footings for either basements or crawl spaces are all similar. They should be below the frost line as in a regular footing. However, they carry a greater weight than do the outside footings. For this reason, they should be 2 to 3 feet square.

Special Strength Needs

Footings for heavier areas of a building such as chimneys, fireplaces, bases for special machinery, and other similar things should be wider and thicker. For chimneys in a one-story building, the footing should project at least 4 inches on each side. The chimneys on two-story buildings are taller and heavier. Therefore, the footing should project 6 to 8 inches on each side of the chimney. Figure 3-10 shows a foundation for a fireplace.

Reinforcement and Strength

Two things are done to the footing to make it stronger. First, it is reinforced with steel rods. Then, the footing is also matched or keyed so that the foundation wall will not shift or slide.

REINFORCEMENT • In most cases the footing should be reinforced with steel rods. These reinforcement rods are called *rebar*. Two or more pieces of rebar are used. The rebar should be located so that at least 3 inches of space for concrete is left around all edges. See Figure 3-11.

KEYED FOOTINGS • The best type of separate footing is *keyed*, as shown in Figure 3-11. This means that the footing has a key or slot formed in the top. The slot is filled when the foundation is formed. The key keeps the foundation from sliding or moving off the footing. Without a key, freezing and thawing of water in the ground could force foundation walls off the footing.

• LAYING OUT THE FOOTINGS •

The procedure for locating the building on the lot was explained in Chapter 2. Batter boards were put up and lines were strung from them to show the location of the walls and corners.

Now the size, shape, and depth of footings must be decided.

Finding Trench Depth

Trenches or ditches must be dug for the footings and foundation. This is called *excavation*. Ground that is extremely rough and uneven should be rough-graded before the excavation is begun. The topsoil that is removed can be piled at one edge of the building site. It should be used later when the ground is smoothed and graded around the building. Before the digging is started, determine how deep it is to be.

The trench at the lowest part of the site must be deep enough for the footing to be below the frost line. If the footing is to be 12 inches below the frost line, the trench at the lowest part must be deep enough for this. Figure 3-12 shows these depths. This lowest point becomes the level line for the entire footing. Elevations are taken at each corner to find out how deep the trenches are at each corner.

Excavating for Deep Footings

Footings must be deep in areas where the frost line is deep. Deep footings are also needed when a basement will be dug.

Figure 3-10. A special footing is used for fireplaces. It supports the extra weight.

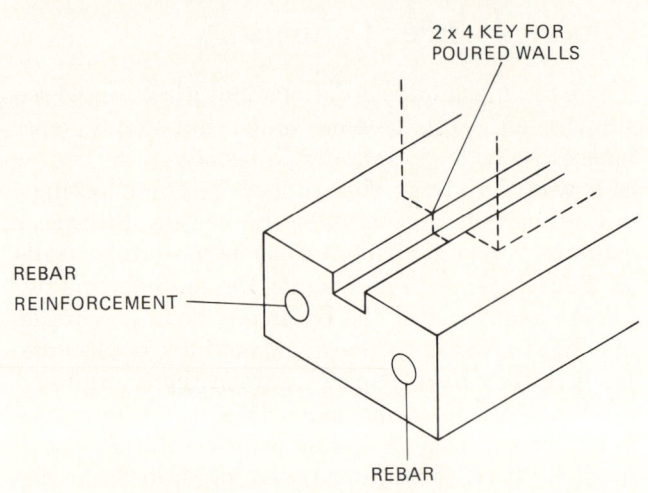

Figure 3-11. Footings may be reinforced. Note the key to keep the foundation from shifting.

44 Carpentry Fundamentals

Rough lines are drawn on the ground. They do not need to be very accurate, but the lines from the batter boards are used as guides. However, the rough line should be about 2 feet outside the line. See Figure 3-13. The trench for the footing is dug much wider than the footing so that there is room to work. Since the footings are made of concrete, the molds (called *forms*) for the concrete must be built. Room is needed to build or put up the forms. Work that must be done after the footing or foundation is formed includes removing the forms, waterproofing the walls, and making proper drainage.

As the trench is dug, the depth is measured. When the trench has been dug to the correct depth, the machinery is removed. The forms are then laid out. For basements, the interior ground is also excavated.

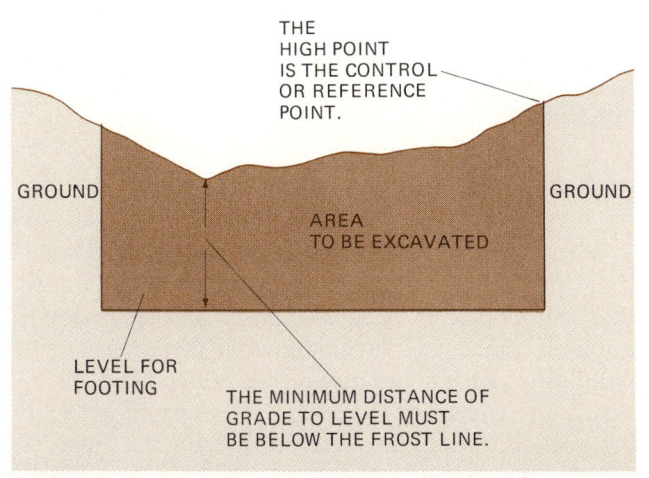

Figure 3-12. Footings must be below the frost line.

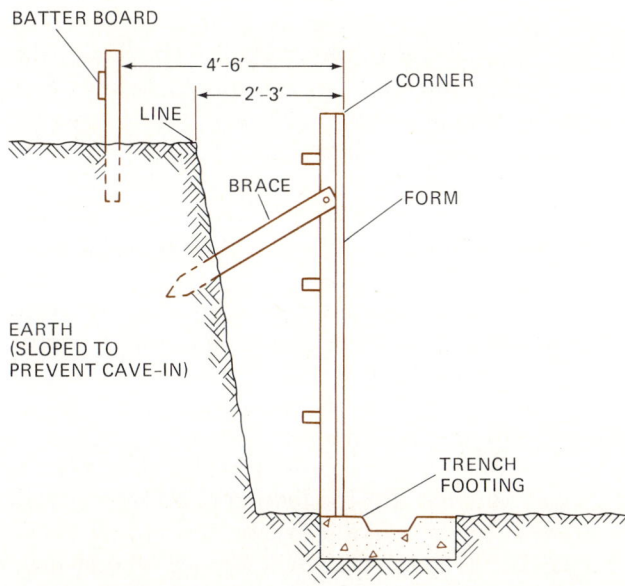

Figure 3-13. The trench is wider than the footing and sloped for safety.

Excavating for Shallow Footings

Rough lines are marked on the ground with chalk or shovel lines. These lines should be marked to show the width of the footing desired. Corner stakes are removed, and lines are taken from the batter boards. A trench is excavated to the correct depth.

The special footings for the interior are also excavated at this point. The excavation for the interior pad footings should be made to the same depth as that for the outside walls.

The concrete is poured directly into these trenches. Any reinforcement is made without forms in the excavation itself. The rebar is suspended with rebar stakes or metal supports called *chairs*. See Figure 3-14.

To form a key in this type of footing, stakes are driven along the edges as in Figure 3-15. The board that forms the key in the footing is suspended in the center of the trench area.

In many areas, concrete block wall is used on this type of footing. The blocks may be secured by inserting rebars into the footing area. The bricks or blocks are laid so that the rebar is centered in an opening in the block. The opening is then filled with mortar or concrete to secure the foundation against slipping.

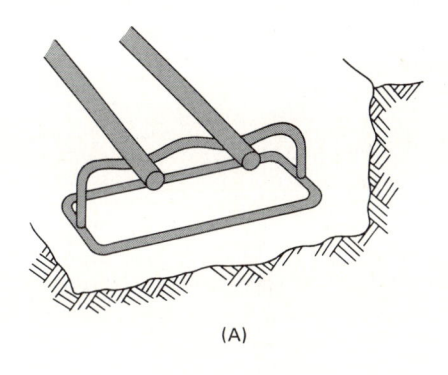

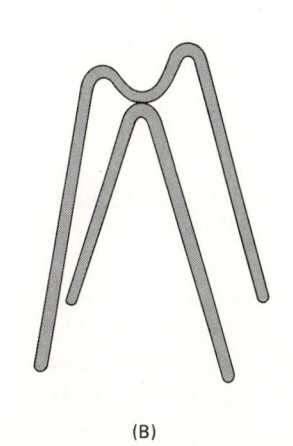

Figure 3-14. Chairs are used to hold rebar in place. *(Richmond Screw Anchor)*

Laying Footings and Foundations **45**

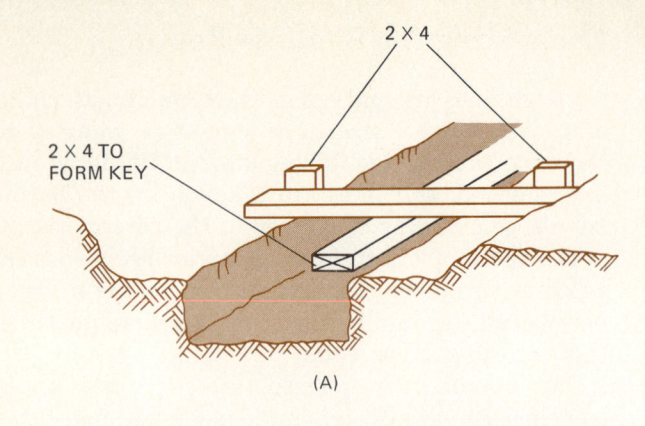

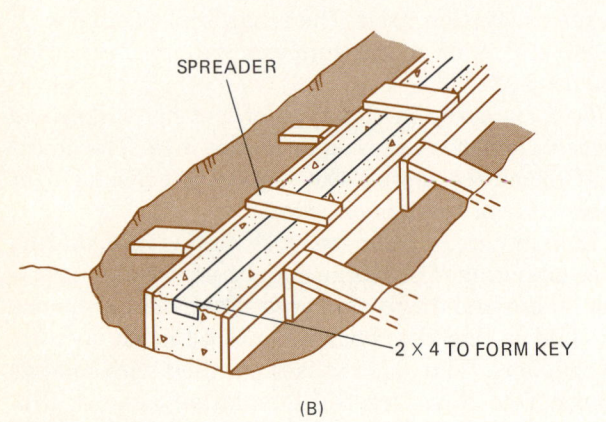

Figure 3-15. Keys are made by suspending 2 x 4s in the form. (A) Keys for trench forms and (B) Keys for board forms.

The important thing to remember is that special forms are not used with shallow footings. Also, they may not be finished smooth. As a result they may appear very rough or unfinished. This is not important if they are the proper shape.

Slab Footings and Basements

Slab footings are used in areas where concrete floors are made. Slabs can combine the concrete floor and the footing as one unit. Slabs, basement floors, and other large concrete surfaces are detailed in another chapter. Basement floors are made separately from the footings, and are done after the footings and basement walls are up.

• BUILDING THE FORMS FOR THE FOOTINGS •

After the excavation has been completed, the corners must be relocated. After the corners are relocated, the forms are built and leveled and the concrete is poured and allowed to harden. Then the forms are removed, and the foundation is erected. In many cases, the footing and the foundation are made as one piece.

Laying Out the Forms

After the excavation is complete, the first step is to relocate the corners and edges for the walls. To do this, the lines from the batter boards are restrung and a plumb bob is used to locate the corner points. The corner points and other reference points are marked with a stake. The stake is driven level for the top of the footing. This level is established by using a transit or a level. Refer to Chapter 2 for this procedure.

Nails

It is best to use double-headed, or duplex, nails for making the forms. Forms should be nailed with the nails on the outside. This means that the nails are not in the space where the concrete will be. This way the nail head does not get embedded in the concrete and is left exposed. The double head lets the nail be driven up tight and still be easy to pull out when the forms are taken apart.

Putting Up the Forms

With the corner stakes used for location and level, the walls of the forms are constructed. The amount that the footing is to project past the wall is determined. Usually this is one-half the thickness of the foundation wall. This dimension is needed because the corner indicates foundation corner and not footing corner. Stakes are driven outside the lines so that the form will be the proper width. The carpenter must allow for the width of the stake and the width of the boards used for the forms. See Figure 3-16. Drive stakes as needed for support. As a rule, the distance between stakes is about twice the width of the footing. Nail the top board to the first stake and level the top board in two directions. For first direction, the top board is leveled with the corner stake. For the second direction, the top board is leveled on its length. See Figure 3-17. After the top board is leveled, nail it to all stakes. Then nail the lower boards to the stakes. Both inside and outside forms are made this way.

If 1-inch-thick boards are used to build the form, stakes should be driven closer together. If boards 2 inches thick are used, the stakes may be 4 to 6 feet apart. In both cases, the stakes are braced as shown in Figure 3-18.

Loose dirt should be removed from under the footing form. It is best for the footing to be deeper than is needed. Never make a footing thinner than the specifications. Never fill any irregular hole or area with loose dirt. Always fill with gravel or coarse sand and tamp it firmly in place.

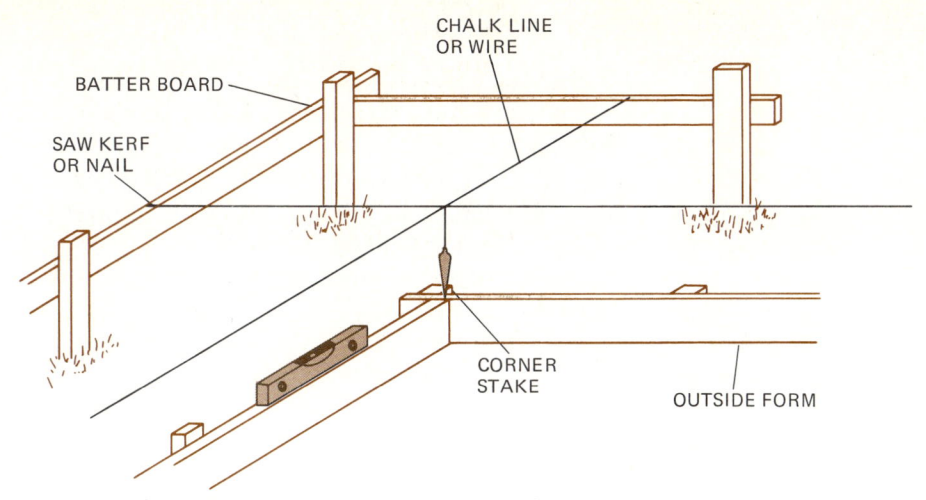

Figure 3-16. The footing corner is located and leveled.

THE KEYED NOTCH • The key or slot in the footing is made with a board. The board is nailed to a brace that reaches across the top of the forms. The brace should be nailed in place at intervals of about 4 feet apart. Refer to Figure 3-15 to see how the key is made.

EXCAVATION FOR DRAINS AND UTILITY LINES • Drain pipes and utility lines are sometimes located beneath the footings in a building. When this is done, trenches are dug underneath the footing forms. These trenches are usually dug by hand underneath the forms. After the drain pipes or utility lines are laid in place, the area is filled with coarse gravel or sand. This gravel or sand is tightly packed in place beneath the form.

SPACING THE WALLS OF THE FORM • The weight of the concrete can make the walls spread apart. To keep the walls straight, braces are used. The braces on the walls provide much support. Special braces called *spreaders* are also nailed across the top. Forms should be braced properly so that the amount of concrete ordered will fill the forms properly. Also, this practice ensures that excess concrete does not add extra weight to the building.

The forms should also be checked to make sure that there are no holes, gaps, or weak areas. These could let the concrete leak out of the form and thus weaken the structure. These leaks are called *blowouts*.

• WORKING WITH CONCRETE •

Before the concrete is ordered and poured, several things are done. The forms should be checked for the proper depth and level. Openings and trenches beneath the footing area for pipes and utility lines are made. These should be properly leveled and

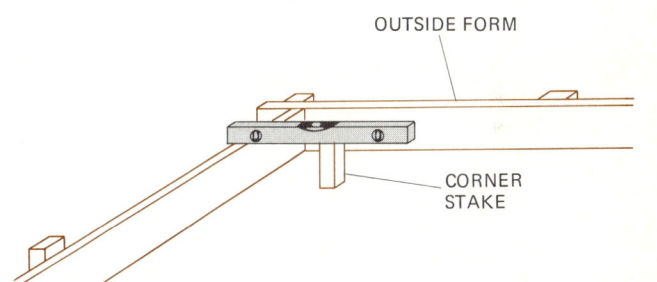

Figure 3-17. The form is leveled all around.

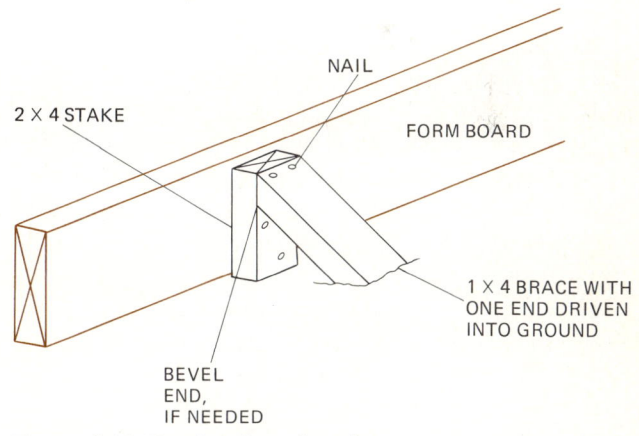

Figure 3-18. Bracing form boards.

filled. The forms should be checked to make sure that they are properly braced and spaced. Finally, chalk lines and corner stakes should be removed from the forms.

Reinforcement

In most cases, reinforcement rods (rebar) are placed in the footing after the forms are finished. The amount of reinforcement is usually given in the plans. As it is laid, the rebar is tied in place.

Laying Footings and Foundations 47

Soft metal wires, called *ties,* are twisted around the rebars. The carpenter must be sure that the footing conforms to the local building codes.

Specifying Concrete

Most concrete used today is made from cement, sand, and gravel mixed with water. The cement is the "glue" that hardens and holds or binds the materials. Most cement used today is *portland* cement. It is made from limestone that is heated, powdered, and mixed with certain minerals. When mixed with *aggregates,* or sand and gravel, it becomes concrete.

Concrete mixes can be denoted by three numbers, such as 1–2–3. This is the volume proportion of cement, sand, and gravel. 1–2–3 is the basic mix, but it is varied for strength, hardening speed, or other factors. However, it is recommended that concrete be specified by the water-to-cement ratio, aggregate size, and bags of concrete per cubic yard. See Table 3-1.

Most concrete today is delivered to the building site. Usually, the concrete is not mixed by the carpenters. It is delivered by concrete trucks from a concrete company. Figure 3-19 shows a transit-mix truck. The concrete is sold in units of cubic yards. The carpenter may need to make the order for concrete to the concrete company. To do so, the carpenter must be able to figure how much concrete to order.

Estimating Concrete Needs

A formula is used to estimate the volume of concrete needed. The basic unit for concrete is the cubic yard. A cubic yard is made up of 27 cubic feet (3×3×3). To convert footing sizes, use the formula:

$$\frac{L'}{3} \times \frac{W''}{36} \times \frac{T''}{36} = \text{cubic yards}$$

where L' = length in feet
W'' = width in inches
T'' = thickness in inches

Example. A footing is 18 inches wide and 8 inches thick. It must support a building 48 feet long and 24 feet wide. The distance around the edges is called the perimeter. The perimeter is $(2 \times 48) + (2 \times 24)$, or 144 feet. This would be:

$$\frac{L'}{3} \times \frac{W''}{36} \times \frac{T''}{36} = \text{cubic yards}$$

$$= \frac{144}{3} \times \frac{18}{36} \times \frac{8}{36}$$

and by cancellation,

$$\frac{48}{1} \times \frac{1}{2} \times \frac{2}{9} = \frac{96}{18} = 5.33 \text{ cubic yards}$$

Figure 3-19. A transit-mix truck delivers concrete to a site. *(Portland Cement)*

TABLE 3-1 CONCRETE USE CHART

Uses	Concrete, Bags per Cubic Yard	Sand, Pounds per Bag of Concrete	Gravel, Pounds per Bag of Concrete	Gravel Size, Average Diameter in Inches	Water, Gallons per Bag of Concrete	Consistency Slump
Footings, basement walls (8-inch), or foundation walls (8-inch thickness)	5.0	265	395	1½"	7	4–6 inches
Slabs, basement floors, sidewalks, etc. (4-inch thickness)	6.2	215	295	1"	6	4–6 inches
Basic 1–2–3 mixture (approximation only)	6.0	190	275	2"	5.5	2–4 inches

NOTES: 1. All figures are for slight to moderate ground water and medium-fineness sand.
2. All figures vary slightly.

The minimum amount that can be ordered is one cubic yard. After the first cubic yard, fractions can be ordered. The estimate is 5⅓ cubic yards. Often a little more is ordered to make sure enough is delivered.

Pouring the Concrete

To be ready, the carpenter sees to two things. First the forms must be done. Then the concrete truck must have a close access. The driver can move the spout to cover some distance. However, it may be necessary to carry the concrete an added distance. This can be done by pumping the concrete or by carrying it. Wheelbarrows, as in Figure 3-20, are sometimes used.

Another method is to use a dump bucket carried by a crane. See Figure 3-21.

The builder must spread, carry, and level the concrete. The truck will only deliver it to the site. The truck driver can remain only a few minutes. The driver is not allowed to help work the concrete. As the concrete is poured, it should be tamped. This is done with a board or shovel that is plunged into the concrete. See Figure 3-22. Tamping helps get rid of air pockets. This makes the concrete solidly fill all the form.

For shallow footings, no smoothing or "finishing" need be done. For deep footings, the surface should be roughly leveled. This is done by resting the ends of a board across the top of the form. The board is then used to scrape the top of the concrete smooth and even with the form.

Figure 3-20. Sometimes the concrete must be carried from the truck to the worksite. *(IRL Daffin)*

• BUILDING THE FOUNDATION FORMS •

The foundation is a wall between the footing and the floor of the building. It is often made of concrete. However, it may also be made of concrete blocks, bricks, or stone. In some regions, foundation walls are made of treated plywood as well.

Figure 3-21. A dump bucket is used to dump concrete into forms that trucks can't reach. *(Universal Form Clamp)*

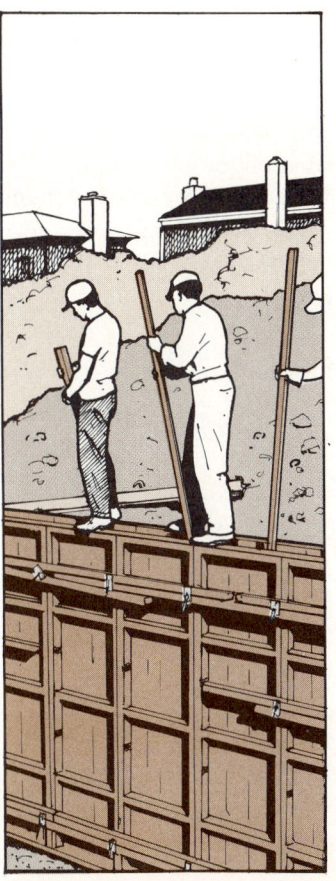

Figure 3-22. Concrete is tamped into forms to get rid of air pockets. *(Portland Cement)*

Laying Footings and Foundations   **49**

When concrete is used, special forms may be used for the foundations. These forms are easily put up and down. Often the footing and the foundation are made in one solid piece.

Builders also make the foundation in much the same way as they make the forms for the footings. After the form is removed, the lumber is used in framing the house.

In either method, the form should be spaced for the correct width. It must also be spaced to prevent the weight of the concrete from spreading the forms. The width of the form is important. A form that is not wide enough will not carry the weight of the building safely. A form that is too wide uses too much concrete. Too much concrete costs more and adds weight to the building. This weight can cause settling problems. However, spreading forms can also cause errors in pouring the concrete. Frequently just enough concrete is ordered to fill the forms. If the forms are allowed to spread, more concrete is used. Thus enough concrete might not be delivered to the site.

Making the Forms

In making forms, several things must be considered. First, sections of a form must fit tightly together. This prevents leaks at the edges. Leaks can cause bubbles and air pockets in the concrete. This is called *honeycomb*. Honeycomb weakens the foundation wall.

When special forms are used and assembled to make the total form, they must be braced properly. Forms up to 4 feet wide are braced on the back side with studs. These forms are made from metal sheets or from plywood sheathing ¾ inch thick or thicker. For building walls higher than 4 feet, special braces called *wales* are used. See Figure 3-23.

The sheathing is nailed to the studs and wales from the inside. The studs are laid out flat on the ground. The sheathing is then laid on the studs and nailed down. The assembled form is then erected and placed into position. It is spaced properly, and wales and braces are added.

Braces are erected every 4 to 6 feet. However, for extra weight or wall height, braces may be closer.

Joining the Forms Together

Edges and corners should be joined tightly so that no concrete leaks occur. When using plywood forms, join the edges together by nailing the plywood sheathing to the studs. Use 16d nails as in Figure 3-24. When nailing the corners together, use the procedure also shown in Figure 3-24. Again, 16d nails are used.

When special metal forms are used, the manufacturer's directions should be carefully followed.

Spreaders

Spreaders are used on all forms to hold the walls apart evenly. Several types may be used. The spreaders may be made of metal at the site. Metal straps may be nailed in between the sections of the forms. However, most builders use spacers that have been made at a factory for that type of form.

After the concrete has hardened, the forms are removed. The spreader is broken off when the form is removed. Special notches are made in the rods to weaken them so that they will break at the required place. Figure 3-25 shows a typical spreader.

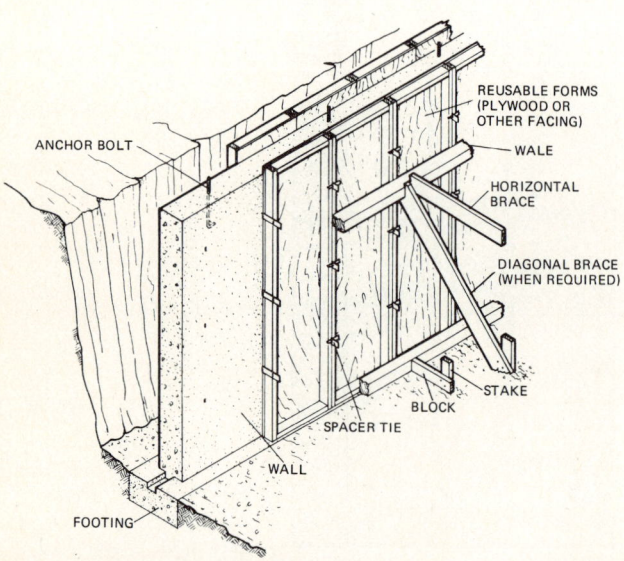

Figure 3-23. Special braces called wales are sometimes needed. *(Forest Products Lab)*

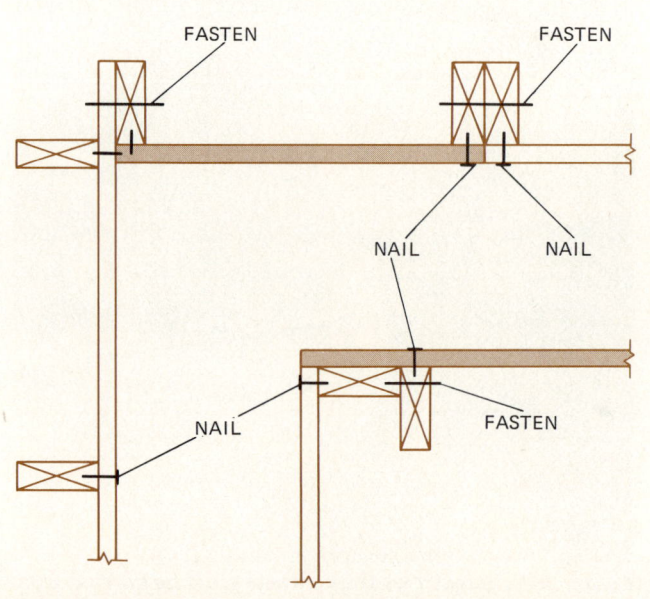

Figure 3-24. Nailing plywood panel forms.

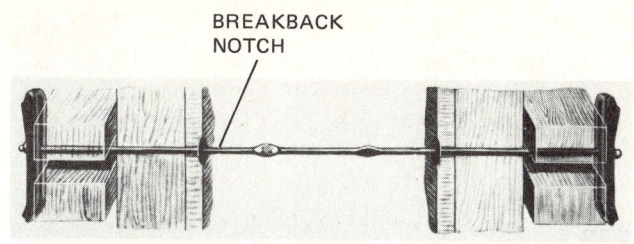

Figure 3-25. Special braces called spreaders keep the forms spaced apart. These rods are later broken off at the notches, called breakbacks. The rest of the spreader remains in the concrete. *(Richmond Screw Anchor)*

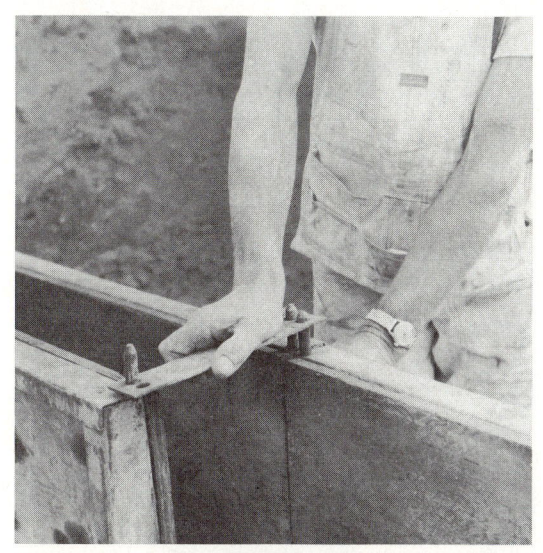

Figure 3-26. Reusable forms are made from panels of plywood or metal. These special panel forms are assembled with special fasteners. *(Proctor Products)*

Figure 3-27. Panel forms can combine the footing and the foundation. *(Proctor Products)*

Using Panel Forms

A panel form is a special form made up in sections. The forms may be used many times. Most are specially made by manufacturers. Each style has special connectors that enables the forms to be quickly and easily erected. By use of standard sizes such as 2x4- or 4x8-ft sections, walls of almost any size and shape can be erected quickly. The forms are made of metal or wood. The advantages are that they are quick and easy to use, they may be used many times, and they may be used on almost any size or shape of wall.

Panel forms must be braced and spaced just as forms constructed on the site are. To the builder making one building there is little advantage to using such forms. They must be purchased, and they are not cheap. However, when they are used many times, the savings in time make them economical. Figure 3-26 shows an example.

One-Piece Forms

When the same style of footing and foundation is often used in the same type of soils, a one-piece form is used. This combines the footings and the foundation as one piece (Figure 3-27). Several versions may be used. Some types allow a footing of any size to be cast with a foundation wall of any thickness. Some incorporate the footing and the foundation wall as a stepped figure. Others, as in Figure 3-28, use a tapered design.

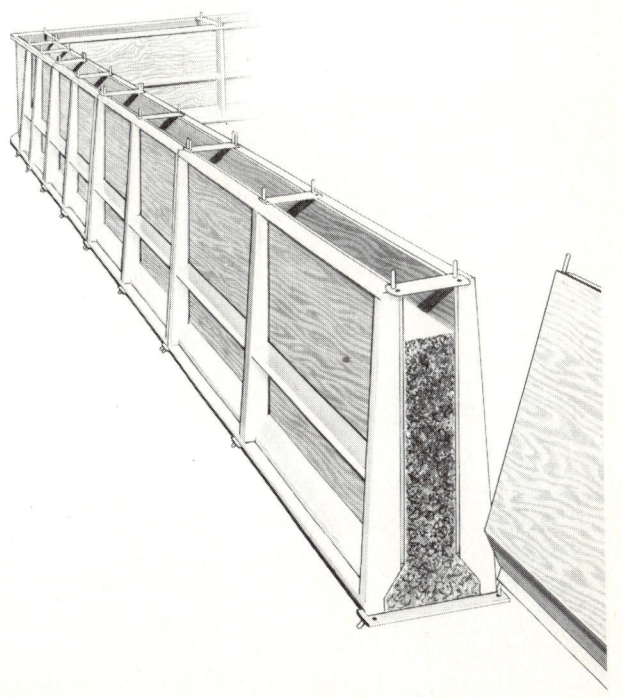

Figure 3-28. Tapered form. *(Proctor Products)*

Laying Footings and Foundations

The one-piece form saves operational steps. Casting is quicker, it is easier, and it is done in one operation. Two-piece forms and the conventional processes require that the footing be cast and allowed to harden. The footing forms are then removed and the foundation forms erected, cast, allowed to harden, and removed. This takes several days and many hours of work. The one-piece form offers many advantages in the savings of time and cost of labor.

Special Forms

Certain types of form are used for shapes that are commonly used. A round form such as that in Figure 3-29 is commonly used. This type of form may be used to cap pilings. It may also be used for the footings under the central foundation pillars.

Other special forms include forms made of steel and cardboard. Steel forms may be used to cast square or round columns. They are normally used in construction of large projects such as bridges and dams. They are also used on large business buildings. The cardboard forms are made of treated paper and fibers. They are used one time and destroyed when they are removed. Figure 3-30 shows a pillar made with a cardboard form.

All such special forms allow time and labor to be saved. Little labor and time is needed to set up special forms. Reinforcement may be added as required. Also, such forms are available in many shapes and sizes.

Openings and Special Shapes

Openings for windows and doors are frequently required in concrete foundation walls. Also, special keys or notches are often needed. These hold the ends of support frames, joists, and girders. At times, utility and sewer lines run through a foundation. Special openings must also be made for these. It is very expensive to try to cut such openings into concrete once it is hardened and cured.

However, if portions of the forms are blocked off, concrete can not enter these areas. This way almost any shape can be built into the wall before it is formed. This shape is called a *block-out* or a *buck*. The concrete is then poured and moves around these blocked-out areas. When the concrete hardens, the shape is part of the wall. This is quicker, cheaper, and easier. It also provides better strength to the wall and makes the forms used more versatile.

Of course, a carpenter can build a block-out of almost any shape in a form. First one wall of the form is sheathed. Then, the shape can then be framed out on that side. The inside of the shaped opening may be used for nails and braces. The outside, next to the concrete, should be kept smooth

Figure 3-29. Pier form. *(Proctor Products)*

Figure 3-30. A round form made of cardboard.

and well finished. However, it is expensive to pay a carpenter to frame special openings if they have to be repeated many times. It is better to use a form that can be used over again. Figure 3-31 shows an example.

When building a buck or block-out, first check the plans. Sometimes bucks are removed and sometimes they are left to form a wooden frame around the opening.

In either case, the size of the opening is the important thing. To determine rough opening sizes for windows in walls, see the chapter on building walls.

Figure 3-31. Openings may be made with special forms that can be reused. *(Proctor Products)*

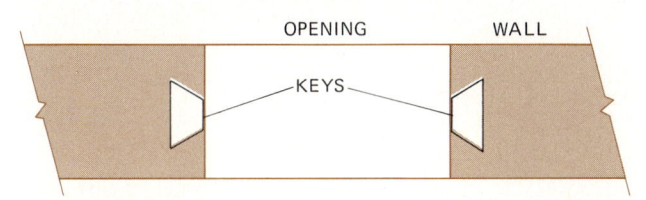

Figure 3-32. Keys are placed along the sides of openings as a nail base. Note undercut so that key cannot be removed.

BUCK KEYS • Strips of wood are used along the sides of openings in concrete. These are used as a nail base to hold frames or units in the opening. These strips are called *keys*. See Figure 3-32. If the buck is removed, the key is left in place. If the buck is left in place, the key holds the buck frame securely. Note how the key is undercut. The undercut prevents the key from being pulled out.

Bucks should be made from 2-inch lumber. The key can be made of either 1- or 2-inch lumber. The key needs to be only 1 or 2 inches wide. Usually, only the sides of the bucks are keyed.

BUCK LEFT AS A FRAME • First find the size of the opening. Next, cut the top and bottom pieces longer than the width. These pieces are usually 3 inches longer than the width. See Figure 3-33. Next, cut the two sides to the same height as the desired opening. Nail with two or three 16d nails as shown. Note that the top piece goes over the sides. The desired size for the opening is the same as the size of the opening in the buck.

BUCK REMOVED • First find the opening size. Next, cut the top and bottom pieces to the exact width of the opening. Then cut the two sides shorter. The amount is usually 3 inches (twice the thickness of the lumber used). See Figure 3-34. Nail the frame together with 16d nails as shown.

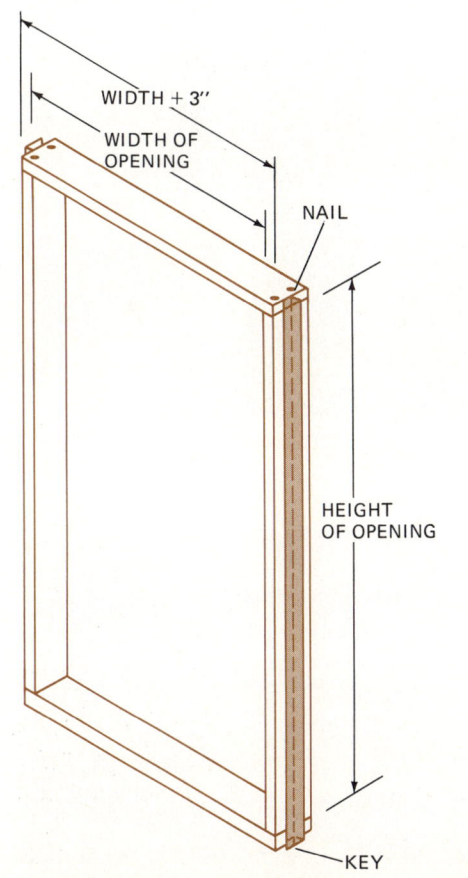

Figure 3-33. The buck may be left in the wall as the frame. In this case the opening in the buck is the desired size.

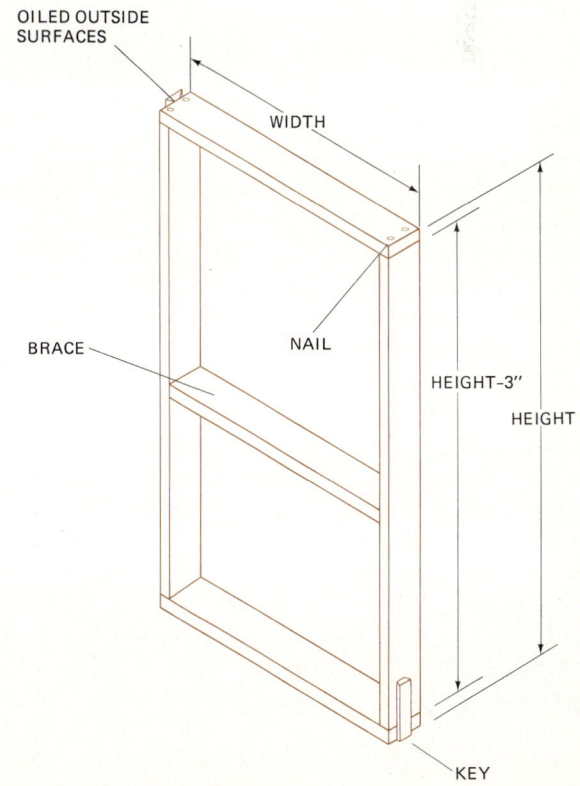

Figure 3-34. The buck may be removed. In this case the buck frame is the size desired.

Laying Footings and Foundations

Note that the opening size is the same as the outside dimensions of the buck. Also, the outside faces are oiled. This keeps the concrete from sticking to the sides of the buck. It also makes it easier to remove the buck.

BUCK BRACES • When the opening is large, the weight of the concrete can bend the boards. If the boards bend, the opening will not be the right size or shape. To prevent this, braces are placed in the opening. See Figure 3-34. Note that the braces can run from side to side or from top to bottom.

Another type of form is used for porches, sidewalks, or overhangs. See Figure 3-35. This allows porch supports to be part of the foundation. The earth is filled in later.

Reinforcing Concrete Foundations

Concrete is very strong when compressed. However, it does not support weight without cracking. It is very brittle although it is very hard. In order to resist shifting soil, concrete should be reinforced. It is not a matter of whether the soil will shift. It is more a matter of how much the soil will shift.

It should be noted that sometimes the reinforce-

Figure 3-35. (A) Special forms used for an overhang on a basement wall.

(B) The overhang after the forms are removed.

(C) Basement walls are coated and waterproofed.

(D) Finally the concrete porch will be poured.

ment is added before the forms are done. When the forms are tall, very narrow, or hard to get at, reinforcement is done first.

Concrete reinforcement is done in two basic ways. The first way is to use concrete reinforcement bars. The second way is to use mesh. Mesh is similar to a large screen made with heavy steel wire. The foundations may be reinforced by running rebars lengthwise across the top and at intervals up and down. Figure 3-36 shows some typical reinforcement.

The amount and type of reinforcement used is determined by the soil and geographic location. The reinforcement is spaced and tied. There should be at least 3 inches of concrete around the reinforcement.

The carpenter should be sure that the foundation and reinforcement conform to the local building codes.

Estimating Concrete Volume

When the forms are complete, the amount of concrete needed is computed. The same formula as for footings is used:

$$\frac{L'}{3} \times \frac{W''}{36} \times \frac{T''}{36} = \text{cubic yards}$$

However, W in inches is replaced by the wall height H in feet, so that

$$\frac{L'}{3} \times \frac{H'}{3} \times \frac{T''}{36} = \text{cubic yards}$$

If the foundation is to be 8 inches thick and 8 feet high and the perimeter is 144 feet, then

$$\frac{144}{3} \times \frac{8}{3} \times \frac{8}{36} = \text{cubic yards}$$

and by cancellation,

$$\frac{48}{1} \times \frac{8}{3} \times \frac{2}{9} = \frac{16}{1} \times \frac{8}{1} \times \frac{2}{9}$$

$$= \frac{256}{9}$$

$$= 28.4 \text{ cubic yards}$$

Delivery and Pouring

Once the needs are estimated and the concrete has been ordered or mixed, the wall should be poured. If a transit-mix truck is used, the concrete is mixed and delivered to the site. The concrete truck

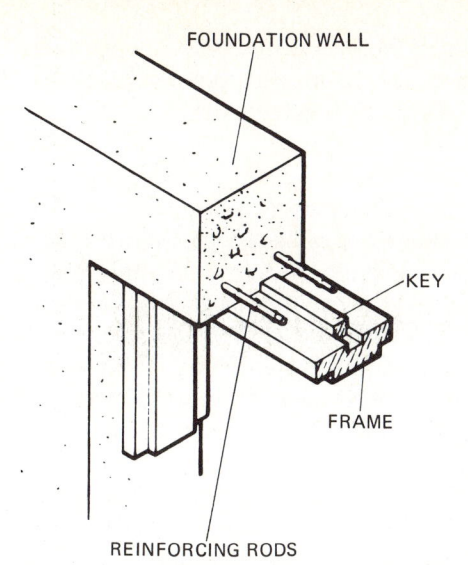

Figure 3-36. Foundations should be reinforced at the top. *(Forest Products Lab)*

Figure 3-37. Concrete is poured or pumped into the finished form. *(IRL Daffin)*

should be backed as close as possible to the forms. As in Figure 3-37, the concrete should be poured into the forms, tamped, and spread evenly. By doing this, air pockets and honeycombs are avoided.

Finishing the Concrete

Two steps are involved in finishing the pouring of the concrete foundation wall. First, the tops of the forms must be leveled. Sometimes the concrete is poured to within 2 or 3 inches of the top. The

Laying Footings and Foundations 55

concrete is then allowed to partially cure and harden. A concrete or grout with a finer mixture of sand may be used to finish out the top of the foundation.

Anchors are embedded in the concrete before it hardens completely. One end of each anchor bolt is threaded and the other is bent. See Figure 3-38. The threaded end sticks up so that a sill plate may be bolted in place. As the concrete begins to harden, the bolts are slowly worked into place by being twisted back and forth and pushed down. Once they are embedded firmly, the concrete around them is troweled smooth. Figure 3-39 shows an anchor embedded in a foundation wall.

The forms are removed after the concrete cures and hardens. Low spots must be filled. A small spot can be shimmed with a wooden shingle. However, a larger area should be filled in with grout or mortar.

• CONCRETE BLOCK WALLS •

Concrete blocks are often used for basement and foundation walls. When the foundation is exposed, it may be faced with brick. See Figure 3-40. Blocks need no formwork and go up more quickly than brick or stone. The most common size is 7⅝ inches high, 8 inches wide, and 15⅝ inches long. The mortar joints are ⅜ inch wide. This gives a finished block size of 8×16 inches, and a wall 8 inches thick.

The footing is rough and unfinished. This is because the mortar for the block is also used to smooth out the rough spots. No key is needed, but reinforcement rod should be used. Figure 3-41 shows a footing for a block wall.

Block walls should be capped with either concrete or solid block. Anchor bolts are mortared in the last row of hollow block. They then pass through the mortar joint of the solid cap. See Figure 3-42.

A special pattern is sometimes used to lay the block. This is done when the wall will also be the visible finish wall. This pattern (see Figure 3-42) is called a *stack bond*. It should be reinforced with small rebar.

• PLYWOOD FOUNDATIONS •

Plywood may be used for foundations or for basement walls in some regions. There are several advantages to using plywood. First, it can be erected in even the coldest weather. It is fast to put up because no forms or reinforcement are required. For the owner, plywood makes a wall that is warmer and easier to finish inside. It also conserves on the energy required to heat the building.

The frame is formed with 2-inch studs located on 12- or 16-inch centers. The frame is then sheathed with plywood. Insulation is placed between the studs. The exterior of the wall is covered with plastic film, and building paper is lapped over the top part. The building paper is laid over the top of the rock fill and helps drainage. See Figures 3-43 and 3-44.

It is important to note that all lumber and plywood must be pressure-treated with preservative.

• DRAINAGE AND WATERPROOFING •

A foundation wall should be drained and waterproofed properly. If a wall is not drained properly, the water may build up and overflow the top of the foundation. Unless the wall is waterproofed, water may seep through it and cause damage to the foundation and footings. It may also cause damage to

Figure 3-38. Anchor bolt.

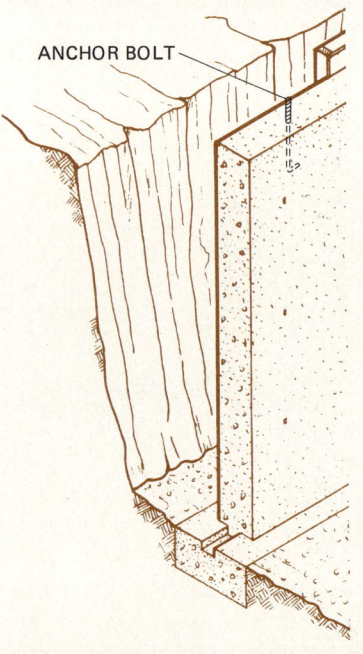

Figure 3-39. Anchor bolts are embedded in foundation walls and protrude from them. *(Forest Products Lab)*

Figure 3-40. Concrete block foundations may be faced with brick or stone.

Figure 3-41. A footing for a concrete block foundation.

TYPICAL PANEL

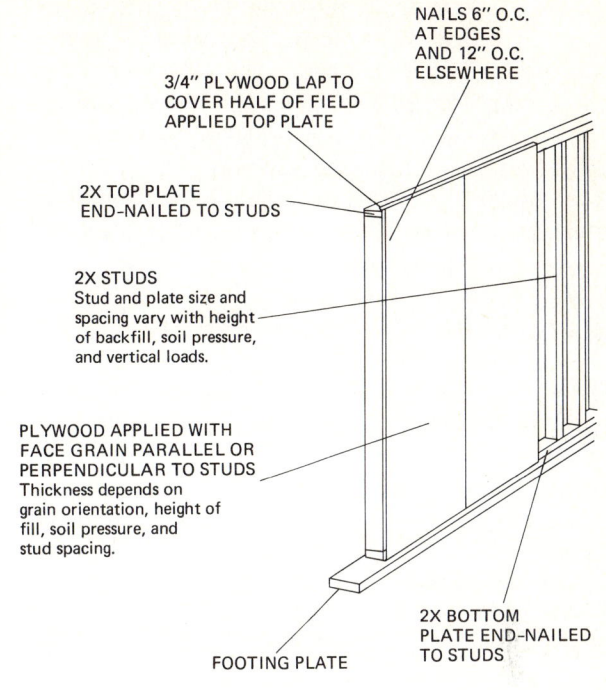

NAILS 6" O.C. AT EDGES AND 12" O.C. ELSEWHERE

3/4" PLYWOOD LAP TO COVER HALF OF FIELD APPLIED TOP PLATE

2X TOP PLATE END-NAILED TO STUDS

2X STUDS
Stud and plate size and spacing vary with height of backfill, soil pressure, and vertical loads.

PLYWOOD APPLIED WITH FACE GRAIN PARALLEL OR PERPENDICULAR TO STUDS
Thickness depends on grain orientation, height of fill, soil pressure, and stud spacing.

2X BOTTOM PLATE END-NAILED TO STUDS

FOOTING PLATE

NOTE: Wood and plywood are treated.

Figure 3-43. A typical plywood foundation panel. *(American Plywood Association)*

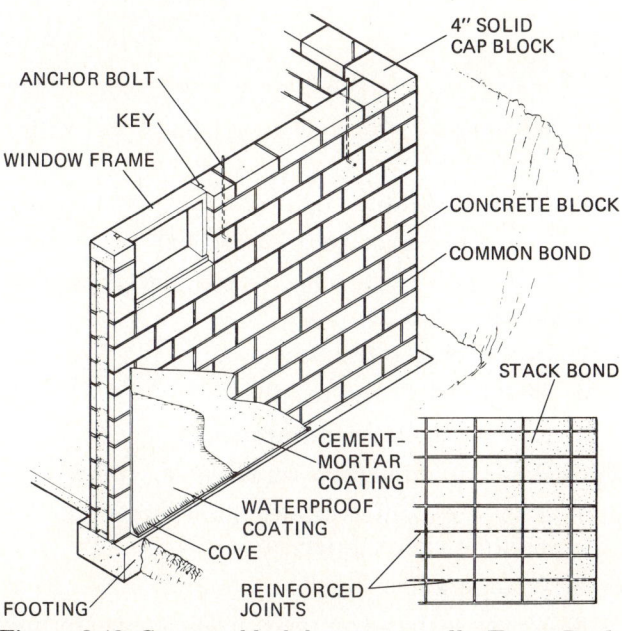

4" SOLID CAP BLOCK
ANCHOR BOLT
KEY
WINDOW FRAME
CONCRETE BLOCK
COMMON BOND
STACK BOND
CEMENT-MORTAR COATING
WATERPROOF COATING
COVE
FOOTING
REINFORCED JOINTS

Figure 3-42. Concrete block basement wall. *(Forest Products Lab)*

BASEMENT WALL

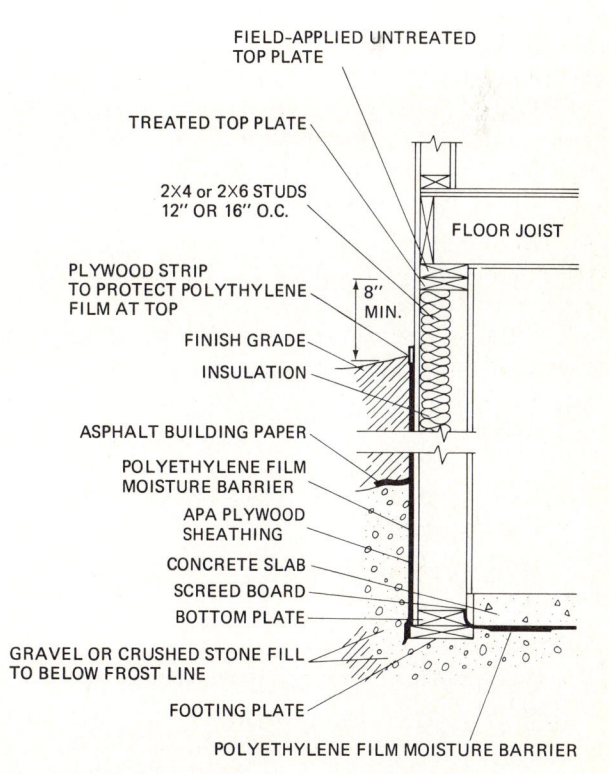

FIELD-APPLIED UNTREATED TOP PLATE
TREATED TOP PLATE
2×4 or 2×6 STUDS 12" OR 16" O.C.
PLYWOOD STRIP TO PROTECT POLYETHYLENE FILM AT TOP
FLOOR JOIST
8" MIN.
FINISH GRADE
INSULATION
ASPHALT BUILDING PAPER
POLYETHYLENE FILM MOISTURE BARRIER
APA PLYWOOD SHEATHING
CONCRETE SLAB
SCREED BOARD
BOTTOM PLATE
GRAVEL OR CRUSHED STONE FILL TO BELOW FROST LINE
FOOTING PLATE
POLYETHYLENE FILM MOISTURE BARRIER

NOTE: Wood and plywood below untreated top plate is treated.

Figure 3-44. Cross section of a plywood basement wall. *(American Plywood Association)*

Laying Footings and Foundations

the interior of a basement. Proper drainage is ensured by placing drain pipes or drain tile around the outside edges of the footing. A gravel fill is used to place the tile slightly below the level of the top of the footing. Figure 3-45 shows the proper location of the drain pipe. The pipe is then covered with loose gravel and compacted slightly.

If the house has a basement, the foundation walls should be waterproofed. If the house has only a crawl space, no waterproofing is needed.

Waterproofing Basement Walls

Three types of walls are commonly used today for basements. The most common is the solid cast-concrete wall. However, concrete blocks are also used and so is plywood.

CONCRETE BLOCK FOUNDATIONS • Concrete blocks should be plastered and waterproofed. Figure 3-42 shows the processes involved. First, the concrete wall is coated with a thin coat of plaster. This is called a scratch coat. After this has hardened, the surface is scratched so that the next coat will adhere more firmly. The second coat of plaster is applied thickly and smoothly over both the wall and the top of the footing. As shown in Figure 3-42, this outside layer is then covered with a waterproof coating. Such a coating could include layers of bitumen, builder's felt, or plastic.

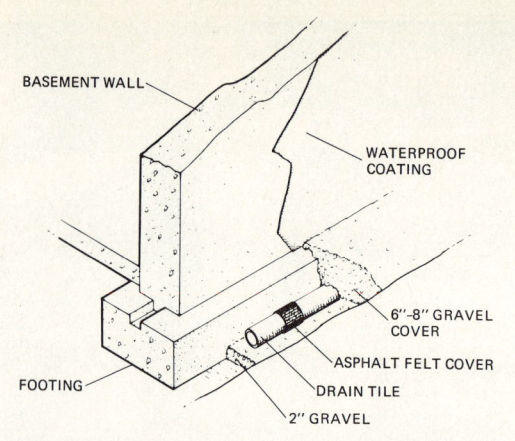

Figure 3-45. Basement walls should be coated and drained. *(Forest Products Lab)*

WATERPROOFING CONCRETE WALLS • No plaster is needed over a cast-concrete wall. The wall may be quickly coated with bitumen. However, plastic sheeting may also be applied to cover both the footing and the foundation in one piece. The most common process, as shown in Figure 3-45, involves a bitumen layer. This bitumen layer is sometimes reinforced by a plastic panel which is then coated with another layer of bitumen.

• WORDS CARPENTERS USE •

footing	pillar	key	concrete	block-out
foundation	pile	rough line	cement	buck
form	excavation	chair	aggregate	anchor bolt
frost line	crawl space	spreader	perimeter	drain tile
stepped footing	rebar	tie	honeycomb	

• STUDY QUESTIONS •

3-1. What is the lowest part of the building?

3-2. How is footing size determined?

3-3. What types of footings are used?

3-4. What is a form?

3-5. How deep should footings be for basements? For crawl spaces?

3-6. How are forms leveled?

3-7. What is a spreader?

3-8. How much concrete would be needed for a one-piece footing and foundation with these dimensions:

   Footing thickness = 8 inches
   Width = 18 inches
   Perimeter = 150 feet
   Foundation thickness = 8 inches
   Height = 9 feet

3-9. What should be used to fill holes for utility lines?

3-10. How is a footing key made?

3-11. How are anchor bolts embedded?

3-12. What two materials are commonly used to make foundation walls?

3-13. How should rebar be spaced?

3-14. How is rebar tied together?

3-15. How are foundations drained?

3-16. How are foundation walls waterproofed?

3-17. Why are double-headed nails preferred for building concrete forms?

3-18. What is a block-out?

3-19. What are the advantages of using panel forms?

3-20. What is a "basic" concrete volume mix?

Carpentry Fundamentals

4
PREPARING CONCRETE SLABS, FLOORS, AND SURFACES

Concrete surfaces are used for many things. Slabs combine footings, foundations, and subfloors in one piece. Concrete floors are common in basements and in baths. Concrete is used outdoors to form stairs, driveways, patios, and sidewalks. Carpenters build the forms and may also help pour and finish the concrete. After this chapter you should be able to:

- Excavate
- Construct the forms
- Prepare the subsurface
- Lay drains and utilities
- Lay reinforcement
- Determine concrete needs
- Ensure correct pouring and surfacing

• CONCRETE SLABS •

Concrete slabs can combine footings, foundations, and subfloors as one piece. Slabs are cheaper to build than basements. In the past, basements were used as storage areas for furnaces, fuels, and ashes. Basements also held cooling and ventilation units and laundry areas. Today, these things are as easily built on the ground. However, in cold climates many people still prefer basements. They keep pipes from freezing and add warmth to the upper floors. They also provide storage space, play areas, and sometimes living areas.

Slabs are best used on level ground and in warm climates. They can be used where ground hardness is uneven. Slabs are good with split-level houses or houses on hills where the slab is used for the lower floor. See Figure 4-1.

In the past, slabs did not make comfortable floors. They were very cold in the winter, and water easily condensed on them. Water could also seep up through the slab from the ground. Today's building methods can solve most of these problems.

Most heat energy is lost at the edges of a slab. This loss is reduced by using rigid insulation. See Figure 4-2. In extreme cases a warmer floor is needed. Heating ducts may be built in the slab flooring itself. This is an efficient method and provides an even temperature in the building. A warmer type of floor can also be built over the concrete floor. This is discussed later in this chapter.

Sequence for Pouring a Slab

Slabs are used for many types of buildings and for several types of outdoor surfaces. In most cases, the general procedure is about the same. After the site is prepared and the building is located, this sequence is common:

1. Excavate
2. Construct forms
3. Prepare subsurface
   a. Spread sand or gravel and level
   b. Install drains, pipes, and utilities
   c. Install moisture barrier
4. Install reinforcement bar

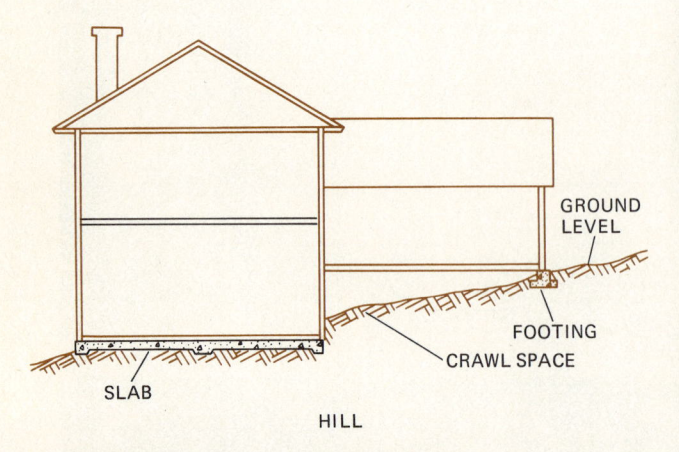

Figure 4-1. A typical split-level home. The lower floor is a slab; the middle floor is a frame floor.

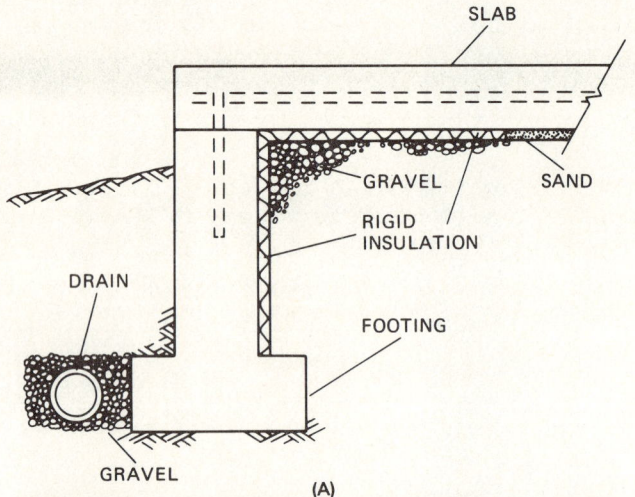

Figure 4-2. (A) Insulation for a slab. (B) A two piece slab. *(Fox and Jacobs)*

60 Carpentry Fundamentals

5. Construct special forms for lower levels, stairs, walks, etc.
6. Level tops of forms
7. Determine concrete needs
8. Pour concrete
9. Tamp, level, and finish
10. Embed anchors
11. Cure and remove forms

Types of Slabs

Slabs are built in two basic ways. Those called *monolithic* slabs are poured in one piece. They are used on level ground in warm climates.

In cold climates, however, the frost line penetrates deeper into the ground. This means that the footing must extend deeper into the ground to be below the frost line. Another method for slabs is often used where the footing is built separately. The two-piece slab is best for cold or wet climates. Figures 4-2 and 4-3 show both types of slabs.

Slab footings must rest beneath the freeze line. This gives the slab stability in the soil. Slabs should be reinforced, but the amount of reinforcement needed varies. It depends on soil conditions and weights to be carried. On dry, stable soil, slabs need little reinforcement. Larger slabs or less stable soils need more reinforcement.

The top of the slab should be 8 inches above ground. If the slab is above the rest of the ground, moisture under the slab can drain away from the building. The ground around the slab should also be sloped for the best drainage. See Figure 4-4.

Excavate

At this point, the building has already been located and batter boards are in place. Lines on the batter boards show the location of corners and walls.

Now, trenches or excavations are made for the footings, drains, and other floor features. These must all be deeper than the rest of the slab. Footings are, of course, around the outside edges. However, a slab big enough for a house should also have central footings.

The locating and digging for slab footings is the same as for foundation footings. There is no difference at all when a two-piece slab is to be made. The main difference for a monolithic slab is how the forms are made.

The trenches can now be dug. Rough lines are used for guides. As a rule, inner trenches are dug first. The trenches for the outside footings are done last. This is easier when machines are moved around the site.

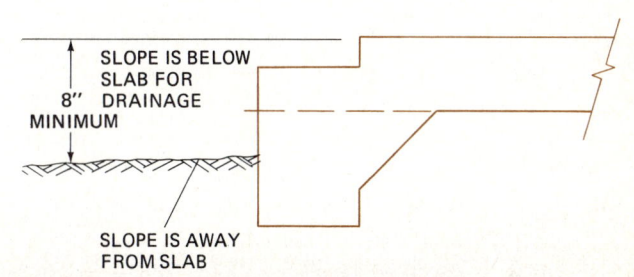

Figure 4-4. Drainage of a slab.

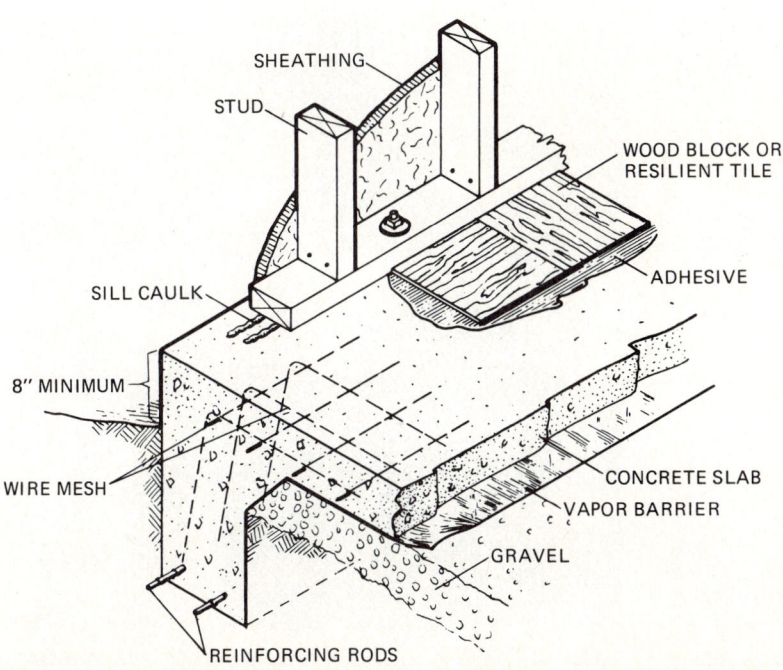

Figure 4-3. Combined slab and foundation (thickened-edge slab). *(Forest Products Lab)*

Preparing Concrete Slabs, Floors, and Surfaces 61

The excavations are then checked for depth level. Remember from Chapter 3 that the lowest point determines the depth. Trenches are also dug for drains and sewer lines inside the slab. Then, trenches are dug from the slab to the main utility lines. Sewer lines must connect the slab to the main sewer line, and so forth.

Construct the Forms

After the excavation is done, the corners are relocated. Lines are restrung on batter boards and the corners are plumbed. See Figure 3-16. The footing forms for two-piece slabs are made like standard footing or foundation forms. Refer to Chapter 3.

Lumber is brought to the place where it is needed, and then forms are constructed. See Figure 4-5. Monolithic forms are made like footing forms. The top board is placed and leveled first. See Figure 3-16. It is leveled with the corner first. Then its length is leveled and the ends are nailed to stakes. As before, double-headed nails should be used from the outside. The remaining form boards are then nailed in place.

Another method may be used for monolithic slabs with shallow footings. It is a very fast and inexpensive way. The form boards are put up before excavating. See Figure 4-6. Be sure to carefully check the plans (Figure 4-7). Next, the sand or gravel is dumped inside the form area. See Figure 4-8. The sand is spread evenly over the form area. As in Figure 4-9, the outside footings are then dug by hand. Chalk lines are strung from the forms. They are then used as guides to dig the central footings. See Figure 4-10. Trenches for drains and sewers are then dug.

Figure 4-5. (Top) Lumber is carried to the site when needed. (Bottom) Forms are erected. *(Fox and Jacobs)*

Figure 4-6. The forms are erected. *(Fox and Jacobs)*

Figure 4-7. Plans must be carefully checked. *(Fox and Jacobs)*

62 Carpentry Fundamentals

Figure 4-8. Sand is dumped and spread in the form area.

Figure 4-10. Excavations are also made for footings in the slab.

Figure 4-9. Excavations are made for outside footings. Plumbing is "roughed in."

Figure 4-11. All openings in pipes are covered. This prevents them from clogging with dirt or concrete.

Prepare the Subsurface

The ground under any type of slab must be prepared for moisture control. Water must be kept from seeping up through a slab. The water must also be drained from under the slab. This preparation is needed because water from rain and snow will seep under the slab.

Outside moisture can be reduced by using good siding methods. The edges of the slab can be stepped for brick. The sheathing can overlap the slab edge on other types of siding. As mentioned before, heat energy can be lost easily from slabs. The main area of loss is around the edges. Rigid foam insulation can be put under the slab's edges. This will reduce the heat energy losses of the slab.

SUBSURFACE PREPARATION • After excavation, drains, water lines, and utilities are "roughed in." These should be placed for areas such as the kitchen, baths, and laundry. Water lines may be run in ceilings or beneath the slab. If they are to be beneath the slab, soft copper should be used. Extra length should be coiled loosely to allow for slab movement. Metal or plastic pipe may be used for the drains. Conduit (metal pipe for electrical wires) should be laid for any electrical wires to go under the slab. Wires should never be laid without the conduit. All openings in the pipes are then capped. See Figure 4-11. This keeps dirt and concrete from clogging them.

The various pipes are then covered with sand or gravel. Sand or gravel is dumped in the slab area

Preparing Concrete Slabs, Floors, and Surfaces **63**

and carefully smoothed and leveled. Chalk lines are strung across the forms to check the level. See Figure 4-12. It is also wise to install a clean-out plug between the slab and the sewer line. See Figure 4-13.

LAY A VAPOR SEAL • Once the sand is leveled, the moisture barrier is laid. The terms vapor barrier, moisture barrier, vapor seal, and membrane mean the same thing. As a rule, plastic sheets are used for moisture barriers. The moisture barrier is laid so that it covers the whole subsurface area. To do this several strips of material will be used. The strips should overlap at least 2 inches at the edges.

INSULATE THE EDGES • Figure 4-2 shows insulation for a slab. Insulation is laid after placement of the vapor barrier. The insulation is placed around the outside edges. This is called *perimeter insulation*. The insulation should extend to the bottom of the footing. It should extend into the floor area at least 12 inches. A distance of 24 inches is recommended.

Rigid foam at least 1 inch thick is used. Perimeter insulation is not always used in warm climates. However, the moisture barrier should always be used.

REINFORCEMENT • Reinforcement should always be used. The amount of reinforcement rods (rebar) used in the footing should conform to local codes. The slab should also be reinforced with mesh. This mesh is made of 10-gage wire. The wires are spaced 4 to 6 inches apart. Figure 4-14 shows the reinforced form ready for pouring. Where the soil

Figure 4-13. Trenches for drains and sewers are also excavated. Cleanout plugs are usually outside the slab.

Figure 4-12. The sand is leveled. Chalk lines are used as guides.

Figure 4-14. The vapor barrier is then laid. The reinforcement is also laid.

64 Carpentry Fundamentals

is unstable, more reinforcement is needed. The amount is usually given on the plans.

As the rebar is laid, it is tied in place. The soft metal ties are twisted around the rebars. See Figure 4-15. The mesh may also be held off the bottom by metal stakes. See Figure 4-16. Mesh may also be lifted into place as the concrete is poured. Chairs hold rebar in place for pouring. See Figure 4-17.

Special Shapes

Special shapes are used for several reasons. A stepped edge, as in Figure 4-18, helps drainage. It prevents rainwater from flowing onto the floor surface. Lower surface areas are also common. They are used for garages, entry ways, and so forth.

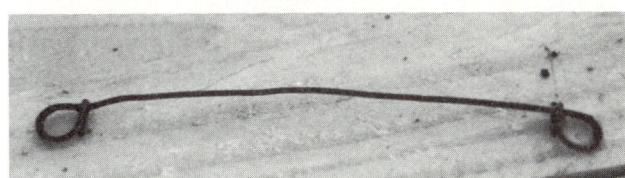

Figure 4-15. Rebar is tied together with soft wire "ties."

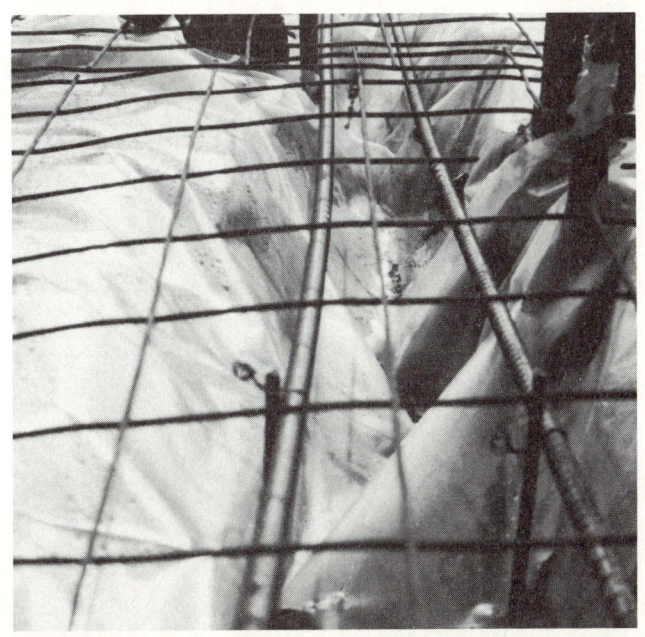

Figure 4-16. Short stakes hold mesh and rebar in place during pouring.

Figure 4-17. Chairs hold rebar in place for pouring. (*Fox and Jacobs*)

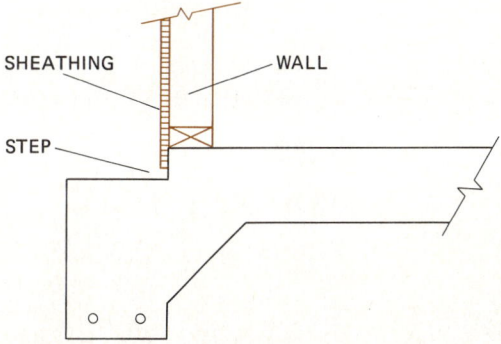

Figure 4-18. A stepped form aids drainage. The step also forms a base for brick siding.

Preparing Concrete Slabs, Floors, and Surfaces **65**

Step edges are easily formed. A 2x4 or 2x6 is nailed to the top of the form. See Figure 4-19A and B. To lower a larger surface, an extension form is used. The first form is used as a nailing base. The extension form is built inside the outer form. The lower area can then be leveled separately. See Figure 4-20.

(A) (B)

Figure 4-19. A stepped edge is formed by nailing a board to the form. *(Fox and Jacobs)*

Figure 4-20. A lower surface is used for a garage. (See top left through bottom right.)

66 Carpentry Fundamentals

Pouring the Slab

First, the corners of the forms are leveled. See Figure 4-21. A transit is used for large slabs, but small areas are leveled with a carpenter's level. See Chapter 2 for the leveling process with a transit.

Diagonals are checked for squareness, and all dimensions are checked. See Figure 4-22.

Most concrete today is delivered to the building site. Usually, the concrete is not mixed by the carpenters. It is delivered by transit-mix trucks from a concrete company. The concrete is sold in units of cubic yards. Before the concrete can be ordered, the amount of concrete needed must be determined.

ESTIMATING VOLUME • A formula is used to estimate the volume of concrete needed. The following formula is used:

$$\frac{L'}{3} \times \frac{W'}{3} \times \frac{T''}{36} = \text{cubic yards}$$

For example, a slab has a footing 1 foot wide. The slab is to be 30 feet wide and 48 feet long. The slab is to be 6 inches thick. From the formula, the amount required is

$$\frac{30}{3} \times \frac{48}{3} \times \frac{6}{36} = \frac{8640}{324}$$

$$\frac{10}{1} \times \frac{16}{1} \times \frac{1}{6} = \frac{160}{6}$$

$$= 26\frac{2}{3}$$

But the perimeter footings are 18 inches deep. They are not just 6 inches deep. Thus, a portion 12 inches thick (18 minus 6) must be added. This additional amount of concrete is calculated as follows. The linear distance around the slab is

$$48 + 48 + 30 + 30 = 156$$

Then

$$\frac{L}{3} \times \frac{W}{3} \times \frac{T}{36} = \frac{1}{3} \times \frac{156}{3} \times \frac{12}{36}$$

$$= \frac{1}{3} \times \frac{52}{1} \times \frac{1}{3} = \frac{52}{9} = 5.7 \text{ cubic yards}$$

Thus, to fill the slab, the two elements are added:

```
  26.7
+  5.7
  32.4 cubic yards
```

POURING. • To be ready, the carpenter sees to two things. First the forms must be done. Then the concrete truck must have a close access.

The builder must spread, carry, and level the con-

Figure 4-21. Corners of forms are leveled before pouring the slab. *(Portland Cement)*

Figure 4-22. Diagonals are measured for squareness and all dimensions are checked. *(Fox and Jacobs)*

Preparing Concrete Slabs, Floors, and Surfaces

crete. The truck will only deliver it to the site. The truck driver can remain only a few minutes. The driver is not allowed to help work the concrete. As the concrete is poured, it should be spread and tamped. This is done with a board or shovel. See Figure 4-23. The board or shovel is plunged into the concrete. Be careful not to cut the moisture barrier. Tamping helps get rid of air pockets. This makes the concrete solidly fill all the form.

After tamping, the concrete is leveled. This is done with a long board called a *strike-off*. See Figure 4-24. The ends rest across the top of the forms. The board is moved back and forth across the top. A short back-and-forth motion is used. If the board is not long enough, special supports are used. These are called *screeds*. A screed may be a board or pipe supported by metal pins. The screed is leveled with the tops of the forms. It is removed after the section of concrete is leveled. Any holes left by the screeds are then patched.

After leveling, the surface is floated. This is done after the concrete is stiff. However, the concrete must not have hardened. A finisher uses a float to tamp the surface gently. During tamping the float is moved across the surface. Large floats called *bull floats* are used. See Figure 4-25. Floating lets the smaller concrete particles float to the top. The large particles settle. This gives a smooth surface to the concrete. Floats may be made of wood or metal.

After floating, the finish is done. A rough, lined surface can be produced. A broom is pulled across the top to make lines. This surface is easier to walk on in bad weather. See Figure 4-26. For most flooring surfaces, a smooth surface is desired. A smooth surface is made by troweling. For small surfaces a hand trowel may be used, as in Figure 4-27. However,

Figure 4-24. As the form is filled, it is leveled. This is done with a long board called a strike-off. *(Goldblatt)*

Figure 4-25. After leveling, the surface is smoothed by "floating" with a bull float. *(Portland Cement)*

Figure 4-26. A broom can be used to make a lined surface. *(Portland Cement)*

Figure 4-23. Concrete is spread evenly in the form.

68 Carpentry Fundamentals

Figure 4-27. Hand trowels may be used to smooth small surfaces. *(Portland Cement)*

Figure 4-28. A power trowel is used to finish large surfaces. *(Portland Cement)*

Figure 4-29. Anchor bolts are imbedded into foundation wall.

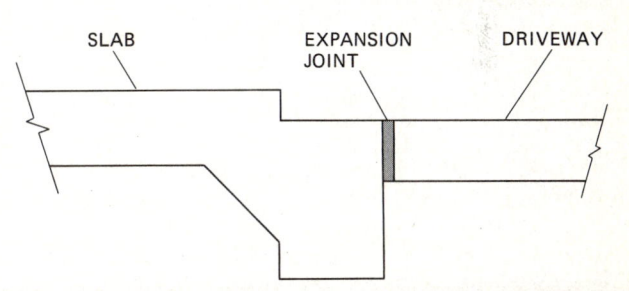

Figure 4-30. Expansion joints are used between large, separate pieces.

for larger areas, a power trowel will be used. See Figure 4-28.

Now the surface has been finished. Next, the anchors for the walls are embedded. Remember, the concrete is not yet hard. Do not let the anchors interfere with joist or stud spacing. The first anchor is embedded at about one-half the stud spacing. This would be 8 inches for 16-inch spacings. Anchors are placed at four to eight foot intervals. Only two or three anchors are needed per wall.

The anchors are twisted deep into the concrete. The anchors are moved back and forth just a little. This settles the concrete around them. After the anchors are embedded, the surface is smoothed. A hand trowel is used. See Figure 4-29.

The concrete is then allowed to harden and cure. Afterward, the form is removed. It takes about 3 days to cure the concrete slab. During this time, work should not be done on the slab. When boards are used for the forms, they are saved. These boards are used later in the house frame. Lumber is not thrown away if it is still good. It is used where the concrete stains do not matter.

Expansion and Contraction

Concrete expands and contracts with heat and cold. To compensate for this movement, expansion joints are needed. Expansion joints are used between sections. The expansion joint is made with wood, plastic, or fiber. Joint pieces are placed before pouring. Such pieces are used between foundations and basement walls or between driveways and slabs. See Figure 4-30. Often, the screed is made of wood. This can be left for the expansion joint.

Joints

Other joints are also used to control cracking. These, however, are shallow grooves troweled or cut into the concrete. They may be troweled in the concrete as it hardens. The joints may also be cut after the concrete is hard. A special saw blade is used. The joints help offset and control cracking. Concrete is very strong against compression. However, it has little strength against bending or twisting. It breaks easily. A concrete sidewalk will break. The builder

Preparing Concrete Slabs, Floors, and Surfaces **69**

only tries to control breaks. They cannot be prevented. The joints form a weak place in concrete. The concrete will break at the weak point. But the crack will not show in the joint. The joint gives a better appearance when the concrete cracks. Figure 4-31 shows a troweled joint. Figure 4-32 shows a cut joint.

• CONCRETE FLOORS •

Concrete is used for floors in basements and commercial buildings. The footings and foundations are built before the floor is made. An entire house may be built before the basement floor is made. See Figure 4-33. The concrete can be poured through windows or floor openings.

A concrete floor is made like a slab. It is also the way to make the two-piece slab. First, the ground is prepared. Drains, pipes, and utilities are positioned and covered with gravel. Next, coarse sand or gravel is leveled and packed. A plastic-film moisture barrier is placed to reach above the floor level. Perimeter insulation is laid as indicated. See Figure 4-34. Reinforcement is placed and tied. Rigid foam insulation can also be used as an expansion joint. Finally, the floor is poured, tamped, leveled, and finished. Separate footings should be used to support beams and girders. Figure 4-35 shows this.

Stairs

Entrances may be made as a part of the main slab. However, stairs used with slabs are poured separately. Separate stairs are used with slabs as an access in steep areas. They are also used with steep

Figure 4-31. Joints may be troweled into a surface. *(Portland Cement)*

Figure 4-32. Some joints may be cut with saws.

Figure 4-33. Footings in a basement later become part of the basement floor.

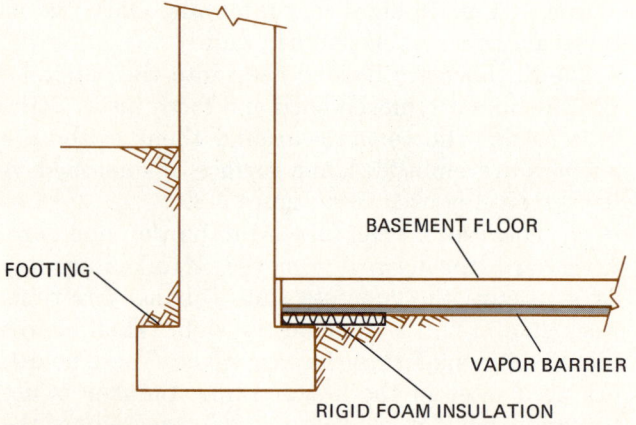

Figure 4-34. Rigid foam should be laid at the edges of the basement floor.

70 Carpentry Fundamentals

lawns. Forms for steps may be made by two methods. The first method uses short parallel boards. See Figure 4-36. The second method uses a long stringer. Support boards are added as in Figure 4-37. The top of the stair tread is left open in both types of forms. This lets the concrete step be finished. The bottoms of the riser forms are beveled. This lets a trowel finish the full surface. The stock used for building supports should be 2-inch lumber. This keeps the weight of the concrete from bulging the forms.

• SIDEWALKS AND DRIVEWAYS •

Sidewalks and driveways should not be a part of a slab. They must float free of the building. Where they touch, an expansion joint is needed. Sidewalks and driveways should slope away from the building. This lets water drain away from the building. Then, special methods are used for drainage. Often, a ledge about 1½ inches high will be used. The floor of the building is then higher than the driveway. One side of the driveway is also raised. Water will then stop at the ledge and will flow off to the side. This keeps water from draining into the building.

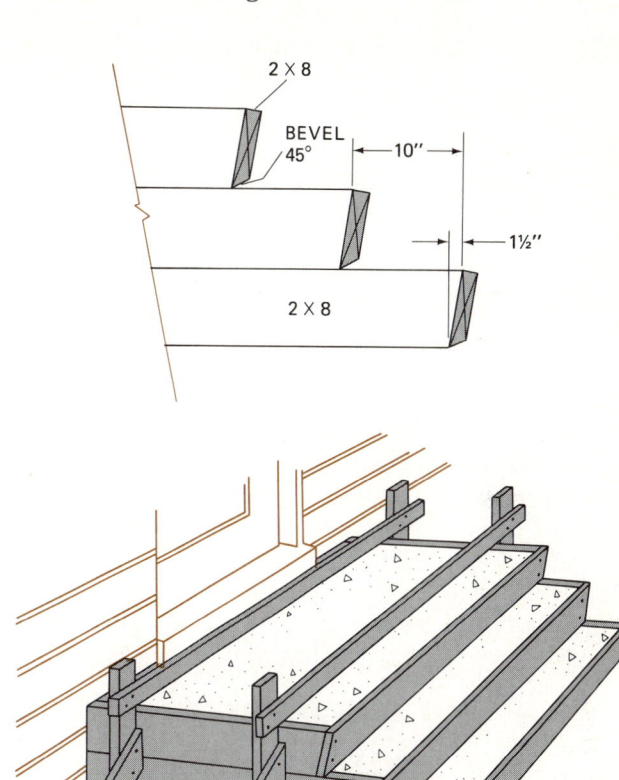

Figure 4-36. A form for steps.

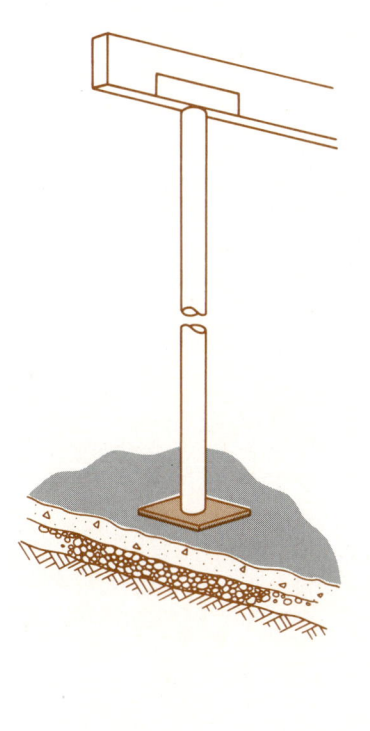

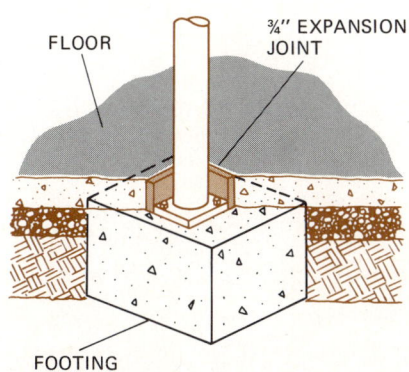

Figure 4-35. Floors should not be major supports. Separate footings should be used.

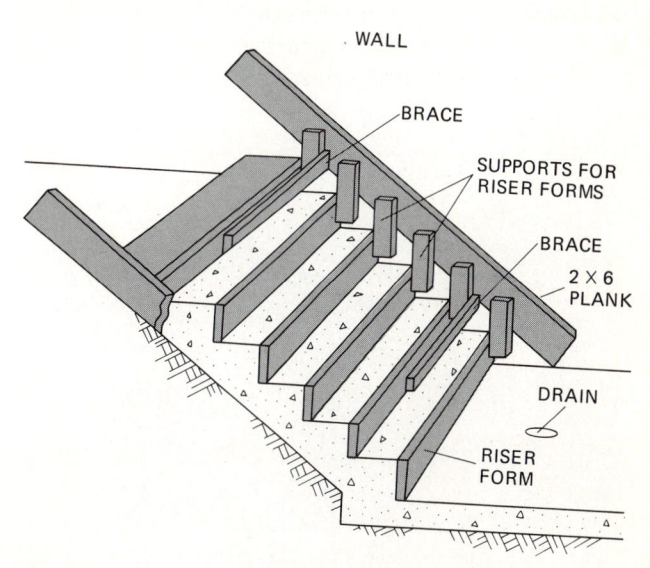

Figure 4-37. A form for steps against a wall.

Preparing Concrete Slabs, Floors, and Surfaces **71**

Two-inch-thick lumber should be used for the forms. It should be four- or six-inches-wide. The width determines slab thickness. Commercial forms may also be used. See Figure 4-38.

Sidewalks

Sidewalks are usually 3 feet wide and 4 inches thick. Main walks or entry ways may be 4 feet or wider. No reinforcement is needed for sidewalks on firm ground. It may be used for sidewalks on soft ground. A sand or gravel fill is used for support of sidewalks on wet ground. The earth and fill should be tamped solid. Figure 4-39 shows a sidewalk form. The slab is poured, leveled, and finished like other surfaces.

Driveways

Driveways should be constructed to handle great weight. Reinforcement rods or mesh should be used in driveways. Slabs 4 inches thick may be used for passenger cars. Six-inch-thick slabs are used where trucks are expected. The standard driveway is 12 feet wide. Double driveways are 20 feet wide. Other dimensions are shown in Figure 4-40. The slab is poured, leveled, and finished like other surfaces.

• SPECIAL FINISHES AND SURFACES •

Concrete may be finished in several ways. Different surface textures are used for better footing or appearance. Also, concrete may be colored. It may also be combined with other materials.

Different Surface Textures

Different surface textures may be used for appearance. However, the most common purpose is for better footing. Driveways are roughened for better tire traction. Better footing and traction are needed in bad weather. Several methods may be used to texture the surface. The surface may be simply floated. This gives a smooth, slightly roughened surface. Floated surfaces are often used on sidewalks.

BRUSHING • The surface may also be brushed. Brushing is done with a broom or special texture brush. The pattern may be straight or curved. For a straight pattern, the brush is pulled across the entire surface. See Figure 4-26 again. For swirls, the brush can be moved in circles.

PEBBLE FINISH • Pebbles can be put in concrete for a special appearance. See Figure 4-41. The pebble finish is not difficult to do. As the concrete stiffens, pebbles are poured on the surface. The pebbles are then tamped into the top of the concrete. The pebbles in the surface are leveled with a board or a float.

Figure 4-39. Sidewalk forms are made with 2x4 lumber.

Figure 4-38. Special forms may be used for sidewalks and driveways. *(Proctor Products)*

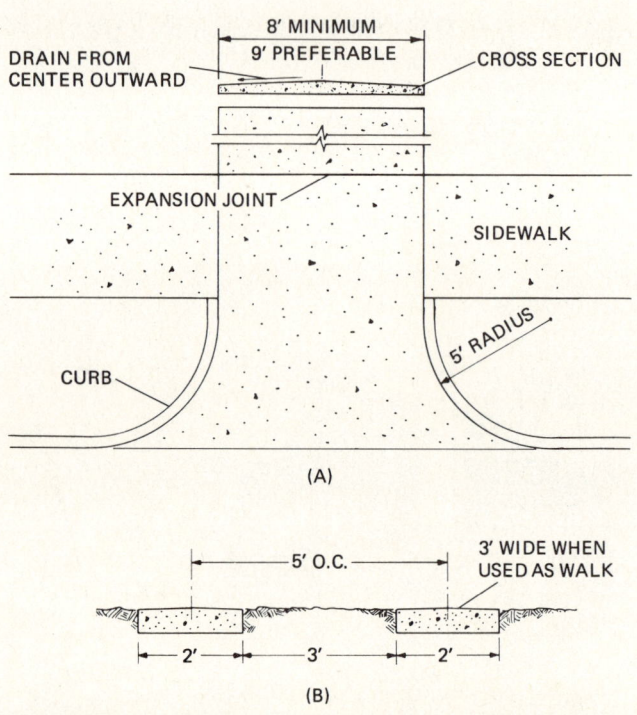

Figure 4-40. Driveway details: (A) Single-slab driveway. (B) Ribbon-type driveway. *(Forest Products Lab)*

Some hours later, the fine concrete particles may be hosed away.

COLOR ADDITIVES • Color may be added to concrete for better appearance. The color is added as the concrete is mixed and is uniform throughout the concrete. The colors used most are red, green, and black. The surface of colored concrete is usually troweled smooth. Frequently, the surface is waxed and polished for indoor use.

TERRAZZO • A terrazzo floor is made in two layers. The first layer is plain concrete. The second layer is a special type of white or colored concrete. Chips of stone are included in the second, or top, layer of concrete. The top layer is usually about ½ to 1 inch thick. It is leveled but not floated. The topping is then allowed to set. After it is hardened, the surface is finished. Terrazzo is finished by grinding it with a power machine. This grinds the surface of both the stones and the cement smooth. Metal strips are placed in the terrazzo to help control cracking. Both brass and aluminum strips are used. Figure 4-42 shows this effect. The result is a durable finish with natural beauty. The floor may be waxed and polished.

Terrazzo will take heavy foot traffic with little wear. It is often used in buildings like post offices or schools. It is also easy to maintain.

CERAMIC TILE, BRICK, AND STONE • Concrete may also be combined with ceramic tile or brick. The result is a better-appearing floor. The floor contrasts the concrete and the brick or tile. Also, the concrete may be finished in several ways. This gives more variety to the contrast. For example, pebble finish on the concrete may be used with special brick. Bricks are available in a variety of shapes, colors, and textures.

Ceramic tile comes in many sizes. The largest is 12×12 inches. Several shapes are also used. Glazed tile is used in bathrooms because the glaze seals water from the tile. Unglazed tile also has many uses, but it is not waterproof.

To set tile, stone, or bricks in a concrete floor, you must have a lowered area. The pieces are laid in the lowered area. The area between the pieces is filled with concrete, grout, or mortar.

Stone, tile, and brick are used. They add contrast and beauty and resist wear. They are easy to clean and resist oil, water, and chemicals.

CONCRETE OVER WOOD FLOORS • Concrete is also used for a surface over wood flooring. Figure 4-43 shows this kind of floor being made. A concrete

Figure 4-41. A pebbled concrete surface.

Figure 4-42. A terrazzo floor is smooth and hard. It can be used in schools and public buildings. *(National Terrazzo and Mosaic Association, Inc.)*

Figure 4.43. A concrete floor is being placed over a plywood floor. The concrete will make the floor more durable. *(American Plywood Association)*

Preparing Concrete Slabs, Floors, and Surfaces

topping on a floor has several advantages. These floors are harder and more durable than wood. They resist water and chemicals and may be used in hallways and rest rooms.

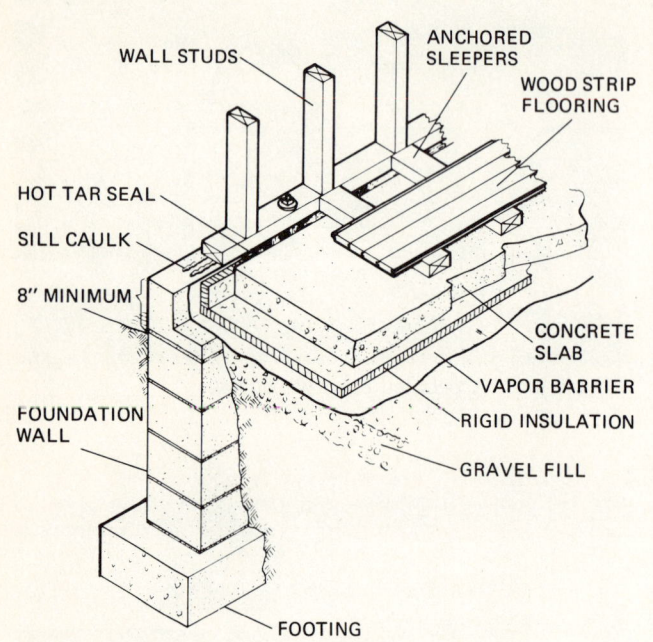

Figure 4-44. (A) Base for wood flooring on a concrete slab with vapor barrier under the slab.

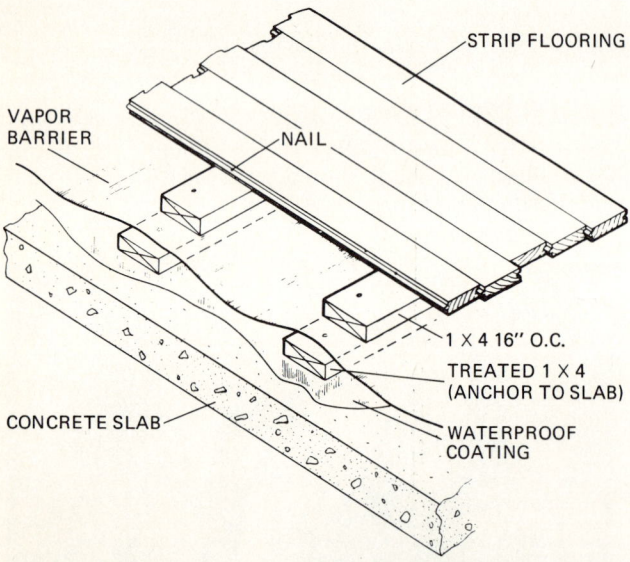

(B) Base for wood flooring on a concrete slab with no vapor barrier under the slab. *(Forest Products Lab)*

WOOD OVER CONCRETE • Wood may be used for a finish floor over a concrete floor. The wooden floor is warmer in cold climates. Because it does not absorb heat as does concrete, energy can be saved. The use of wood can also improve the appearance of the floor.

Two methods are used for putting wood over a concrete floor. The first method is the older. A special glue, called *mastic,* is spread over the concrete. Then strips of wood are laid on the mastic. A wooden floor may be nailed to these strips. Figure 4-44A and B shows a cross section.

For better energy savings, a newer method is used. In this method, rigid insulation is laid on the concrete. Mastic may be used to hold the insulation to the floor. See Figure 4-45. The wooden floor is laid over the rigid insulation. Plywood or chipboard underlayment can also be used. The floor may then be carpeted. If desired, special wood surfaces can be used.

• ENERGY FACTORS •

There are two ways of saving energy that are used with concrete floors. The first is to insulate around the edges of the slab. The second is to cover the floor with insulated material. Both methods have been mentioned previously.

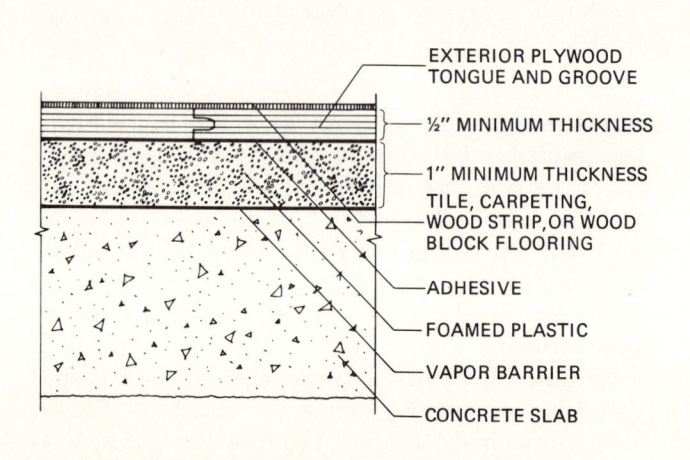

Figure 4-45. A wood floor may be laid over insulation. The insulation is glued to the concrete floor. *(American Plywood Association)*

• WORDS CARPENTERS USE •

moisture barrier	perimeter insulation	extension form	floating	ceramic tile
tamp	reinforcement	transit-mix truck	trowel	mastic
monolithic slab	mesh	strike-off	expansion joint	
vapor barrier		screed	terrazzo	

74 Carpentry Fundamentals

· STUDY QUESTIONS ·

4-1. What does monolithic mean?

4-2. Can you define the terms above?

4-3. What is the sequence for pouring a slab?

4-4. Why are slab forms leveled just before pouring?

4-5. How is moisture beneath a slab drained?

4-6. How is moisture drained away from a building?

4-7. What thickness of lumber is best for making concrete forms?

4-8. What should be the thickness of a sidewalk?

4-9. What should be the thickness of a driveway for trucks?

4-10. How is energy saved when concrete floors are built?

4-11. How are step forms made?

4-12. What is the width of a single driveway?

4-13. Where should footings for a slab be located?

4-14. How is a large surface floated?

4-15. How is a large surface troweled?

4-16. How can concrete be finished for better footing?

4-17. How is a pebble surface made?

4-18. How much concrete would be needed for a plain slab floor 6 inches thick, 21 feet wide, and 36 feet long?

4-19. What happens to lumber used for forms?

4-20. How is cracking controlled? Is it really stopped?

5
BUILDING FLOOR FRAMES

In this unit you will learn how to build frame floors. You will learn how to make floors over basements and crawl spaces. You will also learn how to make openings for stairs and other things. Things you will learn to do are:

- Connect the floor to the foundation
- Place needed girders and supports
- Lay out the joist spacings
- Measure and cut the parts
- Put the floor frame together
- Lay the subflooring
- Build special framing
- Alter a standard floor frame to save energy

• INTRODUCTION •

Floors form the base for the rest of the building. Floor frames are built over basements and crawl spaces. Houses built on concrete slabs do not have floor frames. However, multilevel buildings may have both slabs and floor frames.

First the foundation is laid. Then the floor frame is made of posts, beams, sill plates, joists, and a subfloor. When these are put together they form a level platform. The rest of the building is held up by this platform. The first wooden parts are called the sill plates. The sill plates are laid on the edges of the foundations. Often, additional supports may be needed in the middle of the foundation area. See Figure 5-1. These are called *midfloor supports* and may take several forms. These supports may be made of concrete or masonry. Wooden posts and metal columns are also used. Wooden timbers called *girders* are laid across the central supports. Floor joists then reach (span) from the sill on the foundation to the central girder. The floor joists support the floor surface. The joists are supported by the sill and girder. These in turn rest on the foundation.

Two types of floor framing are used on multistory buildings. The most common is the platform type. In platform construction each floor is built separately. The other type is called the balloon frame. In balloon frames the wall studs reach from the sill to the top of the second floor. Wall frames are attached to the long wall studs. The two differ on how the wall and floor frames are connected. These will be covered in detail later.

• SEQUENCE •

The carpenter should build the floor frame in this sequence:

1. Check the level of foundation and supports
2. Lay sill seals, termite shields, etc.
3. Lay the sill
4. Lay girders
5. Select joist style and spacing
6. Lay out joists for openings and partitions
7. Cut joists to length and shape
8. Set joists
    a. Lay in place
    b. Nail opening frame
    c. Nail regular joists
9. Cut scabs, trim joist edges
10. Nail bridging at tops
11. Lay subfloor
12. Nail bridging at bottom
13. Trim floor at ends and edges
14. Cut special openings in floor

• SILL PLACEMENT •

The sill is the first wooden part attached to the foundation. However, other things must be done before the sill is laid. When the anchors and foundation surface are adequate, a seal must be placed on the foundation. The seal may be a roll of insulation material or caulking. If a metal termite shield is used, it is placed over the seal. Next, the sill is prepared and fitted over the anchor bolts onto the foundation. See Figure 5-2.

Figure 5-1. The piers and foundations walls will help support the floor frame.

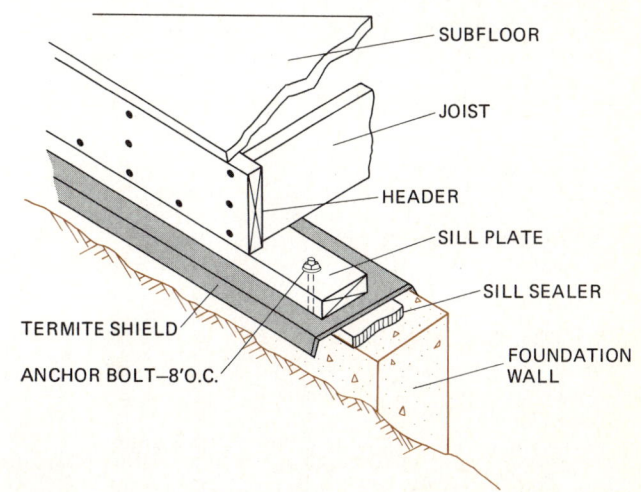

Figure 5-2. Section showing floor, joists, and sill placement.

The seal forms a barrier to moisture and insects. Roll-insulation-type material, as in Figure 5-3, may be used. The roll should be laid in one continuous strip with no joints. At corners, the rolls should overlap about 2 inches.

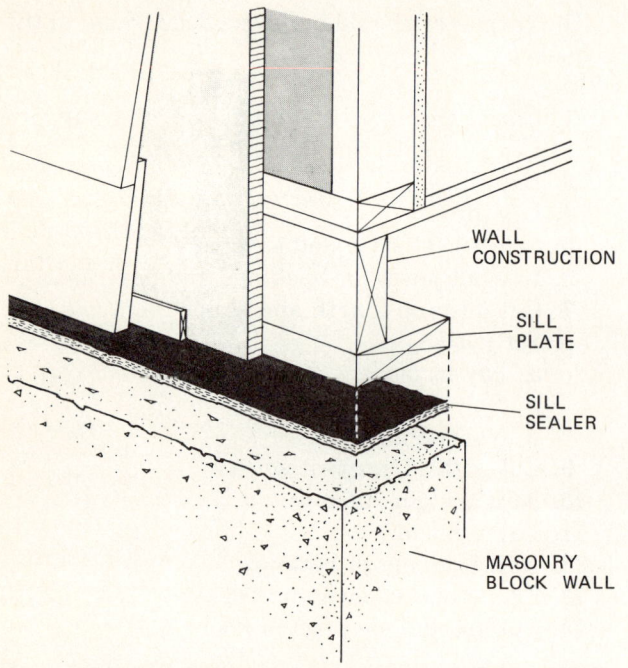

Figure 5-3. A seal fills in between the top of the foundation wall and the sill. It helps conserve energy by making the sill more weather tight. *(Courtesy Conwed)*

To protect against termites, two things are often done. A solid masonry top is used. Metal shields are also used. Some foundation walls are built of brick or concrete block that have hollow spaces. They are sealed with mortar or concrete on the top. A solid concrete foundation provides the best protection from termite penetration.

However, termites can penetrate cracks in masonry. Termites can enter a crack as small as $1/64$ inch in width. Metal termite shields are used in many parts of the country. Figure 5-4 shows a termite shield installed.

Anchor the Sill

The sill must be anchored to the foundation. The anchors keep the frame from sliding from the foundation. They also keep the building from lifting in high winds. Three methods are used to anchor the sill to the foundation. The first uses bolts embedded in the foundation, as in Figure 5-5. Sill straps are also used and so are special drilled bolt anchors. See Figures 5-6 and 5-7. Special masonry nails are also sometimes used. However, they are not recommended for anchoring exterior walls. It is not necessary to use many anchors per wall. Anchors should be used about every 4 feet, depending upon local codes. Anchors may not be required on walls shorter than 4 feet.

The anchor bolts must fit through the sill. The

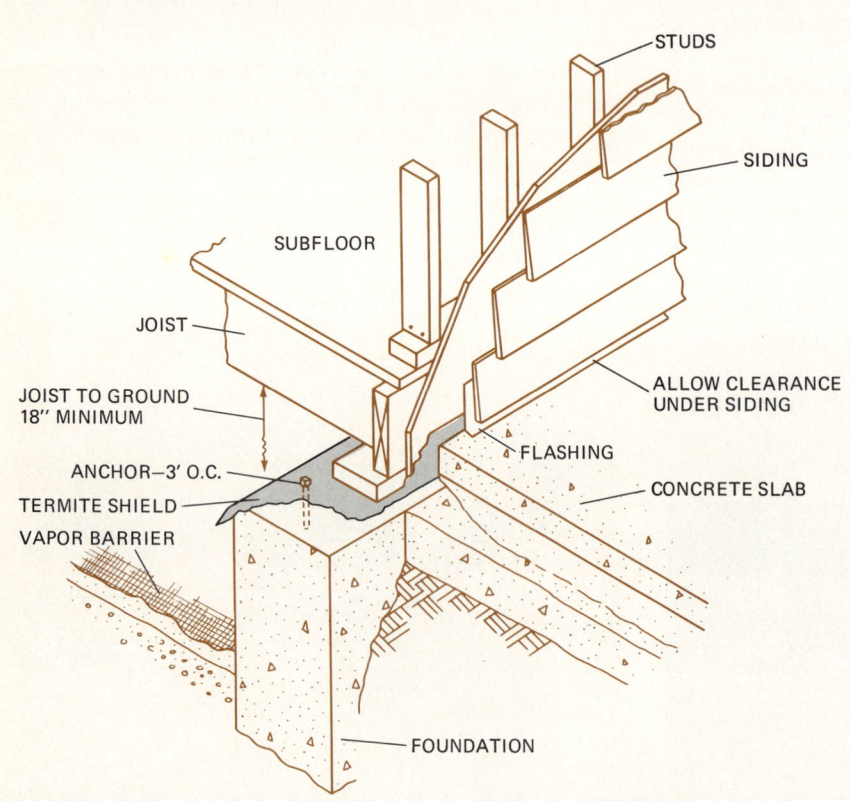

Figure 5-4. Metal termite shield used to protect wood over foundation. *(Forest Products Lab)*

78 Carpentry Fundamentals

holes are located first. Washers and nuts are taken from the anchor bolts. The sill board is laid next to the bolts. See Figure 5-8. Lines are marked using a framing square as a guide. The sheathing thickness is subtracted from one-half the width of the board. This distance is used to find the center of the hole for each anchor. The centers for the holes are then marked. As a rule the hole is bored ¼ inch larger than the bolt. This leaves some room for adjustments and makes it easier to place the sill.

Next, the sill is put over the anchors and the spacing and locations are checked. All sills are fitted and then removed. Sill sealer and termite shields are laid, and the sills are replaced. The washers and nuts are put on the bolts and tightened. The sill is checked for levelness and straightness. Low spots in the foundation can be shimmed with wooden wedges. However, it is best to use grout or mortar to level the foundation.

Special masonry nails may be used to anchor interior walls on slabs. These are driven by sledge hammers or by nail guns. The nail mainly prevents side slippage of the wall. Figure 5-9 shows a nail gun application.

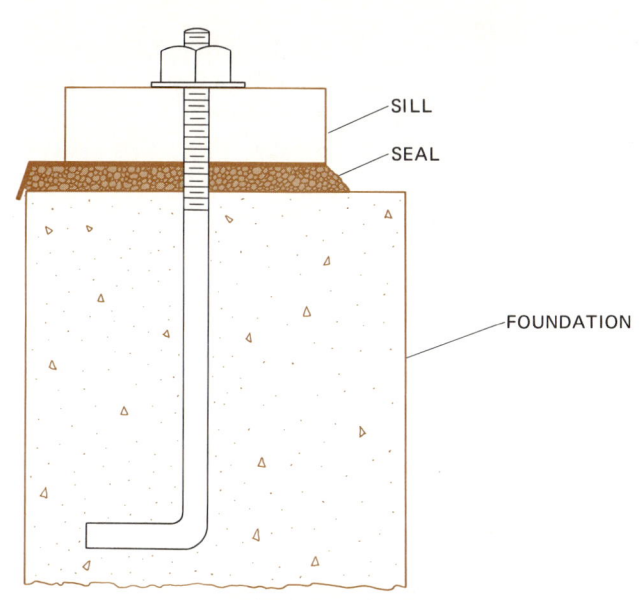

Figure 5-5. Anchor bolt in foundation.

Figure 5-6. Anchor straps or clips can be used to anchor the sill.

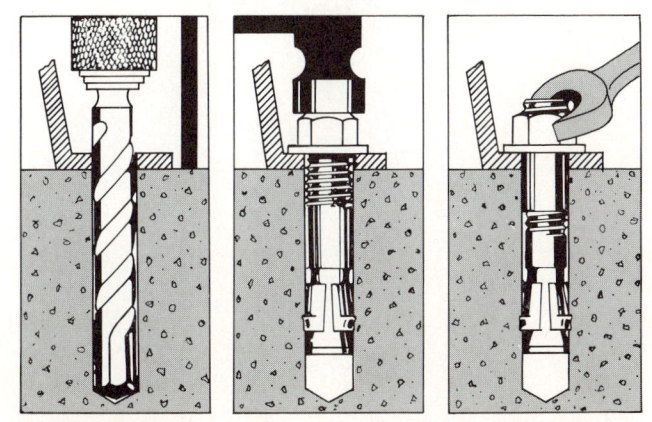

Figure 5-7. Anchor holes may be drilled after the concrete has set. *(Hilti-Fastening Systems)*

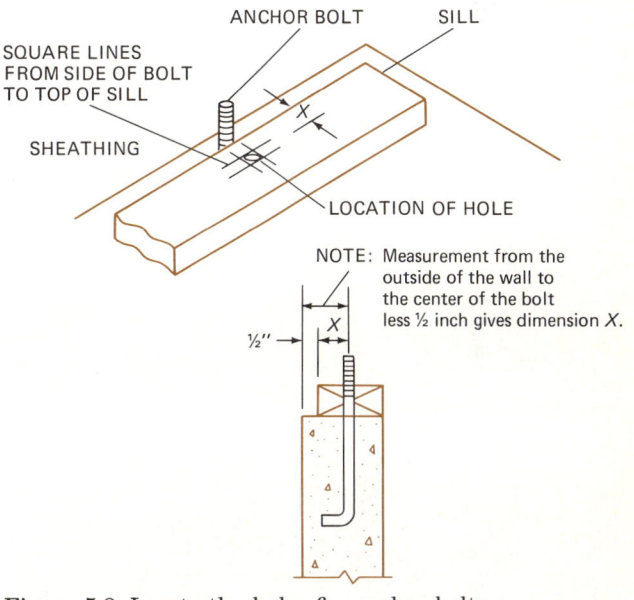

Figure 5-8. Locate the holes for anchor bolts.

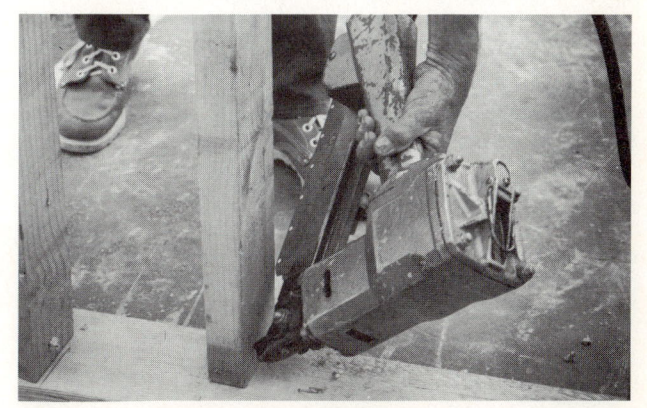

Figure 5-9. A nail gun can be used to drive nails in the slab and for toenailing. *(Duo-Fast)*

Building Floor Frames 79

Setting Girders

Girders support the joists on one end. Usually the girder is placed halfway between the outside walls. The distance between the supports is called the *span*. The span on most houses is too great for joists to reach from wall to wall. Central support is given by girders.

DETERMINE GIRDER LOCATION • Plans give the general spacing for supports and girders. Spans up to 14 feet are common for 2x10-inch or 2x12-inch lumber. The girder is laid across the leveled girder supports. A chalk line may be used to check the level. The support may be shimmed with mortar, grout, or wooden wedges. The supports are placed to equalize the span. They also help lower expense. The piers shown in Figure 5-1 must be leveled for the floor frame.

The girder is often built by nailing boards together. Figure 5-10 shows a built-up girder. Girders are often made of either 2x10-inch or 2x12-inch lumber. Joints in the girder are staggered. The size of the girder and joists is also given on the plans.

There are several advantages to using built-up girders. First, thin boards are less expensive than thick ones. The lumber is more stable because it is drier. There is less shrinkage and movement of this type of girder. Wooden girders are also more fire-resistant than steel girders. Solid or laminated wooden girders take a long time to burn through. They do not sag or break until they have burned nearly through. Steel, on the other hand, will sag when it gets hot. It only takes a few minutes for steel to get hot enough to sag.

The ends of girders can be supported in several ways. Figure 5-11A and B shows two methods. The ends of girders set in walls should be cut at an angle. See Figure 5-12. In a fire, the beam may fall free. If the ends are cut at an angle, they will not break the wall.

Metal girders should have a wooden sill plate on top. This board forms a nail base for the joists. Basement girders are often supported by post jacks. See Figure 5-13. Post jacks are used until the basement

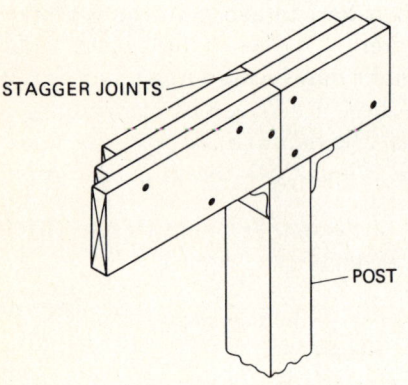

Figure 5-10. Built-up wood girder. *(Forest Products Lab)*

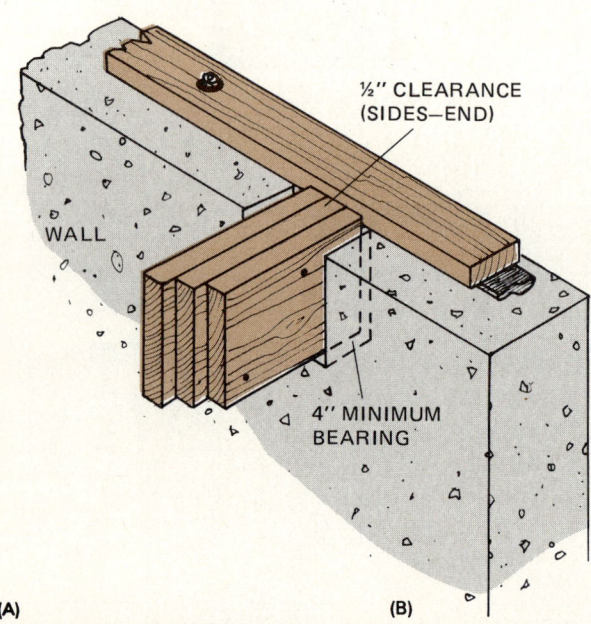

Figure 5-11. Two methods of supporting girder ends: (A) **Projecting** post. (B) **Recessed** pocket. *(Forest Products Lab)*

80 Carpentry Fundamentals

floor is finished. A support post, called a *lolly column*, may be built beneath the girder. It is usually made from 2x4 lumber. Walls may be built beneath the girder. In many areas this is done so that the basement may later be finished out as rooms.

• JOISTS •

Joists are the supports under the floor. They span from the sill to the girder. The subfloor is laid on the joists.

Lay Out the Joists

Joists are built in two basic ways. The first is the *platform* method. The platform method is the more common method today. The other method is called *balloon* framing. It is used for two-story buildings in some areas. However, the platform method is more common for multistory buildings.

JOIST SPACING • The most common spacing is 16 inches. This makes a strong floor support. It also allows the carpenter to use standard sizes. However, 12-inch and 24-inch spacing are also used. The spacing depends upon the weight the floor must carry. Weight comes from people, furniture, and snow, wind, and rain. Local building codes will often tell what joist spacings should be used.

Joist spacings are given by the distance from the center of one board to the center of the next. This is called the distance *on centers*. For a 16-inch spacing, it would be written, *16 inches O.C.*

Modular spacings are 12, 16, or 24 inches O.C. These modules allow the carpenter to use standard-size sheet materials easily. The standard-size sheet is 48x96 inches (4x8 feet). Any of the modular sizes divides evenly into the standard sheet size. By using modules, the amount of cutting and fitting is greatly reduced. This is important since sheet materials are used on subfloors, floors, outside walls, inside walls, roof decks, and ceilings.

JOIST LAYOUT FOR PLATFORM FRAMES • The position of the floor joists may be marked on a board called a *header*. The header fits across the end of the joists. See Figure 5-14.

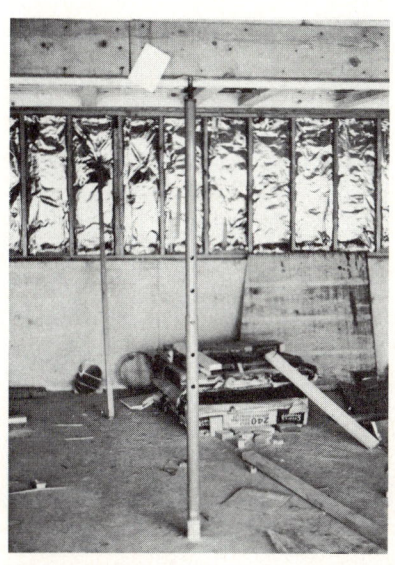

Figure 5-13. A post jack supports the girder until a column or a wall is built.

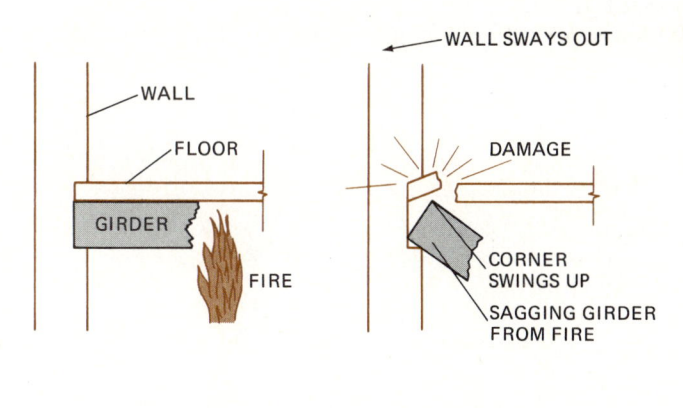

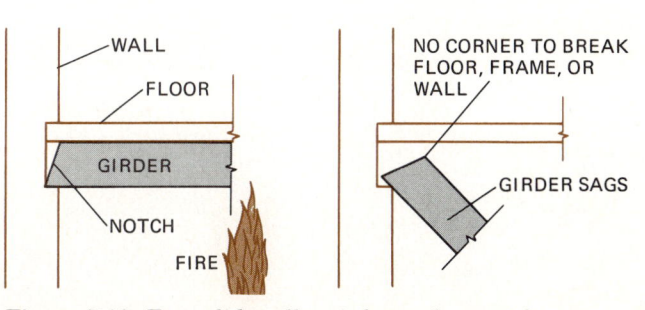

Figure 5-12. For solid walls, girder ends must be cut at an angle.

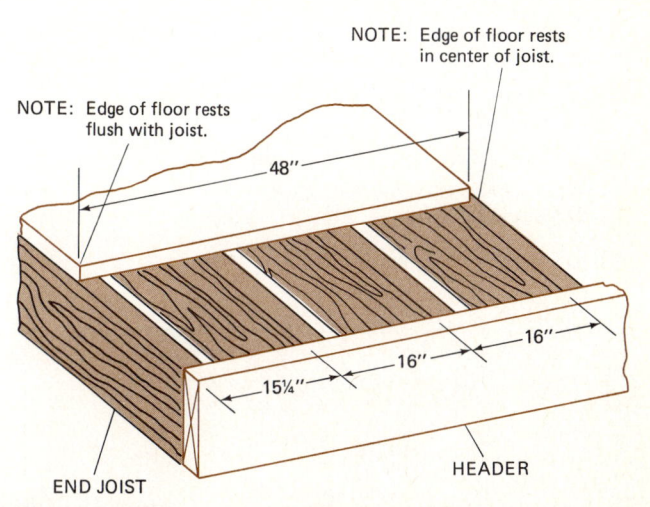

Figure 5-14. The positions of the floor joists may be marked on the header.

Building Floor Frames **81**

Joist spacing is given by the distance between centers. However, the center of the board is a hard mark to use. It is much easier to mark the edge of a board. After all, if the centers are spaced right, the edges will be too!

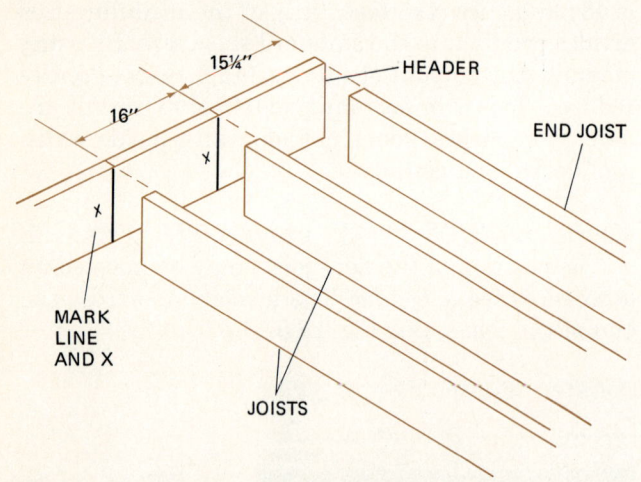

Figure 5-15. The first joist must be spaced ¾" less than the O.C. spacing used. Mark from end of header.

The header is laid flat on the foundation. The end of the header is even with the end joist. The distance from the end of the header to the edge of the first joist is marked. However, this distance is not the same as the O.C. spacing. See Figure 5-15. The first distance is always 3/4 inch less than the spacing. This lets the edge of the flooring rest flush, or even, with the outside edge of the joist on the outside wall. This will make laying the flooring quicker and easier.

The rest of the marks are made at the regular O.C. spacing. See Figure 5-15. As shown, as X indicates on which side of the line to put the joist.

MARK A POLE • It is faster to transfer marks than to measure each one. The spacing can be laid out on a board first. The board can then be used to transfer spacings. This board is called a *pole*. A pole saves time because measurements are done only once. To transfer the marks, the pole is laid next to a header. Use a square to project the spacing from the pole to the header. A square may be used to check the "square" of the line and mark.

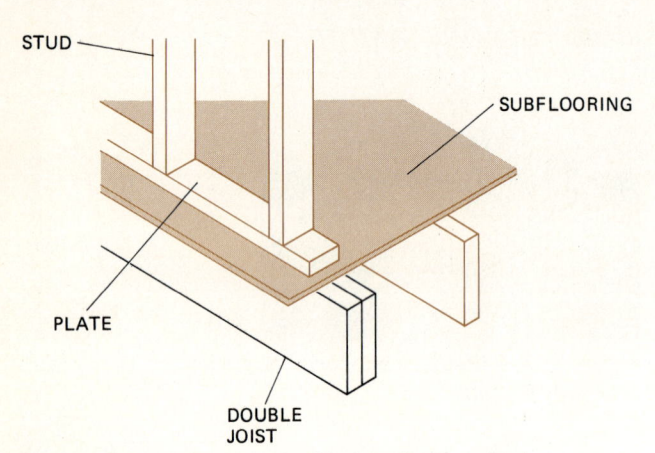

Figure 5-16. Joists are doubled under partitions.

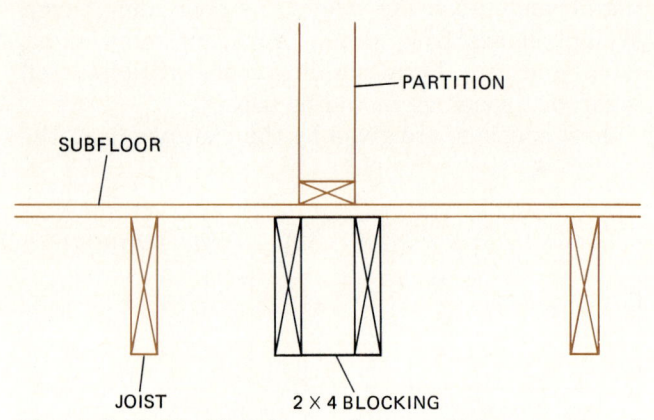

Figure 5-17. Double joists under a partition are spaced apart when pipes must go between them.

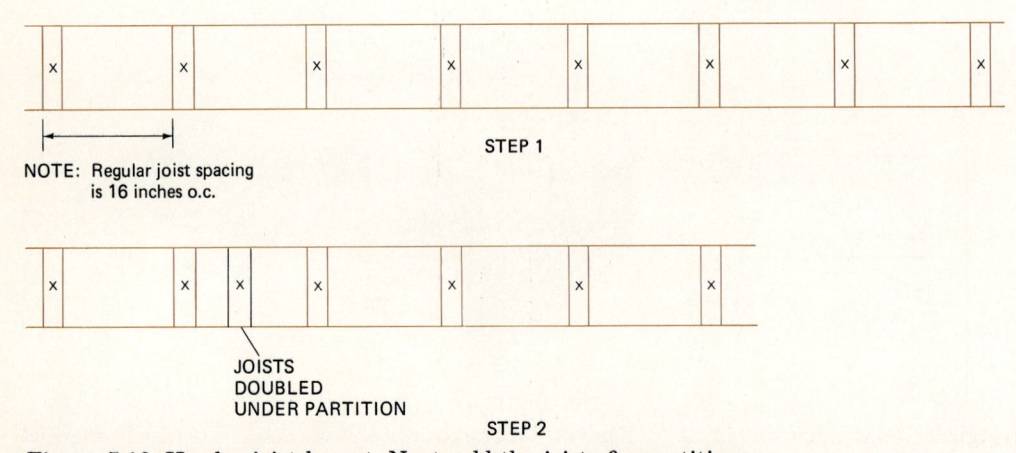

Figure 5-18. Header joist layout. Next, add the joists for partitions.

82 Carpentry Fundamentals

JOISTS UNDER WALLS • Joists under walls are doubled. There are two ways of building a double joist. When the joist supports a wall, the two joists are nailed together. See Figure 5-16. Pipes or vents sometimes go through the floors and walls. Then, a different method is used. See Figure 5-17. The joists are spaced approximately 4 inches apart. This space allows the passage of pipes or vents. Figure 5-18 shows the header layout pole with a partition added. Special blocking should be used in the double joist. Two or three blocks are used. The blocking serves as a fire stop and as bracing.

JOISTS FOR OPENINGS • Openings are made in floors for stairs and chimneys. Double joists are used on the sides of openings. They are called *double trimmers*. Double trimmers are placed without regard for regular joist spacings. Regular spacing is continued on each end of the opening. Short "cripple" joists are used. See Figure 5-19. A pole can show the spacing for the openings. See Figure 5-20.

GIRDER SPACING • The joists are located on the girders also. Remember, marks do not show centerlines of the joists. Centerlines are hard to use, so marks show the edge of a board. These marks are easily seen.

BALLOON LAYOUT • Balloon framing is different. See Figure 5-21. The wall studs rest on the sill. The joists and the studs are nailed together as

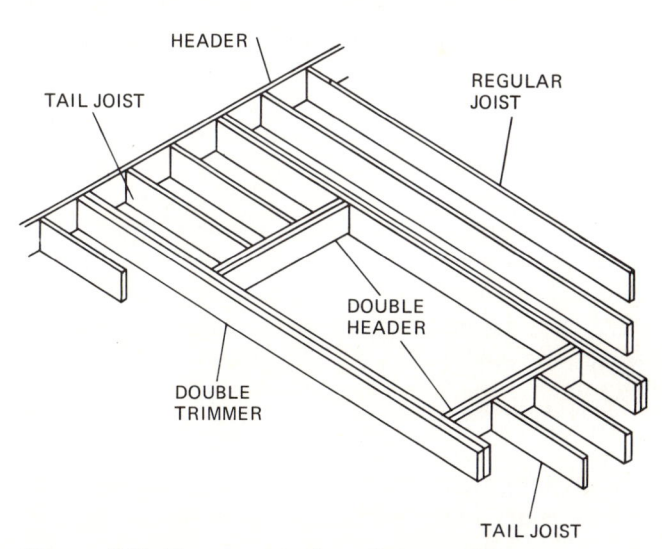

Figure 5-19. Frame parts for a floor opening.

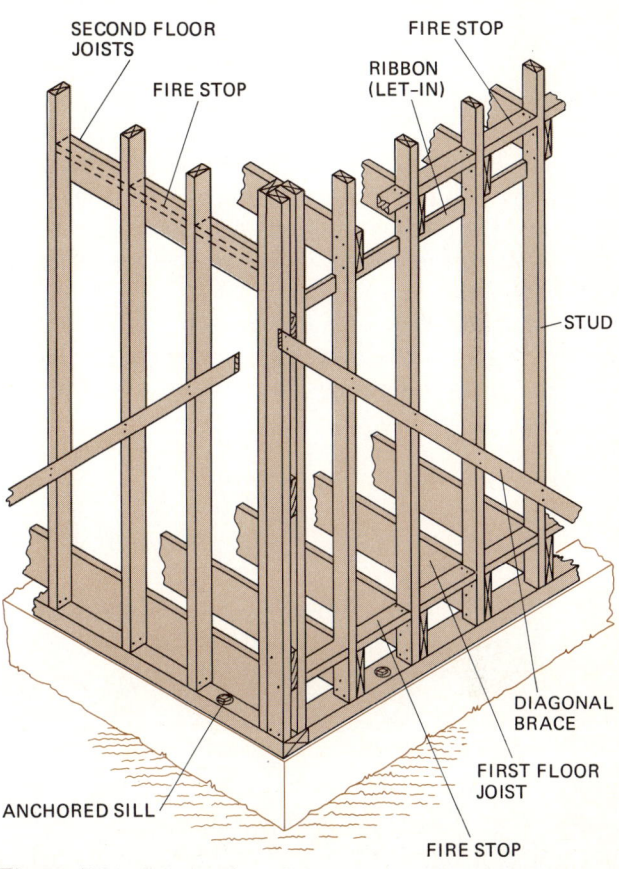

Figure 5-21. Joist and stud framing used in balloon construction. *(Forest Products Lab)*

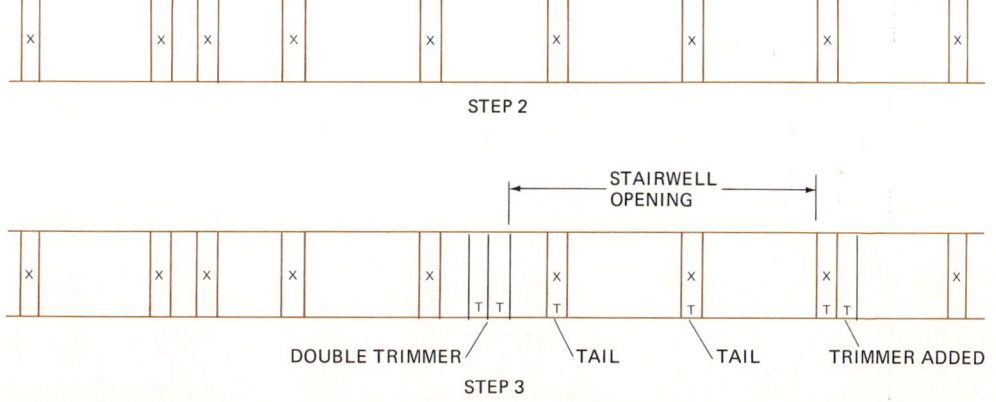

Figure 5-20. Add the trimmers for the opening to the layout pole.

Building Floor Frames 83

shown. However, the end joists are nailed to the end wall studs.

The first joist is located back from the edge. The distance is the same as the wall thickness. The second joist is located by the first wall stud.

The wall stud is located first. The first edge of the stud is ¾ inch less than the O.C. spacing. For 16 inches O.C., the stud is 15¼ inches from the end. A 2-inch stud will be 1½ inches thick. Thus, the edge of the first joist will be 16¾ inches from the edge.

Cut Joists

The joists span, or reach, from the sill to the girder. Note that joists do not cover the full width of the sill. Space is left on the sill for the joist header. See Figure 5-4. For lumber 2 inches thick, the spacing would be 1½ inches. Joists are cut so that they rest on the girder. Four inches of the joist should rest on the girder.

The quickest way to cut joists is to cut each end square. Figure 5-22 shows square-cut ends. The joists overlap across the girder. The ends rest on the sill with room for the joist header. This way the header fits even with the edge.

It is sometimes easier to put the joist header on after the joists are toenailed and spaced. Or, the joist header may be put down first. The joists are then just butted next to the header. But it is very important to carefully check the spacing of the joists before they are nailed to either the sill or the header.

Other Ways to Cut Joists

Ends of joists may be cut in other ways. The ends may be aligned and joined. Ends are cut square for some systems. For others, the ends are notched. Metal girders are sometimes used. Then, joists are cut to rest on metal girders.

END-JOINED JOISTS • The ends of the joists are cut square to fit together. The ends are then butted together as in Figure 5-23. A gusset is nailed (10d) on each side to hold the joists together. Gussets may be made of either plywood or metal. This method saves lumber. Builders use it when they make several houses at one time.

NOTCHED AND LAPPED JOISTS • Girders may be notched and lapped. See Figure 5-24. This connection has more interlocking but takes longer and costs more. First, the notch is cut on the end of the joists. Next a 2x4-inch joist support is nailed (16d) on the girder. Nails should be staggered 6 or 9 inches apart. The joists are then laid in place. The ends overlap across the girder.

JOIST-GIRDER BUTTS • This is a quick method. With it the top of the joist can be even with the top of the girder. To do this the girder must be 4 inches wider than the joist to allow for the ledger.

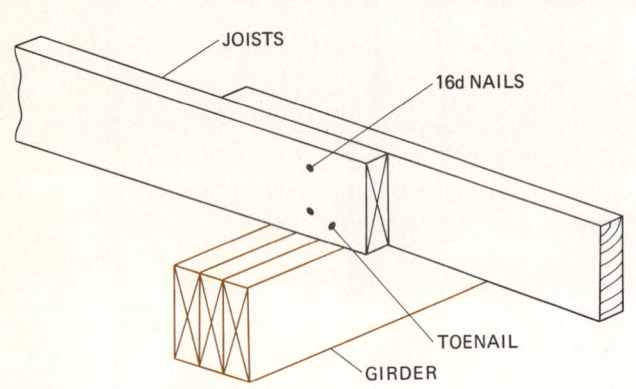

Figure 5-22. Joists may overlap on the girder. The overlap may be long or short.

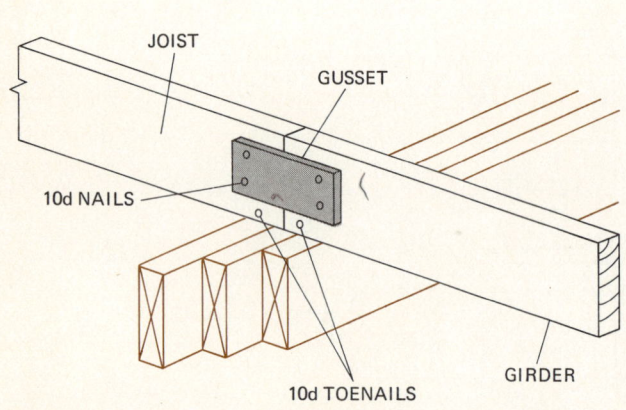

Figure 5-23. Joists may be butt-joined on the girder. Gussets may be used to hold them. Plywood subfloor may also hold the joists.

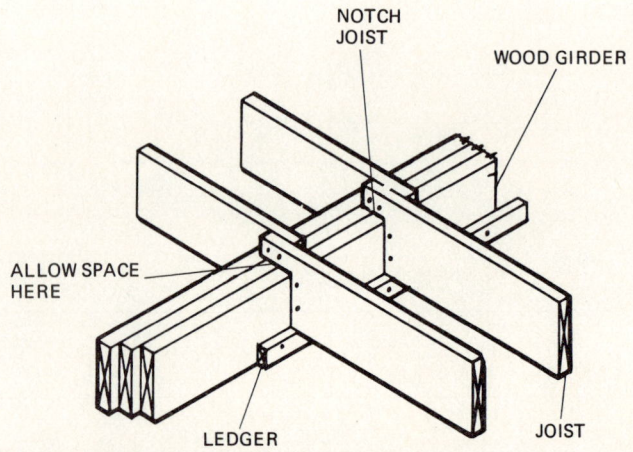

Figure 5-24. Joists can be notched and lapped on the girder. *(Forest Products Lab)*

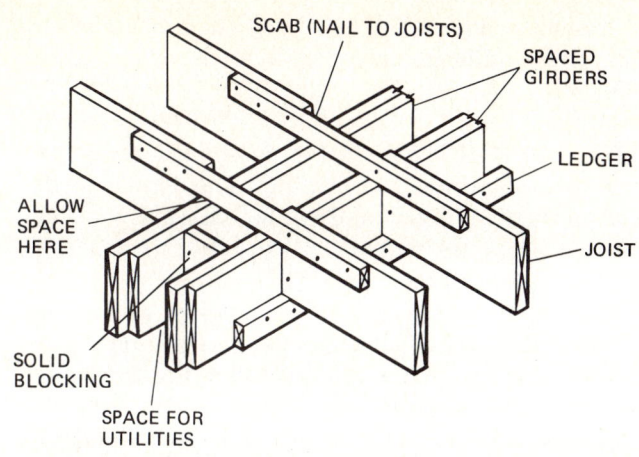

Figure 5-25. Joists may be butted against the girder. Note that the girder may be spaced for pipes, etc. *(Forest Products Lab)*

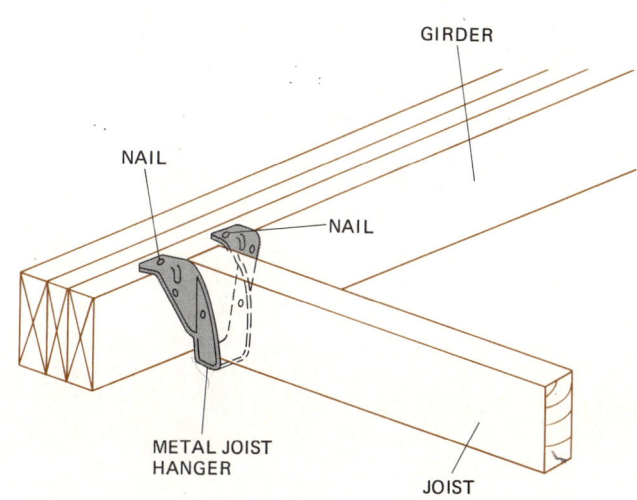

Figure 5-26. Using metal joist hangers saves time.

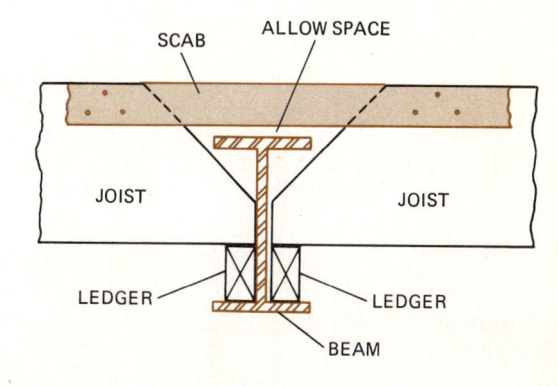

Figure 5-27. Systems for joining joists to metal girders. *(Forest Products Lab)*

A 2x4-inch ledger is nailed to the girder with 16d common nails. See Figure 5-25. The joist rests on the ledger and not on the girder. This method is not as strong.

Also, a board is used to join the girder ends. The board is called a *scab*. The scab also makes a surface for the subfloor. It is a 2x4-inch board. It is nailed with three 16d nails on each end.

JOIST HANGERS • Joist hangers are metal brackets. See Figure 5-26. The brackets hold up the joist. They are nailed (10d) to the girder. The joist ends are cut square. Then, the joist is placed into the hanger. It is also nailed with 10d nails as in Figure 5-26. Using joist hangers saves time. The carpenter need not cut notches or nail up ledgers.

JOISTS FOR METAL GIRDERS • Joists must be cut to fit into metal girders. See Figure 5-27. A 2x4-inch board is first bolted to the metal girder. The ends of the joist are then beveled. This lets the joist fit into the metal girder. The joist rests on the board. The board also is a nail base for the joist. The tops of the joists must be scabbed. The scabs are made of 2x4-inch boards. Three 16d nails are driven into each end.

SETTING THE JOISTS

Two jobs are involved in setting joists. The first is laying the joists in place. The second is nailing the joists. The carpenter should follow a given sequence.

LAY THE JOISTS IN PLACE • First, the header is toenailed in place. Then the full-length joists are cut. Then they are laid by the marks on the sill or

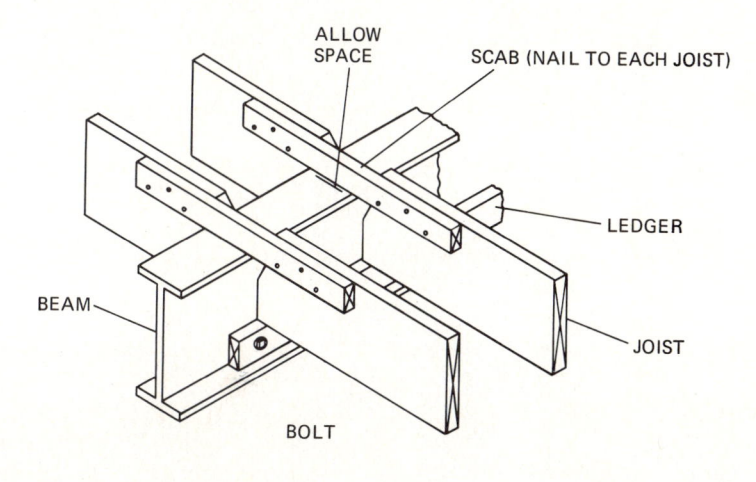

Building Floor Frames **85**

header. Each side of the joist is then toenailed (10d) to the sill. See Figure 5-28. Joists next to openings are not nailed. Next, the ends of the joists are toenailed (10d) on the girder. Then the overlapped ends of the joists are nailed (16d) together. See Figure 5-22. These nails are driven at an angle, as in Figure 5-29.

NAIL OPENING FRAME • A special sequence must be used around the openings. The regular joists next to the opening should not be nailed down. The opening joists are nailed (16d) in place first. These are called *trimmers*.

Then, the first *headers* for openings are nailed (16d) in place. Note that two headers are used. For 2x10-inch joists, three nails are used. For 2x12-inch lumber, four nails are used. Figure 5-30 shows the spacing of the nails.

Next, short tail joists are nailed in place. They span from the first header to the joist header. Three 16d nails are driven at each end. Then, the second header is nailed (16d) in place. See Figure 5-31.

The double trimmer joist is now nailed (16d) in place. These pieces are nailed next to the opening. The nails are alternated top and bottom. See Figure 5-32. This finishes the opening. The regular joists next to the opening are nailed in place. Finally, the header is nailed to the joist ends. Three 16d nails are driven into each joist.

Figure 5-28. The joists are toenailed to the girders.

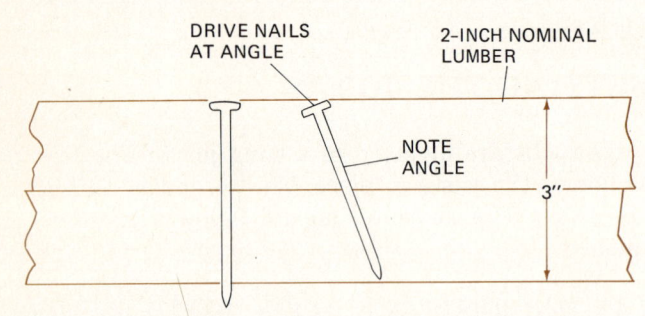

Figure 5-29. When nailing joists together, drive nails at an angle. This holds better and the ends do not stick through.

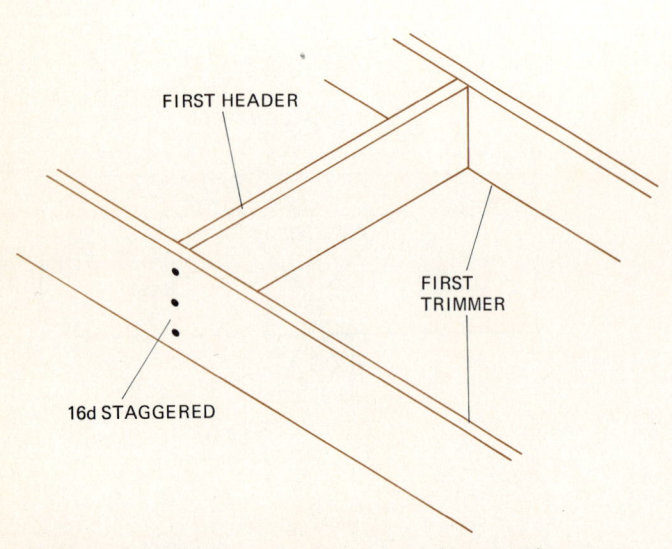

Figure 5-30. Nailing the first parts of an opening.

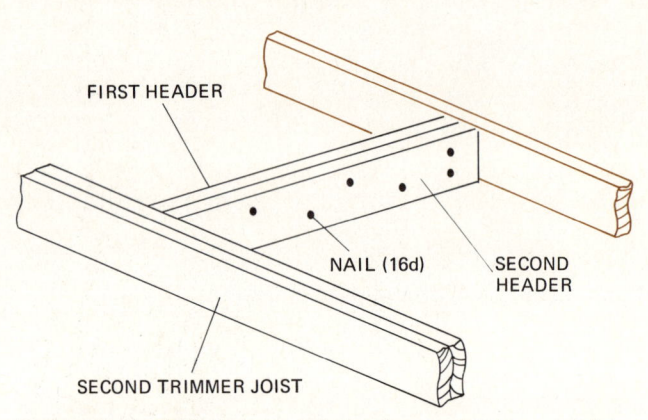

Figure 5-31. Add the second header and trimmer joists.

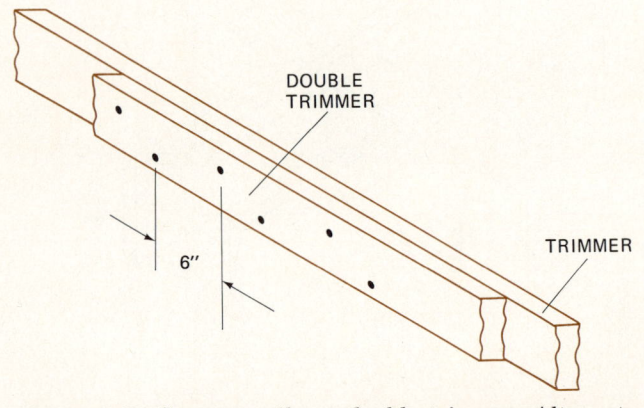

Figure 5-32. Stagger nails on double trimmer. Alternate nails on top and bottom.

86 Carpentry Fundamentals

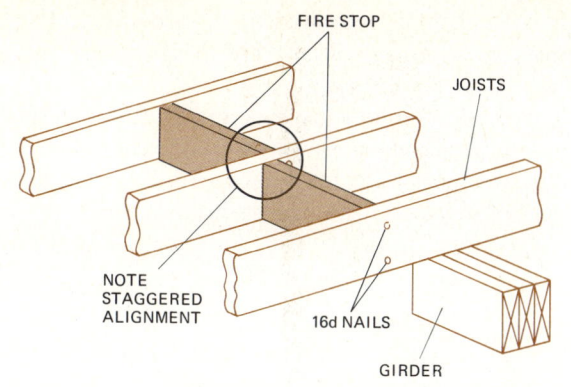

Figure 5-33. Fire stops are nailed in. They keep fire from spreading between walls and floors.

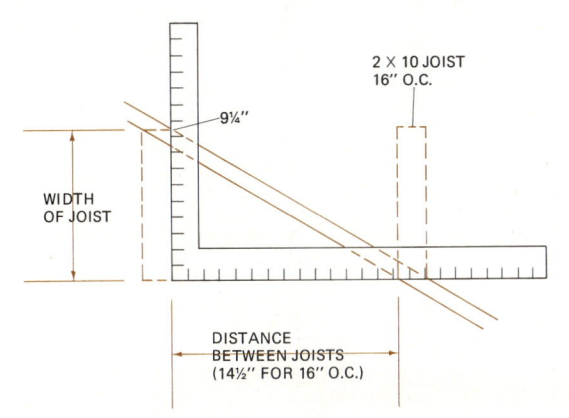

Figure 5-34. Use a square to lay out bridging.

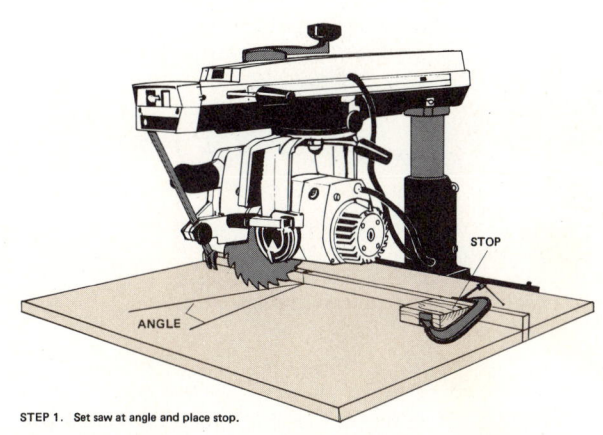

Figure 5-35. Radial arm saw set up for bridging.

FIRE STOPS

Fire stops are short pieces nailed between joists and studs. See Figure 5-33. They are made of the same boards as the joists. Fire stops keep fire from spreading between walls and floors. They also help keep joists from twisting and spreading. Fire stops are usually put at or near the girder. Two 16d nails are driven at each end of the stop. Stagger the boards slightly as shown. This makes it easy to nail them in place.

Bridging

Bridging is used to keep joists from twisting or bending. Bridging is centered between the girder and the header. For most spans, center bridging is adequate. For joist spans longer than 16 feet, more bridging is used. Bridging should be put in every 8 feet. This must be done to comply with most building codes.

Most bridging is cut from boards. It may be cut from either 1-inch or 2-inch lumber. Use the framing square to mark the angles as in Figure 5-34. With this method, the angle may be found.

A radial arm saw may be used to cut multiple pieces. See Figure 5-35. Also, a jig can be built to use a portable power saw. See Figure 5-36.

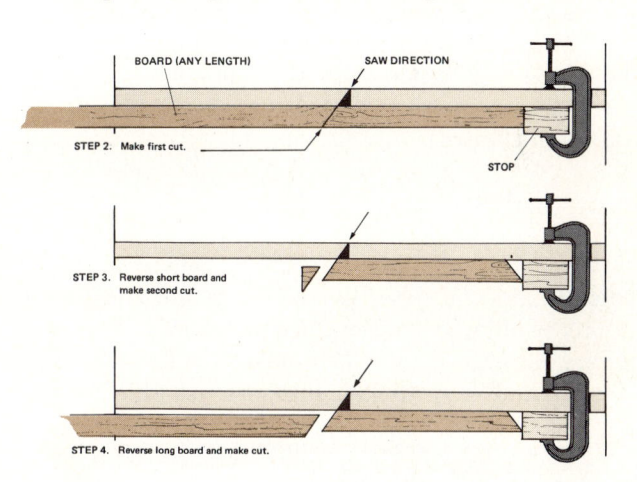

Figure 5-36. A jig may be built to cut bridging.

Building Floor Frames **87**

Special steel bridging is also used. Figure 5-37 shows an example. Often, only one nail is needed in each end. Steel bridging meets most codes and standards.

All the bridging pieces should be cut first. Nails are driven into the bridging before it is put up. Two 8d or 10d nails are driven into each end. Next, a chalk line is strung across the tops of the joists. This gives a line for the bridging.

The bridging is nailed at the top first. This lets the carpenter space the joists for the flooring when it is laid. The bottoms of the bridging are nailed after the flooring is laid.

The bridging is staggered on either side of the chalk line. This prevents two pieces of bridging from being nailed at the same spot on a joist. To nail them both at the same place would cause the joist to split. See Figure 5-38.

• SUBFLOORS •

The last step in making a floor frame is laying the subflooring. Subflooring is also called *underlayment*. The subfloor is the platform that supports the rest of the structure. It is covered with a finish floor material in the living spaces. This may be of wood, carpeting, tile, or stone. However, the finish floor is added much later.

Several materials are commonly used for subflooring. The most common material is plywood. Plywood should be C-D grade with waterproof or exterior glues. Other materials used are chipboard, fiberboard, and boards.

Plywood Subfloor

Plywood is an ideal subflooring material. It is quickly laid and takes little cutting and trimming. It may be either nailed or glued to the joists. Plywood is very flat and smooth. This makes the finished floor smooth and easy to lay. Builders use thicknesses from ½ to ¾ inch plywood. The most common thicknesses are ½ and ⅝ inch. The FHA minimum is ½-inch-thick plywood.

Plywood as subflooring has fewer squeaks than boards. This is because fewer nails are required. The squeak in floors is caused when nails work loose. Table 5-1 shows minimum standards for plywood use.

Figure 5-37. Most building codes allow steel bridging.

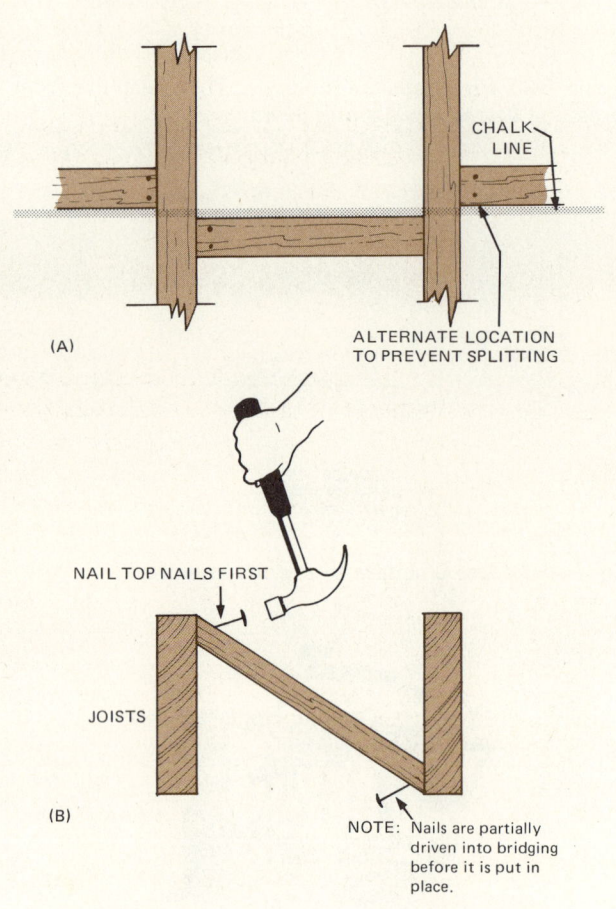

Figure 5-38. Bridging pieces are staggered.

TABLE 5-1 MINIMUM FLOORING STANDARDS

\multicolumn{4}{c}{Single—Layer (Resilient) Floor}
Joists, Inches O.C.	Minimum Thickness, Inches	Common Thickness, Inches	Minimum Index
12	19/32	5/8	24/12
16	5/8	5/8 or 3/4	32/16
24	3/4	3/4	48/24
\multicolumn{4}{c}{Subflooring with Finish Floor Layer Applied}
Joists, Inches O.C.	Minimum Thickness, Inches	Common Thickness, Inches	Minimum Index
12	1/2	1/2 or 5/8	32/16
16	1/2	5/8	32/16
24	3/4	3/4	48/24

NOTES: 1. C-C grade underlayment plywood.
2. Each piece must be continuous over two spans.
3. Sizes can vary with span and depth of joists in some locations.

Chipboard and Fiberboard

As a rule, plywood is stronger than other types of underlayment. However, both fiberboard and chipboard are also used. Chipboard underlayment is used more often. The minimum thickness for chipboard or fiberboard is 5/8 inch. This thickness must also be laid over 16-inch joist spacing. Both chipboard and fiberboard are laid in the same manner as plywood. In any case, the ends of the large sheets are staggered. See Figure 5-39.

Laying Sheets

The same methods are used for any sheet subflooring. Nails are used most often, but glue is also used. An outside corner is used as the starting point. The long grain or sheet length is laid across the joists. See Figure 5-40. The ends of the different courses are staggered. This prevents the ends from all lining up on one joist. If they did, it could weaken the floor. By staggering the end joints, each layer adds strength to the total floor. The carpenter must allow for expansion and contraction. To do this, the sheets are spaced slightly apart. A paper match cover may be used for spacing. Its thickness is about the correct space.

NAILING • The outside edges are nailed first with 8d nails. Special "sinker" nails may be used. The outside nails should be driven about 6 inches apart. Nails are driven into the inner joists about 10 inches apart. See Figure 5-41. Power nailers can be used to save time, cost, and effort. See Figure 5-42.

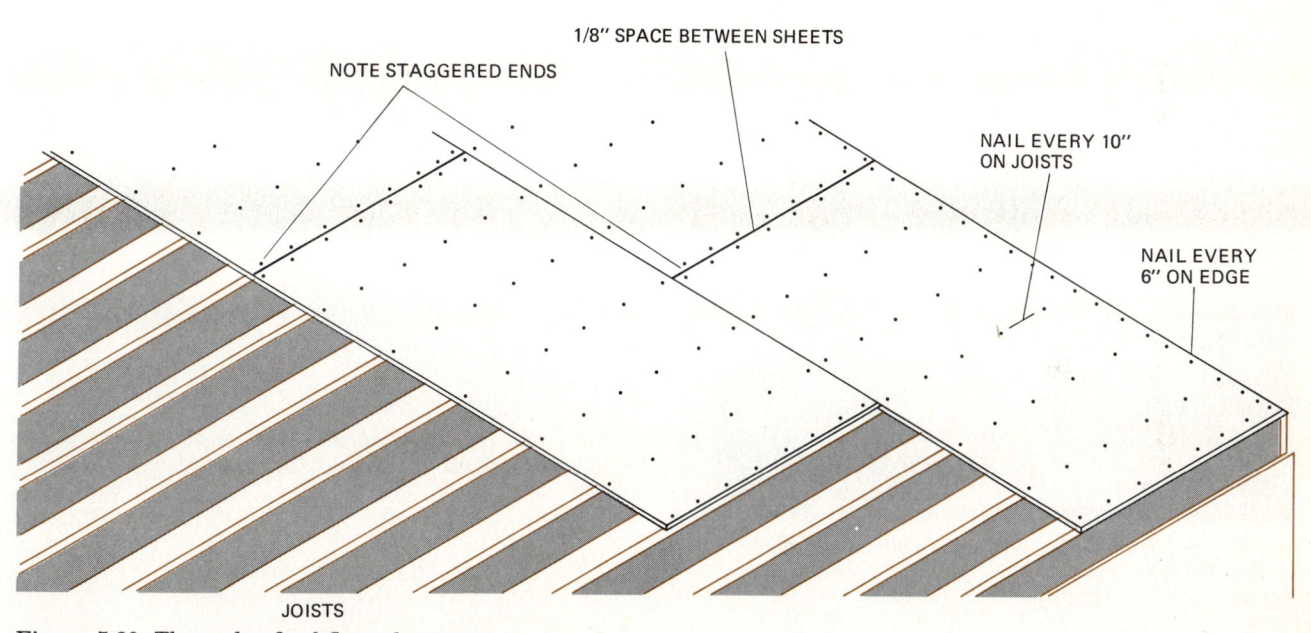

Figure 5-39. The ends of subfloor sheets are staggered.

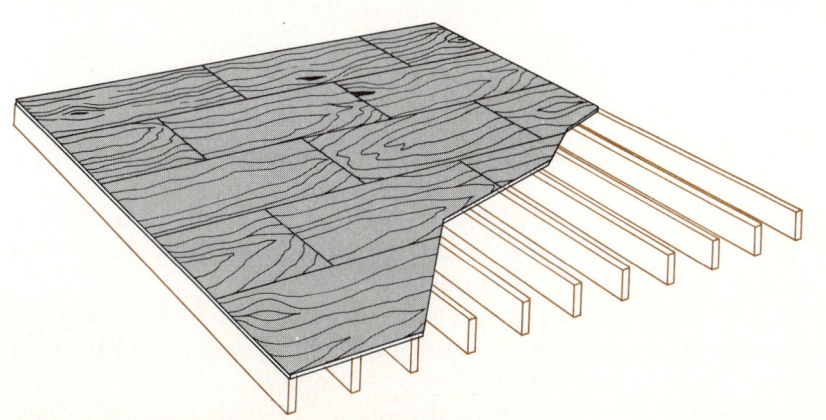

Figure 5-40. The long grain runs across the joists.

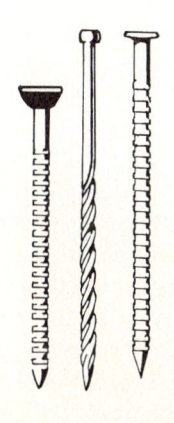

Figure 5-41. Flooring nails. Note the "sinker" head on the first nail.

Building Floor Frames

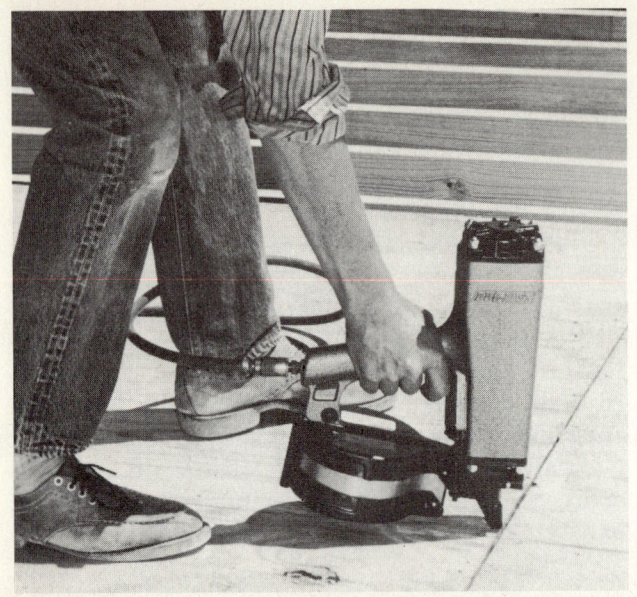

Figure 5-42. Using power nailers saves time and effort. *(Duo-Fast)*

Figure 5-44. Plywood subflooring may also have tongue-and-groove joints. This is stronger. *(American Plywood Association)*

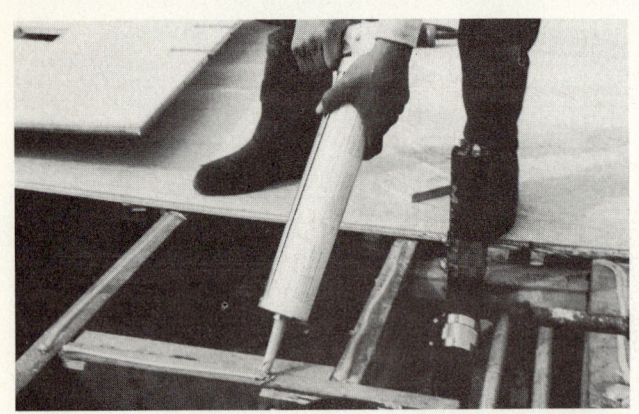

Figure 5-43. Subflooring is often glued to the joists. This makes the floor free of squeaks. *(American Plywood Association)*

Figure 5-45. A buffer board is used to protect the edges of tongue-and-groove panels. *(American Plywood Association)*

GLUING • Gluing is now widely used for subflooring. Modern glues are strong and durable. Glues, also called *adhesives,* are quickly applied. Glue will not squeak as will nails. Figure 5-43 shows glue being applied to floor joists. Floors are also laid with tongue-and-groove joints (Figure 5-44). Buffer boards are used to protect the edges of the boards as the panels are put in place. (See Figure 5-45.)

Board Subflooring

Boards are also used for subflooring. There are two ways of using boards. The older method lays the boards diagonally across the joists. Figure 5-46 shows this. This way takes more time and trimming. It takes a longer time to lay the floor, and more material is wasted by trimming. However, diagonal flooring is still used. It is preferred where wood board finish flooring will be used. This way the finish floor-

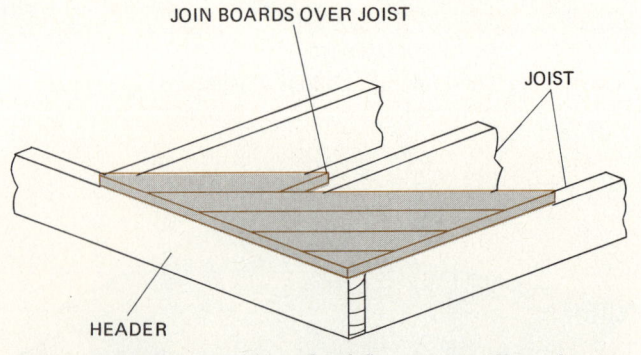

Figure 5-46. Diagonal board subfloors are still used today.

ing may be laid at right angles to the joists. Having two layers that run in different directions gives greater strength.

Today board subflooring is often laid at right angles to the joists. This is appropriate when the finish floor will be sheets of material.

Either way, two kinds of boards are used. *Plain* boards are laid with a small space between the boards. It allows for expansion. End joists must be

90　Carpentry Fundamentals

made over a joist for support. See Figure 5-47. *Grooved* boards are also used. See Figure 5-48. End joints may be made at any point with grooved boards.

NAILING • Boards are laid from an outside edge toward the center. The first course is laid and nailed with 8d nails. Two nails are used for boards 6 inches wide or less. Three nails are used for boards wider than 6 inches.

The boards are nailed down untrimmed. The ends stick out over the edge of the floor. This is done for both grooved and plain boards. After the floor is done, the ends are sawed off. They are sawed off even with the floor edges.

• SPECIAL JOISTS •

The carpenter should know how to make special joists. Several types of joists are used in some buildings. Special joists are used for overhangs and sunken floors. A sunken floor is any floor lower than the rest. Sunken floors are used for special flooring such as stone. Floors may also be lowered for appearance. Special joists are also used to recess floors into foundations. This is done to make a building look lower. This is called the *low-profile* building.

Overhangs

Overhangs are called *cantilevers*. They are used for special effects. Porches, decks, balconies, and projecting windows are all examples. Figure 5-49A shows an example of projecting windows. Figure 5-49B is a different type of bay. But both rest on overhanging floor joist systems. Overhangs are also used

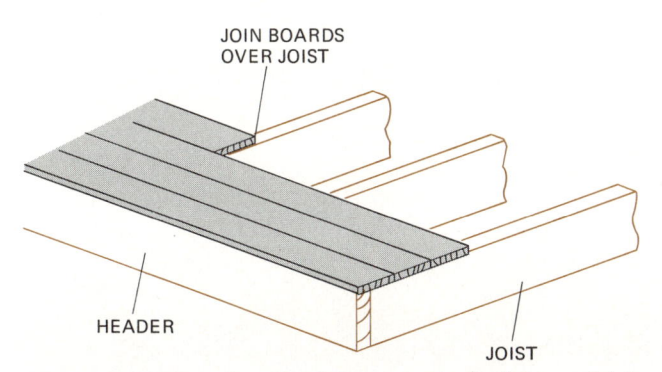

Figure 5-47. Plain board subfloor may be laid across joists. Joints must be made over a joist.

Figure 5-48. Grooved flooring is laid across joists. Joints can be made anywhere.

Figure 5-49. (A) A bay window rests on overhanging floor joists. *(American Plywood Association)*

(B) A different type of bay. Both types rest on overhanging floor joist systems.

Building Floor Frames

for "garrison" style houses. When a second floor extends over the wall of the first, it is called a garrison style. See Figure 5-50A and B.

The longest projection without special anchors is 24 inches. Windows and overhangs seldom extend 24 inches. A balcony, however, would extend more than 24 inches. Thus, a balcony would need special anchors.

OVERHANGS WITH JOIST DIRECTION • Some overhangs project in the same direction as the floor joists. Little extra framing is needed for this. This is the easiest way to build overhangs. In this method, the joists are simply made longer. Blocking is nailed over the sill with 16d nails. Figure 5-51 shows blocking and headers for this type of overhang. Here the joists rest on the sill. Some overhangs extend over a wall instead of a foundation. Then, the double top plate of the lower wall supports the joists.

OVERHANGS AT ANGLES TO JOIST DIRECTION • Special construction is needed to frame this type of overhang. It is similar to framing openings in the floor frame. Stringer joists form the base for the subflooring. Stringer joists must be nailed to the main floor joists. See Figure 5-52. They must

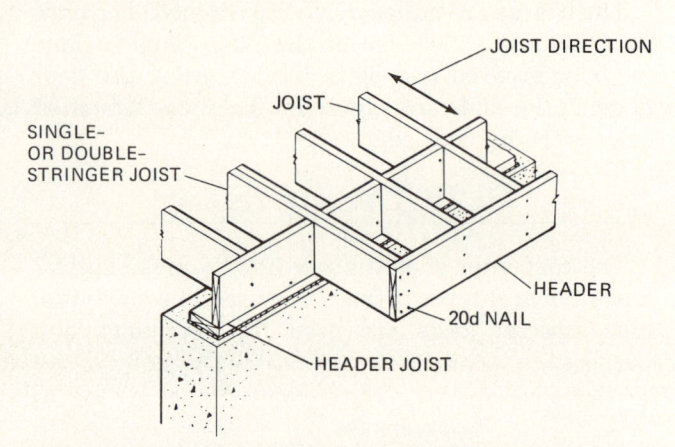

Figure 5-51. Some overhangs simply extend the regular joist. *(Forest Products Lab)*

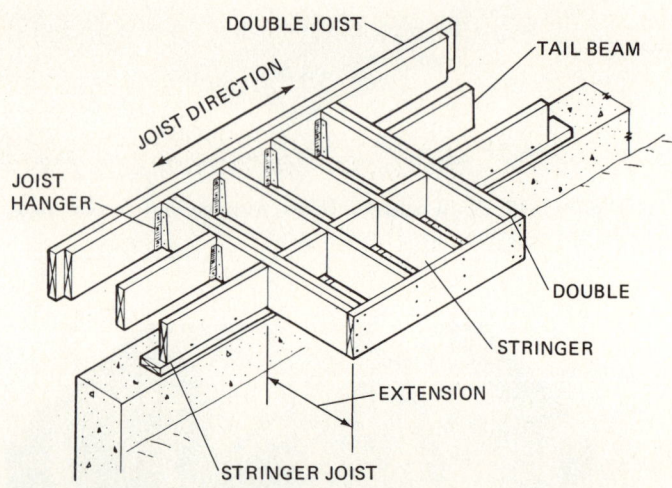

Figure 5-50. (A) Joist protections form the overhang for a garrison type second story.

Figure 5-52. Frame for an overhang at an angle to the joists. *(Forest Products Lab)*

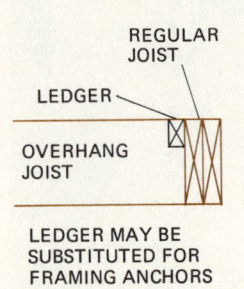

(B) The finished house.

Figure 5-53. A top ledger "let in" is a good anchor.

92 Carpentry Fundamentals

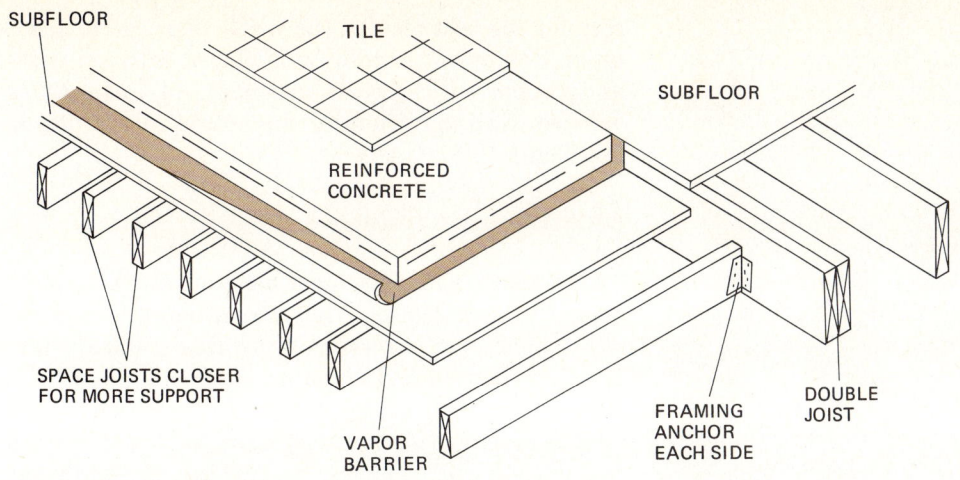

Figure 5-54. Details of frame for a sunken floor.

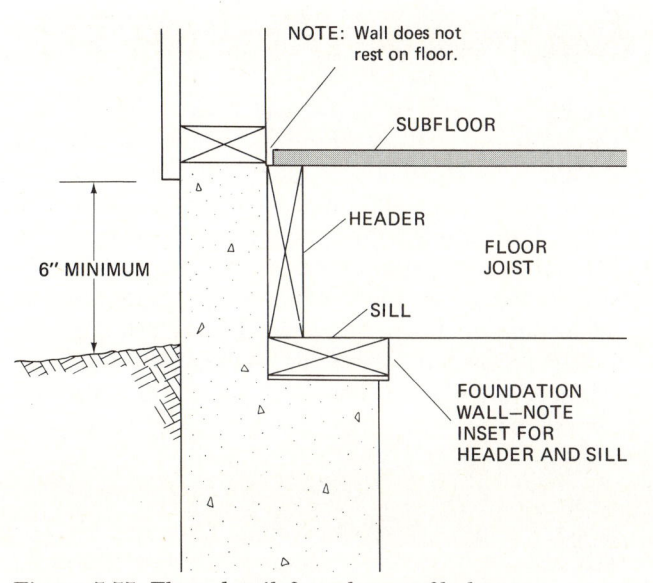

Figure 5-55. Floor detail for a low-profile house.

be inset twice the distance of the overhang. Two methods of attaching the stringers are used. The first method is to use a wooden ledger. However, this ledger is placed on the top. See Figure 5-53. The other method uses a metal joist hanger. Special anchors are needed for large overhangs such as rooms or decks.

Sunken Floors

Subfloors are lowered for two main reasons. A finish floor may be made lower than an adjoining finish floor for appearance. Or the subfloor may be made lower to accommodate a finish floor of a different material. The different flooring could be stone, tile, brick, or concrete. These materials are used for appearance or to drain water. However, they are thicker than most finish floors. To make the floor level, special framing is done to lower the subfloor.

The sunken portion is framed like a special opening. First, header joists are nailed (16d) in place. See Figure 5-54. The headers are not as deep, or wide, as the main joists. This lowers the floor level. To carry the load with thinner boards, more headers are used. The headers are added by spacing them closer together. Double joists are nailed (16d) after the headers.

Low Profiles

The lower profile home has a regular size frame. However, the subfloor and walls are joined differently. Figure 5-55 shows the arrangement. The sill is below the top of the foundation. The bottom plate for the wall is attached to the foundation. The wall is not nailed to the subfloor. This makes the joists below the common foundation level. The building will appear to be lower than normal.

• ENERGY FACTORS •

Most energy is not lost through the floor. The most heat is lost through the ceiling. This is because heat rises. However, energy can be saved by insulating the floor. In the past, most floors were not insulated. Floors over basements need not be insulated. Floors over enclosed basements are the best energy savers.

Floors over crawl spaces should be insulated. The crawl space should also be totally enclosed. The foundation should have ventilation ports. But, they should be closed in winter. The most energy is saved by insulating certain areas. Floors under overhangs and bay windows should be insulated. Floors next to the foundation should also be insulated. The insulation should start at the sill or header. It should extend 12 inches into the floor area. See Figure 5-56. The outer corners are the most critical areas.

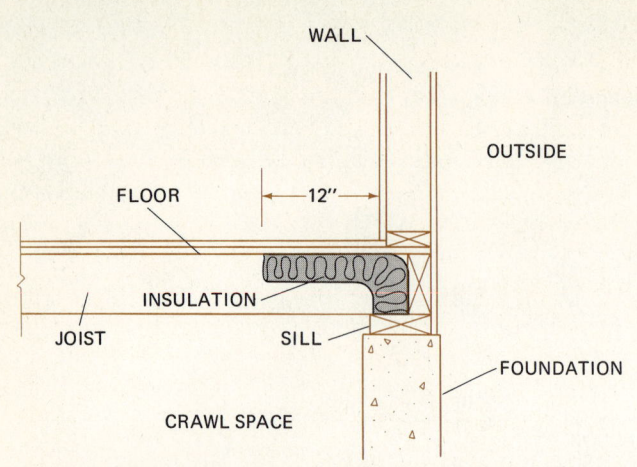

Figure 5-56. Insulate the outside floor edges.

Figure 5-57. Insulation between floor joists should be supported.

But, for the best results, the whole floor can be insulated. Roll or bat insulation is placed between joists and supported. Supports are made of wood strips or wire. Nail (6d) them to the bottom of the joists. See Figure 5-57.

Moisture Barriers

Basements and slabs must have moisture barriers beneath them. Moisture barriers are not needed under a floor over a basement. However, floors over crawl spaces should have moisture barriers. The moisture barrier is laid over the subfloor. See Figure 5-58. A moisture barrier may be added to older floors below the joists. This may be held in place by either wooden strips or wires. Six-mil plastic or builder's felt is used. See Figure 5-59.

Energy Plenums

A plenum is a space for controlled air. The air is pressurized a little more than normal. Plenum systems over crawl spaces allow air to circulate beneath floors. This maximizes the heating and cooling effects. Figure 5-60 shows how the air is circulated. Doing this keeps the temperature more even. Even temperatures are more efficient and comfortable.

The plenum must be carefully built. Insulation is used in special areas. See Figure 5-61. A hatch is needed for plenum floors. The hatch gives access

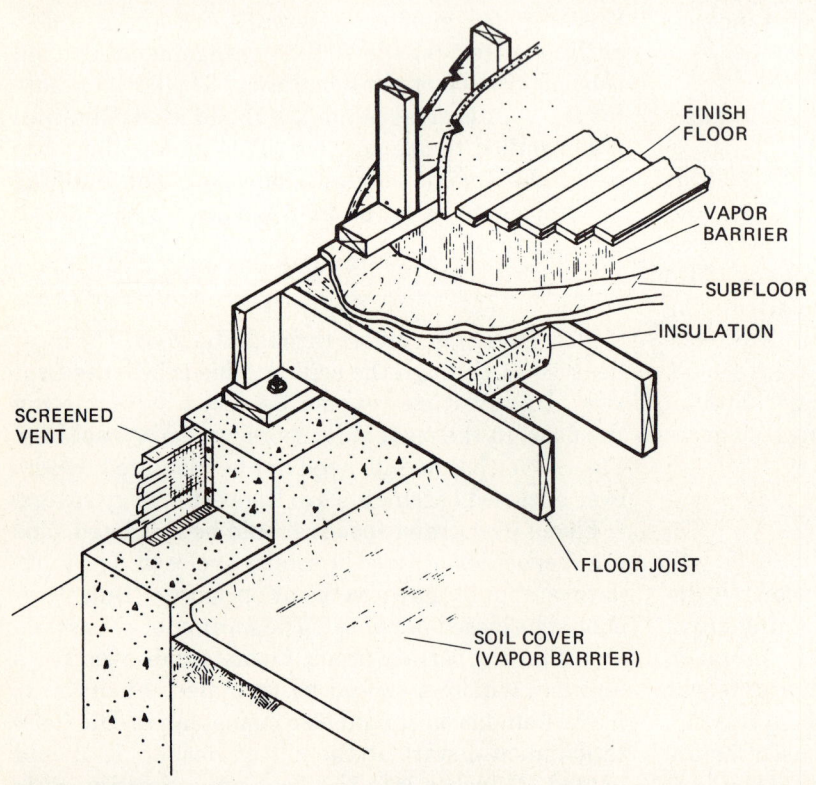

Figure 5-58. A moisture barrier is laid over the subfloor above a crawl space. *(Forest Products Lab)*

94   Carpentry Fundamentals

to the plenum area. Access is needed for inspection and servicing. There are no outside doors or vents to the plenum.

The plenum arrangement offers an advantage to the builder. A plenum house can be built more cheaply. There are several reasons. The circulation system is simpler. No ducts are built beneath floors or in attics. Common vents are cut in the floors of all the rooms. The system forces air into the sealed plenum. Figure 5-62 shows this. The air does not lose energy in the insulated plenum. The forced air then enters the various rooms from the plenum. The blower unit is in a central portion of the house. The blower can send air evenly from a central area. The enclosed louvered space lets the air return freely to the blower.

Rough plumbing is brought into the crawl space first. Then the foundation is laid. Fuel lines and clean-outs are located outside the crawl space. The minimum clearance in the crawl space should be 18 inches. The maximum should be 24 inches. This size gives the greatest efficiency for air movement.

Figure 5-59. A moisture barrier may also be added beneath floors.

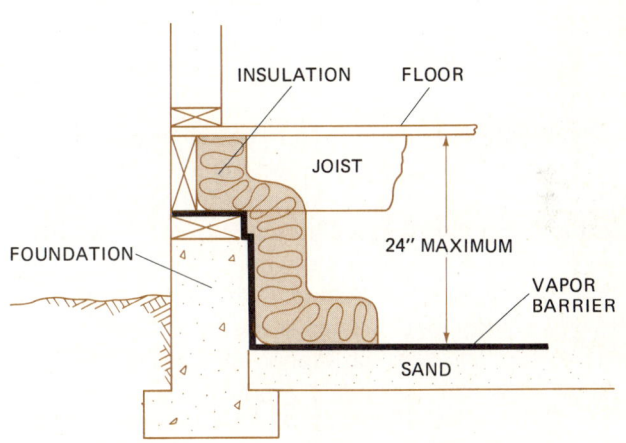

Figure 5-61. Section view of the energy plenum.

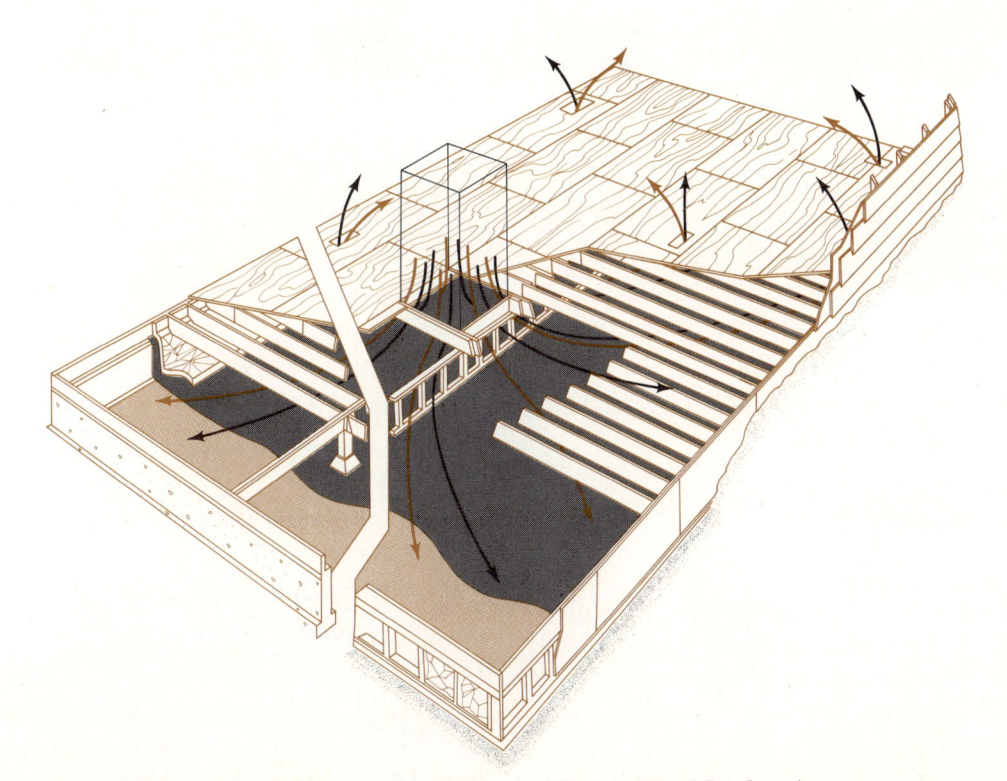

Figure 5-60. The air-circulation system for an energy plenum. *(Western Wood Products)*

Building Floor Frames   95

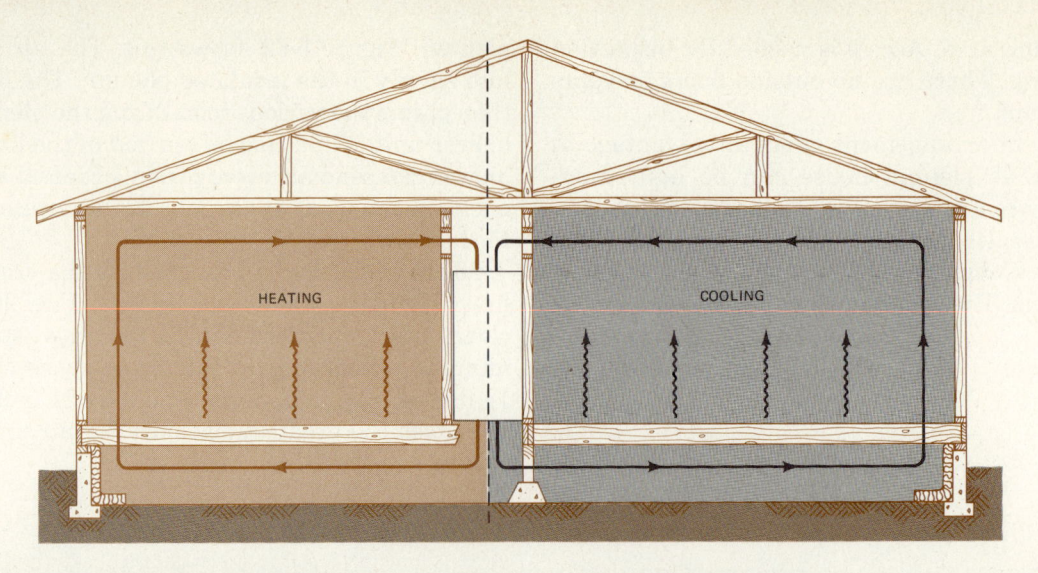

Figure 5-62. Conditioned air is circulated through the plenum. *(Western Wood Products)*

Foundation walls may be masonry or poured concrete. Special treated plywood foundations are also used. Proper drainage is essential. After the foundation is built, the sill is anchored. Standard sills, seals, and termite shields may be used. The plenum area must be covered with sand. Next, a vapor barrier is laid. The vapor barrier is made of 6-mil plastic film. The film must be lapped at least 2 inches at joints. It is laid over the ground and extends up over the sill. See Figure 5-61. This completely seals the plenum area. Then insulation is laid. Either rigid or batt insulation may be used. It should extend from the sill to about 24 inches inside the plenum. The most energy loss occurs at foundation corners. The insulation covers these corners. Then, the floor joists are nailed to the sill. The joist header is nailed (16d) on and insulated. Subflooring is then nailed to the joists. This completes the plenum. After this, the building is built as a normal platform.

• WORDS CARPENTERS USE •

joist	girder	ledger	bridging	cantilever
platform frame	trimmer	scabs	subfloor	insulation
balloon frame	header	joist hanger	underlayment	plenum
anchor bolt	double trimmer	tail joist	chipboard	moisture barrier
sill	joist header	fire stops	sinker nail	

• STUDY QUESTIONS •

5-1. Know the terms above.
5-2. List the procedure for building a floor.
5-3. List the parts of a frame floor.
5-4. Why are regular joists by openings nailed last?
5-5. What does a girder do?
5-6. What are scabs?
5-7. What is a crawl space?
5-8. What holds the wooden frame to the foundation?
5-9. Where is joist spacing marked?
5-10. Why is the first joist spaced differently?
5-11. What is a post jack?
5-12. What are the two joist (framing) methods?
5-13. What are common joist spacings?
5-14. What size lumber is used for joists?
5-15. How are girders made?
5-16. What is the easiest way to cut joists?
5-17. What is a gusset?
5-18. What are the advantages of using joist hangers?
5-19. What sequence is used to nail the opening frames?
5-20. Why are fire stops and bridging used?
5-21. Why are sunken floors used?
5-22. What is done to frame overhangs?
5-23. What is done to frame sunken floors?
5-24. How can floors be changed to save energy?
5-25. What is an energy plenum?

6 FRAMING WALLS

How to build a wall frame is the topic of this unit. How to cut the parts for the wall is covered, then how to connect the wall to the rest of the building. Why the parts are made as they are is also explained. You will learn how to:

- Lay out wall sections
- Measure and cut the parts
- Assemble and erect wall sections
- Join wall sections together
- Change a standard wall frame to save energy and materials

• INTRODUCTION •

There are two ways of framing a building. The most common is called the western platform method. The other is the balloon method. In most buildings today, the western platform method is used.

In the western platform method, walls are put up after the subfloor has been laid. Walls are started by making a frame. The frame is made by nailing boards to the tops and bottoms of other boards. The top and bottom boards are called *plates*. The vertical boards are called *studs*. The frame must be made very strong because it holds up the roof. After the frame is put together, it is raised and nailed in place. The wall frames are put up one at a time. A roof can be built next. Then the walls are covered and windows and doors are installed. Figure 6-1 shows a wall frame in place. Note that the roof is not on the building yet.

Sometimes the first covering for the wall is added before the wall is raised. This first covering is called *sheathing*. It is very easy to nail the sheathing on the frame while the frame is flat on the floor. Doing this also makes the job quicker, and reduces problems with keeping the frame square. Another advantage is that no scaffolds will be needed to reach all the areas. This reduces the possibility of accidents and saves time in moving scaffolds. However, most builders still use the other way.

The wall is attached to the floor in most cases. When the floor is built over joists, the wall is nailed to the floor. See Figure 6-2. When the wall is built on a slab, it is anchored to the slab. See Figure 6-3. As a rule, walls for both types of floors are made the same way.

Walls that help hold up the roof, or the next floor, are made first. As a rule, all outside walls do this. Outside walls are called *exterior* walls. These walls are also called *load-bearing* walls.

Inside walls are called *partitions*. They can be load-bearing walls, too. However, not all partitions carry loads. Interior walls that do not carry loads may be built after the roof is up. Interior walls that do not carry loads are also called *curtain walls*.

After the walls are put up, the roof is built. The walls are then covered. The first cover is the sheathing. Putting on the sheathing after the roof lets a builder get the building waterproof or weatherproof a little sooner. The siding is the wall cover that shows. Siding is put on much later.

Wall sections are made one at a time. The longest outside walls are made first. The end walls are made next. However, the sequence can be changed to fit the job. There are many ways to make walls. This chapter will show the most common one. One or two other ways will also be shown when they are commonly

• SEQUENCE •

The general sequence for making wall frames is:

1. Lay out the longest outside wall section
2. Cut the parts
3. Nail the parts together
4. Raise the wall
5. Brace it in place
6. Lay out the next wall
7. Repeat the process
8. Join the walls
9. Do all outside walls
10. Do all inside walls
11. Build the roof
12. Sheath the outside walls
13. Install outside doors and windows
14. Cut sole plates from inside door openings

• WALL LAYOUT •

All the parts of the wall must be planned. The carpenter must know what parts to cut. The carpenter must also nail the parts together. The plans are called the *wall layout*. The carpenter makes the layout on boards. One layout is done on the top and bottom boards (plates). Another layout is done for the wall studs.

Plate Layout

SOLE PLATES • First, select pieces of 2-inch lumber for the bottom of the wall. The bottom part of the wall is called a *sole plate*. The sole plate is laid along the edges of the floor where the walls will be. No sole plate is put across large door openings. Sometimes, the sole plate can be made across small doors. After the wall is put up, the sole plate is sawed out.

TOP PLATE • After a sole plate has been laid, another piece is laid beside it. It will be the top part of the wall. This piece is called a *top plate*. The sole plate and the top plate are laid next to each other with a flat side up. Figure 6-4 shows how the sole plate and the top plate are laid. The measurements

Figure 6-1. Wall frames are put up after the floor is built.

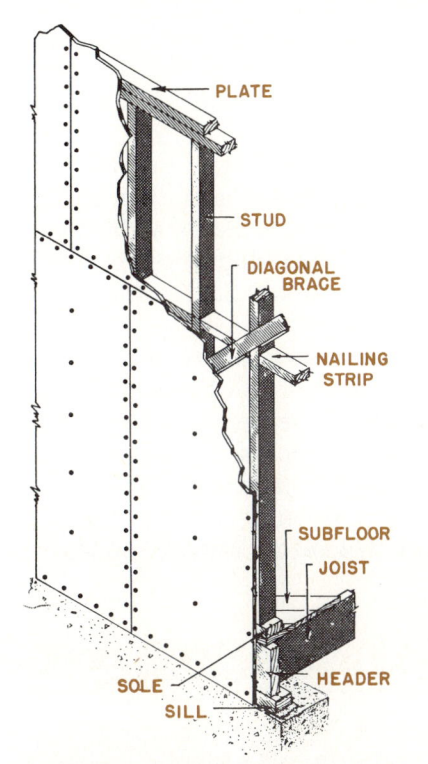

Figure 6-2. Section view of a wall on a frame floor.

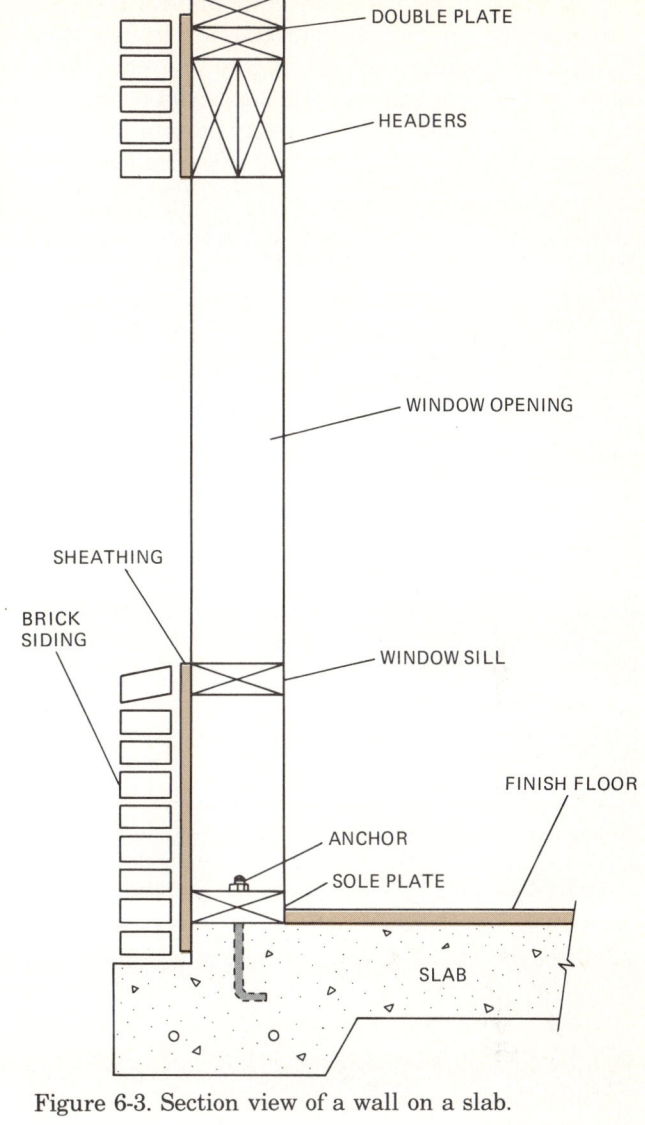

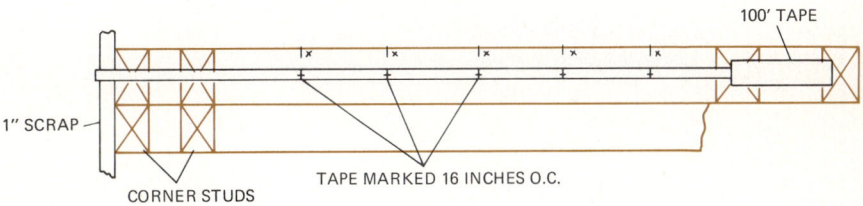

Figure 6-3. Section view of a wall on a slab.

NAIL A PIECE OF 1 INCH SCRAP AT THE END. HOOK THE TAPE OVER THE END. MARK THE INTERVAL. THEN MAKE A SMALL X TO THE RIGHT OF THE MARK.

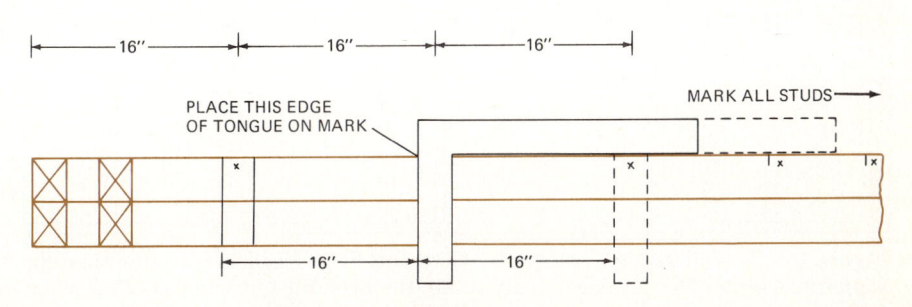

Figure 6-4. Lay sole and top plates for marking.

USE A 2 INCH SCRAP TO MARK WIDTHS. THEN MARK STUD LOCATIONS.

Framing Walls 99

and marks will be made on both plates. This lets you measure and mark both the top and bottom plates at the same time. This way you can check the location of the marks. This keeps you from making mistakes. It keeps both top and bottom locations aligned. Sole and top plates are often spliced. The splice must occur over the center of a full stud. See Figure 6-5. Otherwise the wall section will be weakened.

Stud Layout

Several types of studs are used in walls. The studs that run from the sole plate to the top plate are called *full studs*. Studs that run from the sole plate to the top of a rough opening are called *trimmer studs*. Short studs that run from either plate to a header or a sill are called *cripple studs*. Figure 6-6 shows a part of a wall section.

SPACING FULL STUDS • Studs that run from sole plate to top plate are called full studs. Most full studs are spaced a standard distance apart. This standard distance is an even part of the sizes of plywood, sheathing, and other building materials. The studs act both as roof support and as a nailing base for the sheathing. The most common standard distance is 16 inches. Studs spaced 16 inches apart are said to be 16 inches *on centers* (O.C.). Another common spacing is 24 inches O.C. However, lines on centers are not used to show where boards are placed. An easier way is shown in Figure 6-4. The stud is located by using the left end of the wall as a starting point. Nail a 1-inch scrap piece there. Then the distance between centers is measured from the block on the outside corner of the wall. A mark is made, and the X is made to the right of the mark.

Figure 6-5. Top plates are spliced over a stud.

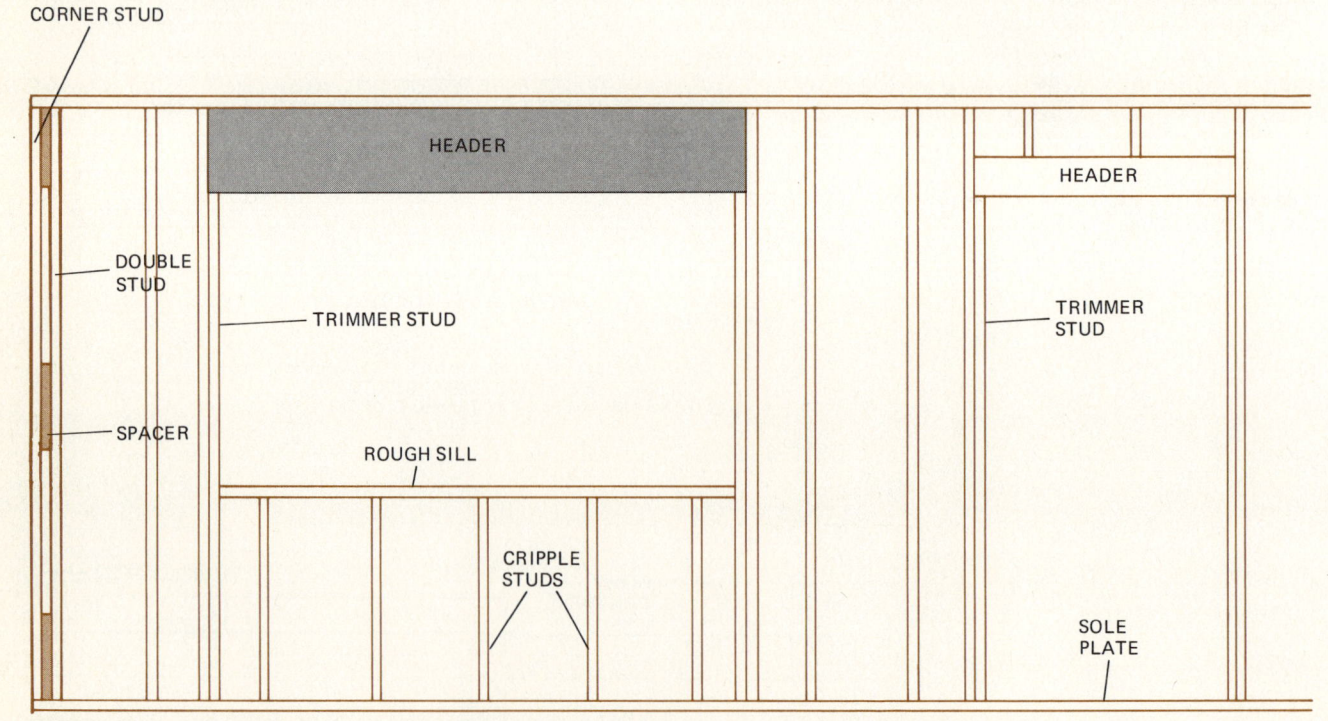

Figure 6-6. A wall section and parts. Note the large header over the opening. It is sometimes larger than needed, to eliminate the need for top cripples. This saves labor costs because it takes longer to cut and nail cripples.

100 Carpentry Fundamentals

All of the marks are made. Then a square is used to line in the locations. This method is easier because the measurement is taken from the side of the stud. The distance is the same whether it is center to center or side to side.

SPACE THE ROUGH OPENINGS • The next step in spacing is to find where the windows and doors are to be made. The openings for windows and doors are framed no matter what the stud spacing is. The openings are called rough openings (R.O.). The size of rough openings may be shown in different ways. The actual size may be shown in the plan. However, many plans show *window schedules*. These often list the window as a number, like "2442." This means that the window sash opening is 2'4" wide and 4'2" high. The rough openings are larger. Wooden windows require larger R.O.s than metal ones. It is best to refer to specifications. However, when they are not available, carpenters can use a rule of thumb. Wooden windows are usually written, for example, 24x42. The R.O. should be 3 inches wider (2'7") and 4 inches higher (4'6"). Metal frames are usually written without the ×. The R.O. is 2 inches wider and 3 inches higher. For a 2442 metal window, the R.O. would be 2'6" wide and 4'5" high.

To find the locations of rough openings, measure the distance from the corner or end of the wall to the center of the rough opening. Make a mark called a centerline on the sole plate. From the centerline measure half of the rough opening width on each side. Make another mark at each side of the centerline for the rough opening. Mark a line, as in Figure 6-7, to show the thickness of a stud. This thickness goes on the outside of the opening. Note that the distance between the lines must be the width of the opening. Mark a T in the regular stud spaces. This tells that a *trimmer* stud is placed there. On the outside of the space for the trimmer stud lay out another thickness. Mark an X from one corner of the opening to the other as shown in the figure. The X is used to show where a full stud is placed. The T is used to show where a trimmer stud is placed. The trimmer stud does not extend from the sole to the top plate. Thus, it is shown only on the sole plate.

Corner Studs

A strong way of nailing the walls together is needed. To make this, a double stud is used for one corner.

CORNER STUDS ON THE FIRST WALL • The spacing of the studs starts at one corner. Another stud space is marked at the second corner. The regular spacing is not used at corners. A stud is laid out at each corner, or end, of the wall.

To make the corner stronger, another stud is placed in the corner section. See Figure 6-8(A). The second corner stud is spaced from the first with spacer blocks. This is called a *built-up corner*. It is done to give the corner greater strength and to make a nail base on the inside. A nail base is needed on the inside to nail the inside wall covering in place. After the end stud is marked on the corner, mark the thickness of one more stud on the plate. Mark it with an S for spacer. Next, lay out the stud as shown in Figure 6-8(B). Mark it with an X to indicate a full stud.

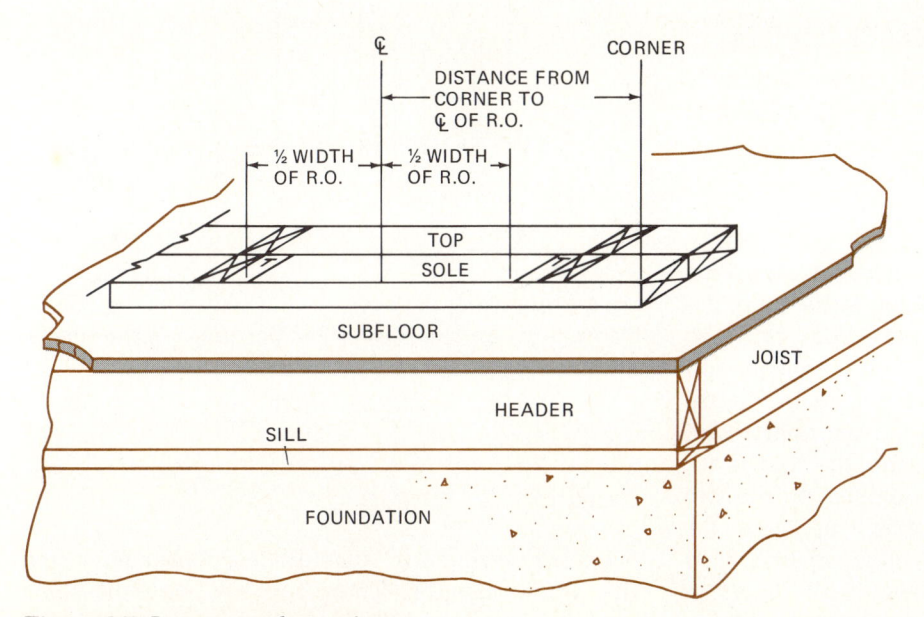

Figure 6-7. Locate rough openings.

Framing Walls 101

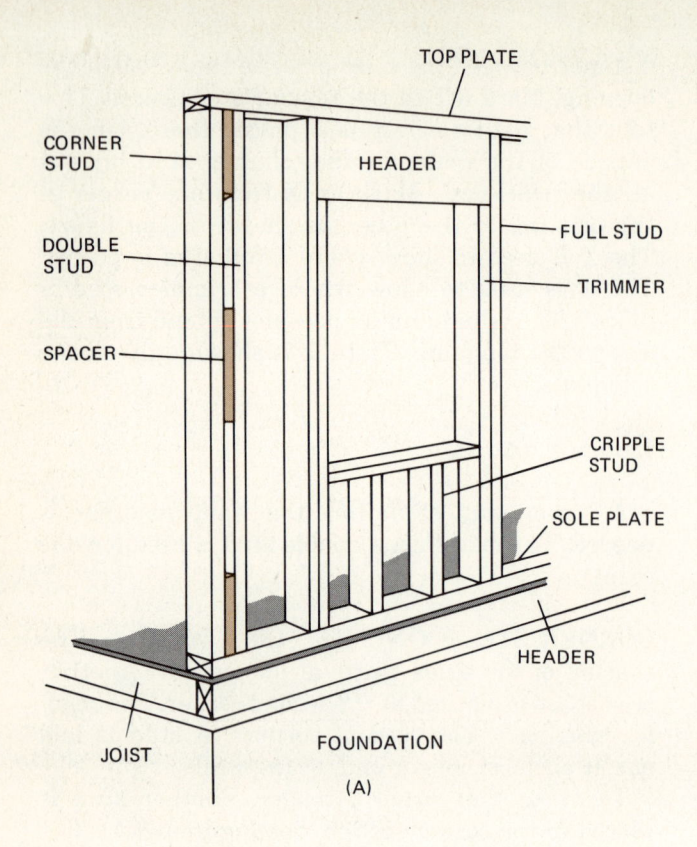

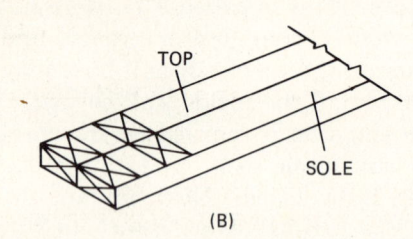

Figure 6-8. Stud layout for corners: (A) the assembly (B) the layout.

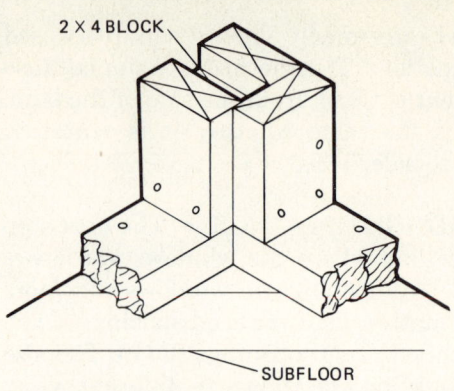

Figure 6-9. The second wall is nailed to the double corner. *(Courtesy Forest Products Lab)*

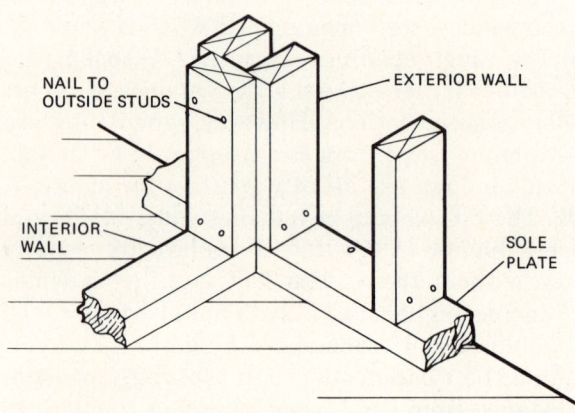

Figure 6-10. Partition walls are nailed to special studs in outside walls. *(Courtesy Forest Products Lab)*

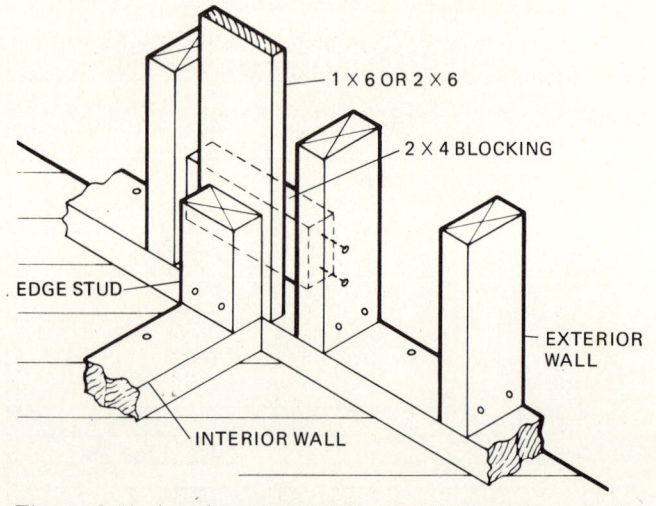

Figure 6-11. Another way to join partitions to the outside walls. *(Courtesy Forest Products Lab)*

CORNER STUDS FOR THE SECOND WALL •

The second wall does not need double studs. It is laid out in the regular way. The walls are nailed together as in Figure 6-9.

Partition Studs

Inside walls are called partitions. The partitions must be solidly nailed to the outside walls. To make a solid nail base, special studs are built into the exterior walls. See Figure 6-10. The most common method is to place two studs as shown in Figure 6-10. The studs are placed 1½ inches apart. The space is made just like corners on the first wall. The corner of the partition can now rest on the two studs. The two special studs act as a nail base for the wall. Also, ¾ inch of each stud is exposed. This makes a nail base for the interior wall covering. Other methods of joining partitions to the outside walls are shown in Figures 6-11 and 6-12.

Find Stud Length

Before cutting any studs, the carpenter must find the proper lengths. There are two ways of doing this. One way lets the carpenter measure the length. In the other way, the carpenter must compute it.

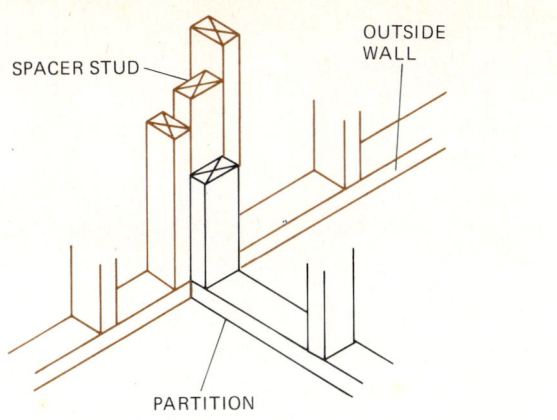

Figure 6-12. A third way to join partitions to the outside walls.

MAKE A MASTER STUD PATTERN • A carpenter can make a master stud pattern. This is the way that stud lengths are measured. A 2-inch board just like the studs is used. A side view of the wall section is drawn. The side view is full size and will show the stud lengths.

The carpenter starts by laying out the distance between the floor and the ceiling height. The distances in between are then shown. Rough openings are added. This then shows the lengths of trimmers and cripples. The pattern also lets the carpenter check the measurements before cutting the pieces. The stud pattern is also called a story pole or rod. As a rule, it is done for only one floor of the house. Figure 6-13 shows a master stud pattern.

COMPUTE STUD LENGTH • Another way to find stud lengths is to compute them. The carpenter must do some arithmetic and check it very carefully. This method is shown in Figure 6-14.

The usual finished floor-to-ceiling distance is 8' ½". The thicknesses of the finish floor and ceiling material are added to this dimension. The floor and ceiling thicknesses are commonly ½ inch each. The distance is written all in inches, and adds up to 97½ inches. The thickness of the sole and top plates is subtracted. For one sole plate and a double top plate, this thickness is 4½ inches (3 times 1½ inches). The remainder is the stud length. For this example, the stud length is 93 inches. This is a commonly used length.

USING PRECUT STUDS • Sometimes the studs are cut to length at the mill and delivered to the site. When this is done, the carpenter does not make

Figure 6-13. Make a master stud pattern by drawing it full size on a board.

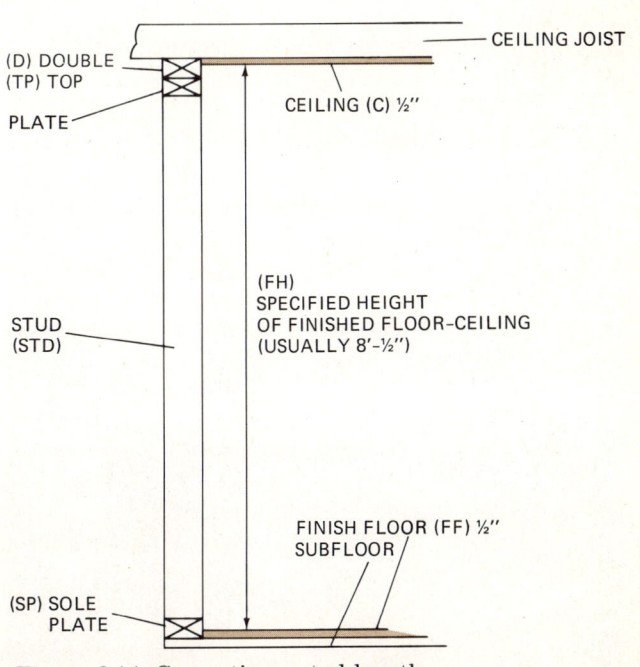

Figure 6-14. Computing a stud length.

Framing Walls 103

any measurements or cuts for full studs. The standard length for such precut studs is 93 inches. The carpenter who orders such materials should be very careful to specify "precision end trim" (P.E.T.) for lumber. See Figure 6-15.

Frame Rough Openings

The locations of full and trimmer studs for the rough opening are shown on the sole plates. The lengths of these can be found from the size of the rough opening. The story pole is used for reference. The width of the R.O. sets the distance between the trimmers. See Figure 6-6. A full stud is used on the outside of the trimmer.

Trimmer Studs

Trimmer studs extend from the sole plate to the top of the rough opening. They provide support for the header. The header must support the wall over the opening. It is important that the header be solidly held. The trimmers give solid support to the ends of the headers. The length of trimmers is the distance from the sole plate to the header.

Header Size

The size of the header is determined by the width of the rough opening. Table 6-1 shows the size for a typical opening width. As in Figure 6-3, two header pieces are used over the rough opening. In some cases, the headers may be large enough to completely fill the space between the rough opening and the top plate. See Figure 6-16. However, doing this will make the wall section around the header shrink and expand at a different rate than the rest of the wall. A solid header section is also harder to insulate.

THE SILL • The bottom of the rough opening is framed in by a sill. The sill does not carry a load. Thus, it can simply be nailed between the trimmer studs at the bottom of the rough opening. The sill does not require solid support beneath the ends.

However, it is common to use another short trimmer for this. See Figure 6-17.

CRIPPLES • Cripple studs are short studs. They join the header to the top plate. They also join the sill to the sole plate. They continue the regular stud spacing. This makes a nail base for sheathing and wallboard.

• CUTTING STUDS TO LENGTH •

After the wall parts are laid out, carpenters must find how many studs are needed. They must do this for each length. A great deal of time can be saved if all studs are cut at one time. Saws are set and full studs are cut first. Then the settings are changed and all trimmer studs are cut. Next, cripple studs are cut. Headers and sills should be cut last.

Cutting Tips

Most cuts to length are done with power saws. Two types are used. Radial arm saws can be moved to a location. See Figure 6-18. Portable circular saws

TABLE 6-1 SIZE OF LUMBER FOR HEADERS

Width of Rough Opening	Minimum Size Lumber	
	One Story	Two or More
3'0"	2 × 4	2 × 4
3'6"	2 × 4	2 × 6
4'0"	2 × 6	2 × 6
4'6"	2 × 6	2 × 6
5'0"	2 × 6	2 × 6
6'0"	2 × 6	2 × 8
8'0"	2 × 8	2 × 10
10'0"	2 × 10	2 × 12
12'0"	2 × 12	2 × 12

Figure 6-15. Precut studs ready at a site.

Figure 6-16. Solid headers may be used instead of cripples.

Carpentry Fundamentals

can be used almost anyplace. See Figure 6-19. With either type of saw, special setups can be used. Pieces can be cut to the right length without measuring each one.

PORTABLE CIRCULAR SAW • Sawing several pieces to the same length is done with a special jig. See Figure 6-20. Two pieces of stud lumber are nailed to a base. Enough space is left between them for another stud. A stop block is nailed at one end. A guide board is then nailed across the two outside pieces. Care should be taken because the guide is for the saw frame. The blade cuts a few inches away.

RADIAL ARM SAW • A different method is used on a radial arm saw. No marking is done. Figure 6-21 shows how to set the stop block. This is quicker and easier. The piece to be cut is simply moved to touch the stop block. This sets the length. The piece is held against the back guide of the saw. The saw is then pulled through the work.

Figure 6-17. A trimmer may be used under both the header and the sill.

Figure 6-18. A radial arm saw can be used on the site.

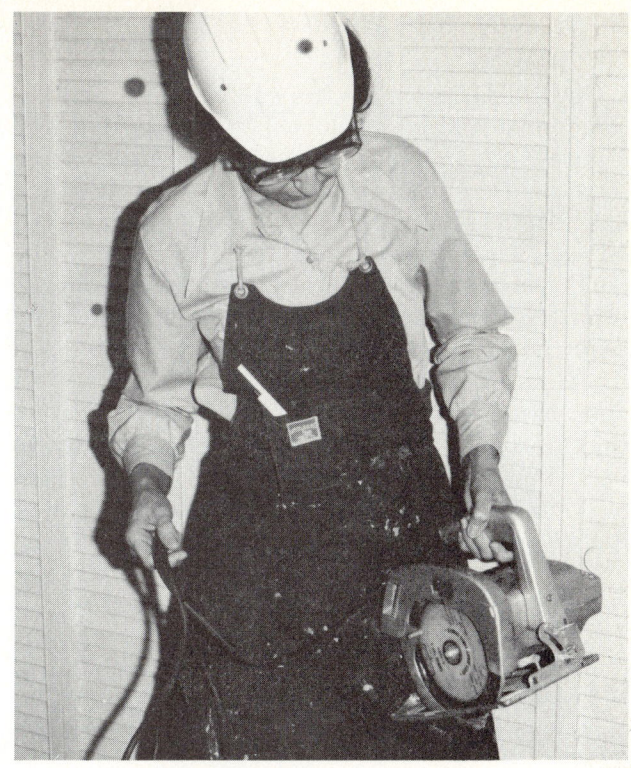

Figure 6-19. Portable circular saws can be used almost anywhere.

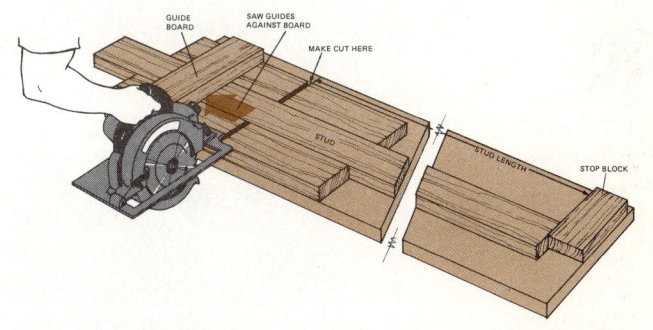

Figure 6-20. A jig for cutting studs with a portable circular saw.

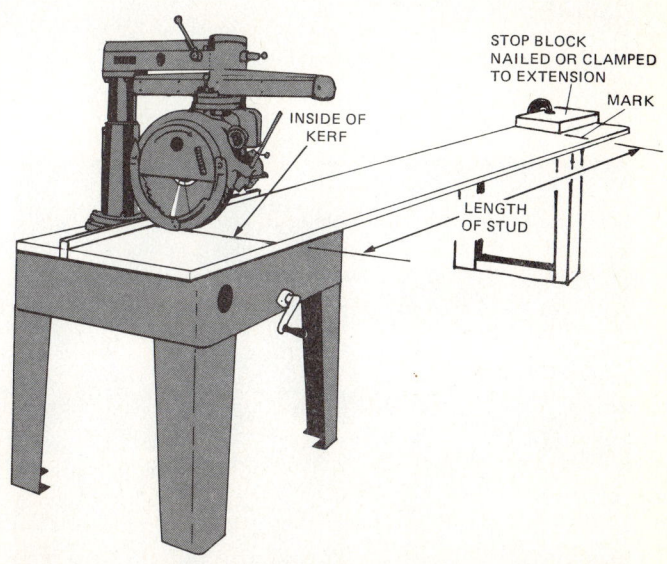

Figure 6-21. A stop block is used to set the length with a radial arm saw. This way all the studs can be cut without measuring. This saves much time and money.

Framing Walls 105

• WALL ASSEMBLY •

Once all pieces are cut, the wall may be assembled. Headers should be assembled before starting. See Figures 6-22 and 6-23. The sole plate is moved about 4 inches away from the edge of the floor. It is laid flat with the stud markings on top. Then it is turned on edge. The markings should face toward the middle of the house. Then the top plate is moved away from the sole plate. The distance should be more than the length of a full stud. The sole plate and top plate markings should be aligned. They must point toward each other. The straightest studs are selected for the corners. They are put at the corners just as they will be nailed. Next, a full stud is laid at every X location between the sole and top plates. Figure 6-24 shows carpenters doing this.

Nailing Studs to Plates

All studs are laid in position. The sole plate and top plate are tapped into position. Make sure that each stud is within the marks.

CORNER STUDS • Before nailing, corner spacers are cut. Three spacers are used at each corner. Each spacer should be about 16 inches long. Spacers are put between the two studs at the corners. One spacer should be at the top, one should be at the bottom, and one should be in the center. Two 16d nails should be used on each side. See Figure 6-25. After the corner studs are nailed together, they are nailed to the plates. Two 16d nails are used for each end of each stud.

FULL STUDS • All the full studs are nailed in place at the X marks. Two 16d nails are used at each end. Figures 6-26 and 6-27 show how.

TRIMMER STUDS • Next, the trimmer studs are laid against the full studs. The spacing of the rough openings is checked. The trimmers are then nailed to the studs from the trimmer side. See Figure 6-28. Use 10d nails. The nails should be staggered

Figure 6-22. Use ½" plywood for spacers between headers.

Figure 6-23. Nailers can be used to assemble headers.

Figure 6-24. Laying the studs in place on the subfloor.

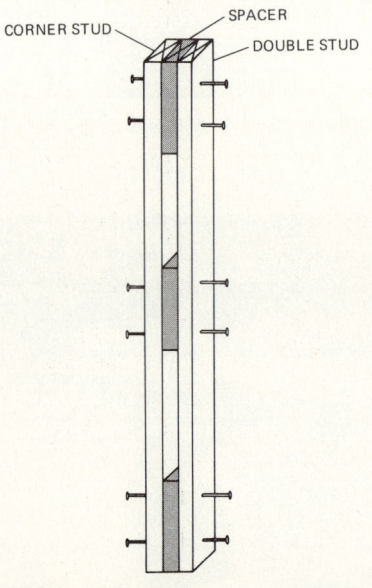

Figure 6-25. Nailing spacers into corner studs. Use two 16d nails on each side of spacer.

106 Carpentry Fundamentals

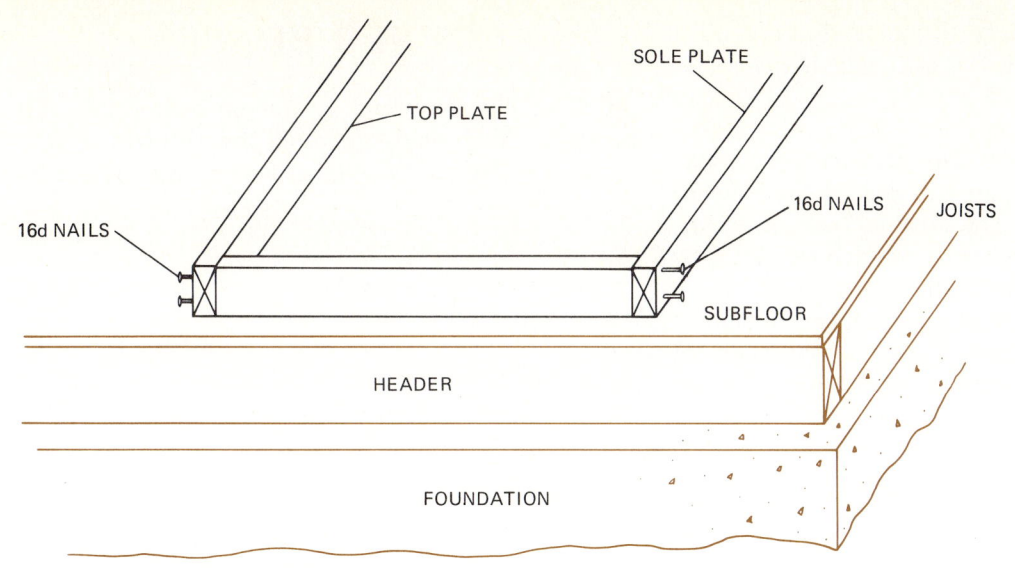

Figure 6-26. Nailing full studs to plates.

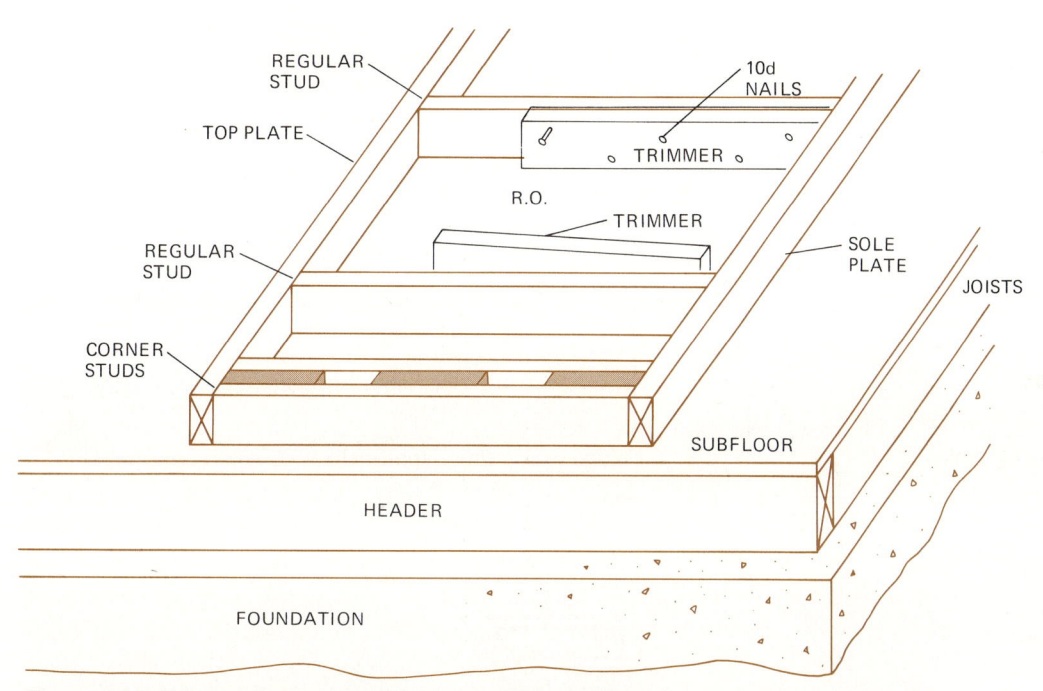

Figure 6-27. Trimmers are nailed into the rough opening.

Figure 6-28. Nailing studs for wall frames.

Framing Walls **107**

and 16 inches apart. Staggered means that one nail is near the top edge and the next nail is near the bottom edge. See Figure 6-29.

HEADERS AND SILLS • The headers are nailed in place next. The headers are placed flush with the edges of the studs. The headers are nailed in place with 16d nails. The nails are driven through the studs into the ends of the headers.

The sills are nailed in place after the headers. Locate the position of the sill. Toenail the ends of the sill to the trimmers. Also, another trimmer may be used. See Figure 6-30.

CRIPPLE STUDS • Next, the cripples are laid out. They are nailed in place from the sole plate with two 16d nails. Then, two 16d nails are used to nail the sill to the ends of the cripples. Finally, the top cripples are nailed into place. See Figure 6-31.

• CORNER BRACES •

Corner braces should be put on before the wall is raised. The bracing prevents the wall frame from swaying sideways. Two methods are commonly used.

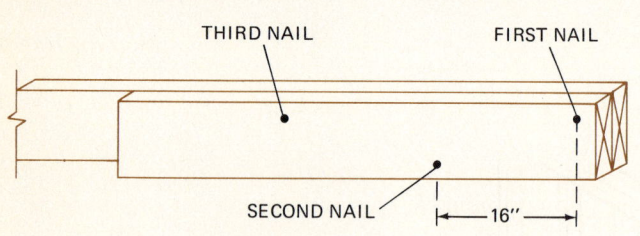

Figure 6-29. In stagger nailing, one nail is near the top and the other is near the bottom.

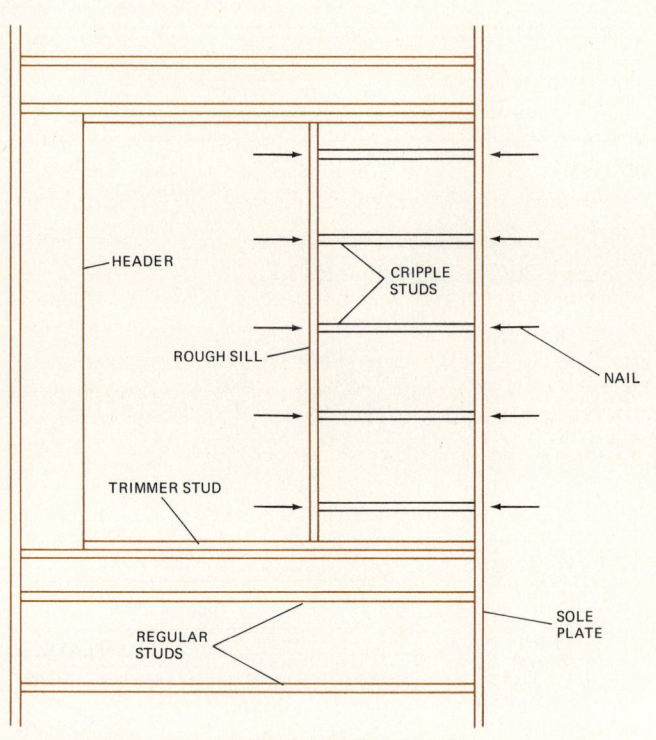

Figure 6-31. Nail cripples through sole plate and sill.

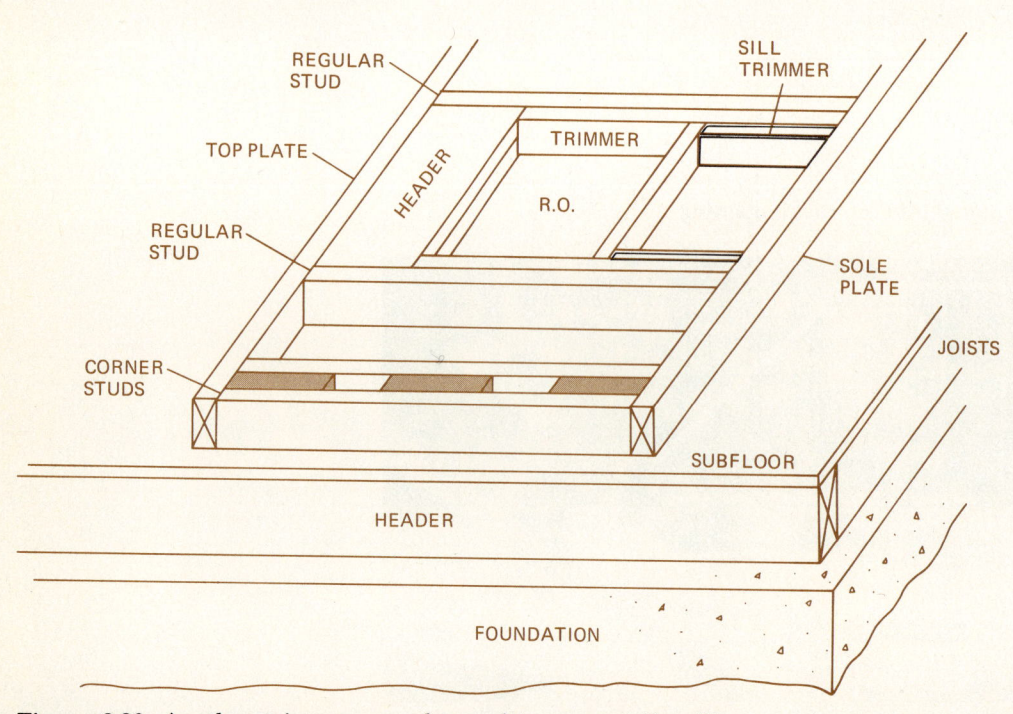

Figure 6-30. Another trimmer may be used to support the sill.

108   Carpentry Fundamentals

Plywood Corner Braces

The first method uses plywood sheets as shown in Figure 6-32. It is best to nail the plywood on before the wall is erected. Plywood bracing costs more but takes little time and effort. If used, it should be the same thickness as the wall sheathing. Plywood should not be used where "energy" sheathing is used. This will be discussed later.

Diagonal Corner Braces

The second method is to make a board brace. The brace is made from 1x4-inch lumber. It is "let in" or recessed into the studs. The angle may be any angle, but 45° is common. Braces are "let in" to the studs so that the outside surface of the wall will stay flat. See Figure 6-33.

To make a diagonal brace, select the piece of wood to be used and nail it temporarily in place across the outside. Mark the layout on each stud. Also mark the angle on the ends of the brace. Remove the brace and gauge the depth of the cut across each stud. Make the cuts with a saw. Knock out the wood between the cuts with a chisel. Then, trim off the ends of the brace. Place the brace in the "let in" slots. Check for proper fit. If the fit is good, nail the brace in place. Use two 8d nails at each stud.

Figure 6-32. Plywood is used for corner bracing.

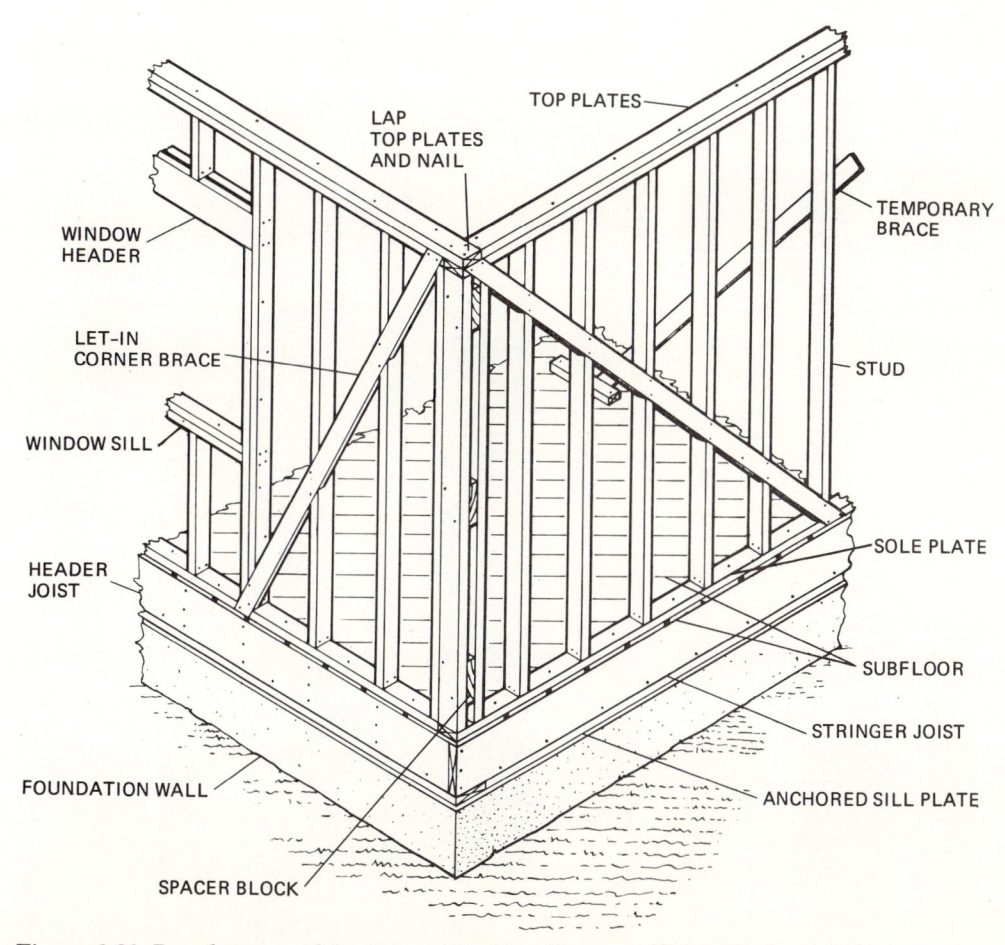

Figure 6-33. Boards are used for corner bracing. *(Courtesy Forest Products Lab)*

Framing Walls 109

• ERECT THE WALLS •

Once the wall section is assembled, it is raised upright. The wall may be raised by hand. See Figure 6-34. Special wall jacks may also be used, as in Figure 6-35. Figure 6-36 shows another method, raising the wall with a fork-lift unit. This is common when a large wall must be placed over anchor bolts on a slab. For slabs, the anchor holes in the sole plate should be drilled before the wall is raised (Figure 6-37). See the section on anchoring sills in Chapter 3. Walls that have been raised must be braced. The brace is temporary. The brace is used to hold the wall upright at the proper angle. Wind will easily blow walls down if they are not braced. Bracing is usually kept until the roof is on. See Figures 6-38 and 6-39.

Wall Sheathing

Wall sheathing may be put on a wall before it is raised. The advantage is that this reduces the amount of lifting and holding. However, it slows

Figure 6-34. Raising a wall section by hand. *(Courtesy American Plywood Association)*

Figure 6-35. Raising a wall with wall jacks. *(Courtesy Proctor Products)*

Figure 6-36. Raising a wall with a fork-lift unit.

Figure 6-37. Wall sections are nailed to slabs after being erected. *(Fox and Jacobs)*

down getting the roof up. It is also harder to make extra openings for vents and other small objects. Wall sheathing may be nailed or stapled. Manufacturer's recommendations should be followed. Sheathing is made from wood, plywood, fiber, fiberboard, plastic foam, or gypsum board.

To Raise the Wall

Before the wall is raised, a line should be marked. The line should show the inside edge of the wall. It is made with the chalk line on the subfloor. This shows the position of the wall.

Figure 6-38. Once raised, walls are held in place with temporary braces.

Before the wall is raised, all needed equipment should be ready. Enough people or equipment to raise the wall should be ready. Also the tools to plumb and brace the wall should be ready.

When all are ready, the wall may be raised. The top edge of the wall is picked up first. Lower parts are grasped to raise it into a vertical position. See Figures 6-34 through 6-36. The wall is held firmly upright. As it is held, the wall is pushed even with the chalk line on the floor.

Some walls are built erect, not raised into position. As a rule, these walls are built on slabs. The layout and cutting are the same.

However, the sole plate is bolted to the slab first. Next studs are toenailed to the sole plate. The top plate can be put on next. Then the wall is braced. Openings and partition bases may then be built.

Put Up a Temporary Brace

After the wall is positioned on the floor, it is nailed in place. See Figure 6-40. Then one end is plumbed for vertical alignment. A special jig can be used for the level. See Figure 6-41. After the wall is plumb, a brace is nailed. See Figure 6-9.

Each wall is treated in the same manner. After the first wall, each added wall will form a corner. The corners are joined by nailing as in Figure 6-47. It is important that the corners be plumb.

Figure 6-39. Nailing a temporary brace. *(Fox and Jacobs)*

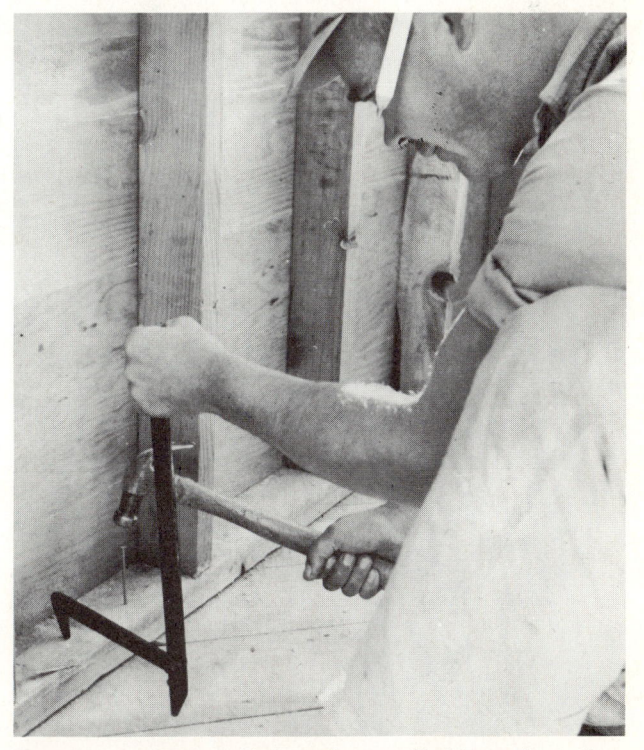

Figure 6-40. Once walls have been pulled into place, they are nailed to the subfloor. *(Courtesy Proctor Products)*

Framing Walls **111**

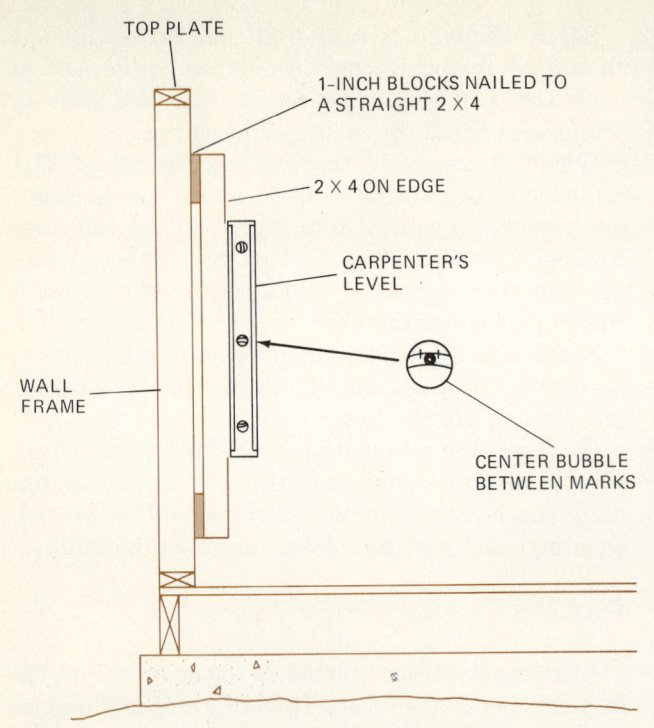

Figure 6-41. Use a spanner jig to plumb corners with the level.

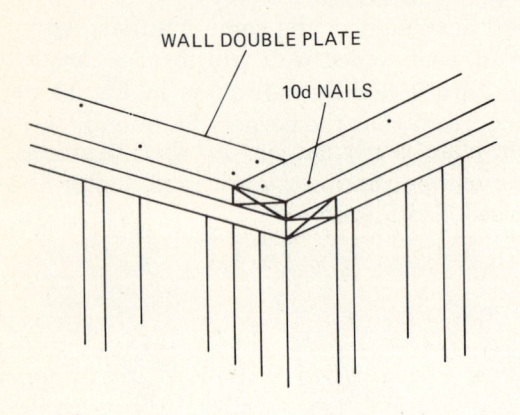

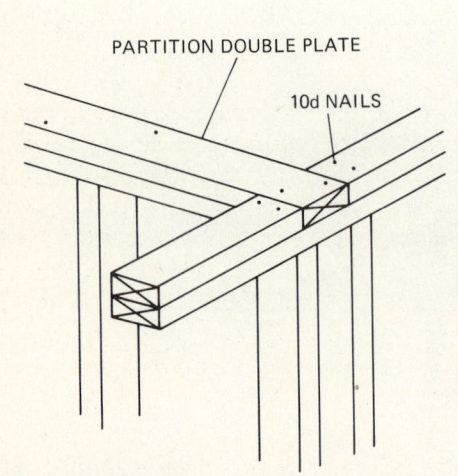

Figure 6-42. Double plates are added after several walls have been erected.

After several walls have been erected, the double plate is added at the top. Figure 6-42 shows a double plate in place. Note that the double plate overlaps on corners to add extra strength.

• INTERIOR WALLS •

Interior walls are called *partitions*. They are made in much the same manner as outside walls. However, the carpenter must remember that studs for inside walls may be longer. This is because inside walls are often *curtain* walls. A curtain wall does not help support the load of the roof. Because of this a double plate might not be used. But, the top of the wall must be just as high.

Most builders wish to make the building weathertight quickly. Therefore the first partitions that are made are load-bearing partitions.

When roof trusses are used, load-bearing partitions may not be necessary. Trusses distribute loads so that inside support is not needed. However, many carpenters make all walls alike. This lets them cut all studs the same. It also lets them use any wall for support.

Locate Sole Plates for Partitions

To locate the partitions, the centerlines are determined from the plans. The centerline is then marked on the floor with the chalk line. Plates are then laid out.

Studs

Studs and headers are cut. The partition walls are done in the same way outside walls. As a rule, the longer partitions are done first. Then the short partition walls are done. Last are the shortest walls for the closets.

Corners

Corners and wall intersections are made just as for outside walls. The size and amount of blocking can be reduced. The purpose of blocking is to provide nail surfaces. These are needed at inside and outside corners. They are a base for nailing wall covering.

Headers and Trimmers

Headers are not required for rough openings in curtain walls. Often, openings are framed with single boards. See Figure 6-43. The header for inside walls is much like a sill. Trimmers are optional. They provide more support. They are recommended when single-board headers are used.

Carpentry Fundamentals

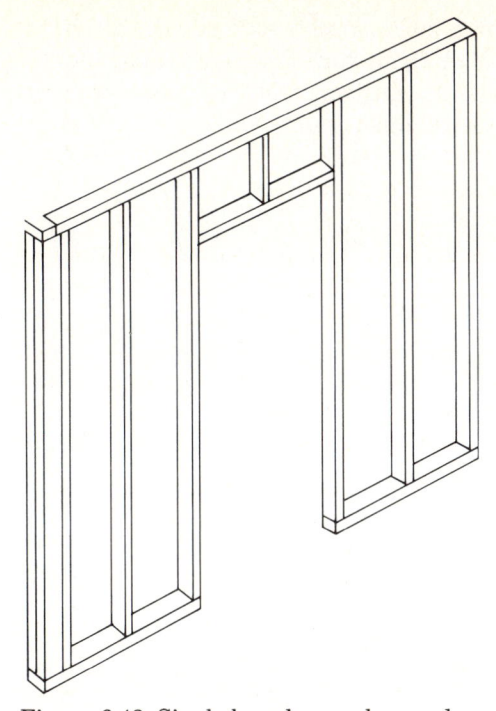

Figure 6-43. Single boards are also used as headers on inside partitions.

WALL DETAIL	DESCRIPTION	STC RATING
	1/2-INCH GYPSUM WALLBOARD	45
	5/8-INCH GYPSUM WALLBOARD (DOUBLE LAYER EACH SIDE)	45
	1/2-INCH GYPSUM WALLBOARD 1 1/2-INCH FIBROUS INSULATION	49
	1/2-INCH SOUND DEADENING BOARD (NAILED) 1/2-INCH GYPSUM WALLBOARD (LAMINATED)	50

Figure 6-45. Sound insulation of double walls. *(Courtesy Forest Products Lab)*

Sole Plate

The sole plate is not cut out for door openings. It is made as one piece. It is cut away after the wall is raised. See Figure 6-44.

Special Walls

Several conditions may call for special wall framing. Walls may need to be thicker to enclose plumbing. Drain pipes may be wider than the standard 3½-inch-thick wall. Thickness may be added in two ways. Wider studs may be used. Or, extra strips may be nailed to the edges of studs. Other special needs are soundproofing and small openings.

SOUNDPROOFING • Soundproofing between rooms requires special framing. The sole and top plates are made wider. Regular studs that are not as wide are used. The studs are staggered from one side to the other. See Figure 6-45. Insulation can now be woven between the studs. Sound is transmitted better through solid objects. But, with staggered studs, there is no solid part from wall to wall. In this way, there is no solid bridge for easy sound passage.

SMALL OPENINGS • Small openings are often needed in walls. They are needed for air ducts, plumbing, and drains. Recessed cabinets, such as bath-

Figure 6-44. At door openings, the partition sole plates are cut out after the wall is up.

Framing Walls 113

room medicine cabinets, also need openings. These openings are framed for strength and support. See Figure 6-46.

• SHEATHING •

Several types of sheathing are used. Sheathing is the first layer put on the outside of a wall. Sheathing makes the frame still and rigid and is insulation. It will also keep out the weather until the building is finished.

There are five main types of sheathing used today. They are fiberboard, gypsum board, boards, plywood, and rigid foam. Corner bracing is needed for all types except plywood and boards.

Fiberboard Sheathing

The most common type of sheathing is treated fiberboard. It should be applied vertically for best bracing and strength characteristics. It is usually ½ inch thick. Plywood ½ inch thick can be used for corner bracing. The ½-inch fiber sheathing and plywood fit flush for a smooth surface. The exterior siding can easily be attached. Fiber sheathing should be fastened with roofing nails. Nails 1½ inches long are spaced 3 to 6 inches apart. Nails should always be driven at least ⅜ inch from the edge. If plywood is not used, diagonal corner braces are required. See Figure 6-47A to C.

Gypsum Sheathing

Sheathing made from gypsum material is also used. See Figure 6-48. This gypsum sheathing is not the same as the sheets used on interior walls. This gypsum sheathing is treated to be weather-resistant. The most common thickness is ½ inch. A ½-inch-thick plywood corner brace may be used. The sheathing should be fastened with roofing nails. Nails 1¾ inches long should be used. The nails should be 4 inches apart.

Plywood Sheathing

Plywood is also used for sheathing. When it is used, no corner bracing is required. Plywood is both strong and fire-resistant. It can be nailed up quickly with little cutting. A moisture barrier must be added when plywood is used. However, plywood and board sheathing are both expensive.

When plywood is used, it should be at least ⁵⁄₁₆ inch thick. Half-inch thickness is recommended. Exterior siding can be nailed directly to ½-inch plywood. Plywood sheets can be applied vertically or horizontally. The sheets should be fastened with 6d or 8d nails. The nails should be 6 to 12 inches apart.

Energy Sheathing

Two types of energy sheathing are commonly used. The first is called *rigid foam*. It is a special plastic foam. See Figure 6-49. Its use can greatly reduce energy consumption. It is roughly equivalent to 3 more inches of regular wall insulation. It is grooved on the sides and ends. It is fitted together horizontally. It is nailed in place with 1¼-inch nails spaced 9 to 12 inches apart. Joints may be made at any point. Rigid foam may also be covered with a shiny foil on one or both sides. See Figure 6-50. The foil surface further reduces energy losses. It does so because the shiny surface reflects heat. Also, the foil prevents air from passing through the foam. However, foam burns easily. A gypsum board interior wall should be used with the foam. The gypsum wall reduces the fire hazard.

The second type is a special fiber. It is also backed on both sides with foil. Its fibers do not insulate as well as foam. It does prevent air movement better

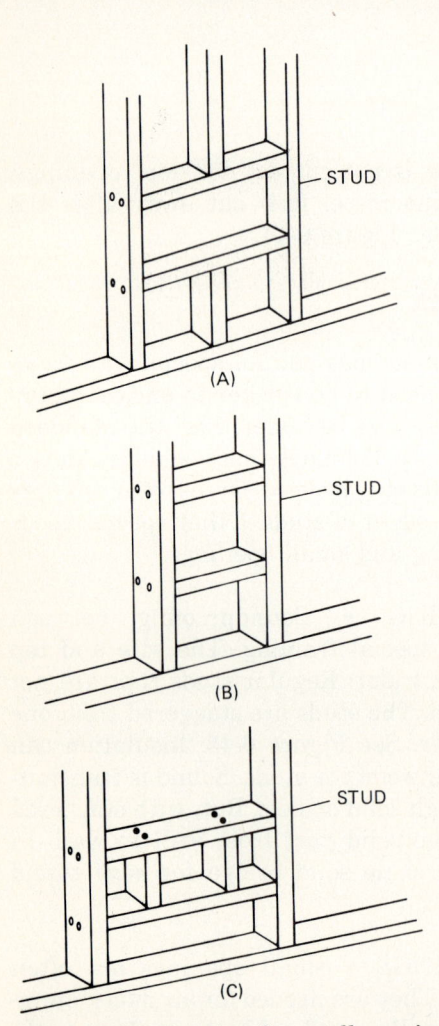

Figure 6-46. Framing for small openings in walls.

114   Carpentry Fundamentals

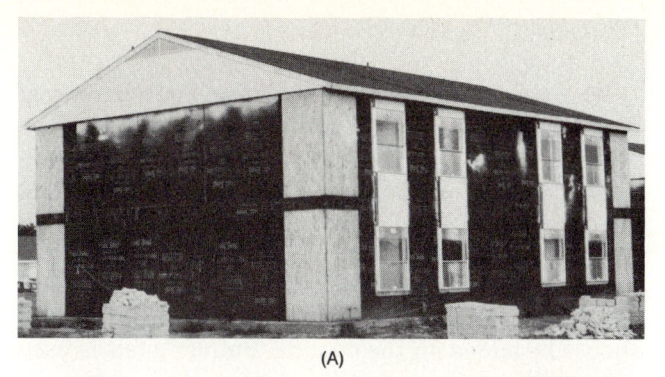

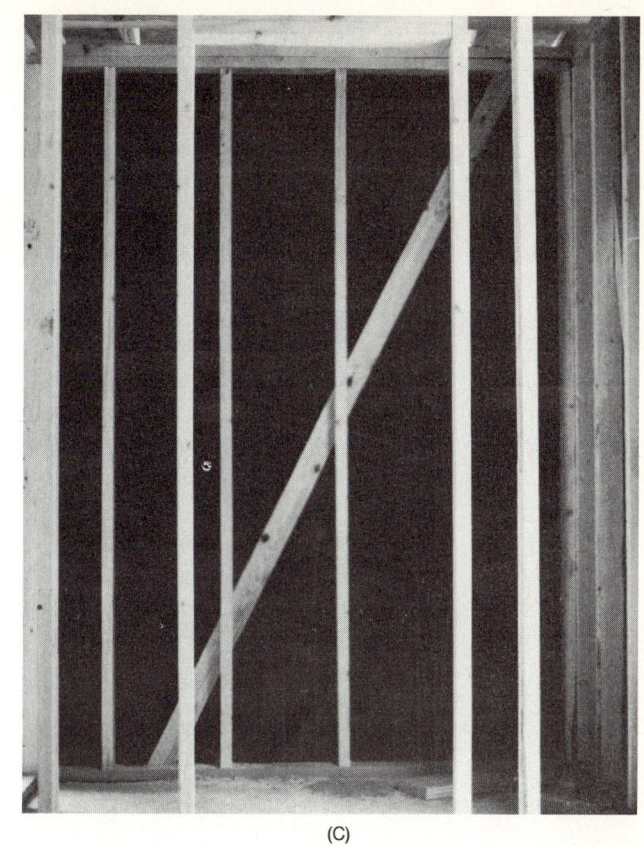

Figure 6-47. Corner bracing:
(A) Fiberboard sheathing with plywood corner braces.
(B) Fiberboard sheathing with inlet board cornerbraces—seen from the outside.
(C) Fiberboard sheathing with inlet board corner braces—seen from the inside.

Figure 6-48. Gypsum sheathing is inexpensive and fire resistive.

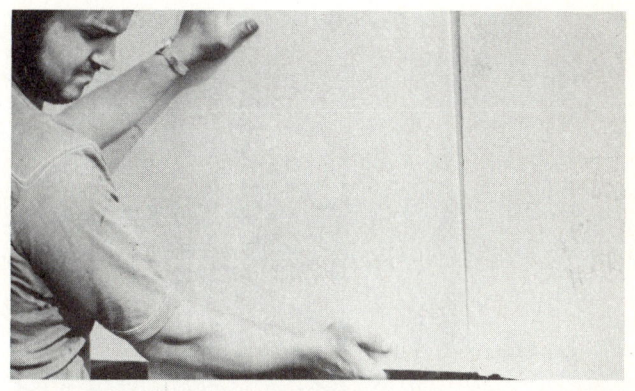

Figure 6-49. Rigid polystyrene foam sheathing provides more insulation. *(Courtesy Dow Chemical)*

Figure 6-50. Rigid polystyrene foam may be coated with reflective foil to increase its effectiveness as an insulation sheathing.

Framing Walls 115

than plain foam. The foil surface is an effective reflector. Also, it is more fire-resistant and costs less. See Figure 6-51.

Figure 6-51. Another style of energy efficient sheathing is similar to hardboard but is coated with reflective foil. *(Courtesy Simplex)*

Boards

Boards are still used for sheathing. However, diagonal boards are seldom used. Plain boards may be used. Boards with both side and end tongue-and-groove joints are used. For grooved boards, joints need not occur over a stud. This saves installation time. Carpenters need not cut boards to make joints over studs. However, for plain boards joints should be made over studs. A special moisture barrier should be added on the outside. Builder's felt is used most often. It is nailed in place with 1-inch nails through metal disks. The bottom layers are applied first. See Figure 6-52.

• FACTORS IN WALL CONSTRUCTION •

A carpenter may learn the procedure for making a wall. However, the carpenter may not know why walls are made as they are. Each part of a wall frame has a specific role. The corner pieces help tie the walls together. The double top plates also add strength to corners. The top plate is doubled to help support the weight of the rafters and ceiling.

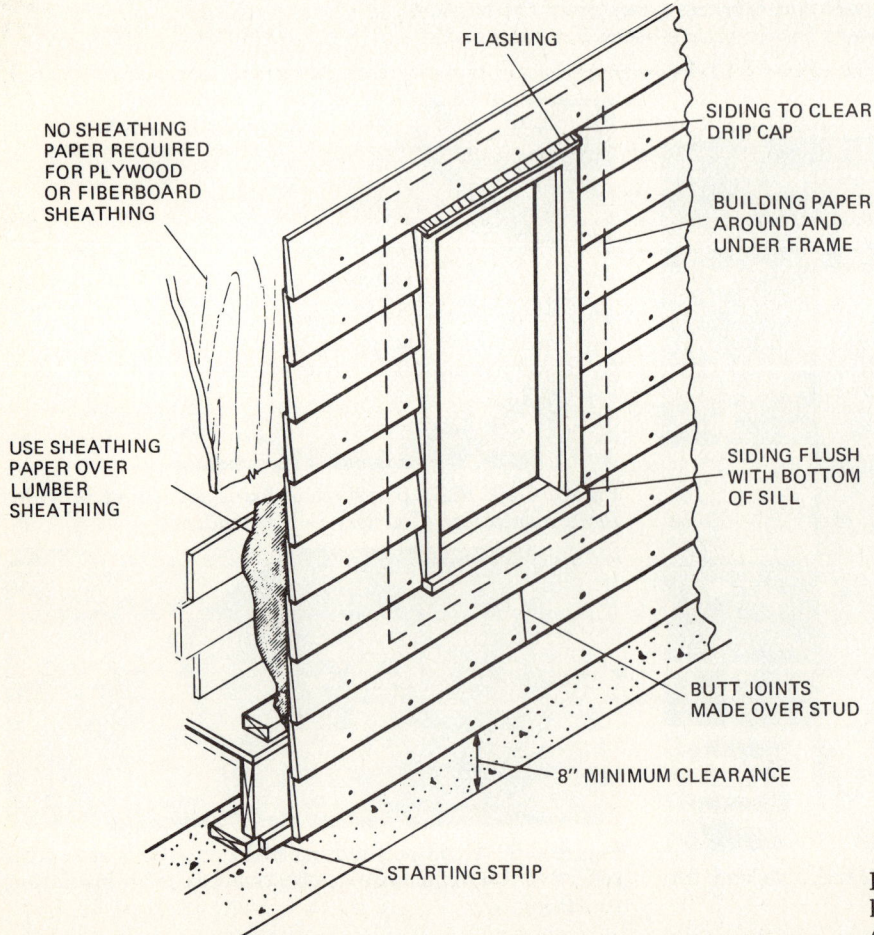

Figure 6-52. Board sheathing must have a separate moisture barrier. *(Courtesy Forest Products Lab.)*

116   Carpentry Fundamentals

A rafter or a ceiling joist between studs is held up by the top plate. A single top plate could eventually sag and bend.

Standard Spacing

The spacing of wall members is also important. Standard construction materials are 4 feet wide and 8 feet long. Standard finish floor-to-ceiling heights are 8′ ½″. The extra ½ inch lets pieces 8 feet long be used without binding.

Further, using even multiples of 4 or 8 means that less cutting is needed. Buildings are often designed in multiple of 4 feet. Two-foot roof overhangs can be used. The overhang at each end adds up to 4 feet. This reduces the cutting done on siding, floors, walls, and ceilings. It reduces time and costs.

Modular Standards

Currently, a new modular system of construction is being used more. The modular system uses stud and rafter systems 24 inches O.C. The wider spacing does not provide as much support for wall sheathing. However, the ability to support the roof is not reduced much. A building with studs on 16-inch centers is very strong. However, the difference in strength between 16- and 24-inch centers is very small. The advantage of using the 24-inch modular system is that it saves on costs. For a three-bedroom house, the cost of wall studs may be reduced about 25 percent. Moreover, the costs of labor are also reduced. This saves the time that would be used for cutting and nailing that many more studs. The savings are even more in terms of money and wages paid. This saving is made because fewer resources are used with almost the same results. In the modular system, several ways of reducing material use are employed. Floor joists are butted and not lapped. Building and roof size are exact multiples of 4 or 8 feet. Also, single top plates are used on partitions.

Energy

Energy is also a matter of concern today. Insulated walls save energy used for heating and cooling. The amount of insulation helps determine the efficiency. A 6-inch-thick wall can hold more insulation than a 4-inch wall. It will reduce the energy used by about 20 percent. However, it takes more material to build such a wall. Canadian building codes often specify 6-inch walls.

The double-wall system of frame buildings is a better insulator than a solid wall. Wood is good insulation when compared with other materials. But the best insulation is a hollow space filled with material that does not conduct heat. A wall should have three layers. An outside, weatherproof wall is needed. This layer stops rain and wind. A thick layer of insulation is next. This helps keep heat energy from being conducted through the wall. The inside wall is the third layer. It helps reduce air movement. It also helps seal and hold the insulation in place.

More insulation effect can also be added by adding a shiny surface. See Figures 6-51 and 6-53.

Energy sheathing adds insulation in two ways. It adds in the same way as regular insulation. However, it also covers the studs. Therefore, the stud does not conduct heat directly through the wall. See Figure 6-54.

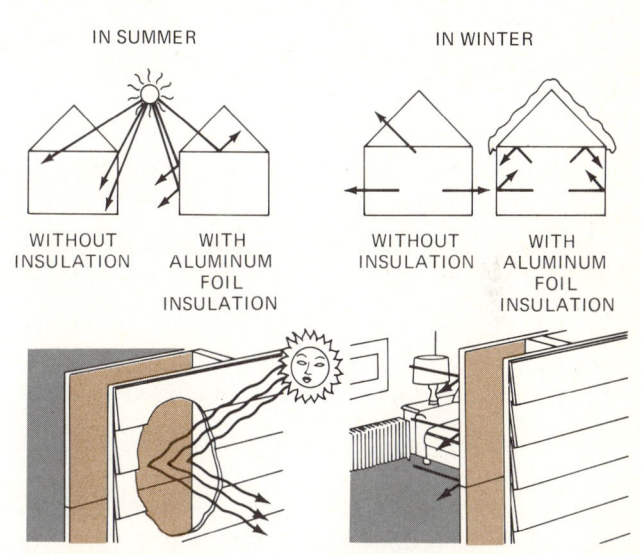

Figure 6-53. Reflected heat makes the house cooler in summer and warmer in winter. *(U.S. Gypsum)*

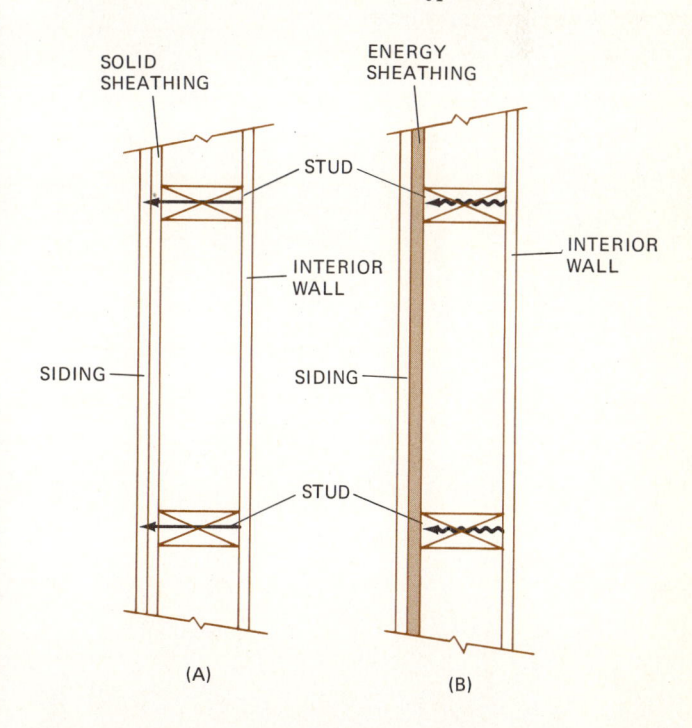

Figure 6-54. (A) Solid sheathing, stud and wall are a solid path. This conducts energy loss straight through a wall. (B) Energy sheathing forms a barrier. There is no solid path for energy.

Framing Walls **117**

The use of headers also affects wall construction. The use of headers reduces time and construction costs. They are also difficult to insulate. The use of truss headers or a single header is better. This allows more insulation to be used.

The builder and the carpenter can alter a wall frame to save energy. Walls can be made thicker. This lets more insulation be used. Thicker walls also let insulation cover studs in the same way as for sound insulation. Energy sheathing can be used to cover the studs. It adds insulation without requiring thick walls. Reflective foil also makes it more efficient. See Figure 6-53.

• WORDS CARPENTERS USE •

stud	partition	trimmer	double plate	moisture barrier
sheathing	curtain wall	rough opening	blocking	
bearing wall	sole plate	sill	header	
load	cripple stud	sway brace	siding	

• STUDY QUESTIONS •

6-1. Which is usually built first, the wall or the floor?

6-2. On what piece are studs spaced?

6-3. How are corners nailed?

6-4. What size nail is used to join studs to plates?

6-5. What size nail is used to join trimmers to studs?

6-6. What size nail is used to join headers to studs?

6-7. How are headers supported?

6-8. How are sills supported?

6-9. Why are top plates doubled?

6-10. When are sole plates cut for interior doors?

6-11. When are headers made?

6-12. What is a curtain wall?

6-13. Why is corner bracing done?

6-14. How are corners braced?

6-15. How are corners plumbed?

6-16. Why are walls braced temporarily?

6-17. What is used for sheathing?

6-18. How can you tell when a wall is in place?

6-19. How can walls be changed to save energy?

6-20. What is a story pole?

7 BUILDING ROOF FRAMES

The framing for a roof is determined by the choice of roof style. There are a number of types of roof lines. Some of them are the gable, mansard, shed, hip, and gambrel. Each has identifying characteristics. Each style presents unique problems. Rafters and sheathing will be shaped according to the roof line desired.

In this unit you will learn how to build roof frames. You will learn how to make roofs of different types. You will also learn how to make openings for chimneys and soil pipes. Things you can learn to do are:

- Design a particular type of roof rafter
- Mark off and cut a roof rafter
- Put roof trusses in place
- Lay sheathing onto a roof frame
- Build special combinations of framing for roofs
- Identify the type of plywood needed for a particular roof

• INTRODUCTION •

The roof is an important part of any building. It is needed to keep out the weather and to control the heat and cooling provided for human comfort inside. There are many types of roofs. Each serves a purpose. Each one is designed to keep the inside of the house warm in the winter and cool in the summer. The roof is designed to keep the house free of moisture. This moisture may be rain, snow, or fog.

The roof is made of rafters and usually supported by ceilings joists. When all the braces and forms are put together, they form a roof. In some cases roofs have to be supported by more than the outside walls. This means some of the partitions must be weight-bearing. However, with truss roofs it is possible to have a huge open room without supports for the roof.

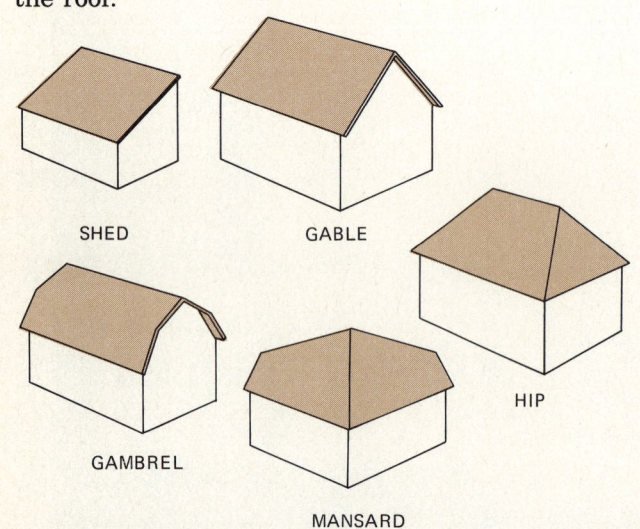

Figure 7-1. Various roof shapes or styles.

Figure 7-1 shows various roof shapes. The shape can affect the type of roofing materials used. The various shapes call for some special details. Roofs have to withstand high winds and ice and snow. The weight of snow can cause a roof to collapse unless it is properly designed. It is necessary to make sure the load on the roof can be supported. This calls for the proper size rafters and decking.

Shingles on a roof protect it from rain, wind, and ice. They have to withstand many years of exposure to all types of weather conditions.

The hip roof with its hip and hip jack rafters is of particular concern. It is one of the most popular types. The common rafter is the simplest in the gable roof. The mansard roof and the gambrel roof present some interesting problems with some very interesting solutions. All this will be explained in detail later in this chapter.

Various valleys, ridge boards, and cripple rafters will be described in detail here.

• SEQUENCE •

The carpenter should build the roof frame in this sequence:

1. Check the plans to see what type of roof is desired.
2. Select the ceiling joist style and spacing.
3. Lay out ceiling joists for openings.
4. Cut ceiling joists to length and shape.
5. Lay out regular rafter spacing.
6. Lay out the rafters and cut to size.
7. Set the rafters in place.
8. Nail the rafters to the ridge board.
9. Nail the rafters to the wall plate.
10. Figure out the sheathing needed for the roof.
11. Apply the sheathing according to specifications.
12. Attach the soffit.
13. Put in braces where needed. See Figure 7-2.
14. Cut special openings in the roof decking.

There may be a need for cutting dormer rafters after the rest of the roof is finished. Structural elements may be installed where needed. Truss roof rafters will need special attention as to spacing and nailing. Double check to make sure they fit the manufacturer's specifications.

• ERECTING TRUSS ROOFS •

Truss Construction

Trussed rafters are commonly used in light frame construction. They are used to support the roof and ceiling loads. Such trusses are designed, fabricated, and erected to meet the design and construction criteria of various building codes. They efficiently use the excellent structural properties of lumber. They are spaced either 16 inches on centers (O.C.) or in some cases 24 inches O.C. See Figure 7-3.

Truss Disadvantages

You should keep in mind that the truss type of construction does have some disadvantages. The attic space is limited by the supports that make up the truss. Truss types of construction may need special equipment to construct them. In some instances it is necessary to use a crane to lift the trusses into position on the job site.

ROOF FRAMING • The roof frame is made up of rafters, ridge board, collar beams, and cripple studs.

See Figure 7-4. In gable roof construction, all rafters are precut to the same length and pattern. Figure 7-5 shows a gable roof. Each pair of rafters is fastened at the top to a ridge board. The ridge board is commonly a 2x8 for 2x6 rafters. This board provides a nailing area for rafter ends. See Figure 7-4.

GETTING STARTED • Getting started with erection of the roof framing is the most complicated part of framing a house. Plan it carefully. Make sure you have all materials on hand. It is best to make a "dry run" at ground level. The erection procedure will be much easier if you have at least two helpers. A considerable amount of temporary bracing will be required if the job must be done with only one or two persons.

Figure 7-2. Braces are put in where they are needed. (Duo-Fast)

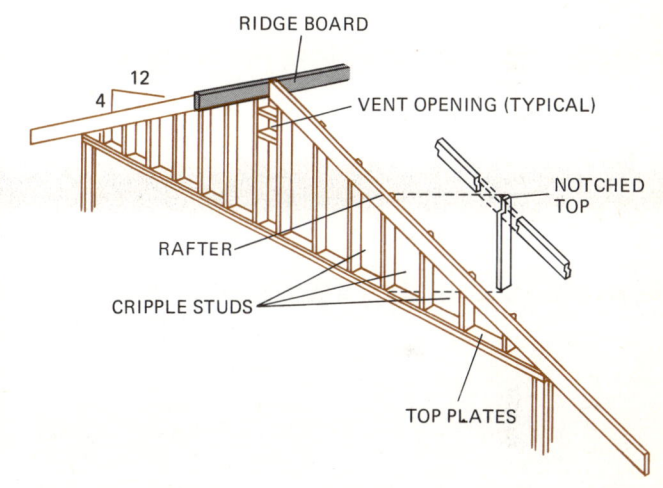

Figure 7-4. Gable-type rafter with cripple studs and ridge board. Note the notched top in the cripple stud. (American Plywood Association)

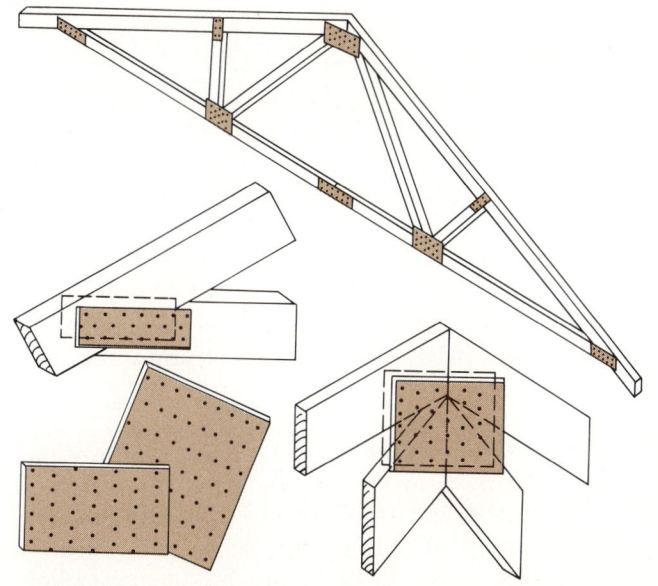

Figure 7-3. W-type truss roof and metal plates used to make the junction points secure.

Figure 7-5. Two-story house with gable roof. Note the bay windows with hip roof.

Building Roof Frames **121**

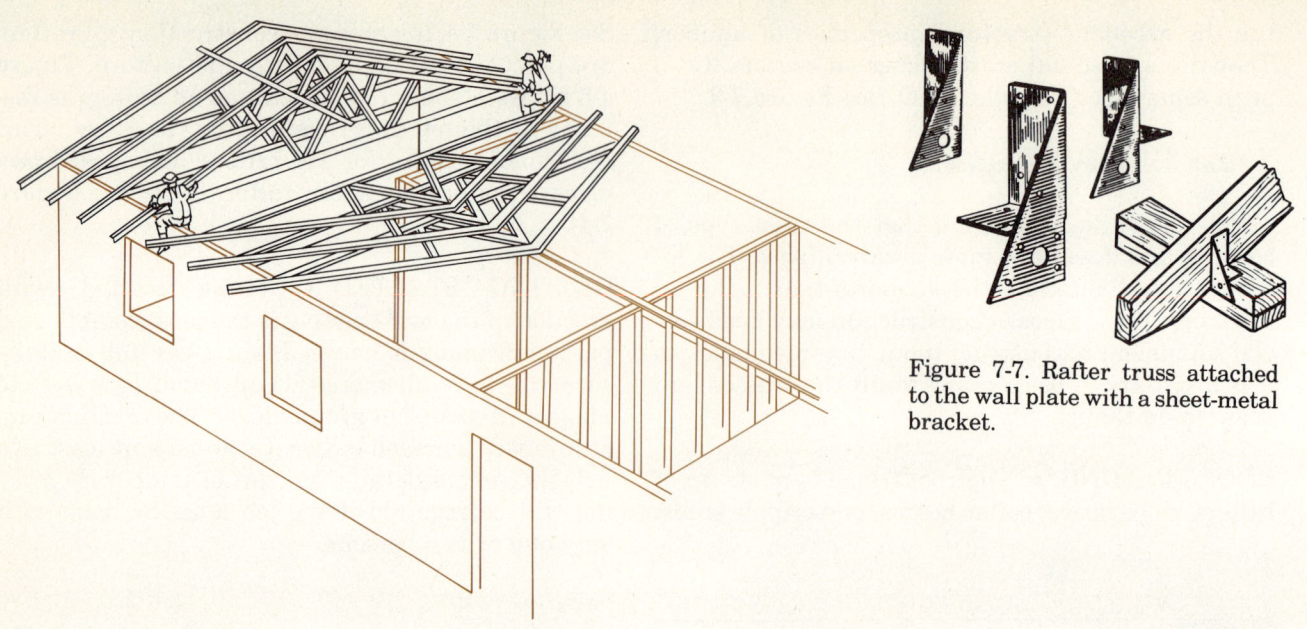

Figure 7-6. Roof trusses are placed on the walls, then tipped up and nailed in place.

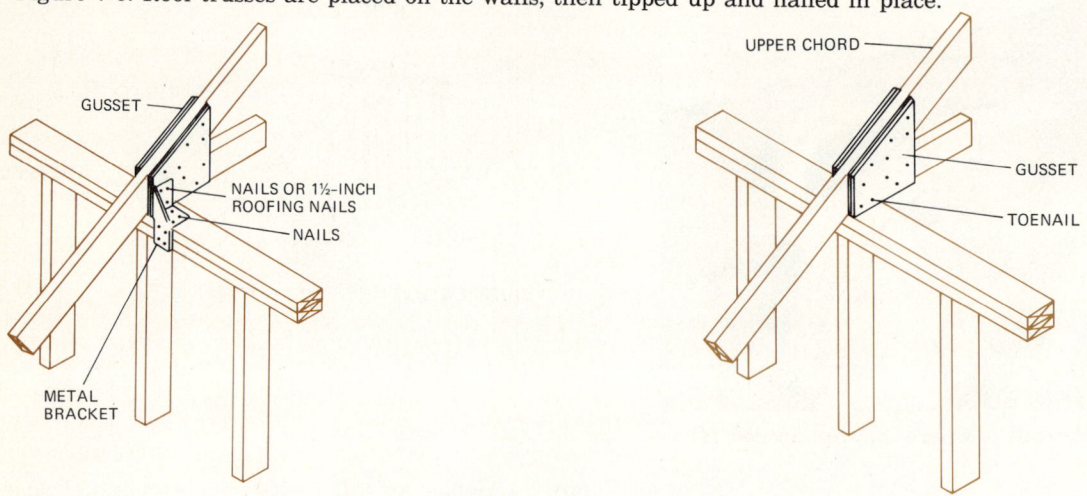

Figure 7-7. Rafter truss attached to the wall plate with a sheet-metal bracket.

Figure 7-8. Using a nailing bracket to attach a truss to the wall plate and nailing into the gusset to attach the rafter.

Figure 7-9. Toenailing the truss rafter to the wall plate.

STEPS IN FRAMING A ROOF • Take a look at Figure 7-6. It shows two persons tipping up trusses. They are tipped up and nailed in place one at a time. Two people, one working on each side, will get the job done quickly. This is one of the advantages of trussed-roof construction. The trusses are made at a lumber yard or some other location. They are usually hauled to the construction site on a truck. Here they are lifted to the roof of the building with a crane or by hand. In some cases the sheet-metal bracket shown in Figure 7-7 holds the truss to the wall plate. Figure 7-8 shows how the metal bracket is used to fasten the truss in place. In Figure 7-9 toenailing is used to fasten the truss to the wall plate.

ADVANTAGES OF TRUSSES • Manufacturers point out that the truss has many advantages. In Figure 7-10 the conventional framing is used. Note how bearing walls are required inside the house. With the truss roof you can place the partitions anywhere. They are not weight-bearing walls. With conventional framing it is possible for ceilings to sag. This causes cracks. The truss has supports to prevent sagging ceilings. See Figure 7-11. Notice how the triangle shape is obvious in all the various parts of the truss. The triangle is a very strong structural form.

DETAILS OF TRUSSES • There are a number of truss designs. A W type is shown in Figure 7-12. Note how the 2x4s are brought together and fastened by plywood that is both nailed and glued. In some cases, as shown in Figure 7-3, steel brackets are used. Whenever plywood is used for the gussets, make sure the glue fits the climate. In humid parts

122   Carpentry Fundamentals

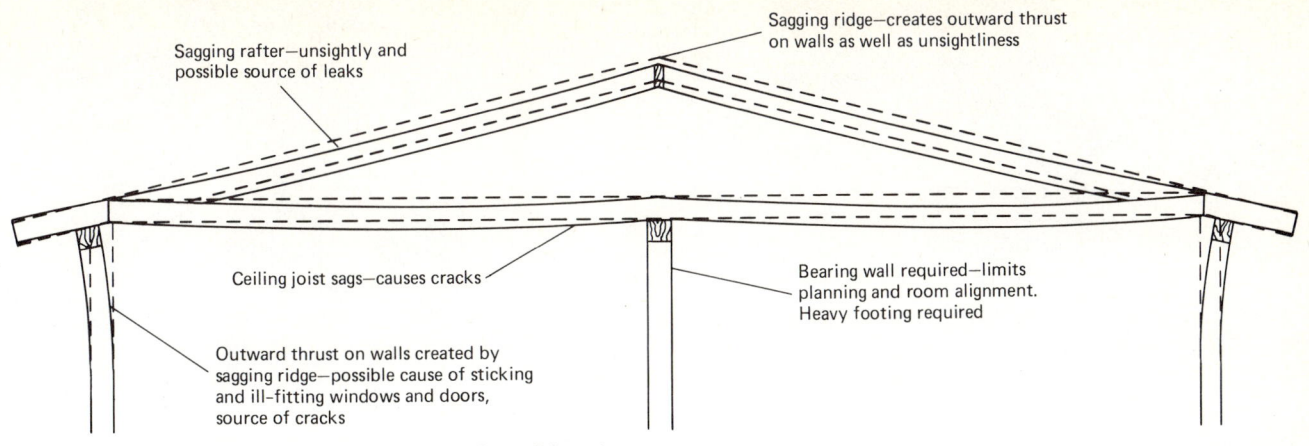

Figure 7-10. Disadvantages of conventional roof framing.

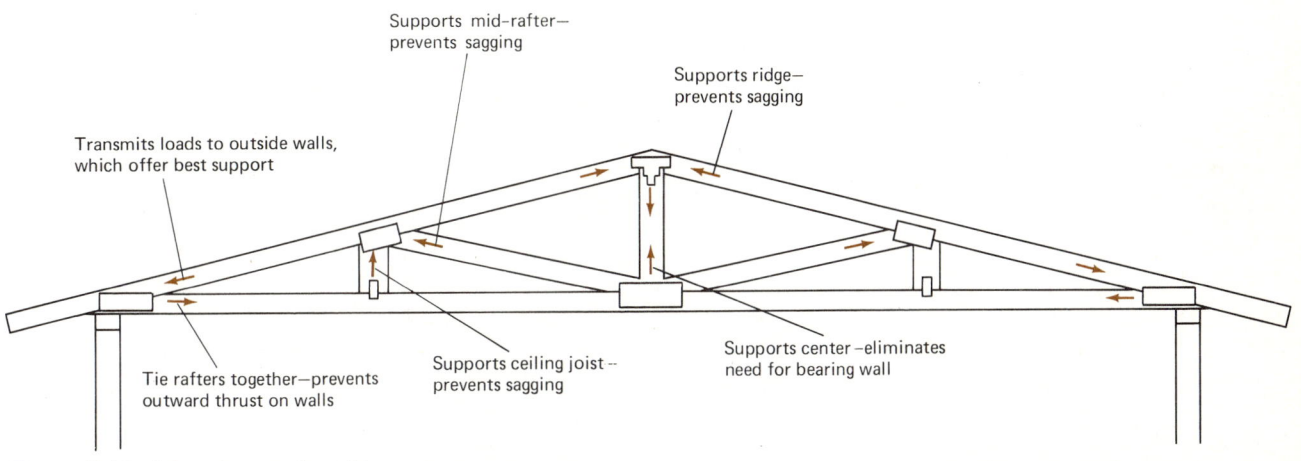

Figure 7-11. Advantages of roof trusses.

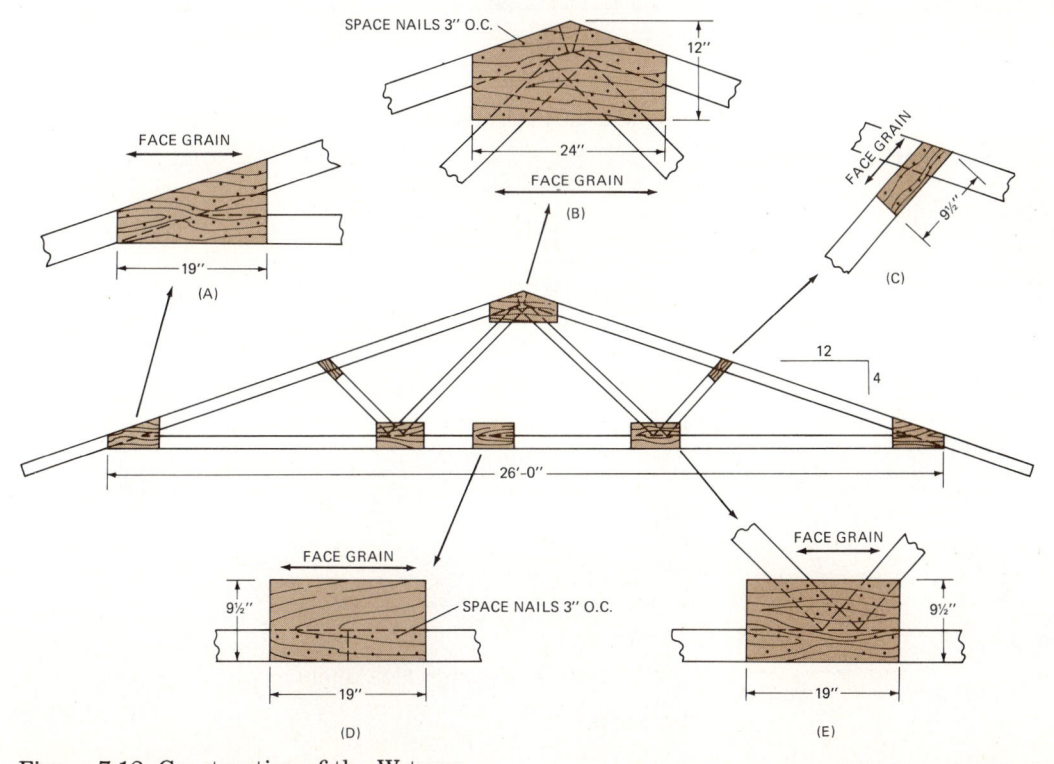

Figure 7-12. Construction of the W truss.

Building Roof Frames **123**

Figure 7-13. Wood trusses. (A) W type. (B) King post type. (C) Scissors type.

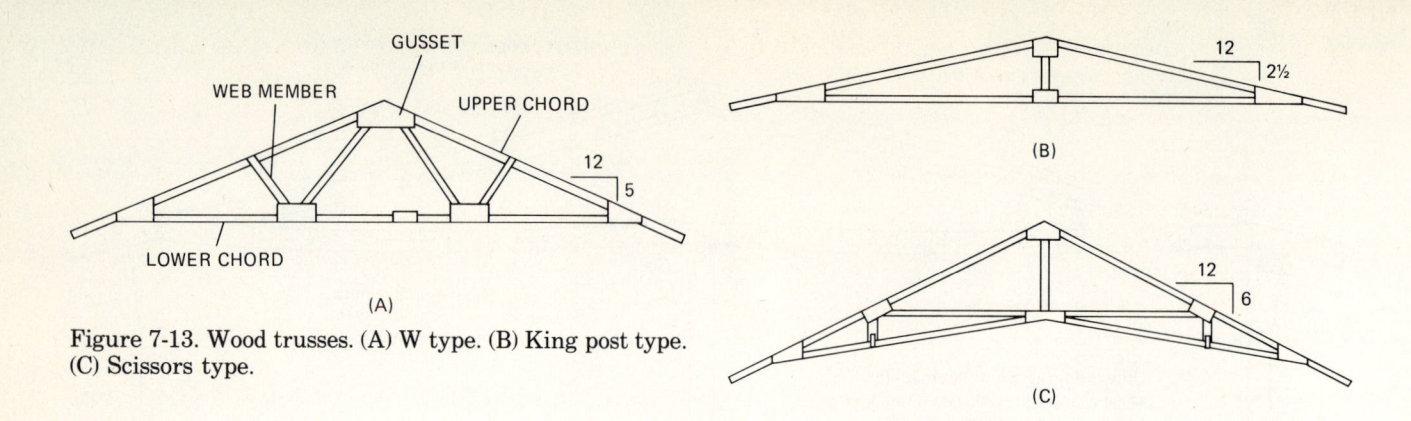

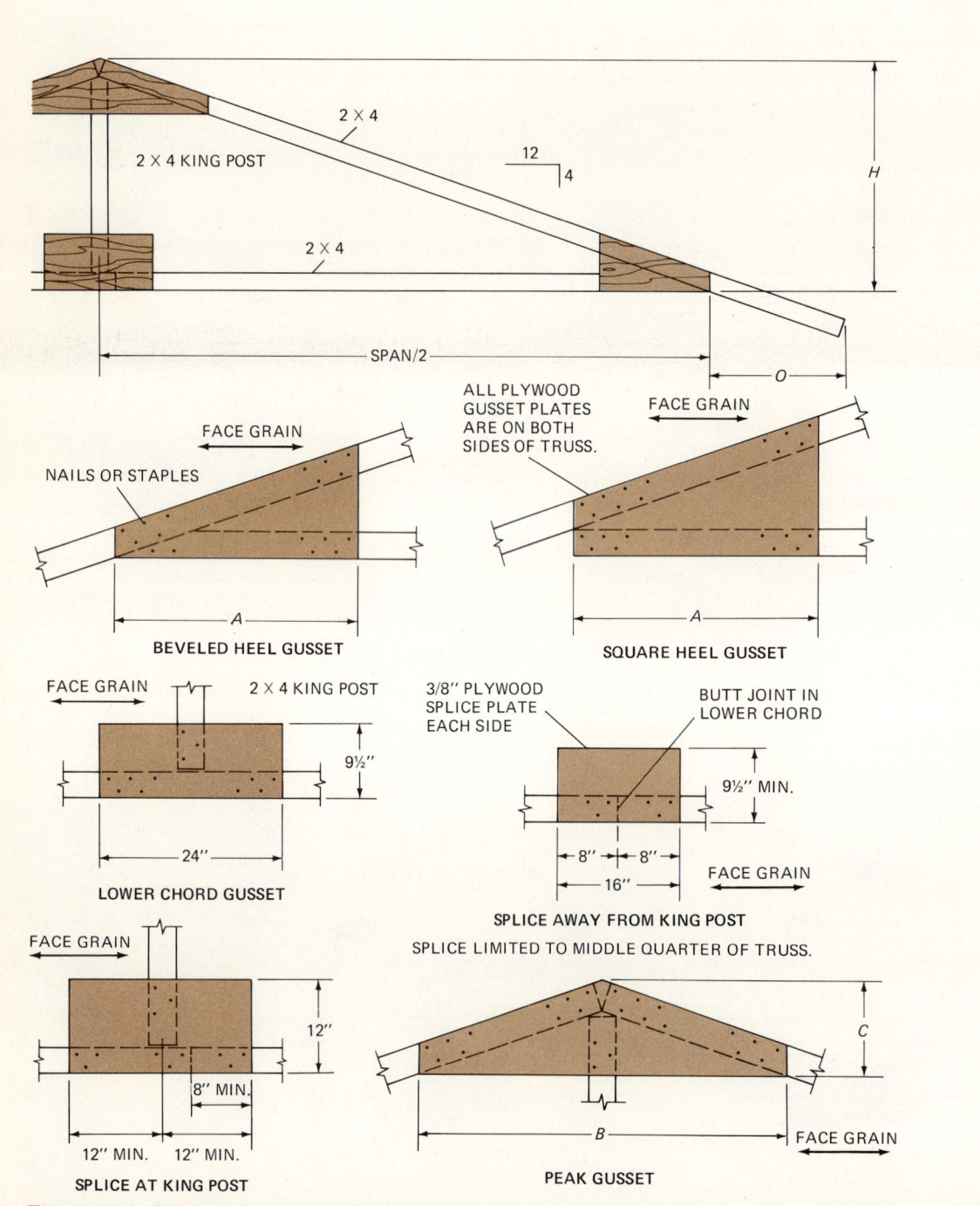

Figure 7-14. Construction of the king post truss.

124 Carpentry Fundamentals

of the country, where the attic may be damp, the glue should be able to take the humidity. The manufacturer will inform you of the glue's use and how it will perform in humid climates. The glue should not lose its strength when the weather turns humid. Most glue containers have this information on them. If not, check with the manufacturer before making trusses. In most instances, the manufacturer of trusses is very much aware of the glue requirements for a particular location.

Figure 7-13 shows how the three suggested designs are different. The truss in Figure 7-13(A) is known as the W type. Note the W in the middle of the truss. In Figure 7-13(B) the king post is simple. It is used for a low-pitch roof. The scissors type is shown in Figure 7-13(C).

The W-type is the most popular. It can be used on long spans. It can also be made of low-grade lumber. The scissor type is used on houses with sloping living room ceilings. It can be used for a cathedral ceiling. This truss is cheaper to make than the conventional type of construction for cathedral ceilings. King-post trusses are very simple. They are used for houses. This truss is limited to about a 26-foot span—that is, if 2x4s are used for the members of the truss. Compare this with a W type, which could be used for a span of 32 feet. King type is economical for use in medium spans. It is also useful in short spans. See Figure 7-14.

The type of truss used depends upon the wind and snow. The weight applied to a roof is an important factor.

Make sure the local codes allow for truss roofs. In some cases the FHA has to inspect them before insuring a mortgage with them in the house.

LUMBER TO USE IN TRUSSES • The lumber used in construction of trusses must be that which is described in Table 7-1. The moisture content should be from 7 to 16 percent. Glued surfaces have to be free of oil, dirt, or any foreign matter. Each piece must be machine-finished, but not sanded.

TABLE 7-1   CHORD CODE TABLE FOR ROOF TRUSSES

Chord Code	Size	Grade and Species Meeting Stress Requirements	Grading Rules	f	t//	c//
1	2 × 4	Select structural light framing WCDF No. 1 dense kiln-dried Southern pine 1.8E	WCLIB SPIB WWPA	1950 2000 2100	1700 2000 1700	1400 1700 1700
2	2 × 4	1500f industrial light framing WCDF 1500f industrial light framing WCH No. 1 2-inch dimension Southern pine 1.4E	WCLIB WCLIB SPIB WWPA	1500 1450 1450 1500	1300 1250 1450 1200	1200 1100 1350 1200
3	2 × 4	1200f industrial light framing WCDF 1200f industrial light framing WCH No. 2 2-inch dimension Southern pine	WCLIB WCLIB SPIB	1200 1150 1200	1100 1000 1200	1000 900 900
4	2 × 6	Select structural J&P WCDF Select structural J&P Western larch No. 1 dense kiln-dried Southern pine 1.8E	WCLIB WWPA SPIB WWPA	1950 1900 2000 2100	1700 1600 2000 1700	1600 1500 1700 1700
5	2 × 6	Construction grade J&P WCDF Construction grade J&P WCH Structural J&P Western larch No. 1 2-inch dimension Southern pine 1.4E	WCLIB WCLIB WWPA SPIB WWPA	1450 1450 1450 1450 1500	1300 1250 1300 1450 1200	1200 1150 1200 1350 1200
6	2 × 6	Standard grade J&P WCDF Standard grade J&P WCH Standard structural Western larch No. 2 2-inch dimension Southern pine	WCLIB WCLIB WWPA SPIB	1200 1150 1200 1200	1100 1000 1100 1200	1050 950 1050 900
7	2 × 4	Select structural light framing WCDF Select structural light framing Western larch No. 1 dense kiln-dried Southern pine 1.8E	WCLIB WWPA SPIB WWPA	1900 1900 2050 2100	1900 1900 2050 1700	1400 1400 1750 1700
8	2 × 4	1500f industrial light framing WCDF Select structural light framing WCH Select structural light framing WH 1500f industrial light framing Western larch No. 1 2-inch dimension Southern pine 1.4E	WCLIB WCLIB WWPA WWPA SPIB WWPA	1500 1600 1600 1500 1500 1500	1500 1600 1600 1500 1500 1200	1200 1100 1100 1200 1350 1200

TABLE 7-1 CHORD CODE TABLE FOR ROOF TRUSSES CONTINUED

Chord Code	Size	Grade and Species Meeting Stress Requirements	Grading Rules	f	t//	c//
9	2 × 4	1200f industrial light framing WCDF	WCLIB	1200	1200	1000
		1200f industrial light framing Western larch	WWPA	1200	1200	1000
		1500f industrial light framing WCH	WCLIB	1500	1500	1000
		1500f industrial light framing WH	WWPA	1500	1500	1000
		No. 2 2-inch dimension Southern pine	SPIB	1200	1200	900
10	2 × 6	Select structural J&P WCDF	WCLIB	1900	1900	1500
		Select structural J&P Western larch	WWPA	1900	1900	1500
		No. 1 dense kiln-dried Southern pine	SPIB	2050	2050	1750
		1.8E	WWPA	2100	1700	1700
11	2 × 6	Construction grade J&P WCDF	WCLIB	1500	1500	1200
		Construction grade J&P Western larch	WWPA	1500	1500	1200
		Select structural J&P WCH	WCLIB	1600	1600	1200
		Select structural J&P WH	WWPA	1600	1600	1200
		No. 1 2-inch dimension Southern pine	SPIB	1500	1500	1350
		1.4E	WWPA	1500	1200	1200
12	2 × 6	Standard grade J&P WCDF	WCLIB	1200	1200	1000
		Standard grade J&P Western larch	WWPA	1200	1200	1000
		Standard grade J&P WCH	WCLIB	1200	1200	1000
		Standard grade J&P WH	WWPA	1200	1200	1000
		No. 2 2-inch dimension dense Southern pine	SPIB	1400	1400	1050

TABLE 7-2 DESIGNS WHEN USING STANDARD SHEATHING AS C-C EXT

Loading Condition, Total Roof Load, psf	Span	Beveled-Heel Gusset Dimensions, Inches A	B	C	H	O	Chord Code Upper	Lower	Square-Heel Gusset Dimensions, Inches A	B	C	H	O	Chord Code Upper	Lower
30 psf (20 psf live load, 10 psf dead load) on upper chord and 10 psf dead load on lower chord. Meets FHA requirements.	20'8"	32	48	12	45⅛	44	2	3	19	32	12	48¾	44	2	3
	22'8"	32	48	12	49⅛	48	1	2	19	32	12	52¾	48	1	3
	24'8"	48	60	16	53⅛	48	1	2	24	48	12	56¾	48	1	3
	26'8"	48	72	16	57⅛	48	1	2	32	60	16	60¾	48	1	2
40 psf (30 psf live load, 10 psf dead load) on upper chord and 10 psf dead load on lower chord.	20'8"	32	48	12	45⅛	43	8	9	19	32	12	48¾	48	7	9
	22'8"	32	60	16	49⅛	48	7	8	19	48	12	52¾	48	7	9
	24'8"								32	60	16	56¾	48	7	9

Lumber with roughness in the gusset area cannot be used. Twisted, cupped, or warped lumber should not be used either. This is especially true if the twist or cup is in the area of the gusset. Keep the intersecting surfaces of the lumber within 1/32 inch.

GLUE FOR TRUSSES • For dry or indoor locations use casein-type glue. It should meet Federal Specification MMM-A-125, Type II. For wet conditions use resorcinol-type glue. Military specifications are MIL-A-46051 for wet locations. If the glue joint is exposed to the weather or used at the soffit, use the resorcinol-type glue.

LOAD CONDITIONS FOR TRUSSES • Table 7-2 shows the loading factors needed in designing trusses. Note the 30 pounds per square foot (psf) and 40 psf columns. Then look up the type of gusset—either beveled-heel or square-heel type. Standard sheathing or C-C Ext-APA grade plywood is the type used here for the gussets and sheathing.

COVERING THE TRUSSES • Once the trusses are in place, cover them with sheathing or plywood. This will make a better structure once the sheathing is applied. The underlayment is applied and the roofing attached properly. The sheathing or plywood makes an excellent nail base for the shingles. See Chapter 8 for applying shingles.

• THE FRAMING SQUARE •

In carpentry it is necessary to be able to use the framing square. This device has a great deal of information stamped on its body and tongue. See Figure 7-15 for what the square can do in terms of a right angle. Figure 7-16 shows the right angle made by a framing square. The *steel square* or *carpenter's square* is made in the form of a right angle. That is, two arms (the body and the tongue of the square) make an angle of 90°. A 90° angle is a right angle.

Note in Figure 7-15 how a right *triangle* is made when points A, B, and C are connected. Figure 7-16 shows the right triangle. A right triangle has one angle which is 90°.

Parts of the Square

The steel square consists of two parts—the body and the tongue. The body is sometimes called the *blade*. See Figure 7-17.

Body The body is the longer and wider part. The body of the Stanley standard steel square is 24 inches long and 2 inches wide.

Tongue The tongue is the shorter and narrower part and usually is 16 inches long and 1½ inches wide.

Heel The point at which the body and the tongue meet on the outside edge of the square is called the *heel*. The intersection of the inner edges of the body and tongue is sometimes also called the *heel*.

Face The face of the square is the side on which the manufacturer's name, Stanley in this case, is stamped, or the visible side when the body is held in the left hand and the tongue is held in the right hand. See Figure 7-17.

Back The back is the side opposite the face. See Figure 7-18.

The modern scale usually has two kinds of marking: scales and tables.

SCALES • The scales are the inch divisions found on the outer and inner edges of the square. The inch graduations are in fractions of an inch. The Stanley square has the following graduations and scales:

Face of body—outside edge	Inches and sixteenths
Face of body—inside edge	Inches and eighths
Face of tongue—outside edge	Inches and sixteenths
Back of tongue—inside edge	Inches and eighths
Back of body—outside edge	Inches and twelfths
Back of body—inside edge	Inches and sixteenths
Back of tongue—outside edge	Inches and twelfths
Back of tongue—inside edge	Inches and tenths

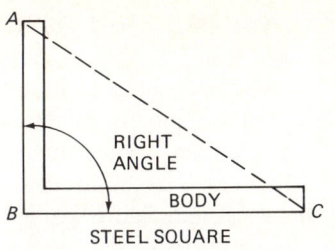

Figure 7-15. The steel square has a 90° angle between the tongue and the body. *(Stanley Tools)*

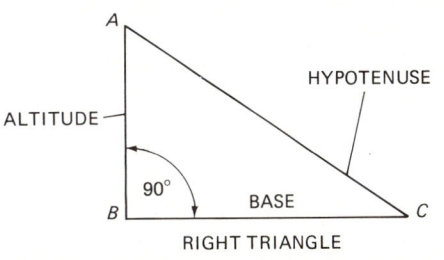

Figure 7-16. The parts of a right triangle. *(Stanley Tools)*

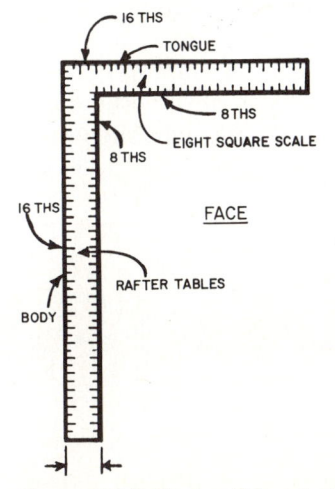

Figure 7-17. Parts of the steel square—face side. *(Stanley Tools)*

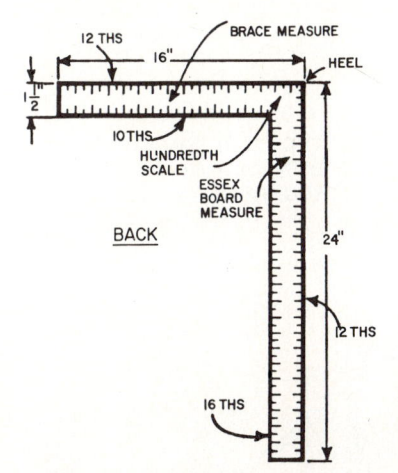

Figure 7-18. Back side of the steel square labeled. *(Stanley Tools)*

Building Roof Frames **127**

Hundredths Scale This scale is located on the back of the tongue, in the corner of the square, near the brace measure. The hundredths scale is 1 inch divided into 100 parts. The longer lines indicate 25 hundredths, while the next shorter lines indicate 5 hundredths. With the aid of a pair of dividers, fractions of an inch can be obtained. See Figure 7-19 for the location of the hundredths scale.

One inch graduated in sixteenths is below the hundredths scale on the latest squares, so that the conversion from hundredths to sixteenths can be made at a glance without the need to use dividers. This comes in handy when you are determining rafter lengths using the figures of the rafter tables, where hundredths are given.

Rafter Scales These tables will be found on the face of the body and will help you to determine rapidly the lengths of rafters as well as their cuts.

The rafter tables consist of six lines of figures. Their use is indicated on the left end of the body. The first line of figures gives the lengths of *common* rafters per foot of run. The second line gives the lengths of *hip and valley* rafters per foot of run. The third line gives the length of the *first jack* rafter and the differences in the length of the others centered at 16 inches. The fourth line gives the length of the *first jack* rafter and the differences in length of the others spaced at 24-inch centers. The fifth line gives the *side cuts of jacks*. The sixth line gives the side cuts of *hip and valley* rafters.

Octagon Scale The octagon or "eight-square" scale is found along the center of the face of the tongue. Using this scale a square timber may be shaped into one having eight sides, or an "octagon."

Brace Scale This table is found along the center of the back of the tongue and gives the exact lengths of *common braces*.

Essex Board Measure This table is on the back of the body and gives the contents of any size lumber.

Steel Square Uses

The steel square has many applications. It can be used as a try square or to mark a 90° line along a piece of lumber. See Figure 7-20.

The steel square can also be used to mark 45° angles and 30–60° angles. See Figure 7-21.

In some instances you may want to use the square for stepping off the length of rafters and braces. See Figure 7-22. Another use for the square is the laying out of stair steps. Figure 7-23 shows how this is done.

However, one of the most important roles the

Figure 7-19. Location of the hundredths scale on a square. *(Stanley Tools)*

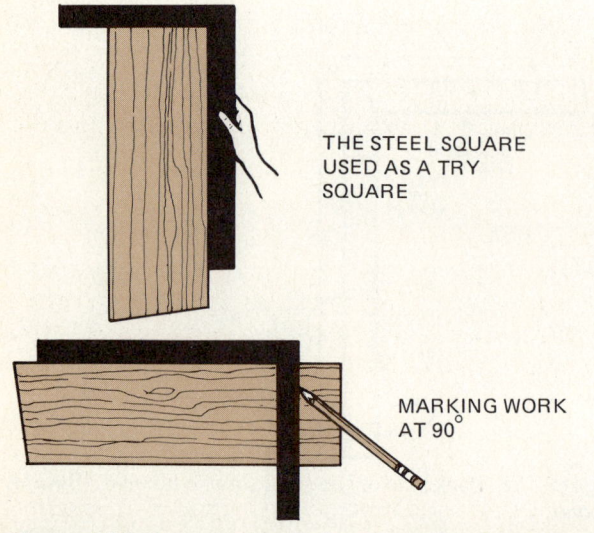

Figure 7-20. Uses of the steel square. *(Stanley Tools)*

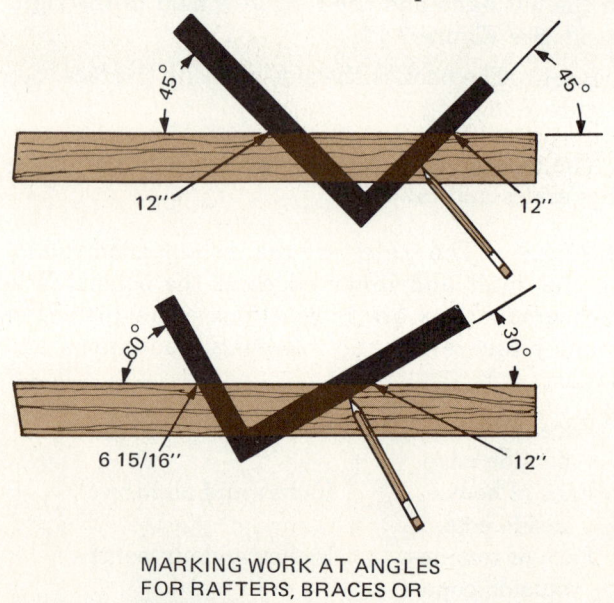

Figure 7-21. Using the steel square for marking angles. *(Stanley Tools)*

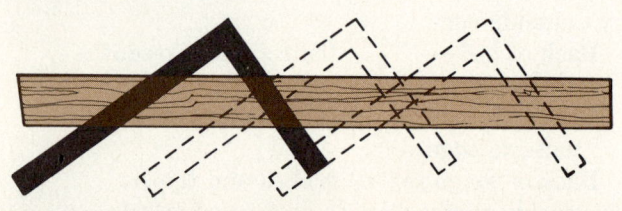

Figure 7-22. Using the steel square to step off the length of rafters and braces. *(Stanley Tools)*

128 Carpentry Fundamentals

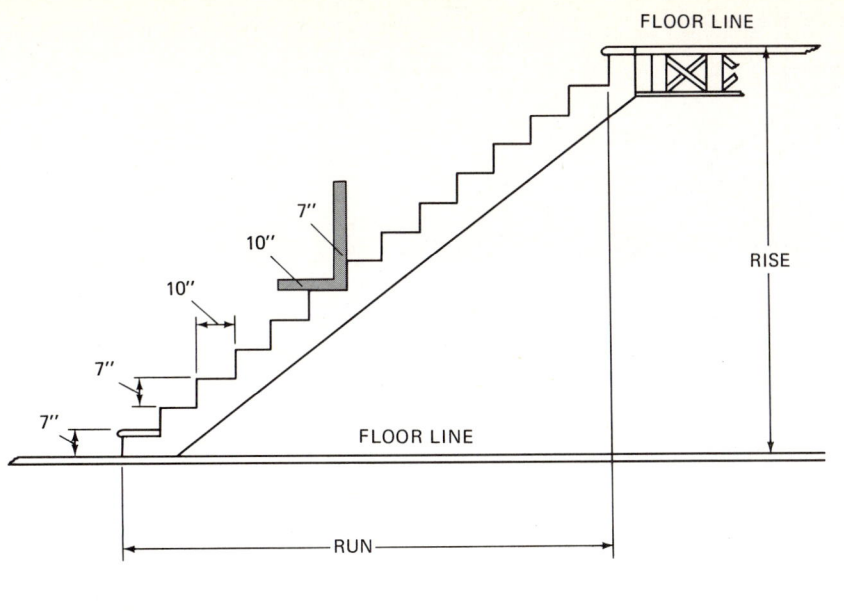

Figure 7-23. Using the square to lay out stairs. *(Stanley Tools)*

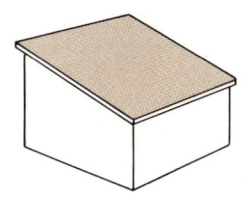

Figure 7-24. Lean-to or shed roof. *(Stanley Tools)*

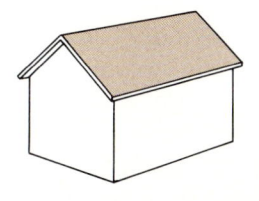

Figure 7-25. Gable roof. *(Stanley Tools)*

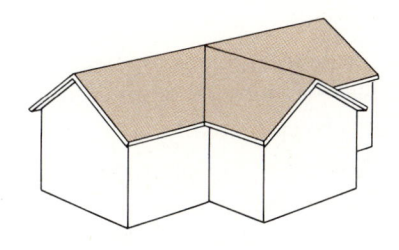

Figure 7-26. Gable-and-valley roof. *(Stanley Tools)*

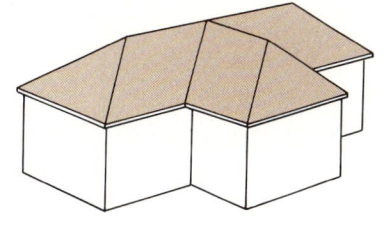

Figure 7-27. Hip-and-valley roof. *(Stanley Tools)*

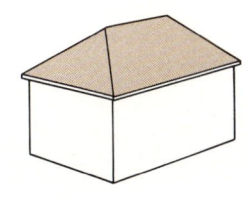

Figure 7-28. Hip roof. *(Stanley Tools)*

square plays in carpentry is the layout of roof framing. Here it is used to make sure the rafters fit the proper angles. The length of the rafters and the angles to be cut can be determined by the use of the framing square. Rafter cuts are shown in the following sections.

• ROOF FRAMING •

There are a number of types of roofs. A great variety of shapes is prevalent, as you can see in any neighborhood. Some of the most common types will be identified and worked with here.

Shed Roof The shed roof is the most common type. It is easy to make. It is also sometimes called the lean-to roof. It has only a single slope. See Figure 7-24.

Gable or Pitch Roof This is another type of roof that is commonly used. It has two slopes meeting at the center or ridge, forming a gable. It is a very simple form of roof, and perhaps the easiest to construct. See Figure 7-25.

Gable-and-Valley or Hip-and-Valley Roof This is a combination of two intersecting gable or hip roofs. The *valley* is the place where two slopes of the roof meet. The roofs run in different directions. There are many modifications of this roof, and the intersections usually are at right angles. See Figures 7-26 and 7-27.

Hip Roof This roof has four sides, all sloping toward the center of the building. The rafters run up diagonally to meet the ridge, into which the other rafters are framed. See Figure 7-28.

Building Roof Frames **129**

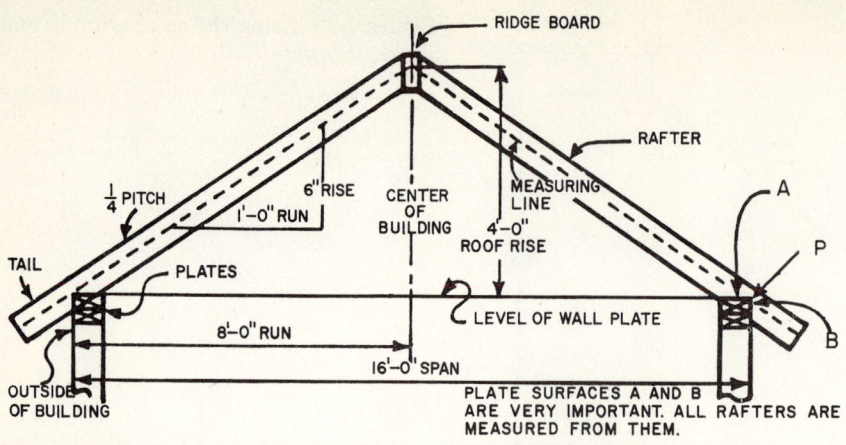

Figure 7-29. Span, run, rise, and pitch. *(Stanley Tools)*

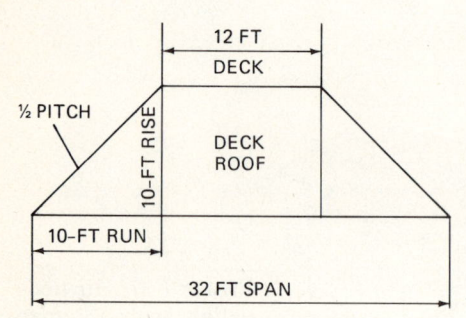

Figure 7-30. Location of the deck roof. *(Stanley Tools)*

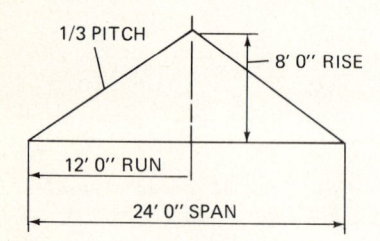

Figure 7-31. One-third pitch. *(Stanley Tools)*

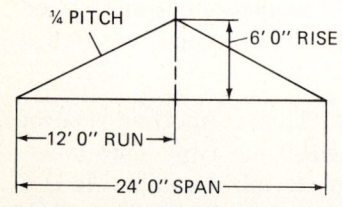

Figure 7-32. One-fourth pitch. *(Stanley Tools)*

Roof Terms

There are a number of terms you should be familiar with in order to work with roof framing. Each type of roof has its own particular terms: however, some of the common terms are:

Span The span of a roof is the distance over the wall plates. See Figure 7-29.

Run The run of a roof is the shortest horizontal distance measured from a plumb line through the center of the ridge to the outer edge of the plate. See Figure 7-29. In equally pitched roofs, the run is always equal to half the span or generally half the width of the building.

Rise The rise of a roof is the distance from the top of the ridge and of the rafter to the level of the foot. In figuring rafters, the rise is considered as the vertical distance from the top of the wall plate to the upper end of the measuring line. See Figure 7-29.

Deck Roof When rafters rise to a deck instead of a ridge, the width of the deck should be subtracted from the span. The remainder divided by 2 will equal the run. Thus, in Figure 7-30 the span is 32 feet and the deck is 12 feet wide. The difference between 32 and 12 is 20 feet, which divided by 2 equals 10 feet. This is the run of the common rafters. Since the rise equals 10 feet, this is a ½-*pitch* roof.

Pitch The pitch of a roof is the slant or the slope from the ridge to the plate. It may be expressed in several ways:

The pitch may be described in terms of the ratio of the *total width* of the building to the *total rise* of the roof. Thus, the pitch of a roof having a 24-foot span with an 8-foot rise will be 8 divided by 24, which equals ⅓ pitch. See Figure 7-31.

The pitch of a roof may also be expressed as so many inches of vertical rise to each foot of horizontal run. A roof with a 24-foot span and rising 8 inches to each foot of run will have a total rise of 8 × 12 = 96 inches or 8 feet. Eight divided by 24 equals ⅓. Therefore, the roof is ⅓ pitch. See Figure 7-31.

Note that in Figure 7-32 the building is 24 feet wide. It has a roof with a 6-foot rise. What is the pitch of the roof? The pitch equals 6 divided by 24, or ¼.

130   Carpentry Fundamentals

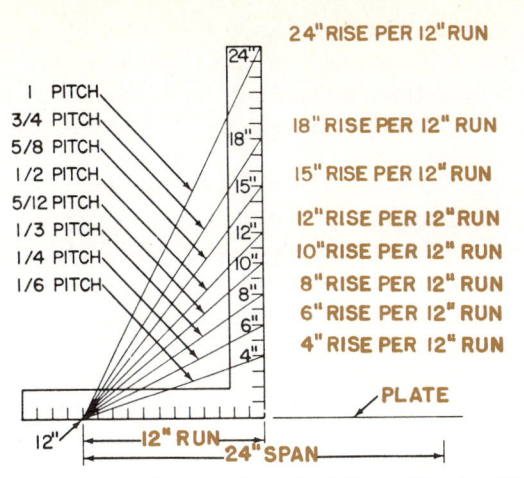

Figure 7-33. Principal roof pitches. *(Stanley Tools)*

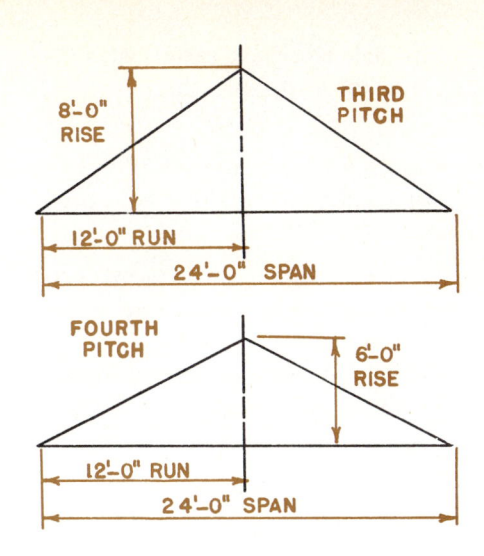

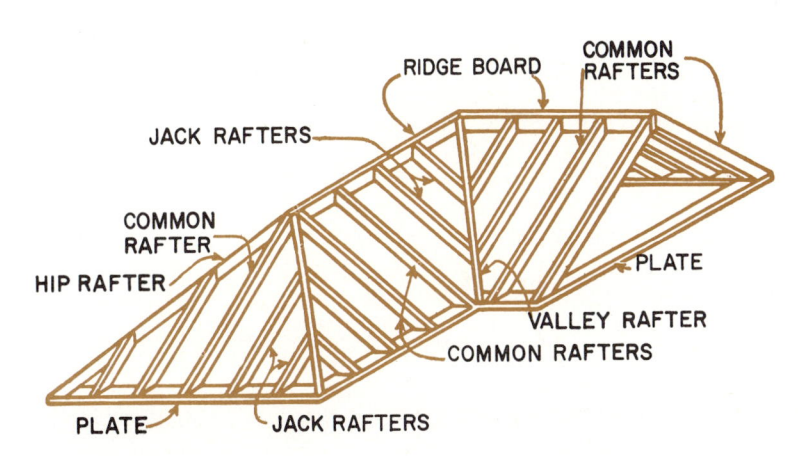

Figure 7-34. Different types of rafters used in a roof frame. *(Stanley Tools)*

PRINCIPAL ROOF PITCHES • Figure 7-33 shows the principal roof pitches. They are called ½ pitch, ⅓ pitch, or ¼ pitch, as the case may be, because the height from the level of the wall plate to the ridge of the roof is one-half, one-third, or one-quarter of the total width of the building.

Keep in mind that roofs of the same width may have different pitches, depending upon the height of the roof.

Take a look at Figure 7-34. This will help you interpret the various terms commonly used in roof construction.

Principal Roof Frame Members

The principal members of the roof frame are the plates at the bottom and the ridge board at the top. To them the various rafters are fastened. See Figure 7-34.

Plate The plate is the roof member to which rafters are framed at their lower ends. The top, *A*, and the outside edge of the plate, *B*, are the important surfaces from which rafters are measured in Figure 7-29.

Ridge Board The ridge board is the horizontal member that connects the upper ends of the rafters on one side to the rafters on the opposite side. In cheap construction the ridge board is usually omitted. The upper ends of the rafters are spiked together.

Common Rafters A common rafter is a member which extends diagonally from the plate to the ridge.

Hip Rafters A hip rafter is a member which extends diagonally from the corner of the plate to the ridge.

Valley Rafters A valley rafter is one which extends diagonally from the plate to the ridge at the line of intersection of two roof surfaces.

Jack Rafters Any rafter that does not extend from the plate to the ridge is called a jack rafter. There are different kinds of jacks. According to the position they occupy, they can be classified as hip jacks, valley jacks, or cripple jacks.

Building Roof Frames 131

Hip Jack A hip jack is a jack rafter with the upper end resting against a hip and the lower end against the plate.

Valley Jack A valley jack is a jack rafter with the upper end resting against the ridge board and the lower end against the valley.

Cripple Jack A cripple jack is a jack rafter with a cut that fits in between a hip and valley rafter. It touches neither the plate nor the ridge.

All rafters must be cut to proper angles so that they will fit at the points where they are framed. These different types of cuts are described below.

Top or Plumb Cut The cut of the rafter end which rests against the ridge board or against the opposite rafter is called the *top* or *plumb cut*.

Bottom or Heel Cut The cut of the rafter end which rests against the plate is called the *bottom* or *heel cut*. The bottom cut is also called the *foot* or *seat cut*.

Side Cuts Hip and valley rafters and all jacks, besides having top and bottom cuts, must also have their sides at the end cut to a proper bevel so that they will fit into the other members to which they are to be framed. These are called *side cuts* or *cheek cuts*. All rafters and their cuts are shown in Figure 7-35.

• RAFTERS •

Layout of a Rafter

This is a temporary line on which the length of the rafter is measured. This line runs parallel to the edge of the rafter and passes through the point P on the outer edge of the plate. This is the point from which all dimensions are determined. See Figure 7-29.

Length The length of a rafter is the shortest distance between the outer edge of the plate and the center of the ridge line.

Tail That portion of the rafter extending beyond the outside edge of the plate is called the tail. In some cases it is referred to as the eave. The tail is figured separately and is not included in the length of the rafter as mentioned above. See Figure 7-29.

Figure 7-36 shows the three variations of the rafter tail. Figure 7-36(A) shows the flush tail or no tail. The rafter butts against the wall plate with no overhang. In Figure 7-36(B), the full tail is shown. Note the overhang. Figure 7-36(C) shows the various shapes possible. This one indicates a separate tail that has been curved. It is nailed onto the no-tail or flush rafter.

All the cuts for the various types of common rafters are made at right angles to the sides of the rafter. Figure 7-37 shows how the framing square is used to lay out the angles. Find the 12-inch mark on the square. Note how the square is set at 12 for laying out the cut.

The distance 12 is the same as 1 foot of run. The other side of the square is set with the edge of the stock to the rise in inches per foot of run. In some cases the tail is not cut until after the rafter is in place. Then it is cut to match the others and aligns better for the fascia board.

Figure 7-38 shows a method of using the square to lay out the bottom and lookout cuts. If there is a ridge board, you have to deduct one-half the thickness of the ridge from the rafter length. Figure 7-38 shows how to place the square for marking the rafter lookout. Scribe the cut line as shown in Figure 7-38(A). The rise is 4 and the run is 12. Then move

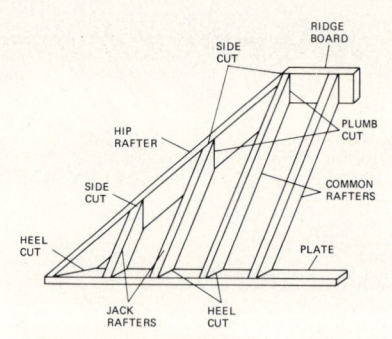

Figure 7-35. Rafter cuts. *(Stanley Tools)*

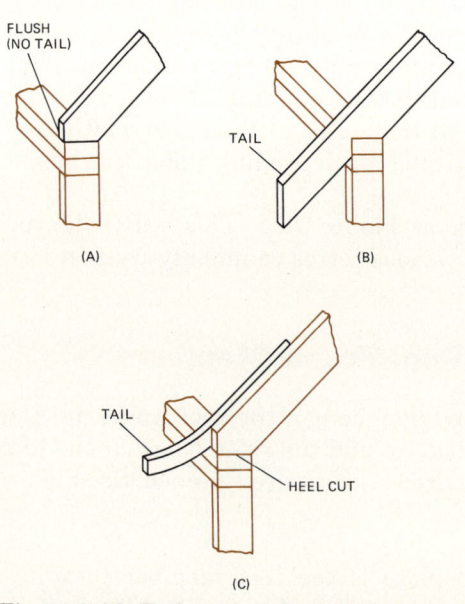

Figure 7-36. Tails or overhangs of rafters. (A) Flush tail (no tail). (B) Full tail. (C) Separate curved tail.

132   Carpentry Fundamentals

the square to the next position and mark from C to E. The distance from B to E is equal to the length of the lookout. Move the square up to E (with the same setting). Scribe line CE. On this line, lay off CD. This is the length of the vertical side of the bottom cut. Now apply the same setting to the bottom edge of the rafter. This is done so that the edge of the square cuts D. Scribe DF. See Figure 7-38(B). This is the horizontal line of the bottom cut. In making this cut, the wood is cut out to the lines CD and DF. See Figure 7-39(A) for an example of rafters cut this way. Note how the portable handsaw makes cuts beyond the marks.

In Figures 7-39(B) and 7-40 you can see the rafters in place. Note the 90° angle. The rafter and the ridge should meet at 90°. The rafter in Figure 7-40 has not had the lookout cut. The overhang is slightly different from the type just shown. Can you find the difference?

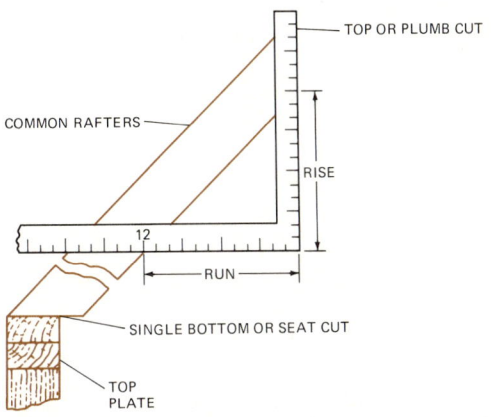

Figure 7-37. Using the steel square to mark off the top or plumb cut.

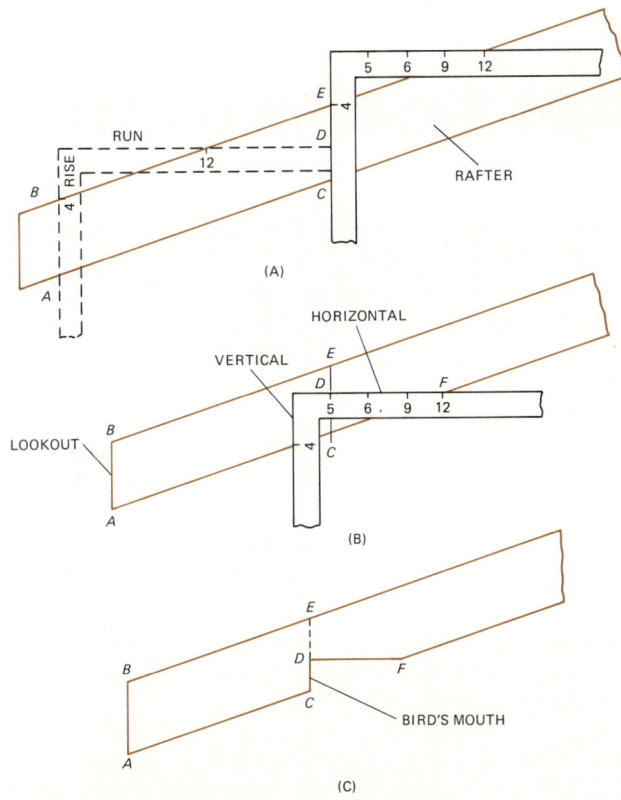

Figure 7-38. Using the steel square to lay out rafter lookout and bird's mouth.

Figure 7-39. (A) Notice the saw cuts past the bird's mouth.

Figure 7-39. (B) Rafters in place. They are nailed to the ridge board, the wall plate, and the ceiling joists. The other side of the roof is already covered with plywood sheathing.

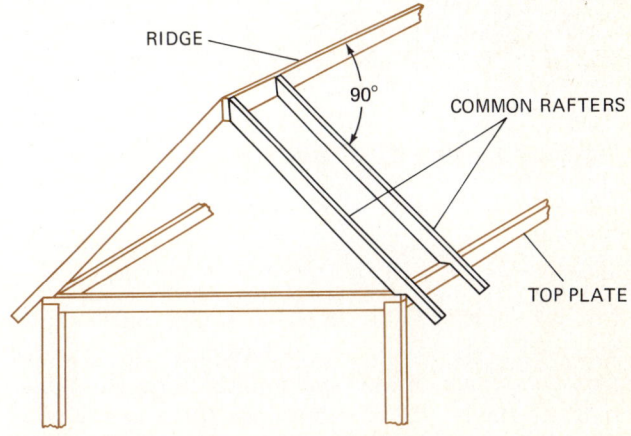

Figure 7-40. Common rafters in place.

Building Roof Frames 133

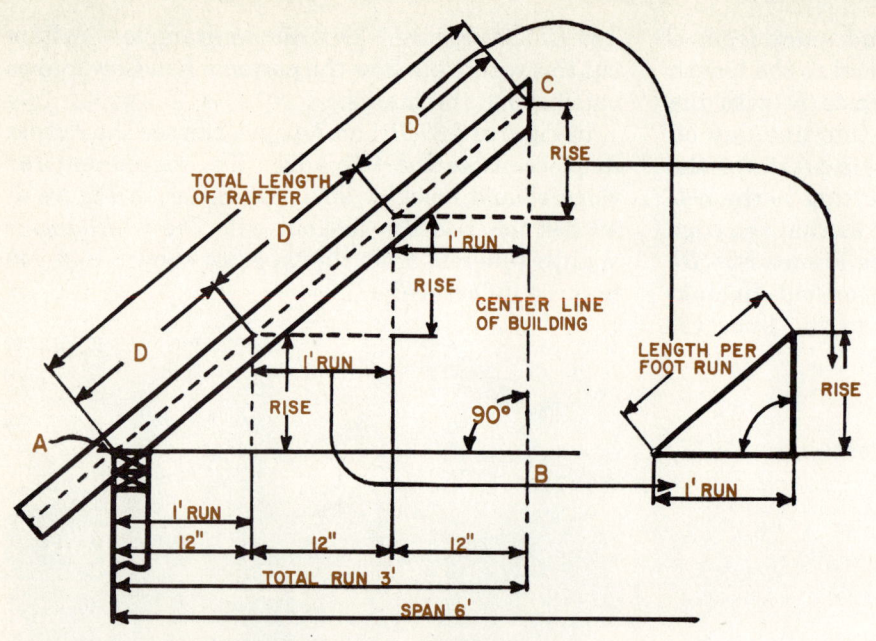

Figure 7-41. Length per foot of run. *(Stanley Tools)*

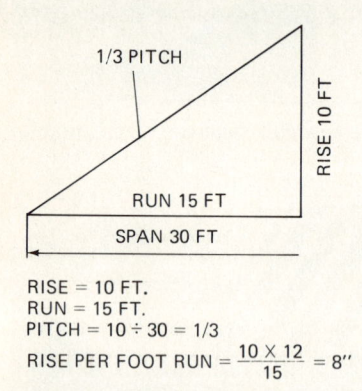

RISE = 10 FT.
RUN = 15 FT.
PITCH = 10 ÷ 30 = 1/3
RISE PER FOOT RUN = $\frac{10 \times 12}{15}$ = 8″

Figure 7-42. Finding the rise per foot of run. *(Stanley Tools)*

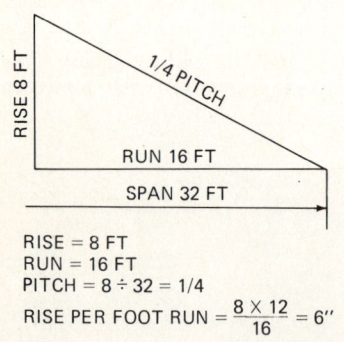

RISE = 8 FT
RUN = 16 FT
PITCH = 8 ÷ 32 = 1/4
RISE PER FOOT RUN = $\frac{8 \times 12}{16}$ = 6″

Figure 7-43. Finding the rise per foot of run. *(Stanley Tools)*

LENGTH PER FOOT OF RUN • The rafter tables on the Stanley squares are based on the *rise per foot run*, which means that the figures in the tables indicate the length of rafters per 1-foot run of common rafters for any rise of roof. This is shown in Figure 7-41.

The roof has a 6-foot span and a certain rise per foot. The figure may be regarded as a right triangle *ABC* having for its sides the run, the rise, and the rafter.

The run of the rafter has been divided into three equal parts, each representing 1-foot run.

It will be noted that by drawing vertical lines through each division point of the run, the rafter also will be divided into three equal parts *D*.

Since each part *D* represents the length of rafter per 1-foot run and the total run of the rafter equals 3 feet, it is evident that the *total length* of rafter will equal the length *D* multiplied by 3.

The reason for using this per-foot-run method is that the length of any rafter may be easily determined for any width of building. The length per foot run will be different for different pitches. Therefore, before you can find the length of a rafter, you must know the rise of roof in inches or the rise per foot of run.

RULE: To find the rise per foot run, multiply the rise by 12 and divide by the length of run.

The factor 12 is used to obtain a value in inches. The rise and run are expressed in feet. See Figures 7-42 and 7-43.

The rise per foot run is always the same for a given pitch and can be easily remembered for all ordinary pitches.

PITCH	½	⅓	¼	⅙
Rise per foot run in inches	12	8	6	4

134 Carpentry Fundamentals

The members of a firmly constructed roof should fit snugly against each other. Rafters that are not properly cut make the roof shaky and the structure less stable. Therefore, it is very important that *all rafters should be the right lengths and their ends properly cut*. This will provide a full bearing against the members to which they are connected.

Correct length, proper top and bottom cuts, and the right side or cheek cuts are the very important features to be observed when framing a roof.

Length of Rafters

The length of rafters may also be obtained in other ways. There are three in particular which can be used:

1. Mathematical calculations
2. Measuring across the square
3. Stepping off with the square

The first method, while absolutely correct, is very impractical on the job. The other two are rather unreliable and quite frequently result in costly mistakes.

The tables on the square have eliminated the need for using the three methods just mentioned. These tables let the carpenter find the exact length and cuts for any rafter quickly, and thus save time and avoid the possibility of errors.

Common Rafters

The common rafter extends from the plate to the ridge. Therefore, it is evident that the rise, the run, and the rafter itself form a right triangle. The length of a common rafter is the shortest distance between the outer edge of the plate and a point on the centerline of the ridge. This length is taken along the *measuring line*. The measuring line runs parallel to the edge of the rafter and is the hypotenuse or the longest side of a right triangle. The other sides of the triangle are the run and the rise. See Figure 7-44.

The rafter tables on the face of the body of the square include the outside edge graduations on both the body and the tongue which are in inches and sixteenths of an inch.

THE LENGTH OF RAFTERS • The lengths of common rafters are found on the first line, indicated as *Length of main rafters per foot run*. There are seventeen of these tables, beginning at 2 inches and continuing to 18 inches. Figure 7-45 shows the square being used.

> **RULE:** To find the length of a common rafter, multiply the length given in the table by the number of feet of run.

For example, if you want to find the length of a common rafter where the rise of roof is 8 inches per foot run, or one-third pitch, and the building is 20 feet wide, you first find where the table is. Then on the inch line on the top edge of the body, find the figure that is equal to the rise of the roof, which in this case will be 8. On the first line under the figure 8 will be found 14.42. This is the length of the rafter in inches per foot run for this particular pitch. Examine Figure 7-46.

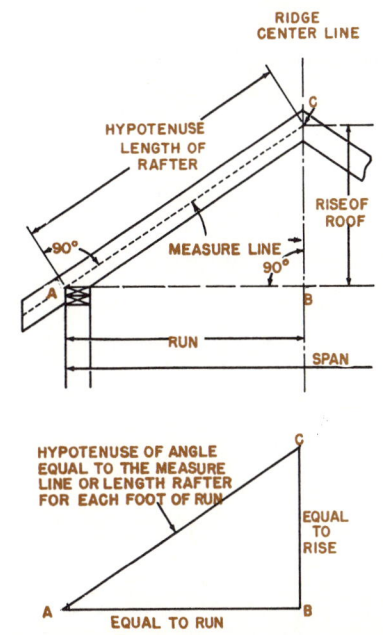

Figure 7-44. Measuring the line for a common rafter. *(Stanley Tools)*

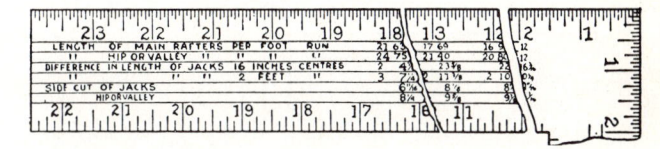

Figure 7-45. Note the labels for tables on the steel square. *(Stanley Tools)*

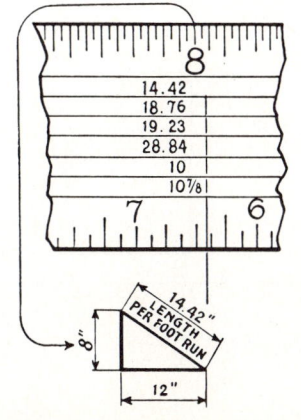

Figure 7-46. Finding the proper number on the square. *(Stanley Tools)*

Building Roof Frames

The building is 20 feet wide. Therefore, the run of the rafter will be 20 divided by 2, which equals 10 feet.

Since the length of the rafter per 1-foot run equals 14.42 inches, the total length of the rafter will be 14.42 multiplied by 10, which equals 144.20 inches, or 144.20 divided by 12, which equals 12.01 feet, or for all practical purposes 12 feet. Check Figure 7-46.

TOP AND BOTTOM CUTS OF THE COMMON RAFTER • The top or plumb cut is the cut at the upper end of the rafter where it rests against the opposite rafter or against the ridge board. The bottom cut or heel cut is the cut at the lower end which rests on the plate. See Figure 7-47.

The top cut is parallel to the centerline of the roof, and the bottom cut is parallel to the horizontal plane of the plates. Therefore, the top and bottom cuts are at right angles to each other.

RULE: To obtain the top and bottom cuts of a common rafter, use 12 inches on the body and the rise per foot run on the tongue. Twelve inches on the body will give the horizontal cut, and the figure on the tongue will give the vertical cut.

To illustrate the rule, we will examine a large square placed alongside the rafter as shown in Figure 7-48. Note that the edge of the tongue coincides with the top cut of the rafter. The edge of the blade coincides with the heel cut. If this square were marked in feet, it would show the run of the rafter on the body and the total rise on the tongue. Line *AB* would give the bottom cut and line *AC* would give the top cut.

However, the regular square is marked in inches. Since the relation of the rise to 1-foot run is the same as that the total rise bears to the total run, we use 12 inches on the blade and the rise per foot on the tongue to obtain the respective cuts. The distance 12 is used as a unit and is the 1-foot run, while the figure on the other arm of the square represents the rise per foot run. See Figures 7-49 and 7-50.

ACTUAL LENGTH OF THE RAFTER • The rafter lengths obtained from the tables on the square

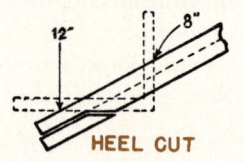

Figure 7-49. Using the steel square to lay out the heel cut. *(Stanley Tools)*

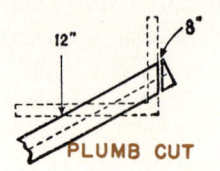

Figure 7-50. Using the steel square to lay out the plumb cut. *(Stanley Tools)*

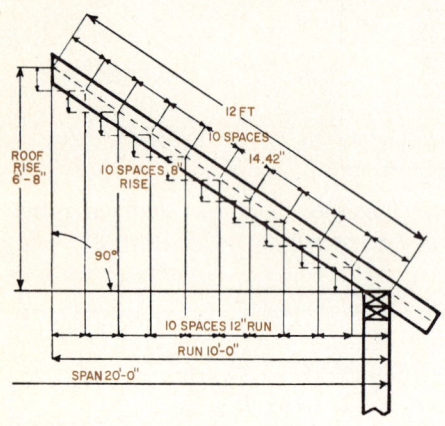

Figure 7-47. Finding the length of the rafter. *(Stanley Tools)*

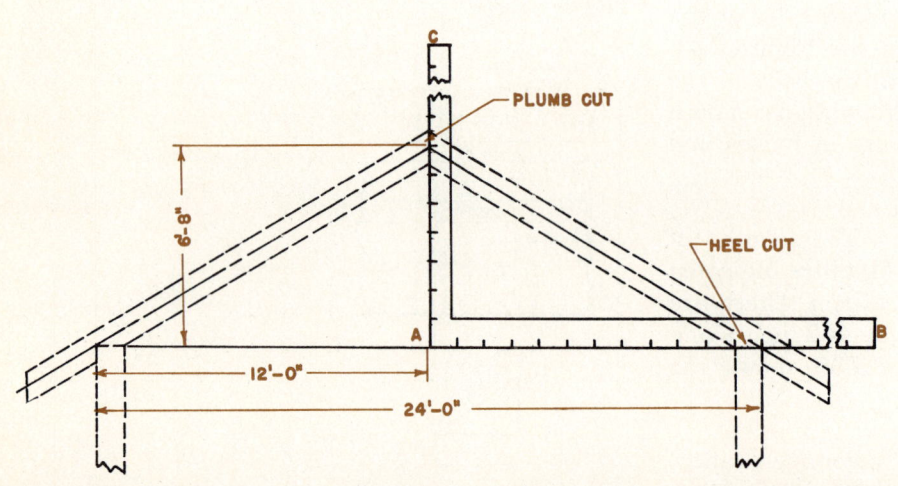

Figure 7-48. Using the steel square to check plumb and heel cuts. *(Stanley Tools)*

are to the centerline of the ridge. Therefore, the thickness of half of the ridge board should always be deducted from the obtained total length before the top cut is made. See Figure 7-51. This deduction of half the thickness of the ridge is measured at right angles to the plumb line and is marked parallel to this line.

Figure 7-52 shows the wrong and right ways of measuring the length of rafters. Diagram D shows the measuring line as the edge of the rafter, which is the case when there is no tail or eave.

CUTTING THE RAFTER • After the total length of the rafter has been established, both ends should be marked and allowance made for a tail or eave. Don't forget to allow for half the thickness of the ridge.

Both cuts are obtained by applying the square so that the 12-inch mark on the body and the mark on the tongue that represents the rise are at the edge of the stock.

All cuts for common rafters are made at right angles to the side of the rafter.

For example, a common rafter is 12'6", and the rise per foot run is 9 inches. Obtain the top and bottom cuts. See Figure 7-53.

Points *A* and *B* are the ends of the rafter. To obtain the bottom or seat cut, take 12 inches on the body of the square and 9 inches on the tongue. Lay the square on the rafter so that the body will coincide with point *A* or the lower end of the rafter. Mark along the body of the square and cut.

To obtain the top cut, move the square so that the tongue coincides with point *B*. This is the upper end of the rafter. Mark along the tongue of the square.

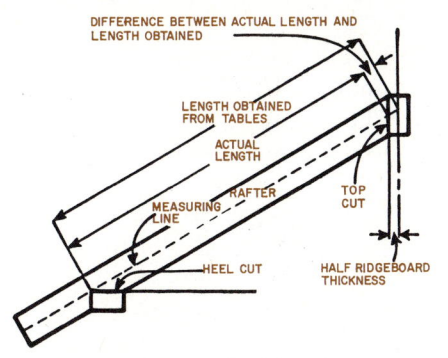

Figure 7-51. How to find the difference between the actual length and the length obtained. *(Stanley Tools)*

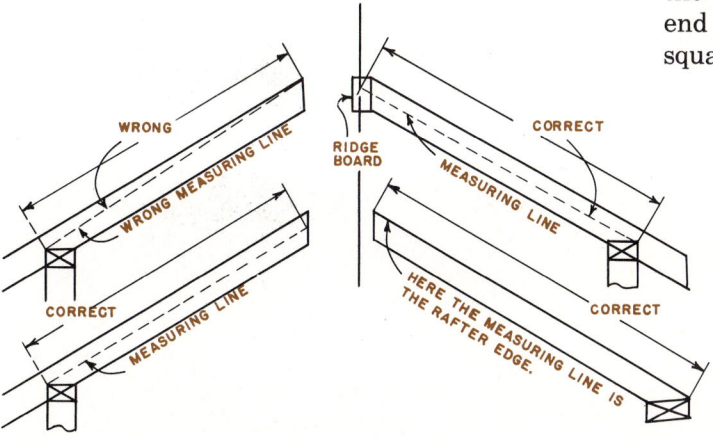

Figure 7-52. Right and wrong ways of measuring rafters. *(Stanley Tools)*

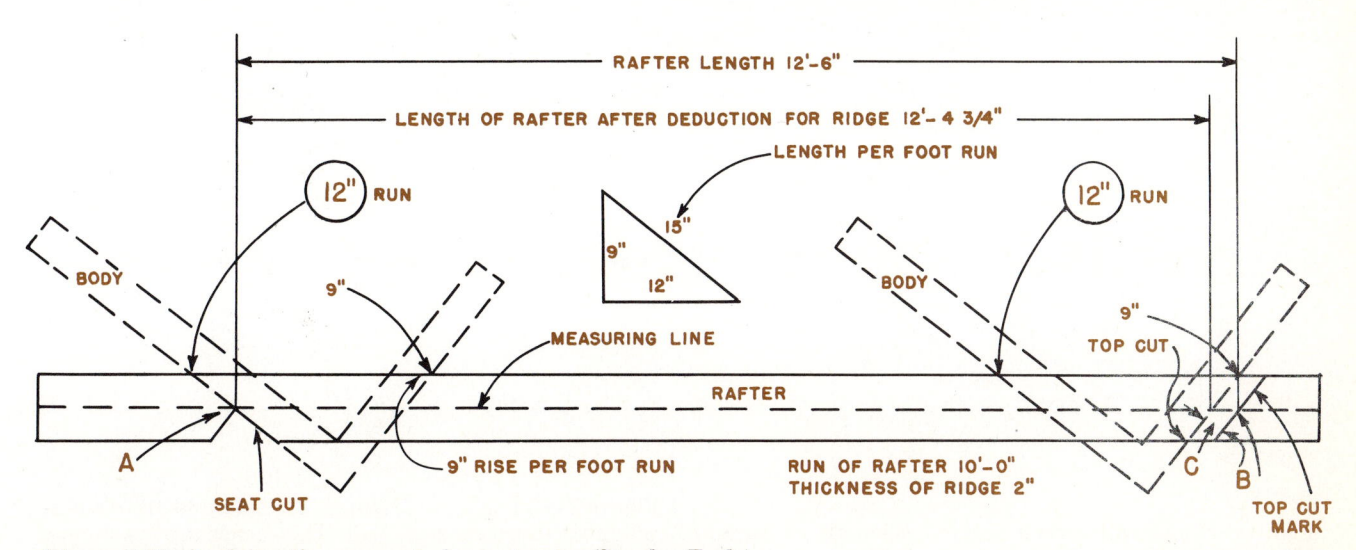

Figure 7-53. Applying the square to lay out cuts. *(Stanley Tools)*

Building Roof Frames

DEDUCTION FOR THE RIDGE • The deduction for half the thickness of the ridge should now be measured. Half the thickness of the ridge is 1 inch. One inch is deducted at right angles to the top cut mark or plumb line, which is point C. A line is then drawn parallel to the top cut mark, and the cut is made. You will notice that the allowance for half the ridge measured along the measuring line is 1¼ inches. This will vary according to the rise per foot run. It is therefore important to measure for this deduction at right angles to the top cut mark or plumb line.

MEASURING RAFTERS • The length of rafters having a tail or eave can also be measured along the back or top edge instead of along the measuring line, as shown in Figure 7-54. To do this it is necessary to carry a plumb line to the top edge from P, and the measurement is started from this point.

Occasionally in framing a roof, the run may have an odd number of inches; for example, a building might have a span of 24'10". This would mean a run of 12'5". The additional 5 inches can be added easily without mathematical division after the length obtained from the square for 12 feet of run is measured. The additional 5 inches is measured at right angles to the last plumb line. See Figure 7-55 for an illustration of the procedure.

Hip and Valley Rafters

The hip rafter is a roof member that forms a hip in the roof. This usually extends from the corner of the building diagonally to the ridge.

The valley rafter is similar to the hip, but it forms a valley or depression in the roof instead of a hip. It also extends diagonally from the plate to the ridge. Therefore, the total rise of the hip and valley rafters is the same as that of the common rafter. See Figure 7-56.

The relation of hip and valley rafters to the common rafter is the same as the relation of the sides of a right triangle. Therefore, it will be well to explain here one of the main features of right triangles.

In a right triangle, if the sides forming the right angle are 12 inches each, the hypotenuse, or the side opposite the right angle, is equal to 16.97 inches. This is usually taken as 17 inches. See Figure 7-57.

The position of the hip rafter and its relation to the common rafter is shown in Figure 7-58, where the hip rafter is compared to the diagonal of a square prism. The prism (as shown in Figure 7-58) has a base of 5 feet square, and its height is 3'4".

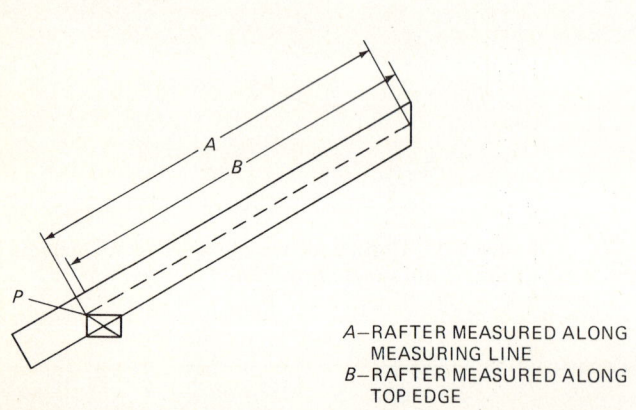

Figure 7-54. Two places to measure a rafter. *(Stanley Tools)*

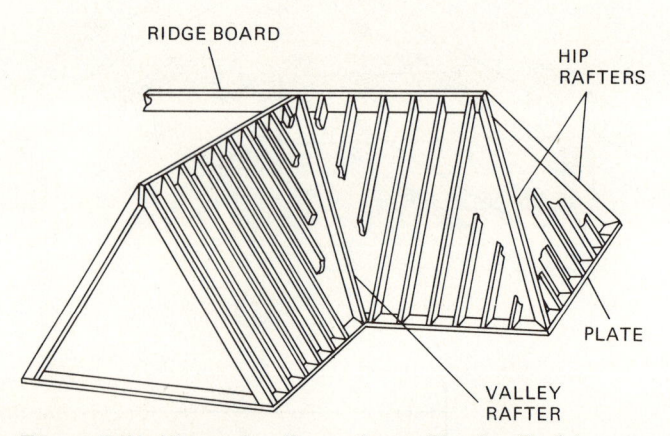

Figure 7-56. Hip and valley rafters. *(Stanley Tools)*

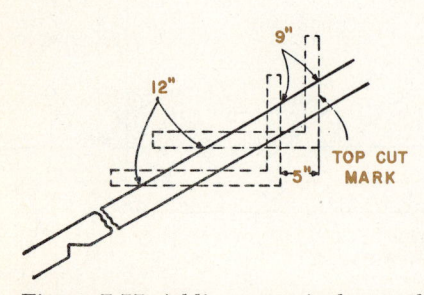

Figure 7-55. Adding extra inches to the length of a rafter. *(Stanley Tools)*

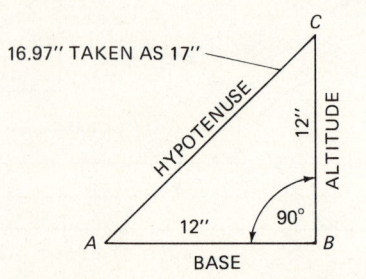

Figure 7-57. In a right triangle, 12" base and 12" altitude produces an isosceles triangle. This means the hypotenuse is 16.97". *(Stanley Tools)*

138 Carpentry Fundamentals

D is the corner of the building
BC is the total rise of the roof
AB is the run of the common rafter
AC is the common rafter
DB is the run of the hip rafter
DC is the hip rafter

It should be noted that the figure *DAB* is a right triangle whose sides are the portion of the plate *DA*, the run of the common rafter *AB*, and the run of the hip rafter *DB*. The run of the hip rafter is opposite right angle *A*. The hypotenuse is the longest side of the right triangle.

If we should take only 1 foot of run of common rafter and a 1-foot length of plate, we would have a right triangle *H*. The triangle's sides are each 12 inches long. The hypotenuse is 17 inches, or more accurately 16.97 inches. See Figure 7-57 and 7-59.

The hypotenuse in the small triangle *H* in Figure 7-59 is a portion of the run of the hip rafter *DB*. It corresponds to a 1-foot run of common rafter. Therefore, the run of hip rafter is always 16.97

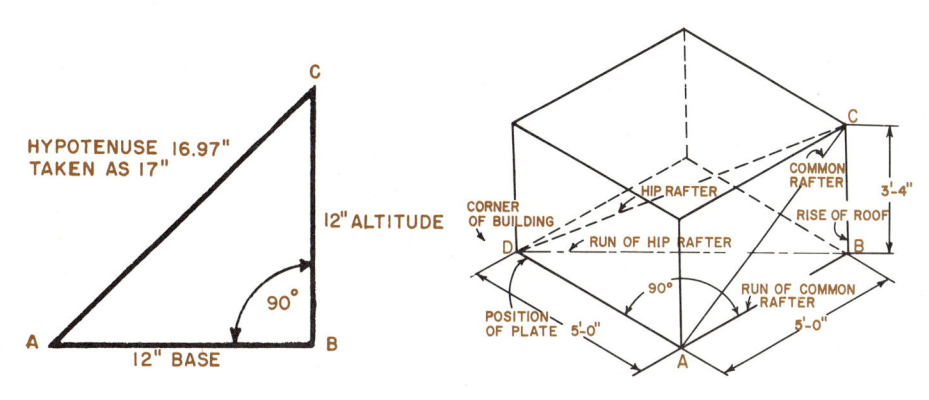

Figure 7-58. Relative position of the hip rafter and the common rafter. *(Stanley Tools)*

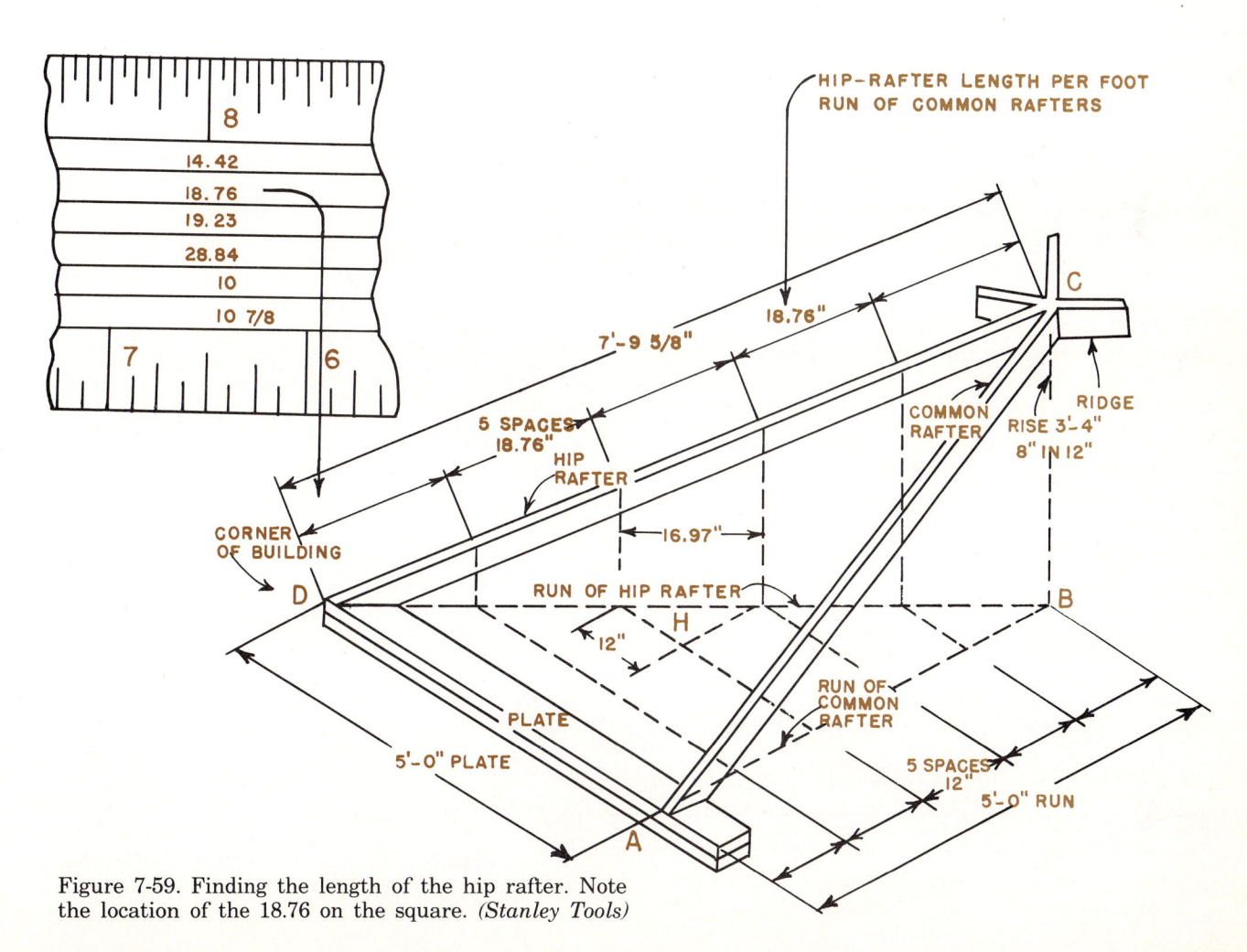

Figure 7-59. Finding the length of the hip rafter. Note the location of the 18.76 on the square. *(Stanley Tools)*

Building Roof Frames   139

inches for every 12 inches of run of common rafter. The total run of the hip rafter will be 16.97 inches multiplied by the number of feet run of common rafter.

LENGTH OF HIP AND VALLEY RAFTERS
• Lengths of the hip and valley rafters are found on the second line of the rafter table. It is entitled *Length of hip or valley rafters per foot run*. This means that the figures in the table indicate the length of hip and valley rafters per foot run of common rafters. See Figure 7-45.

RULE: To find the length of a hip or valley rafter, multiply the length given in the table by the number of feet of the run of common rafter.

For example, find the length of a hip rafter where the rise of the roof is 8 inches per foot of run, or one-third pitch. The building is 10 feet wide. See Figure 7-58.

Proceed as in the case of the common rafters. That is, find on the inch line of the body of the square the figure corresponding to the rise of roof—which is 8. On the second line under this figure is found 18.76. This is the length of hip rafter in inches for each foot of run of common rafter for one-third pitch. See Figure 7-59.

The common rafter has a 5-foot run. Therefore, there are also five equal lengths for the hip rafter, as may be seen in Figure 7-59. We have found the length of the hip rafter to be 18.76 inches per 1-foot run. Therefore, the total length of the hip rafter will be 18.76 × 5 = 93.80 inches. This is 7.81 feet, or for practical purposes 7'9 13/16" or 7'9 5/8".

TOP AND BOTTOM CUTS
• The following rule should be followed for top and bottom cuts.

RULE: To obtain the top and bottom cuts of hip or valley rafters, use 17 inches on the body and the rise per foot run on the tongue. Seventeen on the body will give the seat cut, and the figure on the tongue will give the vertical or top cut. See Figure 7-60.

MEASURING HIP AND VALLEY RAFTERS
• The length of all hip and valley rafters must always be measured along the center of the top edge or back. Rafters with a tail or eave are treated like common rafters, except that the measurement or measuring line is at the center of the top edge.

DEDUCTION FROM HIP OR VALLEY RAFTER FOR RIDGE
• The deduction for the ridge is measured in the same way as for the common rafter (see Figure 7-53), except that half the diagonal (45°) thickness of the ridge must be used.

SIDE CUTS
• In addition to the top and bottom cuts, hip and valley rafters must also have side or check cuts at the point where they meet the ridge.

These side cuts are found on the sixth or bottom line of the rafter tables. It is marked *Side cut hip or valley—use*. The figures given in this line refer to the graduation marks on the outside edge of the body. See Figure 7-45.

The figures on the square have been derived by determining the figure to be used with 12 on the tongue for the side cuts of the various pitches by the following method. From a plumb line, the thickness of the rafter is measured and marked at right angles as at A in Figure 7-61. A line is then squared across the top of the rafter and the diagonal points connected as at B. The line B or side cut is obtained by marking along the tongue of the square.

STEP 1. Shorten for ridge thickness.

STEP 2. Mark a line parallel to the plumb cut, equal in distance to the thickness of the rafter.

STEP 3. Square a line across top of hip rafter. The diagonal is the side cut.

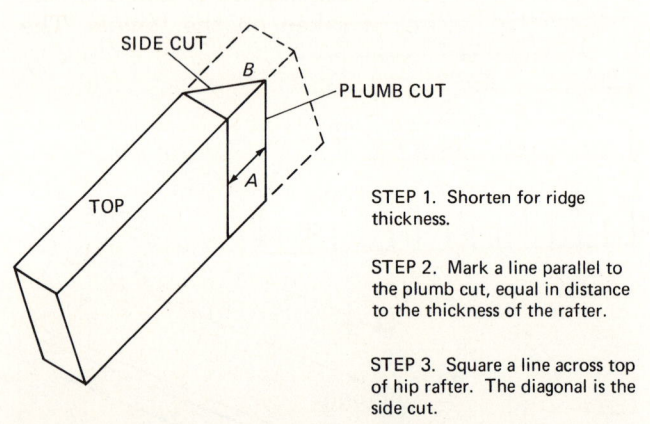

Figure 7-61. Making side cuts so that the hip will fit into the intersection of rafters. *(Stanley Tools)*

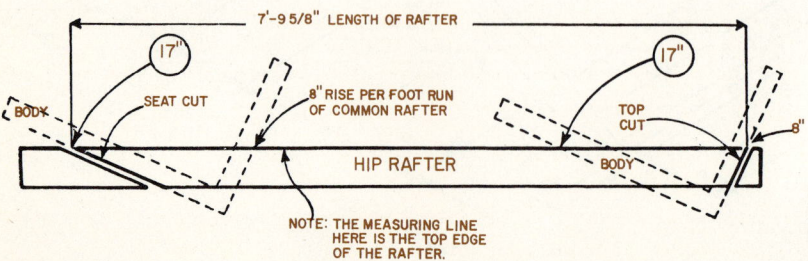

Figure 7-60. Using the square to lay out top and seat cuts on a hip rafter. *(Stanley Tools)*

140 Carpentry Fundamentals

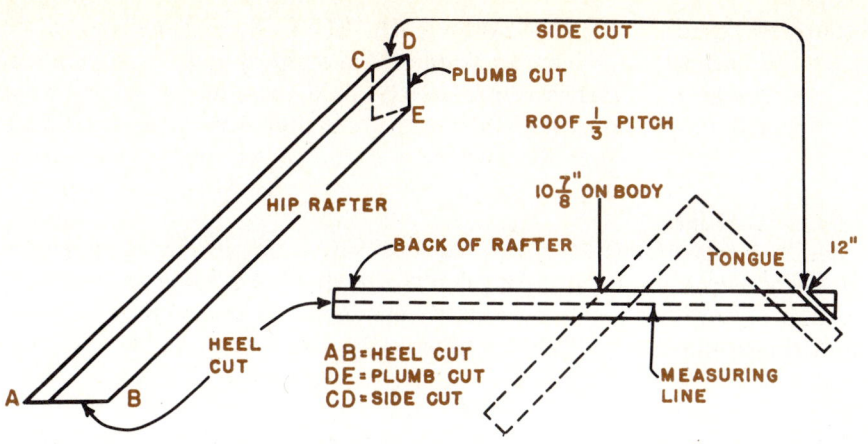

Figure 7-62. Hip rafter cuts. *(Stanley Tools)*

RULE: To obtain the side cut for hip or valley rafters, take the figure given in the table on the body of the square and 12 inches on the tongue. Mark the side cut along the tongue where the tongue coincides with the point on the measuring line.

For example, find the side cut for a hip rafter where the roof has 8 inches rise per foot run or ⅓ pitch. See Figure 7-62.

Figure 7-62 represents the position of the hip rafter on the roof. The rise of the roof is 8 inches to the foot. First, locate the number 8 on the outside edge of the body. Under this number in the bottom line you will find 10⅞. This figure is taken on the body and 12 inches is taken on the tongue. The square is applied to the edge of the back of the hip rafter. The side cut *CD* comes along the tongue.

In making the seat cut for the hip rafter, an allowance must be made for the top edges of the rafter. They would project above the line of the common and jack rafters if the corners of the hip rafter were not removed or backed. The hip rafter must be slightly lowered. Do this by cutting parallel to the seat cut. The distance varies with the thickness and pitch of the roof.

It should be noted that on the Stanley square the 12-inch mark on the tongue is always used in all angle cuts—top, bottom, and side. This leaves the worker with only one number to remember when laying out side or angle cuts. That is the figure taken from the fifth or sixth line in the table.

The side cuts always come on the right-hand or tongue side on rafters. When you are marking boards, these can be reversed for convenience at any time by taking the 12-inch mark on the body and using the body references on the tongue.

Obtain additional inches in run of hip or valley rafters by using the explanation given earlier for common rafters. However, use the diagonal (45°)

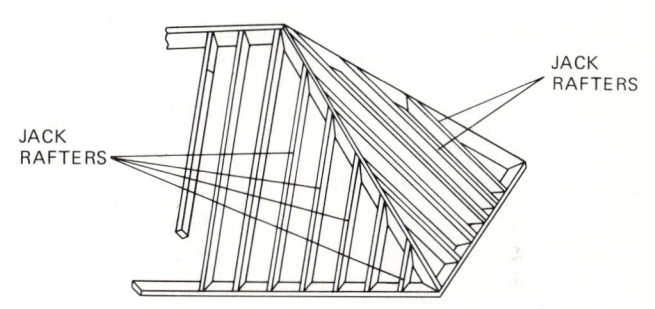

Figure 7-63. Location of jack rafters. *(Stanley Tools)*

of the additional inches. This is approximately 7 1/16 inches for 5 inches of run. This distance should be measured in a similar manner.

Jack Rafters

Jack rafters are *discontinued* common rafters. They are common rafters cut off by the intersection of a hip or valley before reaching the full length from plate to ridge.

Jack rafters lie in the same plane as common rafters. They usually are spaced the same and have the same pitch. Therefore, they also have the same length per foot run as common rafters have.

Jack rafters are usually spaced 16 inches or 24 inches apart. Because they rest against the hip or valley equally spaced, the second jack must be twice as long as the first one, the third 3 times as long as the first, and so on. See Figure 7-63.

LENGTH OF JACK RAFTERS • The lengths of jacks are given in the third and fourth lines of the rafter tables on the square. They are indicated:

Third line: Difference in length of jacks—16 inch centers
Fourth line: Difference in length of jacks—2 foot centers

The figures in the table indicate the length of the first or shortest jack, which is also the difference in length between the first and second jacks, between the second and third jacks, and so on.

Building Roof Frames 141

RULE: To find the length of a jack rafter, multiply the value given in the tables by the number indicating the position of the jack. From the obtained length, subtract half the diagonal (45°) thickness of the hip or valley rafter.

For example, find the length of the second jack rafter. The roof has a rise of 8 inches to 1 foot of run of common rafter. The spacing of the jacks is 16 inches.

On the outer edge of the body find the number 8, which corresponds to the rise of the roof. On the third line under this figure find 19¼. This means that the first jack rafter will be 19¼ inches long. Since the length of the second jack is required, multiply 19¼ by 2, which equals 38½ inches. From this length half the diagonal (45°) thickness of the hip or valley rafter should be deducted. This is done in the same manner that the deduction for the ridge was made on the hip rafter.

Proceed in the same manner when the lengths of jacks spaced on 24-inch centers are required. It should be borne in mind that the second jack is twice as long as the first one. The third jack is 3 times as long as the first jack, and so on.

TOP AND BOTTOM CUTS FOR JACKS • Since jack rafters have the same rise per foot run as common rafters, the method of obtaining the top and bottom cuts is the same as for common rafters. That is, take 12 inches on the body and the rise per foot run on the tongue. Twelve inches will give the seat cut. The figure on the tongue will give the plumb cut.

SIDE CUT FOR JACKS • At the end where the jack rafter frames to the hip or valley rafter, a side cut is required. The side cuts for jacks are found on the fifth line of the rafter tables on the square. It is marked: *Side cut of jacks—use.* See Figure 7-45.

RULE: To obtain the side cut for a jack rafter, take the figure shown in the table on the body of the square and 12 inches on the tongue. Mark along the tongue for side cut.

For example, find the side cut for jack rafters of a roof having 8 inches rise per foot run, or ⅓ pitch. See Figures 7-64 and 7-65. Under the number 8 in the fifth line of the table find 10. This number taken on the outside edge of the body and 12 inches taken on the tongue will give the required side cut.

• BRACE MEASURE •

In all construction there is the need for some braces to make sure certain elements are held securely. See Figure 7-120(A). The brace measure table is found along the center of the back of the tongue of the carpenter's square. It gives the lengths of common braces. See Figure 7-66.

For example, find the length of a brace whose run on post and beam equals 39 inches. See Figure 7-67. In the brace table find the following expression:

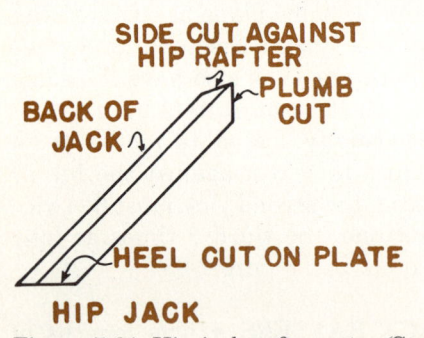

39
     55.15
39

Figure 7-64. Hip jack rafter cuts. *(Stanley Tools)*

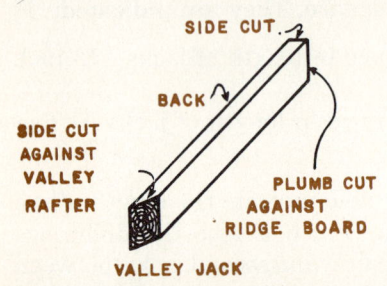

Figure 7-65. Valley jack rafter cuts. *(Stanley Tools)*

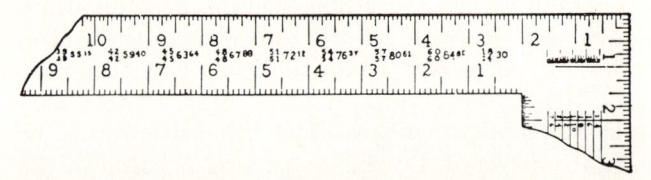

Figure 7-66. Brace measure table on back side of steel square tongue. *(Stanley Tools)*

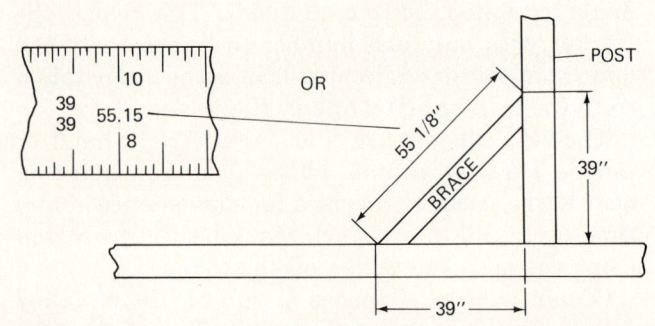

Figure 7-67. Cutting a brace using a square table. *(Stanley Tools)*

142   Carpentry Fundamentals

This means that with a 39-inch run on the beam and a 39-inch run on the post, the length of the brace will be 55.15 inches. For practical purposes, use 55⅛ inches.

Braces may be regarded as common rafters. Therefore, when the brace run on the post differs from the run on the beam, their lengths as well as top and bottom cuts may be determined from the figures given in the tables of the common rafters.

• ERECTING THE ROOF WITH RAFTERS •

Rafters are cut to fit the shape of the roof. Roofs are chosen by the builder or planner. The design of the rafter is determined by the pitch, span, and rise of the type of roof chosen. The gable roof is simple. It can be made easily with a minimum of difficult cuts. In this example we start with the gable type and then look at other variations of roof lines. See Figure 7-68 for an example of the gable roof.

Rafter Layout

One of the most important tools used to lay out rafters is the carpenter's square. All rafters must be cut to the proper angle or bevel. They fasten to the wall plate or to the ridge board. In some cases there is an overhang. This overhang must be taken into consideration when the rafter is cut. Gable siding, soffits, and overhangs are built together (Figure 7-69).

Raising Rafters

Mark rafter locations on the top plate of the side walls. The first rafter pair will be flush with the outside edge of the end wall. See Figure 7-70. Note the placement of the gable end studs. The notch in the gable end stud is made to fit the 2 × 4, or whatever thickness of rafter you are using. Space the first interior rafter 24 inches measured from the end of the building to the center of the rafter. In some cases 16 inches O.C. is used for spacing. All succeeding rafter locations are measured 24 inches center to center. They will be at the sides of ceiling joist ends. See Figure 7-71.

Figure 7-68. Gable roof.

Figure 7-69. Gable siding, soffits, and gable overhangs can be built together. *(Fox and Jacobs)*

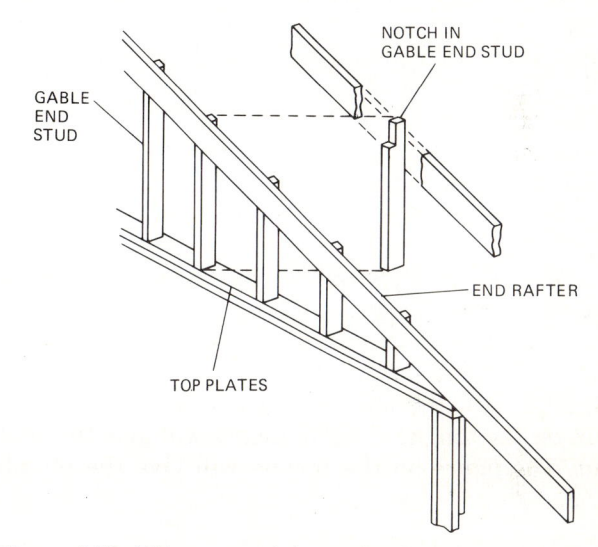

Figure 7-70. Placement of the end rafter. *(American Plywood Association)*

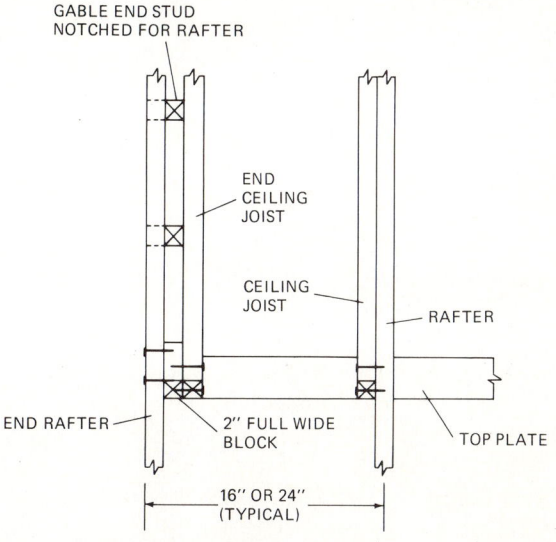

Figure 7-71. Spacing of the first interior rafter. *(American Plywood Association)*

Building Roof Frames 143

Next, mark rafter locations on the ridge board. Allow for the specified gable overhang. To achieve the required total length of ridge board, you may have to splice it. See Figure 7-72. Do not splice at this time. It is easier to erect it in shorter sections, then splice it after it is in place.

Check your house plan for roof slope. For example, a 4-inch rise in 12 inches of run is common. It is usually considered the minimum for asphalt or wood shingles.

Lay out one pair of rafters as previously shown. Mark the top and bottom angles and seat-cut location. Make the cuts and check the fit by setting them up at floor level. Mark this set of rafters for identification and use it as a pattern for the remainder.

Cut the remaining rafters. For a 48-foot house with rafters spaced 24 inches O.C. you will need 24 pairs cut to the pattern (25 pairs counting the pattern). In addition, you will need two pairs of fascia rafters for the ends of the gable overhang. See Figure 7-73. Since they cover the end of the ridge board, they must be longer than the pattern rafters by half the width of the ridge board. Fascia rafters have the same cuts at the top and bottom as the regular rafters. However, they do not have a seat cut.

Build temporary props of 2x4s to hold the rafters and ridge board in place during roof framing installation. The props should be long enough to reach from the top plate to the bottom of the ridge board. They should be fitted with a plywood gusset at the bottom. When the props are installed, the plywood gusset is nailed temporarily to the top plate or to a ceiling joist. The props are also diagonally braced from about midpoint in both directions to maintain true vertical (check with a plumb bob). See Figure 7-74.

Move the ridge board sections and rafters onto the ceiling framing. Lay plywood panels over the ceiling joists for a safe footing. First erect the ridge board and the rafters nearest its ends. See Figure 7-74. If the ridge of the house is longer than the individual pieces of ridge board, you'll find it easier to erect each piece separately, rather than splice the ridge board full-length first. Support the ridge board at both ends with the temporary props. Toenail the first rafter pair securely to the ridge board using at least two 8d nails on each side. Then nail it at the wall. Install the other end rafter pair in the same manner.

Make the ridge board joints, using plywood gussets on each side of the joint. Nail them securely to the ridge board.

Check the ridge board for level. Also check the straightness over the centerline of the house.

After the full length of the ridge board is erected, put up the remaining rafters in pairs. Nail them securely in place. Check them occasionally to make sure the ridge board remains straight. If all rafters are cut and assembled accurately, the roof should be self-aligning.

Toenail the rafters to the wall plate with 10d nails. Use two per side. Also nail the ceiling joists to the rafters. For a 24-foot-wide house, you will need four 16d nails at each lap. In high-wind areas, it is a good idea to add metal-strap fasteners for extra uplift resistance. See Figure 7-75.

Cut and install 1x6 collar beams at every other pair of rafters (4 feet O.C.). See Figure 7-73. Nail each end with four 8d nails. Collar beams should

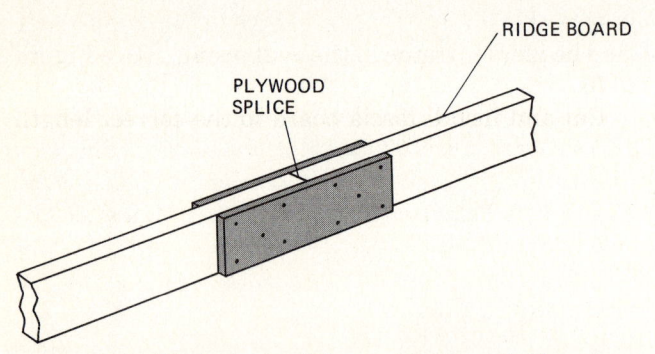

Figure 7-72. Method of splicing a ridge board. *(American Plywood Association)*

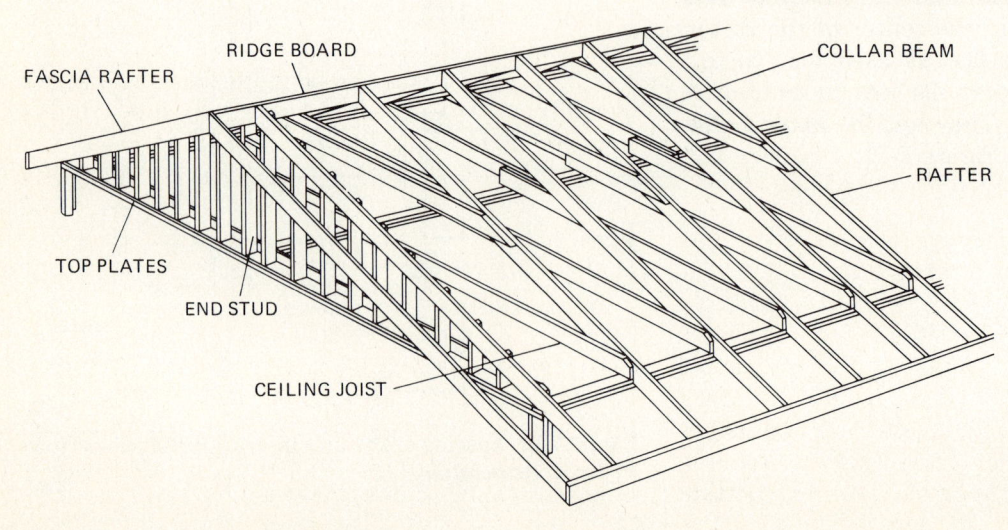

Figure 7-73. Placement of rafters, ridge board, and collar beam. *(American Plywood Association)*

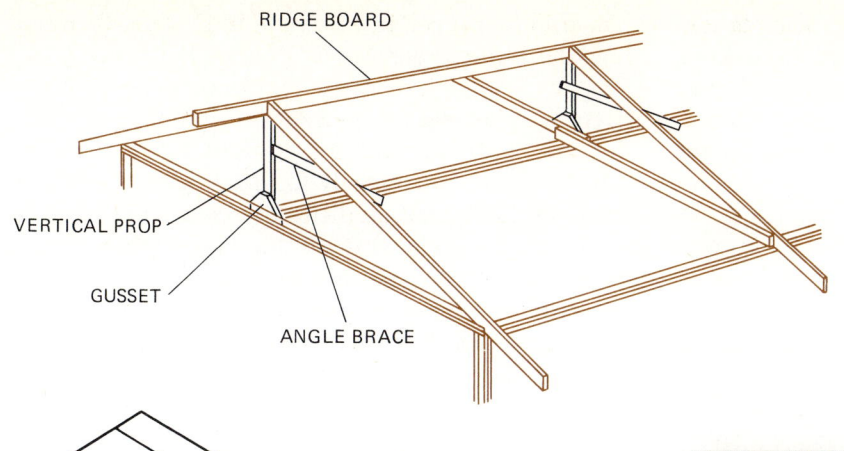

Figure 7-74. Placement of angle braces and vertical props. *(American Plywood Association)*

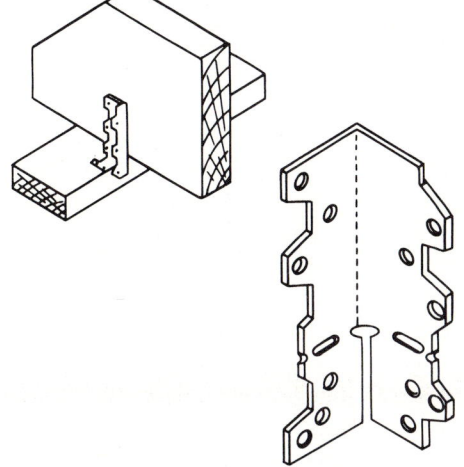

Figure 7-75. Metal framing anchors.

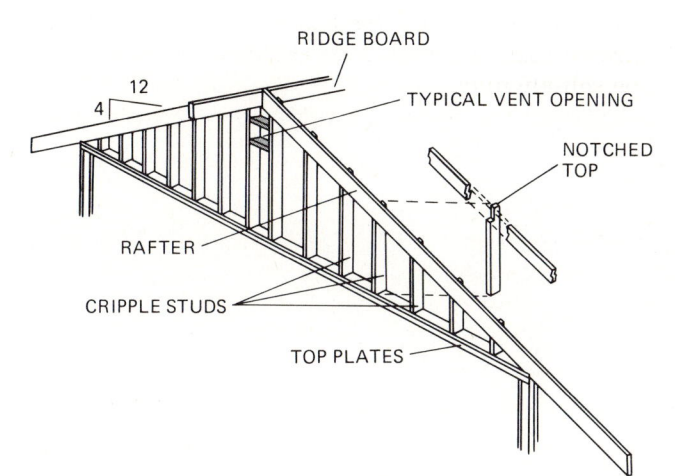

Figure 7-76. Vent openings should be blocked in. *(American Plywood Association)*

Figure 7-77. A dormer.

will be flush with the top plate. Cut the cripple studs and headers to frame in the vent opening. See Figure 7-76.

Cut and install fascia board to the correct length of the ridge board. Bevel the top edge to the roof slope. Nail the board to the rafter ends. Then, install fascia rafters. Fascia rafters cover the end of the ridge board. See Figure 7-73. Where the nails will be exposed to weather, use hot-dipped galvanized or other nonstaining nails.

• SPECIAL RAFTERS •

There are some rafters needed to make special roof shapes. The mansard roof and the hip roof both call for special rafters. Jack rafters are needed for the hip roof. This type of roof may also have valleys and have to be treated especially well. Dormers call for some special rafters, too. For bay windows and other protrusions, some attention may have to be given to rafter construction.

Dormers

Dormers are protrusions from the roof. They stick out from the roof. They may be added to allow light into an upstairs room. Or, they may be added for architectural effect. See Figure 7-77. Dormers may be made in three types. They are:

be in the upper third of the attic crawl space. Remove the temporary props.

Square a line across the end wall plate directly below the ridge board. If a vent is to be installed, measure half its width on each side of the center mark to locate the first stud on each side. Mark the positions for the remaining studs at 16 inches O.C. Then measure and cut the studs. Notch the top end to fit under the rafter so that the stud bottom

Building Roof Frames 145

1. Dormers with flat, sloping roofs that have less slope than the roof in which they are located. This can be called a shed-type dormer (Figure 7-78).
2. Dormers with roofs of the gable type at right angles to the roof (Figure 7-79). No slope in this one.
3. The two types can be combined. This is called the hip-roof dormer.

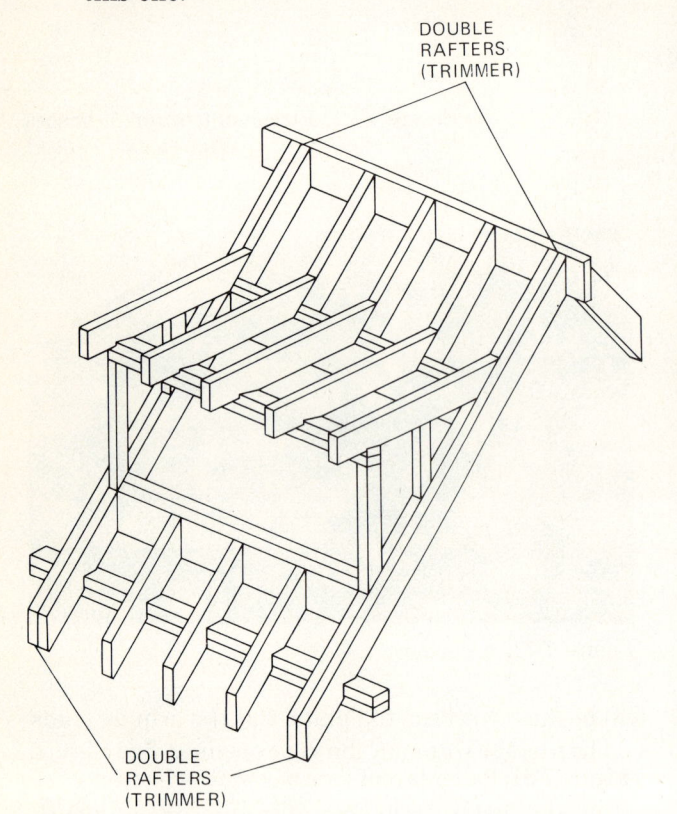

Bay Windows

Bay windows are mostly for decoration. They add to the architectural qualities of a house. They stick out from the main part of the house. This means they have to have special handling. The floor joists are extended out past the foundation the required amount. A *band* is then used to cap off the joist ends. See Figure 7-80. Take a closer look at the ceiling joists and rafters for the bay. The rafters are cut according to the rise called for on the plans. Cuts and lengths have already been discussed. No special problems should be presented by this method of framing. In order to make it easier, it is best to lay out the rafter plan at first so that you know which are the common and which are the hip rafters. In some cases you may need a jack rafter or two, depending upon the size of the bay. See Figure 7-81.

• CEILING JOISTS •

Ceiling joists serve a number of purposes. They keep the wall from falling inward or outward. They anchor the rafters to the top plate. Ceiling joists also hold the finished ceiling in place. The run of the joist is important. The distance between supports for a joist determines its size. In some cases the ceiling joist has to be spliced. Figure 7-82 shows one

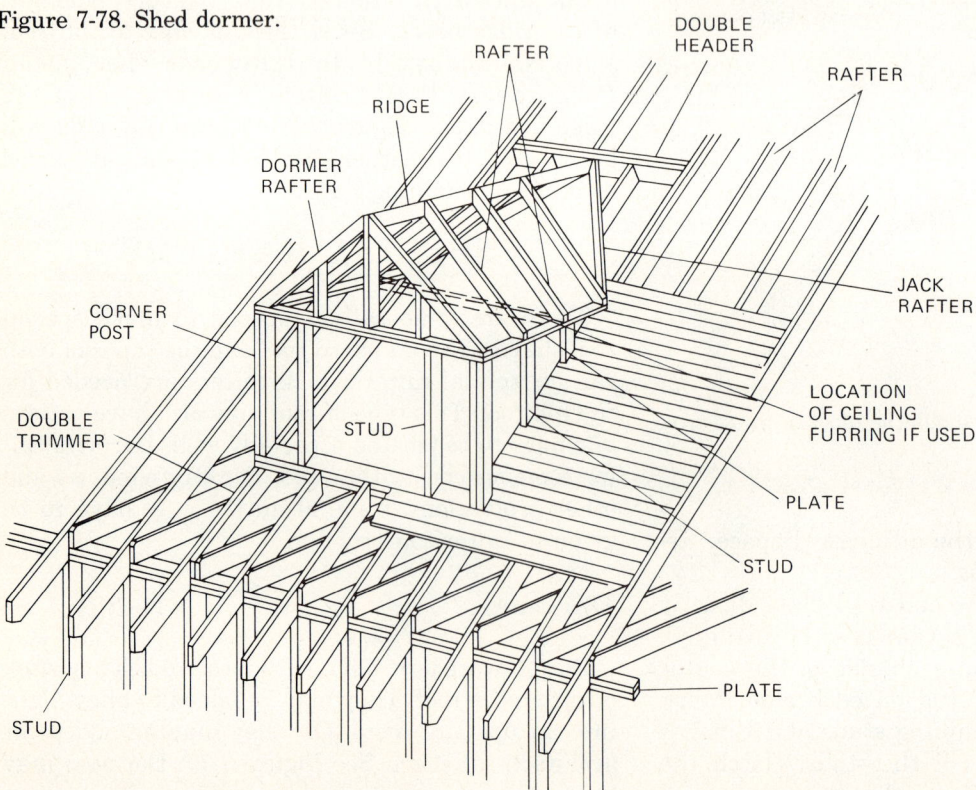

Figure 7-78. Shed dormer.

Figure 7-79. Gable dormer.

146 Carpentry Fundamentals

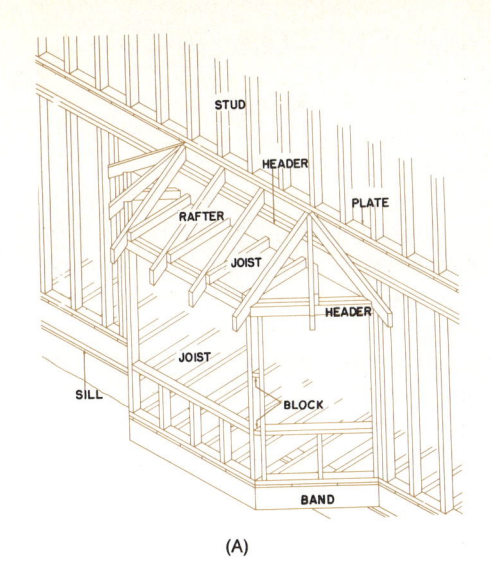

Figure 7-80. (A) Bay window framing. (B) Two bay windows stacked for a two-story house. The metal cover does not require rafters.

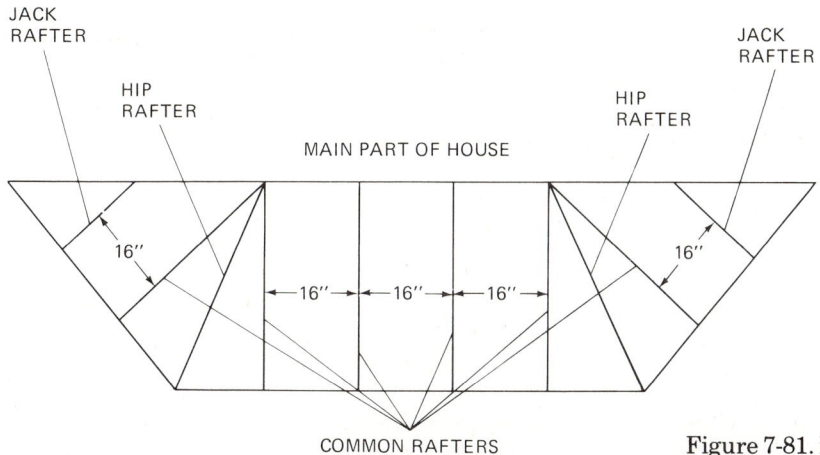

Figure 7-81. Rafter layout for a bay window.

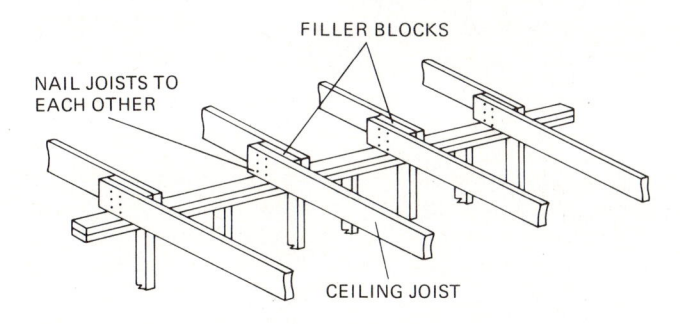

Figure 7-82. Ceiling joist splices are made on a supporting partition.

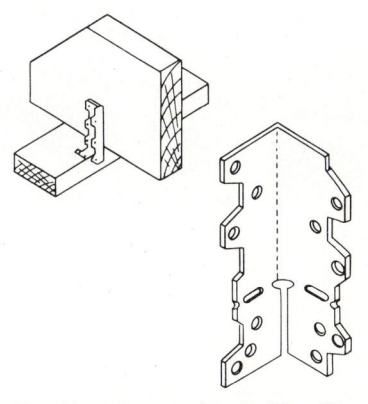

Figure 7-84. Steel bracket used to hold ceiling joist to the top plate. Note how it is bent to fit.

Figure 7-83. Looking up toward the ceiling joists. Notice how these are supported on the wall plate and are not cut or tapered. Two nails are used to toenail the joists to the plate.

method of splicing. Note how the splice is made on a supporting partition. Figure 7-83 shows how the joists fit on top of the plate. In some cases it is best to tie the joist down to the top plate by using a framing bracket. Figure 7-84 shows how one type of bracket is used to hold the joist. This helps in high-wind areas.

Building Roof Frames **147**

The ceiling joists in Figure 7-85 (A) and (B) have been trimmed to take the angle of the rafter into consideration once the rafter is attached to the top plate and joist.

The size of the joist is determined by the local code. However, there are charts that will give you some idea of what size piece of dimensional lumber to use. Table 7-3 indicates some allowable spans for ceiling joists. These are given using non-stress-graded lumber.

There are some special arrangements for ceiling joists. In some cases it is necessary to interrupt the free flow of lines represented by joists. For example, a chimney may have to be allowed for. An attic opening may be called for on the drawings. These openings have to be reinforced to make sure the joists maintain their ability to support the ceiling and some weight in the ceiling at a later time. See Figure 7-86.

Figure 7-87 shows how framing anchors are used to secure the joists to the double header. This method can be used for both attic openings and fireplace openings.

· OPENINGS ·

As mentioned before, fireplaces do come out through the roof. This must be allowed for in the construction process. The floor joists have to be reinforced. The area around the fireplace has to be strong enough to hold the hearth. However, what we're interested in here is how the fireplace opening comes through the rafters. See Figure 7-88. The roofing here has been boxed off to allow the fireplace to come through. The chimney at the top has the flashing ready for installation as soon as the bricks are laid around the flue.

Other openings are for soil pipes. These are used for venting the plumbing system. A hole in the plywood deck is usually sufficient to allow their exit from the inside of the house. See Figure 7-89.

· DECKING ·

A number of types of roof deckings are used. One is plywood. It is applied in 4x8-foot sheets. Plywood adds structural strength to the building. Plywood also saves time, since it can be placed on the rafters rather quickly.

Boards of the 1x6- or 1x8-inch size can be used for sheathing. This decking takes a longer time to

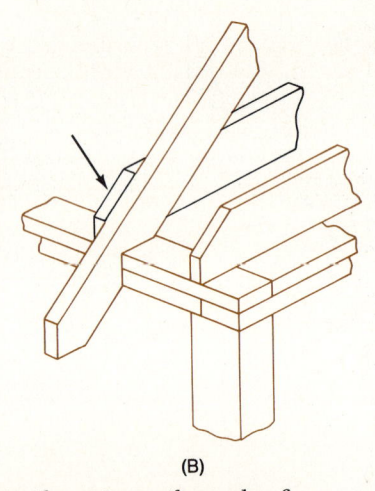

(A)　(B)

Figure 7-85. (A) Ceiling joists in place on the top plate. Note the cuts on the ends of the joists. (B) Set the first ceiling joist on the inside of the end of the wall. This will allow the ceiling drywall to be nailed to it later.

TABLE 7-3　CEILING JOISTS

Size of Ceiling Joists, Inches	Spacing of Ceiling Joists, Inches	Maximum Allowable Span			
		Group I	Group II	Group III	Group IV
2 × 4	12	11'6"	11'0"	9'6"	5'6"
	16	10'6"	10'0"	8'6"	5'0"
2 × 6	12	18'0"	16'6"	15'6"	12'6"
	16	16'0"	15'0"	14'6"	11'0"
2 × 8	12	24'0"	22'6"	21'0"	19'0"
	16	21'6"	20'6"	19'0"	16'6"

Carpentry Fundamentals

apply. Each board has to be nailed and sawed to fit. This type of decking adds strength to the roof also.

Another type of decking is nothing more than furring strips applied to the rafters. This is used as a nail base for cedar shingles.

Plywood Decking

Roof sheathing, the covering over rafters or trusses, provides structural strength and rigidity. It makes a solid base for fastening the roofing material.

A roof sheathing layout should be sketched out first to determine the amount of plywood needed to cover the rafters. See Figure 7-90.

Draw your layout. It may be a freehand sketch, but it should be relatively close to scale. The easiest method is to draw a simple rectangle representing half of the roof. The long side will represent the length of the ridge board. Make the short side equal to the length of your rafters, including the overhangs. If you have open soffits, draw a line inside the ends and bottoms. Use a dotted line as shown in Figure 7-90. This area is to be covered by *exterior*

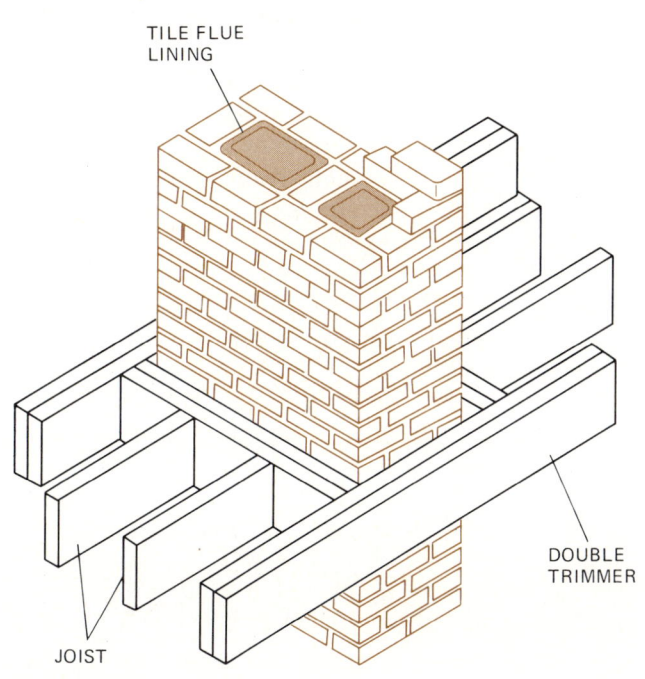

Figure 7-86. Blocking in the joists to allow an opening for a chimney.

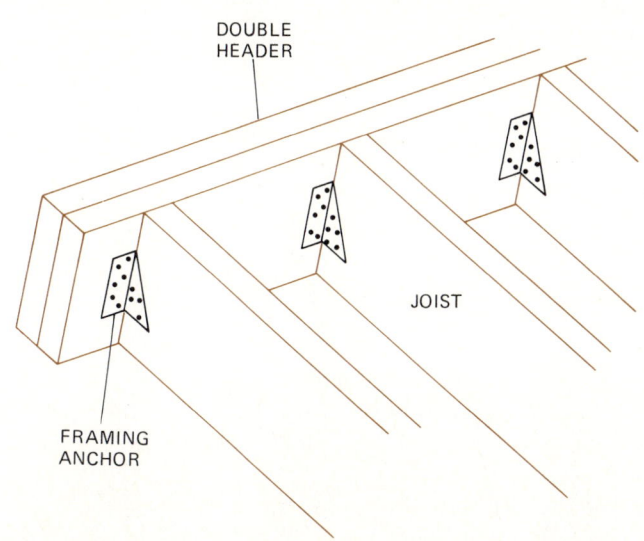

Figure 7-87. Using framing anchors to hold the tail joists to the header.

Figure 7-88. Arrows show the openings in the roof for the chimney.

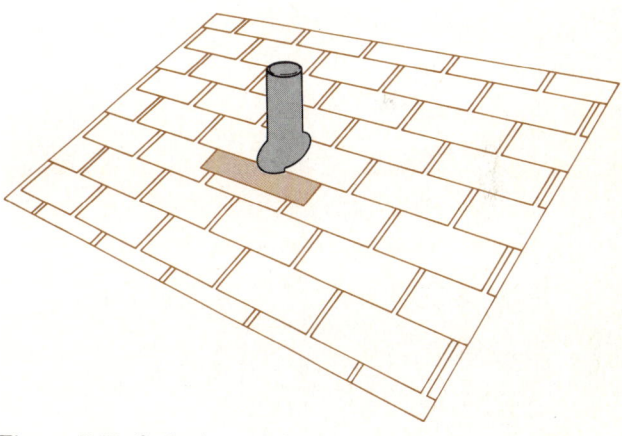

Figure 7-89. Soil pipe stack coming through the roof.

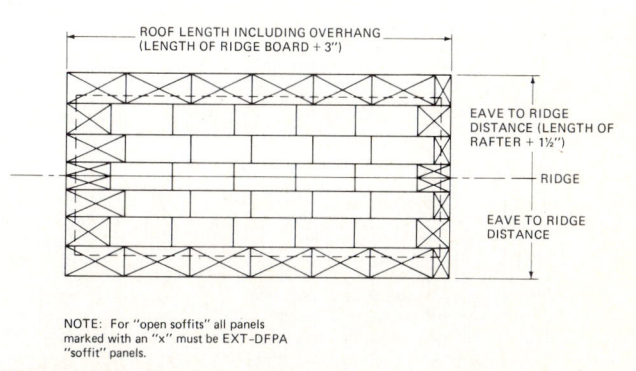

Figure 7-90. Roof sheathing layout for a plywood deck. *(American Plywood Association)*

Building Roof Frames 149

plywood. Remember that this is only half of the roof. Any cutting of panels on this side can be planned so that the cut-off portions can be used on the other side. If your eave overhang is less than 2 feet and you have an open soffit, you may wish to start with a half panel width of soffit plywood. Figure 7-91 shows the open soffit. Figure 7-92 shows the boxed soffit. Otherwise you will probably start with a full 4x8-foot sheet of plywood at the bottom of the roof and work upward toward the ridge. This is where you may have to cut the last row of panels. Stagger panels in succeeding rows.

Complete your layout for the whole roof. The layout shows panel size and placement as well as the number of sheathing panels needed. This is shown in Figure 7-90.

If your diagram shows that you will have a lot of waste in cutting, you may be able to reduce scrap by slightly shortening the rafter overhang at the eave, or the gable overhang.

An example is shown in Figure 7-90, where nearly half of the panels are "soffit" panels. In such a case, rather than use shims to level up soffit and interior sheathing panels, you may want to use interior sheathing panels of the same thickness as your soffit panels, even though they might then be a little thicker than the minimum required.

Cut panels as required, marking the cutting lines first to ensure square corners.

Begin panel placement at any corner of the roof. If you are using special soffit panels, remember to place them best or textured side down.

Fasten each panel in the first course (row), in turn, to the roof framing using 6d common smooth, ring-shank, or spiral-threaded nails. Space nails 6 inches O.C. along the panel ends and 12 inches O.C. at intermediate supports.

Leave a 1/16-inch space at panel ends and 1/8 inch at edge joints. In areas of high humidity, double the spacing.

Apply the second course, using a soffit half panel in the first (overhang) position. If the main sheathing panels are thinner than the soffit sheathing, install small shims to ease the joint transition. See Figure 7-91 for location of the shims.

Apply the remaining courses as above.

Note that if your plans show closed soffits, the roof sheathing will all be the same grade thickness. To apply plywood to the underside of closed soffits, use nonstaining-type nails.

Figure 7-91. Open soffit. *(American Plywood Association)*

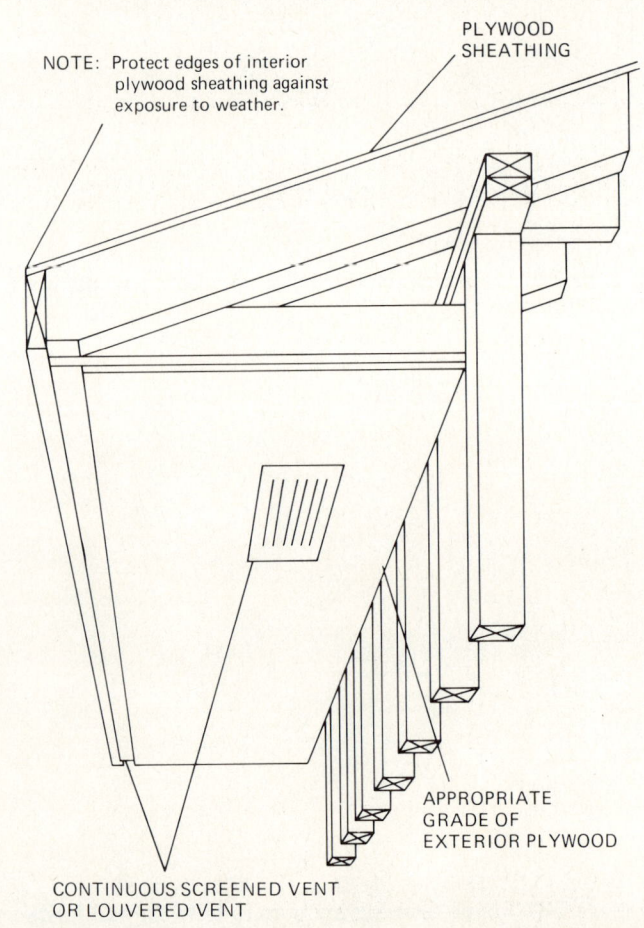

Figure 7-92. Boxed soffit. *(American Plywood Association)*

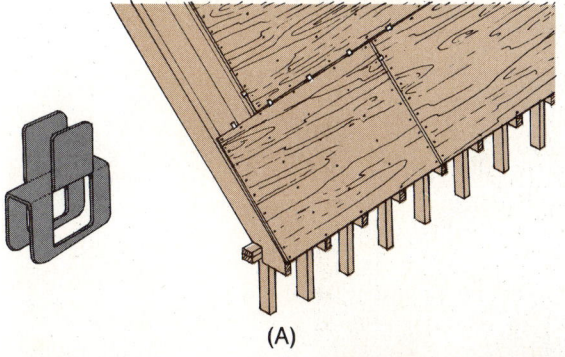

Figure 7-93 shows plywood decking being applied with a stapler. Figure 7-94A and B shows an H clip for plywood support along the long edges. This gives extra support for the entire length of the panels.

Figure 7-95 shows the erection of the sidewalls to a house. In Figure 7-96 the plywood sheathing has been placed on one portion of the rafters. Note the rig (arrow) that allows a sheet of plywood to be passed up from the ground to roof level. The sheet is first placed on the rack. Then it is taken by the person on the roof and moved over to the needed area.

Figure 7-93. Using a stapler to fasten plywood decking to the rafters. (*American Plywood Association*)

Figure 7-95. Erection of sidewalls to a house.

Figure 7-96. Plywood sheathing placed on one portion of rafters. Note the ladder made for plywood lifting.

(A)

(B)

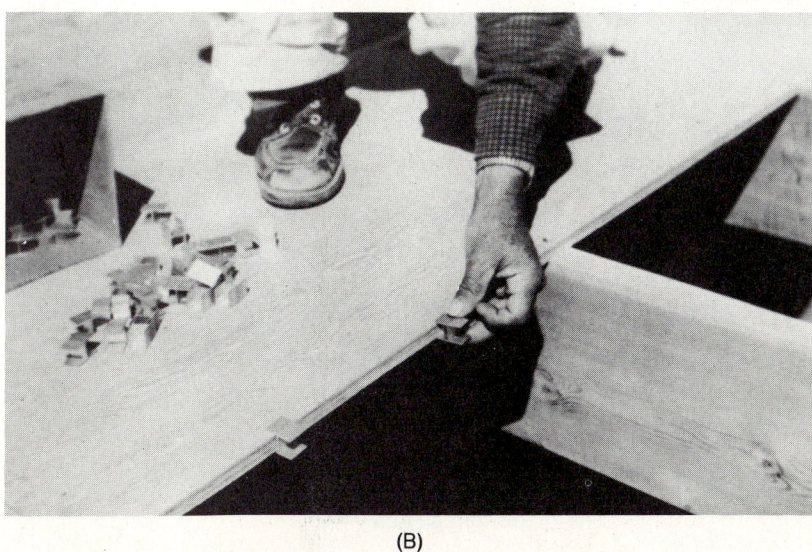

Figure 7-94. Plywood clips reinforce the surface area where the sheathing butts.

Building Roof Frames **151**

Figure 7-97 shows some boxed soffit. Note the louvers already in the board. The temporary supports hold the soffit in place until final nailing is done and it can be supported by the fascia board. The fascia has a groove along its entire length to allow the soffit to slide into it.

Boards for Decking

Lumber may be used for roof decking. In fact, it is necessary in some special-effects ceilings. It is needed where the ceiling is exposed and the underside of the decking is visible from below or inside the room.

Roof decking comes in a variety of sizes and shapes. See Figure 7-98. A 2° angle is cut in the lumber decking ends to ensure a tight joint, (Figure 7-99). This type of decking is usually nailed, and so the nails must be concealed. This requires nailing as shown in Figure 7-100. Eight-inch spikes are usually used for this type of nailing. Note the chimney opening in Figure 7-101.

Figure 7-102 illustrates the application of 1x6 or 1x8 sheathing to the rafters. Note the two nails used to hold the boards down. The common nail is used here. In the concealed nailing it takes a finishing nail to be completely concealed.

Once the decking is in place, it is covered by an underlayment of felt paper. This paper is then covered by shingles as selected by the builder.

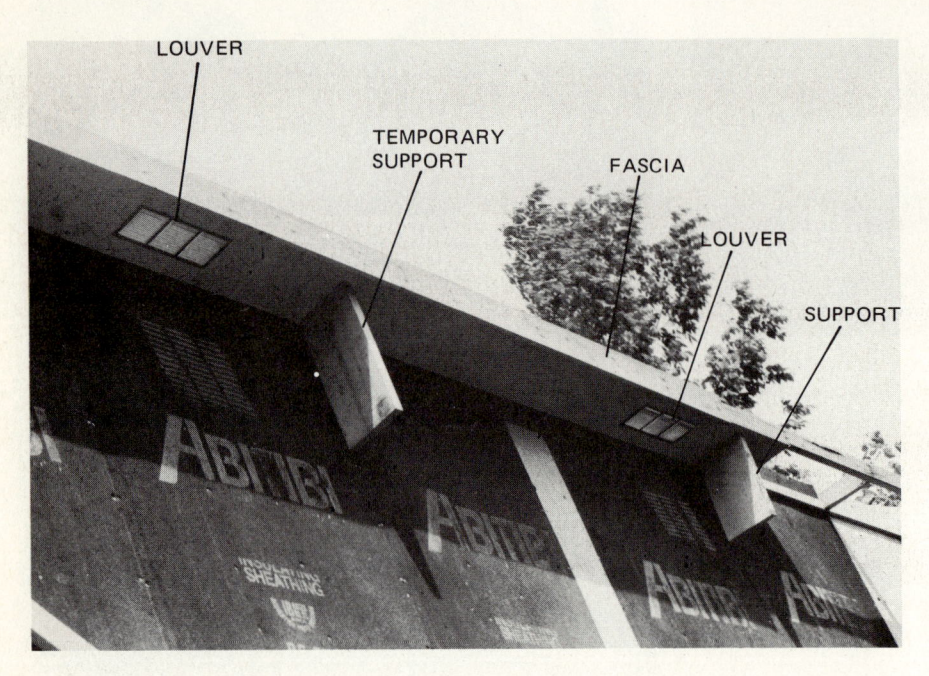

Figure 7-97. Louvered soffit held in place with temporary braces.

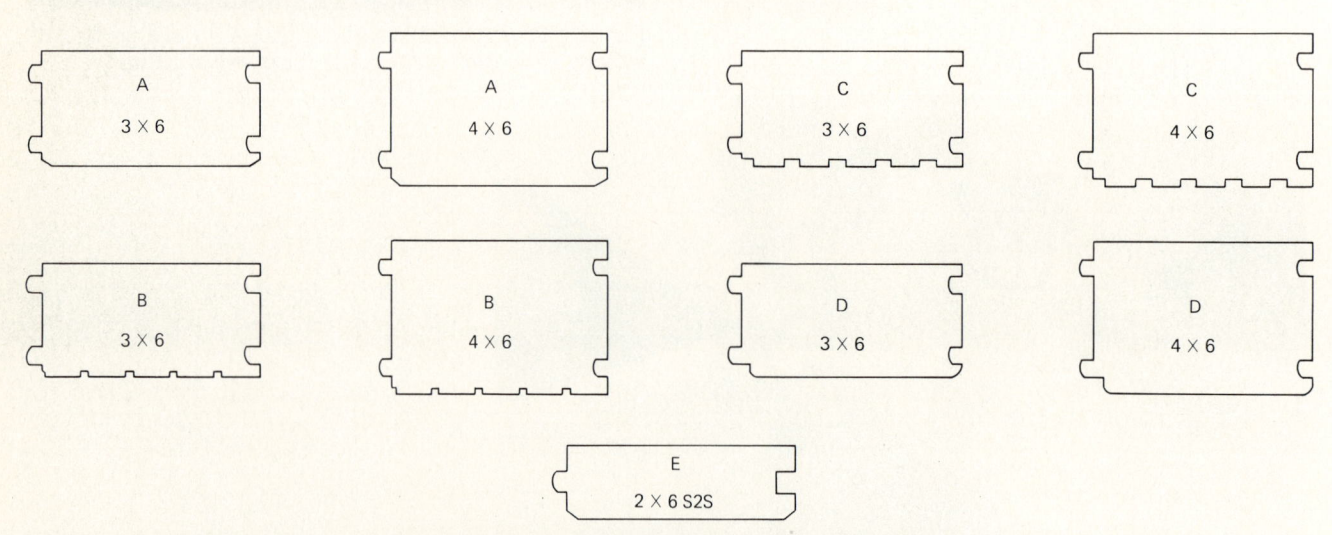

Figure 7-98. Different sizes and shapes of roof decking made of lumber. (A) Shows the regular V-jointed decking. (B) Indicates the straited decking. (C) Shows the grooved type. (D) Illustrates the eased joint or bullnosed type. (E) Indicates single tongue-and-groove with a V joint.

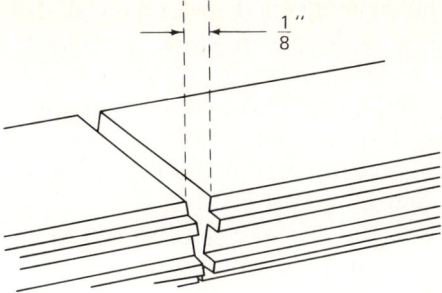

Figure 7-99. A 2° angle is cut in the lumber decking ends. This ensures a tight face joint on the exposed ceiling below.

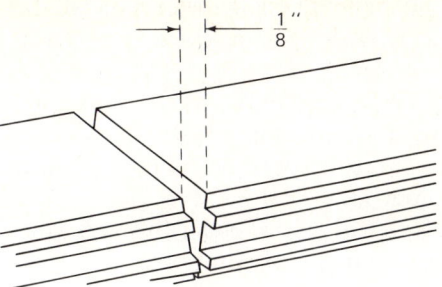

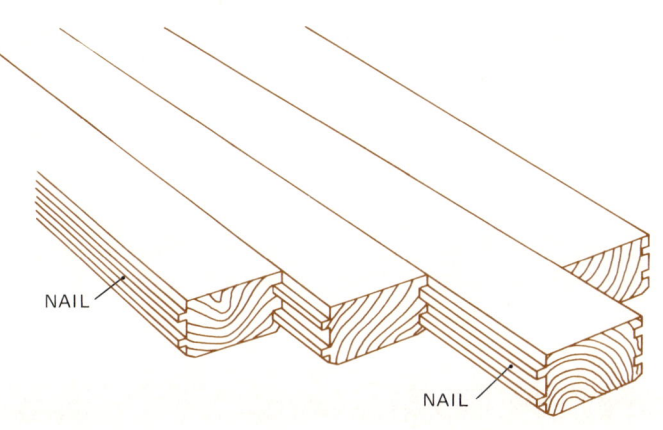

Figure 7-100. Note the location of nails in the lumber decking.

Shingle Stringers

For cedar shingles the roof deck may be either spaced or solid. The climatic conditions determine if it is a solid deck or one that is spaced. In areas where there are blizzards and high humidity, the spaced deck is not used. In snow-free areas, spaced sheathing is practical. Use 1x6s spaced on centers equal to the weather exposure at which the shakes are to be laid. However, the spacing should not be over 10 inches. In areas where wind-driven snow

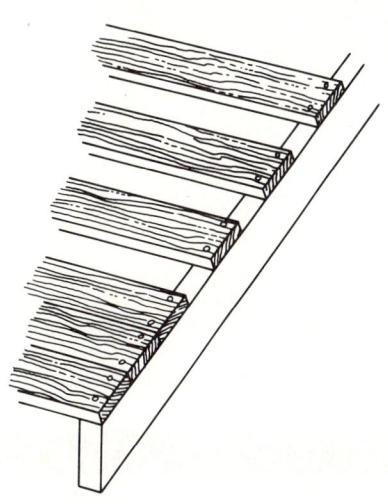

Figure 7-102. Spacing of sheathing for wood shingles. *(Red Cedar Shakes and Hand Split Shingles Bureau)*

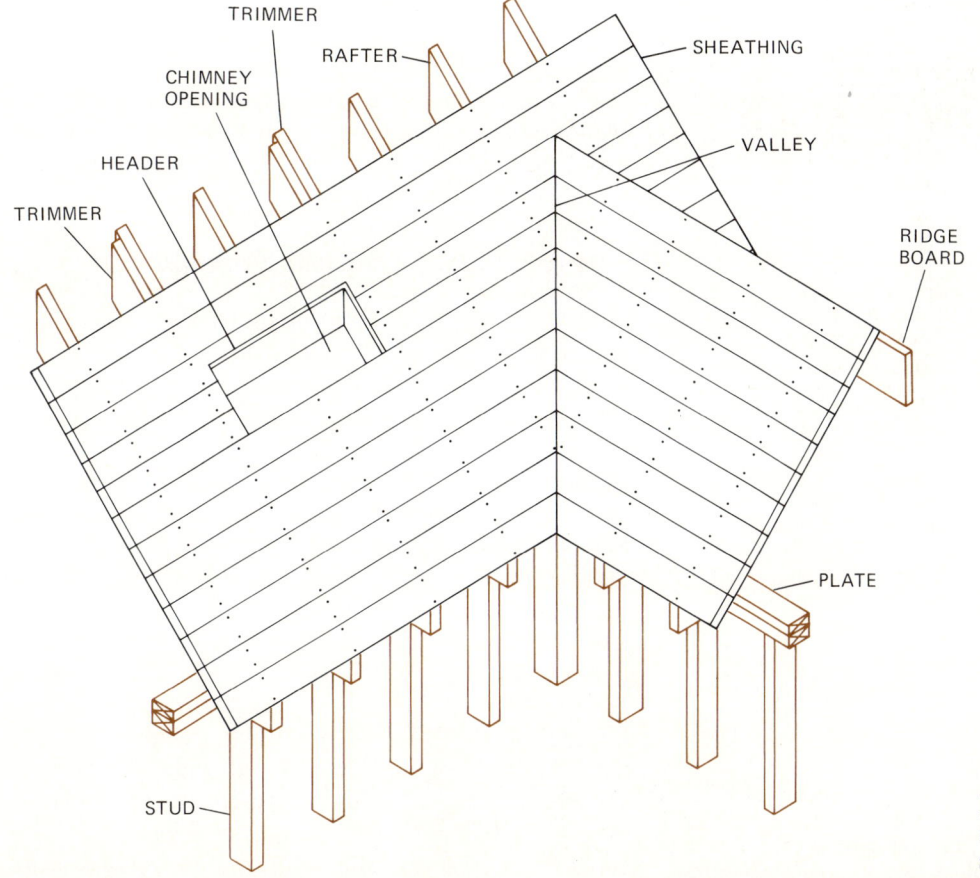

Figure 7-101. Note the chimney opening in this lumber sheathing.

Building Roof Frames 153

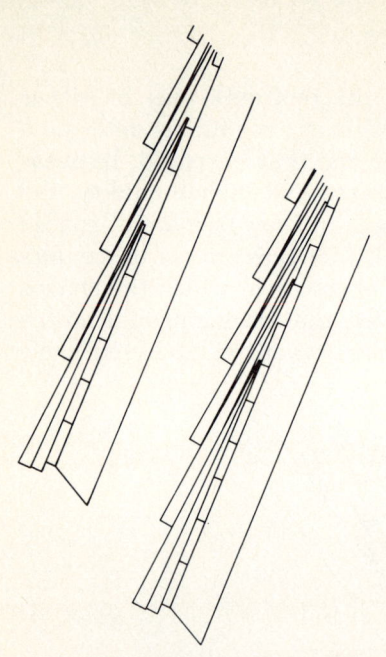

Figure 7-103. Hand split shakes should be used on roofs where the slope is sufficient to ensure good drainage. Two different exposures to the weather are shown. Note the spacing of the sheathing under the shakes. *(Red Cedar Shakes and Hand Split Shingles Bureau)*

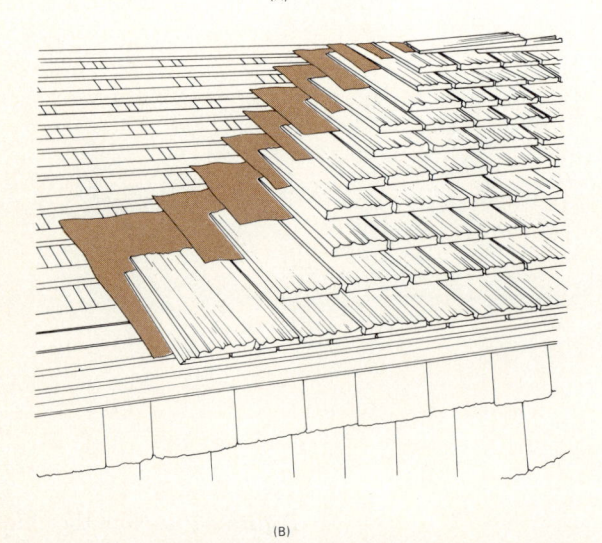

Figure 7-104. (A) Roof ready for application of shingles. (B) Cedar shingles being applied to prepared sheathing. *(Red Cedar Shakes and Hand Split Shingles Bureau)*

is encountered, solid sheathing is recommended. See Figure 7-102 for an example of spaced sheathing.

ROOF PITCH AND EXPOSURE • Hand-split shakes should be used on roofs where the slope or pitch is sufficient to ensure good drainage. Minimum recommended pitch is $\frac{1}{6}$ or 4-in-12 (4-inch vertical rise for each 12-inch horizontal run), although there have been satisfactory installations on lesser slopes—climatic conditions and skill and techniques of application are modifying factors.

Maximum recommended weather exposure is 10 inches for 24-inch shakes and 7½ inches for 18-inch shakes. A superior three-ply roof can be achieved at slight additional cost if these exposures are reduced to 7½ inches for 24-inch shakes and 5½ inches for 18-inch shakes.

Figure 7-103 shows the shakes in place on spaced sheathing. Note how the amount of exposure to the weather makes a difference in the spacing of the sheathing. Figure 7-104(A) and (B) gives a better view of the spaced sheathing and how the roofing is applied to it.

• CONSTRUCTING SPECIAL SHAPES •

The gambrel shape is familiar to most people, since it is the favorite shape for barns. It consists of a double-slope roof. This allows for more space in the attic or upper story. More can be stored there. In modern home designs, this type of roof has been used to advantage. It gives good headroom for an economical structure with two stories. This design was brought to the United States by Germans in the early days of the country.

Gambrel-Shaped-Roof Storage Shed

A storage shed will give you an idea of the simplest way to utilize the gambrel-shaped roof. Examine the details and then obtain the equipment and supplies needed. The bill of materials lists the supplies that are needed. See Figure 7-105. Now take a look at Figures 7-106, 7-107, and 7-108. These show different ways of finishing the shed for different purposes. For instance, the structure can be covered with glass, Plexiglas, or polyvinyl as in Figure 7-107 and made into a greenhouse. Then there is the rustic look shown in Figure 7-106 and the contemporary look shown in Figure 7-108.

FRAME LAYOUT • Note the dimensions of the shed. It is 7 feet high and 8 feet wide. A detail of the framing angle is shown in the frame-to-sill detail (Figure 7-109). Note the spacing of the slopes of the roof. Figure 7-110 indicates how the vertical stud member and the roof member are attached with

| \multicolumn{4}{c}{BILL OF MATERIALS} |
QUAN.	DESCRIPTION	QUAN.	DESCRIPTION
28	2" x 4" x 8 FT. LONG	40	TECO C-7 PLTS.
9	4' x 8' x ½" PLYWD.	30	TECO JOIST HGR.
2	1" x 4" x 6 FT. LONG	12	TECO ANGLES
1 ROLL	ROOFING FELT	30	TECO A-5 PLTS.
1 GAL	ROOF. CEMENT	3	3 BUTT HINGES
1 GAL	BARN-RED PAINT	1	HASP & LOCK
5#	6d COM. NAILS	10 BG.	90# CONC. MIX
2#	12d COM. NAILS		OR
2#	½" ROOF. NAILS	4	6" x 8" x 8' RAIL TIE

Figure 7-105. Bill of materials for a storage shed. *(Courtesy of TECO Products and Testing Corporation, Washington D.C. 20015)*

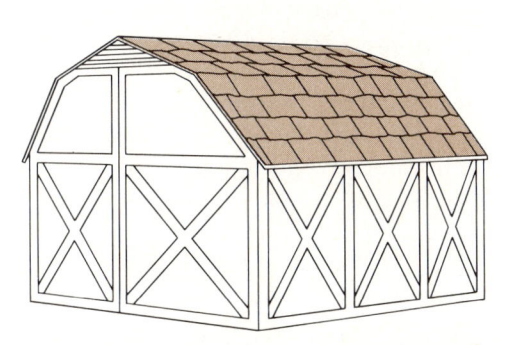

Figure 7-106. Rustic shed design. *(Courtesy of TECO)*

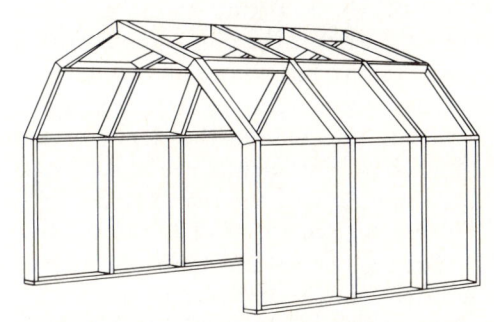

Figure 7-107. A green house can be made by covering the frame with plastic. *(Courtesy of TECO)*

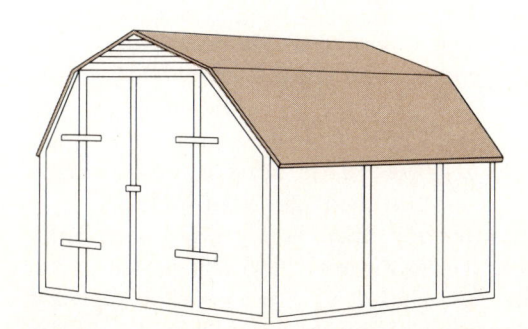

Figure 7-108. Contemporary shed design. *(Courtesy of TECO)*

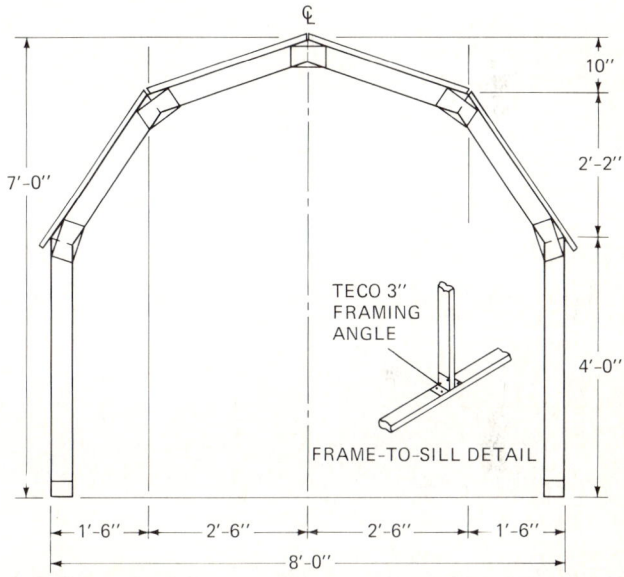

Figure 7-109. Frame layout for the shed. *(Courtesy of TECO)*

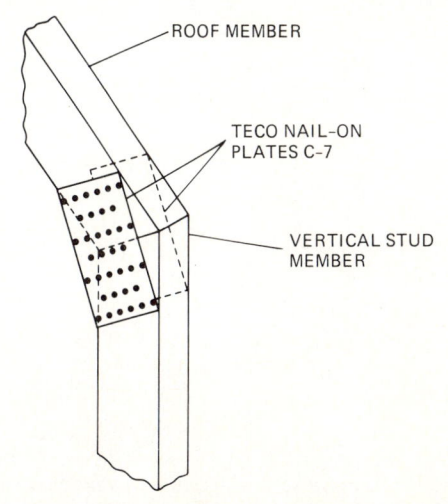

Figure 7-110. Detail of connection of vertical stud member and roof member. *(Courtesy of TECO)*

Building Roof Frames 155

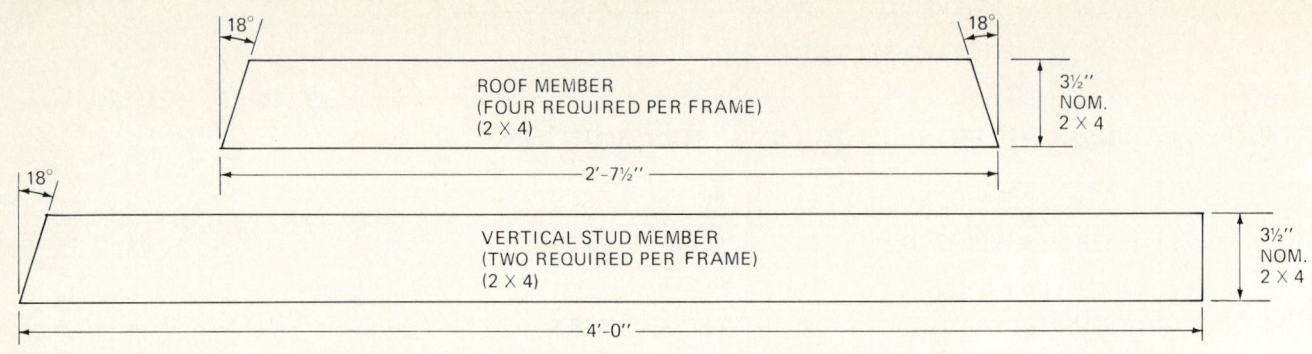

Figure 7-111. Frame cutting instructions. *(Courtesy of TECO)*

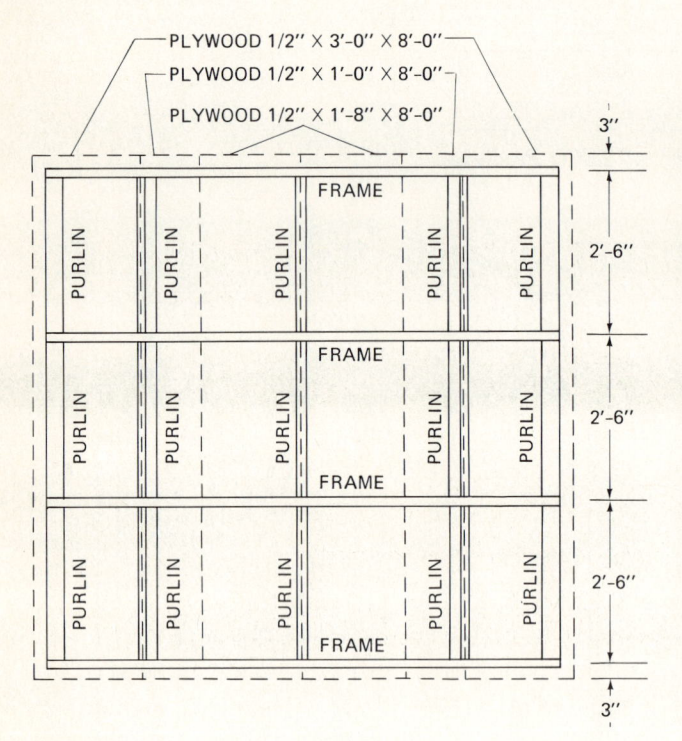

Figure 7-112. Roof-framing plan for the shed. *(Courtesy of TECO)*

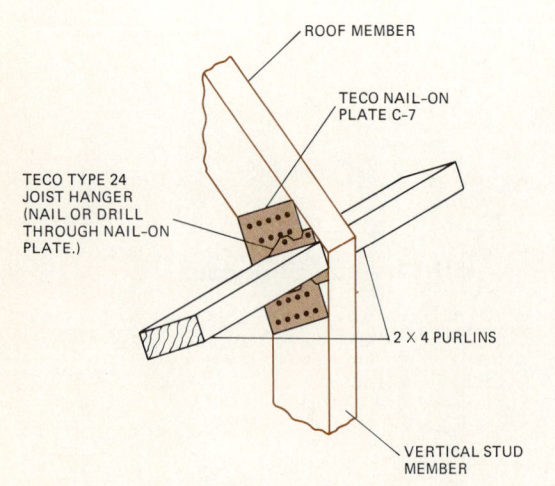

Figure 7-113. Attachment of purlins to vertical stud member and roof member. *(Courtesy of TECO)*

metal plates. Figure 7-111 indicates how an 18° angle is used for cutting the studs and roof members.

FRAME-CUTTING INSTRUCTIONS • Mark off 18° angles on the 2x4s. See Figure 7-111. Cut to length. Note the exact length required for the roof member and the vertical stud. When you have cut the required number of studs and roof members, place the sections on a hard, flat surface. A driveway or sidewalk can be used. Nail the metal plates equally into each member and flip the frame over. Then nail metal plates on this side. Make four frames for a shed with an 8-foot depth. You can change the number of frames to match your length or depth requirements. Just add a frame for every 2'8" of additional depth.

ROOF FRAMING PLAN • Figure 7-112 shows the roof framing layout. Such a layout will eliminate any problems you might have later if you did not properly plan your project. The purlin is extended in length every time you extend the depth of the shed by 2'6". Note how the purlins are attached to the vertical stud member and the roof member. See Figure 7-113. Once you have attached all the purlins, you have the standing frame. It is time to consider other details. Figure 7-114 shows how a vent is built into the rear elevation. It can be a standard window or constructed from 1x2s with a polyvinyl backing.

DOOR DETAILS • You have to decide upon the design of the doors to be used. Figure 7-115 shows how the door is constructed for both rustic and contemporary styles. Cut the angles at 18° when making the door for the rustic style. Use the same template used for the studs and roof members.

The contemporary door is nothing more than squared-off corners. Figure 7-116 shows the placement of the hasps used on the door. The plywood sheathing for the roof and the shingles should follow the instructions in Chapter 8.

156 Carpentry Fundamentals

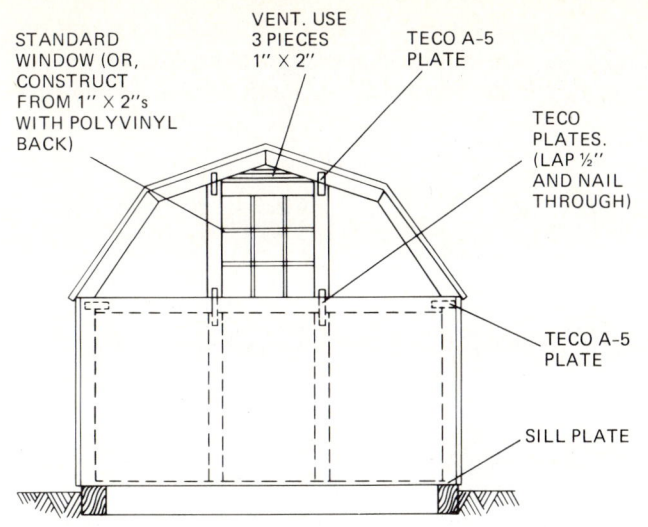

Figure 7-114. Rear elevation of the shed. Note how the vent is built in. *(Courtesy of TECO)*

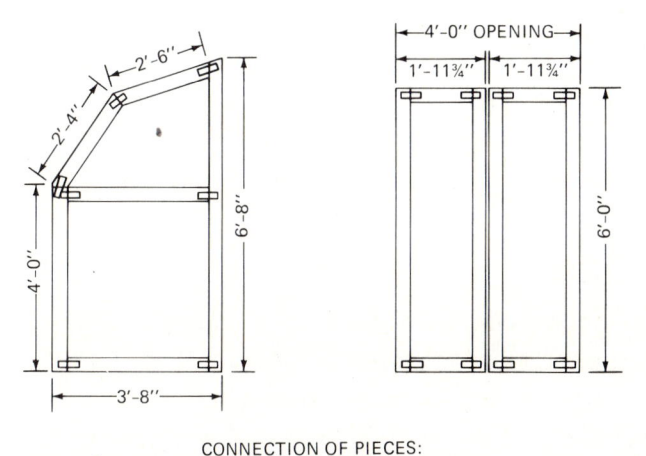

Figure 7-115. Door details for the shed. *(Courtesy of TECO)*

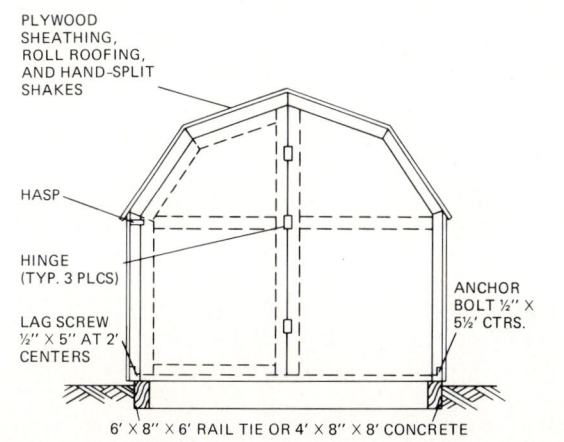

Figure 7-116. Front elevation of the shed. *(Courtesy of TECO)*

Mansard Roofs

The mansard roof has its origins in France. The mansard usually made in the United States is slightly different from the French style. Figure 7-117 shows the American style of mansard. It uses a roof rafter with a steep slope for the side portion and one with a very low slope for the top. Standard 2x4 or 2x6 framing lumber is used to make these rafters. Plywood sheathing is applied over the rafters, and roofing is usually applied over the entire surface. This is supposed to make the house look lower. It effectively lowers the "belt-line" and makes the roof look closer to the ground. Shingles have to be chosen for a steep slope so that they are not blown off by high winds.

Figure 7-118 shows a mansard roof with cedar shakes. The steep slope portion adds to the effect

Figure 7-117. Mansard roof.

Figure 7-118. Hand-split shakes used on a mansard roof. *(Red Cedar Shakes and Hand Split Bureau)*

Building Roof Frames **157**

when covered by cedar shingles. It is best to use the wooden shakes for the low slope as well; however, in some cases the slope may be too low and other shingles may be needed to do the job properly.

Figure 7-119 shows how the French made the mansard roof truss. It was used on hotels and some homes. This style was popular during the nineteenth century. Note the elaborate framing used in those days to hold the various angles together. At that time they used wrought-iron straps instead of today's metal (steel) brackets and plates.

Post-and-Beam Roofs

The post-and-beam type of roof is used for flat or low-slope roofs. This type of construction can use the roof decking as the ceiling below. See Figure 7-120(A). The exposed ceiling or roof decking has to be finished. That means the wood used for the roof deck has to be surfaced and finished on the inside surface. Post-and-beam methods don't use regular rafters. See Figure 7-120(B). The rafters are spaced at a greater distance than in conventional framing. This calls for larger dimensional lumber in the rafters. These are usually exposed also and need to be finished according to plan.

• ROOF LOAD FACTORS •

Plywood roof decking offers builders some of their most attractive opportunities for saving time and money. Big panels really go down fast over large areas. They form a smooth, solid base with a minimum of joints. Waste is minimal, contributing to the low in-place cost. It is frequently possible to cut costs still further by using fewer rafters with a somewhat thicker panel for the decking. For example, use ¾-inch plywood over framing 4 feet O.C. Plywood and trusses are often combined in this manner. For recommended spans and plywood grades, see Table 7-4.

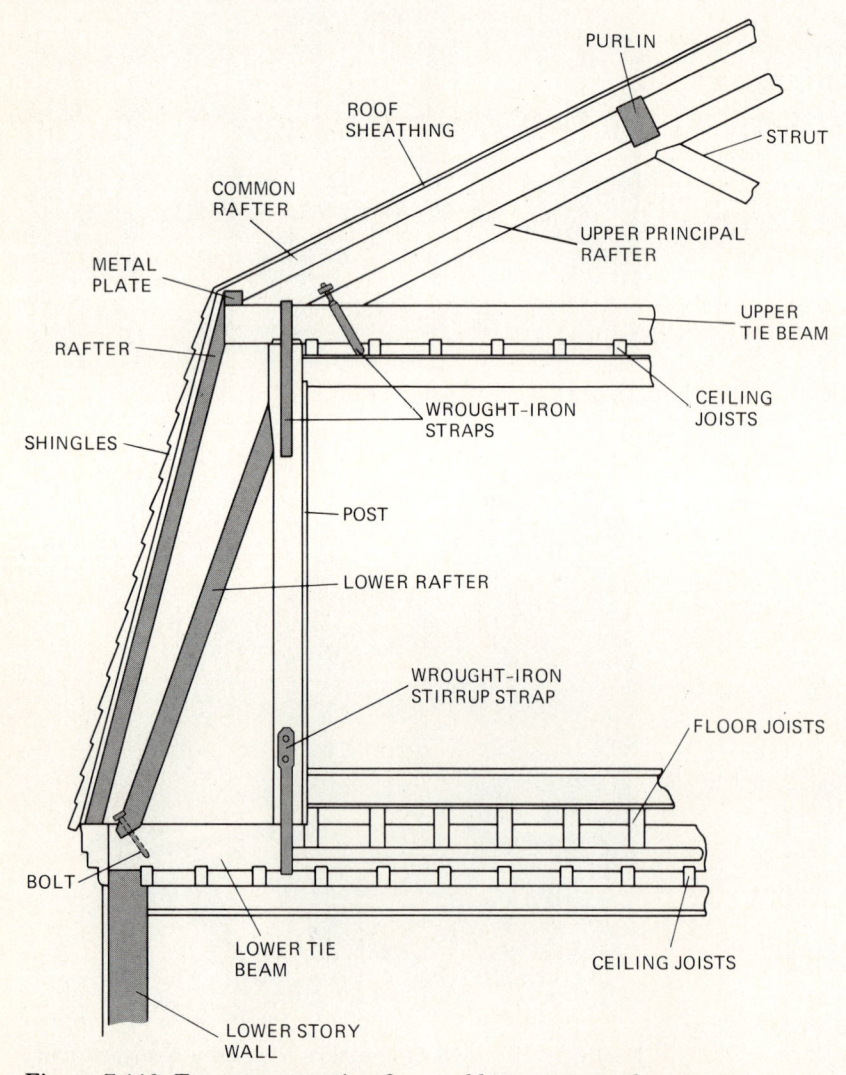

Figure 7-119. Truss construction for an old type mansard roof.

158   Carpentry Fundamentals

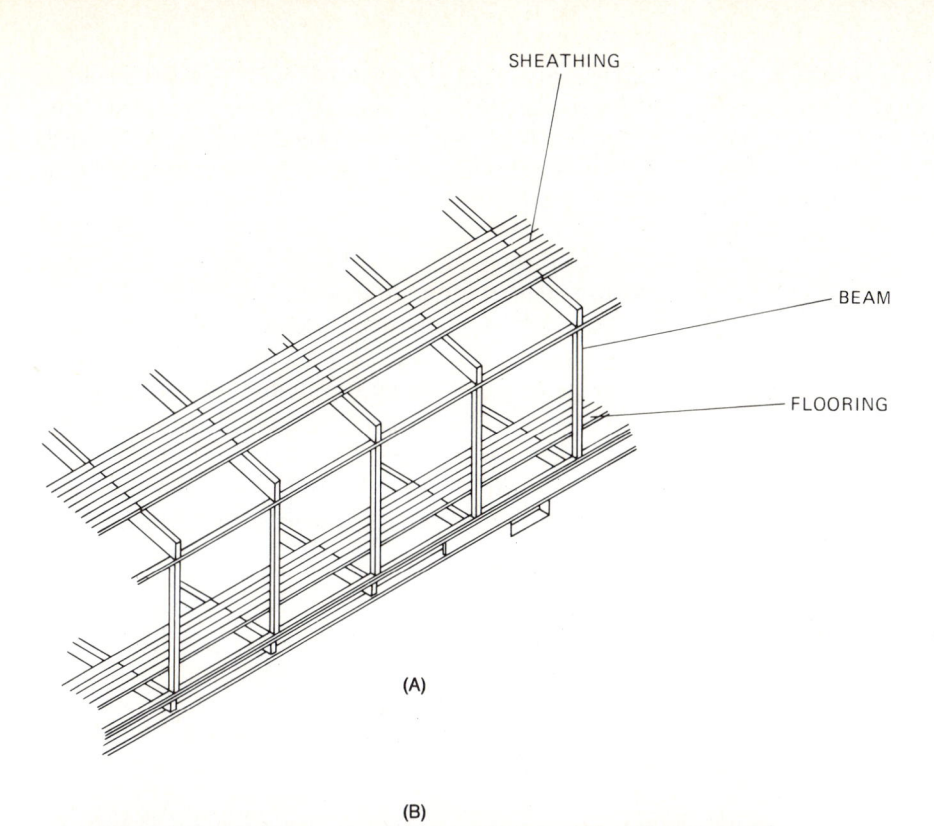

Figure 7-120. (A and B) Post-and-beam framing.

TABLE 7-4 PLYWOOD ROOF DECKING

Identi-fication Index	Plywood Thickness, Inches	Maximum Span, Inches	Unsupported Edge—Max. Length, Inches	Allowable Live Loads, psf Spacing of Supports Center to Center, Inches									
				12	16	20	24	30	32	36	42	48	60
12/0	5/16	12	12	150									
16/0	5/16, 3/8	16	16	160	75								
20/0	5/16, 3/8	20	20	190	105	65							
24/0	3/8, 1/2	24	24	250	140	95	50						
32/16	1/2, 5/8	32	28	385	215	150	95	50	40				
42/20	5/8, 3/4, 7/8	42	32		330	230	145	90	75	50	35		
48/24	3/4, 7/8	48	36			300	190	120	105	65	45	35	
2·4·1	1 1/8	72	48				390	245	215	135	100	75	45
1 1/8" Grp. 1 and 2	1 1/8	72	48				305	195	170	105	75	55	35
1 1/4" Grp. 3 and 4	1 1/4	72	48				355	225	195	125	90	85	40

Building Roof Frames

PLYWOOD ROOF SHEATHING WITH CONVENTIONAL SHINGLE ROOFING • Plywood roof sheathing under shingles provides a tight deck with no wind, dust, or snow infiltration and high resistance to racking. Plywood has stood up for decades under asphalt shingles, and it has performed equally well under cedar shingles and shakes.

Plywood sheathing over roof trusses spaced 24 inches O.C. is widely recognized as the most economical construction for residential roofs and has become the industry standard.

DESIGN • Plywood recommendations for plywood roof decking are given in Table 7-4. They apply for the following grades: C–D INT APA, C–C EXT APA, Structural II and II C–D INT APA, and Structural I and II C–C EXT APA. Values assume 5 pounds per square foot dead load. Uniform load deflection limit is $\frac{1}{180}$ of the span under live load plus dead load, or $\frac{1}{240}$ under live load only. Special conditions, such as heavy concentrated loads, may require constructions in excess of these minimums. Plywood is assumed continuous across two or more spans, and applied face grain across supports.

APPLICATION • Provide adequate blocking, tongue-and-groove edges, or other edge support such as plyclips when spans exceed maximum length for unsupported edges. See Figure 7-121 for installation of plyclips. Use two plyclips for 48-inch or greater spans and one for lesser spans.

Space panel ends $\frac{1}{16}$ inch apart and panel edges $\frac{1}{8}$ inch apart. Where wet or humid conditions prevail, double the spacings. Use 6d common smooth, ring-shank, or spiral-thread nails for plywood $\frac{1}{2}$ inch thick or less. Use 8d nails for plywood to 1 inch thick. Use 8d ring-shank or spiral nails or 10d common smooth for 2·4·1, 1$\frac{1}{8}$ inch and 1$\frac{1}{4}$ inch panels. Space nails 6 inches at panel edges and 12 inches at intermediate supports, except where spans are 48 inches or more. Then space nails 6 inches at all supports.

PLYWOOD NAIL HOLDING • Extensive laboratory and field tests, reinforced by more than 25 years experience, offer convincing proof that even $\frac{5}{16}$-inch plywood will hold shingle nails securely and permanently in place, even when the shingle cover is subjected to hurricane-force winds.

The maximum high wind pressure or suction is estimated at 25 psf except at the southern tip of Florida, where wind pressures may attain values of 40 to 50 psf. Because of shape and height factors, however, actual suction or lifting action even in Florida should not exceed 25 psf up to 30-foot heights. Thus any roof sheathing under shingles should develop at least that much withdrawal resistance in the nails used.

Plywood sheathing provides more than adequate withdrawal resistance. A normal wood-shingled roof will average more than 6 nails per square foot. Each nail need carry no more than 11 pounds. Plywood sheathing only $\frac{5}{16}$ inch thick shows a withdrawal resistance averaging 50 pounds for a single 3d shingle nail in laboratory tests and in field tests of wood shingles after 5 to 8 years' exposure. In addition, field experience shows asphalt shingles consistently tear through at the nail before the nail pulls out of the plywood.

Figure 7-122 shows the markings found on plywood used for sheathing. Note the interior and exterior glue markings. APA stands for the American Plywood Association.

• LAYING OUT A STAIR •

So far you have used the framing square to lay out rafters. There is also another use for this type of instrument. It can be used to lay out the stairs going to the basement or going upstairs in a two-story house.

Much has been written about stairs. Here we only lay out the simplest and most useful of the types

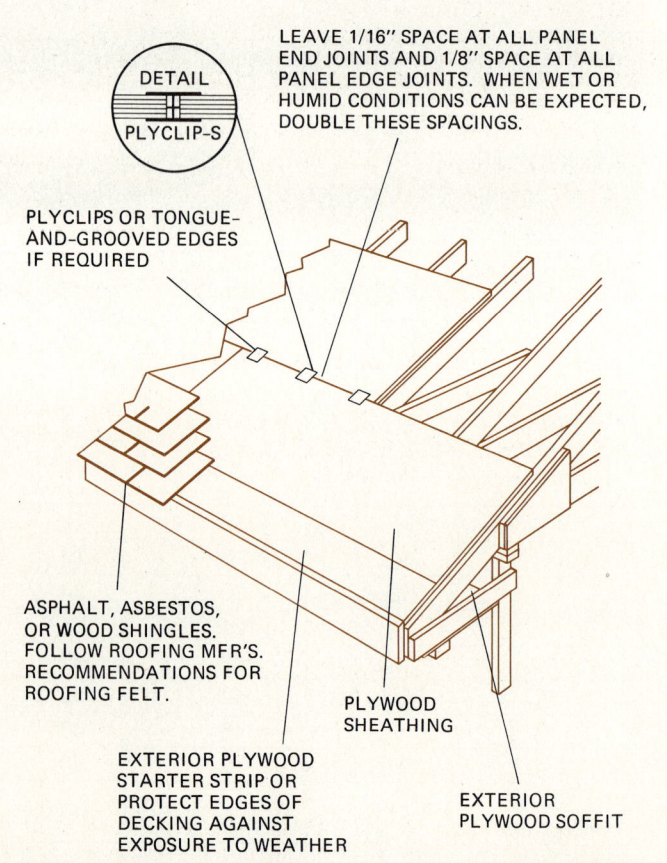

Figure 7-121. Using a plyclip to reinforce plywood decking. *(American Plywood Association)*

160 Carpentry Fundamentals

available. The fundamentals of stair layout are offered here.

1. Determine the height or rise. This is from the top of the floor from which the stairs start, to the top of the floor on which they are to end. See Figure 7-123.
2. Determine the run or distance measured horizontally.
3. Mark the total rise on a rod or a piece of 1x2-inch furring to make a so-called *story pole*. Divide the height or rise into the number of risers desired. A simple method is to lay out the number of risers wanted by spacing off the total rise with a pair of compasses. It is common to have this result in fractions of an inch. For example, a total rise of 8′3¾″ or 99¾ inches divided by 14 = 7.125 or 7⅛-inch riser. This procedure is not necessary in the next step because the horizontal distance, or run, is seldom limited to an exact space as is the case with the rise.
4. Lay out or space off the number of treads wanted in the horizontal distance or run. There is always one less tread than there are risers. If there are 14 risers in the stair, there are only 13 treads. For example, if the tread is 10 inches wide and the riser is 7 inches, the stair stringer would be laid out or "stepped off" with the square, ready for cutting as shown in Figure 7-123. The thickness of the tread should be deducted from the first riser as shown. This is in order to have this first step the same height as all the others.

• ALUMINUM SOFFIT •

So far the soffit has been mentioned as the covering for the underside of the overhang. This has been shown to be covered with a plywood sheathing of ¼ inch thickness or with a cardboard substance about the thickness of the plywood suggested. The cardboard substitute is called Upson board. This is because the Charles A. Upson Company makes it. If properly installed and painted, it will last for years. However, it should not be used in some climates.

One of the better materials for soffit is aluminum. More and more homes are being fitted with this type of maintenance-free material.

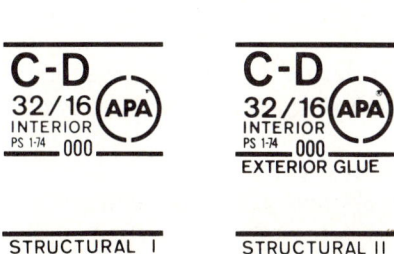

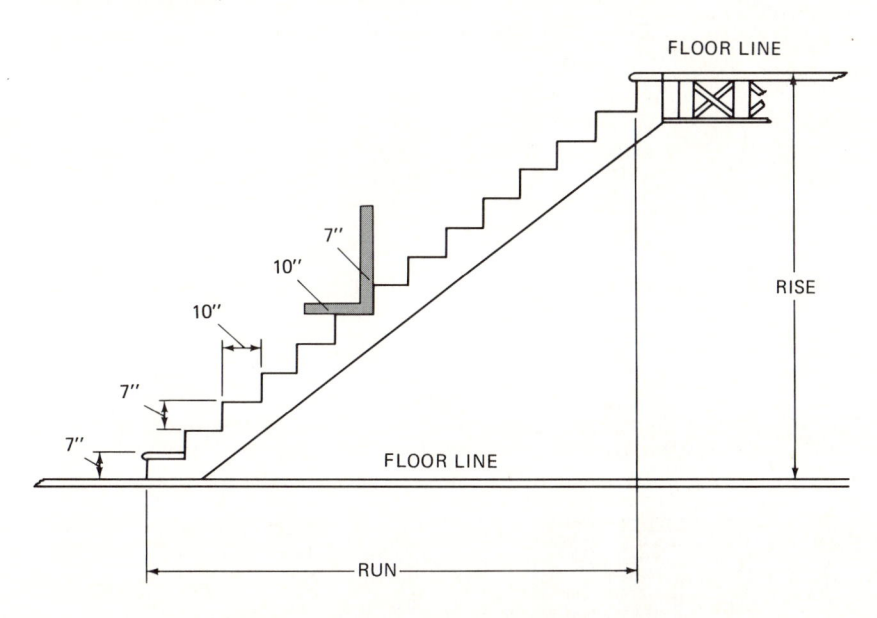

Figure 7-122. Plywood grades identified. *(American Plywood Association)*

Figure 7-123. Laying out stairs with the steel square. *(Stanley Tools)*

Building Roof Frames **161**

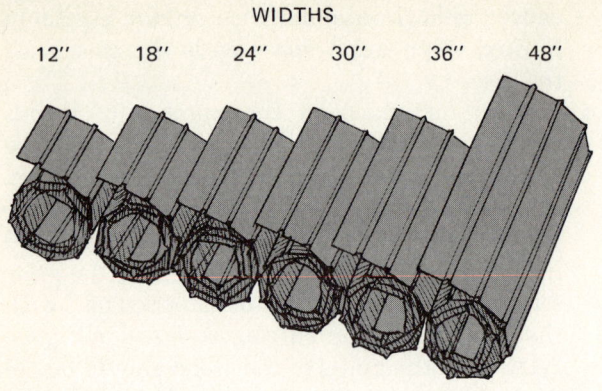

Figure 7-124. Rolls of soffit material. These are made of aluminum.

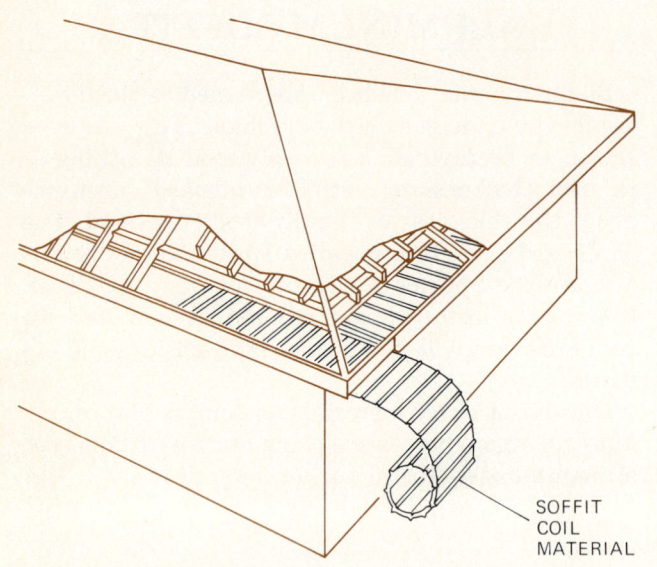

Figure 7-125. Method of inserting the aluminum soffit material. *(Reynolds Metals Products)*

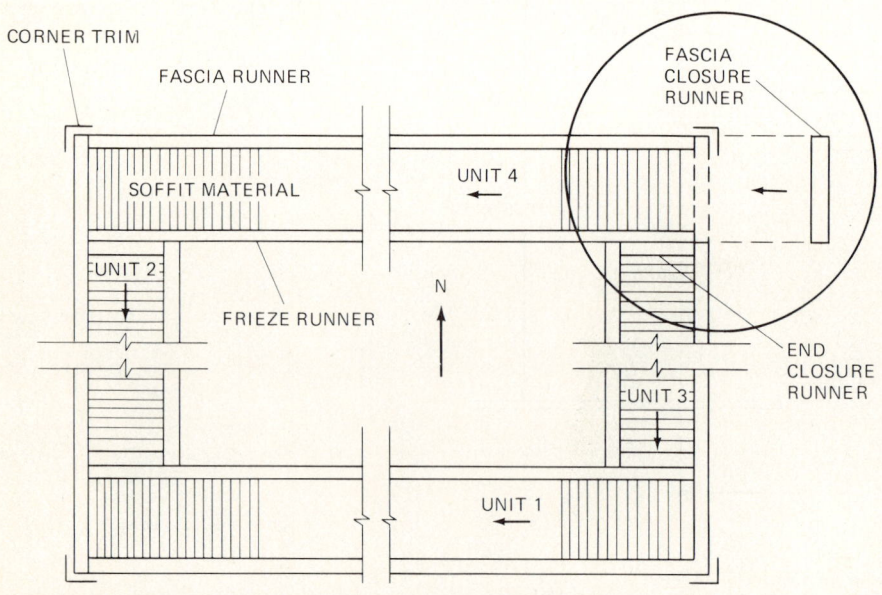

Figure 7-126. Steps in installing aluminum soffit in a hip roof with an overhang all around. *(Reynolds Metals Products)*

Material Availability

Aluminum soffit can be obtained in 50-foot rolls with various widths. They can be obtained (Figure 7-124) in widths of 12, 18, 24, 30, 36, and 48 inches. These are pushed or pulled into place as shown in Figure 7-125. The hip roof with an overhang all around the house would require soffit material pulled in as shown in Figure 7-126. The runners supporting the material are shown in Figure 7-127. Covering the ends is important to making a neat job. Corner trim and fascia closure are available to help give the finished job a look not unlike that of an all-wood soffit.

Figure 7-128 shows how the fascia runner, the frieze runner, and corner trim are located for ease of installation of the soffit coil.

After the material has been put in place and the end cuts have been made, the last step is to insert a plastic liner to hold the aluminum in place. This prevents rattling when the wind blows. The material can be obtained with a series of holes prepunched. This will serve as ventilation for the attic. See Figure 7-129.

Figure 7-130 gives more details on the installation of the runners that support the soffit material. The fascia runner is notched at points *b* for about 1½ inches maximum. Then the tab is bent upward and against the inside of the fascia board. Here it is nailed to the board for support. Take a look at *c* in Figure 7-130 to see how the tab is bent up. Note how the width of the channels is the soffit coil width plus at least ⅜ inch and not more than ⅞ inch. This allows for expansion by the aluminum. Aluminum will expand in hot weather.

162 Carpentry Fundamentals

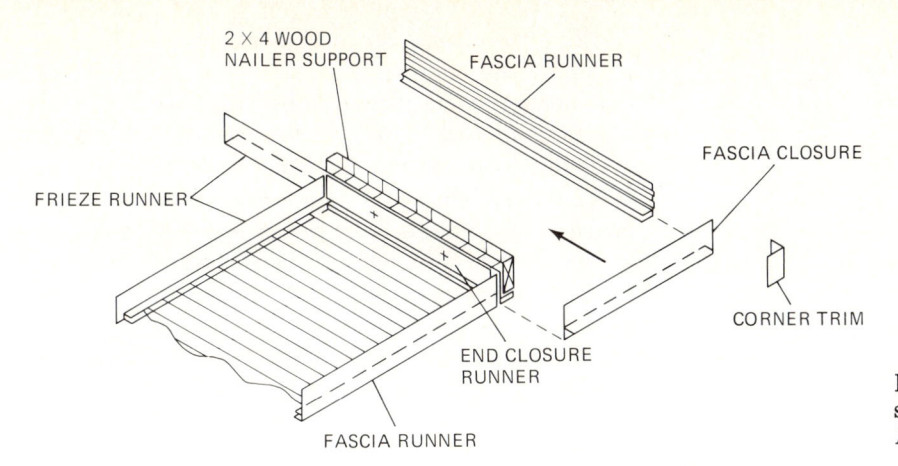

Figure 7-127. Closing off the ends of the soffit with aluminum. *(Reynolds Metal Products)*

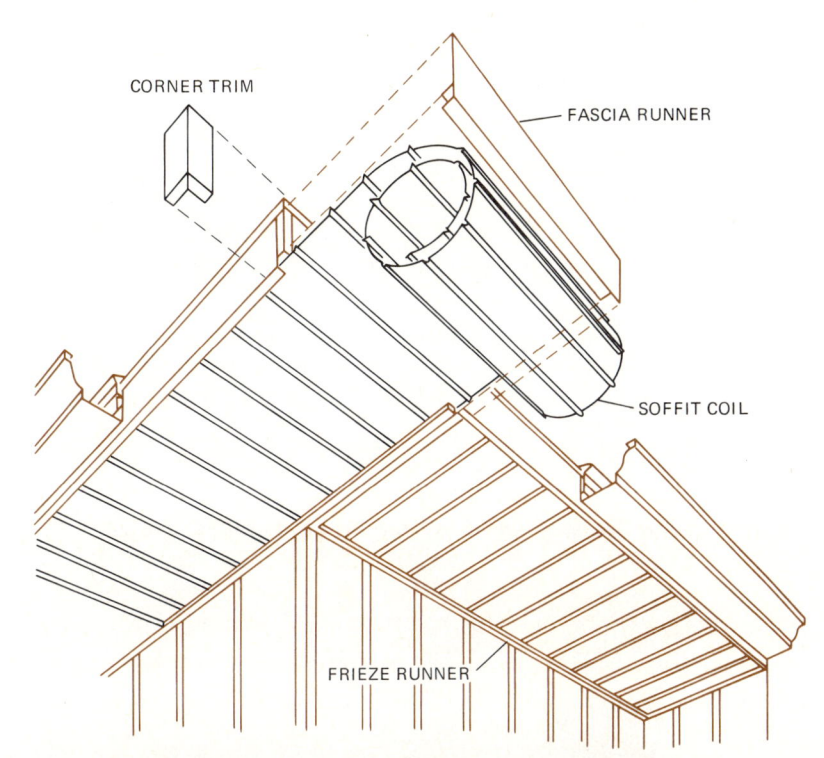

Figure 7-128. Installing soffit coil in the overhang space. Note the corner trim to give a finished appearance. *(Reynolds Metals Products)*

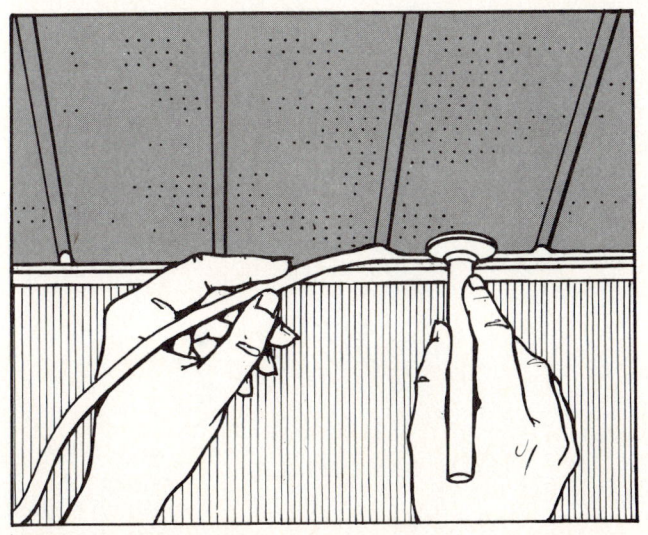

Figure 7-129. Finishing up the job with a vinyl insert to hold the aluminum in place. *(Reynolds Metals Products)*

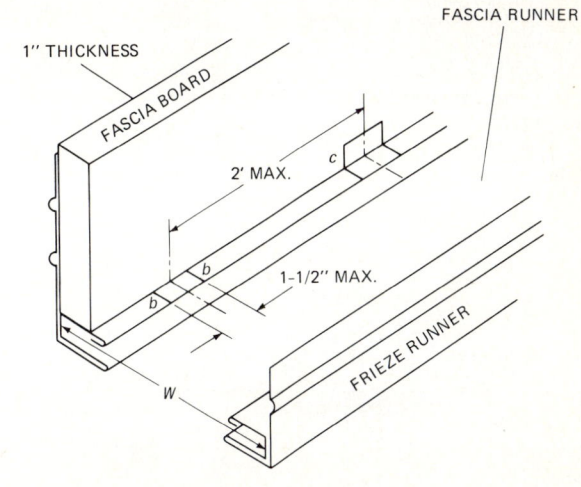

Figure 7-130. Method used to support the runner on the fascia. *(Reynolds Metals Products)*

W = SOFFIT COIL WIDTH + 3/8" TO 7/8"
(3/8" MINIMUM) (7/8" MAXIMUM)

Building Roof Frames **163**

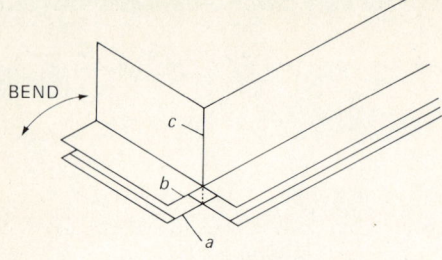

Figure 7-131. Bending and breaking the runner material. *(Reynolds Metals Products)*

Cutting the runner to desired lengths can be done by cutting the channel at *a* and *b* of Figure 7-131. Then bend the metal back and forth along a line such as at *c* until it breaks. Of course you can use a pair of tin snips to make a clean cut.

Figure 7-132 shows how the soffit is installed with a brick veneer house. Part (A) shows how the frieze runner is located along the board. Insert (B) shows how the runner is nailed to the board.

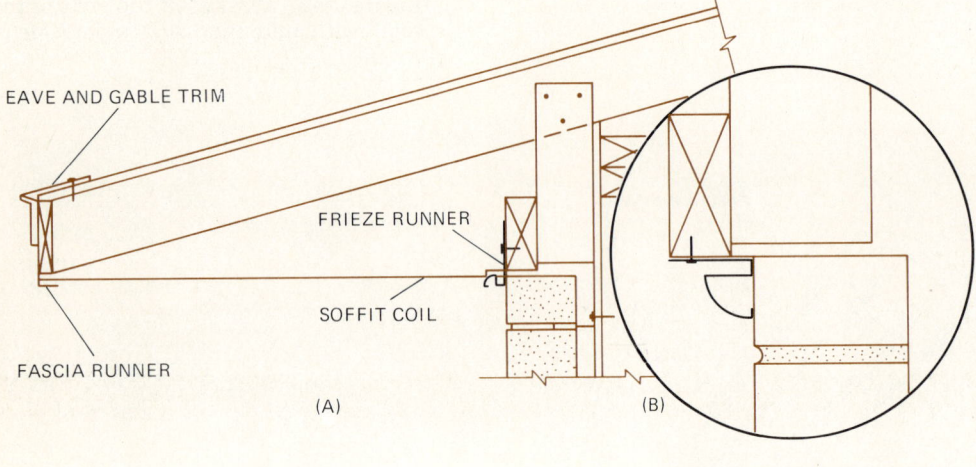

Figure 7-132. Soffit on a brick veneer house. (A) Locating the frieze runner along the board. (B) Using the quarter-round type of frieze runner. *(Reynolds Metals Products)*

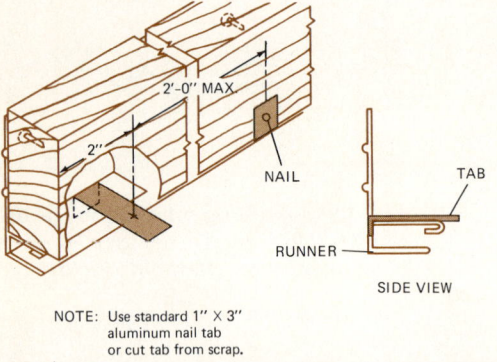

Figure 7-133. Installing a tab to keep the runner free to move as aluminum expands on hot days. *(Reynolds Metals Products)*

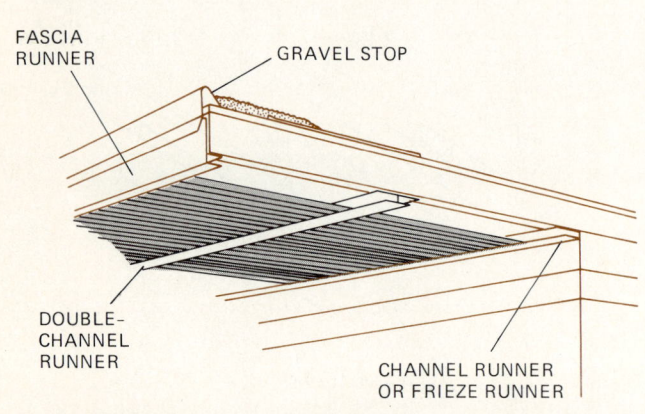

Figure 7-134. Using a double-channel runner and a double row of aluminum soffit material. *(Reynolds Metals Products)*

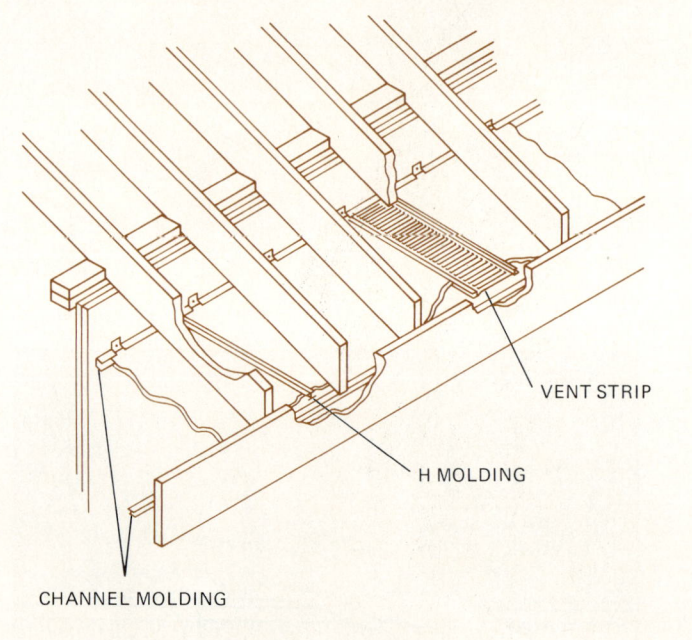

Figure 7-135. Installing the H-molding joint. *(Reynolds Metals Products)*

164 Carpentry Fundamentals

When the fascia board is more than 1 inch wide, it is necessary to place 1-inch aluminum strips as shown in Figure 7-133. The tabs are hooked in between the fold in the runner. Then the runner is brought under the fascia board and bent back and nailed. This will allow the runner to expand when it is hot. Do not nail the overlapping runners to one another.

In some instances it is necessary to use a double-channel runner. This is done so that there will be no sagging of the soffit material. See Figure 7-134. Note how the frieze runner and double-channel runner are located. Note the gravel stop on this flat roof. In some parts of the country more overhang is needed because it gives more protection from the sun.

Figure 7-135 shows how the H-molding joint works to support the two soffit materials as they are unrolled into the channel molding. Note the location of the vent strip, when needed.

As was mentioned before, the aluminum soffit makes for a practically maintenance-free installation. More calls will be made for this type of finish. The carpenter will probably have to install it, since the carpenter is responsible for the exterior finish of the building and the sealing of all the openings. With the advent of aluminum siding, it is only natural that the soffit be aluminum.

• WORDS CARPENTERS USE •

framing	king-post trusses	rafter scales	deck	bottom or heel cut
braces	FHA	octagon scale	pitch	tail
shingles	military specifications	brace scale	plate	fascia
gambrel roof		Essex board measure	common rafters	dormer
ridge boards	load conditions		hip rafters	bay window
cripple rafters	sheathing	valley	valley rafters	shingle stringers
truss	gussets	hip roof	jack rafters	soffit
cripple studs	framing square	span	cripple jack	
cathedral ceiling	steel square	run	plumb cut	

• STUDY QUESTIONS •

7-1. List at least five types of roof lines.

7-2. What is sheathing?

7-3. What is the difference between a hip roof and a mansard roof?

7-4. What are the parts of a roof frame?

7-5. Where are trussed rafters commonly used?

7-6. What is a disadvantage of a truss roof?

7-7. What is an advantage of the truss type of construction?

7-8. What is used to cover trusses?

7-9. What is a framing square?

7-10. Identify the following parts of a steel square:
 a. Body
 b. Tongue
 c. Heel
 d. Face

7-11. How can you make use of the hundredths scale on a square?

7-12. What is the difference between the octagon and brace scales on a square?

7-13. Identify the following terms:
 a. Shed roof
 b. Gable or pitch roof
 c. Valley roof

7-14. What do the following terms mean?
 a. Span
 b. Run
 c. Rise
 d. Pitch

7-15. Identify the following roof frame members:
 a. Plate
 b. Ridge board
 c. Common rafters
 d. Hip rafters
 e. Jack rafters
 f. Valley rafters

7-16. What part of a rafter is the tail?

7-17. What is the difference between a valley rafter and a hip rafter?

7-18. How do you describe a jack rafter?

7-19. What is a dormer?

7-20. What is a bay window?

Building Roof Frames

8 COVERING ROOFS

In this unit you will learn how the carpenter covers roofs. You will learn how the roof is prepared for shingles. How to place the shingles on the prepared surface is discussed. Details for applying an asphalt shingle roof are given. Things you can learn to do are

- Prepare a roof deck for shingles
- Apply shingles to a roof deck
- Estimate shingles needed for a job
- Figure slope of a roof

• INTRODUCTION •

Roofing shingles are made in different sizes and shapes. Roofs have different angles and shapes. Pipes stick up through the roof. Roofing has to be fitted. There are crickets also to be fitted. (A cricket fits between a chimney and the roof.) A number of fine details are presented by roof shapes. All these have to be considered by the roofer.

Some industrial and commercial buildings use roll shingles. They take a slightly different approach. Asphalt shingles are safer than wooden shingles. The asphalt shingles will resist fire longer. Wooden shingles are not allowed in some sections of the country. This is because of fire regulations.

• BASIC SEQUENCE •

The carpenter should apply a roof in this order:

1. Check the deck for proper installation.
2. Decide which shingles to use for the job.
3. Estimate the amount needed for the job.
4. Apply drip strips.
5. Place the underlayment.
6. Nail the underlayment.
7. Start the first course of shingles.
8. Continue other courses of shingles.
9. Cut and install flashings:
    a. valleys
        (1) open
        (2) closed-cut
        (3) woven
    b. soil pipe flashing
    c. chimney flashing
    d. other flashings
10. Cover ridges.
11. Cover all nail heads with cement.
12. Glue down tabs, if needed.

Types of Roofs

There are a number of roof types. They are classified according to shape. Figure 8-1 shows the different types. Each type presents roofing problems. For each, certain methods are used to cover the decking, ridges, and drip areas.

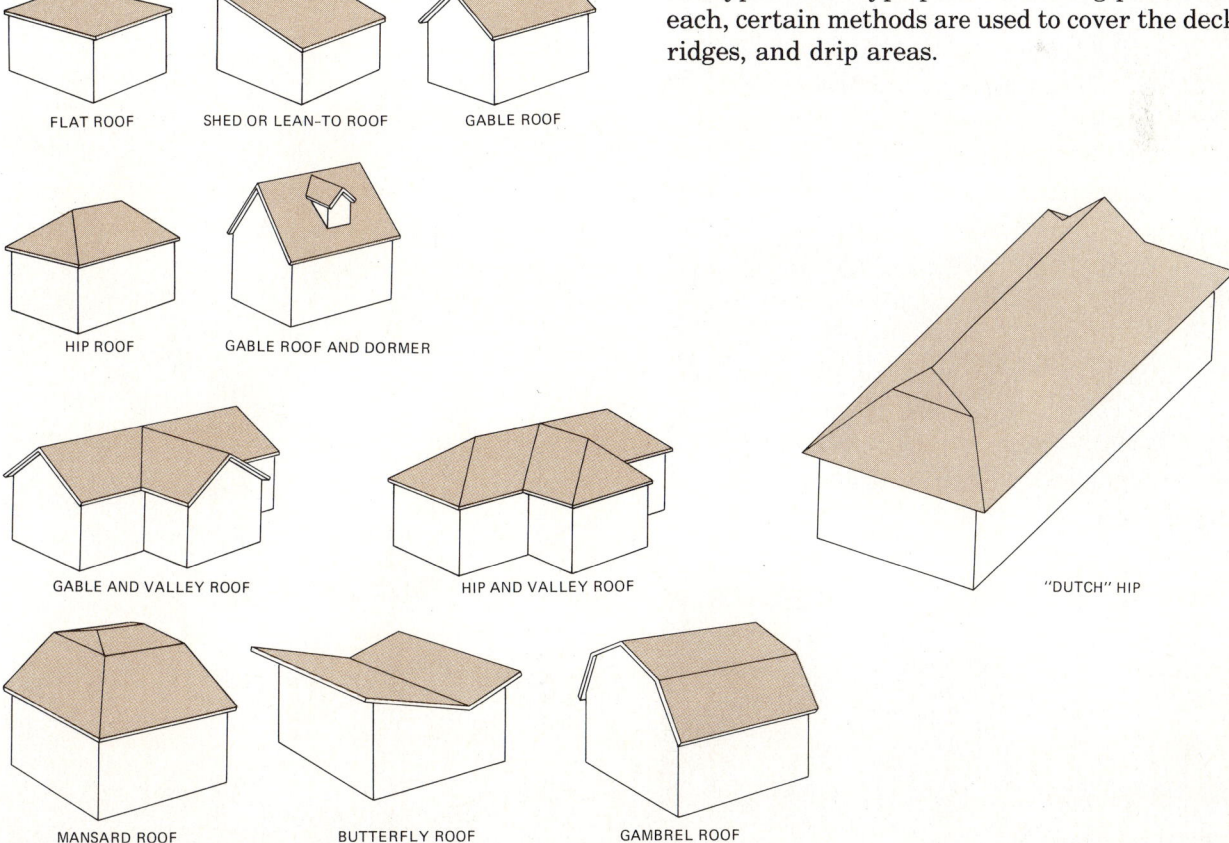

Figure 8-1. Various types of roof shapes.

Covering Roofs **167**

The *mansard* roof presents some problems. Figure 8-2 shows a mansard roof. Note that the dark area is covered with shingles. The attached garage in Figure 8-2 has a hip roof. Figure 8-3 also shows a hip roof. Note how the entrance is also a hip, but shorter.

The *gable* roof is a common type. See Figure 8-4. It is easy to build. This is a simple roof. Figure 8-5 shows a variety of gable roofs. Each is a complete unit. The garage shows the angles of this type of roof very well.

Figure 8-2. Mansard roof with hip on garage.

Figure 8-3. Hip-and-valley roof.

Figure 8-4. Gable roof.

Figure 8-5. Gable roof with add-ons.

Drainage Factors

The main purpose of a roof is to protect the inside of a building. This is done by draining the water from the roof. The water goes onto the ground or into the storm sewer. Some parts of the country allow the water to be dropped onto the earth below. Other sections require the collected water to be moved to a storm sewer. The main idea is to prevent water seepage. Roof water should not seep back into a basement.

Ice is another problem in colder climates. Ice forms and makes a dam for melting snow. See Figure 8-6A. Water backs up under the shingles. This leaks into the ceiling below. It can be caused by insufficient insulation. No soffit ventilation will also cause leaks. These two situations will cause problems. Ice may call for heating cables. Cables can be installed to solve the ice dam problem. Leaking can be prevented by adding ventilation. If insulation can be added, this too should be done. See Figure 8-6B.

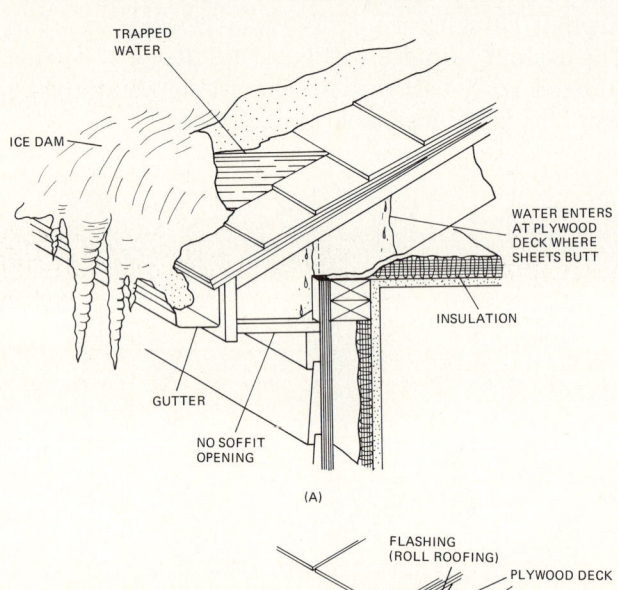

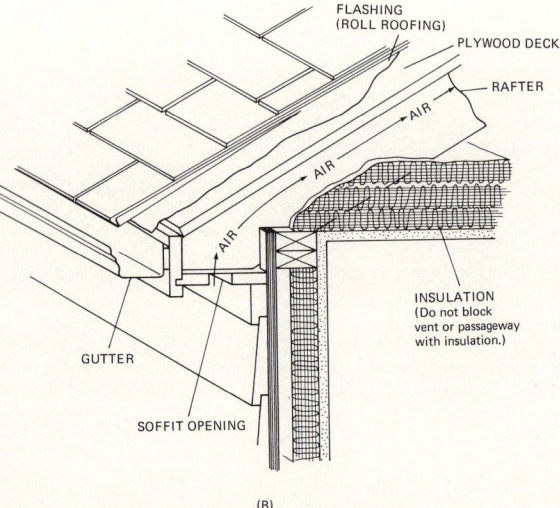

Figure 8-6. (A) Water leakage caused by ice dam. (B) Using ventilation and insulation to prevent leaks.

The point where roof lines come together is called a *valley*. See Figure 8-7A, B, and C. Valleys direct water to the drain. This keeps it out of the house. They need special attention during roofing.

Eave troughs and downspouts drain water from the roof. It is drained into gutters. Downspouts carry the water to the ground. Downspouts connect to other pipes. That piping sometimes goes to the street storm sewer. This eliminates seepage into the basement or under the slab. See Figure 8-8A. In most cases a splash block is located under the downspout. This disperses the water. See Figure 8-8B.

Roofing Terms

There are a number of roofing terms. They are used by roofers and carpenters. This is a special language. You should become familiar with the terms.

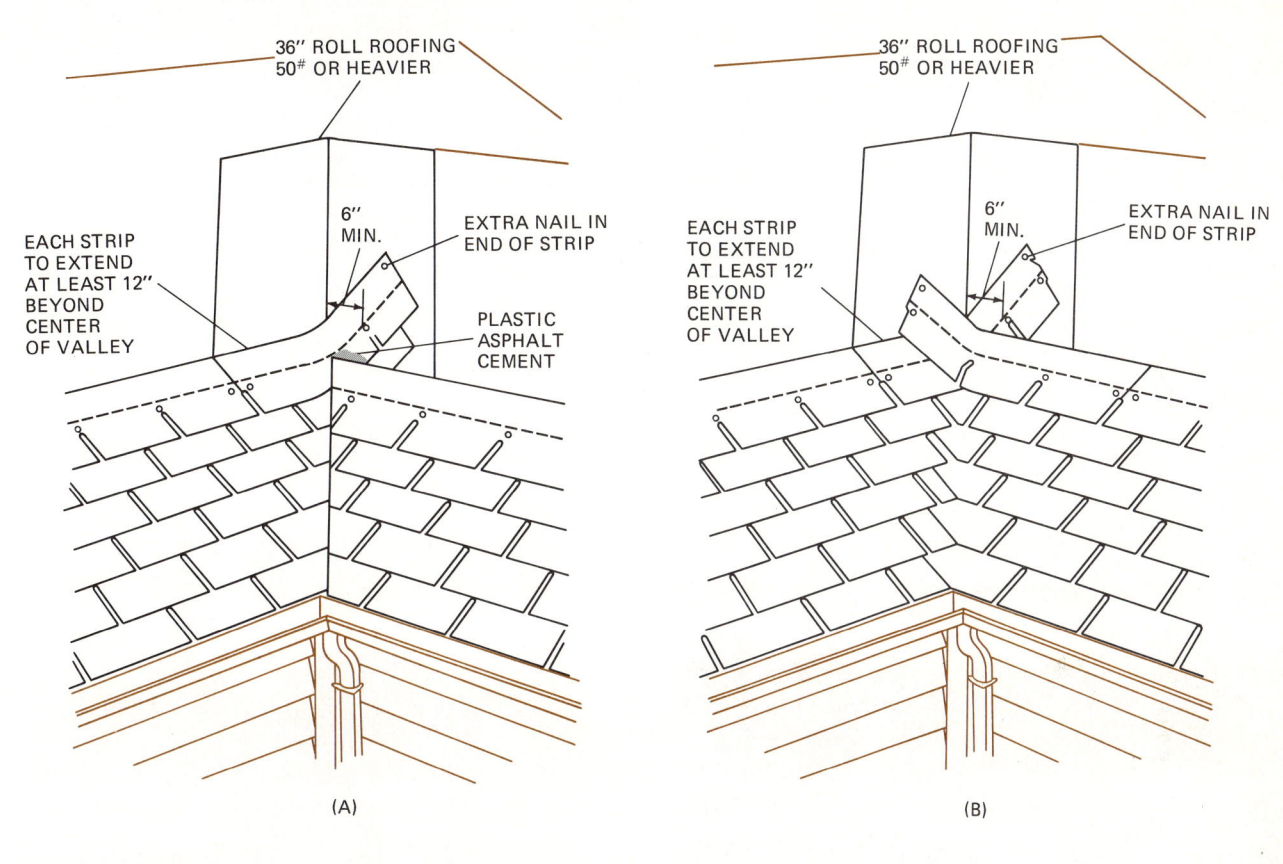

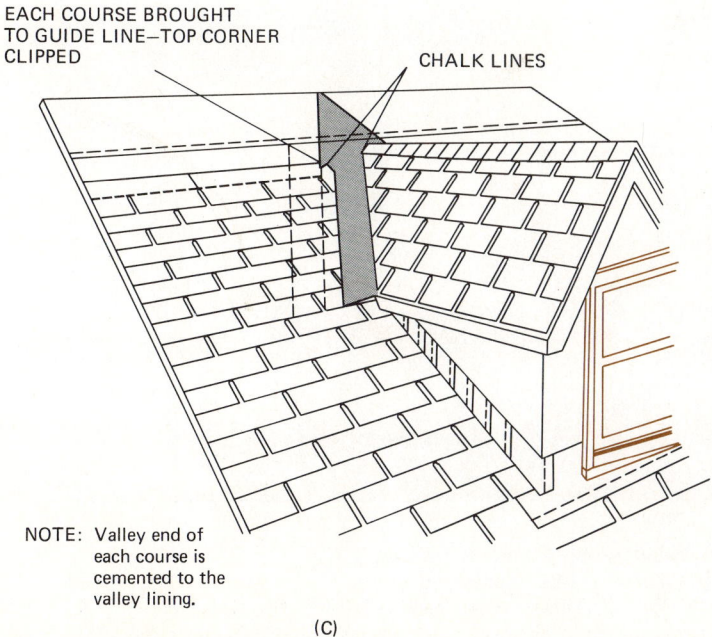

Figure 8-7. (A) Closed-cut valley. (B) Woven valley. (C) Open valley.

Covering Roofs **169**

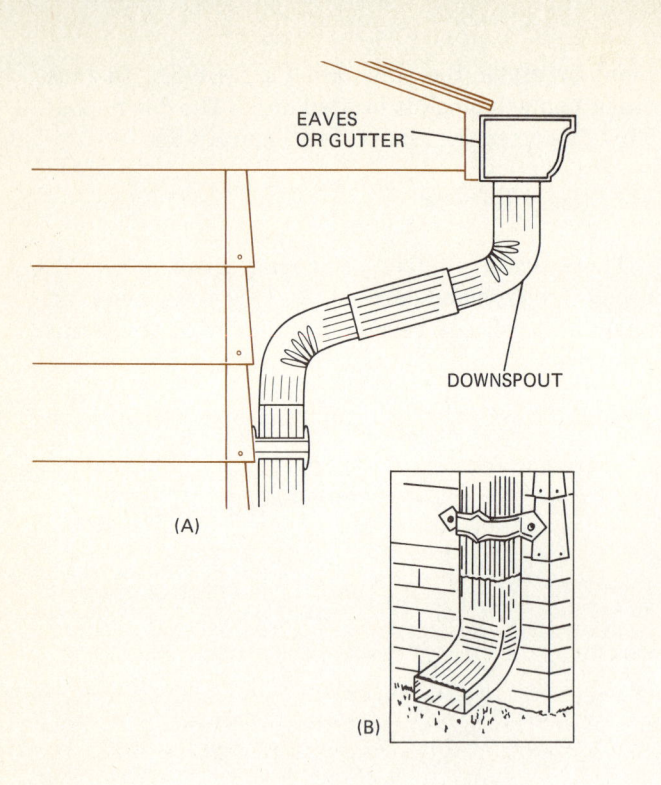

Figure 8-8. (A) Eave trough and downspout. (B) Downspout elbow turns water away from the basement. (C) Preparing a flat roof. Asphalt-saturated felt is mopped down with hot asphalt or tar. (D) Applying a shingle roof. The shingles are packaged so that they can be placed in a location convenient for the roofer.

170　Carpentry Fundamentals

You will then be able to talk with roofing salespersons. Other roofers also use the terms. You will be able to speak their language.

Roofing is part of the exterior building. The carpenter is called upon to place the covering over a frame. This frame is usually covered by plywood. Plywood comes in 4×8-foot sheets. It goes onto the rafters quickly. Sheathing may be 1×6-inch or 1×10-inch boards. Sheathing takes longer to install than plywood. The frame is covered with a tar or felt paper. This paper goes on over the sheathing. The paper allows moisture to move from the wood upward. Moisture then escapes under the shingles. This prevents a buildup of moisture. If the weather is bad, moisture can freeze. This forms a frost under the shingles.

Shingles of wood, asphalt, and asbestos are used for roofing. Fiberglas is now used. Tile and slate were used at one time. They are rather expensive to install. Copper, galvanized iron, and tin are also used as roof coverings.

Commercial buildings may use a built-up roof. This type uses a gravel topping or cap sheet. It has a number of layers. Asphalt-saturated felt is mopped down with hot asphalt or tar. See Figure 8-8C. Choice of roofing is determined by three things. They are cost, slope, and life expected. In some local climates (wind, rain, snow), flat roofs may have to be rejected. Appearance is important in some cases. Shingles are used most frequently for homes. See Figure 8-8D. Shingles are made of asphalt, asbestos, Fiberglas, or cedar. Cedar shingles are not permitted in some places. Wood burns too easily. Once aged, however, it becomes more fire-resistant.

Terms

Square Shingles needed to cover 100 square feet of roof surface. That means 10 feet *square*, or 10 feet by 10 feet.

Exposure Distance between exposed edges of overlapping shingles. Exposure is given in inches. See Figure 8-9. Note the 5-inch and 3-inch exposures.

Head Lap Distance between the top of the bottom shingle and the bottom edge of the one covering it. See Figure 8-10.

Top Lap Distance between the lower edge of an overlapping shingle and the upper edge of the lapping shingle. See Figure 8-10. Top lap is measured in inches.

Side Lap Distance between adjacent shingles that overlap. Measured in inches.

Valley Angle formed by two roofs meeting. The internal part of the angle is the valley.

Rake On a gable roof, the inclined edge of the surface to be covered.

Flashing Metal used to cover chimneys and around things projecting through the roofing. Used to keep the weather out.

Underlayment Usually No. 15 or No. 30 felt paper applied to a roof deck. It goes between the wood and the shingle.

Ridge The horizontal line formed by the two rafters of a sloping roof being nailed together.

Hip The external angle formed by two sides of a roof meeting.

Pitch

Drainage of water from a roof surface is essential. This means that pitch should be considered. The pitch or slope of a roof deck determines the choice of shingle. It also determines drainage.

Pitch limitations are shown in Figure 8-11. Any shingle may safely be used on roofs with normal slopes. Normal is 4 inches rise or more per horizontal foot. An exception exists for square-butt strip shingles. They may be used on slopes included in the shaded area in Figure 8-11.

When the pitch is less than 4 inches per foot, it is best to use roll roofings. In the range of 4 inches down to 1 inch per foot, the following rules apply:

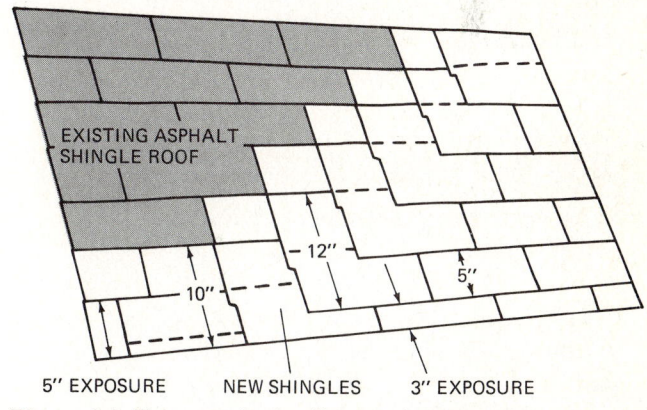

Figure 8-9. Exposure is the distance between the exposed edges of overlapping shingles. *(Bird and Son)*

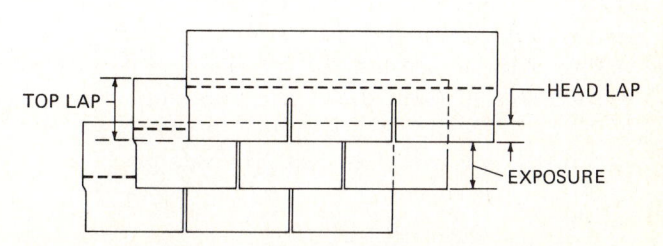

Figure 8-10. Head lap and top lap. *(Bird and Son)*

Covering Roofs 171

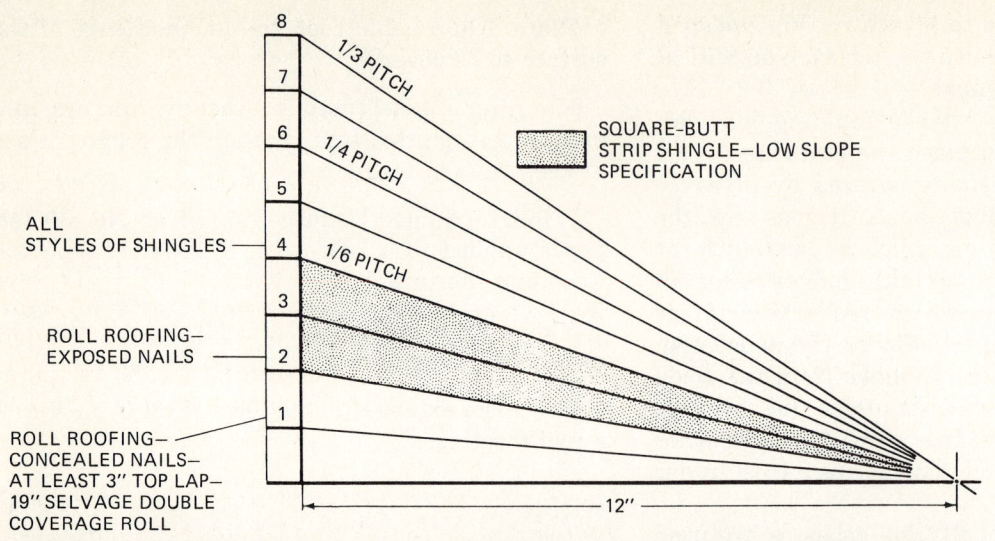

Figure 8-11. Minimum pitch requirements for different asphalt roofing products. *(Bird and Son)*

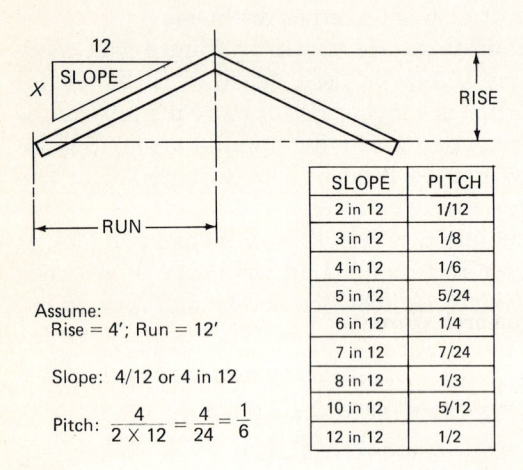

Figure 8-12. Slope, pitch, and run of a roof.

SLOPE	PITCH
2 in 12	1/12
3 in 12	1/8
4 in 12	1/6
5 in 12	5/24
6 in 12	1/4
7 in 12	7/24
8 in 12	1/3
10 in 12	5/12
12 in 12	1/2

Assume:
Rise = 4'; Run = 12'

Slope: 4/12 or 4 in 12

Pitch: $\frac{4}{2 \times 12} = \frac{4}{24} = \frac{1}{6}$

$$\text{Slope} = \frac{\text{rise}}{\text{run}} \qquad \text{Pitch} = \frac{\text{rise}}{2 \times \text{run}}$$

Roll roofing may be applied by the exposed nail method if the pitch is not lower than 2 inches per foot.

Roll roofings applied by the concealed nail method may be used on pitches down to, but not below, 1 inch per foot. This is true if (1) they have at least 3 inches of top lap, and (2) they have double coverage roofing with a top lap of 19 inches.

Any of the above may be applied on a deck with a pitch steeper than the stated minimum. Pitch is given as a fraction. For example, a roof has a *rise* of 8 feet and a *run* of 12 feet. Then its pitch is

$$\frac{8}{2 \times 12} \quad \text{or} \quad \frac{8}{24} \quad \text{or} \quad \tfrac{1}{3}$$

Pitch is equal to the rise divided by 2 times the run. Or

$$\text{Pitch} = \frac{\text{rise}}{2 \times \text{run}}$$

Slope

Slope is how fast the roof rises from the horizontal. See Figure 8-12. Slope is equal to the rise divided by the run. Or

$$\text{Slope} = \frac{\text{rise}}{\text{run}}$$

Slope and pitch are often used with the same meaning. However, you can see there is a difference. Some roofers' manuals use them as if they were the same.

Before a roof can be applied, you have to know how many shingles are needed. This calls for estimating the area to be covered. The number of square feet is needed. Then divide the number of square feet by 100 to produce the number of squares needed.

• HOW TO ESTIMATE QUANTITIES OF ROOFING •

Roofing is estimated and sold in squares. *A square of roofing is the amount required to cover 100 square feet.* To estimate the required amount, you have to compute the total area to be covered. This should be done in square feet. Then divide the amount by 100. This determines the number of squares needed. Some allowance should be made for cutting and waste. This allowance is usually 10 percent. If you use 10 percent for waste and cutting, you will have the correct number of shingles. A simple roof with no dormers will require less than 10 percent.

Complicated roofs will require more than 10 percent for cutting and fitting.

Estimating Area

The areas of simple surfaces can be computed easily. The area of the shed roof in Figure 8-13 is easy. It is the product of the eave line and the rake line ($A \times B$). The area of the simple gable roof in Figure 8-13 equals the sum of the two rakes B and C multiplied by the eave line A. A gambrel roof is estimated by multiplying rake lines A, B, C, and D by eave line E. See Figure 8-13.

Complications arise in roofs such as the one in Figure 8-14. Ells, gables, or dormers can cause special problems. Obtain the lengths of eaves, rakes, valleys, and ridges from drawings or sketches. Simple methods call for dangerous climbing. You may want to estimate without climbing. For that method,

1. The pitch of the roof must be known or determined.
2. The horizontal area in square feet covered by the roof must be computed.

Pitch is shown in Figure 8-15. The pitch of a roof is stated as a relationship between rise and span. If the span is 24'0" and the rise is 8'0", the pitch will be $8/24$ or $1/3$. If the rise were 6'0", then the pitch would be $6/24$ or $1/4$. The $1/3$-pitch roof rises 8 inches per foot of horizontal run. The $1/4$-pitch roof rises 6 inches per foot of run.

You can determine the pitch of any roof without leaving the ground. Use a carpenter's folding rule in the following manner.

Form a triangle with the rule. Stand across the street or road from the building. Hold the rule at

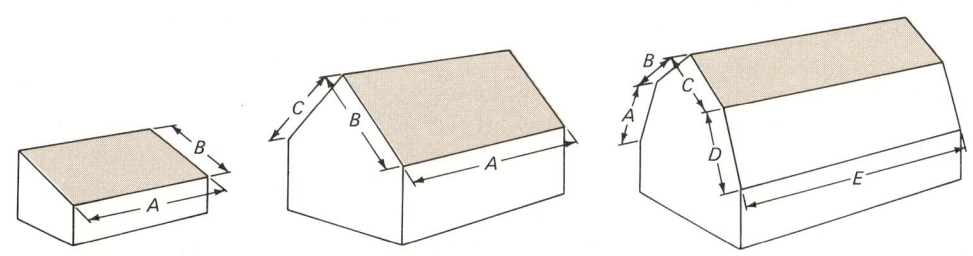

Figure 8-13. Simple roof types and their dimensions: shed, gable, and gambrel. *(Bird and Son)*

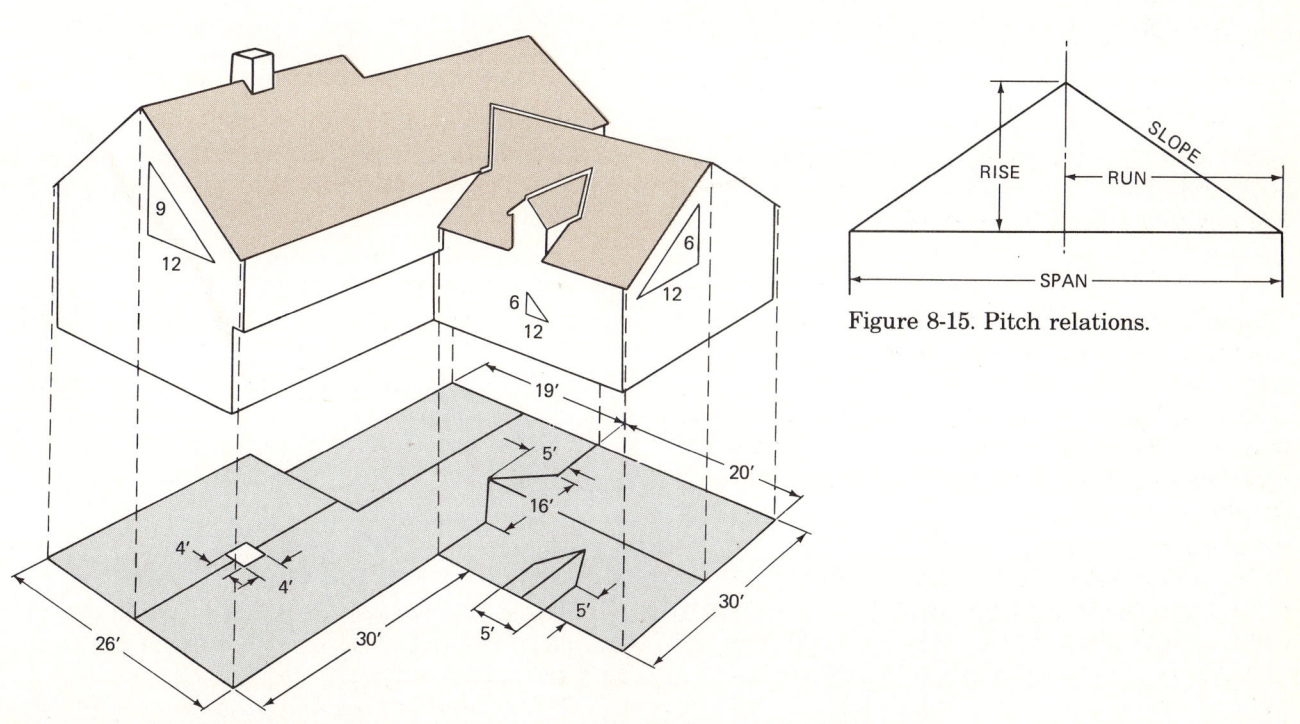

Figure 8-14. Complicated dwelling roof shown in perspective and plan views. *(Bird and Son)*

Figure 8-15. Pitch relations.

Covering Roofs 173

arm's length. Align the roof slope with the sides of the rule. Be sure that the base of the triangle is held horizontal. It will appear within the triangle as shown in Figure 8-16. Take a reading on the base section of the rule. Note the reading point shown in Figure 8-17. Locate in the top line, headed *Rule Reading* in Figure 8-18, the point nearest your reading. Below this point is the pitch and the rise per foot of run. Here the reading on the rule is 22. Under 22 in Figure 8-18, the pitch is designated as 1/3. This is a rise of 8 inches per foot of horizontal run.

Horizontal Area

Figure 8-14 is a typical dwelling. The roof has valleys, dormers, and variable-height ridges. Below the perspective the total ground area is covered by the roof. All measurements needed can be made from the ground. Or they can be made within the attic space of the house. No climbing on the roof is needed.

Computation of Roof Areas

Make all measurements. Draw a roof plan. Determine the pitches of the various elements of the roof. Use the carpenter's rule. The horizontal areas can now be quickly worked out.

Include in the estimate only those areas having the same pitch. The rise of the main roof is 9 inches per foot. That of the ell and dormers is 6 inches per foot.

The horizontal area under the 8-inch-slope roof will be

$26 \times 30 = 780$ square feet
$19 \times 30 = \underline{570}$ square feet
Total   1350 square feet

Less

$8 \times 5 = 40$ (triangular area under ell roof)
$4 \times 4 = \underline{16}$ (chimney)
        56 square feet
$1350 - 56 = 1294$ square feet total

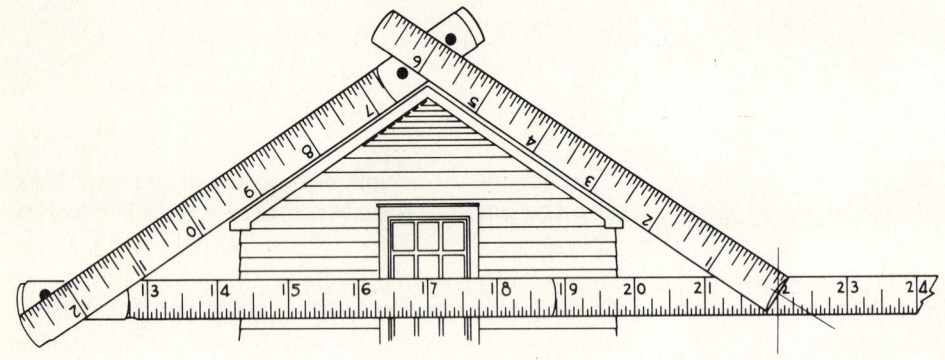

Figure 8-16. Using the carpenter's rule to find the roof pitch. *(Bird and Son)*

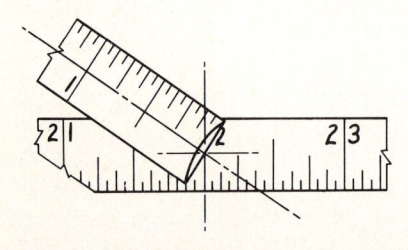

READING POINT

Figure 8-17. Reading the carpenter's rule to find the point needed for the pitch figures. *(Bird and Son)*

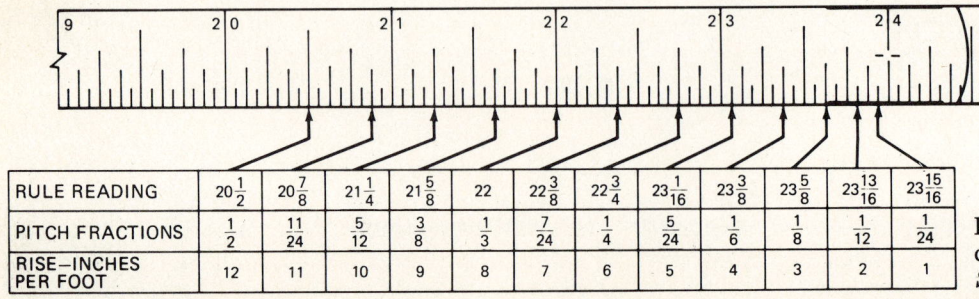

RULE READING	$20\frac{1}{2}$	$20\frac{7}{8}$	$21\frac{1}{4}$	$21\frac{5}{8}$	22	$22\frac{3}{8}$	$22\frac{3}{4}$	$23\frac{1}{16}$	$23\frac{3}{8}$	$23\frac{5}{8}$	$23\frac{13}{16}$	$23\frac{15}{16}$
PITCH FRACTIONS	$\frac{1}{2}$	$\frac{11}{24}$	$\frac{5}{12}$	$\frac{3}{8}$	$\frac{1}{3}$	$\frac{7}{24}$	$\frac{1}{4}$	$\frac{5}{24}$	$\frac{1}{6}$	$\frac{1}{8}$	$\frac{1}{12}$	$\frac{1}{24}$
RISE—INCHES PER FOOT	12	11	10	9	8	7	6	5	4	3	2	1

Figure 8-18. Reading point converted to pitch. *(Bird and Son)*

174 Carpentry Fundamentals

The area under the 6-inch-rise roof will be

20 × 30 = 600 square feet
 8 × 5 = 40 (triangular area projecting over the
       640   main house)

Duplications

Sometimes one element of a roof projects over another. Add duplicated areas to the total horizontal area. If the eaves in Figure 8-14 project only 4 inches, there will be

1. A duplication of 2(7 × ⅓) or 4⅔ square feet under the eaves of the main house. This is where they overhang the rake of the ell section.
2. A duplication under the dormer eaves of 2(5 × ⅓) or 3⅓ square feet.
3. A duplication of 9½ × ⅓ or 3¹⁄₁₆ square feet under the rake of the wide section of the main house. This is where it overhangs the rake of the small section in the rear.

The total is 11⅙ or 12 square feet. Item 2 should be added to the area of the 6-inch-pitch roof. Items 1 and 3 should be added to the 9-inch-pitch roof. The new totals will be 640 + 4 or 644 for the 6-inch pitch and 1294 + 8 or 1302 for the 9-inch pitch.

Converting Horizontal to Slope Areas

Now convert horizontal areas to slope areas. Do this with the aid of the *conversion table*, Table 8-1. Horizontal areas are given in the first column. Corresponding slope areas are given in columns 2 to 12.

The total area under the 9-inch rise is 1302 square feet. Under the column headed 9 (for 9-inch rise) on the conversion table is found:

	Horizontal Area		*Slope Area*
Opposite	1000	is	1250.0
Opposite	300	is	375.0
Opposite	00	is	00.0
Opposite	2	is	2.5
Totals	1302		1627.5

The total area under the 6-inch rise is 644 square feet.

	Horizontal Area		*Slope Area*
Opposite	600	is	670.8
Opposite	40	is	44.7
Opposite	4	is	4.5
Totals	644		720.0

The total area for both pitches is 1627.5 + 720 = 2347.5 square feet.

Now, add a percentage for waste. This should be 10 percent. That brings the total area of roofing required to 2582 square feet. Divide 2582 by 100 and get 25.82 or, rounded, *26 squares*.

One point about this method should be emphasized. The method is possible because of one fact. Over any given horizontal area, at a given pitch, a roof will always contain the same number of square feet regardless of its design. A shed, a gable, and a hip roof, with or without dormers, will each require exactly the same square footage of roofing—that is, if each is placed over the same horizontal area with the same pitch.

Accessories

Quantities of starter strips, edging strips, ridge shingles, and valley strips all depend upon linear measurements. These measurements are taken along the eaves, rake, ridge, and valley. Eaves and ridge are horizontal. The rakes and valleys run on a slope. Quantities for the horizontal elements can be taken off the roof plan. True length of rakes and valleys must be taken from conversion tables.

Length of Rake

Determine the length of the rake of the roof. Measure the horizontal distance over which it extends. In this case the rakes on the ends of the main house span distances are 26 and 19 feet. More rake footage occurs where the higher roof section joins the lower. This amounts to 13 + 3½. The total rake footage is 26 + 19 + 13 + 3½ = 61½ feet.

Refer to Table 8-1 under the 9-inch-rise column. Opposite the figures in column 1 find the length of the rake.

	Horizontal run	Length of rake
	60	75.00
	1	1.3
	0.5	0.6
Totals	61.5	76.9 (actual length of the rake

Use the same method and apply it to the rake of the ell. This will indicate its length, including the dormer, to be 39.1 inches. Add these amounts to the total length of eaves. The figure obtained can be used for an estimate of the amount of edging needed.

Hips and Valleys

Hip and valley lengths can be determined. Use the run of the common rafter. Then refer to the hip and valley table, Table 8-2.

Covering Roofs 175

TABLE 8-1 CONVERSION TABLE

Rise, Inches per Foot of Horizontal Run	1	2	3	4	5	6	7	8	9	10	11	12
Pitch, Fractions	1/24	1/12	1/8	1/6	5/24	1/4	7/24	1/3	3/8	5/12	11/24	1/2
Conversion Factor	1.004	1.014	1.031	1.054	1.083	1.118	1.157	1.202	1.250	1.302	1.356	1.414
Horizontal, area in Square Feet or length in feet												
1	1.0	1.0	1.0	1.1	1.1	1.1	1.2	1.2	1.3	1.3	1.4	1.4
2	2.0	2.0	2.1	2.1	2.2	2.2	3.2	2.4	2.5	2.6	2.7	2.8
3	3.0	3.0	3.1	3.2	3.2	3.2	3.5	3.6	3.8	3.9	4.1	4.2
4	4.0	4.1	4.1	4.2	4.3	4.5	4.6	4.8	5.0	5.2	5.4	5.7
5	5.0	5.1	5.2	5.3	5.4	5.6	5.8	6.0	6.3	6.5	6.8	7.1
6	6.0	6.1	6.2	6.3	6.5	6.7	6.9	7.2	7.5	7.8	8.1	8.5
7	7.0	7.1	7.2	7.4	7.6	7.8	8.1	8.4	8.8	9.1	9.5	9.9
8	8.0	8.1	8.3	8.4	8.7	8.9	9.3	9.6	10.0	10.4	10.8	11.3
9	9.0	9.1	9.3	9.5	9.7	10.1	10.4	10.8	11.3	11.7	12.2	12.7
10	10.0	10.1	10.3	10.5	10.8	11.2	11.6	12.0	12.5	13.0	13.6	14.1
20	20.1	20.3	20.6	21.1	21.7	22.4	23.1	24.0	25.0	26.0	27.1	28.3
30	30.1	30.4	31.0	31.6	32.5	33.5	34.7	36.1	37.5	39.1	40.7	42.4
40	40.2	40.6	41.2	42.2	43.3	44.7	46.3	48.1	50.0	52.1	54.2	56.6
50	50.2	50.7	51.6	52.7	54.2	55.9	57.8	60.1	62.5	65.1	67.8	70.7
60	60.2	60.8	61.9	63.2	65.0	67.1	69.4	72.1	75.0	78.1	81.4	84.8
70	70.3	71.0	72.2	73.8	75.8	78.3	81.0	84.1	87.5	91.1	94.9	99.0
80	80.3	81.1	82.5	84.3	86.6	89.4	92.6	96.2	100.0	104.2	108.5	113.1
90	90.4	91.3	92.8	94.9	97.5	100.6	104.1	108.2	112.5	117.2	122.0	127.3
100	100.4	101.4	103.1	105.4	108.3	111.8	115.7	120.2	125.0	130.2	135.6	141.4
200	200.8	202.8	206.2	210.8	216.6	223.6	231.4	240.4	250.0	260.4	271.2	282.8
300	301.2	304.2	309.3	316.2	324.9	335.4	347.1	360.6	375.0	390.6	406.8	424.2
400	401.6	405.6	412.4	421.6	433.2	447.2	462.8	480.8	500.0	520.8	542.4	565.6
500	502.0	507.0	515.5	527.0	541.5	559.0	578.5	601.0	625.0	651.0	678.0	707.0
600	602.4	608.4	618.6	632.4	649.8	670.8	694.2	721.2	750.0	781.2	813.6	848.4
700	702.8	709.8	721.7	737.8	758.1	782.6	809.9	841.4	875.0	911.4	949.2	989.8
800	803.2	811.2	824.8	843.2	864.4	894.4	925.6	961.6	1000.0	1041.6	1084.8	1131.2
900	903.6	912.6	927.9	948.6	974.7	1006.2	1041.3	1081.8	1125.0	1171.8	1220.4	1272.6
1000	1004.0	1014.0	1031.0	1054.0	1083.0	1118.0	1157.0	1202.0	1250.0	1302.0	1356.0	1414.0

TABLE 8-2 HIP AND VALLEY CONVERSIONS.

Rise, Inches per Foot of Horizontal Run	4	5	6	7	8	9	10	11	12	14	16	18
Pitch { Degrees	18°26'	22°37'	26°34'	30°16'	33°41'	36°52'	39°48'	42°31'	45°	49°24'	53°8'	56°19'
Pitch { Fractions	1/6	5/24	1/4	7/24	1/3	3/8	5/12	11/24	1/2	7/12	2/3	3/4
Conversion Factor	1.452	1.474	1.500	1.524	1.564	1.600	1.642	1.684	1.732	1.814	1.944	2.062
Horizontal, Length in Feet												
1	1.5	1.5	1.5	1.5	1.6	1.6	1.6	1.7	1.7	1.8	1.9	2.1
2	2.9	2.9	3.0	3.0	3.1	3.2	3.3	3.4	3.5	3.6	3.9	4.1
3	4.4	4.4	4.5	4.6	4.7	4.8	4.9	5.1	5.2	5.4	5.8	6.2
4	5.8	5.9	6.0	6.1	6.3	6.4	6.6	6.7	6.9	7.3	7.8	8.2
5	7.3	7.4	7.5	7.6	7.8	8.0	8.2	8.4	8.7	9.1	9.7	10.3
6	8.7	8.8	9.0	9.1	9.4	9.6	9.9	10.1	10.4	10.9	11.7	12.4
7	10.2	10.3	10.5	10.7	10.9	11.2	11.5	11.8	12.1	12.7	13.6	14.4
8	11.6	11.8	12.0	12.2	12.5	12.8	13.1	13.5	13.9	14.5	15.6	16.5
9	13.1	13.3	13.5	13.7	14.1	14.4	14.8	15.2	15.6	16.3	17.5	18.6
10	14.5	14.7	15.0	15.2	15.6	16.0	16.4	16.8	17.3	18.1	19.4	20.6
20	29.0	29.5	30.0	30.5	31.3	32.0	32.8	33.7	34.6	36.3	38.9	41.2
30	43.6	44.2	45.0	45.7	46.9	48.0	49.3	50.5	52.0	54.4	58.3	61.9
40	58.1	59.0	60.0	61.0	62.6	64.0	65.7	67.4	69.3	72.6	77.8	82.5
50	72.6	73.7	75.0	76.2	78.2	80.0	82.1	84.2	86.6	90.7	97.2	103.1
60	87.1	88.4	90.0	91.4	93.8	96.0	98.5	101.0	103.9	108.8	116.6	123.7
70	101.6	103.2	105.0	106.7	109.5	112.0	114.9	117.9	121.2	127.0	136.1	144.3
80	116.2	117.9	120.0	121.9	125.1	128.0	131.4	134.7	138.6	145.1	155.5	165.0
90	130.7	132.7	135.0	137.2	140.8	144.0	147.8	151.6	155.9	163.3	175.0	185.6
100	145.2	147.4	150.0	152.4	156.4	160.0	164.2	168.4	173.2	181.4	194.4	206.2

Covering Roofs

Common rafter run is one-half the horizontal distance which the roof spans. This determines the length of a valley. The run of the common rafter should be taken at the lower end of the valley.

Figure 8-14 shows the portion of the ell roof that projects over the main roof. It has a span of 16 feet at the lower end of the valley. Therefore, the common rafter at this point has a run of 8'0".

There are two valleys at this roof intersection. Total run of the common rafter is 16'0". Refer to Table 8-2. Opposite the figures in the column headed *Horizontal* find the linear feet of valleys. Then check the column under the pitch involved.

One of the intersecting roofs has a rise of 6 inches. The other has a rise of 9 inches. Length for each rise must be found. The average of the two is then taken. This gives a close approximation of the true length of the valley.

Thus,

Horizontal	6-inch rise	9-inch rise
10	15	16
6	9	9.6
16	24	25.6

24 + 25.6 = 49.6
49.6 ÷ 2 = 24.8 length of valleys

Dormer Valleys

The run of the common rafter at the dormer is 2.5 feet. Check Table 8-2. It is found that:

Horizontal	6-inch rise
2.0	3.0
0.5	0.75
2.5	3.75 (length of valley)

Two such valleys will total 7.5 feet.

The total length of valley will be 24.8 + 7.5 = 32.3 feet. Use these figures to estimate the flashing material required.

• ROOFING TOOLS •

Most roofing tools are already in the carpenter's tool box. Tools needed for roofing are shown in Figure 8-19.

Roof Brackets Can be used to clamp onto ladder.

Ladders A pair of sturdy ladders with ladder jacks are needed. Shingles are placed on the roof using a hoist on the delivery truck. These ladders come in handy for side roofing.

Staging This is planks for the ladder jacks. They hold the roofer or shingles. They are very useful on mansard roof jobs.

Apron The carpenter's apron is very necessary. It holds the nails and hammer. Other small tools can fit into it. It saves time in many ways. It keeps needed tools handy.

Hammer The hammer is a necessary device for roofing. It should be a balanced hammer for less wrist fatigue.

Chalk and Line This combination is needed to draw guide lines. Shingles need alignment. The chalk marks are needed to make sure the shingles line up.

Tin Snips Heavy-duty tin snips are needed for trimming flashing. They can also be used for trimming shingles.

Kerosene A cleaner is needed to remove tar from tools. Asphalt from shingles can be removed from tools with kerosene.

Tape Measure A roofer has to make many measurements. This is a necessary tool.

Utility Knife A general-purpose knife is needed for close trimming of shingles.

Putty Knife This is used to spread roofing cement.

Carpenter's Rule This makes measurements and also serves to determine the pitch of a roof. See Figure 8-16.

Stapler Some new construction roofing can use a stapler. This device replaces the hammer and nails.

Safety

Working on a roof can be dangerous. Here are a few helpful hints. They may save you broken bones or pulled muscles.

1. Wear sneakers or rubber-soled shoes.
2. Secure ladders and staging firmly.
3. Stay off wet roofs.
4. Keep away from power lines.
5. Don't let debris accumulate underfoot.
6. Use roofing brackets and planks if the roof slopes 4 inches or more for every 12 inches of horizontal run.

Appearance

How the finished job looks is important. Here are a few precautions to improve roof appearance.

1. Avoid shingling in extremely hot weather. Soft asphalt shingles are easily marred by shoes and tools.
2. Avoid shingling when the temperature is below 40°F. Cold shingles are stiff and may crack.

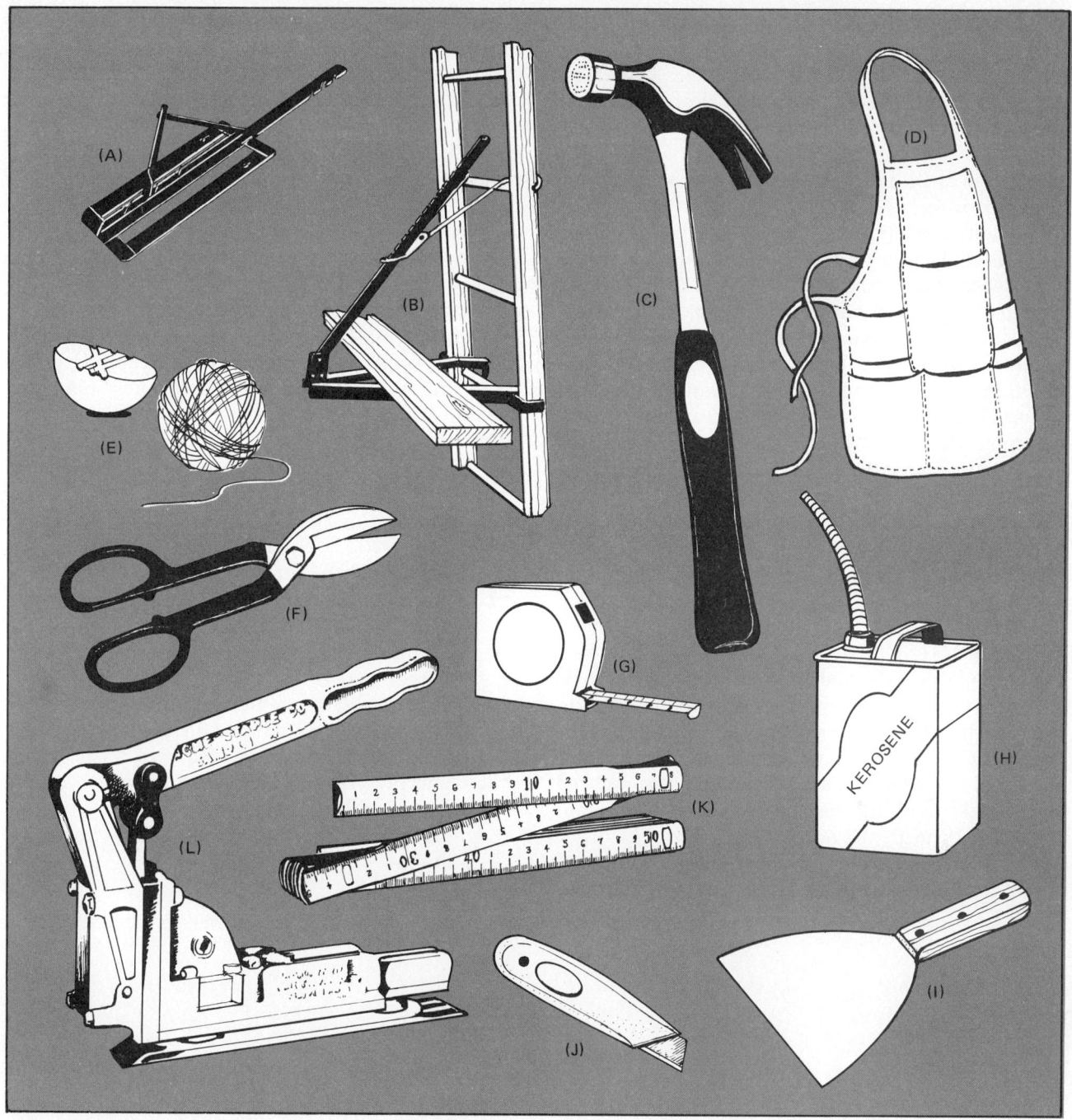

Figure 8-19. Carpenter's tools needed for roofing. *(Bird and Son)* (A) Planking support. (B) Ladder with planking. (C) Claw hammer. (D) Carpenter's apron. (E) Chalk and cord. (F) Snips. (G) Tape measure. (H) Kerosene. (I) Stapler. (J) Carpenter's folding rule. (K) Utility knife. (L) Putty knife.

3. Measure carefully and snap the chalk line frequently. Roof surfaces aren't always square. You'll want to know about problems to come so that you can correct them.
4. Start at the rear of the structure. If you've never shingled before, this will give you a chance to gain experience before you reach the front.

• **APPLYING AN ASPHALT ROOF** •

Asphalt roofing products will serve well when they are correctly applied. Certain fundamentals must be considered. These have to do with the deck, flashing, and application of materials.

Covering Roofs **179**

ROOF PROBLEMS

A number of roof problems are caused by defects in the deck. A nonrigid deck may affect the lay of the roofing. Poorly seasoned deck lumber may warp. This can cause cocking of the shingle tabs. It can also cause wrinkling and buckling of roll roofing.

Improper ventilation can have an effect similar to that of green lumber. The attic area should be ventilated. This area is located directly under the roof deck. It should be free of moisture. In cold weather, be sure the interiors are well ventilated. This applies when plaster is used in the building. A positive ventilation of air is required through the building during roofing. This can usually be provided by opening one or two windows. Windows in the basement or on the first floor can be opened. This can create a positive draft through the house. Open windows at opposite ends of the building. This will also create a flow of air. The moving air has a tendency to dry out the roof deck. It helps to eliminate excess moisture. Condensation under the roofing can cause problems.

DECK CONSTRUCTION • Wood decks should be made from well-seasoned tongue and groove lumber. It should be more than 6 inches wide. Wider sheathing boards are more likely to swell or shrink. This will produce a buckling of the roof material. Sheathing should be tightly matched. It should be secured to the rafter with at least two 8d nails. One should be driven through the edge of the board. The other should be driven through the board face. Boards containing too many resinous areas should be rejected. Boards with loose knots should not be used. Do not use badly warped boards.

Figure 8-20 shows how a wood roof deck is constructed. In most cases today, 4×8-foot sheets of plywood are used as sheathing. The plywood goes over the rafters. C-D grade plywood is used.

UNDERLAYMENT • Apply one layer of no. 15 asphalt-saturated felt over the deck as an underlayment if the deck has a pitch of 4 inches per foot or greater. The felt should be laid horizontally. See Figure 8-20. Do not use no. 30 asphalt felt. Do not use any tar-saturated felt. Laminated waterproof papers should not be used either. Do not use any vapor-barrier-type material. Lay each course of felt over the lower course. Lap the courses 4 inches. Overlap should be at least 2 inches where ends join. Lap the felt 6 inches from both sides over all hips and ridges.

Apply underlayment as specified for low-slope roofs. This is where the roof slope is less than 4 inches per foot and not below 2 inches. Check the maker's suggestions. See Figure 8-21. Felt underlayment performs three functions:

1. It ensures a dry roof for shingles. This avoids buckling and distortion of shingles. Buckling may be caused by shingles being placed over wet roof boards.
2. Felt underlay prevents the entrance of wind-driven rain onto the wood deck. This may happen when shingles are lifted up.
3. Underlay prevents any direct contact between shingles and resinous areas. Resins may cause chemical reactions. These could damage the shingles.

PLYWOOD DECKS • Plywood decks should meet the Underwriters' Laboratories standards. Stan-

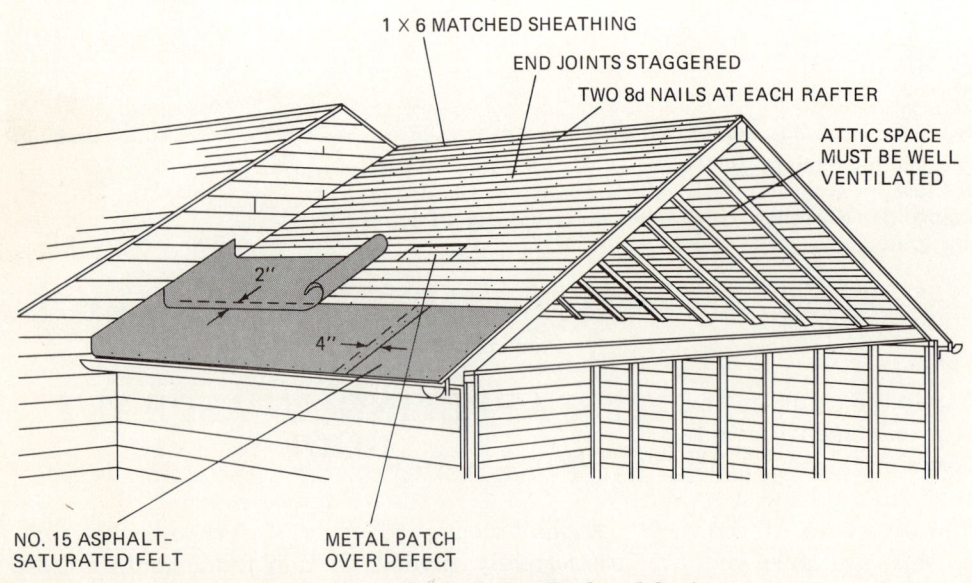

Figure 8-20. Features of a good wood roof deck. *(Bird and Son)*

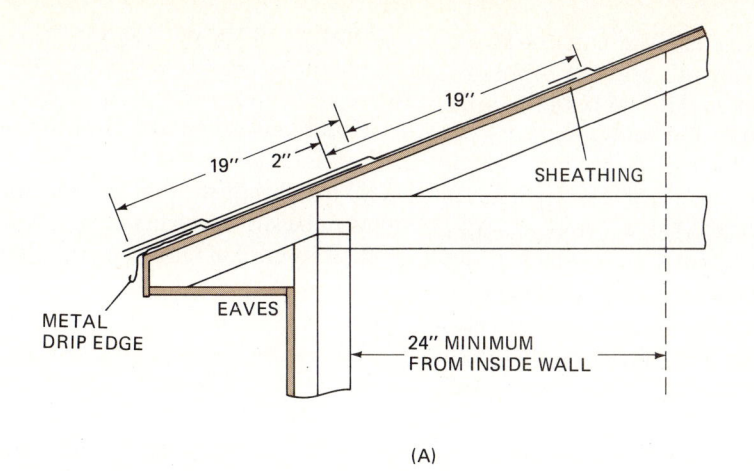

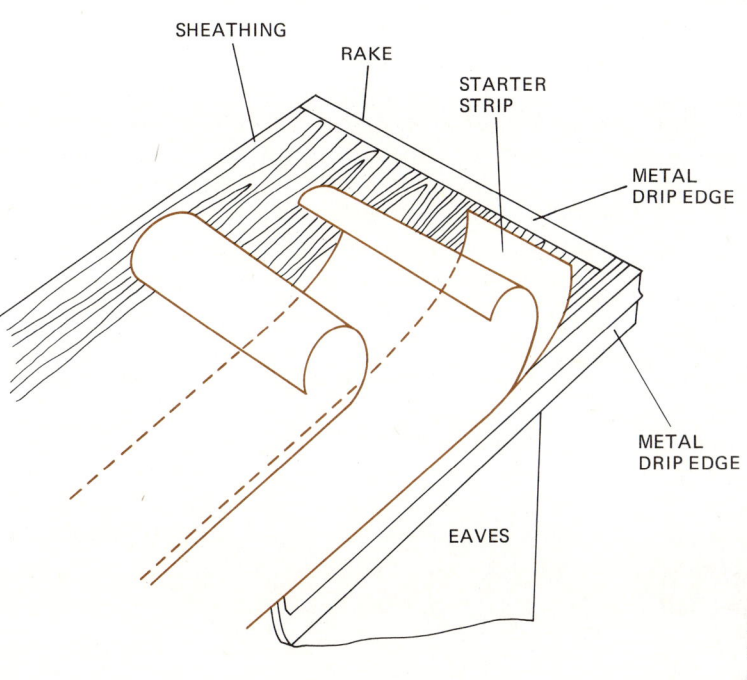

Figure 8-21. (A) Eaves flashing for a low-slope roof. *(Bird and Son)* (B) Placing of sheating and drip edge. *(Bird and Son)* (C) Placement of the underlayment for a shingle roof. *(Bird and Son)*

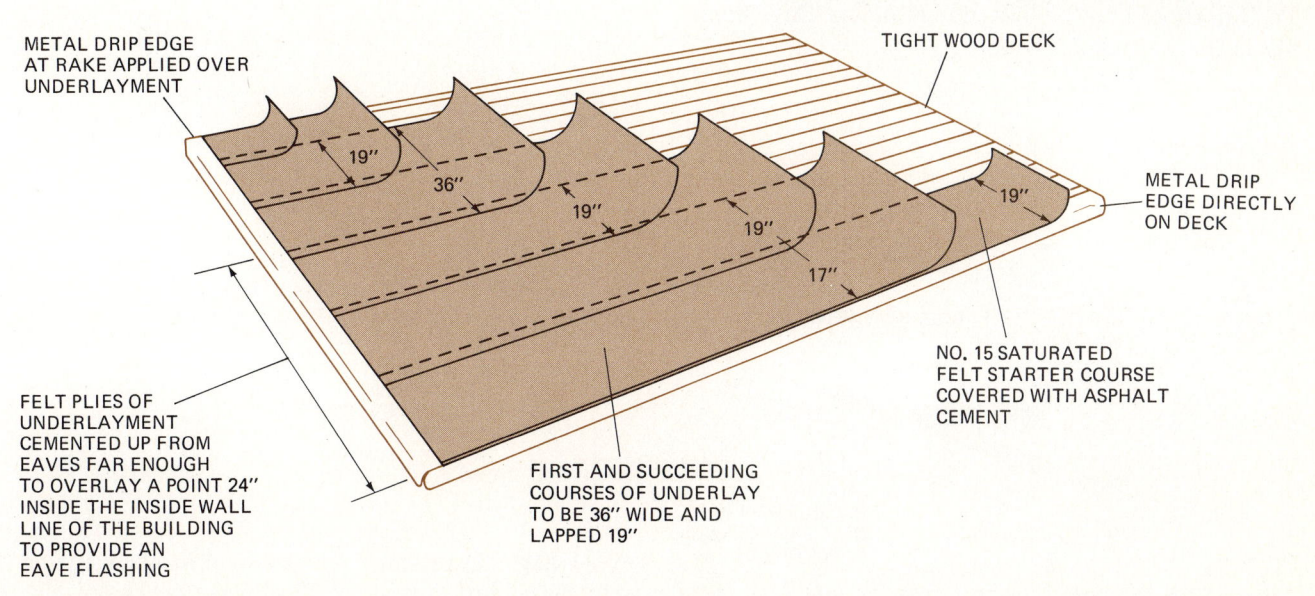

Covering Roofs **181**

dards are set according to grade and thickness. Design the eaves, rake, and ridge to prevent problems. Openings through the deck should be made in such a way that the plywood will not be exposed to the weather. See Figure 8-22.

NONWOOD DECK MATERIALS • Nonwood materials are sometimes used in decks. Such things as fiberboard, gypsum, concrete plank, and tile are nonwoods used for decks. These materials have their own standards. Check with the manufacturer for suggestions.

FLASHINGS • Roofs are often complicated by intersections with other roofs. Some adjoining walls have projections through the deck. Chimneys and soil stacks create leakage problems. Special attention must be given to protecting against the weather here. Such precautions are commonly called *flashing*. Careful attention to flashing is critical. It helps provide good roof performance. See Figure 8-23.

VALLEYS • Valleys exist where two sloping roofs meet at an angle. This causes water runoff toward and along the joint. Drainage concentrates at the joint. This makes the joint an easy place for water to enter. Smooth, unobstructed drainage must be provided. It should have enough capacity to carry away the collected water.

There are three types of valleys. They are called open, woven, and closed-cut. See Figure 8-24.

Each type of valley calls for its own treatment. Figure 8-25 shows felt being applied to a valley. A 36-inch-wide strip of no. 15 asphalt-saturated felt is centered in the valley. It is secured with nails. They hold it in place until shingles are applied. Courses of felt are cut to overlay the valley strip. The lap should not be less than 6 inches. Eave flashing is then applied.

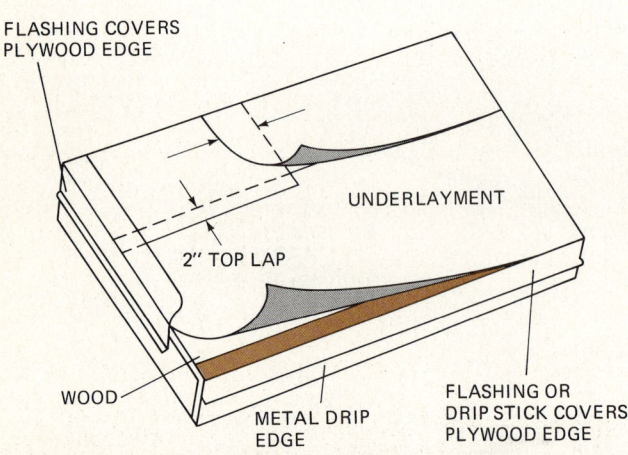

Figure 8-22. Preparing the roof deck for shingling. *(Bird and Son)*

• PUTTING DOWN SHINGLES •

Don't put down the first shingle until you know what is available. Tables 8-3 and 8-4 show typical asphalt shingles. Before you put the shingles down, you need an underlayment. This is shown in Table 8-4. Remember the # symbol means pound or lb.

Nails

Nails used in applying asphalt roofings are large-headed. They are sharp-pointed. Some are hot-galvanized steel. Others are made of aluminum. They may

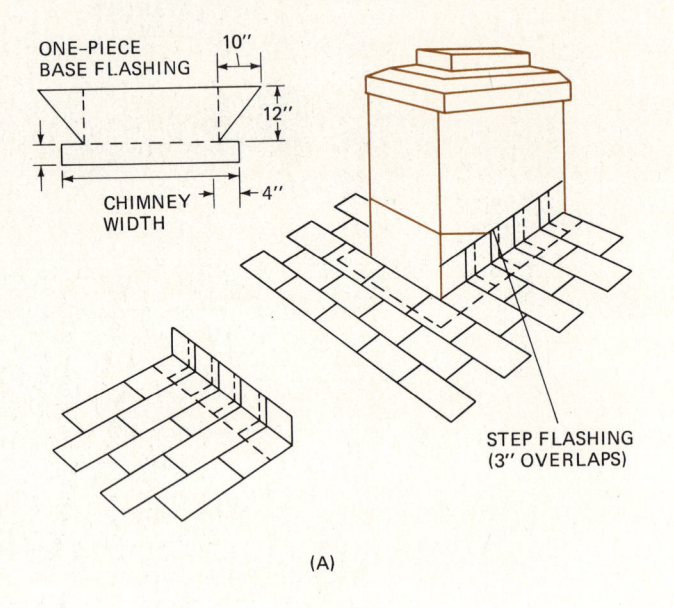

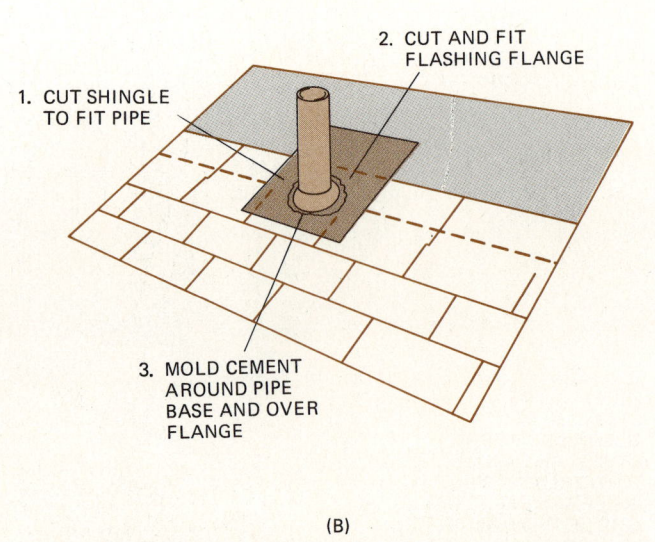

Figure 8-23. (A) Flashing patterns and in place around a chimney. *(Bird and Son)* (B) Flashing around a soil pipe. *(Bird and Son)*

182 Carpentry Fundamentals

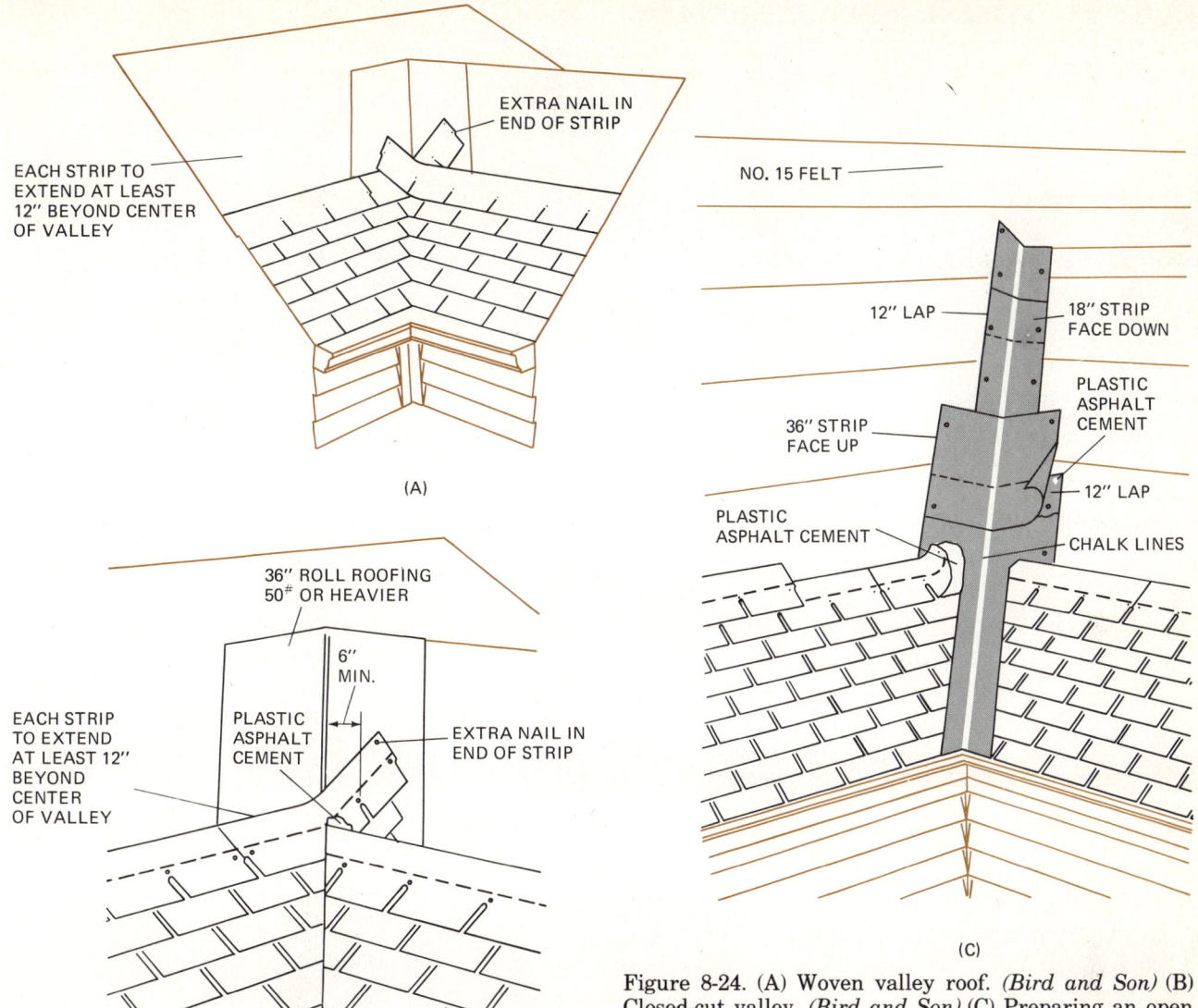

Figure 8-24. (A) Woven valley roof. *(Bird and Son)* (B) Closed-cut valley. *(Bird and Son)* (C) Preparing an open valley. *(Bird and Son)*

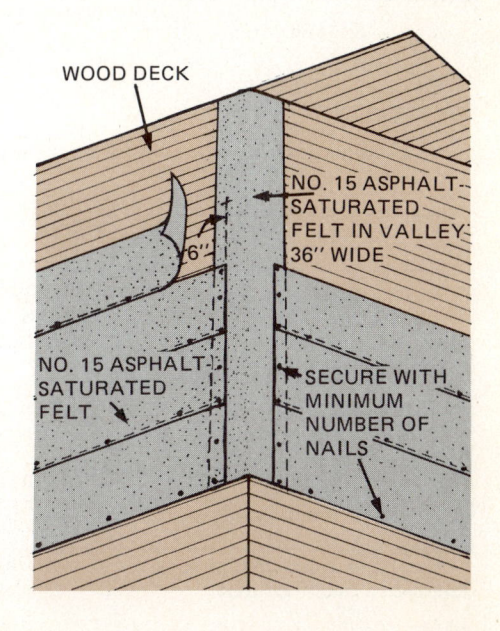

Figure 8-25. Felt underlay centered in the valley before valley linings are applied. *(Bird and Son)*

Covering Roofs 183

TABLE 8-3 TYPICAL ASPHALT SHINGLES.

1	2	3 Per Square			4 Size		5	6
Product*	Configuration	Approximate Shipping Weight	Shingles	Bundles	Width, Inches	Length, Inches	Exposure, Inches	Underwriters' Listing
Wood appearance strip shingle more than one thickness per strip 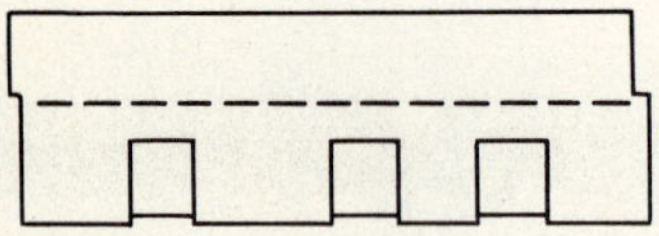 — Laminated or job-applied	Various edge, surface texture, and application treatments	285# to 390#	67 to 90	4 or 5	11½ to 15	36 or 40	4 to 6	A or C, many wind-resistant
Wood appearance strip shingle single thickness per strip 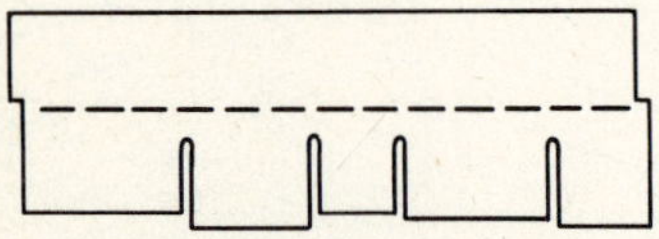	Various edge, surface texture, and application treatments	Various, 250# to 350#	78 to 90	3 or 4	12 or 12¼	36 or 40	4 to 5⅛	A or C, many wind-resistant
Self-sealing strip shingle 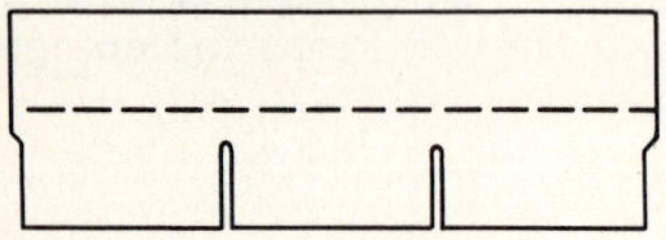	Conventional three-tab	205# to 240#	78 or 80	3	12 or 12¼	36	5 or 5⅛	A or C, all wind-resistant
	Two- or four-tab	Various, 215# to 325#	78 or 80	3 or 4	12 or 12¼	36	5 or 5⅛	
Self-sealing strip shingle  — No cut-out	Various edge and texture treatments	Various, 215# to 290#	78 to 81	3 or 4	12 or 12¼	36 or 36¼	5	A or C, all wind-resistant
Individual lock-down 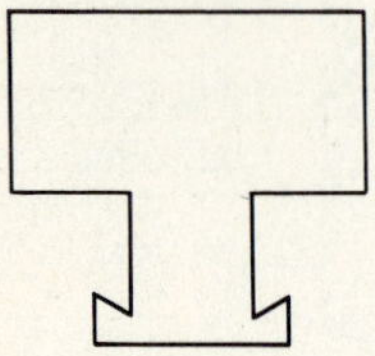 — Basic design	Several design variations	180# to 250#	72 to 120	3 or 4	18 to 22¼	20 to 22½		C, many wind-resistant

* Other types available from some manufacturers in certain areas of the country. Consult your regional Asphalt Roofing Manufacturers Association manufacturer.

TABLE 8-4 TYPICAL ASPHALT ROLLS.

1	2		3	4		5		6	7
Product	**Approximate Shipping Weight**		Squares Per Package	Length, Feet	Width, Inches	Side or End Lap, Inches	Top Lap, Inches	Exposure, Inches	Underwriters' Listing
	Per Roll	Per Square							
Mineral surface roll double coverage	75# to 90#	75# to 90#	1	36 38	36 38	6	2 4	34 32	C
	Available in some areas in 9/10 or 3/4 square rolls.								
Mineral surface roll	55# to 70#	55# to 70#	½	36	36	6	19	17	C
Coated roll	50# to 65#	50# to 65#	1	36	36	6	2	34	None
Saturated felt	60# 60# 60#	15# 20# 30#	4 3 2	144 108 72	36 36 36	4 to 6	2	34	None

be barbed or otherwise deformed on the shanks. Figure 8-26 shows three types of asphalt nails.

Roofing nails should be long enough to penetrate through the roofing material. They should go at least ¾ inch into the wood deck. This requires that they be of the lengths indicated in Table 8-5.

NUMBER OF NAILS • The number of nails required for each shingle type is given by its maker. Manufacturer's recommendations come with each bundle.

Use 2-inch centers in applying roll roofing. This means 252 nails are needed per square. If 3-inch centers are used, then 168 nails are needed per square.

Fasteners for Nonwood Materials

Gypsum products, concrete plank and tile, fiberboard, or unusual materials require special fasten-

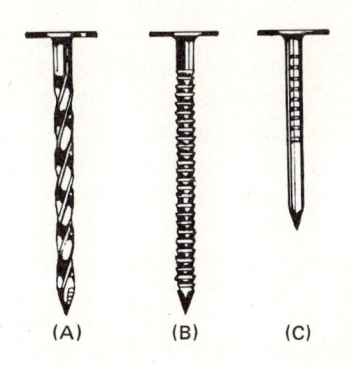

Figure 8-26. (A) Screw-threaded nail. (B) Annular threaded nail. (C) Asphalt shingle nail—smooth.

TABLE 8-5 RECOMMENDED NAIL LENGTH.

Purpose	Nail Length, Inches
Roll roofing on new deck	1
Strip or individual shingles—new deck	1¼
Reroofing over old asphalt roofing	1¼ to 1½
Reroofing over old wood shingles	1¾

Covering Roofs

ers. This type of deck varies with its manufacturer. In such cases follow the manufacturer's suggestions.

Shingle Selection

There are a number of types of shingles available. They may be used for almost any type of roof. Various colors are used to harmonize with buildings. Some are made for various weather conditions. White and light colors are used to reflect the sun's rays. Pastel shingles are used to achieve a high degree of reflectivity. They still permit color blending with siding and trim. Fire and wind resistance should be considered. Simplicity of application makes asphalt roofings the most popular in new housing. They are also rated high for reroofing.

FARM BUILDINGS • There is no one kind of asphalt roofing for every job. Building types are numerous. The style of roof on the farm house may affect the choice of roof on other buildings. They probably should have the same color roofing. This would make the group harmonize. A poultry laying house or machine storage shed near the house calls for a roof like the farm house. An inexpensive roll roofing might be used for an isolated building.

The main idea is to select the right product for the building. The main reason for selecting a roofing material is protection of the contents of a building. The second reason for selecting a roofing is low maintenance cost.

STAPLES • Staples may be used as an alternative to nails. This is only for new buildings. Staples must be zinc-coated. They should be no less than 16 gauge. A semiflattened elliptical cross section is preferred. They should be long enough to penetrate ¾ inch into the wood deck. They must be driven with *pneumatic* (air-driven) staplers. The staple crown must bear tightly against the shingle. However, it must not cut the shingle surface. Use four staples per shingle. See Figure 8-27. The crown of the staple must be parallel to the tab edge. Position it as shown in Figure 8-27. Figure 8-28 shows how shingles are nailed. Figure 8-29 shows how the shingles are overlapped.

Cements

Six types of asphalt coatings and cements are:

1. Plastic asphalt cements
2. Lap cements
3. Quick-setting asphalt adhesives
4. Asphalt water emulsions
5. Roof coatings
6. Asphalt primers

METHODS OF SOFTENING • The materials are flammable. Cement should be applied to a dry, clean surface. It should be troweled or brushed vigorously. This removes air bubbles. The material should flow freely. It should be forced into all cracks and openings. An emulsion may be applied to damp or wet surfaces. It should not be applied in an exposed location. It should not be rained on for at least 24 hours. Emulsions are water soluble.

USES • Plastic asphalt cements are used for flashing cements. They are so processed that they will not flow at summer temperatures. They are elastic after setting. This compensates for normal expansion and contraction of a roof deck. They will not become brittle at low temperatures.

Lap cements come in various thicknesses. Follow the manufacturer's suggestions. Lap cement is not as thick as plastic cement. It is used to make a watertight bond. The bond is between lapping elements of roll roofing. It should be spread over the entire

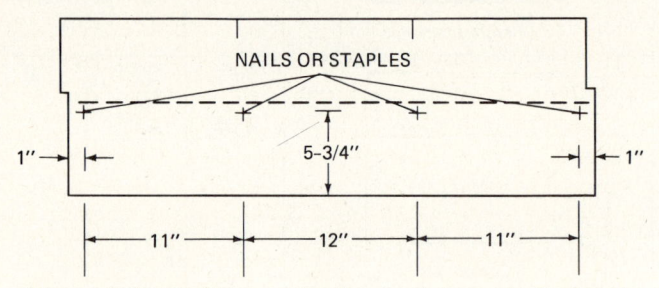

Figure 8-27. Nailing or stapling a strip asphalt shingle. *(Certain-Teed)*

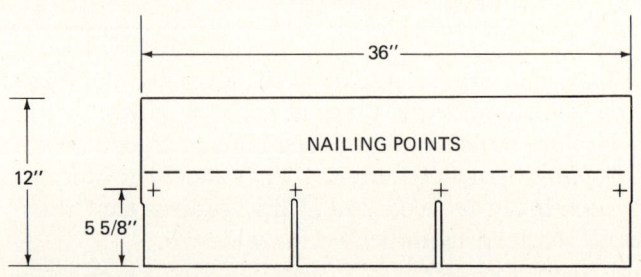

Figure 8-28. Nailing points on a strip shingle. *(Bird and Son)*

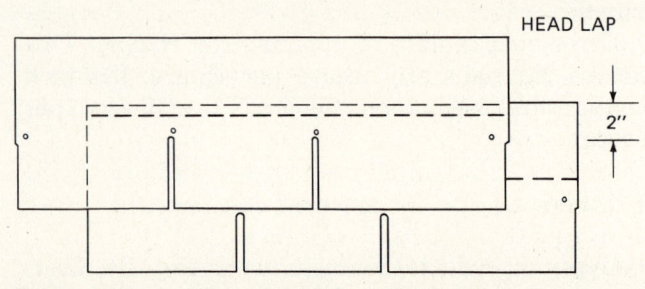

Figure 8-29. Overlap of shingles. *(Bird and Son)*

186 Carpentry Fundamentals

lapped area. Nails used to secure the roofing should pass through the cement. The shank of the nail should be sealed where it penetrates the deck material.

Seal down the free tabs of strip shingles with quick-setting asphalt adhesive. It can also be used for sealing laps of roll roofing.

Quick-setting asphalt adhesive is about the same thickness as plastic-asphalt cement. However, it is very adhesive. It is mixed with a solvent. The solvent evaporates quickly. This permits the cement to set up rapidly.

Roof coatings are used in spray or brush thickness. They are used to coat the entire roof. They can be used to resurface old built-up roofs. Old roll roofing or metal roofs can also be coated.

Asphalt water emulsions are a special type of roof coating. They are made with asphalt. Sometimes they are mixed with other materials. They are emulsified with water. This means they can freeze. Store them in a warm location. They should not be rained on for at least 24 hours.

Masonry primer is very fluid. Apply it with a brush or by spray. It must be thin enough to penetrate rapidly into the surface pores of masonry. It should not leave a continuous surface film. Thin, if necessary, by following instructions on the can.

Asphalt primer is used to prepare the masonry surface. It should make a good bond with other asphalt products. These are found on built-up roofs. Other products are plastic-asphalt cement or asphalt coatings.

As a carpenter you will be using these products.

Starter Course

Putting down the shingles isn't too hard—that is, if you have the roof deck in place. It should be covered by the proper underlayment. The eaves should be properly prepared. See Figure 8-30.

STARTING AT THE RAKE • Use only the upper portion of the asphalt shingle. Cut off the tabs. Position it with the adhesive dots toward the eaves. The starter course should overhang the eaves and rake edges by ¼ inch. Nail it in a line 3 to 4 inches above the eaves. See Figure 8-31.

Start the first course with a full strip. Overhang the drip edges at the eaves and rake by ¼ inch. Nail the strip in place. Drive the nails straight. The heads should be flush with the surface of the shingle.

Snap a chalk line along the top edge of the shingle. The line should be parallel with the eaves. Snap several others parallel with the first. Make them 10 inches apart. Use the lines to check alignment at every other course. Snap lines parallel with the rake at the shingle cutouts. Use the lines to check cutout alignment.

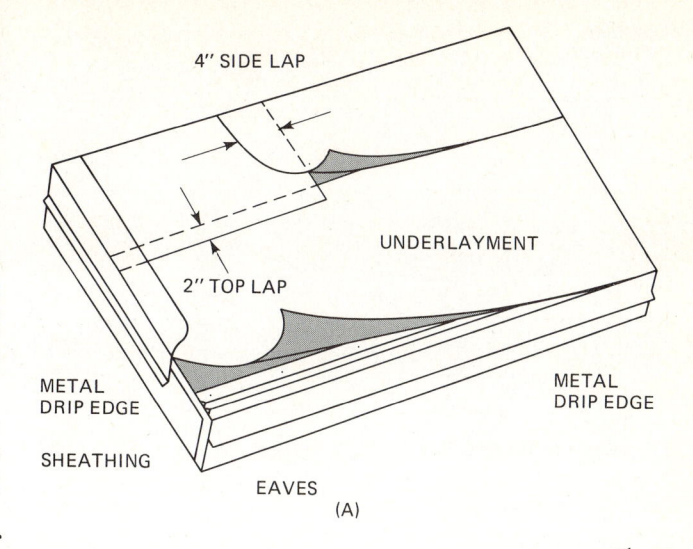

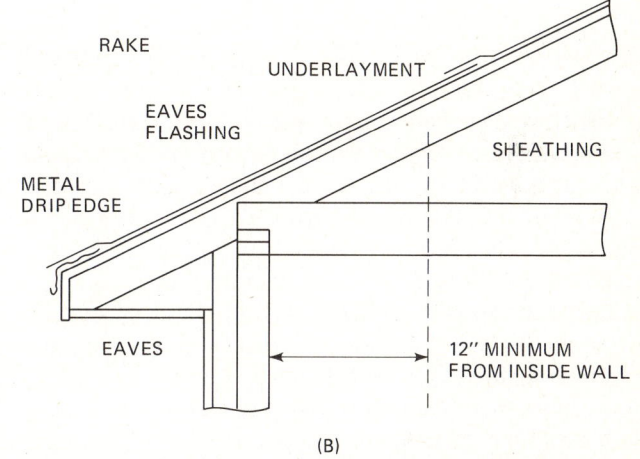

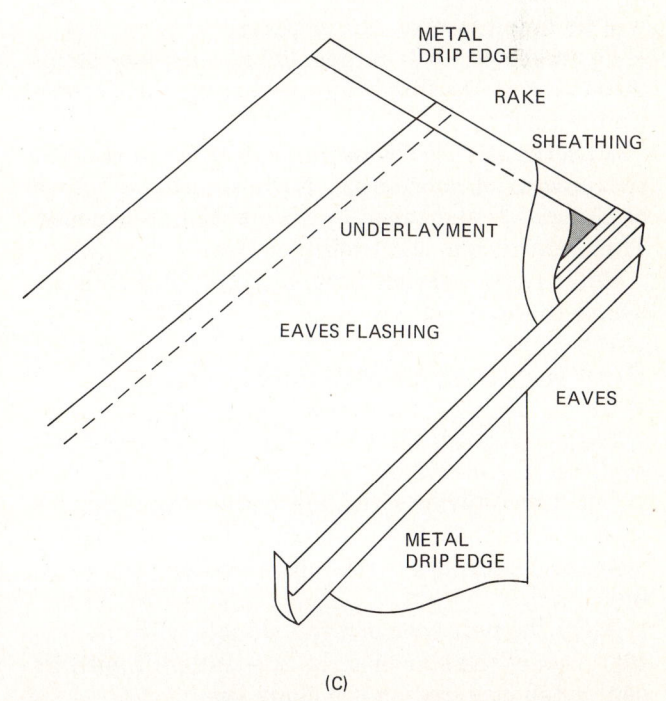

Figure 8-30. (A) Underlayment and drip edge. *(Bird and Son)* (B) Eaves flashing for a roof. *(Bird and Son)* (C) Underlayment, eaves flashing, and metal drip edge. *(Bird and Son)*

Covering Roofs **187**

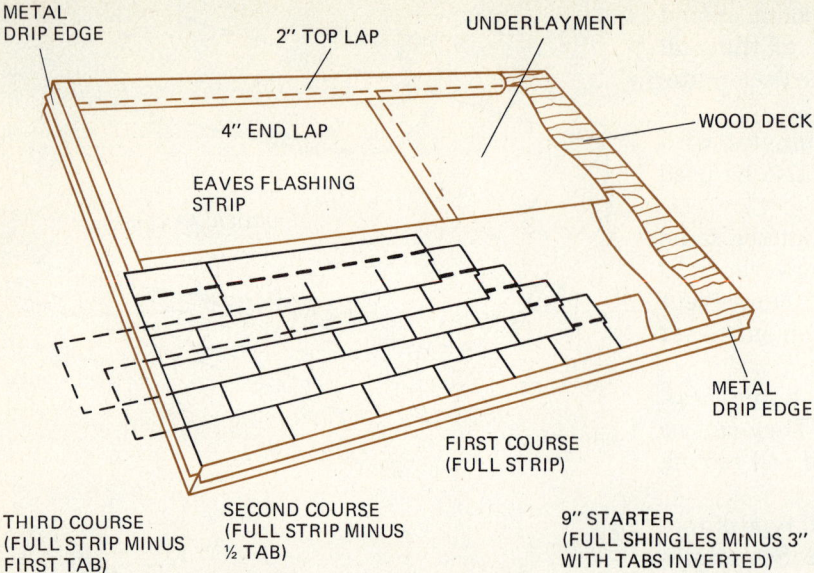

Figure 8-31. Starting asphalt shingles at the rake. *(Bird and Son)*

Start the second course with a full strip less 6 inches. This means half a tab is missing. Overhang the cut edge at the rake. Nail the shingle in place.

Start the third course with a full strip less a full tab.

Start the fourth course with half a strip.

Continue to reduce the length of the first shingle in each course by an additional 6 inches. The sixth course starts with a 6-inch strip.

Return to the eaves. Apply full shingles across the roof, finishing each course. Dormer, chimney, or vent pipe instructions are another matter. They will be found later in this chapter.

For best color distribution, lay at least four strips in each row. Do this before repeating the pattern up the roof.

Start the *seventh course* with a full shingle. Repeat the process of shortening. Each successive course of shingles is shortened by an additional 6 inches. This continues to the twelfth course.

Return to the seventh course. Apply full shingles across the roof.

Starting at the Center (Hip Roof)

Snap a vertical chalk line at the center of the roof. See Figure 8-32.

Put down starter strips along the eaves. Do this in each direction from the chalk line. Go slightly over the centerlines of the hips. Overhang the eaves by ¼ inch.

Align the butt edge of a full shingle with the bottom edge of the starter strip. Also, align it with its center tab centered on the chalk line.

Snap a chalk line along the top of the shingle parallel with the eaves. Snap several others parallel with the first. They should be 10 inches apart. Use the lines to check the alignment of alternate courses.

Finish the first course with full shingles. Extend the shingles part way over the hips.

Finish the remaining courses with full shingles.

Valleys

There are three types of valleys. One is the open type. Here, the saturated felt can be seen after the shingles are applied. Another type is the woven valley. This one has the shingles woven. There is no obvious valley line. The other is the closed-cut valley. This one has a straight line where the roofs intersect.

OPEN VALLEYS • Use mineral-surfaced-material roll roofing for this valley. Match or contrast the color with that of the roof covering. The open valley method is shown in Figure 8-33. The felt underlay is centered in the valley before shingles are applied. See Figure 8-34.

Center an 18-inch-wide layer of mineral-surfaced roll roofing in the valley. The surfaced side goes down. Cut the lower edge to conform to and be flush with the eave flashing strip. The ends of the upper segments overlap the lower segments in a splice. The ends are secured with plastic asphalt cement. See Figure 8-33. Use only enough nails, 1 inch in from each edge, to hold the strip smoothly. Press the roofing firmly in place into the valley as you nail. Place another strip 36 inches wide on top of the first strip. This is placed surfaced side up. Center it in the valley. Nail it in the same manner as the underlying 18-inch strip.

Do this before the roofing is applied. Snap two chalk lines the full length along the valley, one line on each side of the valley. They should be 6 inches

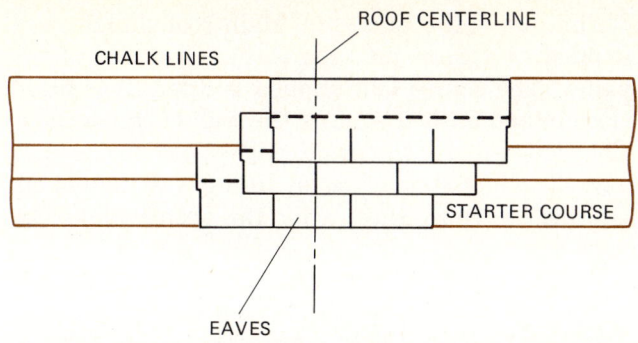

Figure 8-32. Starting asphalt shingles at the center. (Bird and Son)

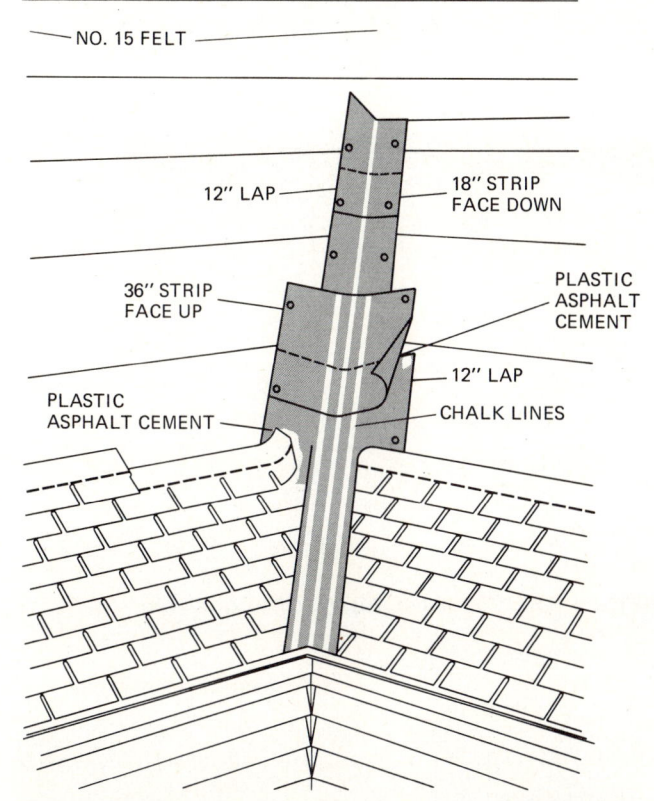

Figure 8-33. Use of roll roofing for open valley flashing. (Bird and Son)

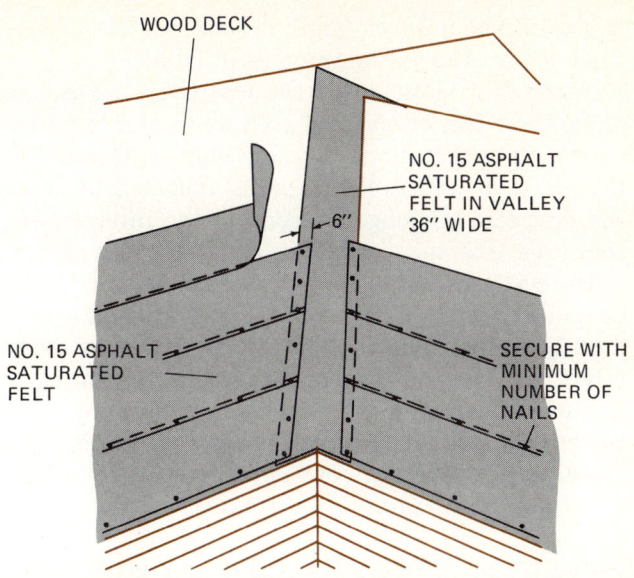

Figure 8-34. Felt underlay centered in the valley before valley linings are applied. (Bird and Son)

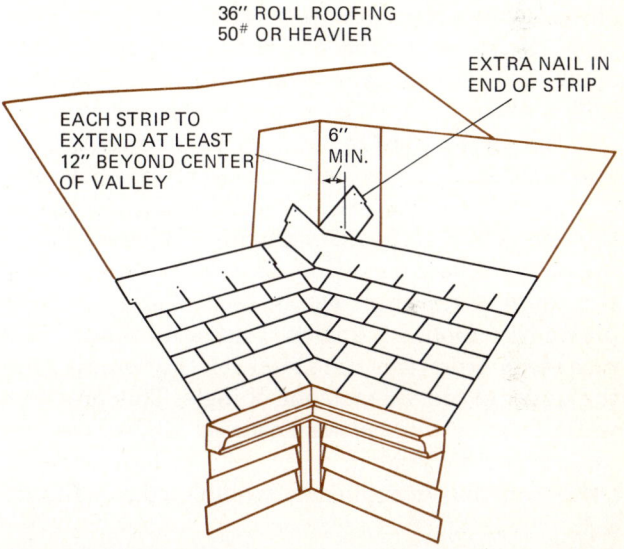

Figure 8-35. Weaving each course in turn to make a woven valley. (Bird and Son)

apart at the ridge. This means 3 inches when measured from the center of the valley. The marks diverge at the rate of 1/8 inch per foot as they approach the eaves. A valley of 8 ft in length will be 7 inches wide at the eaves. One 16 feet long will be 8 inches wide at the eaves. The chalk line serves as a guide in trimming the last unit to fit the valley. This ensures a clean, sharp edge. The upper corner of each end shingle is clipped. See Figure 8-33. This keeps water from getting in between the courses. The roofing material is cemented to the valley lining. Use plastic asphalt cement.

WOVEN AND CLOSED-CUT VALLEYS • Some roofers prefer woven or closed-cut valleys. These are limited to strip-type shingles. Individual shingles cannot be used. Nails may be required at or near the center of the valley lining. Avoid placing a nail in an overlapped shingle too close to the center of the valley. It may sometimes be necessary to cut a strip short. That is done if it would otherwise end near the center. Continue from this cut end over the valley with a full strip. These methods increase the coverage of the shingles throughout the length of the valleys. This adds to the weather resistance of the roofs at these points.

WOVEN VALLEYS • There are two methods of weaving the shingles. See Figure 8-35. They can be applied on both roof areas at the same time. This means you weave each course, in turn, over the valley. Or, you may cover each roof area first. Do this

Covering Roofs 189

to a point about 3 feet from the center of the valley. Then weave the valley shingles in later.

In the first method, lay the first course. Place it along the eaves of one roof area up to and over the valley. Extend it along the adjoining roof area. Do this for a distance of at least 12 inches. Then lay the first course along the eaves of the intersecting roof area. Extend it over the valley. It goes on top of the previously applied shingle. The next courses go on alternately. Lay along one roof area and then along the other. Weave the valley shingles over each other. See Figure 8-35. Make sure that the shingles are pressed tightly into the valley. Nail them in the normal manner. No nails are located closer than 6 inches to the valley centerline. Two nails are located at the end of each terminal strip. See Figure 8-35.

CLOSED-CUT VALLEYS • For a closed-cut valley, lay the first course of shingles along the eaves of one roof area up to and over the valley. Extend it along the adjoining roof section. The distance is at least 12 inches. Follow the same procedure when applying the next courses of shingles. See Figure 8-36. Make sure that the shingles are pressed tightly into the valley. Nail in the normal manner, except that no nail is to be located closer than 6 inches to the valley centerline. Two nails are located at the end of each terminal strip. See Figure 8-37.

Apply the first course of shingles. Do this along the eaves of the intersecting roofs. Extend it over previously applied shingles. Trim a minimum of 2 inches up from the centerline of the valley. Clip the upper corner of each end shingle. This prevents water from getting between courses. Embed the end in a 3-inch-wide strip of plastic asphalt cement. Other courses are applied and completed. See Figure 8-37.

AN OPEN VALLEY FOR A DORMER ROOF • A special treatment is needed where an open valley occurs at a joint between the dormer roof and the main roof through which it projects. See Figure 8-38.

Figure 8-36. Worker installing shingles. Note that the first strip is at the bottom. Also note the felt strip in the valley. *(Fox and Jacobs)*

First apply the underlay. Main roof shingles are applied to a point just above the lower end of the valley. The course last applied is fitted. It is fitted close to and flashed against the wall of the dormer. The wall is under the projecting edge of the dormer eave. The first strip of valley lining is then applied. Do this the same way as for the open valley. The

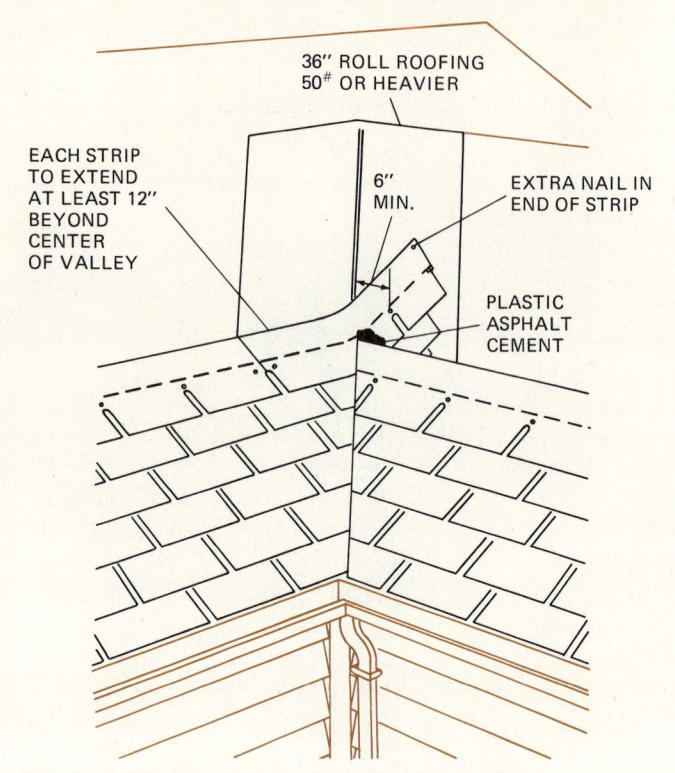

Figure 8-37. Closed-cut valley. *(Bird and Son)*

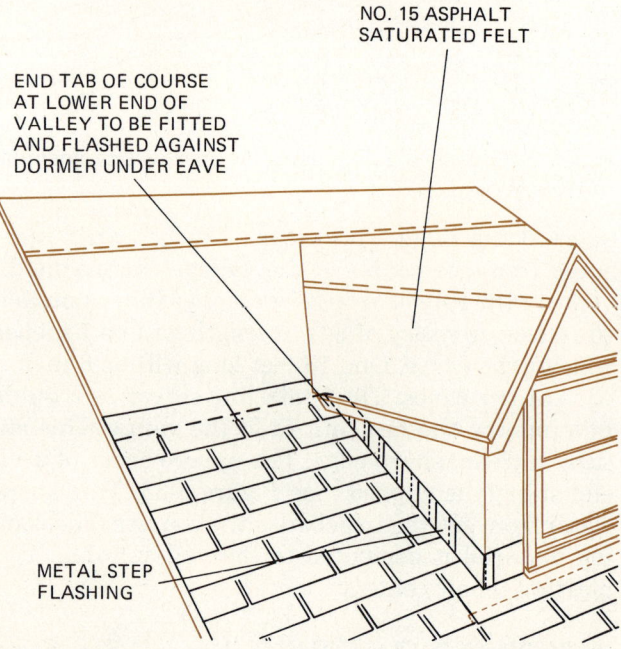

Figure 8-38. An open valley for a dormer roof. Shingles have been laid on main roof up to lower end of the valley. *(Bird and Son)*

190 Carpentry Fundamentals

bottom end is cut so that it extends ¼ inch below the edge of the dormer deck. The lower edge of the section lies on the main deck. It projects at least 2 inches below the joining roofs. Cut the second or upper strip on the dormer side. It should match the lower end of the underlying strip. Cut the side that lies on the main deck. It should overlap the nearest course of shingles. This overlap is the same as the normal lap of one shingle over another. It depends on the type of shingle being applied. In this case it extends to the top of the cutouts. This is a 12-inch-wide three-tab square-butt strip shingle.

The lower end of the lining is then shaped. See Figure 8-39. It forms a small canopy over the joint between the two decks.

Apply shingles over the valley lining. The end shingle in each course is cut. It should conform to the guidelines. Bed the ends in a 3-inch-wide strip of plastic asphalt cement. Valley construction is completed in the usual manner. See Figure 8-40.

Flashing against a Vertical Wall

Step flashing is used when the rake of a roof abuts a vertical wall. It is best to protect the joint by using metal flashing shingles. They are applied over the end of each course of shingles.

The flashing shingles are rectangular in shape. They are from 5 to 6 inches long. They are 2 inches wider than the exposed face of the roofing shingles. When used with strip shingles laid 5 inches to the weather, they are 6 to 7 inches long. They are bent so as to extend 2 inches out over the roof deck. The remainder goes up the wall surface. Each flashing shingle is placed just uproof from the exposed edge of the single which overlaps it. It is secured to the wall sheathing with one nail in the top corner. See Figure 8-41. The metal is 7 inches wide. The roof shingles are laid 5 inches to the weather. Each ele-

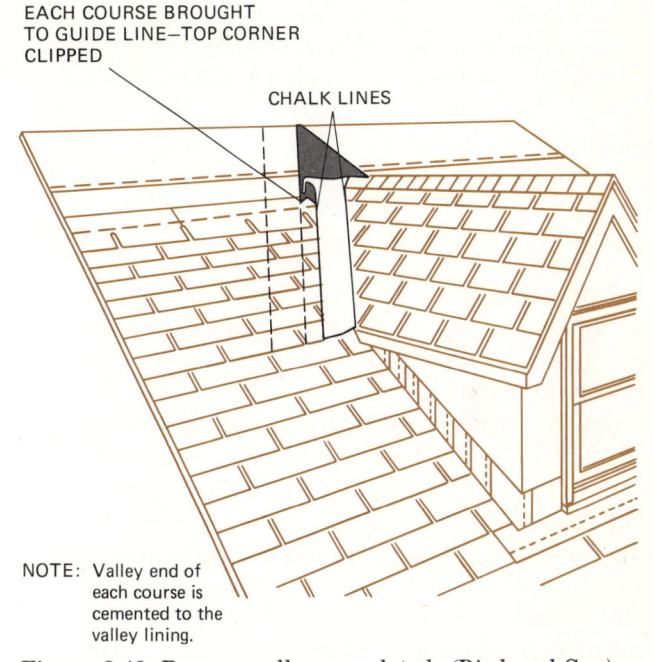

Figure 8-40. Dormer valley completed. *(Bird and Son)*

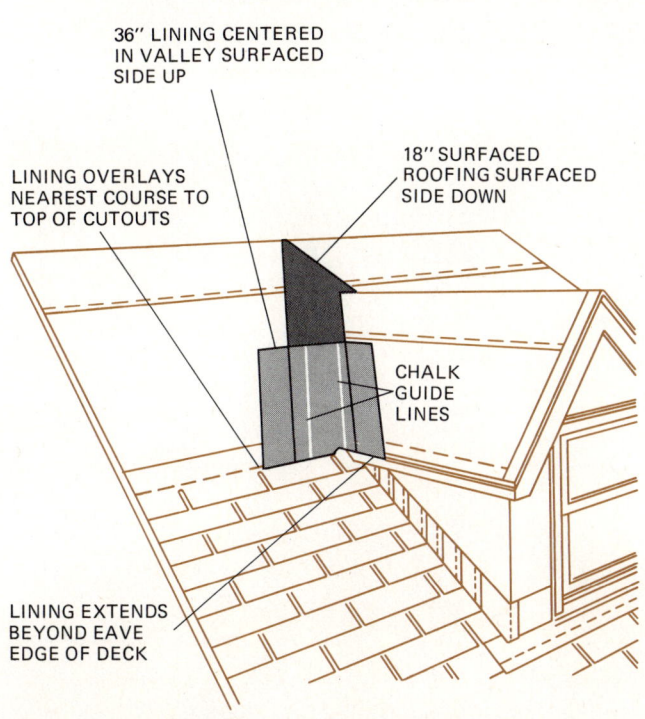

Figure 8-39. Application of valley lining for a dormer roof. *(Bird and Son)*

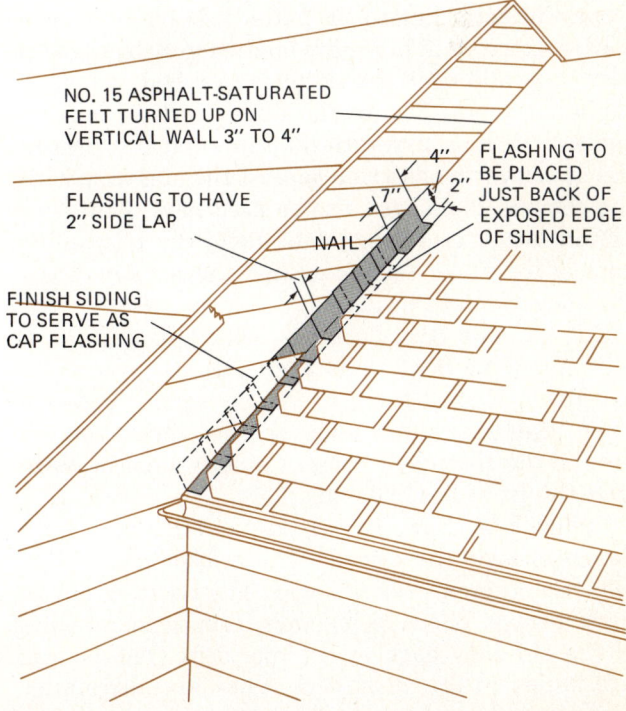

Figure 8-41. Use of metal flashing shingles to protect the joint between a sloping roof and a vertical wall. *(Bird and Son)*

Covering Roofs

ment of flashing will lap the next by 2 inches. See Figure 8-41.

The finished siding is brought down over the flashing. It serves as a cap flashing. However, it is held far enough away from the shingles. This allows the ends of the boards to be painted. Paint excludes dampness and prevents rot.

Chimneys

Chimneys are usually built on a separate foundation. This avoids stresses and distortions due to uneven settling. It is subject to differential settling. Flashing at the point where the chimney comes through the roof calls for something that will allow movement without damage to the water seal. It is necessary to use base flashings. They should be secured to the roof deck.

The counter or cap flashings are secured to the masonry. Figures 8-42 through 8-46 show how roll roofing is used for base flashing. Metal is used for cap flashing.

Apply shingles over the roofing felt up to the front face of the chimney. Do this before any flashing are placed. Make a saddle or cricket. See Figure 8-42. This sits between the back face of the chimney and the roof deck. The cricket keeps snow and ice from piling up. It also deflects downflowing water around the chimney.

Apply a coat of asphalt primer to the brickwork. This seals the surface. This is where plastic cement will later be applied. Cut the base flashing for the front. Cut according to the pattern shown in Figure 8-43 (pattern A). This one is applied first. The lower section is laid over the shingles in a bed of plastic asphalt cement. Secure the upper vertical section against the masonry with the same cement. Secure the upper vertical section against the masonry with the cement. Nails can also be used here. Nails are driven into the mortar joints. Bend the triangular ends of the upper section around the corners of the chimney. Cement in place.

Cut the side base flashings next. Use pattern B of Figure 8-43. Bend them to shape and apply them as shown. Embed them in plastic asphalt cement. Turn the triangular ends of the upper section around the chimney corners. Cement them in place over the front base flashing.

Figure 8-44 shows the cutting and fitting of base flashings over the cricket. The cricket consists of two triangular pieces of board. The board is cut to form a ridge. The ridge extends from the centerline of the chimney back to the roof deck. The boards are nailed to the wood deck. They are also nailed to one another along the ridge. This is done before felt underlayment is applied. Cut the base flashing. See Figure 8-45A. Bend it to cover the entire cricket. Extend it laterally to cover part of the side base

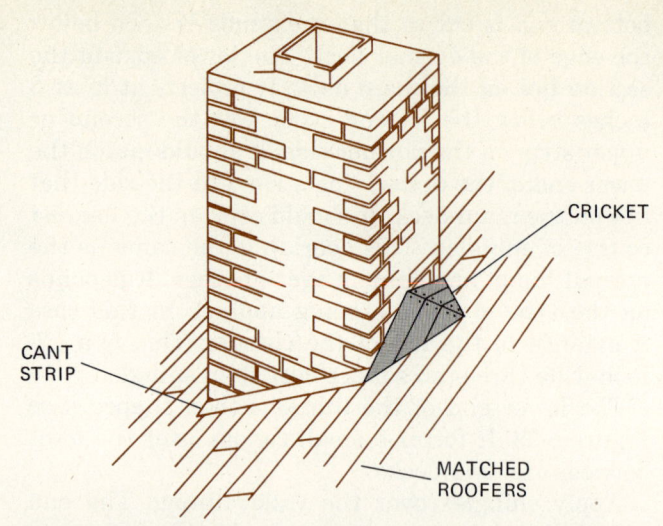

Figure 8-42. Cricket or saddle built behind the chimney. *(Bird and Son)*

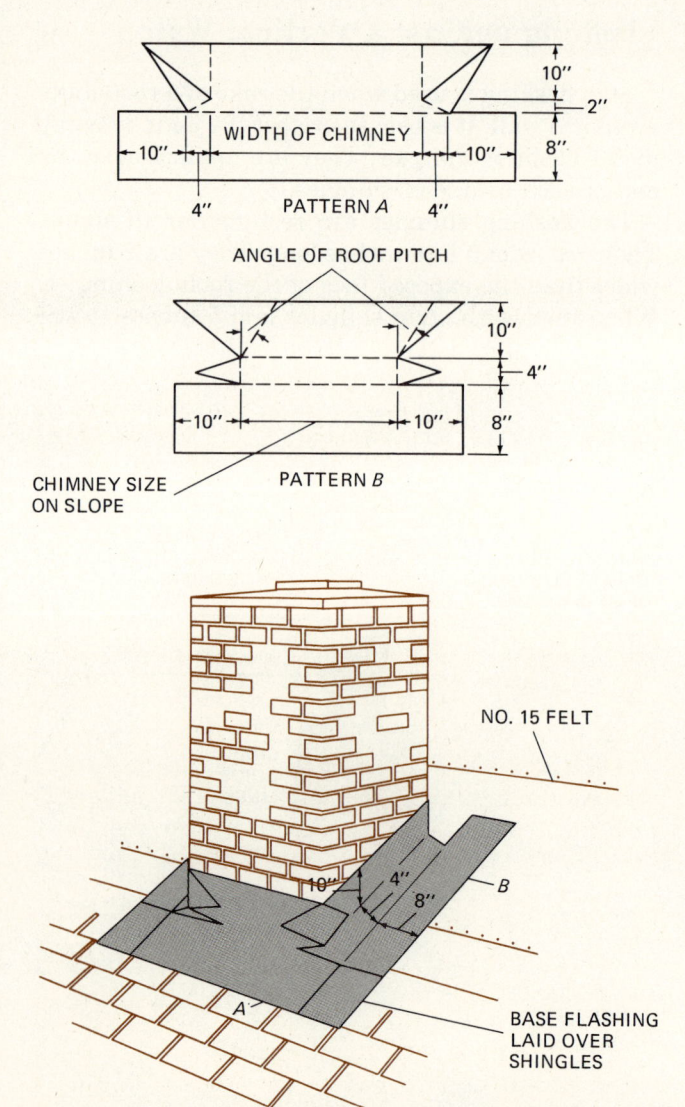

Figure 8-43. Base flashings cut and applied. *(Bird and Son)*

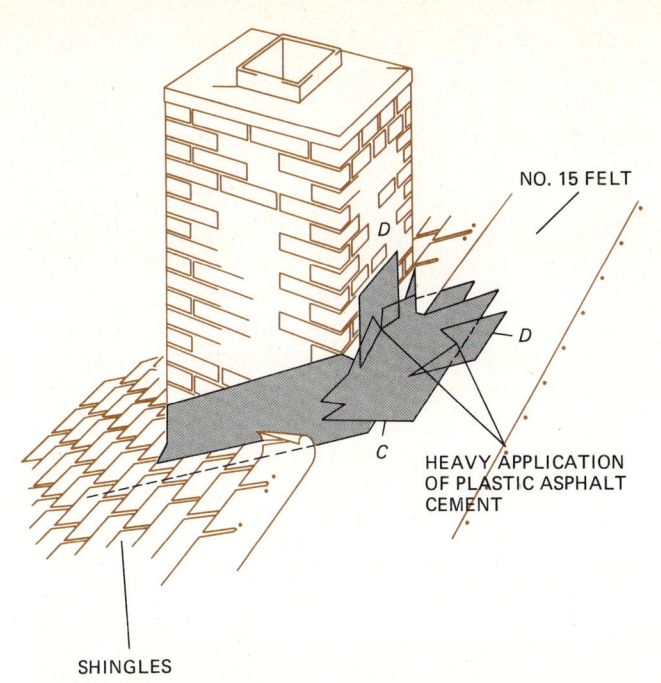

Figure 8-44. Flashing over the cricket in the rear of the chimney. *(Bird and Son)*

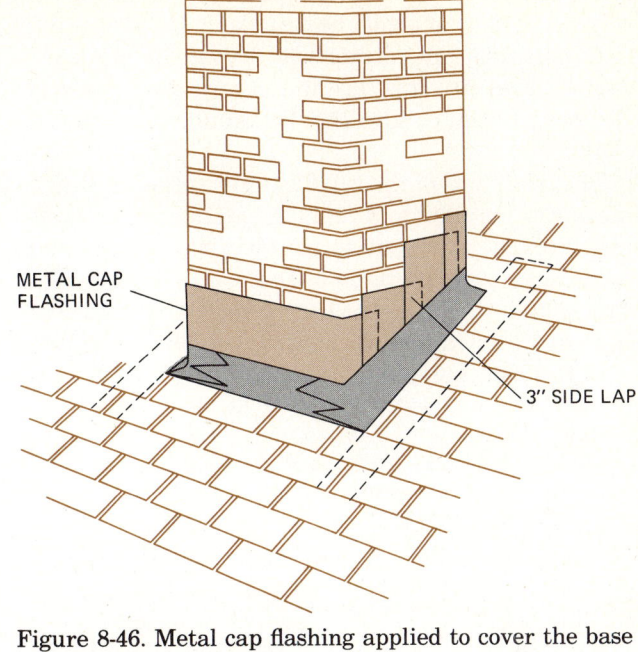

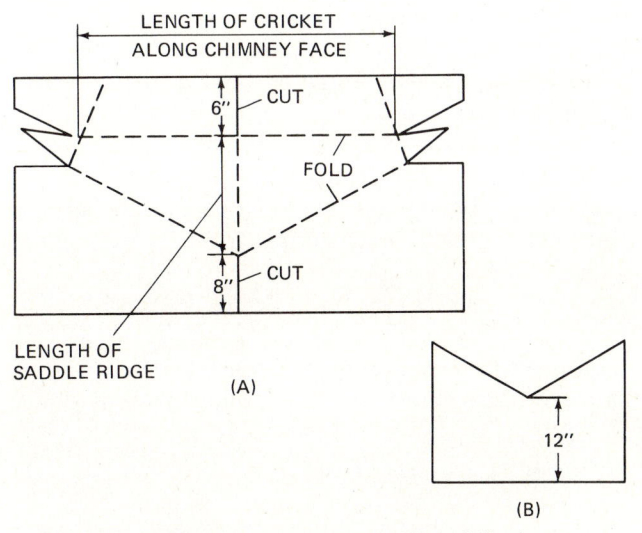

Figure 8-45. Flashing patterns. *(Bird and Son)*

Figure 8-46. Metal cap flashing applied to cover the base flashing. *(Bird and Son)*

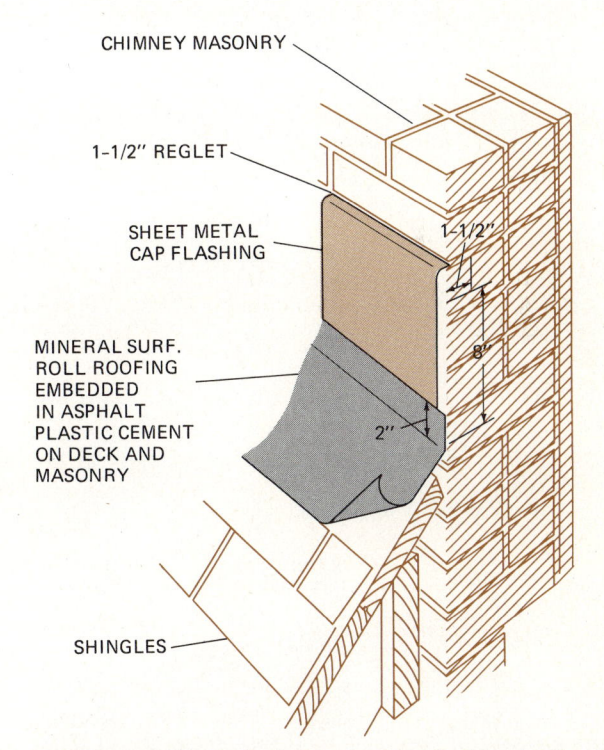

Figure 8-47. Method of securing cap flashing to the chimney. *(Bird and Son)*

flashing. Cut a second rectangular piece of roofing. See Figure 8-45B. Make a cutout on one side to conform to the rear angle of the cricket. Set it tightly in plastic asphalt cement. Center it over that part of the cricket flashing extending up to the deck. This piece provides added protection where the ridge of the cricket meets the deck. Cut a second similar rectangular piece of flashing. Cut a V from one side. It should conform to the pitch of the cricket. Place it over the cricket ridge and against the chimney. Embed it in plastic asphalt cement.

Use plastic asphalt cement generously. Use it to cement all standing portions of the base flashing to the brickwork.

Cap flashings are shown in Figure 8-46. They are made of sheet copper. It should be 16-ounce or heavier copper. You can also make the caps of 24-gage galvanized steel. If steel is used, it should be painted on both sides.

Brickwork is secured to the cap flashing in Figures 8-46 and 8-47. These drawings show a good method. Rake the mortar joint to a depth of 1½ inches. Insert the bent back edge of the flashing into the cleared

Covering Roofs **193**

space between the bricks. It is under slight spring tension. It cannot easily be dislodged. Refill the joint with portland cement mortar. Or you can use plastic asphalt cement. This flashing is bent down to cover the base flashing. The cap lies snugly against the masonry.

The front unit of the cap flashing is one continuous piece. On the sides and rear, the sections are of similar size. They are cut to conform to the locations of brick joints. The pitch of the roof is also needed. The side units lap each other. See Figure 8-46. This lap is at least 3 inches. Figure 8-48 shows another way to flash a sloping roof abutting a vertical masonry wall. This is known as the *step flashing* method. Place a rectangular piece of material measuring 8×22 inches over the end tab of each course of shingles. Hold the lower edge slightly back of the exposed edge of the covering shingle. Bend it up against the masonry. Secure it with suitable plastic asphalt cement. Drive nails through the lower edge of the flashing into the roof deck. Cover the nails with plastic asphalt cement. Repeat the operation for each course. Flashing units should be wide enough to lap each other at least 3 inches. The upper one overlays the lower one each time.

Asphalt roofing can be used for step flashing. It simply replaces the base flashing already shown. Metal cap flashings must be applied in the usual manner. The metal cap completes a satisfactory job.

Soil Stacks

Most building roofs have pipes or ventilators through them. Most are circular in section. They call for special flashing methods. Asphalt products may be successfully used for this purpose. Figures 8-49 through 8-54 show a step-by-step method of flashing for soil pipes. A soil pipe is used as a vent for plumbing. It is made of cast iron or copper. The pipe gets its name from being buried in the soil as a sanitary sewer pipe.

An alternative procedure for soil stacks can be used. Obtain noncorrodible metal pipes. They should have adjustable flanges. These flanges can be applied as a flashing to fit any roof pitch.

• STRIP SHINGLES •

Prepare the deck properly before starting to apply strip shingles. First place an underlayment down.

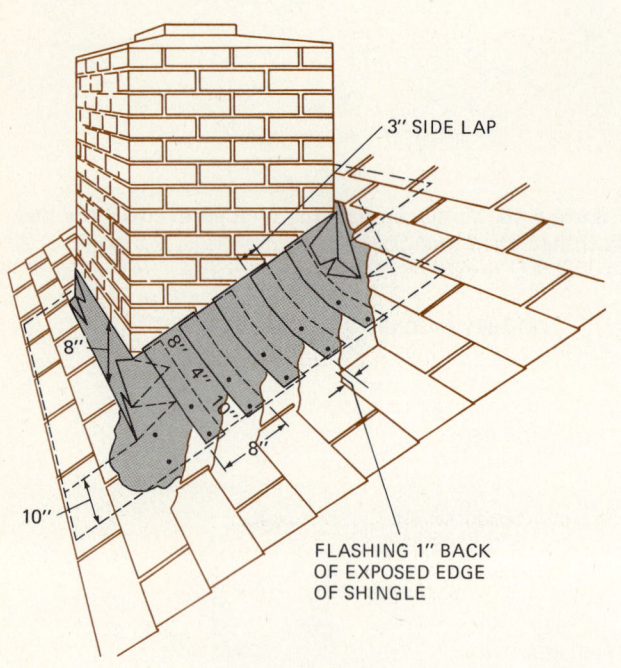

Figure 8-48. Alternative base flashing method. *(Bird and Son)*

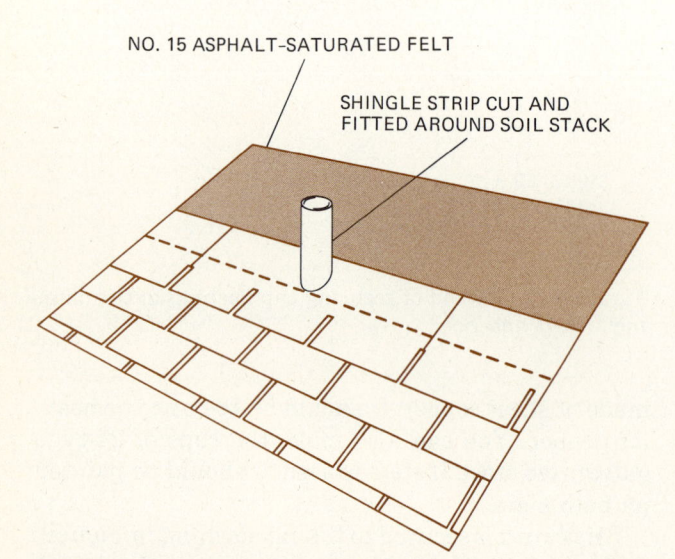

Figure 8-49. Roofing is first applied up to the soil pipe and fitted around it. *(Bird and Son)*

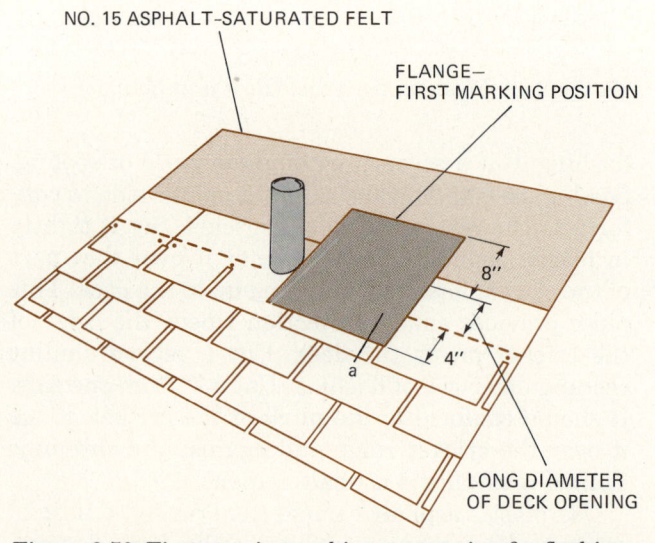

Figure 8-50. First step in marking an opening for flashing. *(Bird and Son)*

194 Carpentry Fundamentals

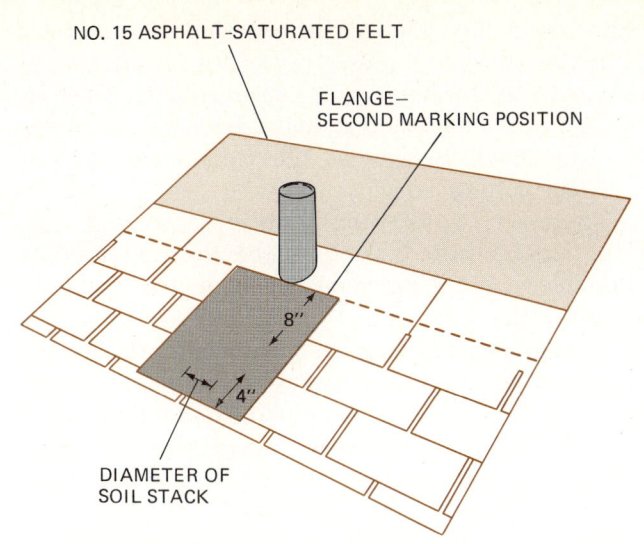

Figure 8-51. Second step in marking an opening for flashing. *(Bird and Son)*

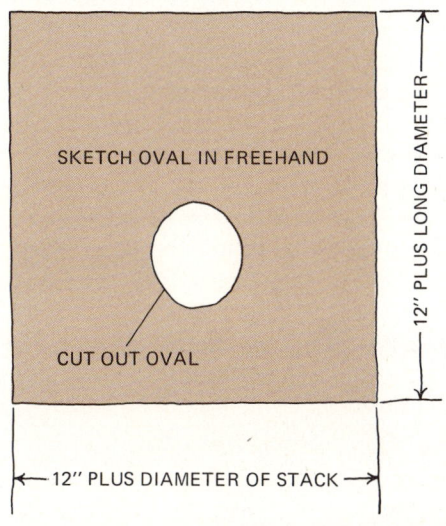

Figure 8-52. Cut oval in the flange. *(Bird and Son)*

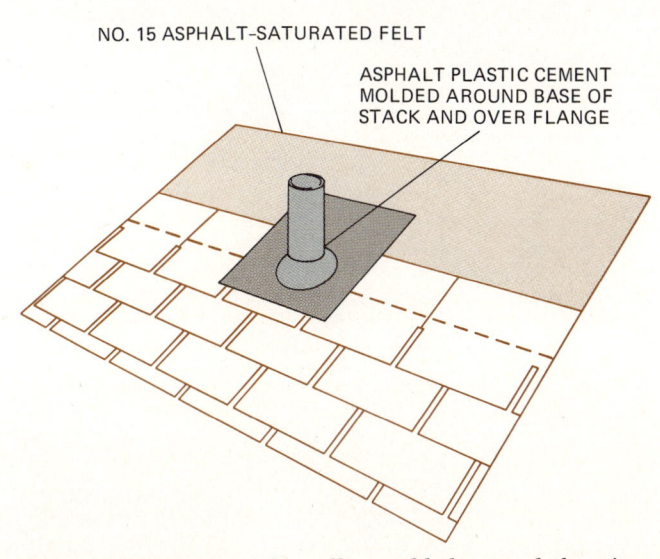

Figure 8-53. Cement the collar molded around the pipe. *(Bird and Son)*

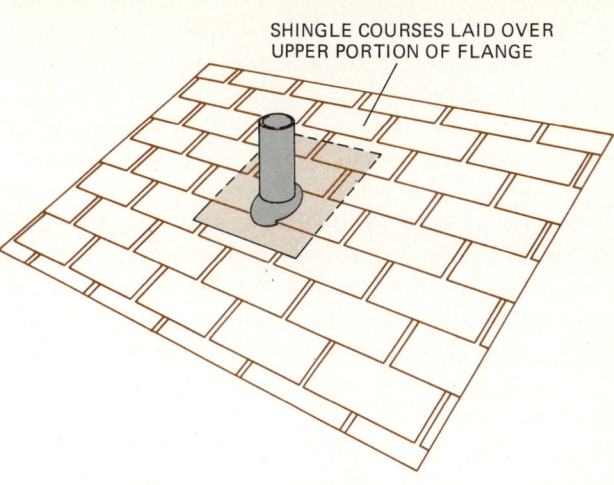

Figure 8-54. Shingling completed past and above the pipe. *(Bird and Son)*

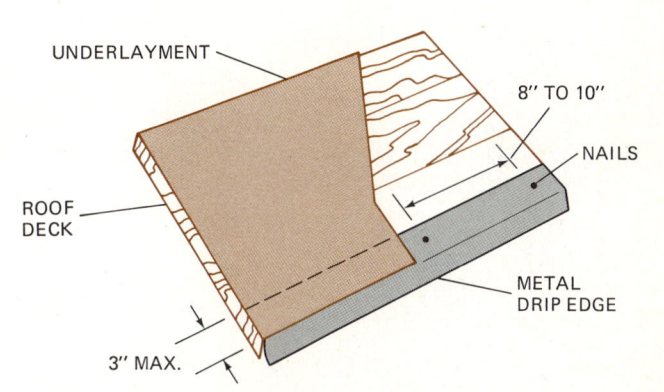

Figure 8-55. Application of the metal drip edge at eaves directly onto the deck. *(Bird and Son)*

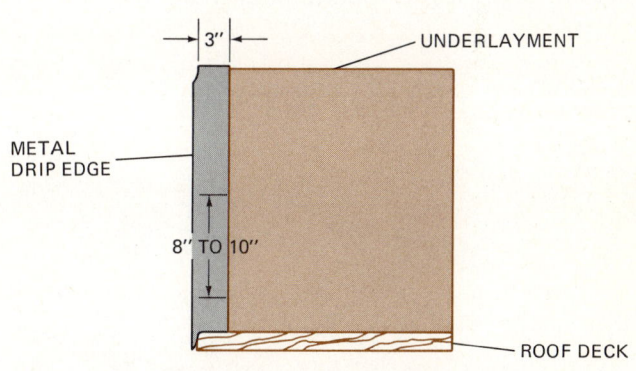

Figure 8-56. Application of the metal drip edge at the rakes over the underlayment. *(Bird and Son)*

Deck Preparation

METAL DRIP EDGE • Use a metal drip edge, made of noncorrodible, nonstaining metal. Place it along the eaves and rakes. See Figures 8-55 and 8-56. The drip edge is designed to allow water runoff to drip free into a gutter. It should extend back from the edge of the deck not more than 3 inches. Secure it with nails spaced 8 to 10 inches apart. Place the nails along the edge. Drip edges of other materials may be used. They should be of approved types.

Covering Roofs 195

UNDERLAYMENT • Place a layer of no. 15 asphalt-saturated felt down. Cover the entire deck as shown. See Figure 8-20.

CHALK LINES • On small roofs, strip shingles may be laid from either rake. On roofs 30 feet or longer, it is better to start them at the center. Then work both ways from a vertical line. This ensures better vertical alignment. It also provides for meeting and matching above a dormer or chimney. Chalk lines are used to control shingle alignment.

EAVES FLASHING • Eaves flashing is required in cold climates. (January daily average temperatures of 25°F or less call for eaves flashing.) In cold climates there is a possibility of ice forming along the eaves. If this happens, it causes trouble. Flashing should be used if there is doubt. Ice jams and water backup should be avoided. They can cause leakage into the ceiling below.

There are two flashing methods to prevent leakage. The methods depend on the slope of the roof. Possible severe icing conditions is another factor in choice of method.

Normal slope is 4 inches per foot or over. Install a course of 90-pound mineral-surfaced roll roofing. Or, apply a course of smooth roll roofing. It should not be less than 50 pounds. Install it to overhang the underlay and metal drip edge from ¼ to ⅜ inch. It should extend up to the roof. Cover a point at least 12 inches inside the building's interior wall line. For a 36-inch eave overhang, the horizontal lap joint must be cemented. It should be located on the roof deck extending beyond the building's exterior wall line. See Figure 8-57.

First and Succeeding Courses

Start the first course with a full shingle. Succeeding courses are started with full or cut strips. It depends upon the style of shingles being applied. There are three major variations for square-butt strip shingles.

1. Cutouts break at joints on the thirds. See Figure 8-58.

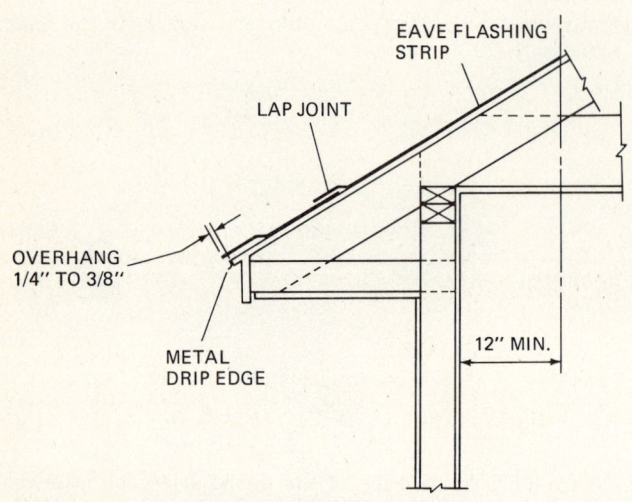

Figure 8-57. Eave flashing strip for normal-slope roof. This means 4″ per foot or over. *(Bird and Son)*

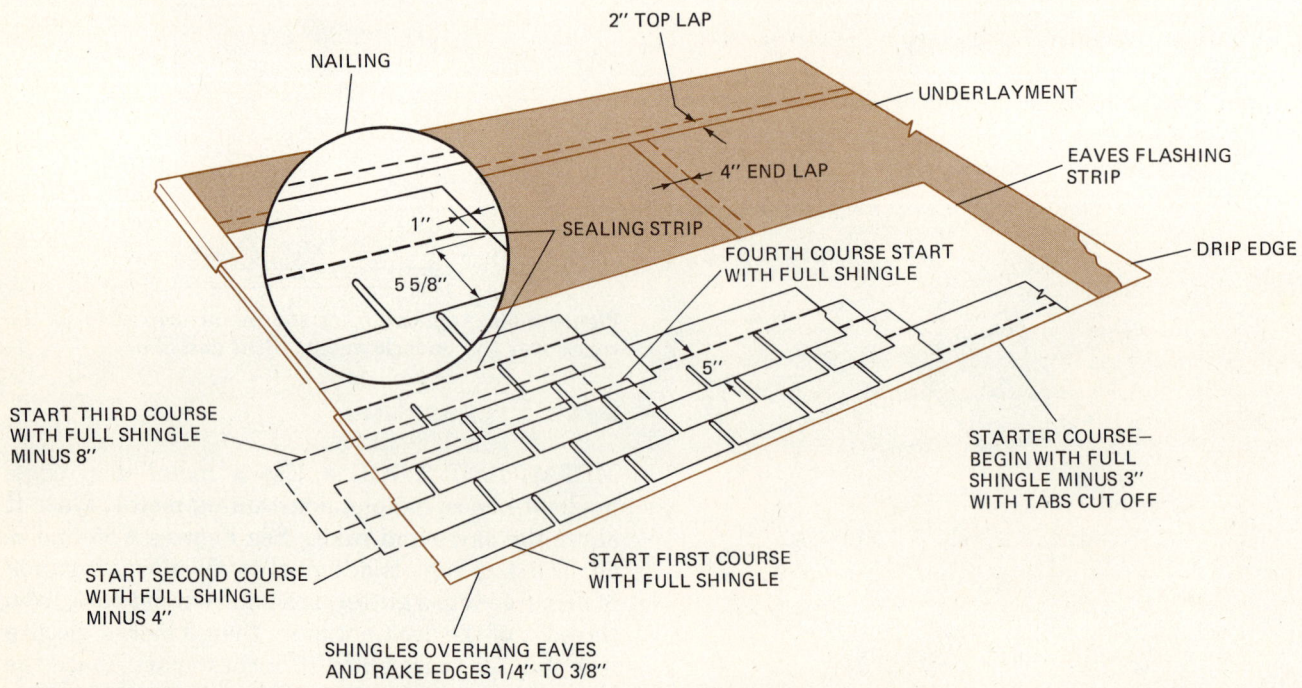

Figure 8-58. Applying three-tab square butt strips so that cutouts break the joints at thirds. *(Bird and Son)*

196 Carpentry Fundamentals

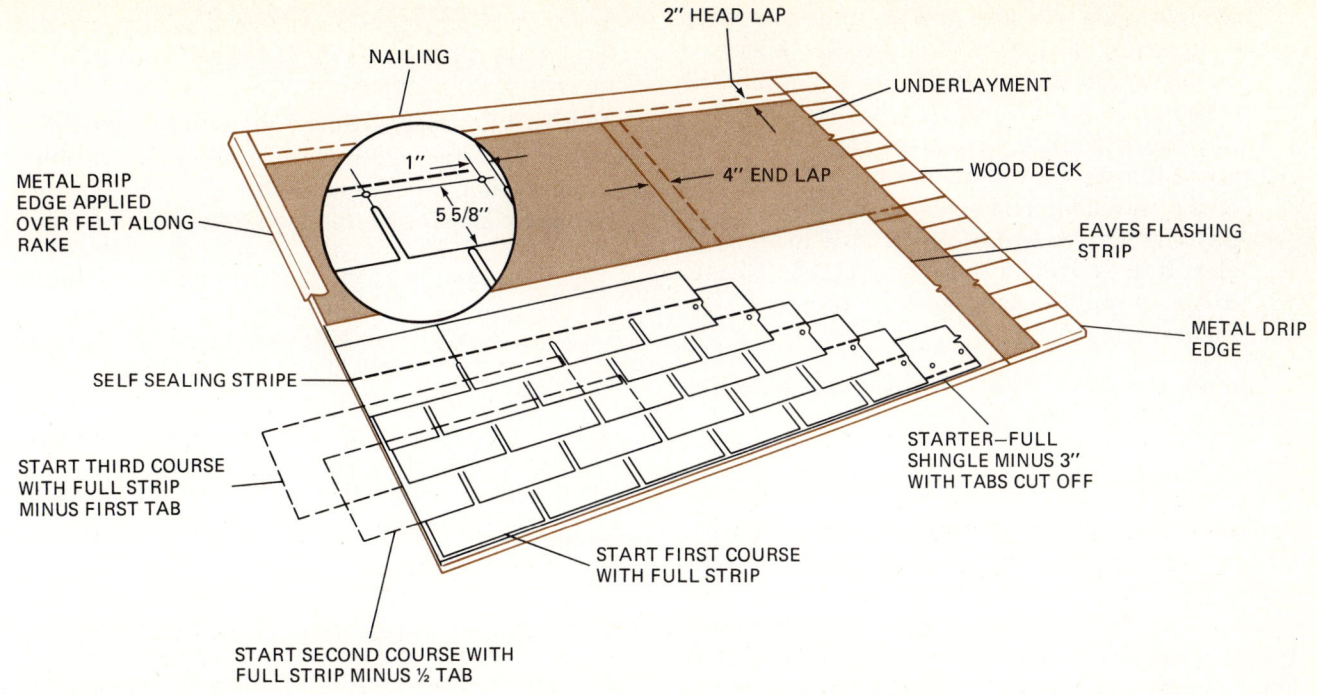

Figure 8-59. Applying three-tab square butt strips so that cutouts are centered over the tabs in the course below. *(Bird and Son)*

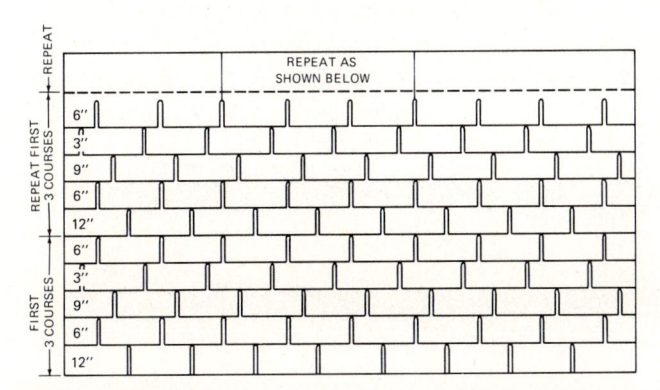

Figure 8-60. Random spacing of three-tab square butt strips. *(Bird and Son)*

2. Cutouts break at joints on the halves. See Figure 8-59.
3. Random spacing. See Figure 8-60.

Random spacing can be done by removing different amounts from the rake tab of succeeding courses. The amounts are removed according to the following scheme:

1. The width of any rake tab should be at least 3 inches.
2. Cutout centerlines should be located at least 3 inches laterally from other cutout centerlines. This means both the course above and the course below.
3. The rake tab widths should not repeat closely enough to cause the eye to follow a cutout alignment.

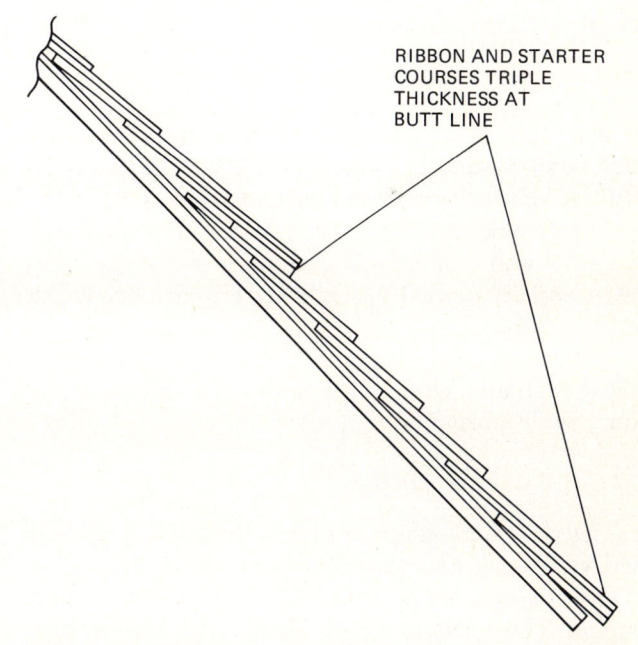

Figure 8-61. Ribbon courses. Side view. *(Bird and Son)*

Ribbon Courses

Use a ribbon course to strengthen the horizontal roof lines. It adds a massive appearance. Some people like this. See Figure 8-61.

One method involves special starting procedures. This is repeated every fifth course. Some people prefer this method.

1. Cut 4 inches off the top of a 12-inch-wide strip

Covering Roofs 197

shingle. This will give you an unbroken strip 4 inches by 36 inches. You also get a strip 8 inches by 36 inches. Both strips contain the cutouts.

2. Lay the 4×36-inch strip along the eave.
3. Cover this with the 8×36-inch strip. The bottom of the cutouts is laid down to the eave.
4. Lay the first course of full (12×36-inch) shingles. It goes over layers *B* and *C*. The bottom of the cutouts is laid down to the eave. See Figure 8-62.

Cutouts should be offset. This is done according to thirds, halves, or random spacing.

Wind Protection

High winds call for specially designed shingles. Cement the free tabs for protection against high winds. See Figure 8-63 and 8-64.

Use a spot of quick-setting cement on the underlaying shingle. Use a putty knife or caulking gun. The cement should be about the size of a half-dollar. Press the free tab against the spot of cement. Do not squeeze the cement beyond the edge of the tab. Don't skip or miss any shingle tabs. Don't bend tabs back farther than needed.

Two- and Three-Tab Hex Strips

Nail two- and three-tab hex strips with four nails per strip. Locate the nails in a horizontal line 5¼ inches above the exposed butt edge.

Figure 8-65 shows how the two-tab strip is applied. Use one nail 1 inch back from each end of the strip. One nail is applied ¾ inch back from each angle of the cutouts.

Three-tab shingles require one nail 1 inch back from each end. One nail is centered above each cutout. See Figure 8-66.

Hips and Ridges

Use hip and ridge shingles to finish hips and ridges. They are furnished by the makers of shingles. You can cut them from 12×36-inch square-butt strips. They should be at least 9×12 inches. One method of applying them is shown in Figure 8-67.

Bend each shingle lengthwise down the center. This gives equal exposure on each side of the hip or ridge. Begin at the bottom of a hip. Or you can begin at one end of a ridge. Apply shingles over the hip or ridge. Expose them by 5 inches. Note the direction of the prevailing winds. This is important when you are placing ridge shingles. Secure each shingle with one nail on each side. The nail should be 5½ inches back from the exposed end. It should be 1 inch up from the end. Never use metal ridge roll with asphalt roofing products. Corrosion may discolor the roof.

• STEEP-SLOPE AND MANSARD ROOFS •

New roof lines have caused changes recently. The mansard roof line requires some variations in shingle application. See Figure 8-68.

Excessive slopes give reduced results with factory

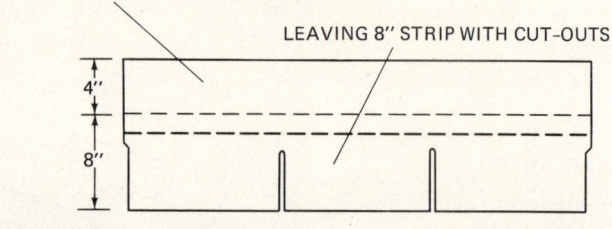

Figure 8-62. Laying the ribbon courses. *(Bird and Son)*

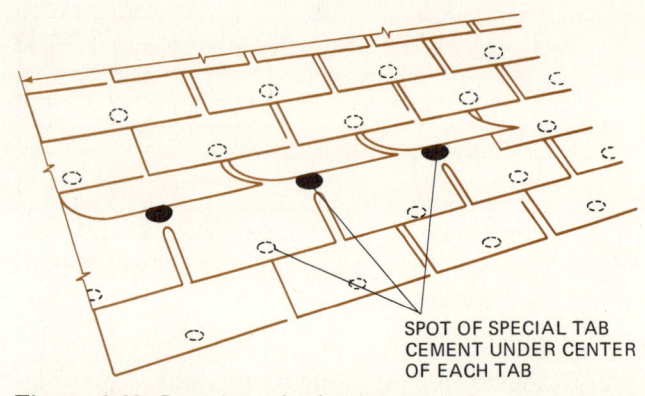

Figure 8-63. Location of tab cement under square butt tabs. *(Bird and Son)*

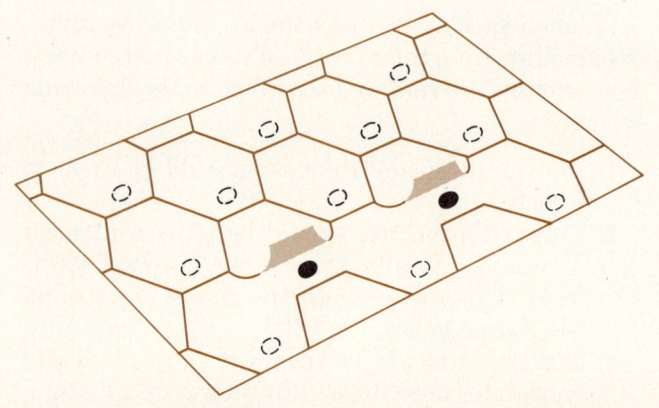

Figure 8-64. Location of tab cement under hex tabs. *(Bird and Son)*

Figure 8-65. Application of two-tab hex strips. *(Bird and Son)*

Figure 8-66. Application of three-tab hex strips. *(Bird and Son)*

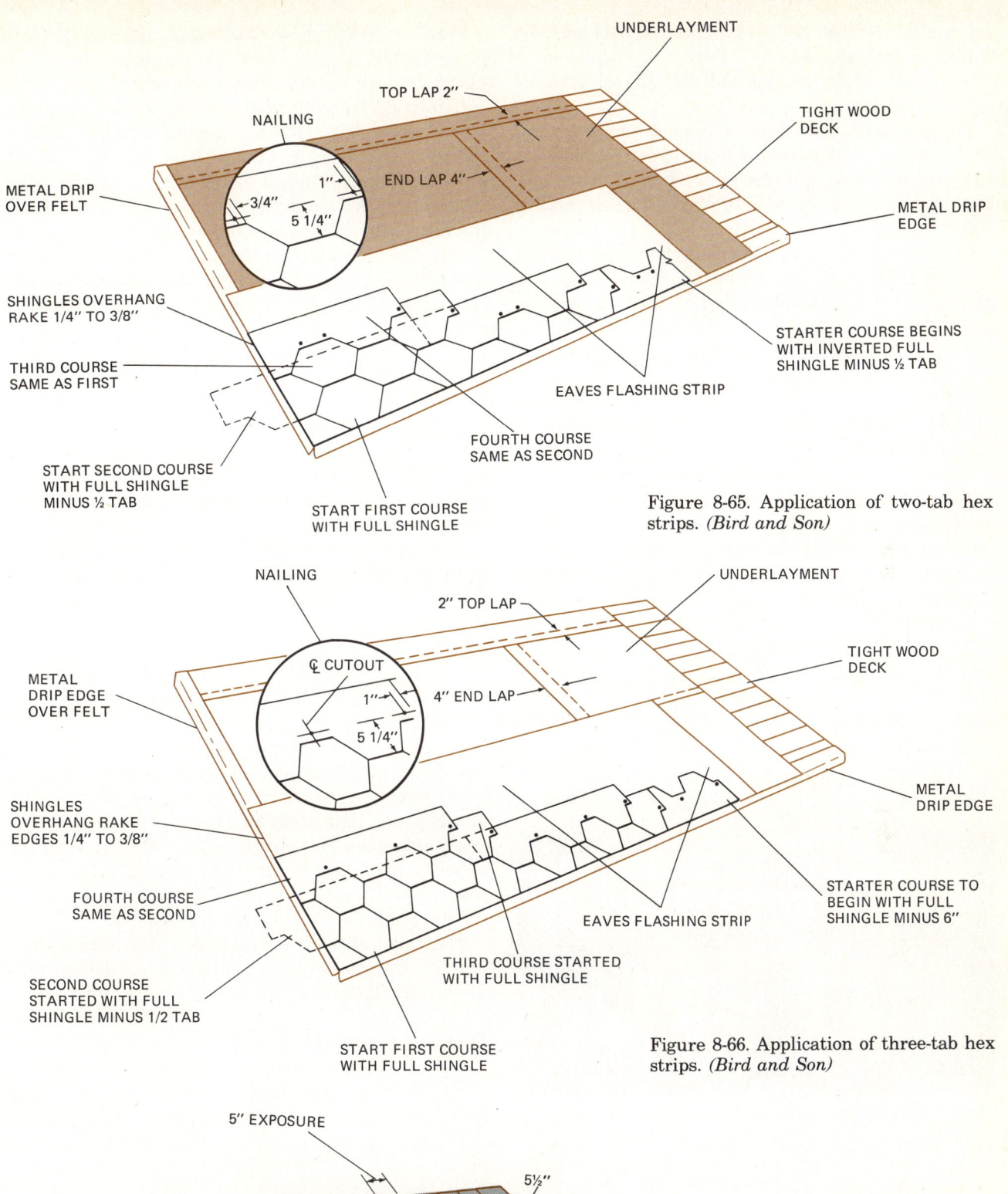

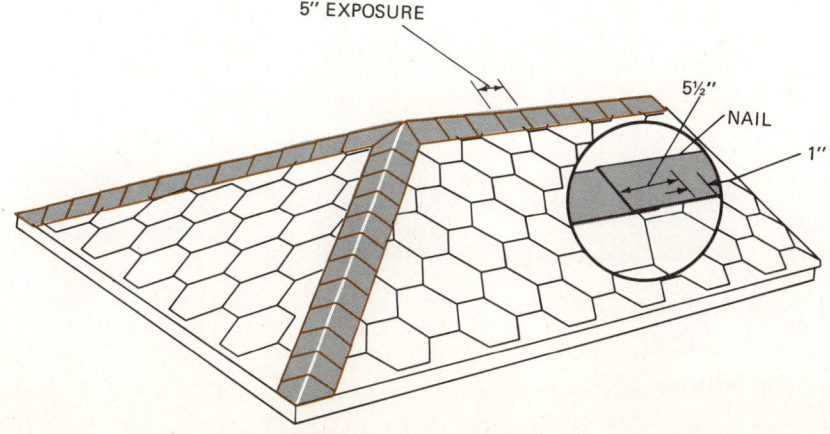

Figure 8-67. Hip and ridge shingles applied with hex strips. *(Bird and Son)*

Covering Roofs 199

self-sealing adhesives. This becomes obvious in colder or shaded areas.

Maximum slope for normal shingle application is 60° or 21 inches per foot.

Shingles on a steeper slope should be secured to the roof deck with roofing nails. See the maker's suggestions. Nails are placed 5⅝ inches above the butt. They should not be in or above the self-sealing strip.

Quick-setting asphalt adhesives should be applied as spots. The spots should be about the size of a quarter. They should be applied under each shingle tab. Do this immediately upon installation. Ventilation is needed to keep moisture-laden air from being trapped behind sheathing.

• INTERLOCKING SHINGLES •

Lockdown shingles are designed for windy areas. They have high resistance to high winds. They have an integral locking device. They can be classified into five groups. See Figure 8-69.

These shingles generally do not require the use of adhesives. They may require a restricted use of cement. This is needed along the rakes and eaves. That is where the locking device may have to be removed.

Lock-type shingles can be used for both new and old buildings. Roof pitch is not too critical with these shingles. The designs can be classified into types 1, 2, 3, 4, and 5. Type 5 is a strip shingle with two tabs per strip.

Figure 8-68. Mansard roof.

Figure 8-70 shows how locking shingles work. Nail placement is suggested by the manufacturer. Nail placement is important for good results.

Figure 8-71 shows how the drip edge should be placed. The drip edge is designed to allow water runoff to drip free into gutters. The drip edge should not extend more than 3 inches back from the edge of the deck. Secure it with appropriate nails. Space the nails 8 to 10 inches apart along the inner edge.

Use the manufacturer's suggestions for starter course placement. Chalk lines will be very useful. This type of shingle, since it is very short, needs chalk lines for alignment.

Hips and Ridges

Hips and ridges use shingles made for that purpose. However, you can cut them from standard shingles. They should be 9x12 inches. Either 90-pound mineral-surfaced roofing or shingles can be used. See Figure 8-72.

In cold weather, warm the shingles before you use them. This keeps them from cracking when they are bent to lock. Do not use metal hip or ridge materials. Metal may become corroded and discolor the roof.

• ROLL ROOFING •

Roofing in rolls in some instances is very economical. Farm buildings and sheds are usually covered with inexpensive roll roofing. It is easier to apply and cheaper than strip shingles.

Do not apply roll roofing when the temperature is below 45°F. If it is necessary to handle the material at lower temperatures, warm it before unrolling it. Warming avoids cracking the coating.

The sheet should be cut into 12-foot and 18-foot lengths. Spread them in a pile on a smooth surface until they flatten.

Windy Locations

Roll roofings are recommended for use in windy locations. Apply them according to the maker's sug-

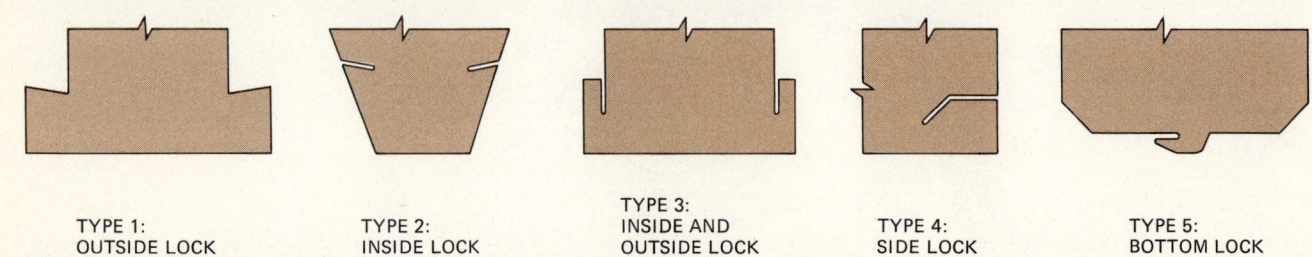

TYPE 1: OUTSIDE LOCK
TYPE 2: INSIDE LOCK
TYPE 3: INSIDE AND OUTSIDE LOCK
TYPE 4: SIDE LOCK
TYPE 5: BOTTOM LOCK

Figure 8-69. Locking devices used in interlocking shingles. *(Bird and Son)*

200 Carpentry Fundamentals

TYPE 1: OUTSIDE LOCK

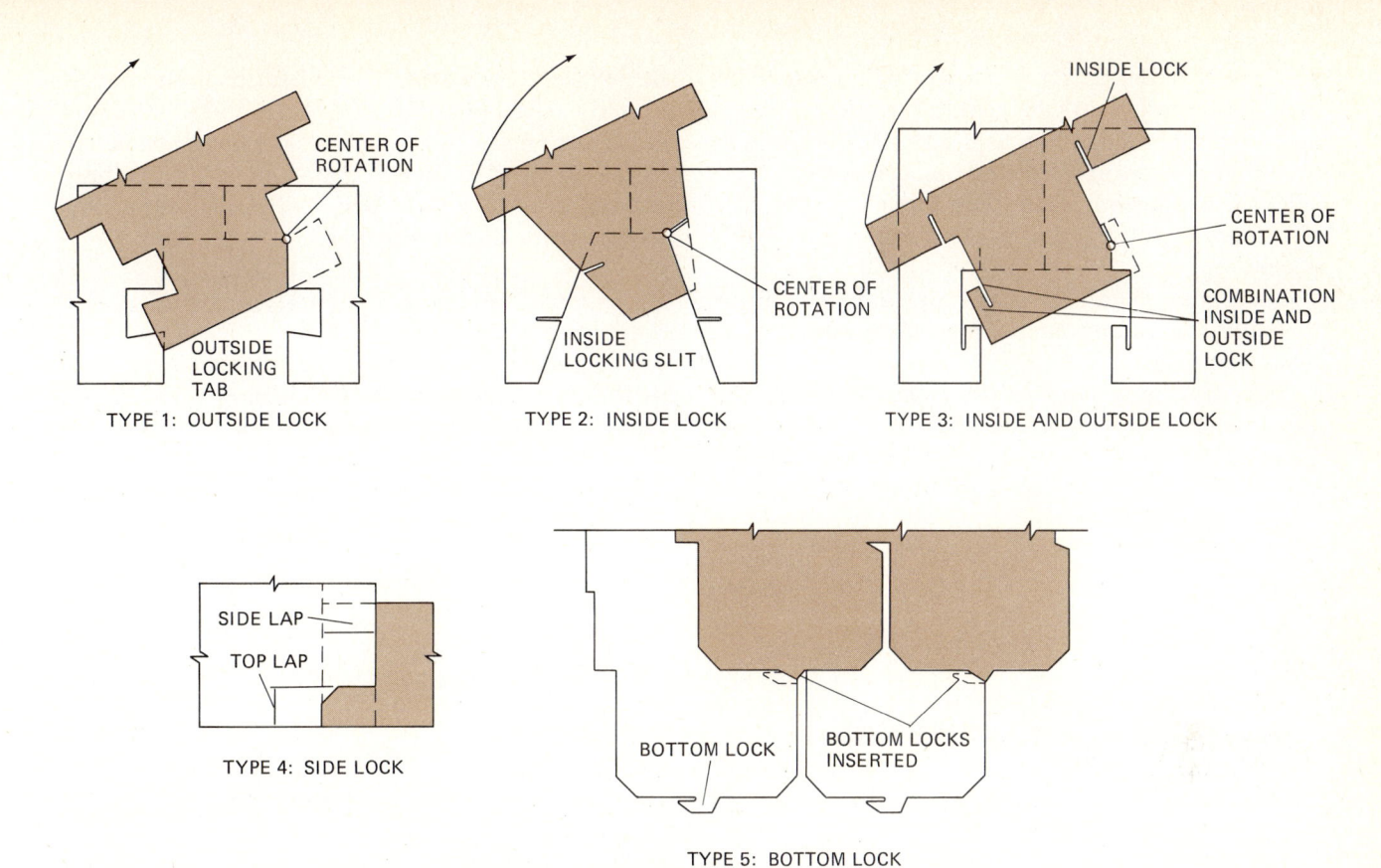

TYPE 2: INSIDE LOCK

TYPE 3: INSIDE AND OUTSIDE LOCK

TYPE 4: SIDE LOCK

TYPE 5: BOTTOM LOCK

Figure 8-70. Methods of locking shingles types 1 through 5. In each case only the locking device is shown. *(Bird and Son)*

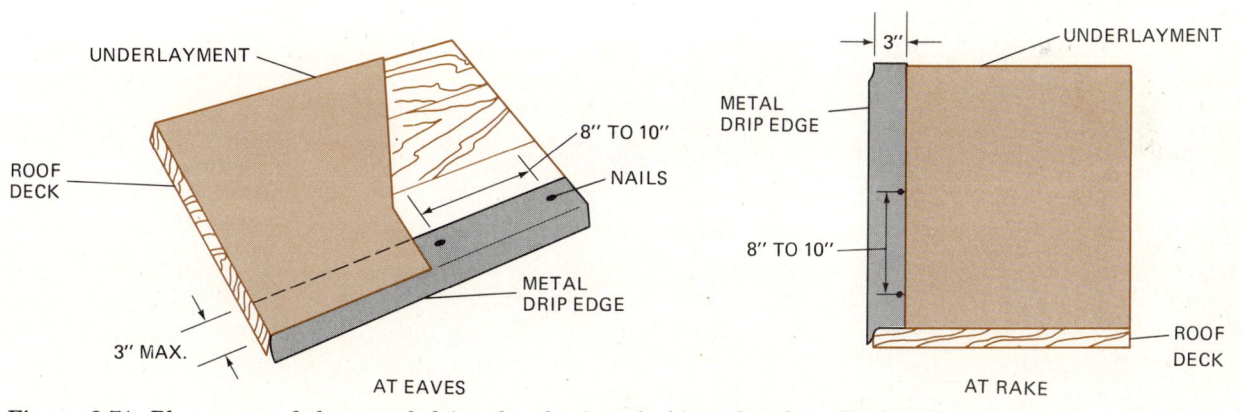

Figure 8-71. Placement of the metal drip edge for interlocking shingles. *(Bird and Son)*

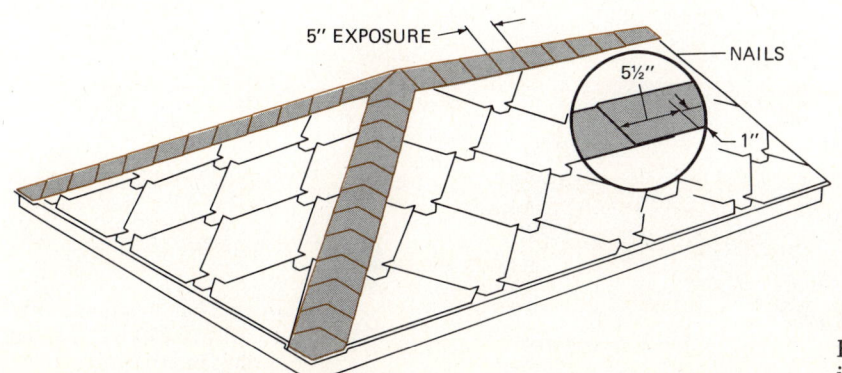

Figure 8-72. Hip and ridge application of interlocking shingles. *(Bird and Son)*

Covering Roofs **201**

gestions. Use the pattern edge and *blind nail.* This means the nails can't be seen after the roofing is applied. Blind nailing can be used with 18-inch-wide mineral-surfaced or 65-pound smooth roofing. This can be used in windy areas. The 19-inch selvage double-coverage roll roofing is also suited for windy places.

Use concealed nailing rather than exposed nailing to apply the roll roofing. This ensures maximum life in service.

Use only lap cement or quick-setting cement. It should be cement the maker of the roofing suggests. Cements should be stored in a warm place until you are ready to use them. Place the unopened container in hot water to warm. Never heat asphalt cements directly over a fire. Use 11- or 12-gage hot-dipped galvanized nails. Nails should have large heads. This means at least ⅜-inch-diameter heads. The shanks should be ⅞ to 1 inch long. Use nails long enough to penetrate the wood below.

Exposed Nails—Parallel to the Rake

Exposed nailing, parallel to the eaves, is shown in Figure 8-73. Figure 8-74 also shows the exposed nail method. It is parallel to the rake in this case. The overhang is ¼ to ⅜ inch over the rake. End laps are 6 inches wide and cemented down. Stagger

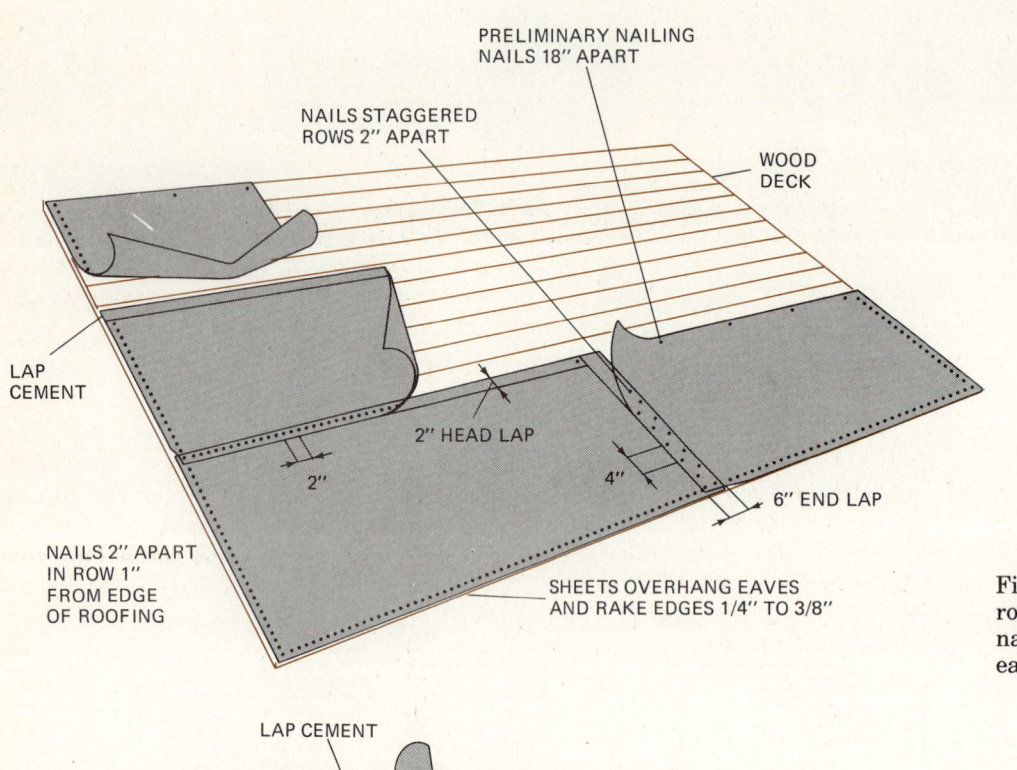

Figure 8-73. Application of roll roofing by the exposed nail method (parallel to the eaves). *(Bird and Son)*

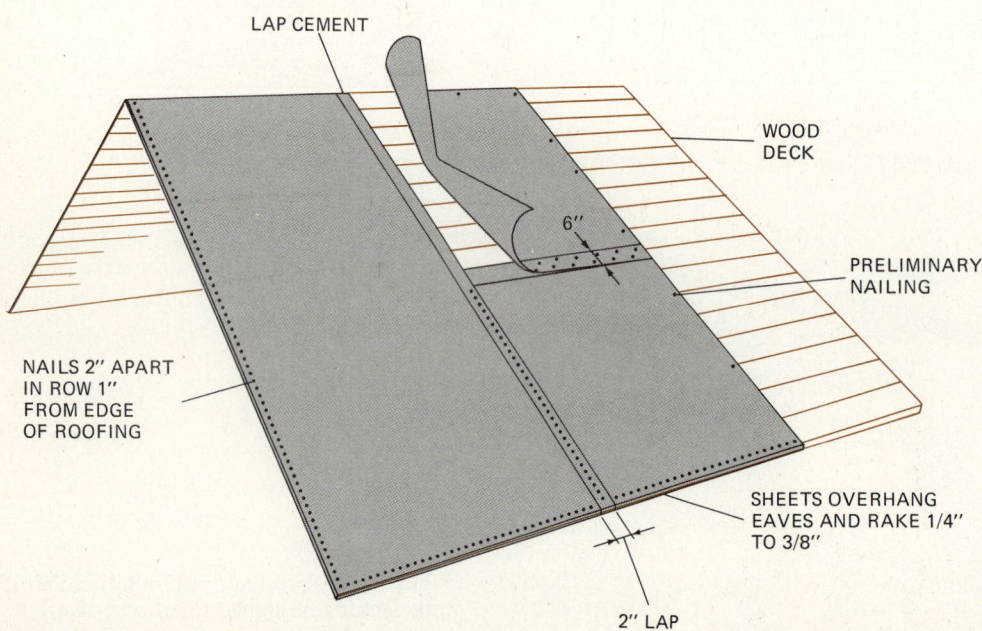

Figure 8-74. Application of roll roofing by the exposed nail method (parallel to the rake). *(Bird and Son)*

202 Carpentry Fundamentals

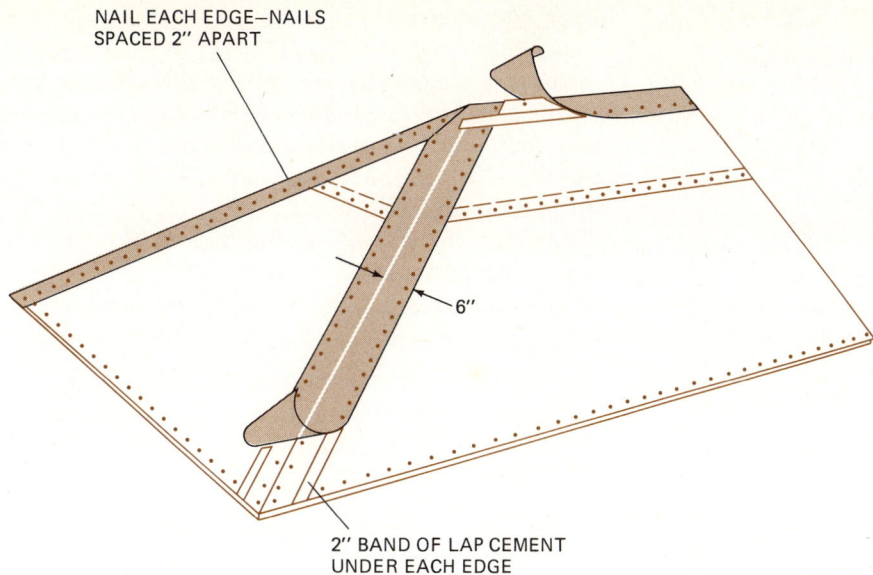

Figure 8-75. Hip and ridge application of roll roofing. *(Bird and Son)*

Hips and Ridges

For the method used to place a cap over the hips and ridge, see Figure 8-75. Butt and nail sheets of roofing as they come up on either side of a hip or ridge. Cut strips of roll roofing 12 inches wide. Bend them lengthwise through their centers. Snap a chalk line guide parallel to the hip or ridge. It should be 5½ inches down on each side of the deck.

Cement a 2-inch-wide band on each side of the hip or ridge. The lower edge should be even with the chalk line. Lay the bent strip over the hip or ridge. Embed it in asphalt lap cement.

Secure the strip with two rows of nails. One row is placed on each side of the hip or ridge. The rows should be ¾ inch above the edges of the strip. Nails are spaced on 2-inch centers. Be sure that the nails penetrate the cemented portion. This seals the nail hole with some of the asphalt.

• WOOD SHINGLES •

Wood shingles are the oldest method of shingling. In early U.S. history, pine and other trees were used for shingles. Then the western United States discovered the cedar shingle. It is resistant to water and rot. If properly cared for, it will last at least 50 years. Application of this type of roofing material calls for some different methods.

Sizing Up the Job

You need to know a few things before ordering these shingles. First, find the pitch of your roof. See Figure 8-76. Simply measure how many inches it rises for every foot it runs.

Remember that a square contains four bundles. It will cover 100 square feet of roof area. See Figure 8-77.

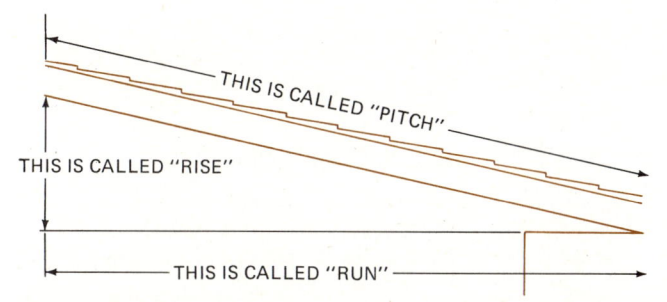

Figure 8-76. Figuring the pitch of a roof. *(Red Cedar Shingle and Handsplit Shake Bureau)*

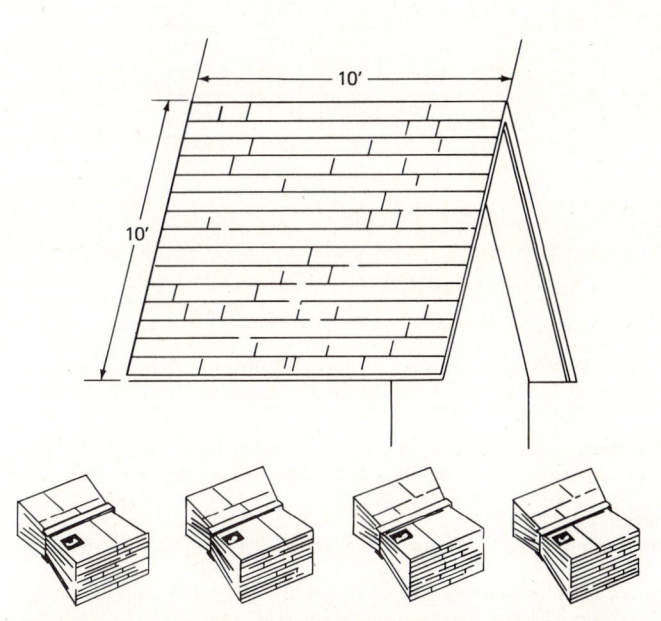

Figure 8-77. A square of shingles contains four bundles. *(Red Cedar Shingle and Handsplit Shake Bureau)*

the nails in rows, 1 inch apart. Space the nails on 4-inch centers in each row. Stagger all end laps. Do not have an end lap in one course over or adjacent to an end lap in the preceding course.

Covering Roofs **203**

Roof Exposure

Exposure means the area of the shingle that contact the weather. See Figure 8-78. Exposure depends upon roof pitch. A good shingle job is never less than three layers thick. See Table 8-6 for important information about shingles.

There are three lengths of shingles: 16 inches, 18 inches, and 24 inches.

If the roof pitch is 4 inches in 12 inches or steeper *(three-ply roof):*

For 16-inch shingles, allow a 5-inch exposure
For 18-inch shingles, allow a 5½-inch exposure
For 24-inch shingles, allow a 7½-inch exposure

If the roof pitch is less than 4 inches in 12 inches but not below 3 inches in 12 inches *(four-ply roof)*

For 16-inch shingles, allow a 3¾-inch exposure

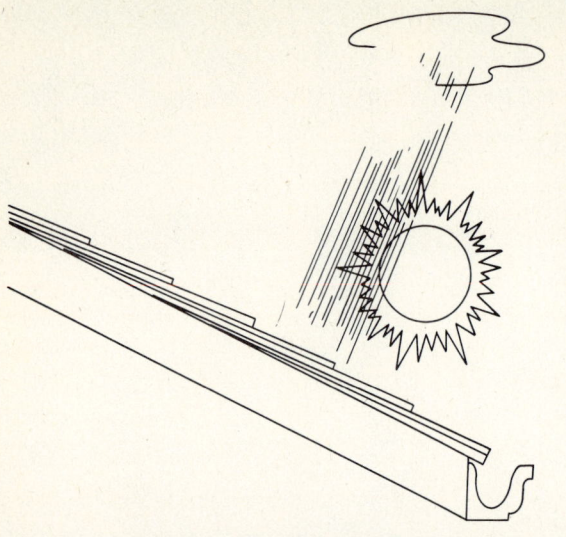

Figure 8-78. Shingle exposure to the weather. *(Red Cedar Shingle and Handsplit Shake Bureau)*

TABLE 8-6 SUMMARY OF SIZES, PACKING, AND COVERAGE OF WOOD SHINGLES

Shake Type, Length, and Thickness, Inches	No. of Courses per Bundle	No. of Bundles per Square	Approximate Coverage (in Square Feet) of One Square, When Shakes Are Applied with ½-Inch Spacing, at Following Weather Exposures (in Inches):								
			5½	6½	7	7½	8½	10	11½	14	16
18 × ½ medium resawn	9/9[a]	5[b]	55[c]	65	70	75[d]	85[e]	100[f]			
18 × ¾ heavy resawn	9/9[a]	5[b]	55[c]	65	70	75[d]	85[e]	100[f]			
24 × ⅜ handsplit	9/9[a]	5		65	70	75[g]	85	100[h]	115[i]		
24 × ½ medium resawn	9/9[a]	5		65	70	75[c]	85	100[j]	115[i]		
24 × ¾ heavy resawn	9/9[a]	5		65	70	75[c]	85	100[j]	115[i]		
24 × ½ to ⅝ tapersplit	9/9[a]	5		65	70	75[c]	85	100[j]	115[i]		
18 × ⅜ true-edge straight-split	14[k] straight	4								100	112[l]
18 × ⅜ straight-split	19[k] straight	5	65[c]	75	80	90[j]	100[i]				
24 × ⅜ straight-split	16[k] straight	5		65	70	75[c]	85	100[j]	115[i]		
15 starter-finish course	9/9[a]	5	Use supplementary with shakes applied with not over 10-inch weather exposure.								

[a] Packed in 18-inch wide frames.
[b] Five bundles will cover 100 square feet of roof area when used as starter-finish course at 10-inch weather exposure; six bundles will cover 100 square feet wall area when used at 8½-inch weather exposure; seven bundles will cover 100 square feet roof area when used at 7½-inch weather exposure.[m]
[c] Maximum recommended weather exposure for three-ply roof construction.
[d] Maximum recommended weather exposure for two-ply roof construction; seven bundles will cover 100 square feet of roof area when applied at 7½-inch weather exposure.[m]
[e] Maximum recommended weather exposure for sidewall construction; six bundles will cover 100 square feet when applied at 8½-inch weather exposure.[m]
[f] Maximum recommended weather exposure for starter-finish course application; five bundles will cover 100 square feet when applied at 10-inch weather exposure.[m]
[g] Maximum recommended weather exposure for application on roof pitches between 4 in 12 and 8 in 12.
[h] Maximum recommended weather exposure for application on roof pitches of 8 in 12 and steeper.
[i] Maximum recommended weather exposure for single-coursed wall construction.
[j] Maximum recommended weather exposure for two-ply roof construction.
[k] Packed in 20-inch-wide frames.
[l] Maximum recommended weather exposure for double-coursed wall construction.
[m] All coverage based on ½-inch spacing between shakes.

For 18-inch shingles, allow a 4¼-inch exposure
For 24-inch shingles, allow a 5¾-inch exposure

If the roof pitch is less than 3 inches in 12 inches, cedar shingles are not recommended. These exposures are for no. 1 grade shingles. In applying no. 3 shingles, make sure you check with the manufacturer.

Estimating Shingles Needed

Determine the ground area of your house. Include the eaves and cornice overhang. Do this in square feet. If the roof pitch found previously:

Rises 3 in 12, add 3 percent to the square foot total

Rises 4 in 12, add 5½ percent to the square foot total

Rises 5 in 12, add 8½ percent to the square foot total

Rises 6 in 12, add 12 percent to the square foot total

Rises 8 in 12, add 20 percent to the square foot total

Rises 12 in 12, add 42 percent to the square foot total

Rises 15 in 12, add 60 percent to the square foot total

Rises 18 in 12, add 80 percent to the square foot total

Divide the number you have found by 100. The answer is the number of shingle "squares" you should order to cover your roof if the pitch is 4 inches in 12 inches or steeper. If the roof is of lesser pitch, allow one-third more shingles to compensate for reduced exposure.

Also, add 1 square for every 100 linear feet of hips and valleys.

Tools of the Trade

A shingler's hatchet speeds the work. See Figure 8-79. Sneakers or similar traction shoes make the job safer. A straight board keeps your rows straight and true.

• APPLYING THE SHINGLE ROOF •

Begin with a double thickness of shingles at the bottom edge of the roof. See Figure 8-80. Let the shingles protrude over the edge to assure proper spillage into the eaves-trough or gutter. See *A* on Figure 8-80.

Figure 8-81 shows how the nails are placed so that the next row above will cover the nails by not more

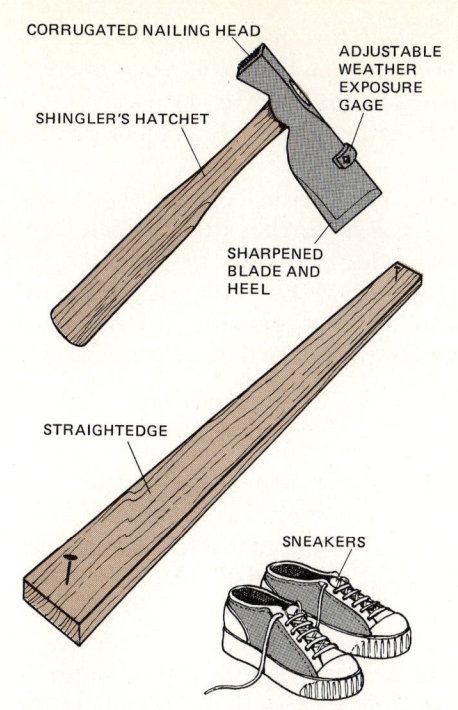

Figure 8-79. Tools of the trade. *(Red Cedar Shingle and Handsplit Shake Bureau)*

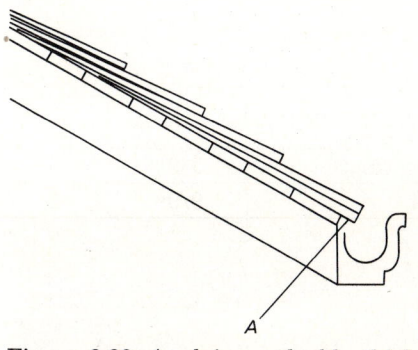

Figure 8-80. Applying a double thickness of shingles at the bottom edge of the roof. *(Red Cedar Shingle and Handsplit Shake Bureau)*

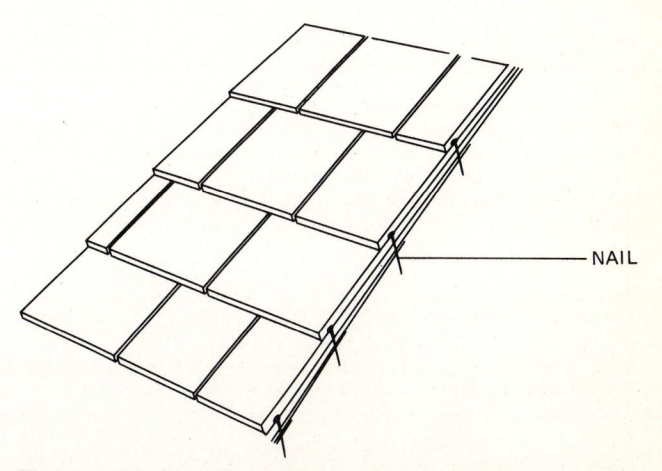

Figure 8-81. Covering nails in the previous course. *(Red Cedar Shingle and Handsplit Shake Bureau)*

Covering Roofs 205

than 1 inch. Use the board as shown in Figure 8-82. Use it as a straightedge to line up rows of shingles. Tack the board temporarily in place as a guide. It makes the work faster and the results look professional.

In Figure 8-83 you can see the location of the nails. They should be placed no further than ¾ inch from the edge of the shingle.

Figure 8-84 shows how the shingles are spaced ¼ inch apart to allow for expansion. Other simple rules are also shown in Figure 8-84.

Valleys and Flashings

Extend the valley sheets beneath shingles. They should extend 10 inches on either side of the valley center. This is the case if the roof pitch is less than 12 inches in 12 inches. For steeper roofs, the valley sheets should extend at least 7 inches. See Figure 8-85.

Most roof leaks occur at points where water is channeled for running off the roof. Or they occur where the roof abuts a vertical wall or chimney. At these points, use pointed metal valleys and flashings to assist the shingles in keeping the roof sound and dry. Suppliers will provide further information on which of the various metals to use. Figure 8-86 shows the flashing installed around a chimney.

Shingling at Roof Junctures

Apply the final course of shingles at the top of the wall. Install metal flashing (26-gage galvanized

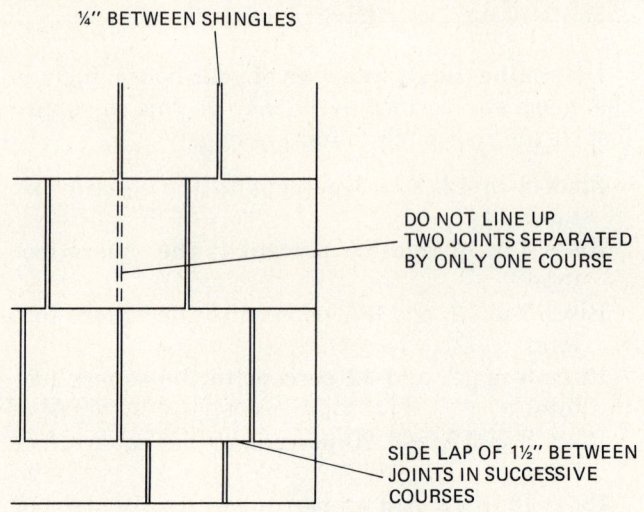

Figure 8-84. Spacing of shingles between courses. *(Red Cedar Shingle and Handsplit Shake Bureau)*

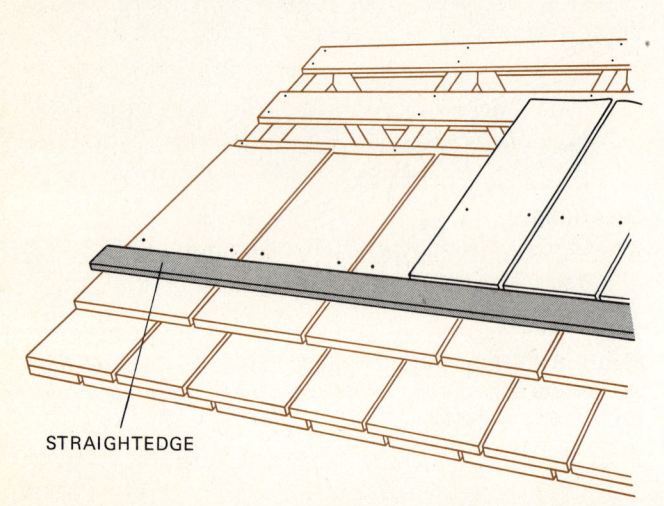

Figure 8-82. Using a straightedge to keep the ends lined up. *(Red Cedar Shingle and Handsplit Shake Bureau)*

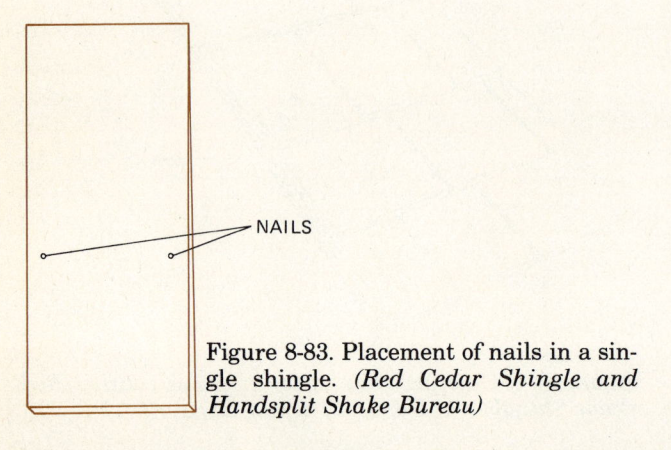

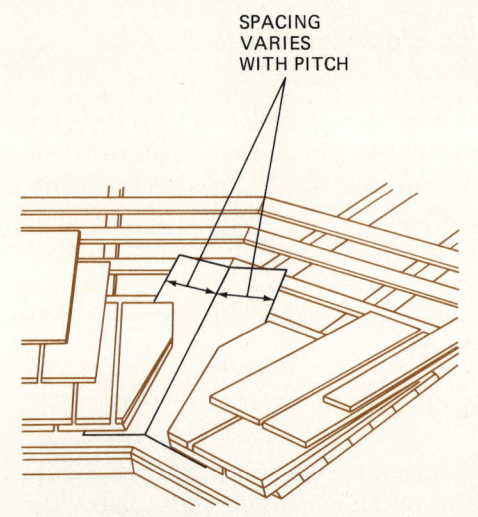

Figure 8-85. Spacing for valleys. *(Red Cedar Shingle and Handsplit Shake Bureau)*

Figure 8-83. Placement of nails in a single shingle. *(Red Cedar Shingle and Handsplit Shake Bureau)*

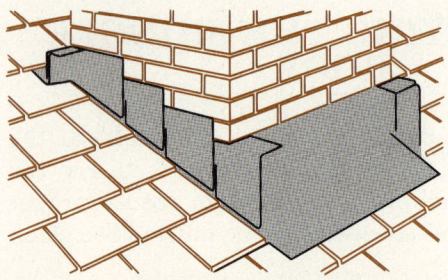

Figure 8-86. Flashing around a chimney. *(Red Cedar Shingle and Handsplit Shake Bureau)*

iron, 8 inches wide). Cover the top 4 inches of the roof slope. Bend the flashing carefully. Avoid fracturing or breaking it. Make sure the flashing covers the nails that hold the final course. Apply a double starter course at the eave. Allow for a 1½-inch overhang of the wall surface. Complete the roof in the normal manner. See Figure 8-87 for the *convex* juncture.

For the concave juncture, apply the final course of shingles as shown in Figure 8-88. Install the metal flashing to cover the last 4 inches of roof slope and bottom 4 inches of wall surface. Make sure the flashing covers the nails that hold the final course. Apply a double starter course at the bottom of the wall surface. Complete the shingling in the normal manner.

Before applying the final course of shingles, install 12-inch-wide flashing. This is to cover the top 8 inches of roof. Bend the remaining 4 inches to cover the top portion of the wall. See Figure 8-89 for the treatment of apex junctures. Complete the roof shingling to cover the flashing. Allow the shingle tips to extend beyond the juncture. Complete the wall shingling. Trim the last courses to fit snugly under the protecting roof shingles. Apply a molding strip to cover the topmost portion of the wall. Trim the roof shingles even with the outer surface of the molding. Apply a conventional shingle "ridge" across the top edge of the roof. This is done in a single strip without matching pairs.

Applying Shingles to Hips and Ridges

The alternative overlap-type hip and ridge can be built by selecting uniform-width shingles. Lace them as shown in Figure 8-90.

Time can be saved if factory-assembled hip and ridge units are used. These are shown in Figure 8-91.

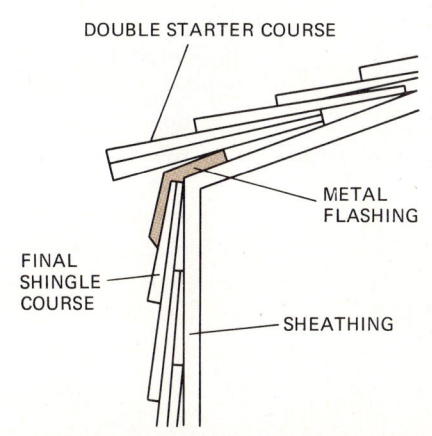

Figure 8-87. Convex roof juncture shingling. *(Red Cedar Shingle and Handsplit Shake Bureau)*

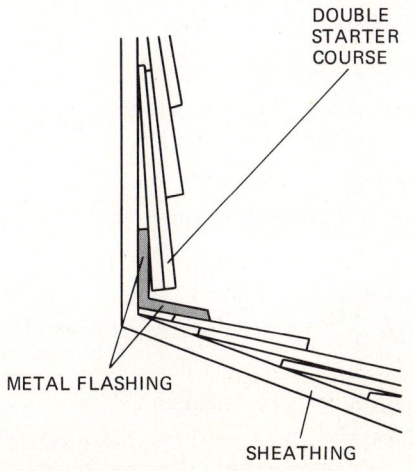

Figure 8-88. Concave roof juncture shingling. *(Red Cedar Shingle and Handsplit Shake Bureau)*

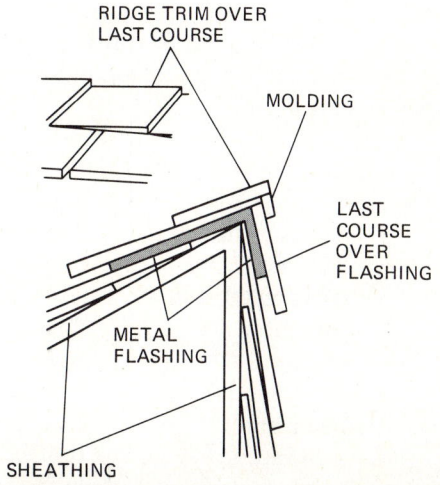

Figure 8-89. Apex roof juncture shingling. *(Red Cedar Shingle and Handsplit Shake Bureau)*

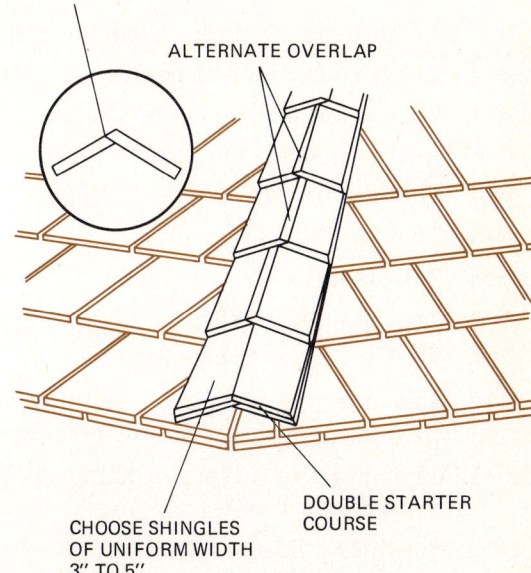

Figure 8-90. Applying overlap-type roofing ridges. *(Red Cedar Shingle and Handsplit Shake Bureau)*

Covering Roofs **207**

Figure 8-91. Applying a factory-assembled hip or ridge unit. *(Red Cedar Shingle and Handsplit Shake Bureau)*

Nails for Wooden Shingles

Rust-resistant nails are very important. Zinc-coated or aluminum nails can be used. Don't skimp on nail quality. (See Figure 8-92.)

	FOR NEW ROOF CONSTRUCTION			OVER-ROOFING CONSTRUCTION		DOUBLE COURSING
	3d	3d	4d	5d	6d	5d
	FOR 16" AND 18" SHINGLES		FOR 24" SHINGLES	FOR 16" AND 18" SHINGLES	FOR 24" SHINGLES	FOR ALL SHINGLES
	1¼" LONG	1¼" LONG #14½ GAGE	1½" LONG #14 GAGE	1¾" LONG #14 GAGE	2" LONG #13 GAGE	1¾" LONG #14 GAGE
	APPROX. 376 NAILS TO LB	APPROX. 515 NAILS TO LB	APPROX. 382 NAILS TO LB	APPROX. 310 NAILS TO LB	APPROX. 220 NAILS TO LB	APPROX. 380 NAILS TO LB

Figure 8-92. Nails used for wood shingles. *(Red Cedar Shingle and Handsplit Shake Bureau)*

• WORDS CARPENTERS USE •

Dutch hip	asphalt shingle	underlayment	chalk line	emulsions
mansard	shakes	ridge	tin snips	primers
hip	galvanized iron	pitch	utility knife	course
gable	saturated felt	slope	stapler	saddle
valley	square	rise	tongue and groove	cricket
eaves	exposure	shed	deck	random spacing
gutter	head lap	dormer	soil stack	hex strips
downspouts	side lap	roof brackets	open	interlocking
sheathing	rake	ladder	woven	
shingles	flashing	staging	closed-cut	

• STUDY QUESTIONS •

8–1. Sketch a mansard type of building roof.

8–2. Sketch a gable roof shape on a simple building.

8–3. What is the main purpose of a roof?

8–4. How does soffit ventilation affect ice dams?

8–5. What is a valley?

8–6. What is a hip roof?

8–7. What is a ridge?

8–8. What is a hip ridge?

8–9. What are eaves troughs used for?

8–10. How are plywood and sheathing different when used on a roof to provide a deck?

8–11. What are three types of shingles that may be used on a roof?

8–12. How many shingles are there in a square?

8–13. What is meant by shingle exposure?

8–14. What is flashing?

8–15. What is meant by underlayment?

8–16. What is the difference between pitch and slope?

8–17. What is a dormer?

8–18. What is a woven valley?

8–19. How does a woven valley differ from a closed-cut valley?

8–20. What is staging?

8–21. What grade of plywood is used for roofing decks?

8–22. How long should roofing nails be?

8–23. Can one type of asphalt roofing be used for every job?

8–24. What are asphalt-water emulsions?

8–25. Why is plastic asphalt cement used in roofing?

8–26. How many courses of shingles are used with wood shingles?

9
INSTALLING WINDOWS AND DOORS

Windows and doors play an important part in any type of house or building. They allow the air to circulate. They also allow passage in and out of the structure. Doors open to allow traffic in a planned manner. Windows are closed or open in design. There may be open and closed combinations, too. The design of a window or door is dictated by the use to be made of the building.

Buildings like those in Figure 9-1 use the window as part of the design. The shape of a window may add to or detract from the design. The designer must be able to determine which is the right window for a building. The designer must also be able to choose a door which is architecturally compatible. Doors and windows come in many designs. However, they are limited in their function. This means some standards are set for the design of both. Most residential doors, for instance, are 6′ 8″ in height. If you are taller than that, you have to duck to pass through. Since most people are shorter than this, it is a safe height to use for doors.

Windows should be placed so that they will allow some view from either a standing or a sitting position. In most instances the window top and the door top are even. This way they look better from the outside.

In this unit you will learn how the carpenter installs windows and doors. You will learn how windows and doors are used to enhance a design. You will learn the various sizes and shapes of windows. You will learn the different sizes and hinging arrangements of doors. Locks will be presented so that you can learn how to install them.

Details for the placement of windows and doors are given. Things you will learn to do are

- Prepare the window for installation
- Shim the window if necessary
- Secure the window in its opening
- Level and check for proper operation of the window
- Prepare a door for installation
- Shim the door if necessary
- Secure the door in its opening
- Level and check for proper operation

• BASIC SEQUENCE •

The carpenter should install a *window* in this order:

1. Check for proper window opening.
2. Uncrate the prehung window.
3. Remove the braces, if called for by the manufacturer.
4. Place builder's paper or felt between window and sheathing.
5. Place the window in the opening and check for level and plumb.
6. Attach the window at the corner or place one nail in the casing or flange (depending on the window design).
7. Check for proper operation of the window.
8. Secure the window in its opening.

The carpenter should install a *door* in this order:

1. Check the opening for correct measurements.
2. Uncrate the prehung door. If the door is not prehung, place the molding and trim in place first. Attach the door hinges by cutting the gains and screwing in the hinges.
3. Check for plumb and level.
4. Temporarily secure the door with shims and nails.

Figure 9-1. Windows can add to or detract from a piece of architecture. *(Western Wood Products)*

210 Carpentry Fundamentals

5. Check for proper operation.
6. Secure the door permanently.
7. Install the lock and its associated hardware.

Types of Windows

There are many types of windows. They hinge or swing in different directions. They may be classified as:

1. Horizontal sliding windows
2. Awning picture windows
3. Double-hung windows
4. Casement windows

Most windows are made in factories today. They are ready to be placed into a rough opening when they arrive at the site. There are standards for windows. For instance, Commercial Standard 190 is shown in Figure 9-2. The window requires a number of features which are pointed out in the drawing. It must be weather-stripped to prevent air infiltra-

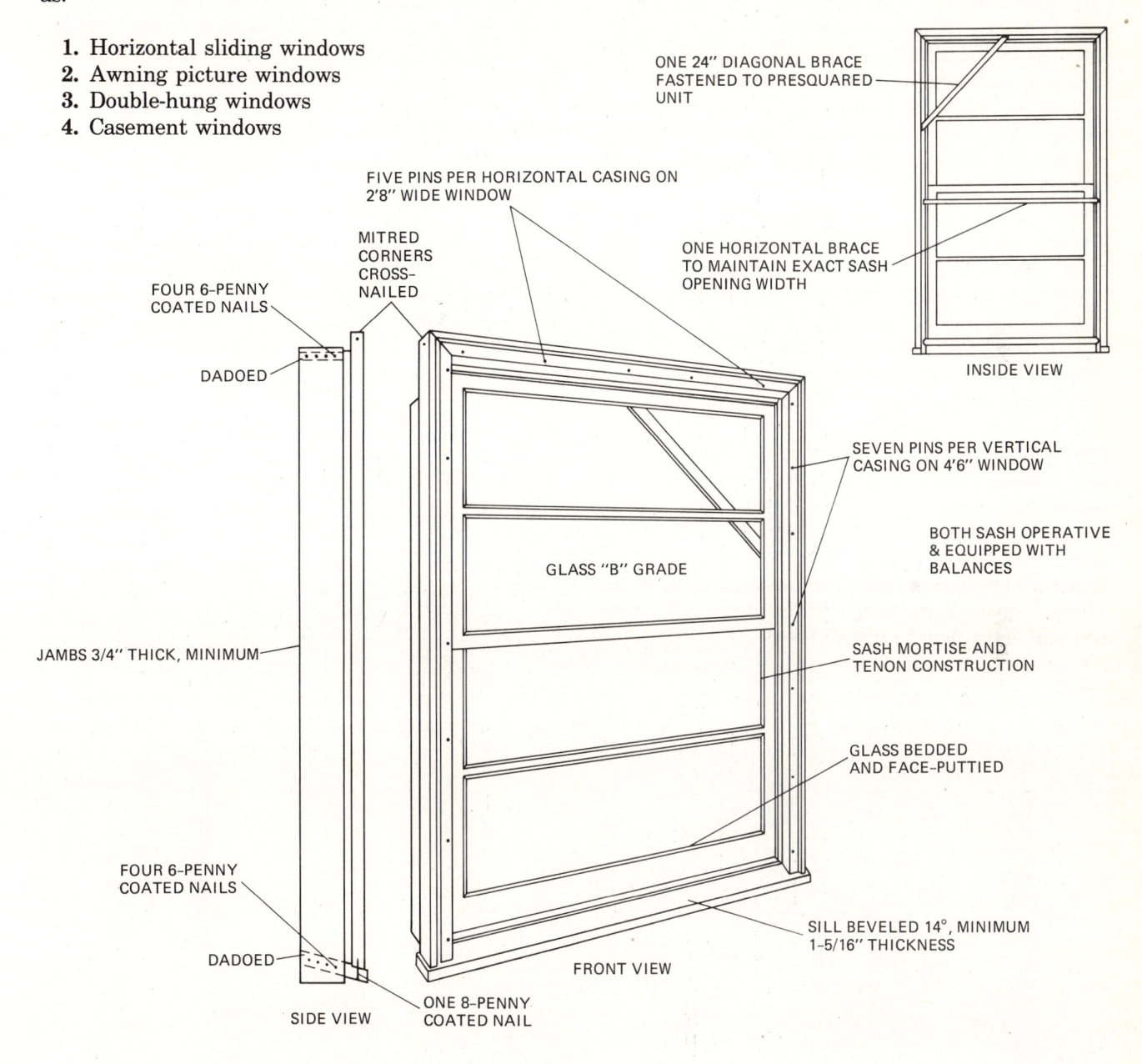

1. Weather-stripped to prevent air infiltration in excess of 0.75 cubic feet per minute per perimeter foot, under 25 MPH wind pressure.
2. No more than two species of wood per unit.
3. Chemically treated in accordance with NWMA minimum standards.
4. Finger-jointing permitted. See Commercial Standard 190, Par. 3.1.7.
5. Sash manufactured under Commercial Standard 163.
6. Ease of operation. See Commercial Standard 190, Par. 3.1.5

Figure 9-2. Standard features of a window required to meet Commercial Standard 190. *(C. Arnold & Sons)*

Installing Windows and Doors

tion in excess of 0.75 cubic feet per minute per perimeter foot. This is under a 25 mph wind.

No more than two species of wood can be used in a unit. The wood used is ponderosa pine or a similar type of pine. Spruce, cedar, redwood, and cypress may also be used in the window frame, sill, and sash.

The wood has to be chemically treated in accordance with NWMA (National Window Manufacturers Association) minimum standards. Finger jointing is permitted. The sash has to be manufactured under Commercial Standard 163. Ease of operation is also spelled out in the written standard. Note how braces are specified to hold the window square and equally distant at all points. These braces are removed once the window has been set in place.

HORIZONTAL SLIDING WINDOWS • This type of window is fitted with a vinyl coating. The wood is not exposed at any point. This means less maintenance in the way of painting or glazing. The window is trimmed in vinyl so that it can be nailed into place on the framing around the rough opening. Figure 9-3 shows the window and details of its operation.

Figure 9-4 shows how the rough opening is made for a window. In this case the framing is on a 24-inch O.C. spacing. The large timber over the opening has to be large enough to support the roof without a stud where the window is placed. This prevents the window from buckling, which would stop the window from sliding. The cripple studs under the window opening are continuations of the studs that would be there normally. They are placed there to properly support the opening and to remove any weight from the window frame.

Sliding windows are available in a number of sizes. Figure 9-5 shows the possibilities. To find the overall unit dimension for a window which has a nonsupporting mullion, add the sum of the unit dimensions and subtract 2 inches. The *mullion* is the vertical bar between windows in a frame which holds two or more windows. The overall rough-opening dimension is equal to ¾ inch less than the overall unit dimension.

DOUBLE-HUNG WINDOW • This window gets its name from the two windows that slide past one another. In this case they slide vertically. See Figure 9-6. This is the most common type used today. The window shown is coated with plastic (vinyl) and can be easily attached to a stud through holes already drilled into the plastic around the frame. This plastic is called the *flashing* or the *flange*.

(A)

(B)

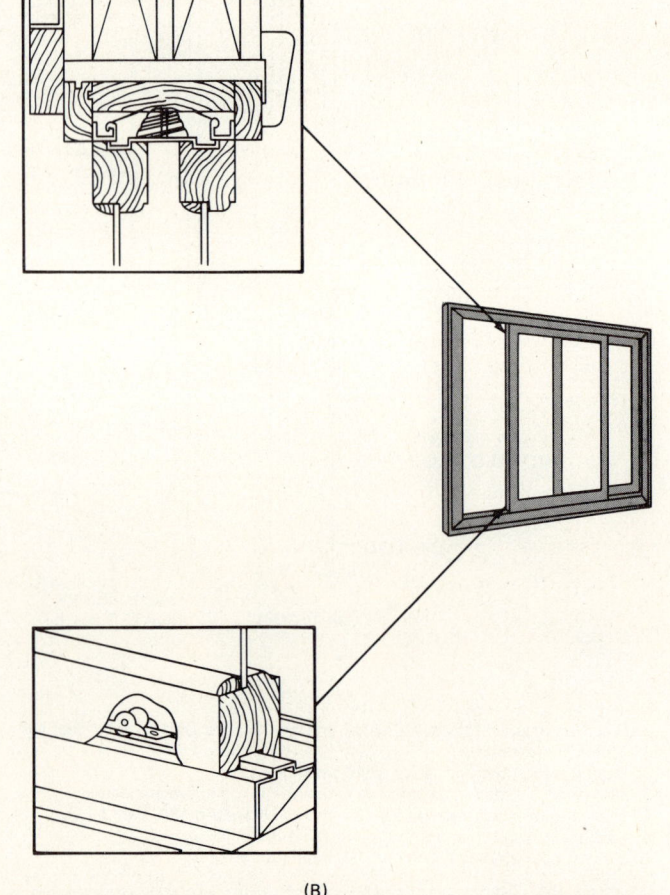

Figure 9-3. (A) Horizontal sliding window. (B) Details. *(Andersen)*

212 Carpentry Fundamentals

WINDOW ON MODULE

6' WALL SECTION

STUDS 24 L.F.
JACKS 14 L.F.
CRIPPLES 3 L.F.
41 L.F.

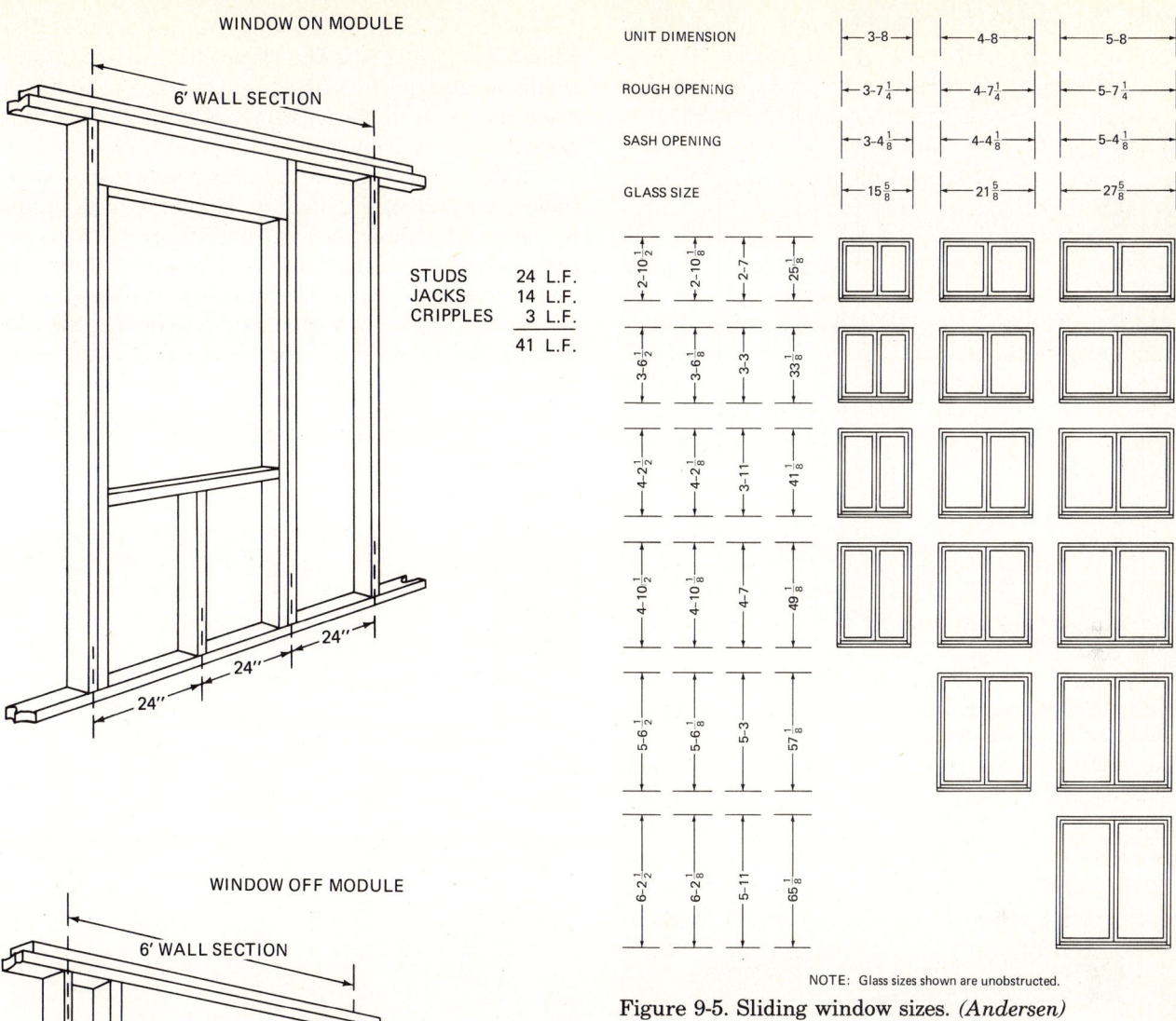

24" 24" 24"

WINDOW OFF MODULE

6' WALL SECTION

STUDS 32 L.F.
JACKS 14 L.F.
CRIPPLES 6 L.F.
52 L.F.

(23% MORE FRAMING REQUIRED)

L.F. = LINEAR FEET

24" 24" 24"

Figure 9-4. Window openings in a house frame. Window on module and off module with 24" centers. *(American Plywood Association)*

UNIT DIMENSION	3-8	4-8	5-8
ROUGH OPENING	3-7 1/4	4-7 1/4	5-7 1/4
SASH OPENING	3-4 1/8	4-4 1/8	5-4 1/8
GLASS SIZE	15 5/8	21 5/8	27 5/8

2-10 1/2	2-10 1/8	2-7	25 5/8
3-6 1/2	3-6 1/8	3-3	33 1/8
4-2 1/2	4-2 1/8	3-11	41 1/8
4-10 1/2	4-10 1/8	4-7	49 1/8
5-6 1/2	5-6 1/8	5-3	57 1/8
6-2 1/2	6-2 1/8	5-11	65 1/8

NOTE: Glass sizes shown are unobstructed.

Figure 9-5. Sliding window sizes. *(Andersen)*

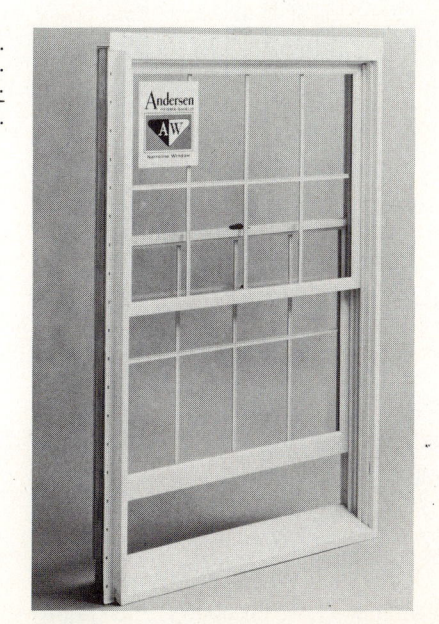

Figure 9-6. Double-hung window. *(Andersen)*

Installing Windows and Doors **213**

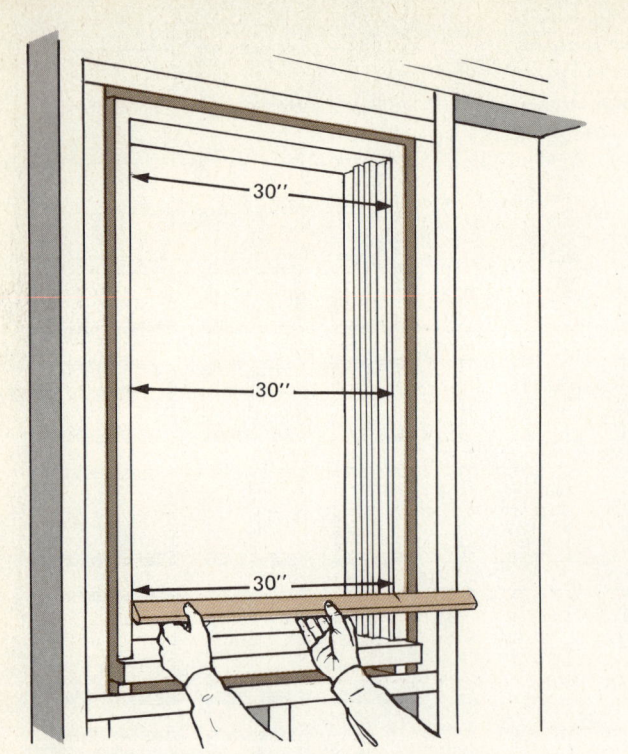

Figure 9-7. Measure the distance in at least three places to make sure the window is square. *(Andersen)*

The double-hung window can be installed rather easily. See Figure 9-7. The distance between the side jambs is checked to make sure they are even. Once the window is in the opening, place a shim where necessary. See Figure 9-8. Note the placement of the nails here. Notice that this window does not have a vinyl flange around it. That is why the nails are placed as shown in Figure 9-8. Figure 9-9 shows how shims are placed under the raised jamb legs and at the center of the long sills of a double window. Figure 9-10 shows a sash out of alignment. See the arrow. The sashes will not be parallel if the unit is out of square.

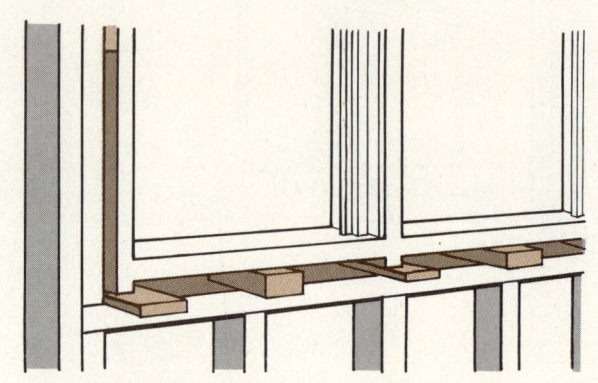

Figure 9-9. Shims raise the jamb legs and keep the window square. *(Andersen)*

Figure 9-8. Shim the window where necessary. Slide jambs are nailed through the shims. *(Andersen)*

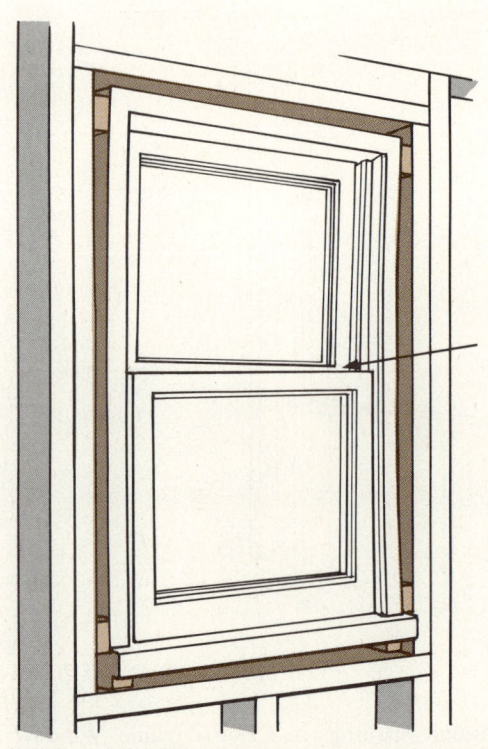

Figure 9-10. If the unit is not square, the sash rails will not be parallel. This can be spotted by eye. *(Andersen)*

214　Carpentry Fundamentals

Once the window is in place and properly seated, you can pack insulation between the jambs and the trimmer studs. See Figure 9-11. Figure 9-12 shows how 1¾-inch galvanized nails are placed through the vinyl anchoring flange. This flange is then covered by the outside wall covering. The nails are not exposed to the weather.

Double-hung windows can be bought in a number of sizes. Figure 9-13 shows some of the sizes. In Figure 9-14 the windows are all in place. The upper-story windows will be butted by siding. The downstairs windows are sitting back inside the brick. The upper windows will have part of the trim sticking out past the exterior siding.

Once the windows are in place and the house completed, the last step is to put the window dividers in place. They snap into the small holes in the sides of the window. The design may vary. See Figure 9-15. The plastic grill patterns can be changed to meet the needs of the architectural style of the house. They can be removed by snapping them out.

Figure 9-11. Loose insulation batting can be placed around the window to prevent drafts. *(Andersen)*

Figure 9-12. Installing the plastic flange around a prehung window with 1½" galvanized nails. *(Andersen)*

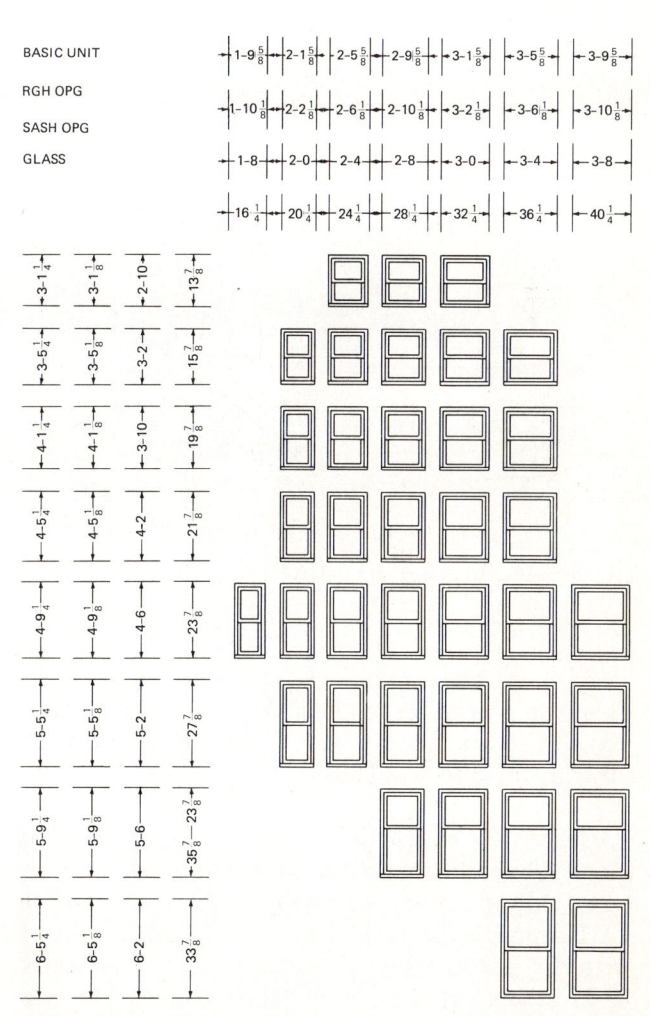

Figure 9-13. Various sizes of double-hung windows. *(Andersen)*

Installing Windows and Doors **215**

Figure 9-14. Double-hung windows in masonry (bottom floor) and set for conventional siding (top floor).

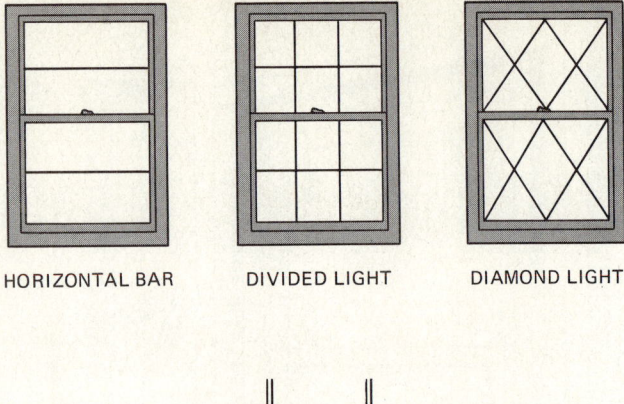

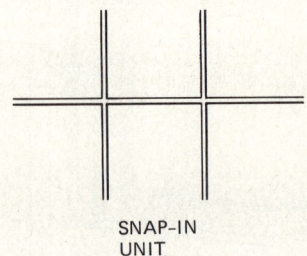

Figure 9-15. Plastic dividers make possible different windowpane treatments. *(Andersen)*

Figure 9-16. Diamond light pattern installed with plastic dividers.

This way it is easier to clean the window pane. Figure 9-16 shows a house with the diamond light pattern installed in windows that swing out.

CASEMENT WINDOW • The casement window is hinged so that it swings outward. The whole window opens, allowing for more ventilation. See Figure 9-17. This particular type has a vinyl flange for nailing it to the frame opening. It can be used as a single or in groups. This type of window can more easily be made weatherproof if it opens outward instead of inward.

Plastic muntins can be added to give a varied effect. They can be put into the windows as shown in Figure 9-16. The diamond light muntins divide the glass space into small diamonds which resemble individual panes of glass.

The crank is installed so that the window opens outward with a twist of the handle. Figure 9-18 shows some of the ways this type of window may be operated.

Screens are mounted on the inside. Storm windows are mounted on the outside as in Figure 9-19. In most cases, however, there is a thermopane

216 Carpentry Fundamentals

Figure 9-17. Casement window. *(Andersen)*

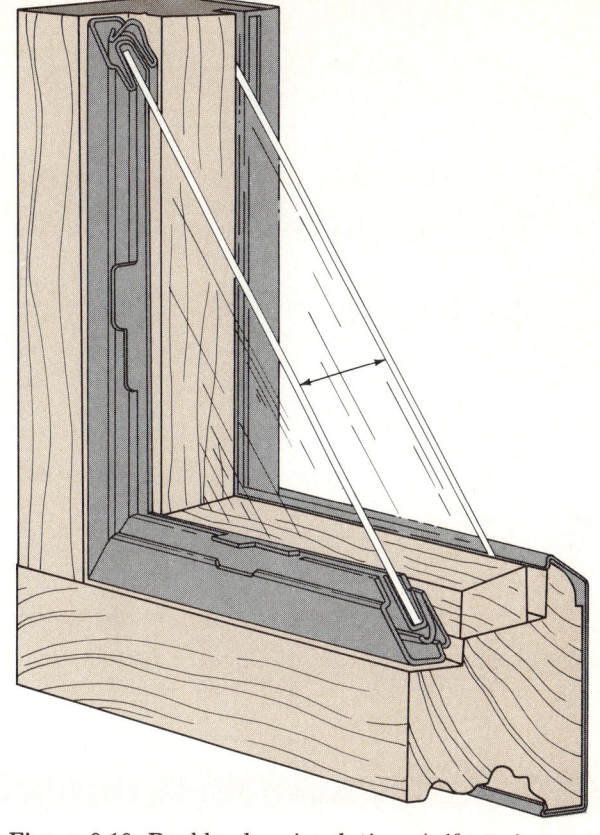

Figure 9-19. Double glass insulation. A $^{13}/_{16}''$ air space is placed between panes. *(Pella)*

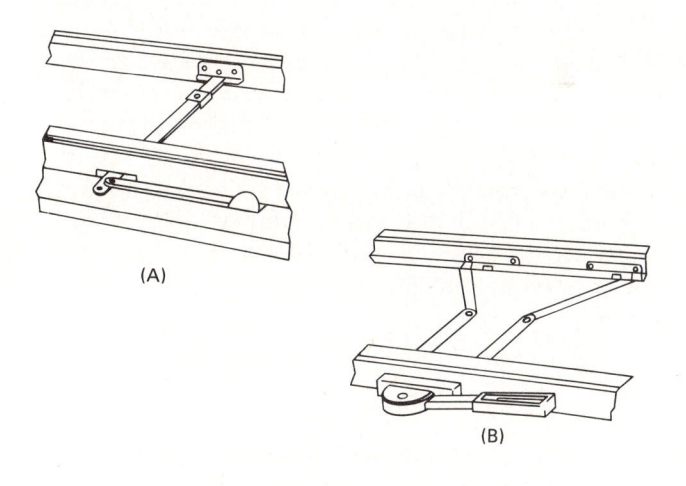

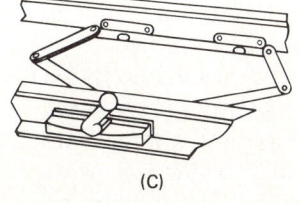

Figure 9-18. Methods of operating casement windows. (A) Standard push bar. (B) Lever lock. (C) Rotary gear. *(Andersen)*

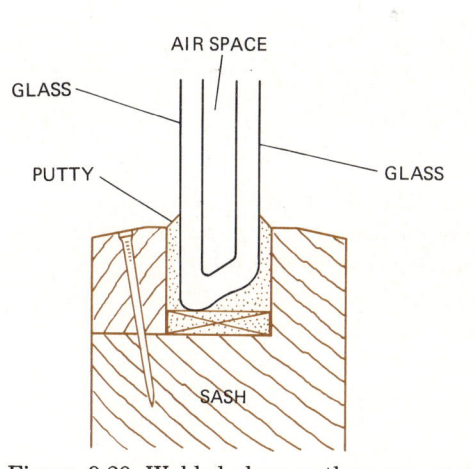

Figure 9-20. Welded glass or thermopane windows.

used for insulation purposes. The thermopane is a double sheet of glass welded together with an air space between. This is then set into the sash and mounted as one piece of glass. See Figure 9-20.

Multiple units are available. They may be movable or stationary. They may have one stationary part in the middle and two movable parts on the ends. Various combinations are available, as shown in Figure 9-21.

Installing Windows and Doors **217**

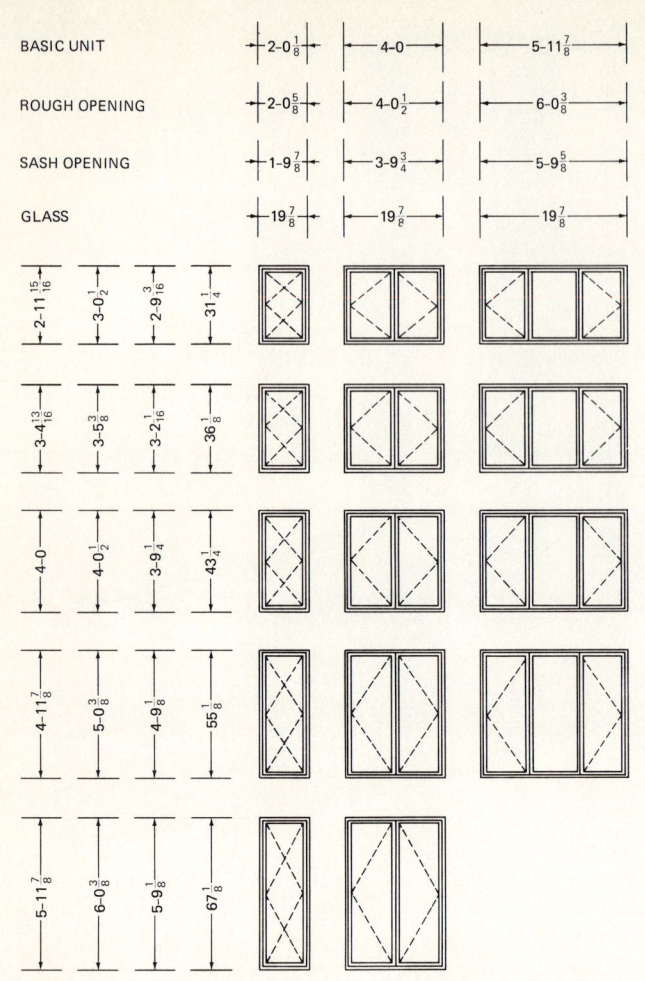

Figure 9-21. Various sizes of casement windows. *(Andersen)*

Figure 9-22. Picture window with awning bottom. *(Andersen)*

AWNING PICTURE WINDOW • This type of window has a large glass area. It also has a bottom panel which swings outward. A crank operates the bottom section. As it swings out, it has a tendency to form an awning effect—thus the name for this type. See Figure 9-22. A number of sizes and combinations are available in this type of window. See Figure 9-23. The fixed sash with an awning sash is also available in multiple units. Glass sizes are given in Figure 9-23. If you need to find the overall basic unit dimension, add the basic unit to 2⅞ inch. The rough opening is the sum of the basic units plus ½ inch.

Figure 9-24 shows the inswinging hopper type and the outswinging awning type of casement window. The awning sash type, shown in Figure 9-22, has a bottom sash that swings outward. You have to specify which type of opening you want when you order. Specifying bottom-hinged or top-hinged is the quickest way to order.

Figure 9-25 shows the various sizes of hopper- and awning-type casements available. They can be stacked vertically. If this is done, the overall unit dimension for stacked units is the sum of the basic units plus ¾ inch for two units high, plus ¼ inch for three units high, and less 1¼ inches for four units high. To find the overall basic unit width of multiple units, add the basic unit dimensions plus 2⅞ inches to the total. To find the rough opening width, add the basic unit width plus ½ inch.

Figure 9-26 shows how a number of units may be stacked vertically. All units in this case open for maximum ventilation.

• PREPARING THE ROUGH OPENING FOR A WINDOW •

It is important that you consult the window manufacturer's specifications before you make the rough opening.

Installation techniques, materials, and building codes vary according to area. Contact the window dealer for specific recommendations.

The same rough opening preparation procedures are used for wood and Perma-Shield windows. Figures 9-27 and 9-28 show the primed wood window and the Perma-Shield windows made by Andersen. These will be the windows discussed here. The instructions for installation will show how a manufactured window is installed.

218 Carpentry Fundamentals

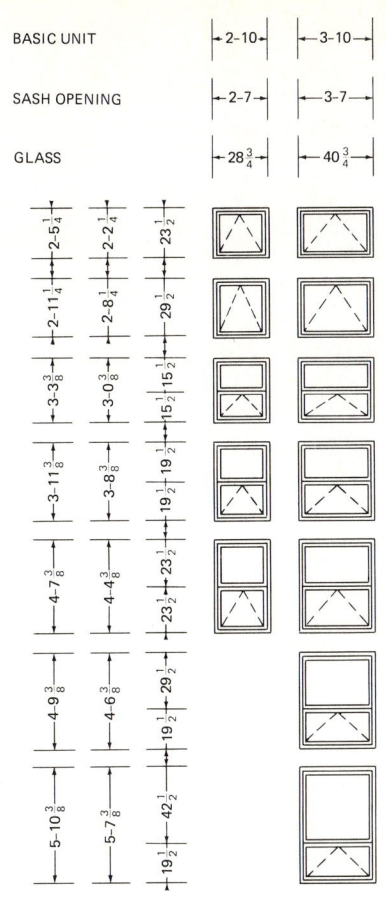

Figure 9-23. Various sizes of the fixed sash with awning sash windows. *(Andersen)*

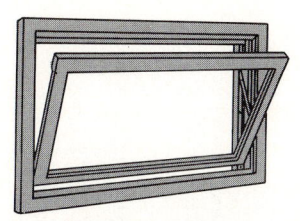

Figure 9-24. In swinging hopper and the out swinging awning types of window. These are casement windows.

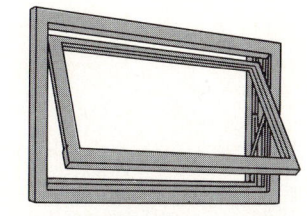

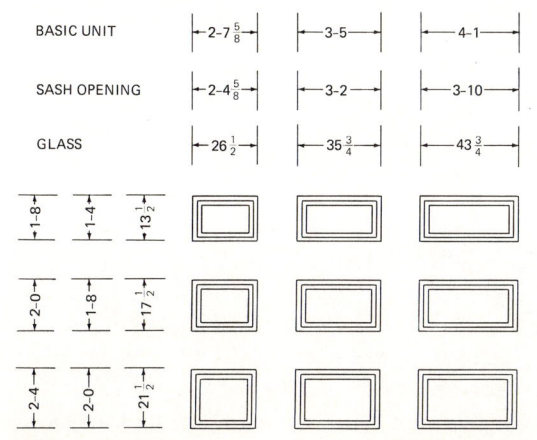

Figure 9-25. Various sizes of out swinging and in swinging windows. *(Andersen)*

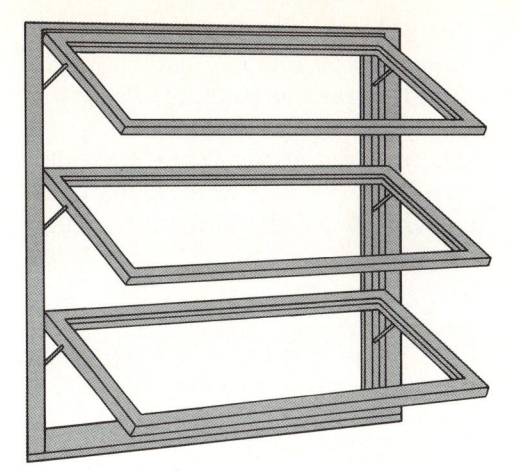

Figure 9-26. Vertical stacking of out swinging windows.

Figure 9-27. Primed wood window. *(Andersen)*

ANDERSEN WOOD CASEMENT WINDOW

Figure 9-28. Perma-Shield window. *(Andersen)*

PERMA-SHIELD® NARROLINE® WINDOW

Installing Windows and Doors **219**

Figure 9-29 shows how the wood casement window operates. Note the operator and its location. Figure 9-30 gives the details of the Perma-Shield Narroline window.

Figure 9-31 shows some of the possible window arrangements available from a window manufacturer. There is a window for almost any use. Select the window needed and follow the instructions or similar steps for your window.

Brick veneer with a frame backup wall is similar in construction to the frame wall in the following illustrations.

When the opening must be enlarged, make certain the proper size header is used. Contact the dealer for the proper size header. To install a smaller size window, frame the opening as in new installation.

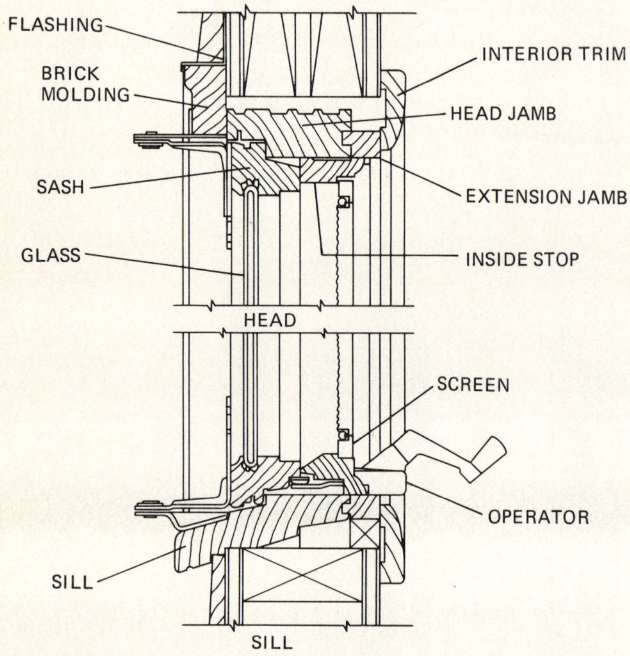

Steps in Preparing the Rough Opening

1. Layout the window-opening width between regular studs to equal the window rough opening width plus the thickness of two regular studs. See Figure 9-32.
2. Cut two pieces of window header material to equal the rough opening of the window plus

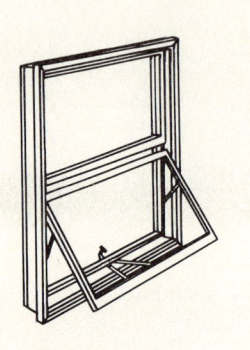

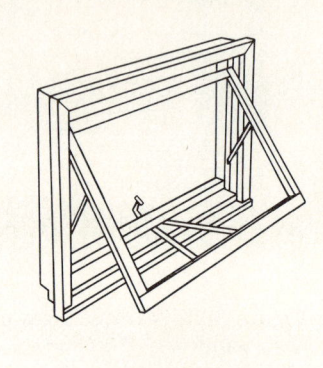

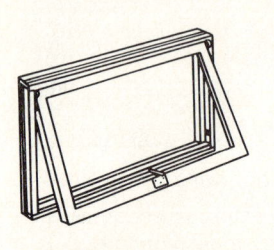

Figure 9-29. Details of the primed wood casement window. *(Andersen)*

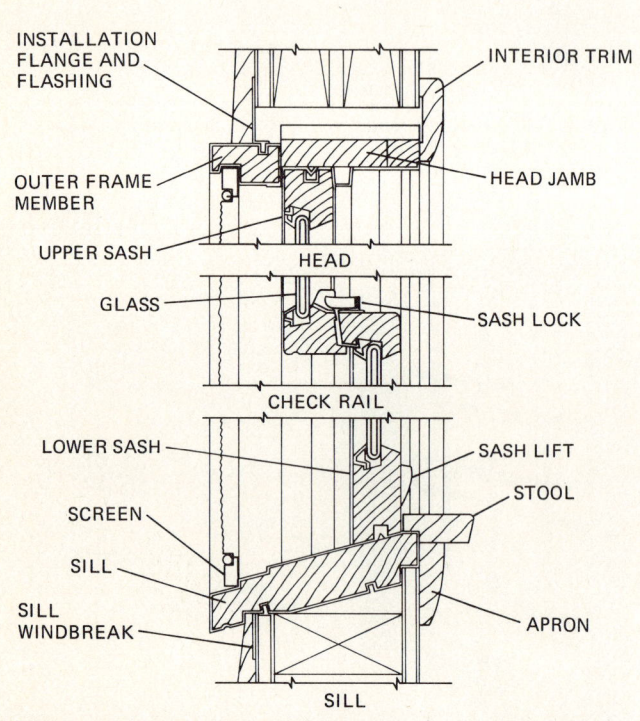

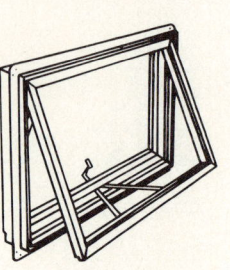

Figure 9-30. Details of the Perma-Shield window. *(Andersen)*

Figure 9-31. Various types of manufactured windows ready for quick installation. *(Andersen)*

220   Carpentry Fundamentals

the thickness of two jack studs. Nail the two header members together using an adequate spacer so that the header thickness equals the width of the jack stud. See Figure 9-33.
3. Position the header at the desired height between the regular studs. Nail through the regular studs into the header to hold the header in place until the next step is completed. See Figure 9-34.
4. Cut the jack studs to fit under the header for support. Nail the jack studs to the regular studs. See Figure 9-35.
5. Measure the rough opening height from the bottom of the header to the top of the rough sill. Cut 2×4-inch cripples and the rough sill to the proper length. See Figure 9-36. The rough sill length is equal to the rough opening width of the window. Assemble the cripples by nailing the rough sill to the ends of the cripples.
6. Fit the rough sill and cripples between the jack studs. See Figure 9-37. Toenail the cripples to the bottom plate and the rough sill to the jack studs at the sides. See the round insert in Figure 9-37.
7. Apply the exterior sheathing (fiberboard, plywood, etc.) flush with the rough sill, header, and jack stud framing members. See Figure 9-38.

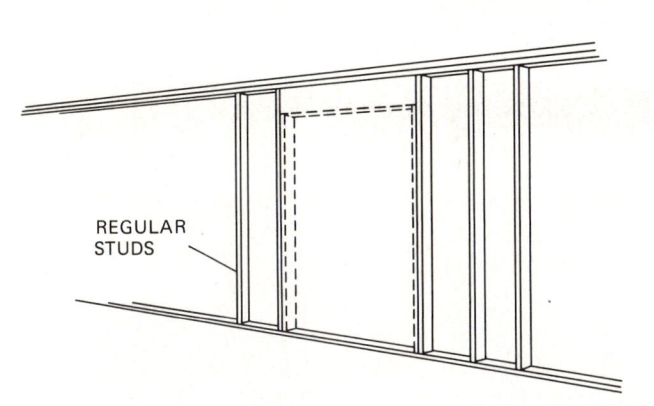

Figure 9-32. How to locate a window rough opening. *(Andersen)*

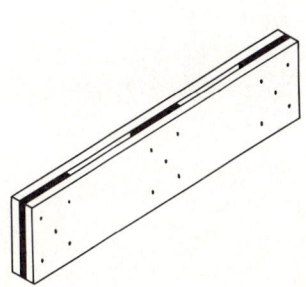

Figure 9-33. Making the header outside of the window opening. *(Andersen)*

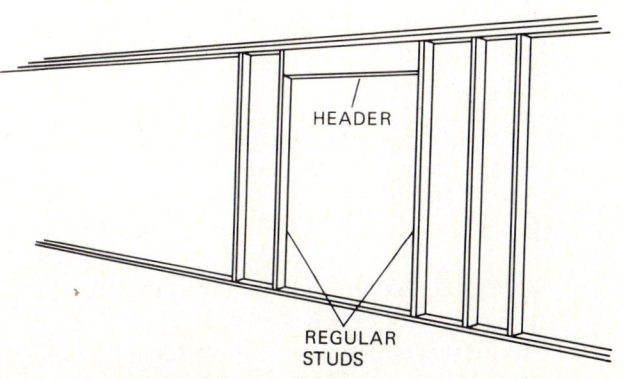

Figure 9-34. Placing the header where it belongs. *(Andersen)*

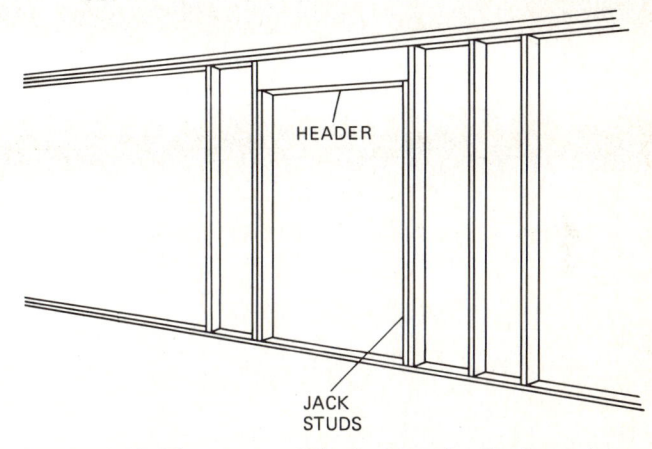

Figure 9-35. Placement of the jack studs. *(Andersen)*

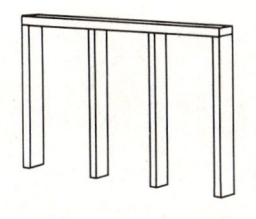

Figure 9-36. Assembling the cripples for easy placement. *(Andersen)*

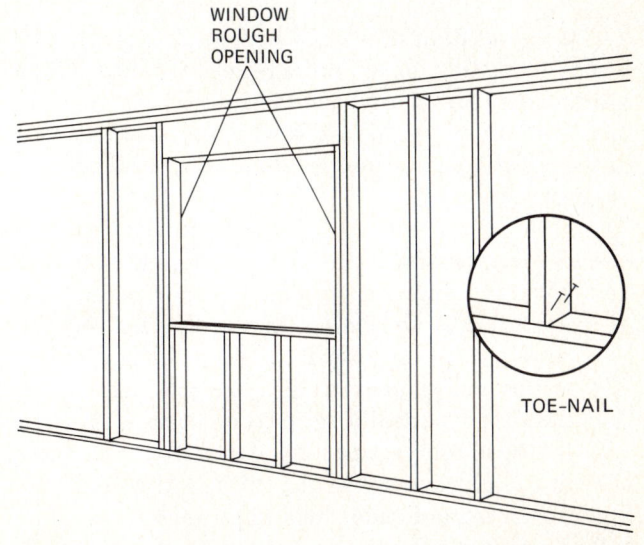

Figure 9-37. Placing the rough sill and cripples between the jack studs. Note the insert showing the toenailing. *(Andersen)*

Installing Windows and Doors

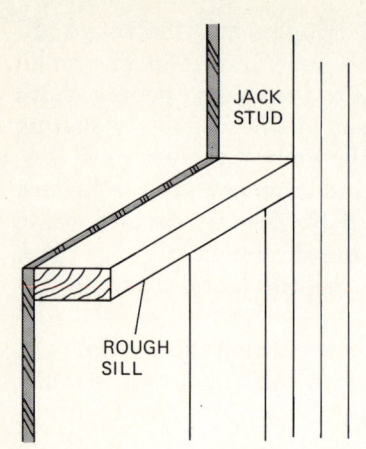

Figure 9-38. Applying the exterior sheathing. *(Andersen)*

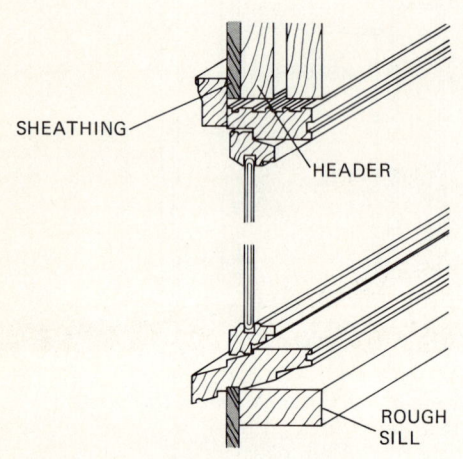

Figure 9-39. Placing the window in the rough opening. *(Andersen)*

• INSTALLATION OF A WOOD WINDOW •

The installation of a wood window is slightly different from that of a Perma-Shield window. However, there are many similarities. The following steps will show you how the windows are installed in the rough opening you just made from the preceding instructions.

1. Set the window in the opening from the outside with the exterior window casing overlapping the exterior sheathing. Locate the unit on the rough sill and center it between the side framing members (jack studs). Use 3½-inch casing nails and partially secure one corner through the head casing. See Figure 9-39. Drive the nail at a slight upward angle, through the head casing into the header. See Figure 9-40.
2. Level the window across the casing and nail it through the opposite corner with a 3½-inch casing nail. It may be necessary to shim the

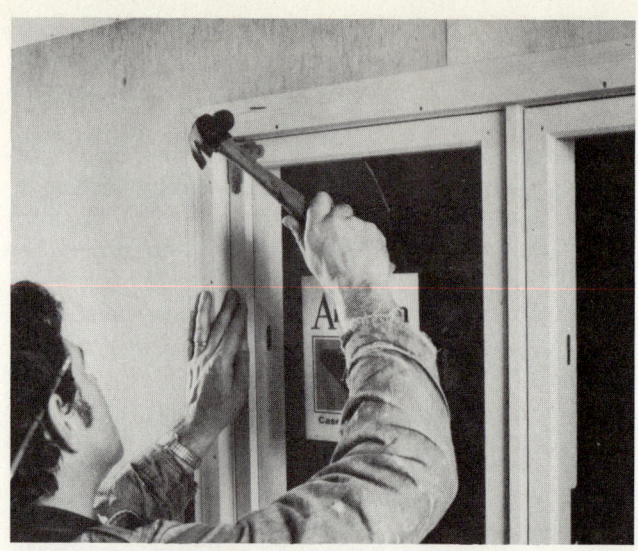

Figure 9-40. Uses 3½" nails to partially secure one corner through the head casing. *(Andersen)*

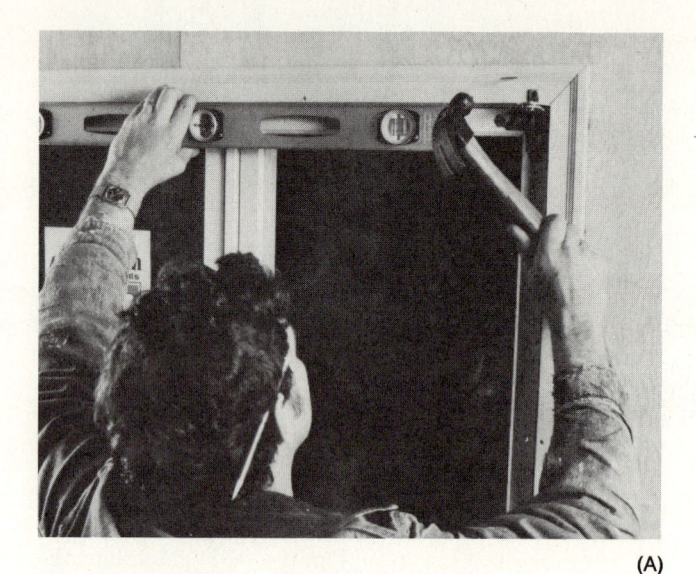

Figure 9-41. (A) Level the window across the casing and nail through the opposite corner. *(Andersen)* (B) Location of the shim that holds the window level. *(Andersen)*

window under the side jambs at the sill to level it. See Figure 9-41A and B.
3. Plumb (check the vertical of) the side jamb on the exterior window casing and drive a nail into the lower corner. See Figure 9-42. Com-

222   Carpentry Fundamentals

Figure 9-42. Check for plumb and nail the lower corner. *(Andersen)*

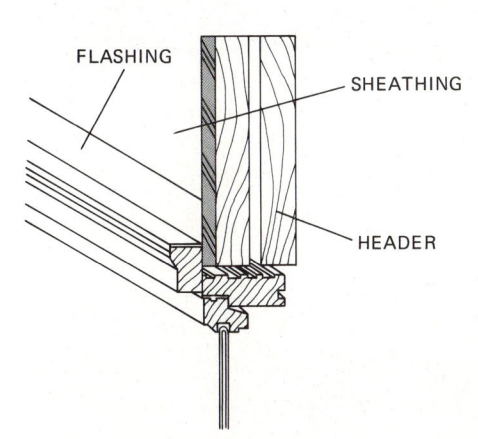

Figure 9-43. Apply the flashing with the rigid part over the head casing. *(Andersen)*

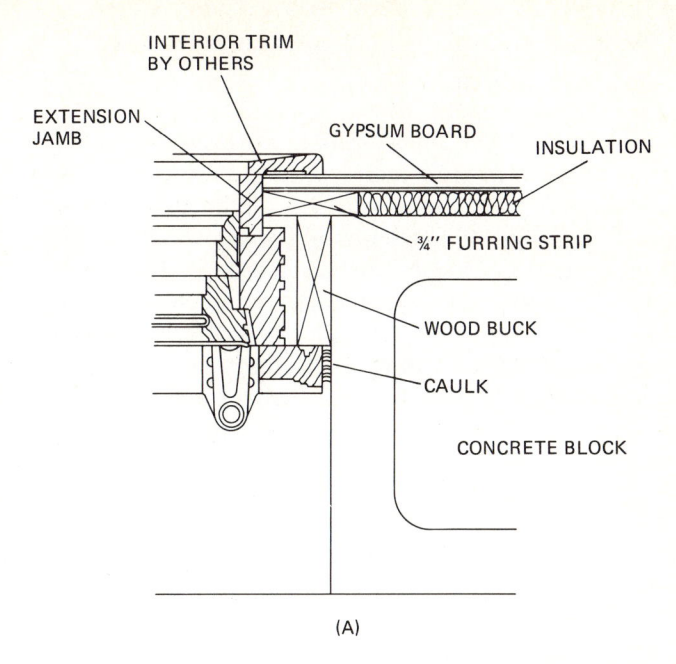

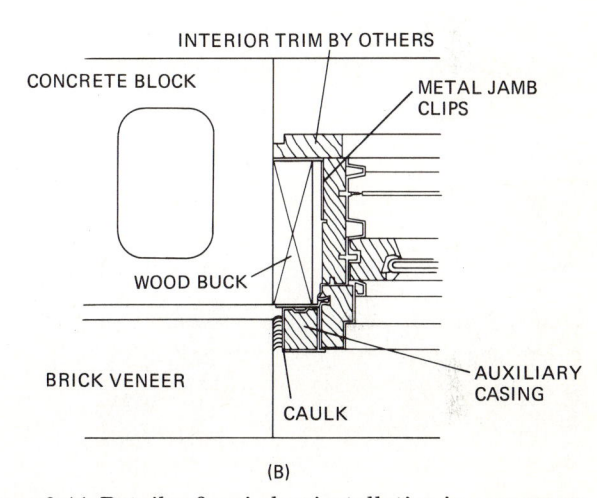

Figure 9-44. Details of a window installation in a masonry or brick veneer wall. (A) Wood window; (B) Perma-Shield window. *(Andersen)*

plete the installation by nailing through the exterior casing with 3½-inch nails. Space the nails about 10 inches apart.
4. Before you finally nail in the window, make sure you check the sash to see that it operates easily.
5. Apply a flashing with the rigid portion over the head casing. See Figure 9-43. Secure this flashing with 1-inch nails through the flexible vinyl into the sheathing. *Do not nail into the head casing.*
6. Caulk around the perimeter of the exterior casing after the exterior siding or brick is applied.

MASONRY OR BRICK VENEER WALL • This type of window can be installed in masonry wall construction. Fasten the wood buck to the masonry wall and nail the window to the wood buck using the procedures just shown for frame wall construction.

In Figure 9-44A you see the wood window installed in a masonry wall. Figure 9-44B shows a Perma-Shield window installed with metal jam clips in a masonry wall with brick veneer. The metal jam clips and the auxiliary casing are available when specified.

Keep in mind that when brick veneer is used as an exterior finish, adequate clearance must be left for caulking between the window sill and the masonry. This will prevent damage and bowing of the sill. The bowing is caused by the settling of the structural member. Shrinkage will also cause damage. Shrinkage takes place as the rough-in lumber dries after enclosure and the heat is turned on in the house.

Installing Windows and Doors 223

Installation of the Andersen Perma-Shield Window

To install the Perma-Shield window, you follow the installation procedure just described. There are some exceptions, however. See Figure 9-45.

The design of Perma-Shield windows eliminates the need for exterior casing to secure the window to the wall. Just apply nails (1⅞-inch×10 galvanized) through the predrilled installation flange/flashing. This installation eliminates the need for separate head flashing. See Figures 9-46 and 9-47.

Figure 9-48 shows a house with windows installed using the methods just described.

TERMS USED IN WINDOW INSTALLATION • Now is a good time to review the terms associated with the installation of a window. This will make it possible for you to understand the terminology when you work with a crew installing windows.

Plumb The act of checking the vertical line of a window when installing it in a rough opening.

Level The act of checking the horizontal line of a window when installing it in a rough opening.

Figure 9-46. Installing the window with the plastic shield around it. Note how easy it is to place the window into position and secure it.

Figure 9-45. Installing the Perma-Shield window. It eliminates the need for exterior casing to secure the window to the wall.

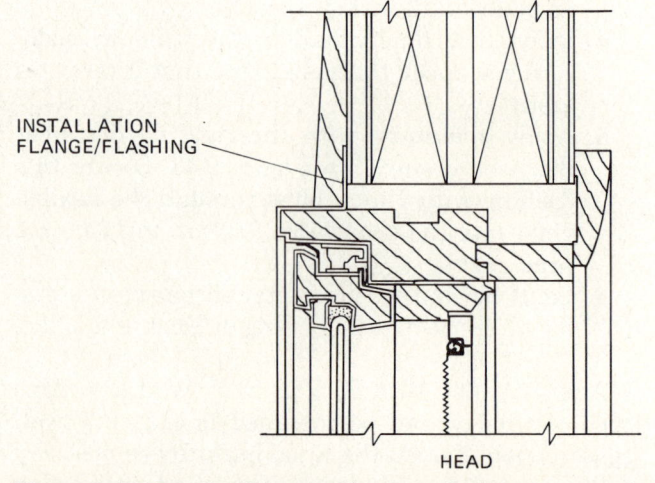

Figure 9-47. Details of how the flange or flashing is covered by the exterior wall covering.

224 Carpentry Fundamentals

Figure 9-48. Windows help make the house attractive. *(Andersen)*

Figure 9-50. A prehung exterior door with three panels of glass. Note how the trim sticks out sufficiently for the siding to butt against it.

construction, or steel I beams in heavier masonry construction.

Rough Sill A horizontal framing member, usually a single 2×4, located across the bottom of the window rough opening. The window unit rests on the rough sill.

Cripples Short vertical framing members spaced approximately 16 inches O.C., located below the rough sill across the width of the rough opening. Also used between the header and the top plate, depending upon the size of the headers required.

Shim An angled wood member—wedge shaped—used as a filler at the jamb and sill. (Wood shingle makes a good shim.)

Wood Buck A structural wood member secured to a masonry opening to provide an installation frame for the window unit.

• PREHUNG DOORS •

Figure 9-49. Various door designs. *(National Woodwork Manufacturers)*

Regular Stud A vertical frame member that runs from the bottom plate on the floor to the top plate at the ceiling. In normal construction, this is a 2×4 approximately 8 feet long.

Jack Stud A vertical frame member that forms the window rough opening at the sides. It runs from the bottom plate at the floor to the underside of the header. It supports the header.

Header A horizontal framing member located over the window rough opening supported by the jack studs. Depending upon the span, headers usually are double 2×6s, 2×8s, or 2×10s in frame wall

Types of Doors

Exterior doors are made in many sizes and shapes. See Figure 9-49. They may be solid with a glass window. They may have an X shape at the bottom, in which case they are referred to as a cross-buck. These doors are made in a factory and crated and shipped to the site. There they are unpacked and placed in the proper opening. There is little to do with them other than level them and nail them in place. The hardware is already mounted on the door.

Figure 9-50 shows a door that was prehung and shipped to the site. Note how it sticks out from the

Installing Windows and Doors **225**

Figure 9-51. The proper door can do much to improve the looks of the house.

sheathing so that the siding can be applied and butted to the side jamb. Doors are chosen for their contribution to the architecture of the building. They must harmonize with the design of the house. Figure 9-51 shows a door that adds to the design of the house. The door facing or trim adds to the column effect of the porch.

FLUSH DOORS • Flush doors are made of plywood or some facing over a solid core. The core may be made of a variety of materials. In some instances where the door is used inside, the inside of the door is nothing more than a mesh or strips. See Figure

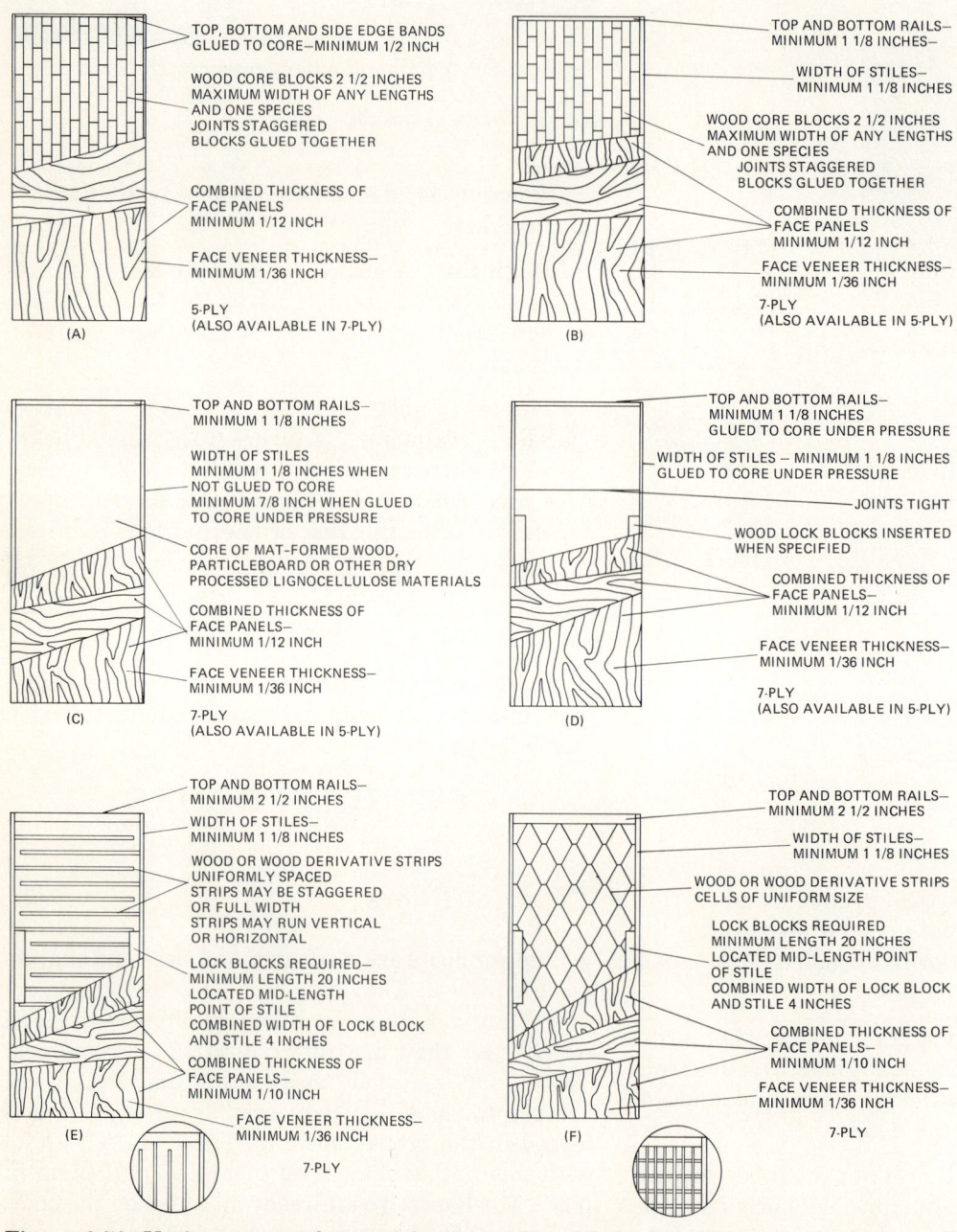

(A)
- TOP, BOTTOM AND SIDE EDGE BANDS GLUED TO CORE—MINIMUM 1/2 INCH
- WOOD CORE BLOCKS 2 1/2 INCHES MAXIMUM WIDTH OF ANY LENGTHS AND ONE SPECIES JOINTS STAGGERED BLOCKS GLUED TOGETHER
- COMBINED THICKNESS OF FACE PANELS MINIMUM 1/12 INCH
- FACE VENEER THICKNESS— MINIMUM 1/36 INCH
- 5-PLY (ALSO AVAILABLE IN 7-PLY)

(B)
- TOP AND BOTTOM RAILS— MINIMUM 1 1/8 INCHES
- WIDTH OF STILES— MINIMUM 1 1/8 INCHES
- WOOD CORE BLOCKS 2 1/2 INCHES MAXIMUM WIDTH OF ANY LENGTHS AND ONE SPECIES JOINTS STAGGERED BLOCKS GLUED TOGETHER
- COMBINED THICKNESS OF FACE PANELS MINIMUM 1/12 INCH
- FACE VENEER THICKNESS— MINIMUM 1/36 INCH
- 7-PLY (ALSO AVAILABLE IN 5-PLY)

(C)
- TOP AND BOTTOM RAILS— MINIMUM 1 1/8 INCHES
- WIDTH OF STILES MINIMUM 1 1/8 INCHES WHEN NOT GLUED TO CORE MINIMUM 7/8 INCH WHEN GLUED TO CORE UNDER PRESSURE
- CORE OF MAT-FORMED WOOD, PARTICLEBOARD OR OTHER DRY PROCESSED LIGNOCELLULOSE MATERIALS
- COMBINED THICKNESS OF FACE PANELS— MINIMUM 1/12 INCH
- FACE VENEER THICKNESS— MINIMUM 1/36 INCH
- 7-PLY (ALSO AVAILABLE IN 5-PLY)

(D)
- TOP AND BOTTOM RAILS— MINIMUM 1 1/8 INCHES GLUED TO CORE UNDER PRESSURE
- WIDTH OF STILES – MINIMUM 1 1/8 INCHES GLUED TO CORE UNDER PRESSURE
- JOINTS TIGHT
- WOOD LOCK BLOCKS INSERTED WHEN SPECIFIED
- COMBINED THICKNESS OF FACE PANELS— MINIMUM 1/12 INCH
- FACE VENEER THICKNESS— MINIMUM 1/36 INCH
- 7-PLY (ALSO AVAILABLE IN 5-PLY)

(E)
- TOP AND BOTTOM RAILS— MINIMUM 2 1/2 INCHES
- WIDTH OF STILES— MINIMUM 1 1/8 INCHES
- WOOD OR WOOD DERIVATIVE STRIPS UNIFORMLY SPACED STRIPS MAY BE STAGGERED OR FULL WIDTH STRIPS MAY RUN VERTICAL OR HORIZONTAL
- LOCK BLOCKS REQUIRED— MINIMUM LENGTH 20 INCHES LOCATED MID-LENGTH POINT OF STILE COMBINED WIDTH OF LOCK BLOCK AND STILE 4 INCHES
- COMBINED THICKNESS OF FACE PANELS— MINIMUM 1/10 INCH
- FACE VENEER THICKNESS— MINIMUM 1/36 INCH
- 7-PLY

(F)
- TOP AND BOTTOM RAILS— MINIMUM 2 1/2 INCHES
- WIDTH OF STILES— MINIMUM 1 1/8 INCHES
- WOOD OR WOOD DERIVATIVE STRIPS CELLS OF UNIFORM SIZE
- LOCK BLOCKS REQUIRED MINIMUM LENGTH 20 INCHES LOCATED MID-LENGTH POINT OF STILE COMBINED WIDTH OF LOCK BLOCK AND STILE 4 INCHES
- COMBINED THICKNESS OF FACE PANELS— MINIMUM 1/10 INCH
- FACE VENEER THICKNESS— MINIMUM 1/36 INCH
- 7-PLY

Figure 9-52. Various types of materials are used to fill the interior space in flush doors. *(National Woodwork Manufacturers)*

9-52. Wood is usually preferred to metal for exterior doors of homes. Wood is nature's own insulator. While metal readily conducts heat and cold, wood does not. Wood is 400 times more effective an insulator than steel and 1800 times more effective than aluminum.

Stock wood flush doors come in a wide variety of sizes, designs, and shapes. Standard wood door frames will accommodate wood combination doors, storm doors, and screen doors without additional framing expense.

PANEL DOORS • This type of door has solid vertical members, rails, and panels. Many types are available. See Figures 9-53 and 9-54. The amount of wood and glass varies. Many people want a glass section in the front door. The four most popular types are clear, diamond obscure, circle obscure, and amber Flemish. Figure 9-54 shows some of the decorative variations in doors. Note how the type of door can improve the architecture. The main entrance may be highlighted with sidelights. See Figure 9-55. These are 12 or 14 inches wide and may be used

Figure 9-53. Add-on panels and lights give designers options with flush doors. *(General Products)*

CHATEAU 8-PANEL

WILLIAMSBURG 6-PANEL

COLONIAL 9-LIGHT

CROSS BUCK 9-LIGHT

Figure 9-54. The four most popular door styles. They have clear, diamond obscure, circle obscure, and amber flemish safety glass. *(General Products)*

Installing Windows and Doors **227**

Figure 9-55. Sidelights are designed for fast installation as integrated units in wood or plastic models. This insulated safety glass comes in 12" or 14" widths. *(General Products)*

on one or both sides of the door. The panels of glass are varied to meet different requirements.

SLIDING DOORS • Sliding doors are just what the name suggests. They usually have tempered or safety glass. They can slide to the left or to the right. You have to specify a right- or left-sliding door when ordering. Most have insulating construction. They have two panes of glass with a dead-air space in between. See Figure 9-56. In Figure 9-57 you can see the sizes available. Also note the arrow which indicates the direction in which the door slides.

FRENCH DOORS • This is usually two or more doors grouped to open outward onto a patio or ve-

Figure 9-56. Sliding door. *(Andersen)*

randa. They have glass panes from top to bottom. They may be made of metal or wood. Later in this chapter you will see two- and three-window groupings mounted step by step.

• INSTALLING AN EXTERIOR DOOR •

The door frame has to be installed in an opening in the house frame before the door can be hung. Figure 9-58 shows the parts of the door frame. Note the way it goes together. Figure 9-59 shows the frame in position with a wire on the right. It has been pulled through for the installation of the doorbell push button. Note the spacing of the hinges. The door in the background is a six-panel type that is already hung. The siding has not yet been butted against the door casing. It will be placed as close as possible and then caulked to prevent moisture from damaging the wood over a period of time.

Figure 9-60A shows the general information you need to be able to identify the parts of a door. It is important that the door be fitted so that there is a uniform ⅛-inch clearance all around to allow for free swing. Allow ½ to ¾ inch at the bottom for a better fit with carpet.

There are nine steps to installing an exterior door:

228 Carpentry Fundamentals

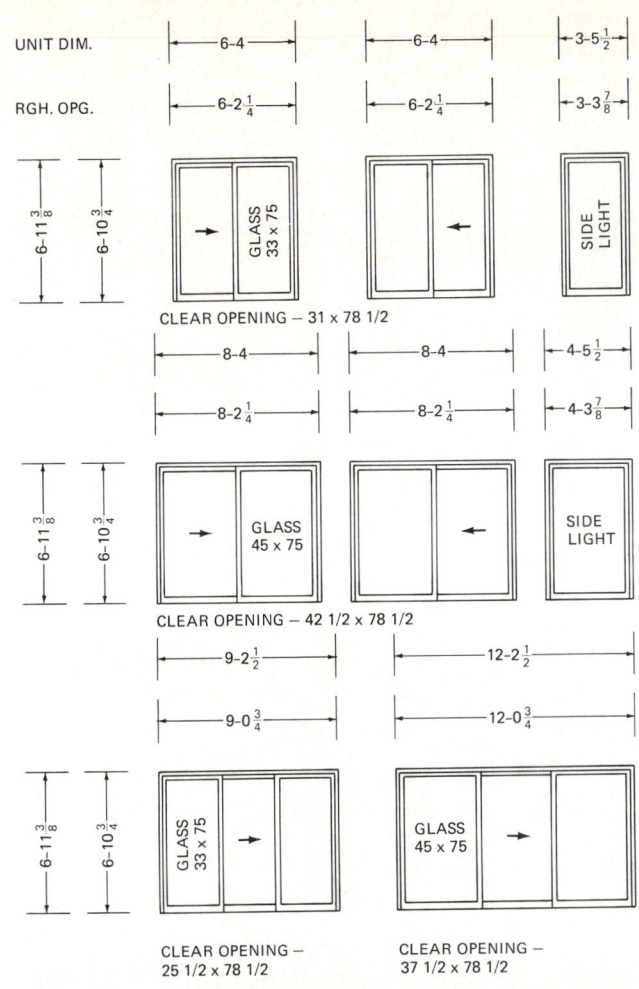

Figure 9-57. Various sizes of sliding doors. *(Andersen)*

Figure 9-59. Prehung door in place. Note the wires hanging out on the left side. They indicate where an outside light will go.

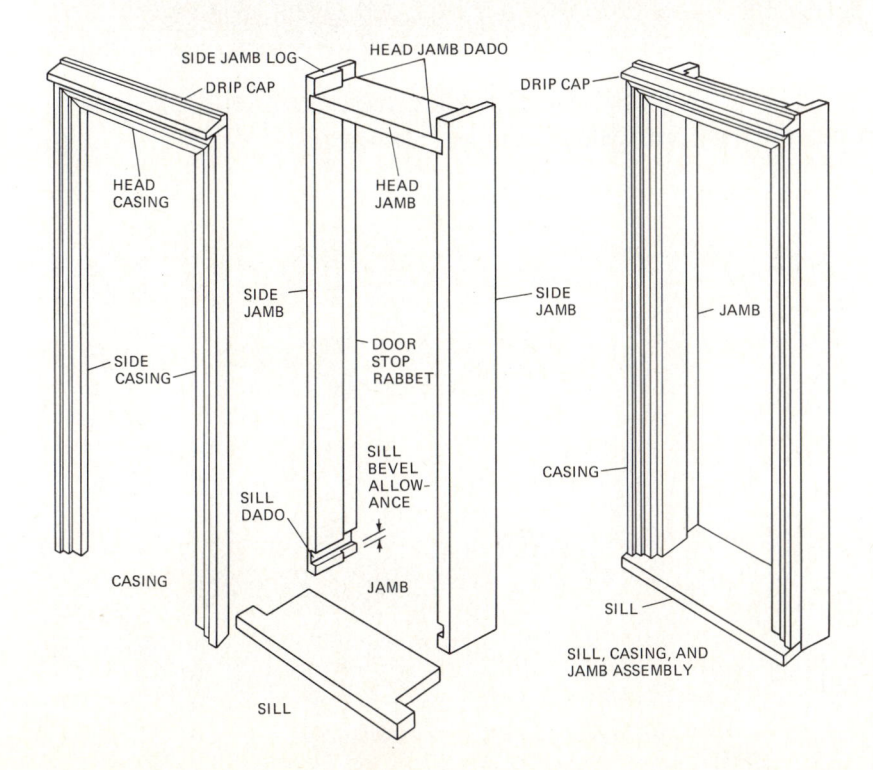

Figure 9-58. Assembling a door frame (left to right).

Installing Windows and Doors **229**

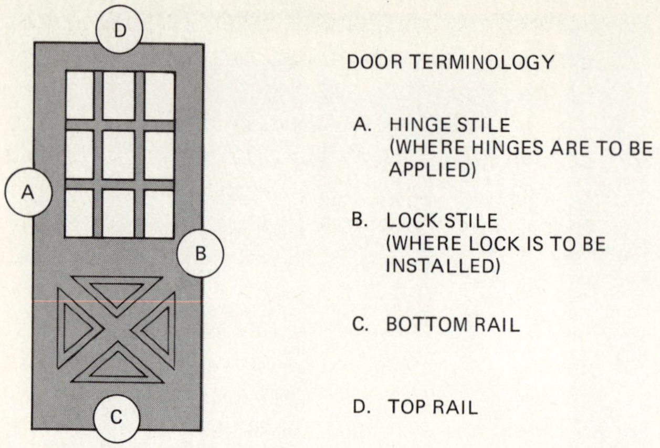

DOOR TERMINOLOGY

A. HINGE STILE
   (WHERE HINGES ARE TO BE APPLIED)

B. LOCK STILE
   (WHERE LOCK IS TO BE INSTALLED)

C. BOTTOM RAIL

D. TOP RAIL

Figure 9-60. (A) Terms used with doors. *(Grossman Lumber)*

(B) Trim the door height.

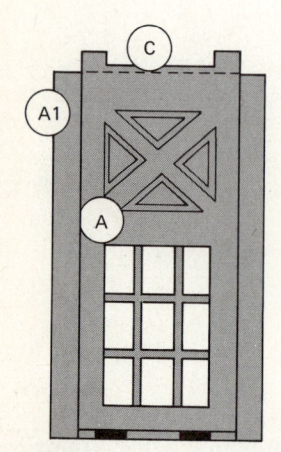

(C) Mark the door to fit the frame.

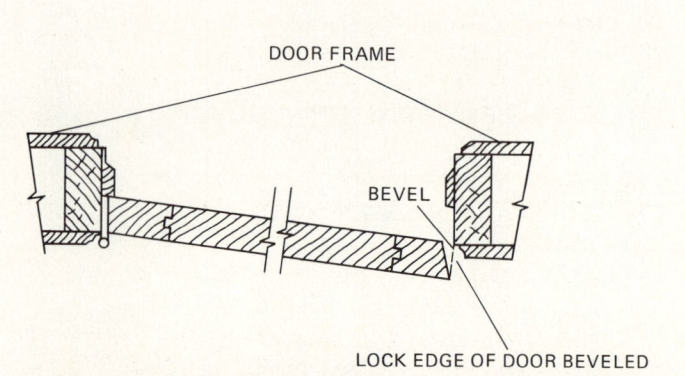

(D) Mark the other end of the door.

(E) Trim the width of the door.

(F) Bevel the lock stile of the door.

230 Carpentry Fundamentals

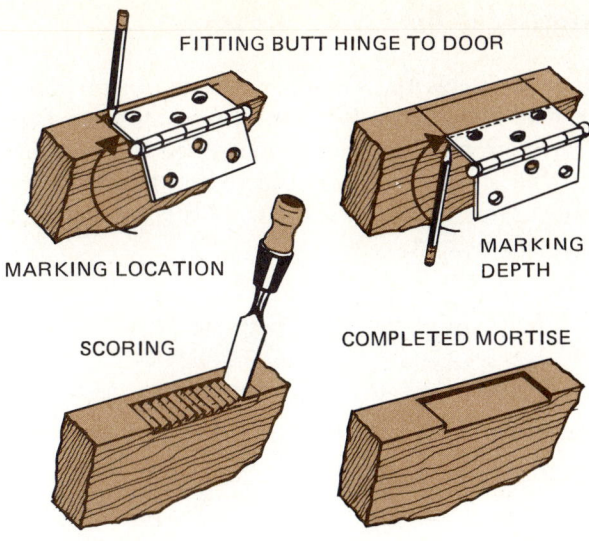

FITTING BUTT HINGE TO DOOR

MARKING LOCATION

MARKING DEPTH

SCORING

COMPLETED MORTISE

(G) Install the hinges.

(H) Check the height of the hinges.

1. Figure 9-60B. Trim the height of the door. Many doors are made with extra long stiles or horns A. Before proceeding, cut the horns off the *top* of the door, even with the top of the top rail. When cutting, start with the saw at the outside edge to avoid splintering the edges of the door.
2. See Figure 9-60C. On the inside of the room, place the door into the opening upside down, and against the door jamb. Keep the hinge stile tight against the hinge jamb A1 of the door frame. This edge must be kept straight to ensure that hinges will be parallel when installed. Place two ¼-inch blocks under the door. These will raise the door to allow for ⅛-inch clearance at both top and bottom when cut. Mark the door at C, the top of the door frame opening, for cutting.
3. Figure 9-60D. After cutting the door to the proper height, place the door into the opening again—this time right side up, with ⅛-inch blocks under the door for clearance. With the door held tightly against the hinge jamb of the door frame, have someone mark a pencil line along the lock stile of the door from outside the door opening (line 1 to 1A), holding a ⅛-inch block between the pencil and the door frame. This will automatically allow for the necessary ⅛-inch clearance needed.
4. See Figure 9-60E. Trim the width of the door. If the amount of wood to be removed from the door (line 1 to 1A) is more than ¼ inch, it will be necessary to trim both edges of the door. Trim one-half the width of the wood to be removed from each edge of the door. Use a smooth or jack plane.
5. See Figure 9-60F. Bevel the lock stile of the door. The lock stile of the door should be planed to about a 3° angle so that it will clear the door frame when the door is swung shut.
6. See Figure 9-60G. Install the hinges. Be sure that markings for hinges are uniform for all hinges used. The mortises (cutouts) for hinges should be of uniform depth (thickness of the hinge). Measure 7 inches down from the top of the door and 7⅛ inches down from the underside of the door frame top. Mark the locations of the top edge of the upper hinge. Placement of hinges will be unnecessary if the door is pre-hung.
7. See Figure 9-60H. The bottom hinge is 9 inches up from the door bottom (9⅛ inches up from the threshold). The middle hinge is centered in the door height. Attach the hinge leafs to the door and door frame. Hang the door in the opening. If the mortises are cut properly but the hinges still bind, the frame jamb may be distorted or bowed. It may be necessary to place a thin shim under one edge of the frame hinge leaf to align it parallel and relieve the binding.
8. Install the lockset. Because of the variety of styles of locksets, there is no one way to install them. This subject will be discussed in detail later in this chapter. Each lockset comes with a complete set of instructions. The best advice is to follow these instructions.
9. Finish the door. Care must be taken to paint or seal all door surfaces. Top, bottom, edges and faces should be sealed and painted. Weather and moisture can hurt a door and decrease its performance. A properly treated door will give many years of satisfactory service.

Installing Windows and Doors

Hanging a Two-door System

In some cases the two-door system is the main entrance. See Figure 9-61. In other cases the two-door system may be French doors. See Figure 9-62. This type of door comes in pieces and has to be assembled. The details for hanging of this type of door are shown in Figure 9-63.

Figure 9-61. Double doors installed. The one on the left is the active door.

Figure 9-62. Double door ready for assembly. *(General Products)*

Step 1 — Attach Astragal to Inactive Door

Remove plastic filler plates from deadbolt and lock locations. Remove appropriate metal knock-out at deadbolt location.
Place door on edge, place astragal with *notch to door top* as shown. Compress outer flange against face of door, install bolt retainer spring at top and bottom of astragal, and secure astragal with five (5) self-tapping screws using power driver or drill.
Place two nylon screw bosses under the strike route and secure strike with 2 No. 8 screws provided.
Strike has tab that can be adjusted to assure proper closing while hanging door.
Place the bolt assembly in top and bottom of astragal. Adjust the bolt retainer spring to proper position and secure bolt retainer with Allen wrench provided.
Snap in 2 ea. channel closures in bolt recess above and below strike location. (proper lengths provided).
Tape one pile pad to interior face of inactive door for installation on astragal after door is installed in opening.

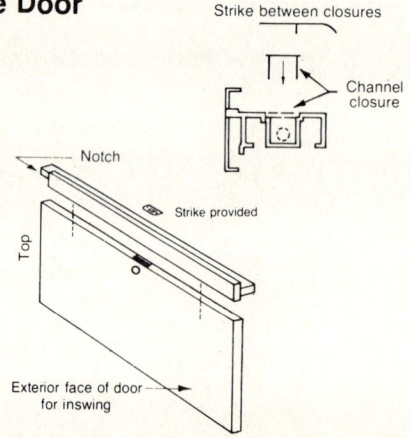

Step 2 — Attach Sweep

Place inactive door on prehang table, <u>interior</u> face up. Pick up sweep with flush bolt hole, place door bottom spacer on latch end of sweep as shown.
Place sweep on bottom of door, flush latch end with latch edge of astragal.
Compress tightly against door bottom and drive self drilling screws into door skin at extreme bottom of slot. Tighten screws moderately to hold sweep in up position.

Check operation of bolt assembly.
Turn door over (Interior face down)
Place active door on table (interior face up) and install sweep in same manner as first door.
Turn door over and proceed with frame prehanging. (Step 3)

232 Carpentry Fundamentals

Step 3

Attach Hinge Jambs

Place hinge jambs beside edge of doors as shown and attach hinges to doors with No. 10 machine screws.

Apply caulking tape to jambs at threshold locations. Make sure it follows contour of vinyl threshold part.

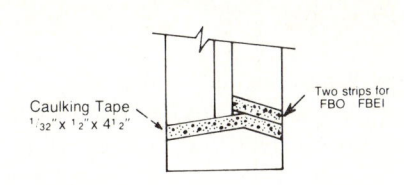

Step 4

Installing Header and Threshold

Stick 1/8" PAK-WIK spacers—2 ea. on header jambs, and 3 ea. on the astragal side at locations marked "X", to maintain 1/8" clearance between doors and jambs. Place header jamb against header stop block. Line up doors with header jamb and press doors firmly against it. Raise jambs up, make sure they are flush with header jamb at top corners. Drive 3 ea. 2¼" long staples in each corner of frame.

Assemble vinyl and aluminum threshold parts. Place threshold in frame and secure with #10 x 1½" screws through pre-drilled holes. Back edge to be flush with frame. (Make sure pile is firmly in contact with threshold and weatherstripping.) If not, remove and reposition.

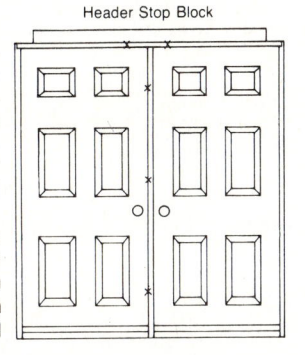

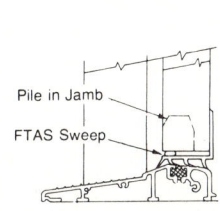

Step 5

Attaching Brickmold

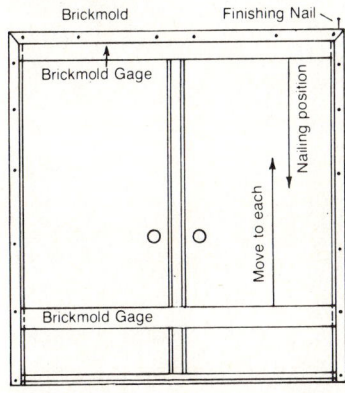

NOTE: If door is to be outswing, proceed with bracing shown in Step 6. Then turn unit upside down and install bolt strike as in Step 6. Proceed with brickmold in Step 5. NOTE: Outswing frame requires 5/8" longer header brickmold than inswing. Place brickmold gage at top and bottom of jambs. Position miter joints of jamb and header brickmold. Align fit of mitered corners and properly space reveal. Tack each corner, nail header brickmold, move gage down jambs and nail brickmold. Use 6 ea. No. 10 x 2½" finishing nails as shown, drive 2 ea. No. 10 x 2½" finishing nails in two corners as shown.

Step 6

Installing Flush Bolt Strike and Bracing Frame

Turn unit upside down and replace on table. Place pencil mark at flush bolt locations on header and threshold. Open inactive door, place thin dab of putty at bolt pencil marks on header and threshold. Close inactive door. Move bolts to mark the putty. Open inactive door. Center punch top and bottom bolt locations with nail. Remove putty. Drill 5/8" hole in header and threshold. Place bolt strike on header and align with drilled hole. Install 2 ea. No. 6 x 1" screws provided. Close inactive door and secure bolts. Close active door. Tack corner braces as shown. Tack strip of wood across frame, approximately 12" above threshold as shown. Use 8d coated box nails. On outswing unit, cut bracing to fit between brickmold.

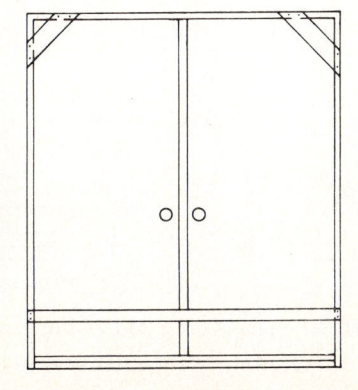

Installing Windows and Doors   233

Figure 9-64 shows how the active and inactive doors are identified first. In most cases both doors do not open. This is especially the case when the entrance door is involved. In the case of French doors, both doors can open.

Handing Instructions

With the assembly kit furnished, one of the first things you will want to check is the "hand" of the door. Hand of doors is always determined by the *outside*. Inswinging doors are more common. See Figure 9-65. The right-hand symbol is RH, and it means the door swings on hinges that are mounted on the right. If the door swings out, it is a right-hand reverse door and the symbol is RHR. It is still hinged on the right looking at it from the outside.

The left-hand symbol is LH. That means that on an inswinging door the hinges are on the left. This is most convenient for persons who are right-handed. If the door is outswinging, the left-hand reverse symbol is LHR. Doors usually swing into a wall where they can rest against the adjoining wall. They usually swing back only 90°. The traffic pattern also determines the way a door swings. Doors swing out in most commercial, industrial, and school buildings. This lets a person open the door outward so that it will be easy to leave the building if there is a fire. Safety is the prime consideration in this case.

Figure 9-66 shows how energy conservation has entered the picture. The figure shows how the top and bottom of the door are fitted to make sure air does not leak through the door. The door may be made of metal. Because metal conducts heat and cold, the door is insulated. See Figure 9-67.

Figure 9-64. Active and inactive door designation. *(General Products)*

Figure 9-65. Handling chart. *(General Products)*

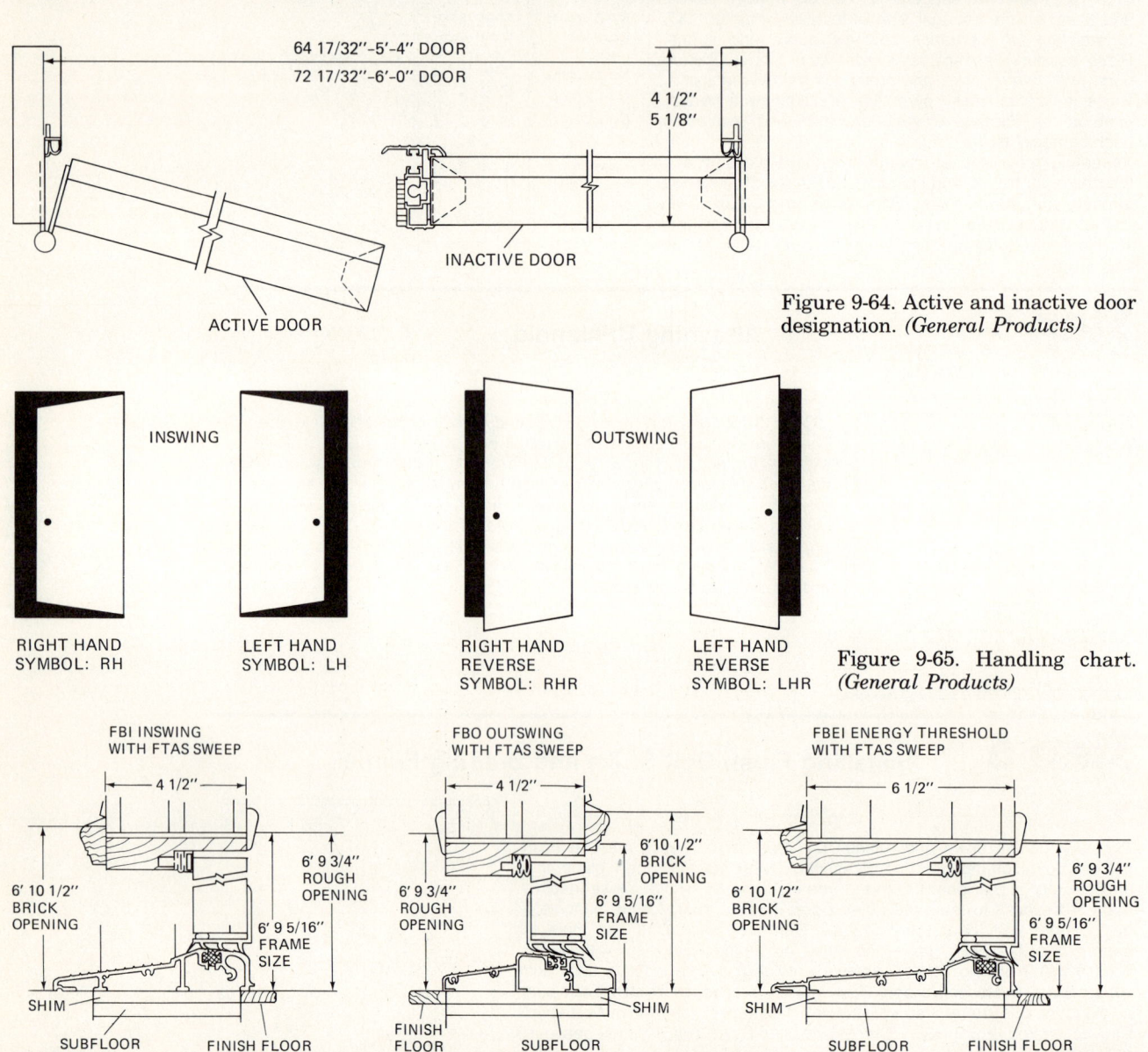

Figure 9-66. Finished dimensions on a double-hung door. *(General Products)*

234 Carpentry Fundamentals

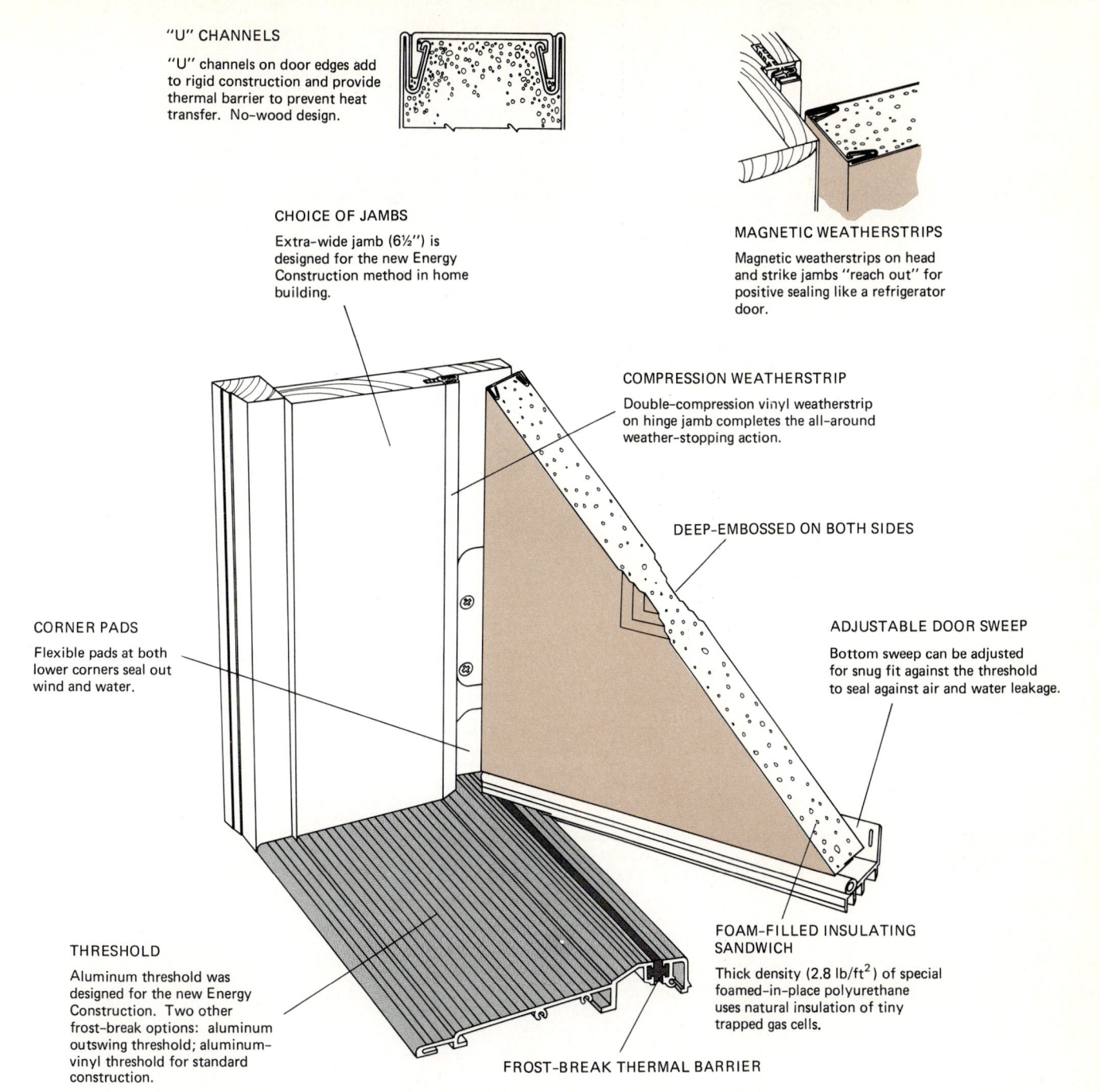

"U" CHANNELS
"U" channels on door edges add to rigid construction and provide thermal barrier to prevent heat transfer. No-wood design.

CHOICE OF JAMBS
Extra-wide jamb (6½") is designed for the new Energy Construction method in home building.

MAGNETIC WEATHERSTRIPS
Magnetic weatherstrips on head and strike jambs "reach out" for positive sealing like a refrigerator door.

COMPRESSION WEATHERSTRIP
Double-compression vinyl weatherstrip on hinge jamb completes the all-around weather-stopping action.

DEEP-EMBOSSED ON BOTH SIDES

CORNER PADS
Flexible pads at both lower corners seal out wind and water.

ADJUSTABLE DOOR SWEEP
Bottom sweep can be adjusted for snug fit against the threshold to seal against air and water leakage.

FOAM-FILLED INSULATING SANDWICH
Thick density (2.8 lb/ft^2) of special foamed-in-place polyurethane uses natural insulation of tiny trapped gas cells.

THRESHOLD
Aluminum threshold was designed for the new Energy Construction. Two other frost-break options: aluminum outswing threshold; aluminum-vinyl threshold for standard construction.

FROST-BREAK THERMAL BARRIER

Figure 9-67. Insulated metal door for commercial, industrial, or residential use. Note the energy-saving features. *(General Products)*

Metal Doors

Metal doors may also be used for residential houses. In some cases they are used to replace old or poorly fitting wooden doors. Figure 9-68 shows how doors with metal frames are designed for ease of installation. In Figure 9-69 you can see how the metal frame is attached to concrete, wood, and concrete blocks.

One of the advantages of metal doors is their fire resistance. Frames for 2'8" and 3'0" doors carry a 1½-hour label. The frames for 3'6" and double doors are not labeled as to fire resistance.

In the case of metal frames, the frame has to be installed before the walls are constructed. The frame requires a rough opening 4½ inches wider and 2¼ inches higher than nominal. The stock frame is usually 5¾ inches wide.

Installing Windows and Doors 235

• INSTALLING FOLDING DOORS •

Folding doors are used to cover closets with any number of interesting patterns. They may be flush, and plain or mirrored. They may have two panels or four panels. The sizes and door widths vary to suit the particular application. See Table 9-1.

Figure 9-70 shows the openings and the details of fitting the metal bifold door.

Figure 9-71 shows the details of the four panels to be installed and different panel styles available. Note the names given to the parts so you can follow the installation instructions.

1. Carefully center the top track lengthwise in the finished opening. This should suit both flush and recessed mountings. Attach the track with No. 10x1¼-inch screws in the provided holes. The bifold track is assembled for four-panel door installation. The track may be separated at the center without the use of tools when a two-panel door is installed. See Figure 9-72. Knobs, screws, and rubber stops are packaged for two-panel installation.

2. Place the bottom track, either round edge or square edge toward the room. Plumb the groove with the top track. See Figure 9-73. Screw the track to the floor with ½-inch screws or fasten it to a clean floor with 3M double-coated tape no. 4432.

3. On all two-panel sections, lower the bottom pivot rod until it projects ¾ inch below the edge of the door. Make it 1¼ inch if the carpet is under the door.

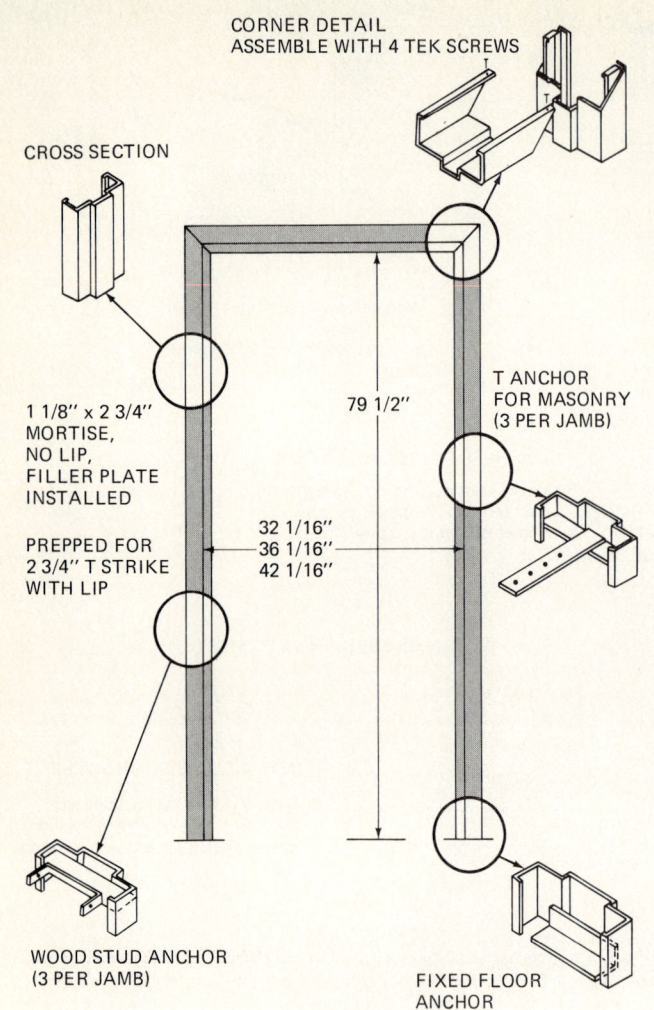

Figure 9-68. Putting together a metal frame for a door. *(General Products)*

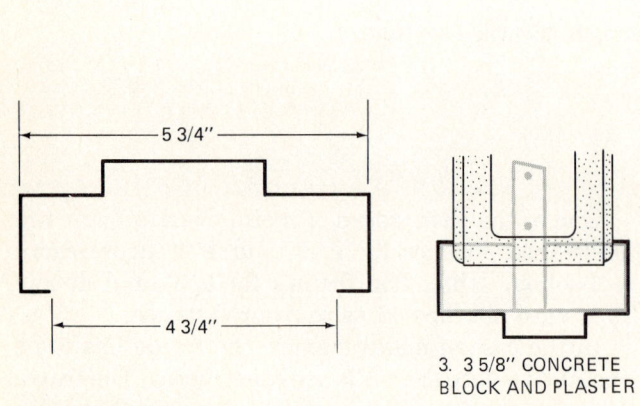

1. 3 1/4" STEEL STUD 3/8" SHEET ROCK AND PLASTER
2. 3 1/2" WOOD STUD 1/2" DRYWALL
3. 3 5/8" CONCRETE BLOCK AND PLASTER
4. 5 5/8" CONCRETE BLOCK WALL
5. 7 5/8" CONCRETE BLOCK WALL

Figure 9-69. Typical installations with metal door frames. *(General Products)*

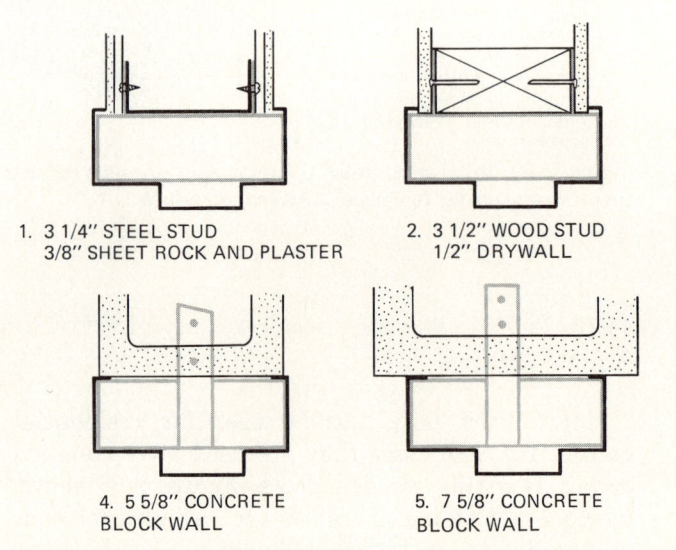

236 Carpentry Fundamentals

TABLE 9-1 FINISHED OPENING SIZES FOR BIFOLD DOORS.

Door Width	Number of Panels	Door opening, Inches* 6'8"	Door opening, Inches* 8'0"	Actual Door Width, Inches
1'6"	2	18½ × 80¾	18½ × 95¼	17 7/16
2'0"	2	24½ × 80¾	24½ × 95¼	23 7/16
2'6"	2	30½ × 80¾	30½ × 95¼	29 7/16
3'0"	2	36½ × 80¾	36½ × 95¼	35 7/16
3'0"	4	36 × 80¾	36 × 95¼	35
3'6"	2	42½ × 80¾	42½ × 95¼	41 7/16
4'0"	4	48 × 80¾	48 × 95¼	47
5'0"	4	60 × 80¾	60 × 95¼	59
6'0"	4	72 × 80¾	72 × 95¼	71
7'0"	4	84 × 80¾	84 × 95¼	83

* Finished opening width shown provides ½ inch clearance each side of door. Finished opening width may be reduced by ½ inch provided finished opening is square and plumb. This will require cutting track. Finished opening heights shown provide ⅜ inch clearance between door and track—top and bottom. (This makes ⅞ inch between door and floor.) Doors can be raised to have 1⅛ inch clearance door to floor without increasing opening height.

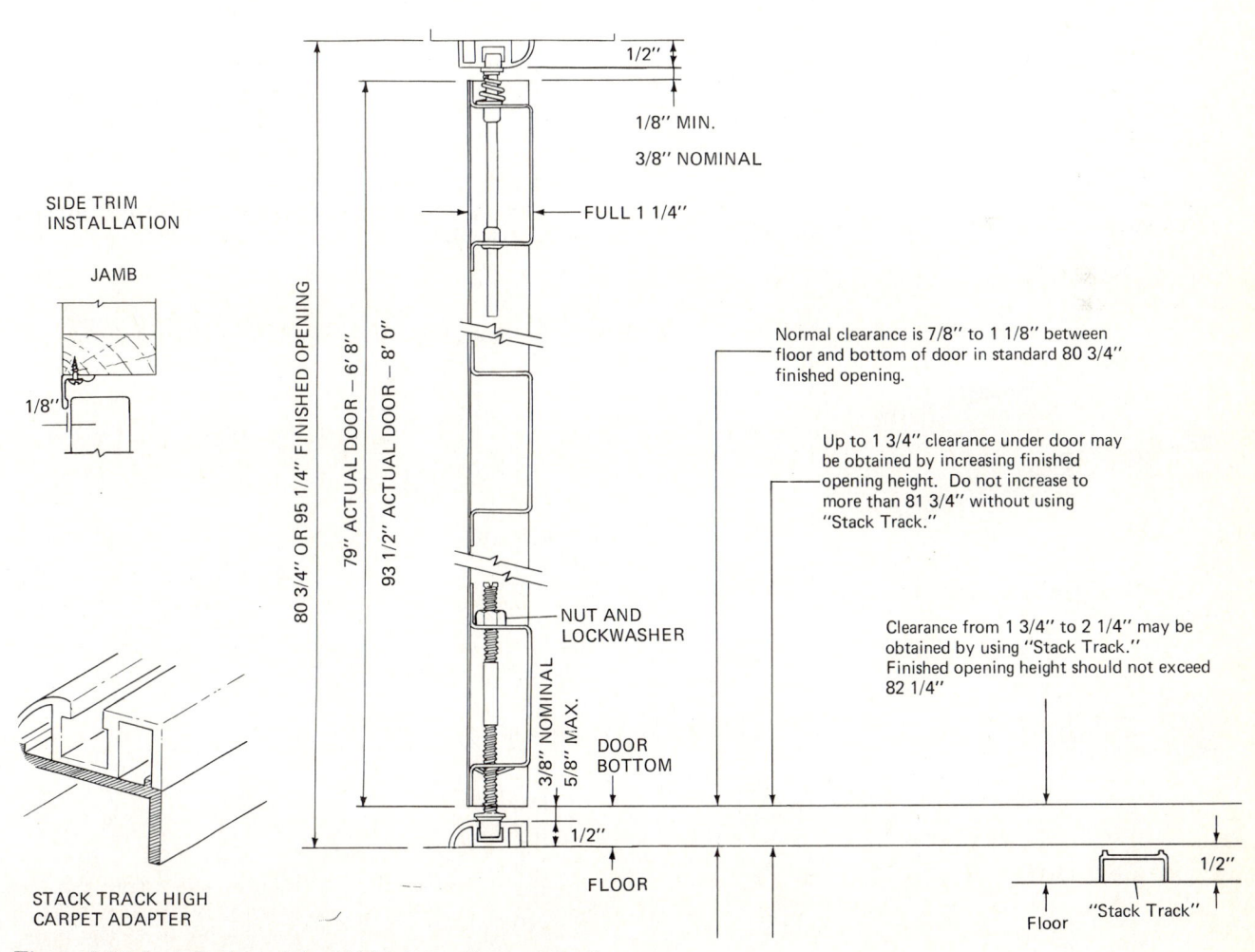

Figure 9-70. Installation of the bifold door. *(General Products)*

Installing Windows and Doors 237

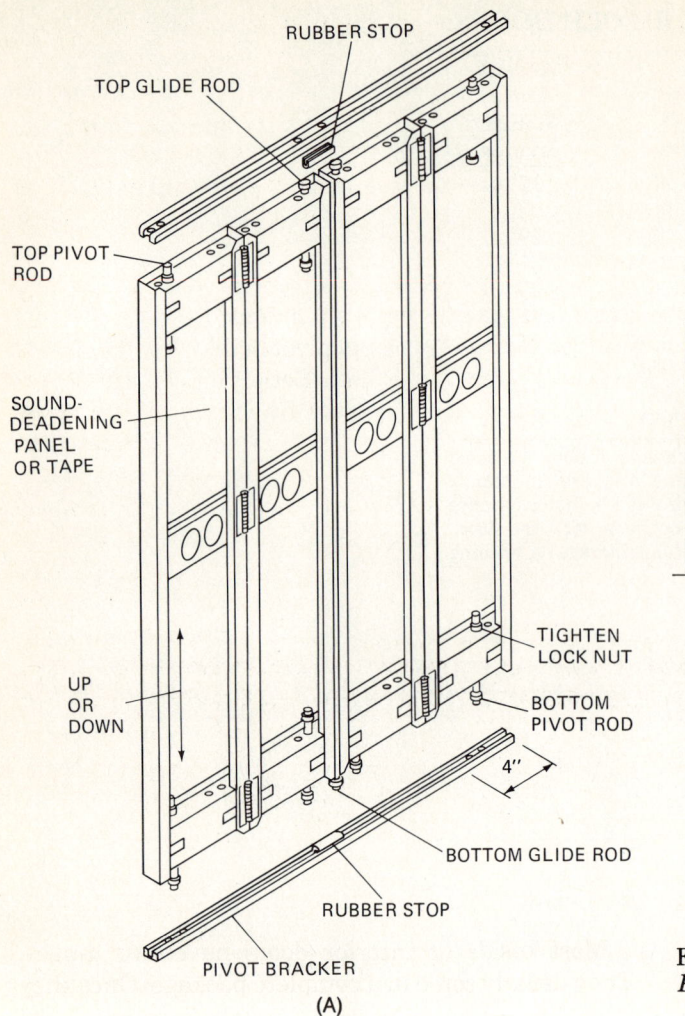

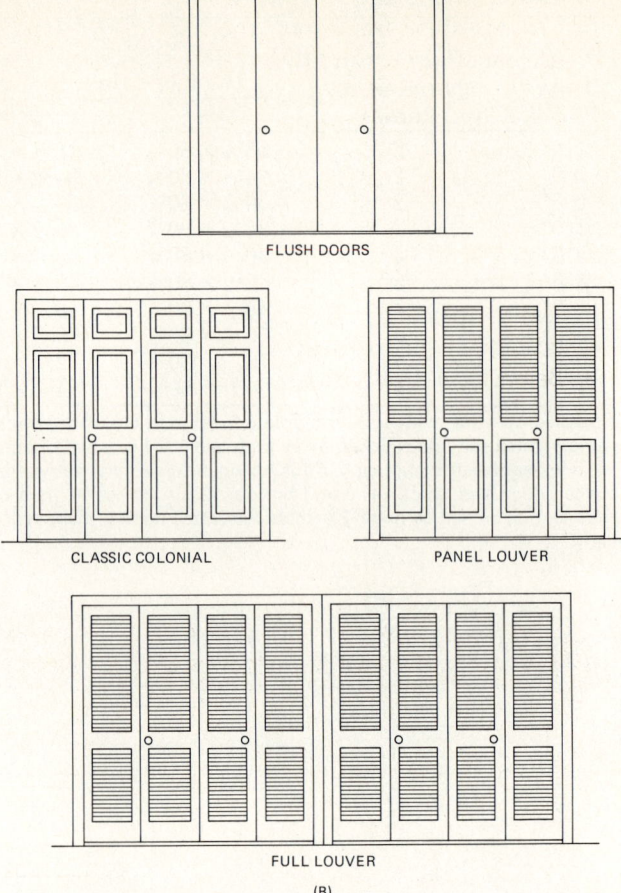

Figure 9-71. (A) Details of the metal bifold door. *(General Products)* (B) Different designs for bifold doors.

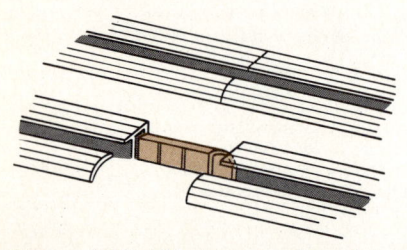

Figure 9-72. Vinyl connectors let you snap apart four-panel track instantly to install two-panel bifolds. *(General Products)*

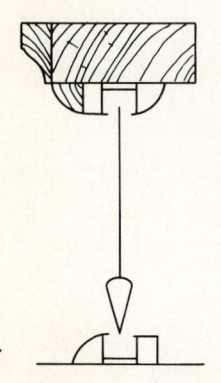

Figure 9-73. Using a plumb bob to make sure the tracks line up. *(General Products)*

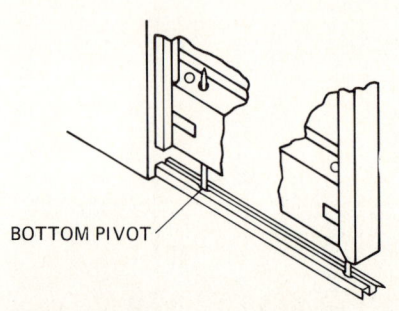

Figure 9-74. Pivoting the bottom pin in the track. *(General Products)*

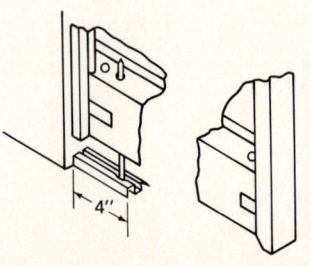

Figure 9-75. You can remove most of the bottom track if you don't want it on the floor. This lets carpeting run straight through to the closet wall. *(General Products)*

238 Carpentry Fundamentals

4. Attach the doorknobs.
5. Lift one door set. Insert the bottom pivot rod (threaded) into the bottom pivot bracket. Pull down the top spring-loaded pivot rod. Insert it into the pivot bracket in the top rack. Insert the top and bottom nylon glide rod tips into the track. See Figure 9-74.
6. Install the second door set the same way.
7. Insert the rubber stop in the center of the top and bottom tracks. Make sure that the stop seats firmly in the track. For a two-panel installation, cut the rubber stop to the proper length.
8. Because of their design, the bifold doors are rigid enough to operate smoothly without a full bottom track. This permits better carpeting in the closet. Saw off a 4-inch section of the bottom track. Place this on the floor. Use a plumb bob to pivot the bottom points with the top pivot. See Figure 9-75. Fasten the section to the floor with two ½-inch screws. Remove the bottom glide rods from the doors. Single-track installation is not recommended for 8'0"-high doors, 7'0"-wide four-panel doors, or 3'6"-wide two-panel doors.

FINAL ADJUSTMENTS • To raise or lower the doors to the desired height, turn the threaded bottom pivot rod with a screwdriver. Make sure the doors are even and level across the top. Tighten the locknut.

Doors should close snugly against the rubber stop. For horizontal (lateral) alignment, loosen the screw holding the top or bottom pivot brackets in the track. Adjust the door in or out. Retighten the screw.

Keep all glides and track free from paint and debris. The aluminum track is already lubricated to ensure smooth operation. Occasionally repeat the lubrication with silicon spray, paraffin, or soap. This keeps door operation free and easy.

These instructions are for a particular make of door. However, most manufacturers' instructions are basically the same. There are some minor adjustments you will have to make for each manufacturer. Make sure you follow the manufacturer's recommendations.

• DOOR AND WINDOW TRIM •

Interior Door Trim

Most inside or interior doors have two hinges. They usually come in a complete package. Once they are set in place, the casing has to be applied. See Figure 9-76 for the location of the casing around an interior door. Note how the jamb is installed. The stop is attached with nails and has a bevel cut at the bottom of the door. It prevents the door from swinging forward more than it should.

In Figure 9-77 you will find two of the most commonly used moldings applied to the trim of a door. These two are colonial and ranch casing moldings. These are the names you use when ordering them. They are ordered from a lumberyard or mill.

Figure 9-78 shows how a molded casing is mitered at the corner. It is secured with a nail through the 45° cut. In the other part of this figure you see the butt joint. This is where the casing meets at the

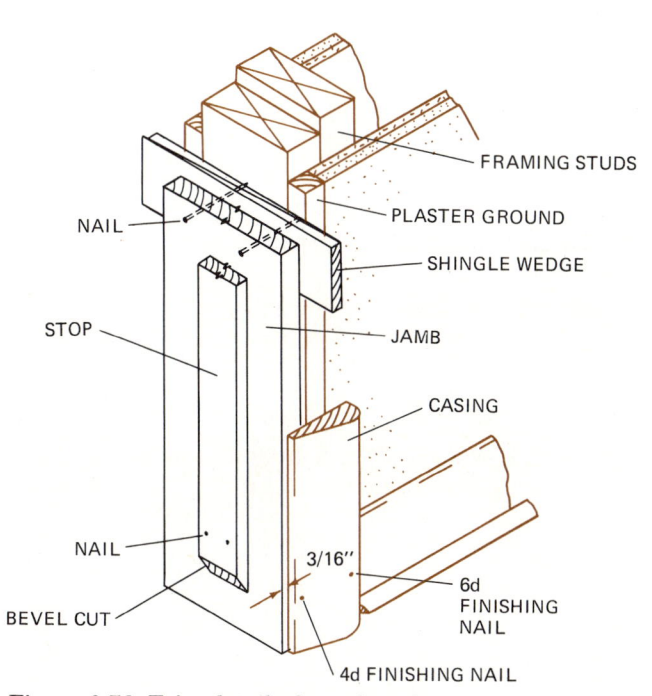

Figure 9-76. Trim details for a door frame.

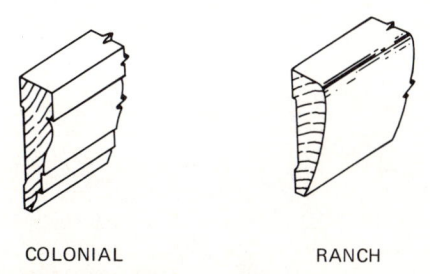

Figure 9-77. Two popular types of molding used for trim around a door.

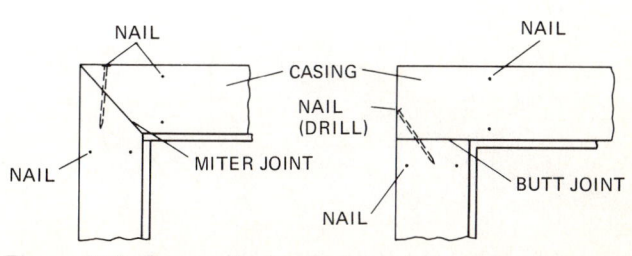

Figure 9-78. Two methods of joining trim over a door.

Installing Windows and Doors **239**

side and top. Notice the way the nail is placed to hold the two pieces securely. Also notice the other nail locations. Why do you need to drill the nail hole for the toenailed side?

Installation of the strike plate in the side jamb is shown in Figure 9-79. It has to be routed or drilled out. This allows the door locking mechanism to move into the hole. Figure 9-80 shows how the strike plate is mounted onto the door jamb.

Window Trim

Windows have to be trimmed. This completes the installation job. See Figure 9-81. There are a couple of ways to trim a window. Take a look at Figure 9-82 and see the difference. It shows a trimmed window with casing at the bottom instead of a stool and apron. This is a quicker and simpler method of finishing a window. There is no need for a stool to overlap the apron or casing in some instances. This is the choice of the architect or the owner of the home. There are problems with the apron and stool method. The apron and stool will pull away

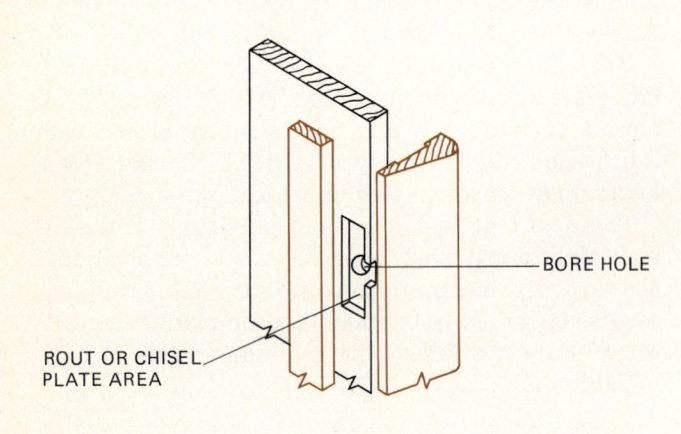

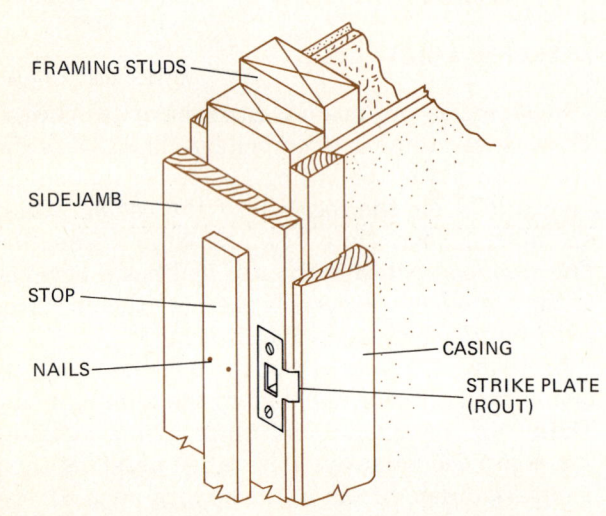

Figure 9-79. Installing a strike plate for the lockset.

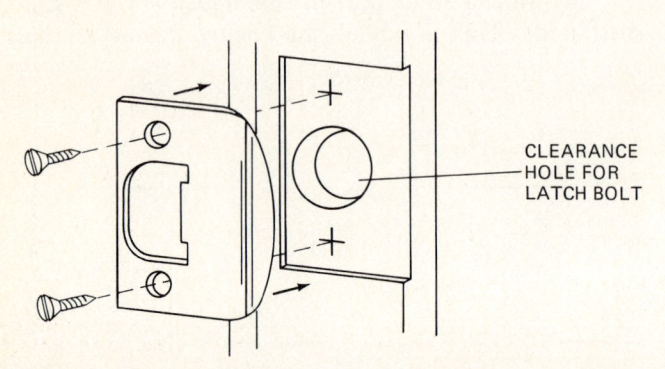

Figure 9-80. Installing the strike plate on the door jamb.

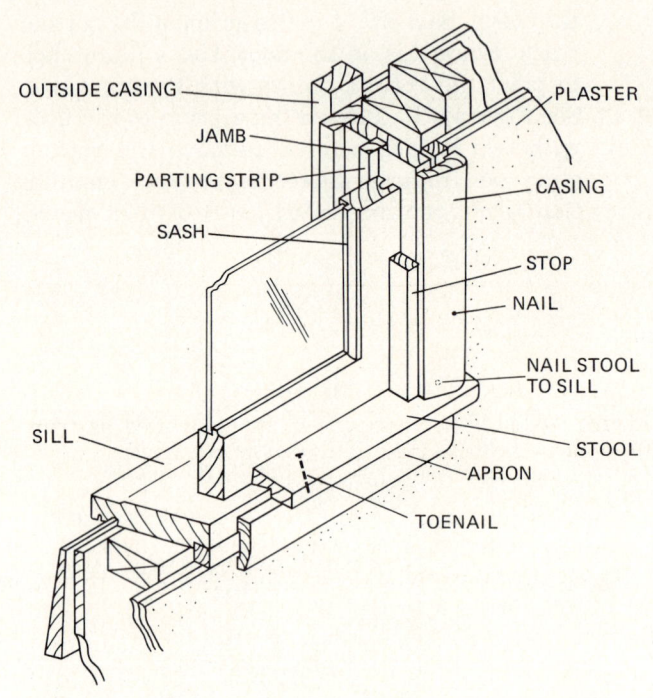

Figure 9-81. Window trim. Note the apron and stool.

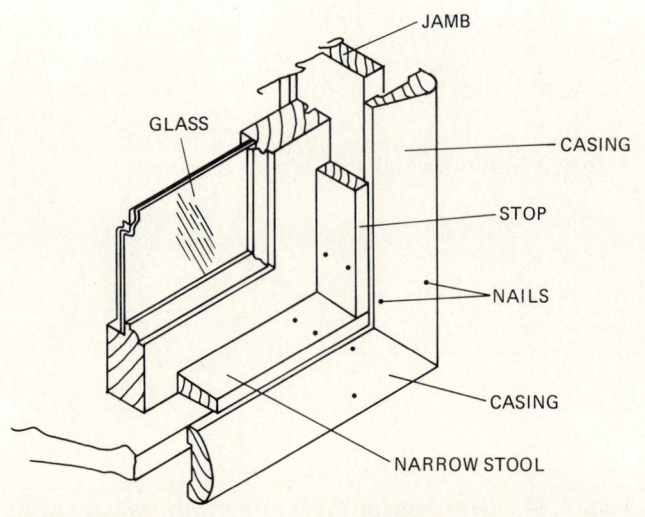

Figure 9-82. Window trim. Note the absence of the apron.

240 Carpentry Fundamentals

from the inside casing. This leaves a gap of up to ¼ inch. It can become unsightly in time.

Figure 9-83 shows some of the moldings which can be used in trimming windows, doors, or panels. These moldings are available in prepared lengths of 8 feet and 12 feet. Generally speaking, the simpler the molding design, the easier it is to clean. Many depressions or designs in a piece of wood can allow it to pick up dust. Some are very difficult to clean.

• INSTALLING LOCKS •

Installing a Lock

There are seven simple steps to installing a lock in a door. Figure 9-84 shows them in order.

In some cases you may want to reverse the lock. This may be the case when you change the lock

WP 444   11/16″ × 3-1/2″

WP 412   11/16″ × 3-1/2″
WP 432   9/16″ × 3-1/2″
WP 433   9/16″ × 3-1/4″

WP 452   11/16″ × 2-1/2″
WP 472   9/16″ × 2-1/2″

STOPS

WP 816   7/16″ × 1-3/8″
WP 818   7/16″ × 1-1/8″
WP 820   7/16″ × 7/8″

WP 846   7/16″ × 1-3/8″
WP 848   7/16″ × 1-1/8″
WP 850   7/16″ × 7/8″

WP 876   7/16″ × 1-3/8″
WP 878   7/16″ × 1-1/8″
WP 880   7/16″ × 7/8″

WP 620   9/16″ × 4-1/4″
WP 622   9/16″ × 3-1/2″
WP 623   9/16″ × 3-1/4″

WP 662   9/16″ × 3-1/2″
WP 663   9/16″ × 3-1/4″
WP 664   9/16″ × 3″

WP 712   9/16″ × 3-1/2″
WP 713   9/16″ × 3-1/4″
WP 714   9/16″ × 3″

WP 906   7/16″ × 1-3/8″
WP 908   7/16″ × 1-1/8″
WP 910   7/16″ × 7/8″

WP 936   7/16″ × 1-3/8″
WP 938   7/16″ × 1-1/8″
WP 940   7/16″ × 7/8″

MULLION CASING

WP 978   3/8″ × 1-3/4″

WP 983   3/8″ × 1-3/4″

Figure 9-83. Different designs of molding used as trim for windows, doors, and other parts of the house.

ADJUSTABLE LOCKSETS
FACTORY PRESET LOCKSETS

1. Mark hole centers with template. Drill for bolt and lock mechanism.

2. Mortise for face plate and strike plate.

3. Position lock and latch by slot engagement.

4. Fasten face plate with 2 wood screws.

3. Insert latch in 7/8″ hole and fasten face plate with 2 wood screws.

4. Insert lockset into 1 7/8″ hole so lock case "slot-engages" the latch.

5. Put on clamp plate by turning to engage keyway. Tighten screws.

6. Snap on rose. Depress spring retainer and apply knob.

7. Fasten strike plate with 2 wood screws.

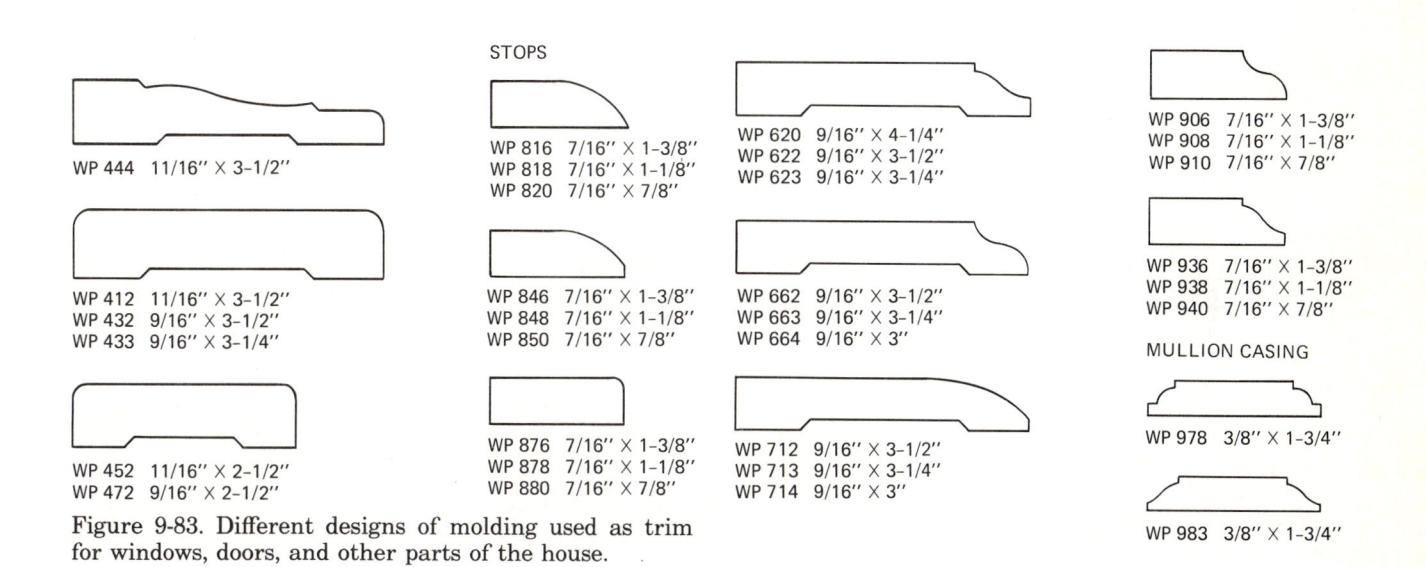

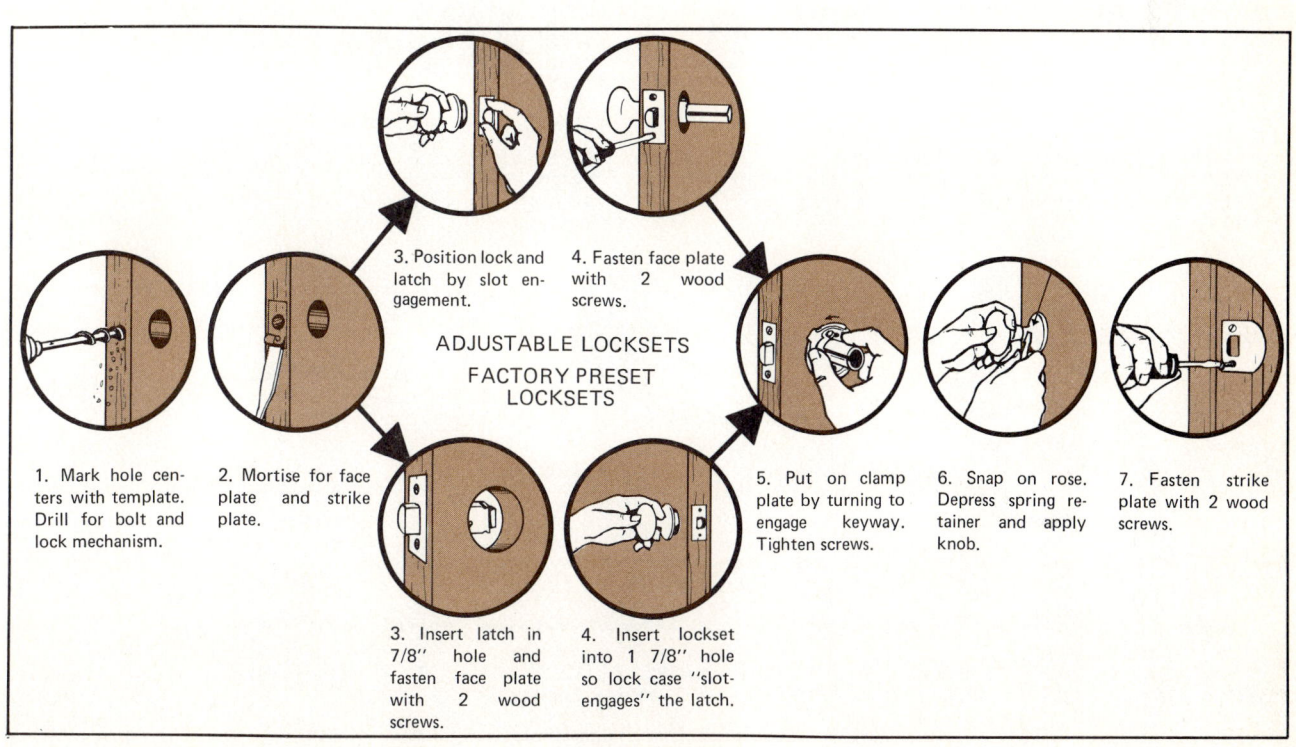

Figure 9-84. Seven simple steps to install a lockset. *(National Lock)*

Installing Windows and Doors

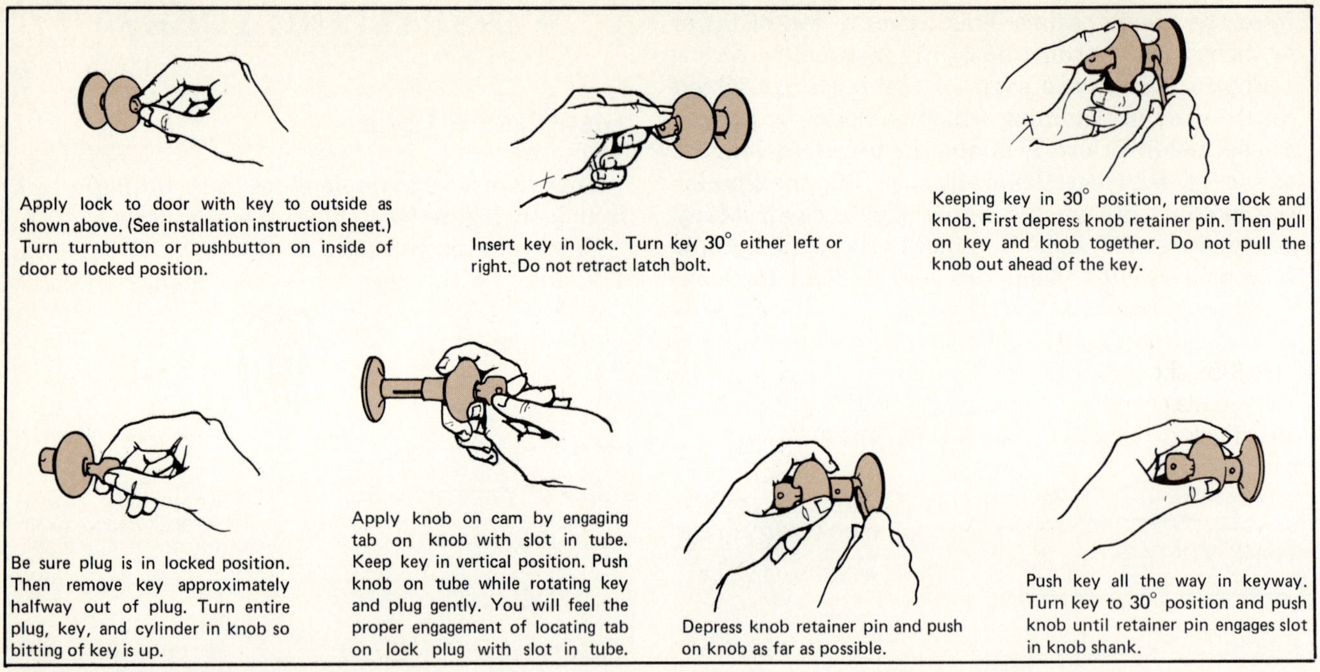

Apply lock to door with key to outside as shown above. (See installation instruction sheet.) Turn turnbutton or pushbutton on inside of door to locked position.

Insert key in lock. Turn key 30° either left or right. Do not retract latch bolt.

Keeping key in 30° position, remove lock and knob. First depress knob retainer pin. Then pull on key and knob together. Do not pull the knob out ahead of the key.

Be sure plug is in locked position. Then remove key approximately halfway out of plug. Turn entire plug, key, and cylinder in knob so bitting of key is up.

Apply knob on cam by engaging tab on knob with slot in tube. Keep key in vertical position. Push knob on tube while rotating key and plug gently. You will feel the proper engagement of locating tab on lock plug with slot in tube.

Depress knob retainer pin and push on knob as far as possible.

Push key all the way in keyway. Turn key to 30° position and push knob until retainer pin engages slot in knob shank.

Figure 9-85. Reversing the hand of a lock. *(National Lock)*

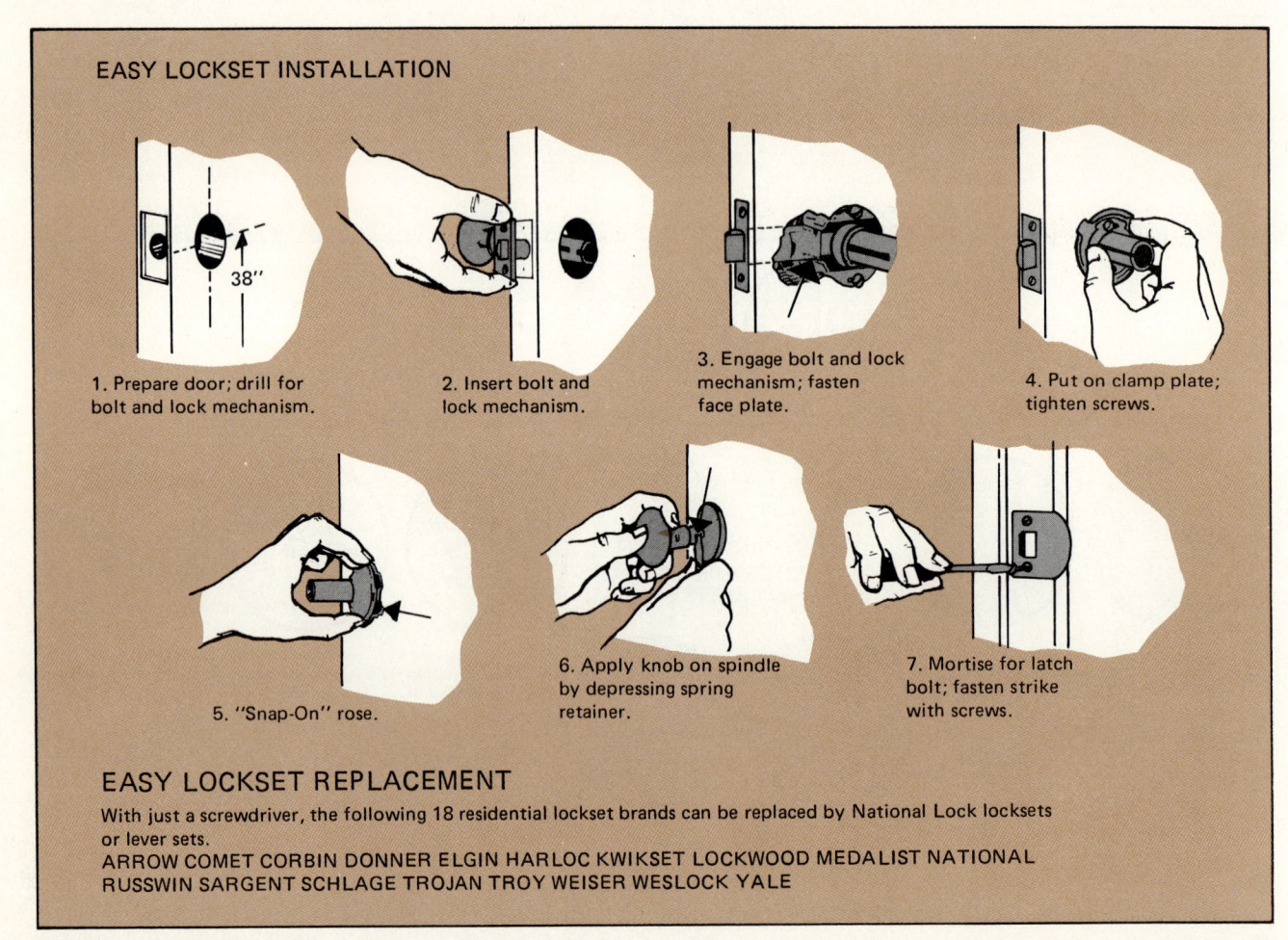

EASY LOCKSET INSTALLATION

1. Prepare door; drill for bolt and lock mechanism.
2. Insert bolt and lock mechanism.
3. Engage bolt and lock mechanism; fasten face plate.
4. Put on clamp plate; tighten screws.
5. "Snap-On" rose.
6. Apply knob on spindle by depressing spring retainer.
7. Mortise for latch bolt; fasten strike with screws.

EASY LOCKSET REPLACEMENT

With just a screwdriver, the following 18 residential lockset brands can be replaced by National Lock locksets or lever sets.
ARROW COMET CORBIN DONNER ELGIN HARLOC KWIKSET LOCKWOOD MEDALIST NATIONAL RUSSWIN SARGENT SCHLAGE TROJAN TROY WEISER WESLOCK YALE

Figure 9-86. Replacement of a lockset. *(National Lock)*

from one door to another. The hand of the door may be different. In Figure 9-85 you can see how simple it is to change the hand of the lock. In some cases you may have bought the lock without noticing how it should fit. This way you are able to make it fit in either direction.

There are a number of locks available. Figure 9-86 shows how 18 different locksets can be replaced by National Lock's locksets or lever sets. Figure 9-87 shows some of the designs available for strikes. The strike is always supplied with the lockset. Figure 9-88 shows the latch bolts. They may have round or square corners. They may have or may not have deadlock capability.

ENTRANCE HANDLE LOCKS • Most homes have an elaborate front door handle. In Figure 9-89 you can see two of the types of escutcheons used to decorate the doorknob.

Door handles also become something of a decorative item. They come in a number of styles. Each lock manufacturer offers a complete line. See Figure 9-90 for an illustration of two such handles. These handles are usually cast brass.

A number of lockset designs are available for entrance and interior doors. They may lock, then require a nail or pin to be opened. Or they may require a key. See Figure 9-91.

AUXILIARY LOCKS • Auxiliary locks are those placed on exterior doors to prevent burglaries. They are called deadlocks. They usually have a 1-inch bolt that projects past the door. It fits into the door jamb (Figure 9-92). In Figure 9-93 you can see three of the various deadlock designs used. The key side, of course, goes on the outside of the door.

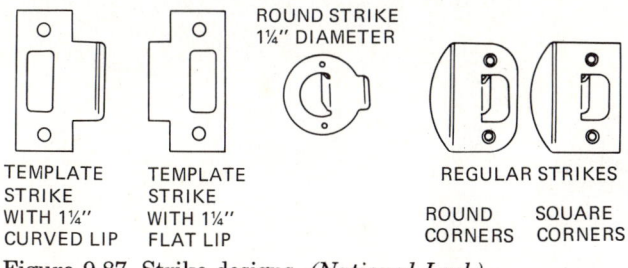

Figure 9-87. Strike designs. *(National Lock)*

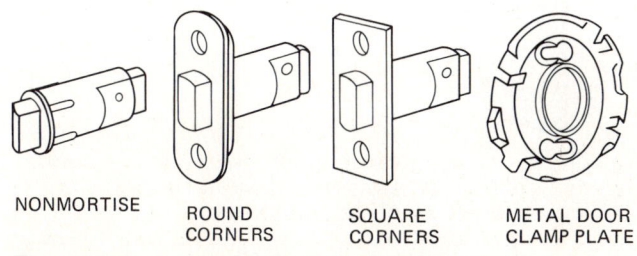

Figure 9-88. Latch bolt designs. *(National Lock)*

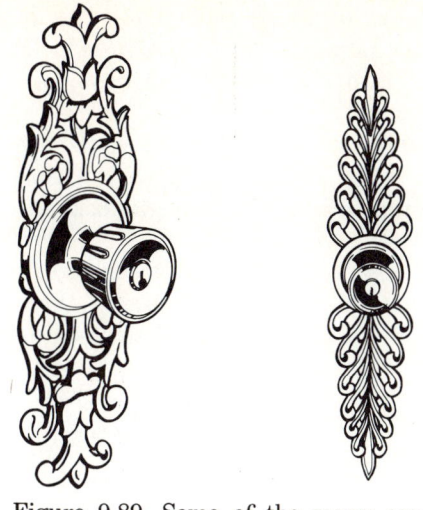

Figure 9-89. Some of the many escutcheons for locks. *(Weiser Lock)*

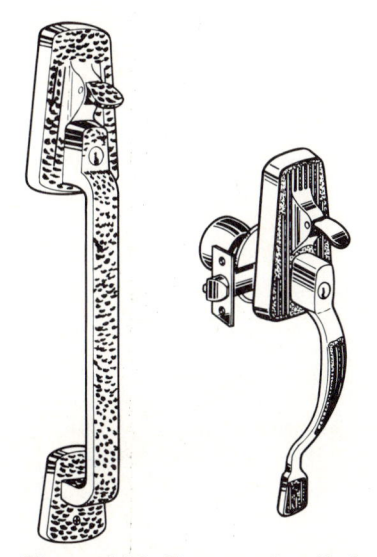

Figure 9-90. Entrance handle locks. *(Weiser Lock)*

Figure 9-91. Locksets for interior and exterior doors. *(Weiser Lock)*

Figure 9-92. A deadlock projects out from the door and fits into the door jamb to make a secure door. *(Weiser Lock)*

Installing Windows and Doors 243

A standard type of lock is exploded for you in Figure 9-94. Note the names of the parts. These locks have keys that fit into a cylinder. The keys lock or unlock the latch bolt.

Exposed brass, bronze, or aluminum parts are buffed or brushed. They are protected with a coat of lacquer. Aluminum is brushed and anodized.

CONSTRUCTION KEYING • There are a couple of methods used for keying locksets. One of them makes it possible for a construction supervisor to get into a number of buildings with one key. Once the building is occupied, the lock is converted. The lock's key and no other will then operate it. See Figure 9-95.

Builders use a short four-pin tumbler key to operate the lock during the construction period. A nylon wafer is inserted in the keyway at the factory. This blocks operation of the fifth and sixth tumblers. Accidental deactivation of the builder's key is unlikely. This is because a conscious effort to apply a 10- to 15-pound force is required to dislodge the nylon wafer the first time the five-tumbler key is used. The owner's keys are packed in specially marked, sealed envelopes.

When construction is completed, the unit is ready for occupancy. The homeowner inserts the regular five-pin tumbler key to move the nylon wafer. This makes the fifth tumbler operative. It automatically deactivates the four-pin tumbler arrangement. This makes the builder's key useless. Now the locksets can be operated only by the owner's keys.

Other lockmakers have different methods for this key operation. See Figure 9-96. Figure 9-97 shows another method of key operation of locks. In this case the whole cylinder of the new lock is removed.

Figure 9-93. Auxiliary locks. Chain and bolt, and two types of deadbolts. *(Weiser Lock)*

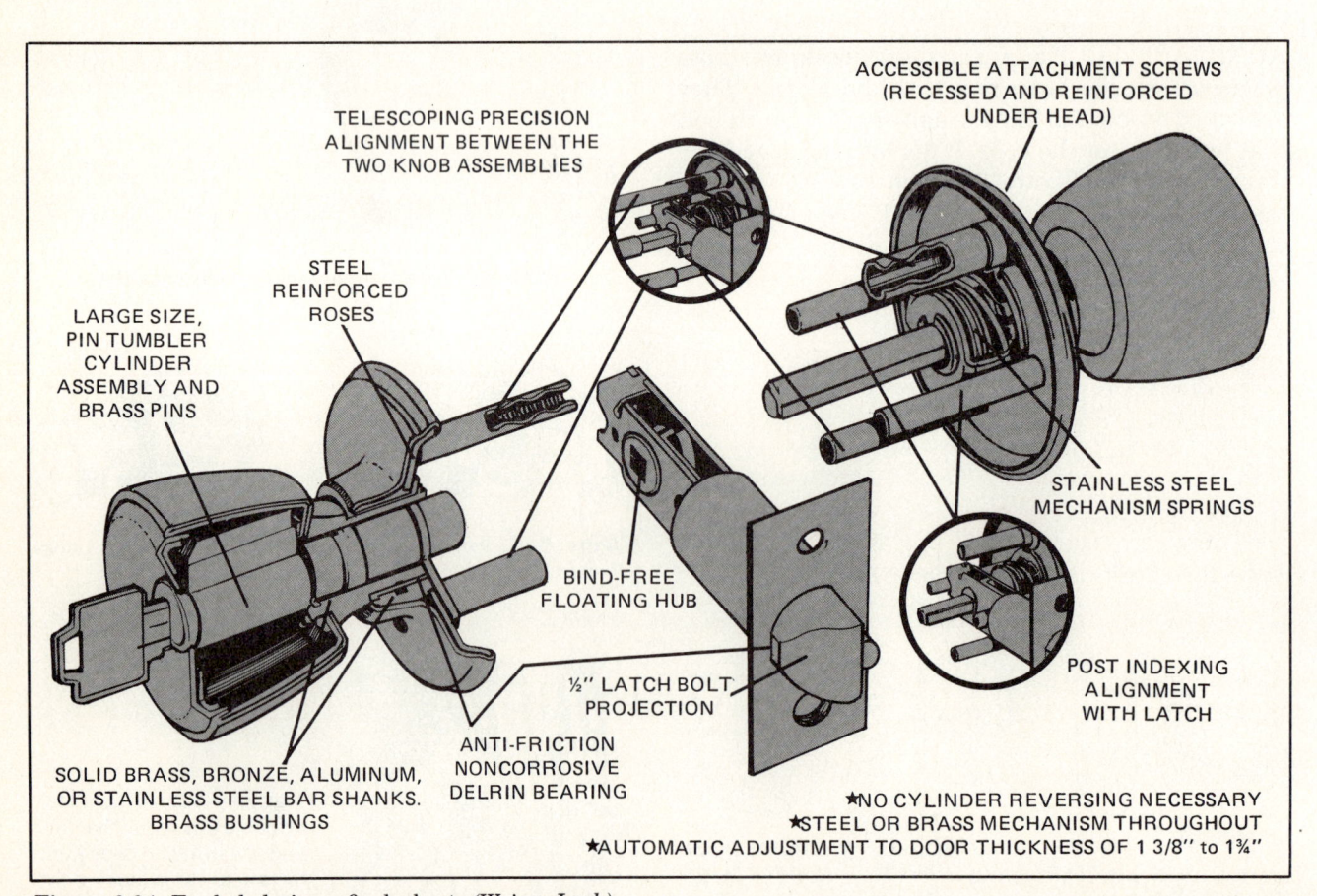

Figure 9-94. Exploded view of a lockset. *(Weiser Lock)*

244 Carpentry Fundamentals

The construction worker inserts a different cylinder. This cylinder works with the master key. Once the job is finished, the original cylinder is reinserted. The construction worker's key no longer operates this lock. Only the homeowner can operate the lock.

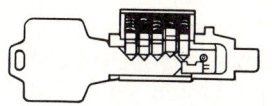

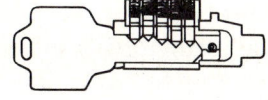

DURING CONSTRUCTION AFTER CONSTRUCTION

Figure 9-95. Construction keying of locksets. *(National Lock)*

Storm Doors and Windows

In most windows thermopane is used. It consists of two pieces of glass welded together with an air space. (See Figure 9-20.) Some windows have the space evacuated so that a vacuum exists inside. This cuts down on the transfer of heat from the inside of a heated building to the cold outside, or the reverse during the summer. In some cases, as shown in Chapter 15, another piece of glass or plastic is added. It fits over the twin panes of glass. See Figure 9-98.

Metal transfers heat faster than wood. Wood is 400 times more effective than steel as an insulator. It is 1800 times more effective than aluminum.

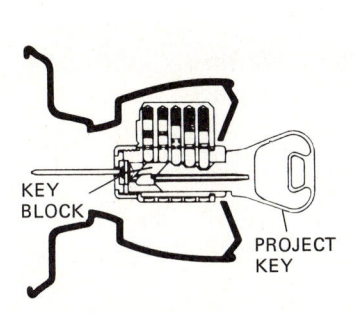

KEY BLOCK — PROJECT KEY

Lock cylinder is operated by the special "project key." The last two pins in the cylinder are held inoperative by the key block.

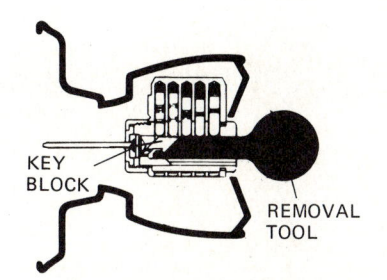

KEY BLOCK — REMOVAL TOOL

The special "project key" is canceled out by removal of the key block. A key block removal tool is furnished with the master keys for the locks. Simply push the removal tool into the keyway. Upon withdrawal, the key block will come out of the keyway. Thereafter, the "project key" no longer will operate the lock cylinder.

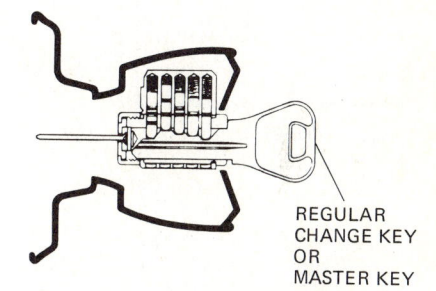

REGULAR CHANGE KEY OR MASTER KEY

Cylinder is now only operable by the regular change key or master key.

Figure 9-96. Another method of construction keying. *(Weiser Lock)*

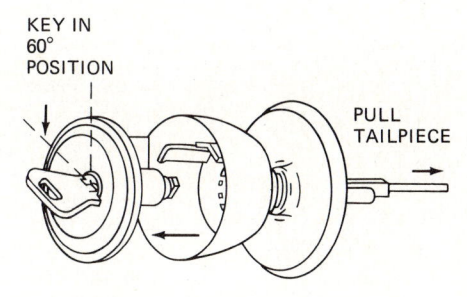

KEY IN 60° POSITION — PULL TAILPIECE

REMOVE REGULAR KEY CYLINDER

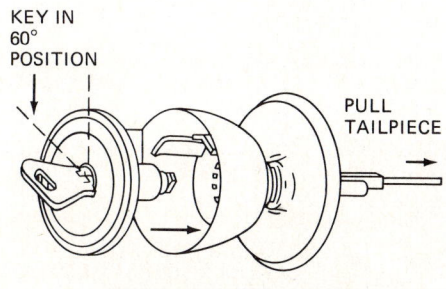

KEY IN 60° POSITION — PULL TAILPIECE

INSERT CONSTRUCTION CYLINDER

Figure 9-97. Some types of construction cylinders have to be removed. *(Weiser Lock)*

Figure 9-98. Examples of thermopane windows. *(Andersen)*

Installing Windows and Doors   **245**

Storm doors come in a wide variety of shapes and designs. They usually have a combination of screen and glass. The glass is removed in the summer and a screen wire panel is inserted in its place. This way the storm door serves year-round. See Figure 9-99. They are delivered prehung and ready for installation. All that has to be done to install them is to level the door and add screws in the holes around the edges. A door closer is added to make sure the door closes after use. In some cases a spring adjustment device is added to the top so that the door closer and the door are protected from wind gusts. Most storm doors are made of metal. However, they are available in wood or plastic.

Standard sizes are for openings from 35¾ to 36⅜ inches wide and from 79¾ to 81¼ inches high. There is an extender Z bar available for openings up to 37⅛ inches.

According to the company's testing lab, this plastic (polypropylene) door has 45 percent more heat retention than an aluminum door. See Figure 9-100.

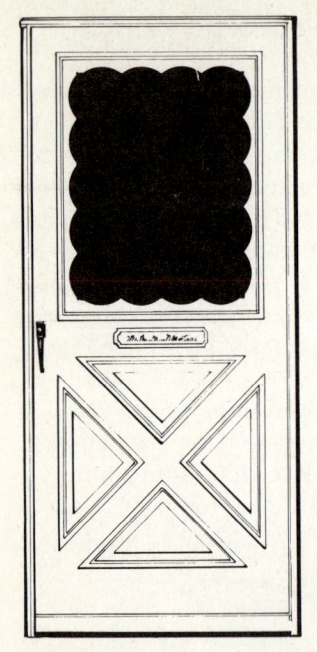

Figure 9-99. Storm door. *(EMCO)*

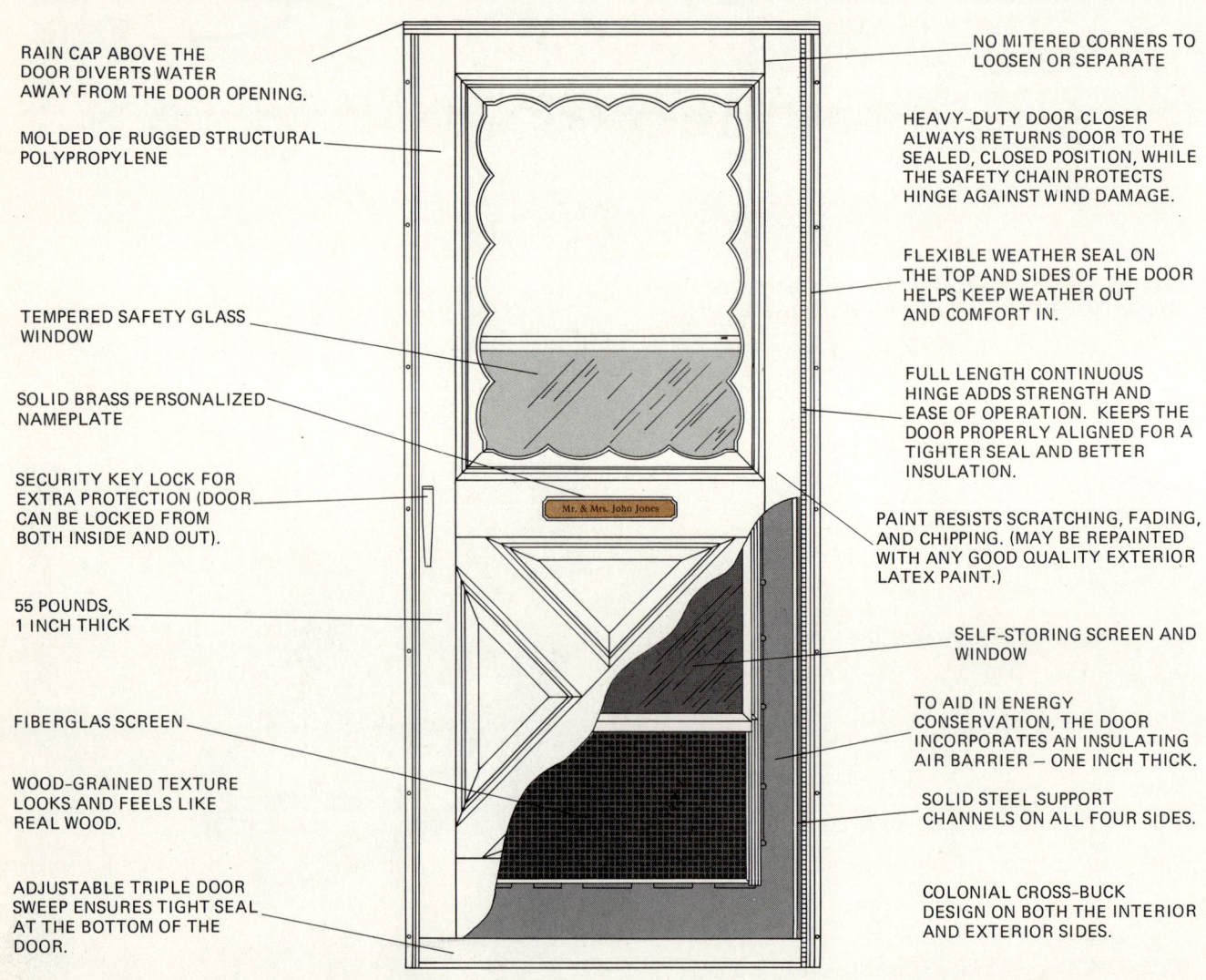

Figure 9-100. Storm door features. *(EMCO)*

Callouts:
- RAIN CAP ABOVE THE DOOR DIVERTS WATER AWAY FROM THE DOOR OPENING.
- MOLDED OF RUGGED STRUCTURAL POLYPROPYLENE
- TEMPERED SAFETY GLASS WINDOW
- SOLID BRASS PERSONALIZED NAMEPLATE
- SECURITY KEY LOCK FOR EXTRA PROTECTION (DOOR CAN BE LOCKED FROM BOTH INSIDE AND OUT).
- 55 POUNDS, 1 INCH THICK
- FIBERGLAS SCREEN
- WOOD-GRAINED TEXTURE LOOKS AND FEELS LIKE REAL WOOD.
- ADJUSTABLE TRIPLE DOOR SWEEP ENSURES TIGHT SEAL AT THE BOTTOM OF THE DOOR.
- NO MITERED CORNERS TO LOOSEN OR SEPARATE
- HEAVY-DUTY DOOR CLOSER ALWAYS RETURNS DOOR TO THE SEALED, CLOSED POSITION, WHILE THE SAFETY CHAIN PROTECTS HINGE AGAINST WIND DAMAGE.
- FLEXIBLE WEATHER SEAL ON THE TOP AND SIDES OF THE DOOR HELPS KEEP WEATHER OUT AND COMFORT IN.
- FULL LENGTH CONTINUOUS HINGE ADDS STRENGTH AND EASE OF OPERATION. KEEPS THE DOOR PROPERLY ALIGNED FOR A TIGHTER SEAL AND BETTER INSULATION.
- PAINT RESISTS SCRATCHING, FADING, AND CHIPPING. (MAY BE REPAINTED WITH ANY GOOD QUALITY EXTERIOR LATEX PAINT.)
- SELF-STORING SCREEN AND WINDOW
- TO AID IN ENERGY CONSERVATION, THE DOOR INCORPORATES AN INSULATING AIR BARRIER — ONE INCH THICK.
- SOLID STEEL SUPPORT CHANNELS ON ALL FOUR SIDES.
- COLONIAL CROSS-BUCK DESIGN ON BOTH THE INTERIOR AND EXTERIOR SIDES.

246 Carpentry Fundamentals

Figure 9-101. Primed-wood gliding door. *(Andersen)*

Figure 9-103. Perma-Shield® gliding door. *(Andersen)*

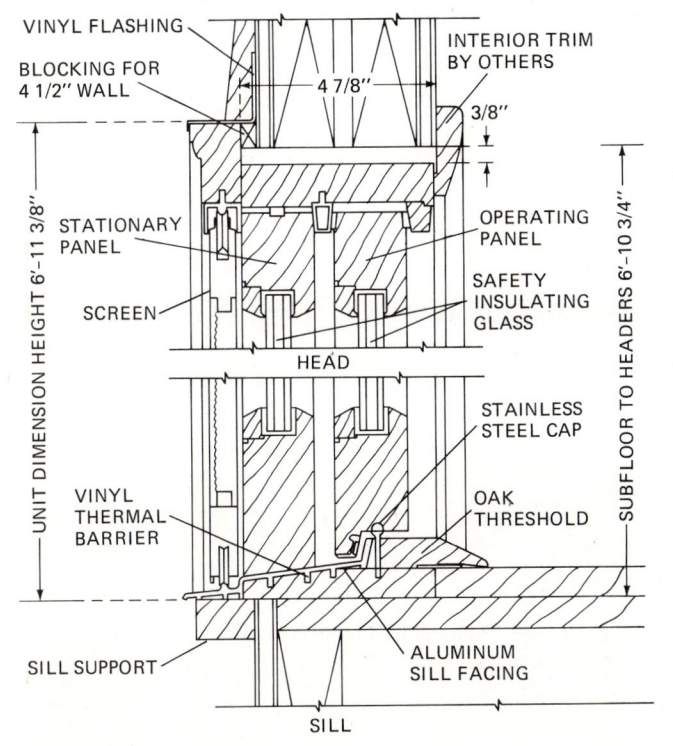

Figure 9-102. Details of the primed-wood gliding door. *(Andersen)*

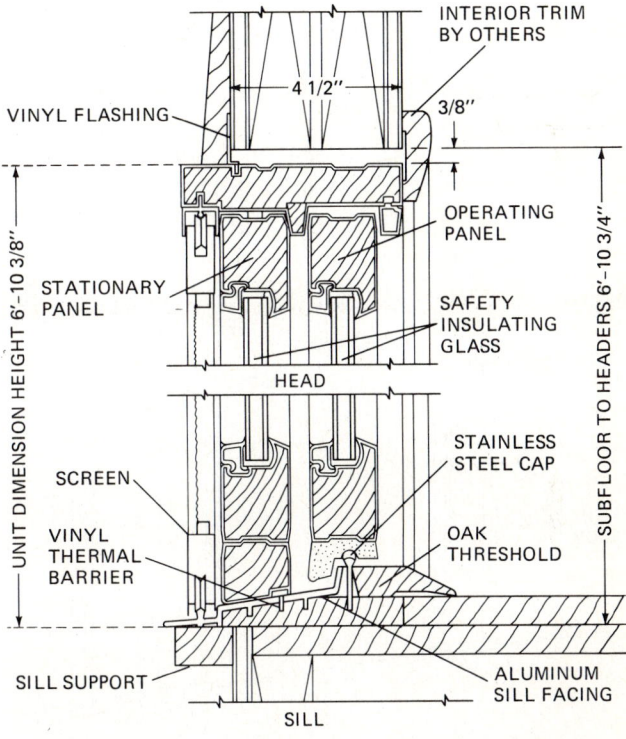

Figure 9-104. Details of the Perma-Shield® gliding door. *(Andersen)*

• INSTALLING A SLIDING DOOR •

Sliding doors are a common addition to a house today. The doors slide open so that the patio can be reached easily. It takes some special precautions to make sure the sliding doors will operate correctly. Most of these doors are made by a manufacturer like Andersen. They require a minimum of effort on the part of the carpenter. However, some very special steps are required. This portion of the chapter will deal primarily with the primed-wood type of gliding door and the Andersen Perma-Shield type of gliding door.

The primed-wood gliding door (see Figure 9-101) does not have a flange around it for quick installation. It requires some special attention. You will get an idea of how it fits into the rough opening from Figure 9-102.

The Perma-Shield gliding door is slightly different from the primed-wood type. See Figure 9-103. In Figure 9-104 you will find the details of the Perma-Shield door so that you can see the differences between the two types.

Installing Windows and Doors 247

Installation of both types of doors requires a rough opening in the frame structure of the house or building. The rough opening is constructed the same way for both types of doors.

Preparation of the Rough Opening

Installation techniques, materials, and building codes vary from area to area. Contact your local material supplier for specific recommendations for your area.

The same rough opening procedures can be used for both doors. There are, however, variations to note in gliding door installation procedures. These will be looked at more fully as we go along here.

If you need to enlarge the opening, make sure you use the proper-size header. Header size is usually given by the manufacturer of the door, or you can obtain it from your local supplier.

Preparation of the rough opening should follow these steps:

1. Lay out the gliding-door opening width between the regular studs to equal the gliding-door rough opening width plus the thickness of two regular studs. See Figure 9-105.
2. Cut two pieces of header material to equal the rough opening width of the gliding door plus the thickness of two jack studs. Nail two header members together using an adequate spacer so that the header thickness equals the width of the jack stud. See Figure 9-106.
3. Position the header at the proper height between the regular studs. Nail through the regular studs into the header to hold the header in place until the next step is completed. See Figure 9-107.
4. Cut the jack studs to fit under the header. This will support the header. Nail the jack studs to the regular studs. See Figure 9-108.
5. Apply the exterior sheathing (fiberboard, plywood, etc.) flush with the header and jack stud members. See Figure 9-109.

Figure 9-105. Layout of a rough opening for a gliding door. *(Andersen)*

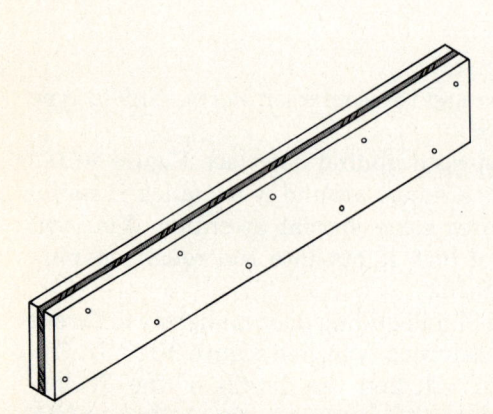

Figure 9-106. Header for a gliding-door opening. *(Andersen)*

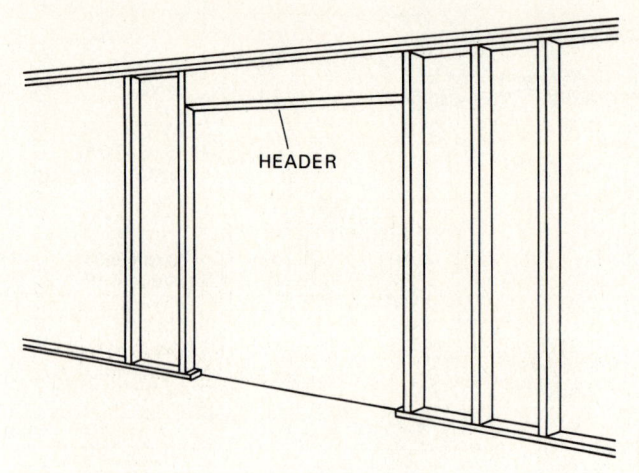

Figure 9-107. Placement of the header in the rough opening. *(Andersen)*

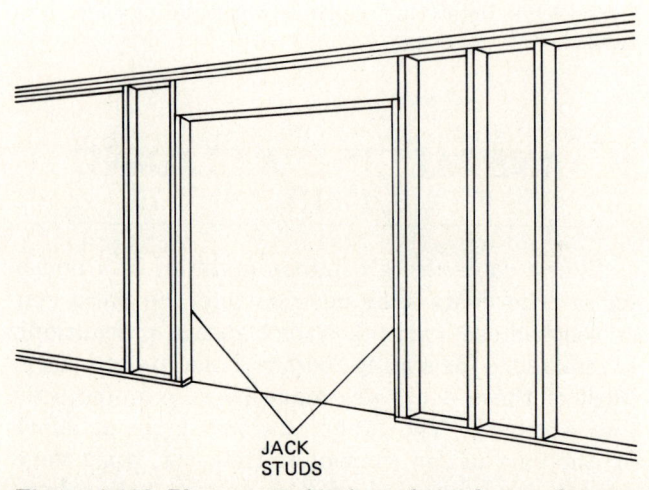

Figure 9-108. Placement of jack studs in the rough opening. *(Andersen)*

Installation of the Wood Gliding Door

Keep in mind that all these illustrations are as viewed from the outside. Be sure the subfloor is level and the rough opening is plumb and square before installing the gliding-door frame. If you follow these steps closely, the door should be properly installed.

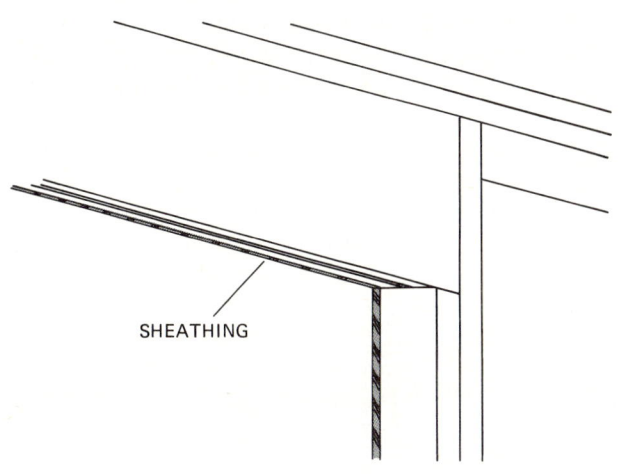

Figure 9-109. Application of exterior sheathing over the header. *(Andersen)*

1. Run caulking compound across the opening to provide a tight seal between the door sill and the floor. Remove the shipping skids from the sill of the frame if the gliding door has been shipped set up. Follow the instructions included in the package if the frame is *not* set up. See Figure 9-110.
2. Position the frame in the opening from the outside. See Figure 9-111. Apply pressure to the sill to properly distribute the caulking compound. The sill must be level. Check carefully and shim if necessary.
3. After leveling the sill, secure it to the floor by nailing along the inside edge of the sill with 8d coated nails spaced approximately 12 inches apart. See Figure 9-112.
4. The jamb must be plumb and straight. Temporarily secure it in the opening with 10d casing nails through each side casing into the frame members. Using a straightedge, check the jambs for bow and shim. *Shim solidly (five per jamb) between side jambs and jack studs.*
5. Complete the exterior nailing of the unit in the opening by nailing through the side and head casings into the frame members with 10d casing nails. See Figure 9-113.

Figure 9-110. Running a bead of caulking to seal the sill and door for a sliding door.

Figure 9-112. Securing the sill to the floor by nailing.

Figure 9-111. Leveling the sill.

Figure 9-113. Exterior nailing of the unit. *(Andersen)*

Installing Windows and Doors **249**

6. Position the flashing on the head casing and secure it by nailing through the vertical leg. The vertical center brace may now be removed from the frame. Be sure to remove and save the head and sill brackets. See Figure 9-114.
7. Apply the treated wood sill support under the protruding metal sill facing. See Figure 9-115. Install it tight to the underside of the metal sill with 10d casing nails.
8. Position the stationary door panel in the outer run. Be sure the bottom rail is straight with the sill. Force the door into the run of the side jamb with a 2×4 wedge. See Figure 9-116. Check the position by aligning the screw holes of the door bracket with the holes in the sill and head jamb. Repeat the above procedure for stationary panels of the triple door (if one is used here). Before the left-hand stationary panel is installed in a triple door, be sure to remove the screen bumper on the sill. Keep in mind, however, that only a double door is shown here.
9. Note the mortise in the bottom rail for a bracket. Secure the bracket with No. 8 one-inch flat head screws through the predrilled holes. See Figure 9-117 for details. Align the bracket with the predrilled holles in the head jamb and secure it with No. 8 one-inch flat head screws. See Figure 9-118. Repeat the procedure for the stationary panels of a triple door. The head stop is now removed if the unit has been shipped set up.
10. Apply security screws. Apply the two 1½-inch No. 8 flat head painted head screws through

Figure 9-114. Securing the flashing on the head casing by nailing.

Figure 9-116. Positioning the stationary door panel in the outer run by using a 2 x 4 wedge.

Figure 9-115. Applying the sill support under the metal sill facing.

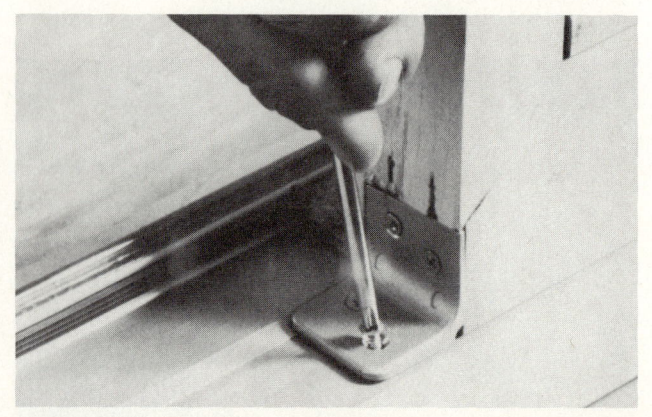

Figure 9-117. Securing the bottom bracket with a screw. (Andersen)

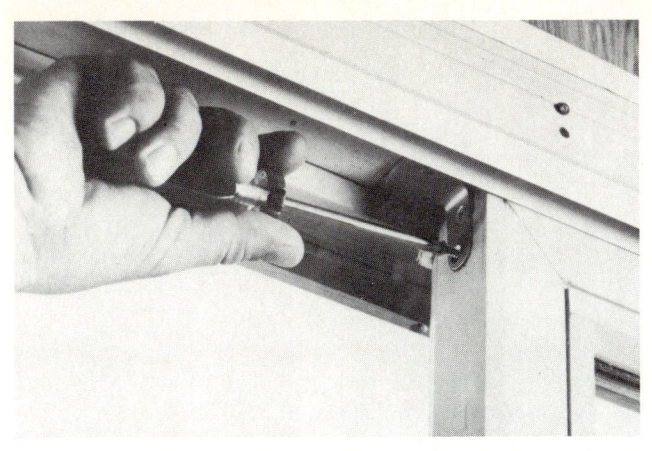

Figure 9-118. Securing the top bracket with a screw. *(Andersen)*

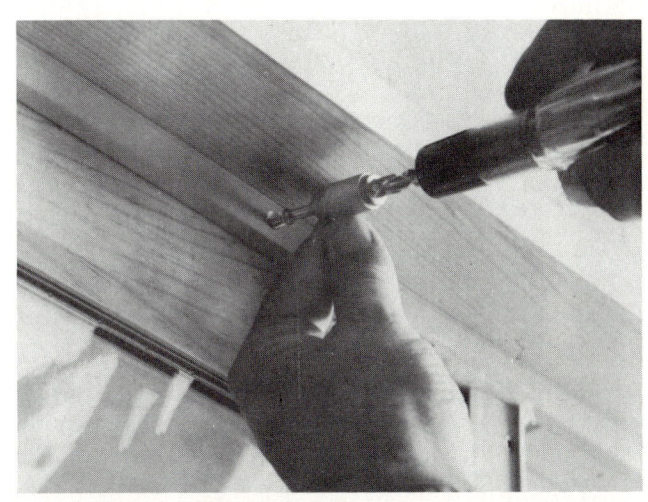

Figure 9-119. Applying the security screws. *(Andersen)*

Figure 9-120. Placing the operating door on the rib of the metal sill facing. *(Andersen)*

the parting stop into the stationary door top rail. See Figure 9-119. Repeat for the stationary panels of a triple door.

11. Place the operating door on the rib of the metal sill facing, and tip the door in at the top. See Figure 9-120. Position the head stop and apply with 1 9/16-inch No. 7 screws. See Figure 9-121.
12. Check the door operation. If the door sticks or binds or is not square with the frame, locate the two adjustment sockets on the outside of the bottom rail. See Figure 9-122. Simply remove the caps, insert the screwdriver, and turn to raise or lower the door. Replace the caps firmly.
13. If it is necessary to adjust the "throw" of the latch on two-panel doors, turn the adjusting screw to move the latch in or out. See Figure 9-123. The lock may be adjusted on triple doors by loosening the screw to move the lock plate. See Figure 9-124.

Figure 9-121. Positioning the head stop. *(Andersen)*

Figure 9-122. Adjusting the door for square. *(Andersen)*

Installing Windows and Doors **251**

Figure 9-123. The throw of the door is adjusted by this screw. *(Andersen)*

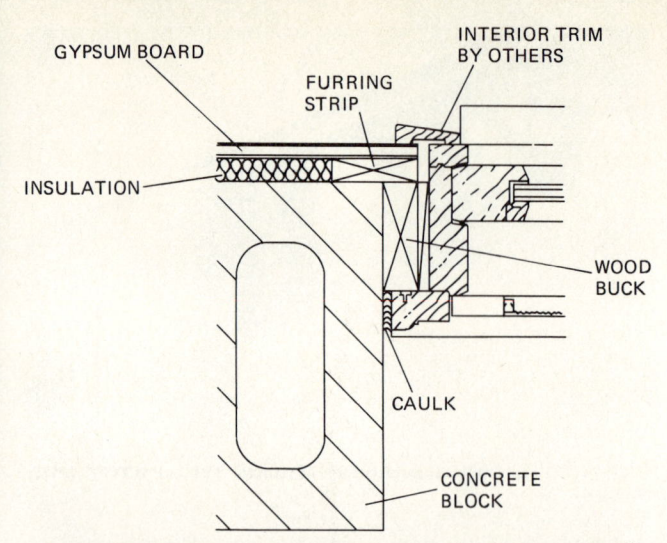

Figure 9-125. Wood gliding door installed in masonry wall. *(Andersen)*

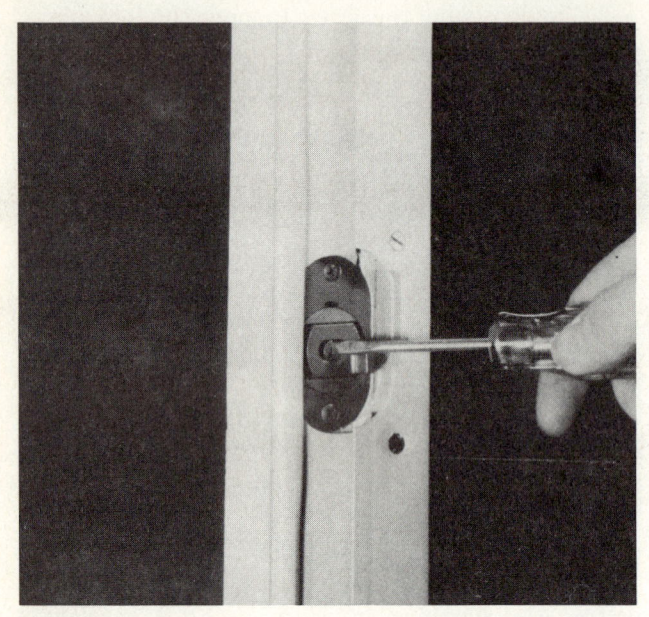

Figure 9-124. Lock adjustment on a triple door. *(Andersen)*

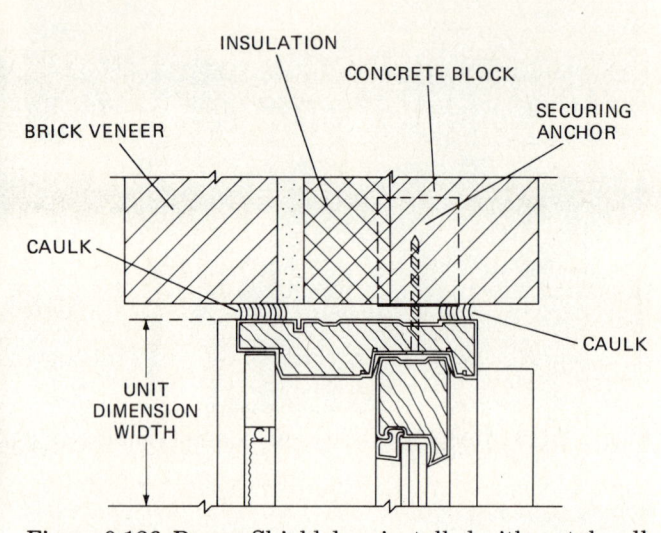

Figure 9-126. Perma-Shield door installed with metal wall plugs or extender plugs in masonry wall with a brick veneer. *(Andersen)*

Masonry or Brick-Veneer Wall Installation of a Gliding Door

Gliding doors can be installed in masonry wall construction. Fasten a wood buck to the masonry wall and nail the sliding door to the wood buck using the procedures shown for frame wall construction.

In Figure 9-125 a wood gliding door is installed in a masonry wall. Figure 9-126 shows a Perma-Shield door installed with metal wall plugs or extender plugs in the masonry wall with brick veneer. Metal wall plugs or extender plugs and auxiliary casing can be specified when the door is ordered.

Keep in mind that when brick veneer is used as an exterior finish, adequate clearance must be left for caulking between the frame and the masonry. This will prevent damage and bowing caused by shrinkage and settling of the structural lumber.

Installation of the Perma-Shield Gliding Door

The Perma-Shield type of door is installed in the same way as has just been described, with some exceptions. The exceptions follow:

1. Note that wide vinyl flanges which provide flashing are used at the head and side jambs. See Figure 9-127A. Locate the side member flush with the bottom of the sill with an offset leg pointing toward the inside of the frame. Tap with the hammer using a wood block to

252 Carpentry Fundamentals

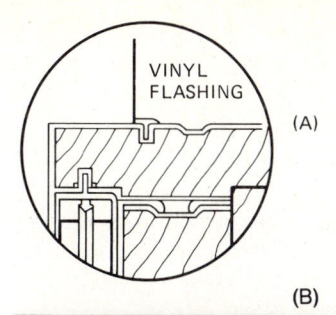

Figure 9-127. (A) Location of the vinyl flashing after it has been applied. *(Andersen)* (B) Using a wooden block to apply the vinyl flashing. *(Andersen)*

Figure 9-128. Clamps draw the flanges tightly against the sheathing. *(Andersen)*

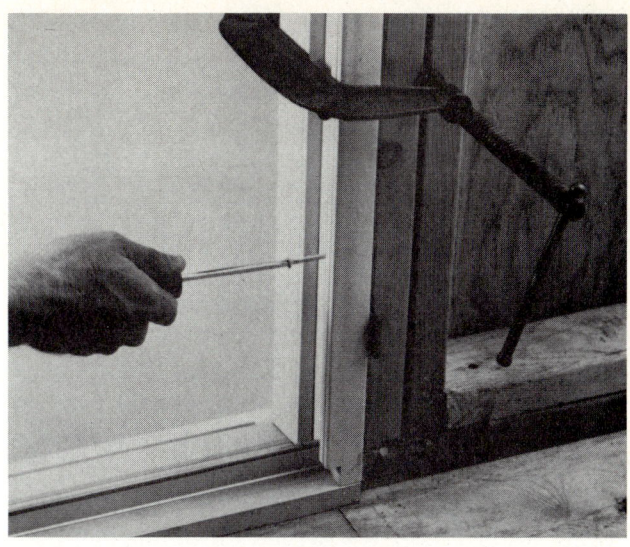

Figure 9-129. Securing the door frame to the studs with screws. *(Andersen)*

Figure 9-130. Placing screws in predrilled holes in the door header. *(Andersen)*

firmly seat the flashing in the groove. See Figure 9-127B. Apply the head member similarly. Overlap the side flange on the outside.
2. After securing the frame to the floor with nails through the sill, apply clamps to draw the flanges tightly against the sheathing. See Figure 9-128.
3. Temporarily secure the door in the opening with 10d casing nails through each side casing into the frame members. Using a straightedge, check the jambs for bow and shim. The jamb must be plumb and straight. Shim solidly using five shims per jamb between the side jambs and the jack studs.
5. Side members on the Perma-Shield gliding doors have predrilled holes to receive 2½-inch No. 10 screws. See Figure 9-129. Shim at all screw holes between the door frame and the studs. Drill pilot holes into the studs. Secure the door frame to the studs with screws.
6. The head jamb also has predrilled holes to receive the 2½-inch No. 10 screws. Shim at all screw holes between the door frame and header. Drill the pilot holes into the header. Insert the screws through the predrilled holes and draw up tightly. Do not bow the head jamb. See Figure 9-130.

Installing Windows and Doors

Figure 9-131 shows Perma-Shield door completely installed. Interior surfaces of the panel and frame should be primed before or immediately after installation for protection.

• ENERGY FACTORS •

There are some coatings for windows and door glass that reflect heat. A thin film is placed over the glass area. This thin film reflects up to 46 percent of the solar energy. Only 23 percent is admitted. That means 77 percent of the solar energy is reflected or turned away. Look at Figure 9-132 to see the extent of this ability to absorb energy and reflect it.

A thin vapor coating of aluminum prevents solar radiation from passing through glass. It does this by reflecting it back to the exterior. The temperature of the glass is not raised significantly. The coating minimizes undesirable secondary radiation from the glass. Visible light is reduced. However, the level of illumination remains acceptable. During the winter this coating reflects long-wave radiation and keeps the heat in the room.

The film is easily applied to existing windows. See Figure 9-132. Step 1 calls for spraying the entire window with cleaner. Scrape every square inch of the window with a razor blade. This removes the paint, varnish, and excess putty. Wipe the window clean and dry. Step 2 calls for respraying the window with cleaner. Squeegee the entire window. Use vertical strokes from top to bottom of the window. Use paper towels to dry the edges and corners.

Step 3 calls for clear water to be sprayed onto the glass area. Remove the separator film. See Figure 9-133. Test for the tacky side of film by folding over and touching to itself. The tacky side will stick slightly. Position the tacky side of the film against the top of the glass. Smooth the film by carefully pulling on the edges, or gently press the palms of your hands against the film. Slide the sheet of film easily to remove large wrinkles.

In step 4, spray the film, which is now on the glass, with water. Standing at the center, use a squeegee in gentle, short vertical and horizontal strokes to flatten the film. Slowly work out the wrinkles. Work out the bubbles and remove excess water from under the film. Be careful not to crease or wrinkle the film. Trim the edges with a razor blade. Wipe the excess water from the edges. The film will appear hazy for about two days until the excess water evaporates.

This is shown here to illustrate the possibilities of making your home an energy saver. Films and other devices will be forthcoming as we look forward to conserving energy.

Figure 9-131. Finished installation of a Perma-Shield gliding door. (Andersen)

Figure 9-132. Energy-saving film for windows. (Kelly-Stewart)

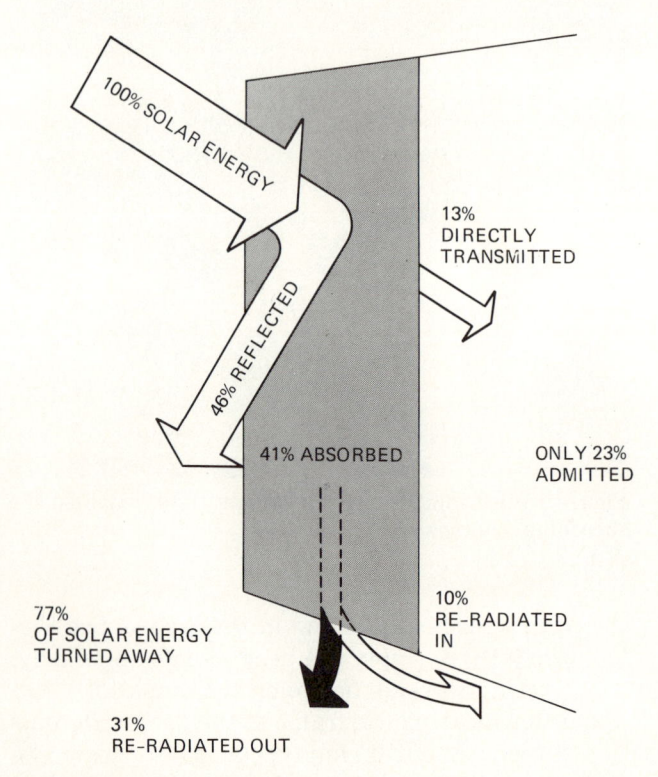

*Applicable to glass surfaces only.

Figure 9-133. Step-by-step application of energy-saving film. *(Kelly-Stewart)*

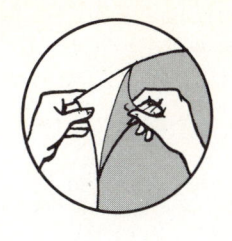

• WORDS CARPENTERS USE •

prehung	casement	sliding door	left-hand door	auxiliary locks
sliding window	jamb	French doors	bifold	contractor's key
awning picture	shim	stiles	track	storm door
window	flush	lockset	pivot	
double hung	panel	right-hand door	moldings	

• STUDY QUESTIONS •

9–1. What are the purposes of windows?

9–2. What purpose does a door serve?

9–3. How high should a window be?

9–4. What is the height of a door?

9–5. Where are windows made today?

9–6. What is meant by a prehung door?

9–7. What is a double-hung window?

9–8. What is a casement window?

9–9. What is an awning-type window?

9–10. What is meant by a hopper-type window?

9–11. What is a flush door?

9–12. What is a bifold?

9–13. Why does a door need a 3° angle on the lock stile?

9–14. What determines the hand of a door?

9–15. What does LHR mean?

9–16. What is a construction key?

9–17. How are the locks in a house changed after the house is complete?

9–18. What is an auxiliary lock?

9–19. How are storm doors installed?

9–20. Where are storm windows installed?

Installing Windows and Doors **255**

10
FINISHING THE EXTERIOR WALLS

Exterior walls are finished by two basic processes. The first is by covering the wall with a wood or wood product material called *siding*. The other is to cover the wall with a masonry material such as brick, stone, or stucco. The carpenter will install the exterior siding and will sometimes prepare the exterior wall for the masonry materials. However, people employed in the "trowel trades" install the masonry materials.

Exterior siding is applied over the wall sheathing. It adds protection against weather, strengthens the walls, and also gives the wall its final appearance or beauty. Siding may be made of many materials. Often, more than one type of material is used. Wood and brick could be combined for a different look. Other materials may also be combined. This chapter will help you learn these new skills:

- Prepare the wall for the exterior finish
- Estimate the amount of siding needed
- Select the proper nails for the procedure
- Erect scaffolds
- Install flashing and water tables to help waterproof the wall
- Finish the roof and edges
- Install exterior siding
- Trim out windows and doors

• INTRODUCTION •

Before the exterior is finished, windows and doors must be installed. Also, the roof should be up and the sheathing should be on the walls. Then, to finish the exterior, three things are done. First, the cornices and rakes around the roof are enclosed. Next, the siding is applied. Finally, the finish trim for windows and doors is installed.

The *cornice* is the area beneath the roof overhang. This area is usually enclosed or boxed in. Figure 10-1 shows a typical cornice. In many areas, the cornice is boxed in as part of the roofing job. The cornice is often painted before siding is installed. This is particularly likely when brick or stone is used.

Carpenters install several types of exterior siding. Most types of siding are made of wood, plywood, or wood fibers. Some types of siding are made of plastic and metal. These are usually formed to look like wood siding.

Siding is the outer part of the house that people see. Beauty and appearance are important factors. However, siding is also selected for other reasons. The builder may want a siding that can be installed quickly and easily. This means that it costs less to install. The owner may want a siding that requires little maintenance. Both will want a siding that will not rot or warp. Whether insects will attack the siding is another factor. The type of siding selected will depend upon many factors.

• TYPES OF SIDING •

Siding is made of plywood, wood boards, wood fibers, various compositions, metal, or plastics. See Figure 10-2A to D. However, the shape is the main

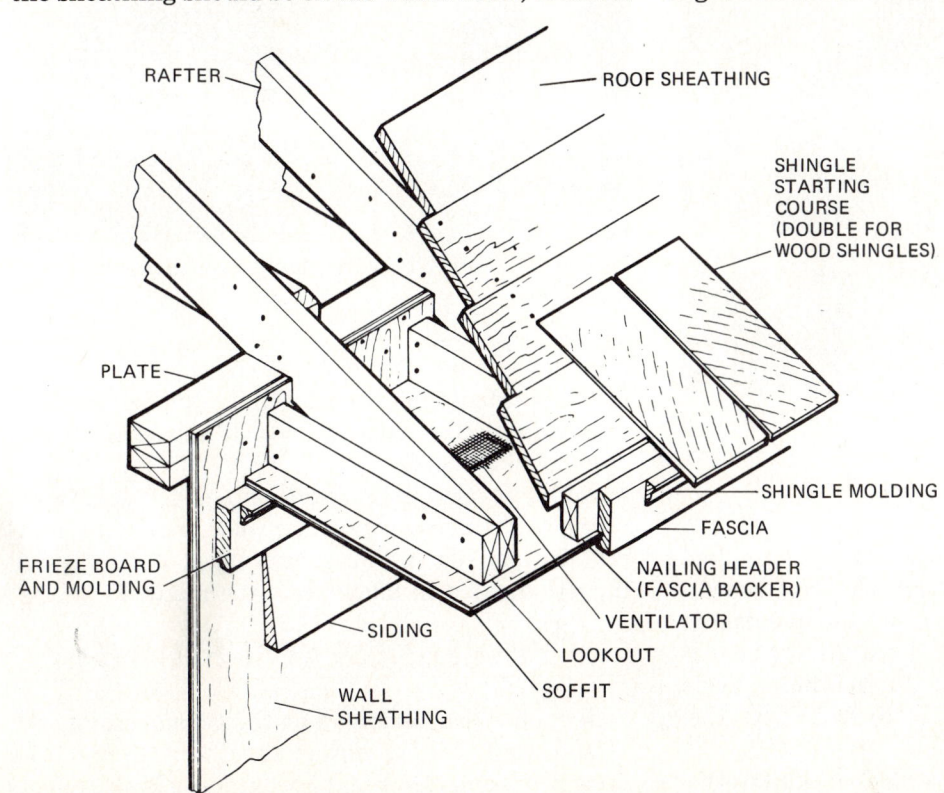

Figure 10-1. Typical box cornice framing *(Courtesy Forest Products Laboratory)*

Finishing the Exterior Walls

Figure 10-2 (A). Cedar boards are used both horizontally and diagonally for this siding. *(Potlatch)*

Figure 10-2 (B). Plywood siding has many "looks" and styles. This looks like boards. *(American Plywood Association)*

Figure 10-2 (C). This siding combines rough brick, smooth stucco, and boards.

Figure 10-2 (D). Here stone siding is applied over frame construction.

factor that determines how the siding is applied. Wood fiber is made in two main shapes, "boards" and sheets or panels, which often look like boards. The board-shaped fiber siding is put up like boards. However, the sheets are put up as whole sheets. Fiber panels can also be made to look like other types of siding.

Plywood panels can also be made to look like other types of siding. Figure 10-2B shows a house with plywood siding that looks like individual boards. Plywood can also be made in "board" strips. These are applied like boards.

Siding is also made from shingles and shakes. Shingles are made from either wood or asbestos mineral compounds. Actually, both types of shingle siding are applied in much the same manner.

• SEQUENCE FOR SIDING •

Sequence is determined by the type of siding, the type of roof, and the type of sheathing (if any) used on the building. How high the building is also affects

258 Carpentry Fundamentals

the sequence. To work on high places, carpenters erect platforms. These let them stand at the right level to work. These platforms are called *scaffolds*.

In most cases, board siding is put on from bottom to top. However, if scaffolds have to be nailed to the wall, the siding on the bottom could be damaged. In that case the sequence can be changed. The scaffolds can be put up and the siding can be put on the upper areas. Then the scaffolds can be removed and the siding can be put on the bottom. This way the siding is not damaged by scaffold nails or bolts.

However, many scaffolds can stand alone. They need not be nailed to the wall. The common sequence for finishing an exterior wall is

1. Prepare for the job. Make sure that the windows are installed, the vapor barrier is in place, the nails are selected, and the amount of siding is estimated.
2. Erect necessary scaffolds.
3. Install the flashing and water tables.
4. Finish the roof edges.
5. Install siding on the upper gable ends and on upper stories.
6. Install siding on the sides.
7. Finish the corners.
8. Trim windows and doors.

• PREPARE FOR THE JOB •

Several things should be done before siding is put on the wall. Windows and doors should be properly installed. Rough openings should have been moisture-shielded. Then any spaces between window units and the wall frame are blocked in. A moisture barrier should be present. Some types of sheathing are also good moisture barriers. However, other types are not. These must have a moisture barrier installed. Figure 10-3 shows a window that has been blocked properly so that siding may now be installed.

Vapor Barrier

Some types of sheathing need no vapor barrier. However, board and plywood sheathing must have a separate vapor barrier. Also, for some types of siding, no sheathing is used. Then, a vapor barrier is also needed. The vapor barrier is applied to the wall frame. Also, remember that several terms are used, but they mean the same thing.

A vapor barrier can help save energy. Air can leak through the wall without a vapor barrier. It is important to remember that air can move both ways. This can result in the loss of either heated or cooled air. This air movement can lessen the efficiency of the air treatment in the building. Also, this movement can damage the exterior siding.

Figure 10-3. A window unit blocked for brick siding.

Figure 10-4. Foil surfaces reflect energy.

Heated air contains moisture. If it travels from inside the building to the outside, damage can result. As the moisture goes through a siding, it will cause paint on the outside to blister and bubble. This will also trap damp air. Damp air can increase the susceptibility to rot and damage.

Reflective vapor barriers are also used. See Figure 10-4. They help keep energy on one side of a wall.

Finishing the Exterior Walls 259

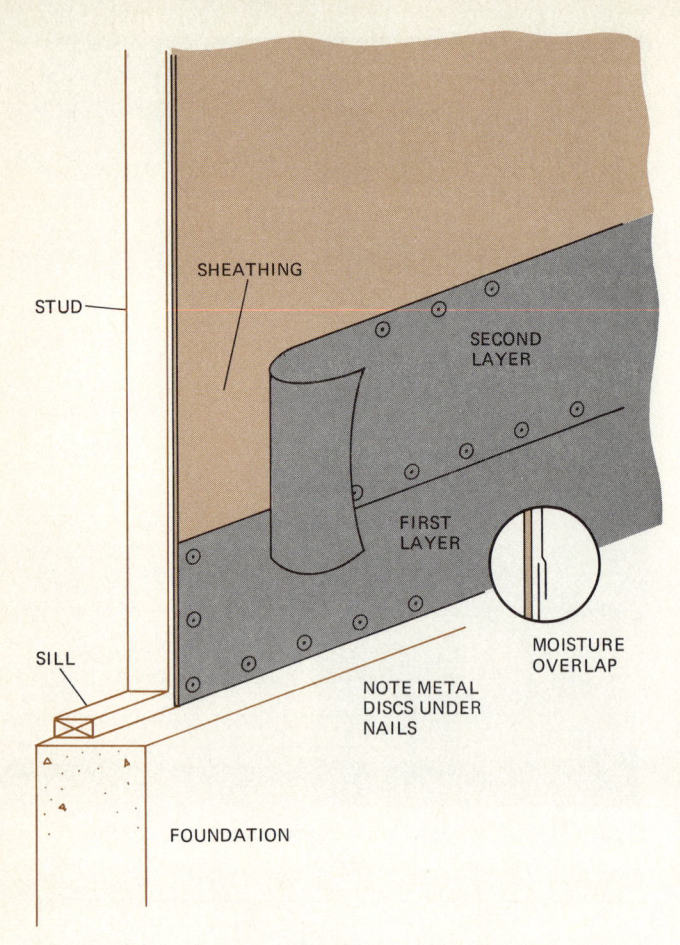

Figure 10-5. Moisture barrier is applied from the bottom to the top. The overlap helps shed moisture.

They are warmer in winter and cooler in summer. However, the foil should be put next to the heat source.

To put up a vapor barrier, start at the bottom. Nail the top part in place. Most vapor-barrier materials will be applied in strips. The strips should shed water to the outside. See Figure 10-5. To do this, the bottom strip is installed first. Each added strip is overlapped. Moisture barriers should also be lapped into window openings. See Chapter 9 on installing windows for this procedure.

Vapor barriers are made from plastic films, from metal foil, and from builder's felt (also called tar paper). Builder's felt and plastic films are used most frequently.

Nail Selection

Several methods of nailing can be used. Some siding may be put up by any of several methods. Other siding should be put up with special fasteners. The important thing is that the nail must penetrate into something solid in order to hold. For example, nails driven into fiber sheathing will not hold. Siding over this type of sheathing must be nailed at the studs. If the studs are placed on 16-inch centers, the nails are driven at 16-inch intervals. Likewise, splices in the siding should be made over studs. Also, the nail must be long enough to penetrate into the stud.

However, if plywood or hardboard sheathing is used, then nails can be driven at any location. The plywood will hold even a short nail.

Strips of wood are also used to make a nail base. These strips are placed over the sheathing. Then they are nailed to the studs through the sheathing. Nail strips are used for several types of shingled siding. See Figure 10-6.

The nail should enter the nail base for at least ½ inch. A nonstructural sheathing such as fiber sheathing cannot be the nail base. With fiber or foam sheathing, the nail must be longer. The nail must go through siding and the sheathing and ½ inch into a stud.

Also, the type of nail should be considered. When natural wood is exposed, a finishing or casing-head nail would look better. The head of the nail can be set beneath the surface of the wood. The head will not be seen. This way the nail will not detract from the appearance. However, if the wood is to be painted, a common or a box-head nail may be used. Coated nails are preferred for composition and mineral siding. These nails are coated with zinc to keep them from rusting. This makes them more weather-resistant.

For some siding, the nail is driven at an angle. This means that it must be longer than the straight-line distance. This type of nailing is frequently done in grooved and edged siding. The nails are driven in the grooves and edges. This way the siding is put on so that the nails are not exposed to the weather or the eye of the viewer.

Estimate the Amount of Siding Needed

Many people think that "one by six" (1x6-inch) boards are 1 inch thick and 6 inches wide. However, a 1x6 board is actually ¾ inch thick by 5½ inches wide. Also, most board siding overlaps. Rabbet, bevel, and drop sidings all overlap. When overlapped, each board would not expose 5½ inches. Each board would expose only about 5 inches.

Several things must be known to estimate the amount of siding needed. First, the height and width of the wall must be found. Then, find the type of siding and the sizes of windows or doors. Consider the following example:

Siding: 1x8-inch bevel siding
Overlap: 1½ inches
Wall height: 8 feet
Wall length: 40 feet

260 Carpentry Fundamentals

Windows and Doors: Two windows, each 2x4 feet
One door, 8x3 feet

First, find the total area to be covered. To do this, multiply the length of the wall times the height.

Area = 40x8
= 320 square feet

Next, subtract the area of the doors and windows from the area of the wall. To do this:

Area of windows = width × length
= 2 × 4
= 8 square feet × number of windows
= 16 square feet
Door area = width × length
= 3 × 8
= 24 square feet
Total opening area = 16 square feet
+ 24 square feet
Total   40 square feet

In order to get enough siding to cover this area, the percentage of overlap must be considered. Also, there is always some waste in cutting boards so that they join at the proper place. When slopes and corners are involved, there is more waste. The amount of overlap for 8-inch siding lapped 1¼ inches is approximately 17 percent. However, it is best to add another 15 percent to this for waste. Thus, about 32 percent would be added to the total requirements. Thus, to side the wall would require 320 − 40 = 280 square feet. However, enough siding should be ordered to cover 280 square feet plus 32 percent (90 board feet). A total of 370 board feet of siding should be ordered. Table 10-1 shows allowances for different types of siding.

To estimate the area for gable ends, the same procedure is used. Find the length and the height of the gable area. Then multiply the two dimensions for the total. However, since slope is involved, as shown in Figure 10-7, only one-half of this figure is required. The allowance for waste and overlap is based upon this halved figure.

The amount of board siding required for an entire building may be estimated. First, the total area of all the walls is found. Then the areas of all the gable sections are found. These areas are added together. The allowances are made on the total figure.

TABLE 10-1 ALLOWANCES THAT MUST BE ADDED TO ESTIMATE SIDING NEEDS.

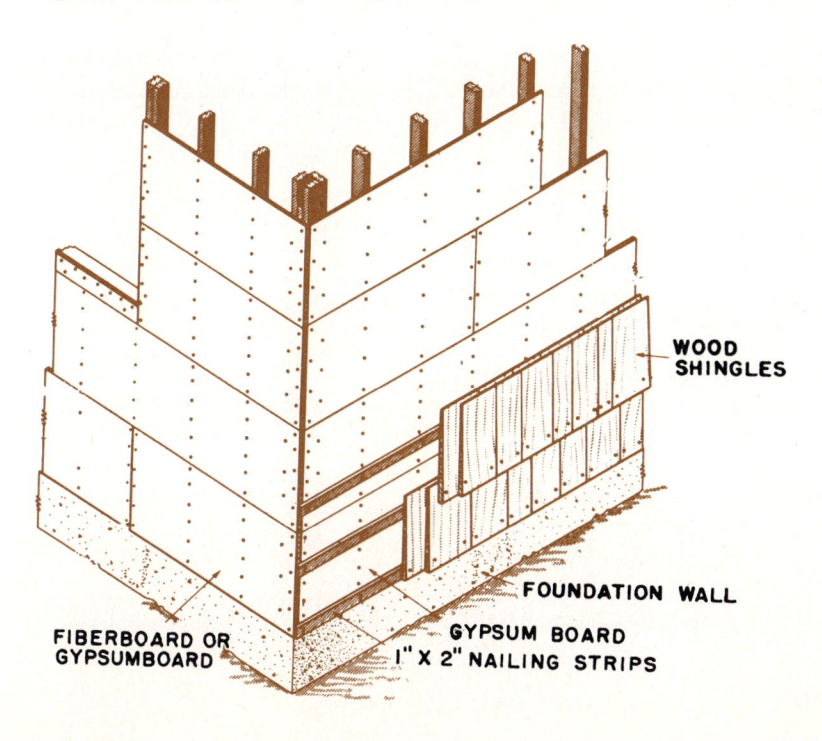

Type	Size, Inches	Amount of Lap, Inches	Allowance, Percent (Add)
Bevel siding	1 × 4	¾	45
	1 × 6	1	35
	1 × 8	1¼	32
	1 × 10	1½	30
	1 × 12	1½	25
Drop siding (shiplap)	1 × 4		30
	1 × 6		20
	1 × 8		17
Drop siding (matched)	1 × 4		25
	1 × 6		17
	1 × 8		15

Figure 10-6. Nail strips can be used over sheathing. (*Courtesy Forest Products Laboratory*)

Finishing the Exterior Walls

ORDERING PANELED SIDING • Paneled or plywood siding is sold in sheets. The standard sheet size is 4 feet wide and 8 feet long. The standard height of the wall is 8 feet. Thus a panel fully covers a 4-foot length of wall. To estimate the amount needed, find the length of the wall. Then divide by 4, the width of a panel. For example, a wall is 40 feet long. The number of panels is 40 divided by 4.

Thus, it would take 10 panels to side this wall with plywood paneling. If, however, the length of the wall is 43 feet, then 11 panels would be required. Only whole panels may be ordered. No allowances for windows and doors are made. Sections of panels are cut out for windows and doors.

ESTIMATING SHINGLE COVERAGE • To find the amount of shingles needed, the actual wall area should be found. The areas of the doors and windows should be deducted from the total area. Shingles vary in length and width. The size of the shingle must be considered when estimating. Also, shingles come in packages called *bundles*. Generally, four bundles are approximately 1 "square" of coverage. This means that four bundles will cover about 100 square feet of surface area. However, the amount of shingle exposed determines the actual coverage. If only 4 inches of a 16-inch-long shingle is exposed, the coverage will be much less. Also, if more of the shingle is exposed, a greater area can be covered. For walls, more of the shingle can be exposed than for roofs. Table 10-2 shows the coverage of 1 square of shingles when different lengths are exposed.

• ERECTING SCAFFOLDS •

Many areas on a building are high off the ground. When carpenters must work up high, they stand on platforms called *scaffolds*. Figure 10-8 shows a typical scaffold. Scaffolds are also sometimes called *stages*.

A scaffold must be strong enough to hold up the weight of everything on it. This includes the people, the tools, and the building materials. There are several types of scaffolds that can be used. The type used depends upon the number of workers involved and the weight of the materials. Also, how high the scaffold will be is a factor. How long the scaffold will remain up must be considered.

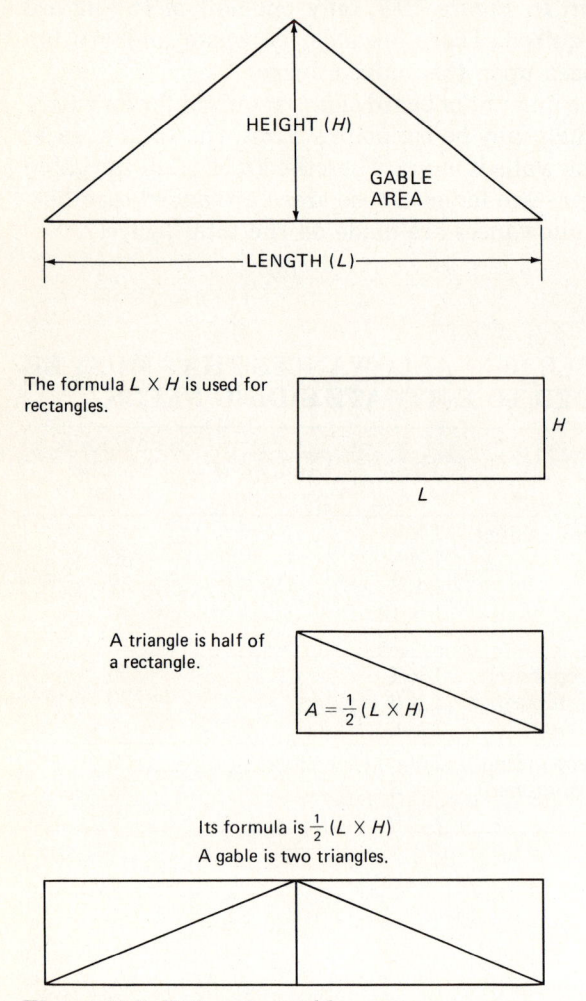

The formula L × H is used for rectangles.

A triangle is half of a rectangle.

$A = \frac{1}{2}(L \times H)$

Its formula is $\frac{1}{2}(L \times H)$
A gable is two triangles.

Figure 10-7. Estimating gable areas.

TABLE 10-2 COVERAGE OF WOOD SHINGLES FOR VARYING EXPOSURES

Length and Thickness[b]	Approximate Coverage of Four Bundles or One Carton,[a] Square Feet								
	Weather Exposure, Inches								
	5½	6	7	7½	8	8½	9	10	11	11½
Random-width and dimension 16″ × 5/2″	110	120	140	150[c]	160	170	180	200	220	230
18″ × 5/2¼″	100	109	127	136	145½	154½[c]	163½	181½	200	209
24″ × 4/2″	—	80	93	100	106½	113	120	133	146½	153[c]

[a] Nearly all manufacturers pack four bundles to cover 100 square feet when used at maximum exposures for roof construction; rebutted-and-rejointed and machine-grooved shingles typically are packed one carton to cover 100 square feet when used at maximum exposure for double-coursed sidewalls.
[b] Sum of the thickness, e.g. 5/2″ means 5 butts equal 2″.
[c] Maximum exposure recommended for single-coursed sidewalls.

Figure 10-8. A typical job-made scaffold for low, light work.

Job-Built Scaffolds

Scaffolds are often built from job lumber. The maximum distance above the ground should be less than 18 feet. Three main types of job scaffolds can be built. Supports for the platform should be no more than 10 feet apart, and less is desirable.

DOUBLE-POLE SCAFFOLDS • This style of scaffold is shown in Figure 10-9. The poles form each support section. They should be made from 2×4-inch pieces of lumber. A 1×6 board is nailed near the bottom of both poles for a bottom brace. Three 12d nails should be used in each end of the 1×6-inch board. The main support for the working platform is called a *ledger*. It is nailed to the poles at the desired height. It should be made of 2×4-inch or 2×6-inch lumber. It is nailed with three 16d common nails in each end of the ledger. Sway braces should be nailed between each end pole as shown in Figure 10-9. The sway braces can be made of 1-inch lumber. They are nailed as needed.

After two or more of the pole sections have been formed on the ground, they can be erected. Place a small piece of wood under the leg of the pole to provide a bearing surface. This keeps the ends of the pole from sinking into the soft earth. The sections should be held in place by the carpenters. Then another carpenter nails braces between each pole section. These are nailed diagonally, as are sway braces. They may be made of 1×6-inch lumber. They are nailed with three 14d common nails at each end.

When the ledgers are more than 8 feet above the ground, a guard rail should be used. The guard rail is nailed to the pole about 36 inches above the platform. Note that the guard rail is nailed to the inside of the pole. This way a person can lean against the guard rail without pushing it loose.

The double-pole type of scaffold can be free-standing. That is, it need not be attached to the building. However, carpenters often nail a short board between the scaffold and the wall. This holds the scaf-

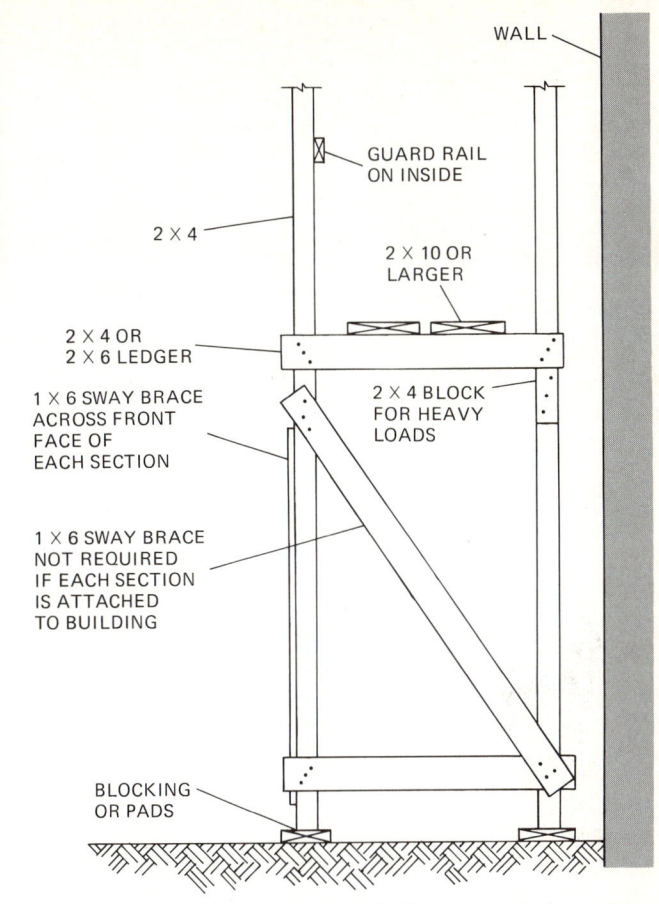

Figure 10-9. A double-pole scaffold can stand free of the building. Guard rails are needed above 10'-0".

fold safely. If this is done, the board should be nailed near the top of the scaffold.

SINGLE-POLE SCAFFOLDS • Single-pole scaffolds are similar to double-pole scaffolds. However, the building forms one of the "poles." Blocks are nailed to the building. They form a nail base for the brace and ledgers. Figure 10-10 shows a single-pole scaffold. Sway braces are not needed on the pole sections. However, sway braces should be used between each section of the scaffold.

WALL BRACKETS • Wall brackets are often made on the job. See Figure 10-11. Wall brackets are most often used when the distance above the ground is not very great. They are quicker and easier to build than other scaffold types. Of course, they are also less expensive.

On the ends, the wall brackets are nailed to the outside corners of the wall. On intersections, nail blocks are used, similar to the single-pole scaffold blocks. Wall brackets must be nailed to solid wall members. It is best to use 20d common nails to fasten them in place. Figure 10-12 shows another type of wall bracket. This type is sturdier than the other.

Finishing the Exterior Walls

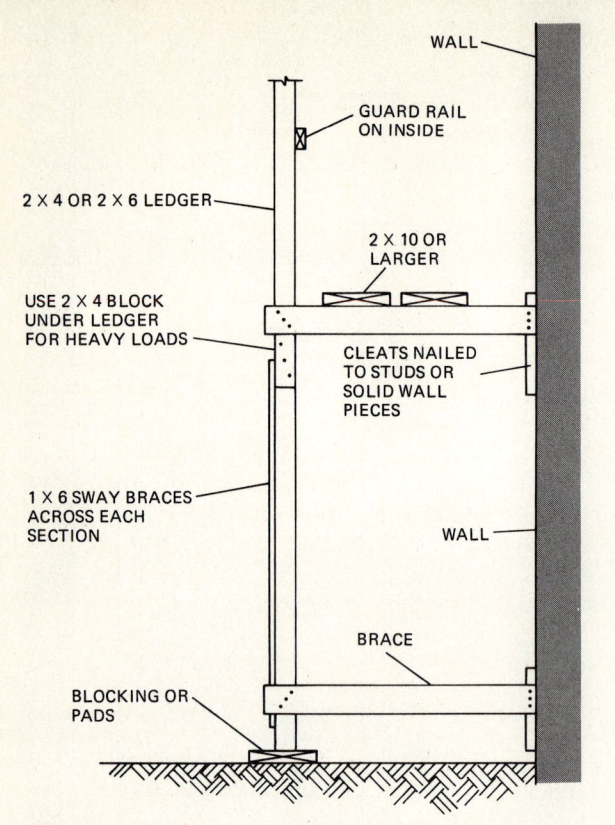

Figure 10-10. A single-pole scaffold must be attached to the building. Guard rails are needed above 10'-0".

Figure 10-12. A job-made wall-bracket scaffold. It is dangerous to work on this kind of scaffold in high places without safety equipment.

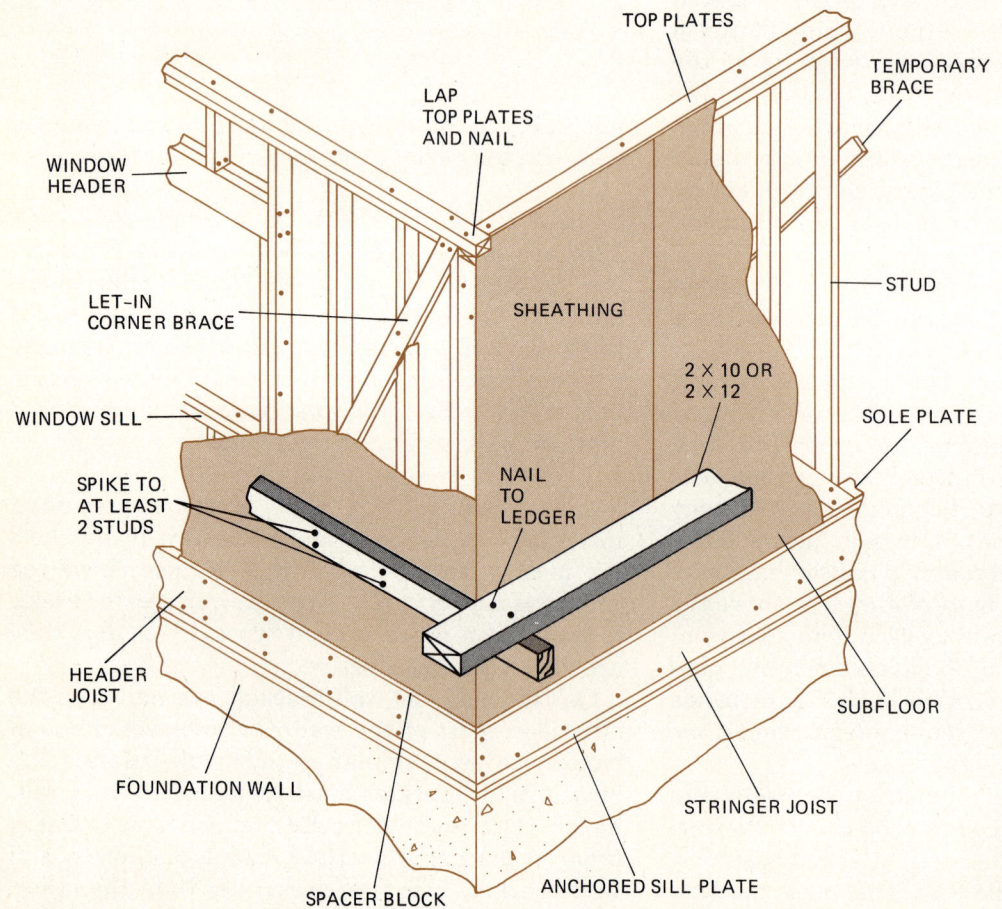

Figure 10-11. A typical low-level wall bracket.

264 Carpentry Fundamentals

Factory Scaffolds

Today, many builders use factory-made scaffolds. These have several advantages for a builder. They are quick and easy to erect. They are strong and durable and can be easily taken down and reused.

No lumber is used. Also, no cutting and nailing is needed to erect the scaffold. Thus, no lumber is ruined or made unusable for the building. The metal scaffold parts are easily stored and carried. They are not affected by weather and will not rot. There are several different types of scaffolds.

DOUBLE-POLE SCAFFOLD SECTIONS • The double-pole type of scaffold features a welded steel frame. It includes sway braces, base plates, and leveling jacks. See Figures 10-13 and 10-14. Two or more sections can be used to gain greater distance above the ground. Special pieces provide sway bracing and leveling. Guard rails and other scaffold features may also be included.

WALL BRACKETS • Wall brackets, as in Figures 10-15 through 10-17, are common. No nail blocks are needed on the walls for these metal brackets. These brackets are nailed or bolted directly to the wall. Be sure that they are nailed to a stud or another structural member. Use 16 or 20d common nails. After the nails are driven in place, be sure to check the heads. If the nail heads are damaged, remove them and renail the bracket with new nails. Remember that it is the nail head that holds the bracket to the wall. A damaged nail head may break. A break could let the entire scaffold fall.

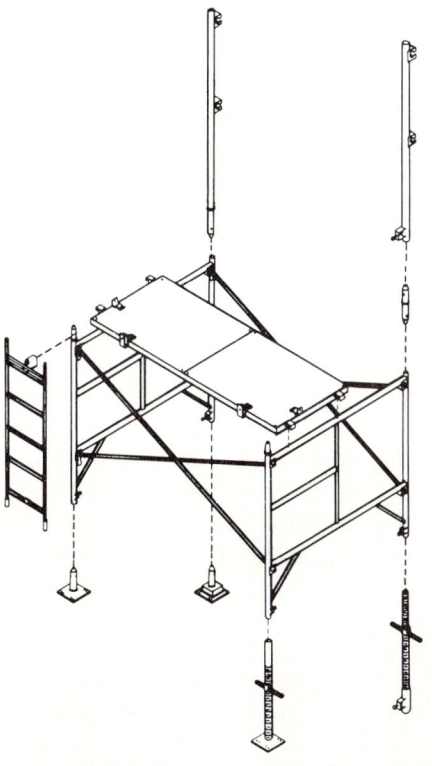

Figure 10-13. Metal double-pole scaffolds are widely used. *(Beaver-Advance)*

Figure 10-15. Metal wall brackets are quick to put up and take down. *(Richmond Screw Anchor)*

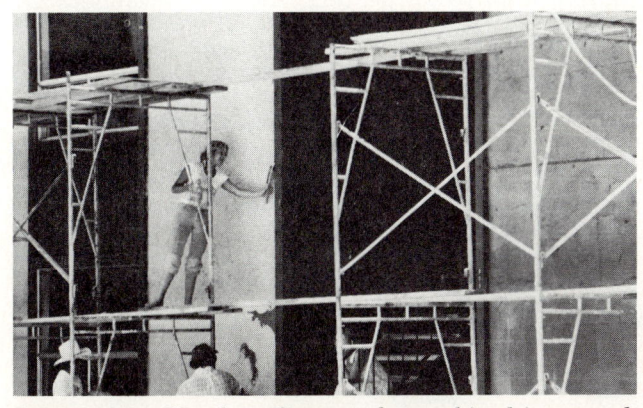

Figure 10-14. Metal sections can be combined in several ways to get the right height and length.

Figure 10-16. Wall brackets can be used on concrete forms.

Finishing the Exterior Walls 265

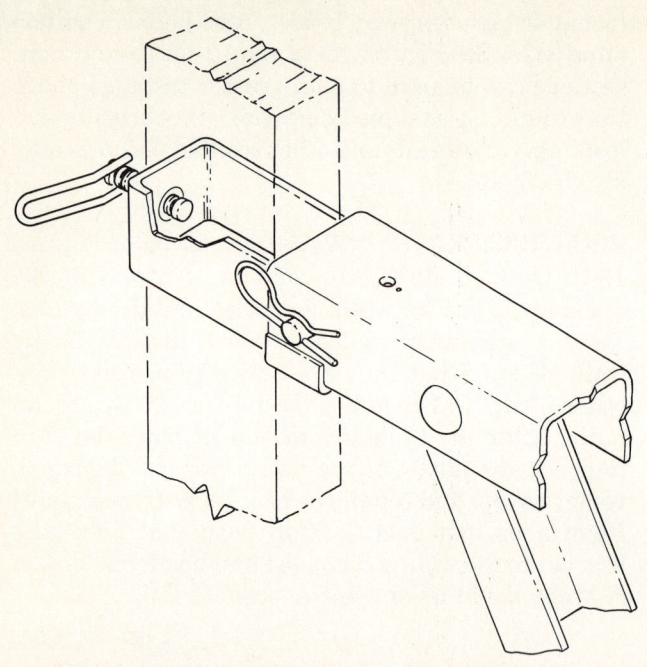

Figure 10-17. Guard rails may be added through an end bracket. *(Richmond Screw Anchor)*

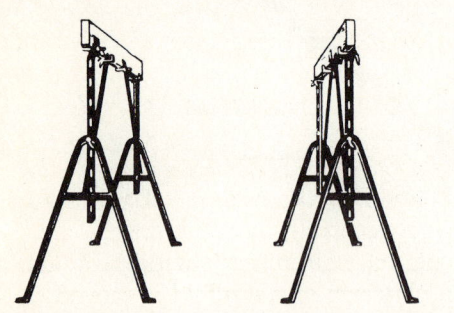

Figure 10-18. Two trestle jacks can form a base for scaffold planks. *(Patent Scaffolding Co.)*

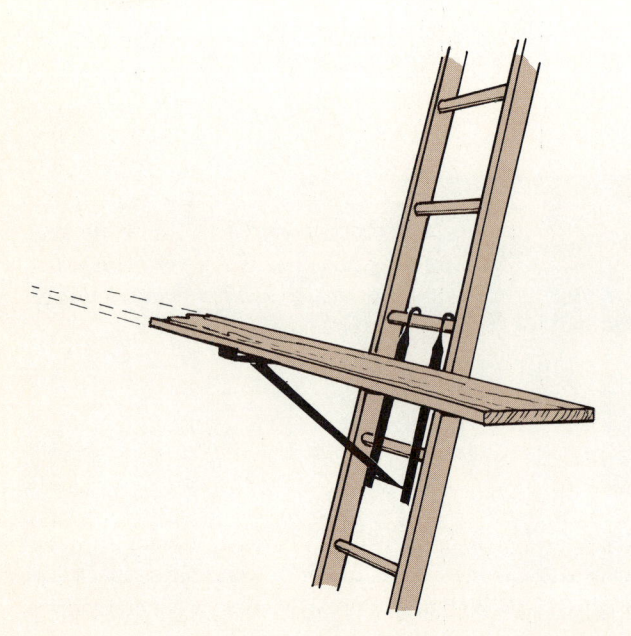

TRESTLE JACK • Trestle jacks are used for low platforms. They are used both inside and outside. They can be moved very easily. Trestle jacks are shown in Figure 10-18. Two trestle jacks should be used at each end of the section. A ledger, made of 2×4-inch lumber, is used to connect the two trestle jacks. Platform boards are then placed across the two ledgers. Platform boards should always be at least 2 inches thick.

As you can see, it takes four trestle jacks for a single section. Trestle jacks can be used on uneven areas, but they provide for a platform height of only about 24 inches. However, this is ideal for interior use.

LADDER JACKS • Ladder jacks, as in Figure 10-19, hang a platform from a ladder. They are most suitable for repair jobs and for light work where only one carpenter is on the job. Two types of jacks are used. The type shown in Figure 10-19 puts the platform on the outside of the ladder. The type shown in Figure 10-20 places the platform below or on the inside of the ladder.

Ladder Use

Using ladders safely is an important skill for a carpenter. A ladder to be erected is first laid on the ground. The top end of the ladder should be near the building. The top end is raised and held overhead. The carpenter gets directly beneath the ladder. The hands are then moved from rung to rung. As the top end of the ladder is raised, the carpenter walks away from the building. Thus, the ladder is raised higher and higher with every step. Care is taken to watch where the end of the ladder is going. The top of the ladder is guided to the proper position. Then the ladder is leaned firmly against some part of the building. The base of the ladder is made secure on solid ground or concrete. Both ends of the bottom should be on a firm base.

Both wooden and aluminum ladders are commonly used by carpenters. Aluminum ladders are

Figure 10-19. Outside ladder jacks.

266 Carpentry Fundamentals

Figure 10-20. Inside ladder jacks.

light and easy to manage. However, they have a tendency to sway more than wooden ladders. They are very strong and safe when used properly. Wooden ladders do not sway or move as much, but they are much heavier and harder to handle.

Wooden ladders are sometimes made by carpenters. A ladder is made from clear, straight lumber. It should have been well seasoned or treated. The sides are called *rails* and the steps are *rungs*. Joints are cut into the rails for the rungs. Boards should never be just nailed between two rails.

Ladder Safety

1. Ladder condition should be checked before use.
2. The ladder should be clean. Grease, oil, or paint on rails or rungs should be removed.
3. Fittings and pulleys on extension ladders should be tight. Frayed or worn ropes and lines should be replaced.
4. The bottom ends of the ladder must rest firmly and securely on a solid footing.
5. Ladders should be kept straight and vertical. Never climb a ladder that is leaning sideways.
6. The bottom of the ladder should be one-fourth the height from the wall. For example, if the height is 12 feet, the bottom should be 3 feet from the wall.

Scaffold Safety

1. Scaffolds should be checked carefully before each use.
2. Design specifications from the manufacturer should be followed. State codes and local safety rules should also be followed.
3. Pads should be under poles.
4. Flimsy steps on scaffold platforms or ladders should never be used. Height should only be increased with scaffold of sound construction.
5. For platforms, planking that is heavy enough to carry the load and span should be used.
6. Platform boards should hang over the ledger at least 6 inches. This way, when boards overlap, the total overlap should be at least 12 inches.
7. Guard rails and toe boards should be used.
8. Scaffolds should never be put up near power lines without proper safety precautions. The electric service company can be consulted for advice when a procedure is not known.
9. All materials and equipment should be taken off before a platform or a scaffold is moved.

• FINISHING ROOF EDGES •

Most roofs have an overhang. This is the part of the roof that hangs over the wall. This portion is called the *eave* of the roof. If eaves are enclosed, they are also called a *cornice*. See Figure 10-21. Usually, the edges of roofs are finished when the roof is sheathed. However, building sequences do vary from place to place. Two methods of finishing the eaves are commonly employed. These are the open method and the closed method. There are several versions of the closed method.

Open Eaves

A board is usually nailed across the ends of the rafters when the roof is sheathed. See Figure 10-22. This board is called the *fascia*. The fascia helps to brace and strengthen the rafters. Fascia should be joined or spliced as shown in Figure 10-23. However, the fascia is not needed structurally. Some

Finishing the Exterior Walls

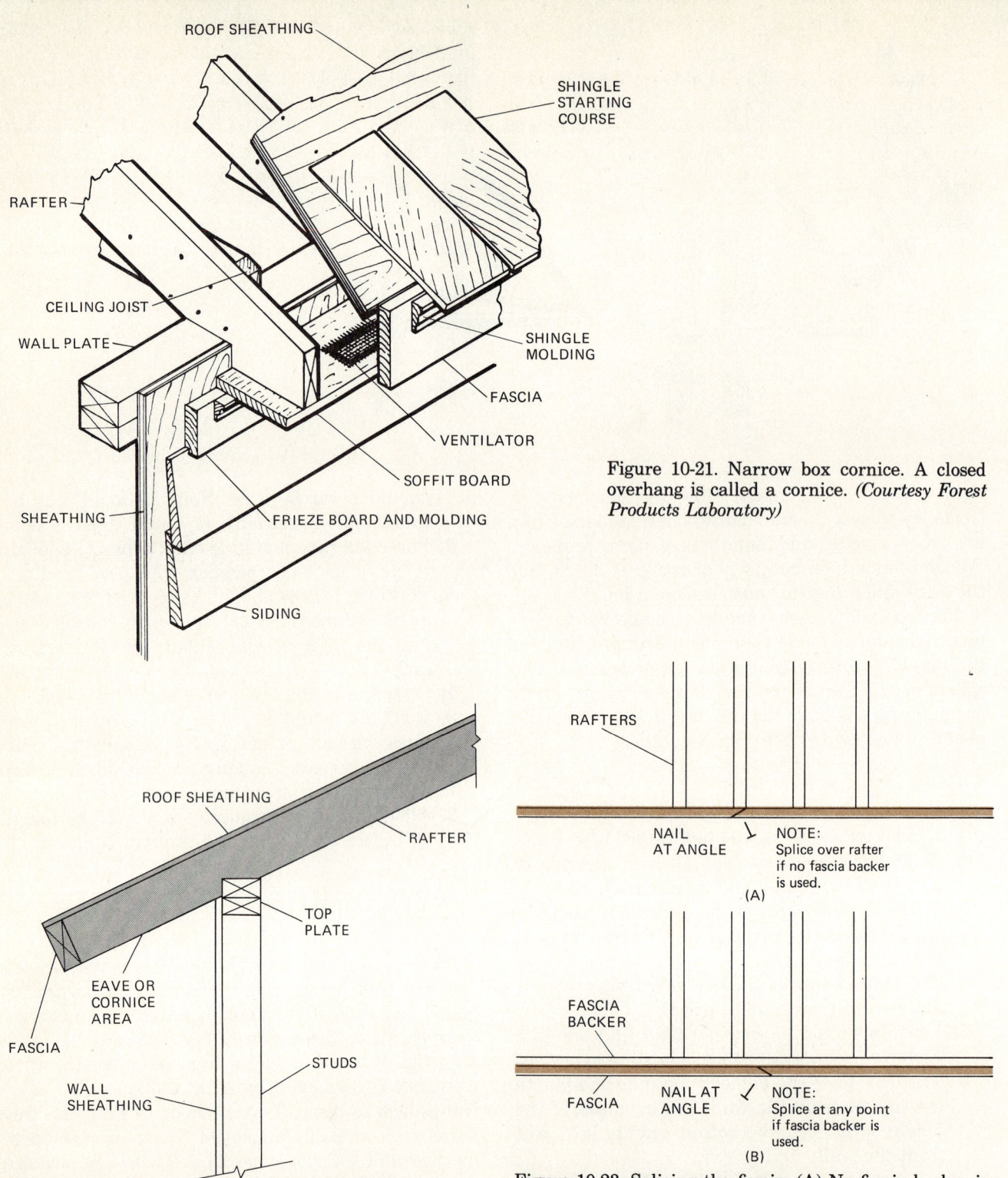

Figure 10-21. Narrow box cornice. A closed overhang is called a cornice. *(Courtesy Forest Products Laboratory)*

Figure 10-22. Open overhangs are called eaves.

Figure 10-23. Splicing the fascia. (A) No fascia backer is used. (B) A fascia backer is used.

types of open-eave construction do not use fascia.

Open eaves expose the area where the rafters and joists rest on the top plate of a wall. This area should be sealed either by a board or by the siding. See Figure 10-24. Sealing this area helps prevent air currents from entering the wall. It also helps keep insects and small animals from entering the attic area. However, to completely seal this restricts airflow. This airflow is important to keep the roof members dry. It also prevents rotting from moisture. Airflow will also cool a building. To allow airflow, vents should be installed. These vents are backed with screen or hardware cloth to keep animals and insects from entering.

Enclosed Cornices

Eaves are often enclosed for a neater appearance. Enclosed eaves are called *cornices*. There are several ways of enclosing cornices. Also, some houses do not have eaves or cornices. These are called *close* cornices. See Figure 10-25. However, the two most common types are the standard slope cornice and the flat cornice. Both types seal the cornice area with panels. These panels are called *soffits*. The panels are usually made of plywood or metal. Note that both types should have some type of ventilation. Special cornice vents are used. However, when plywood is used for the soffits, an opening is often left as in Figure 10-26. This provides a continuous vent strip. This strip is covered with some type of grill or screen.

Figure 10-24. Open eaves should be sealed and vented properly.

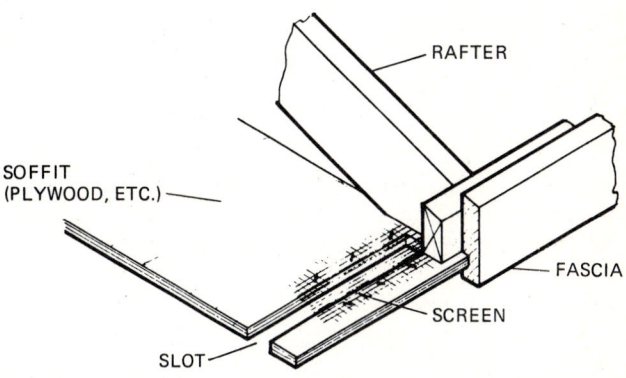

Figure 10-26. A strip opening may be used for ventilation instead of cornice vents. *(Courtesy Forest Products Laboratory)*

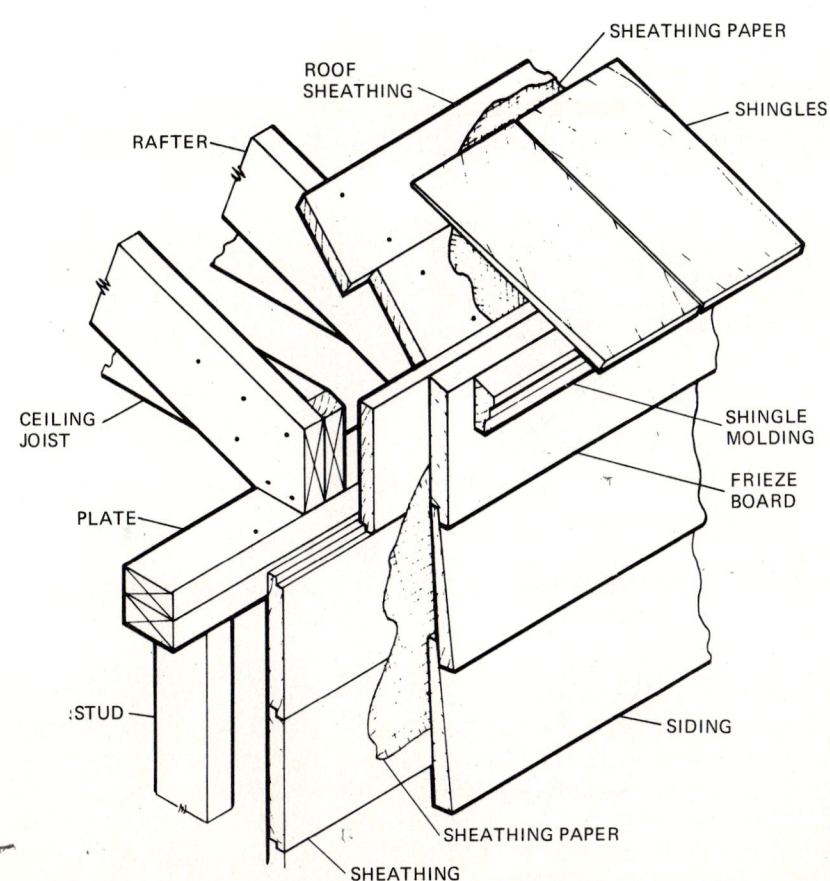

Figure 10-25. With close cornices, the roof does not project over the walls. *(Courtesy Forest Products Laboratory)*

Finishing the Exterior Walls **269**

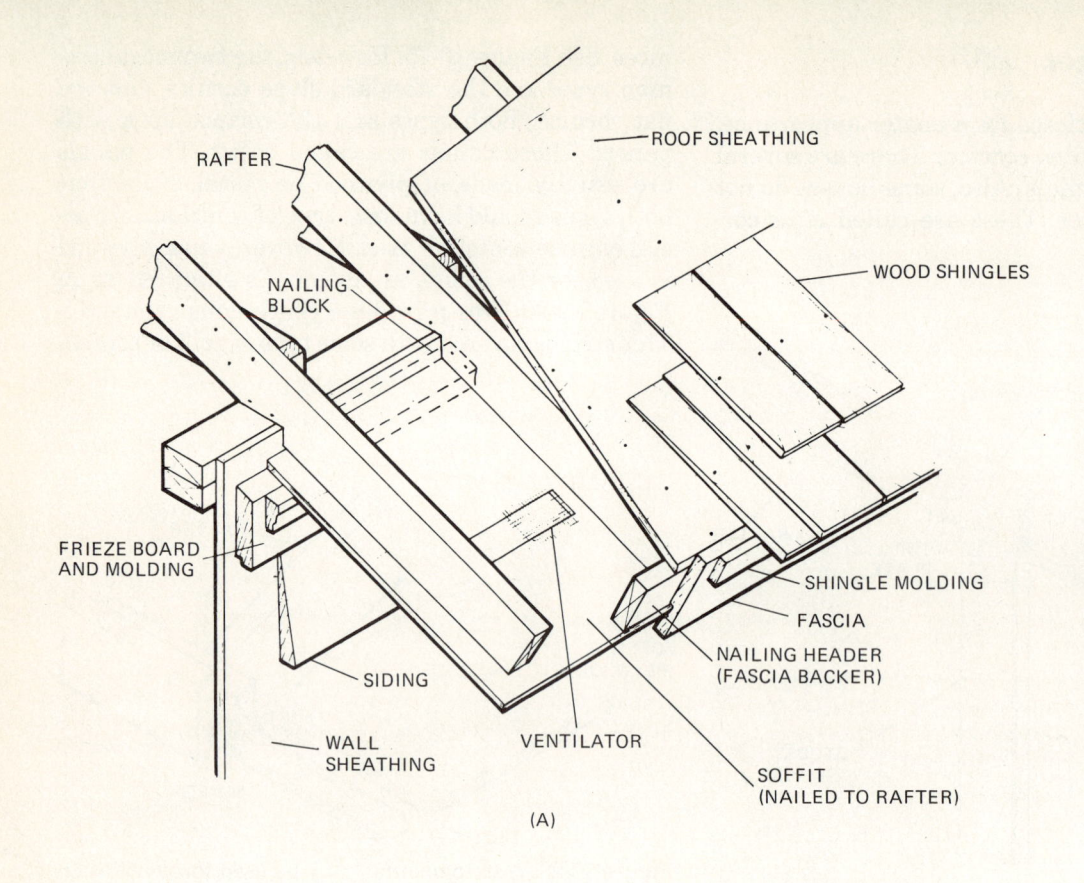

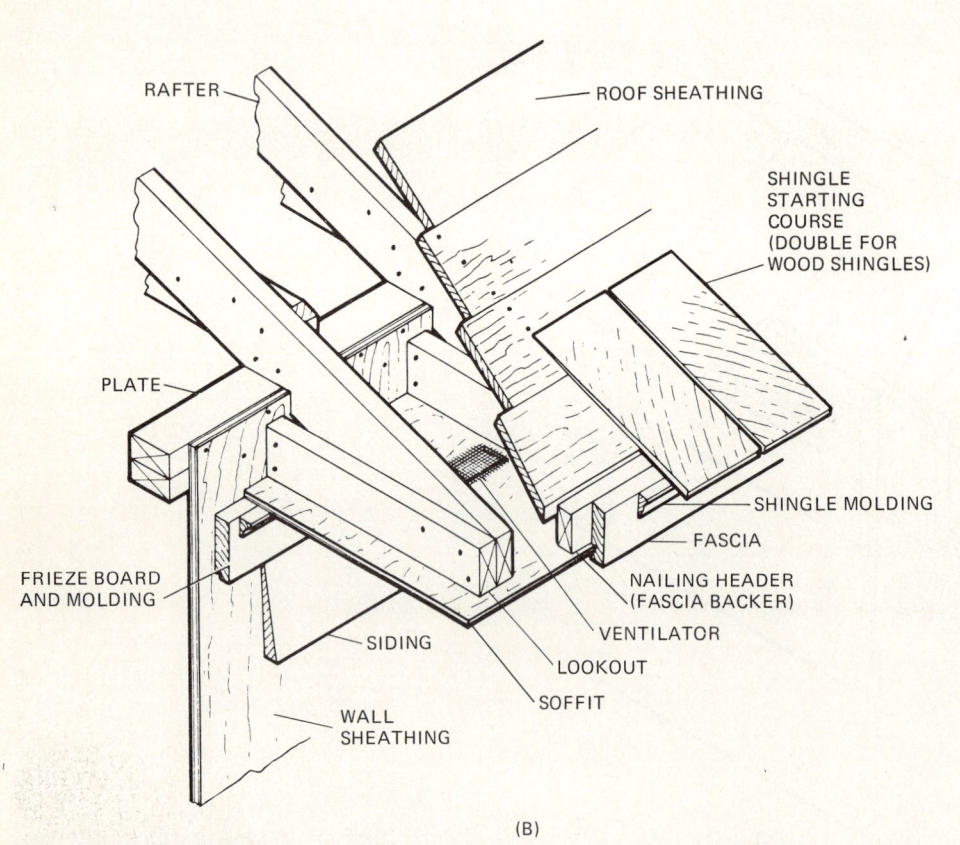

Figure 10-27. (A) Standard slope cornice. (B) Standard flat cornice. *(Courtesy Forest Products Laboratory)*

270 Carpentry Fundamentals

STANDARD SLOPE CORNICE • The standard slope cornice is the simplest and quickest to make. It is shown in Figure 10-27A. The soffit panel is nailed directly to the underside of the rafters. If ¼-inch paneling is used, a 6d box nail is appropriate. Rust-resistant nails are recommended. Casing nails, when used, should be set and covered before painting. Panels should join on a rafter. In this way, each end of each panel has firm support. One edge of the panel is butted against the fascia. The edge next to the wall is closed with a piece of trim called a *frieze*.

STANDARD FLAT CORNICES • The standard flat cornice has a soffit that is flat or horizontal with the ground. It is not sloped with the angle of the roof.

A nail base must be built for the soffit. Special short joists are constructed. These are called *lookouts*. See Figure 10-27B. The flat cornice should be vented, as are other types. Either continuous strips or stock vents can be used.

The lookouts need a nail base on both the wall and the roof. They are nailed to the tail of the rafter for the roof support. On the wall, they can be nailed to the top plate or to a stud. More often than not, neither the top plate nor the stud can be used. Then a ledger is nailed to the studs through the sheathing. See Figure 10-28.

Either 16d or 20d common nails should be used to nail the ledger to the wall. The bottom of the ledger should be level with the bottom of the rafters.

Next, find the correct length needed for the lookouts. Cut the lookouts to length with both ends cut square. Drive nails into one end of the lookout. The nails are driven until the tips just barely show through the lookout. Then they can be held against the rafter and driven down completely. Butt the other end of the lookout against the ledger. Then toenail the lookout on each side using an 8d or 12d common nail.

Soffit panels should be cut to size. Cornice vents should be cut or spaced next. Cornice vents can be attached to soffits before mounting. After the soffits are ready, they are nailed in place. For ¼-inch panels, a 6d common nail or box nail can be used.

Soffit panels should be joined on a solid nail base. The ends should join at the center of a lookout. One edge is butted against the fascia. The inner edge is sealed with a frieze board or a molding strip.

Many builders use prefabricated soffit panels. Sometimes fascia is grooved for one edge of the soffit panel. When prefabricated soffits are used, a vent is usually built into the panel.

CLOSED RAKES • The *rake* is the part of the roof that hangs over the end of a gable. See Figure 10-29. When a cornice is closed, the rake should also be closed. However, most rakes are not vented. This is because they are not connected with the attic space as are the cornices.

It is common to add the gable siding before the rake soffit is added. Then a 2×4-inch nailing block is nailed to the end rafter. See Figure 10-30. Use 16d common nails. A fly rafter is added and supported by the fascia and the roof sheathing.

The soffit is then nailed to the bottom of the nail block and the fly rafter. Frieze boards or bed molding are then added to finish the soffit.

SOLID RAKE • Today many roofs do not extend over the ends of the gables. In this type of roof, a fascia is nailed directly onto the last rafter. The roof is then finished over the fascia. This becomes what is called a solid rake. See Figure 10-31.

Siding the Gable Ends

On many buildings, cornices are painted before any siding is installed. In other cases the gable siding is installed first. Then, both the cornice and gables are painted. After these are done, siding is installed. This is common when two different types of siding are used. For example, wood siding is put on the

Figure 10-28. Lookouts are nailed first to the rafter tail. Then they are toenailed to a solid part of the wall.

Finishing the Exterior Walls

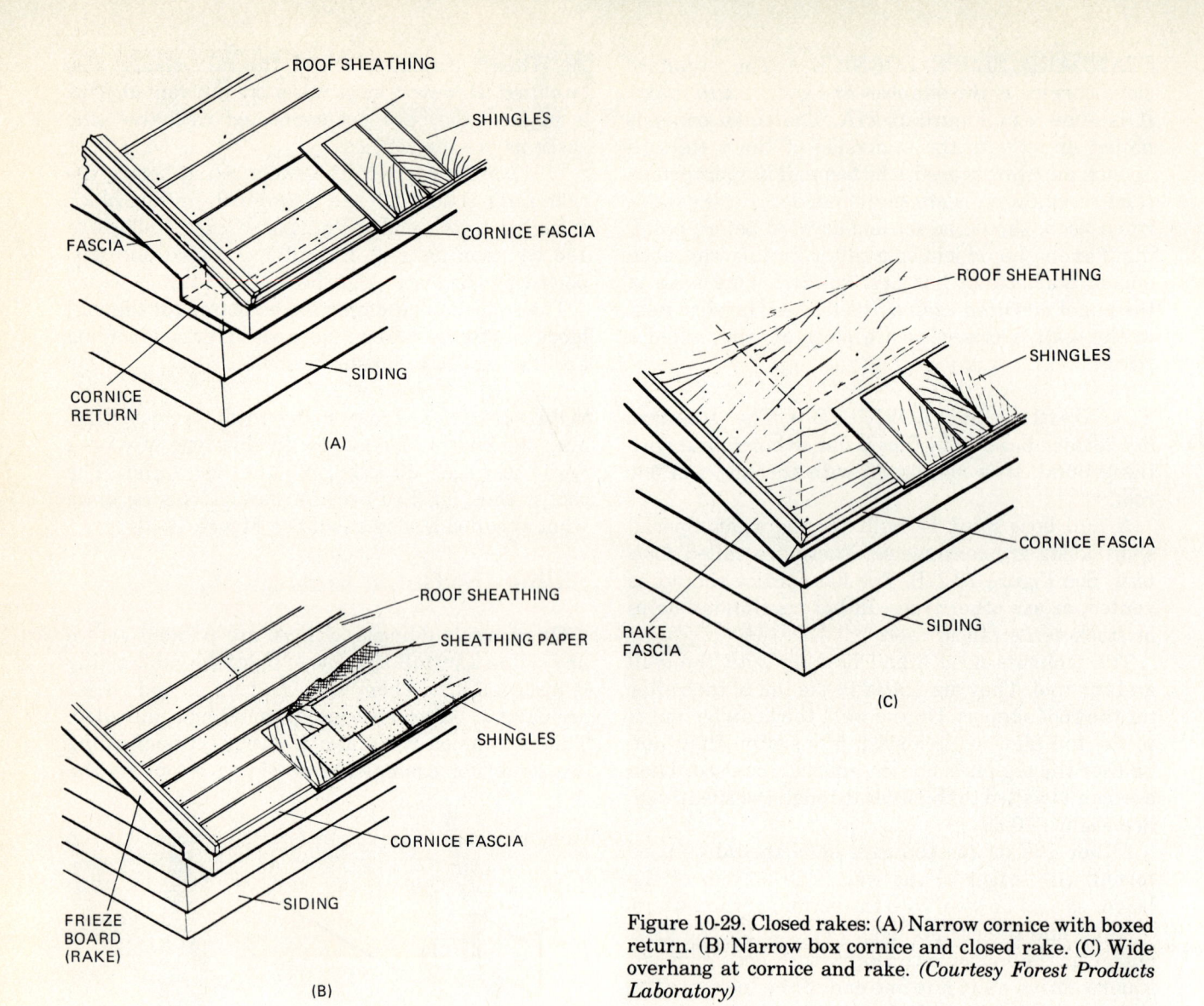

Figure 10-29. Closed rakes: (A) Narrow cornice with boxed return. (B) Narrow box cornice and closed rake. (C) Wide overhang at cornice and rake. *(Courtesy Forest Products Laboratory)*

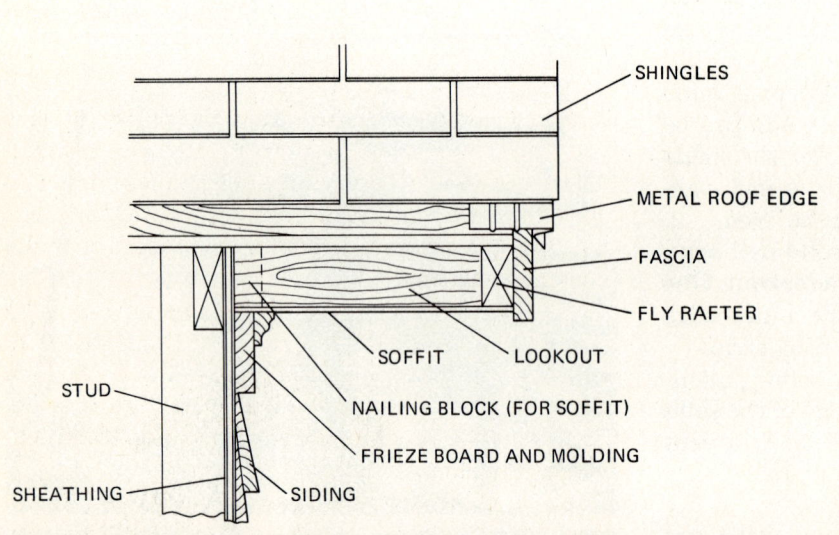

Figure 10-30. A nailing block is nailed to the end rafter for the soffit. *(Courtesy Forest Products Laboratory)*

272 Carpentry Fundamentals

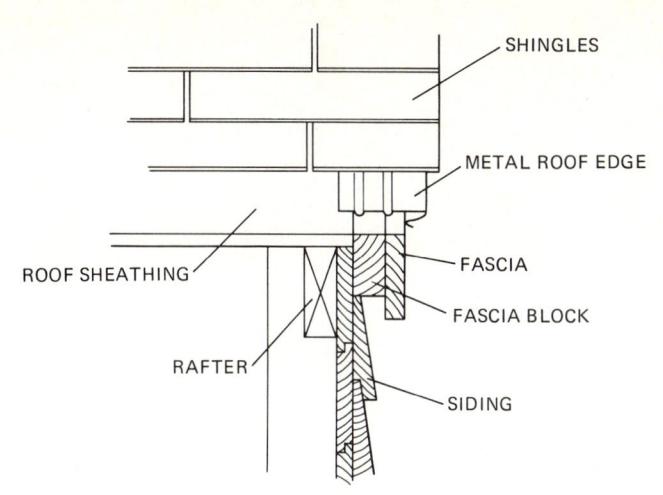

Figure 10-31. Detail for a solid rake. *(Courtesy Forest Products Laboratory)*

Figure 10-33. A wall and gable prepared for a combination of diagonal siding and brick veneer. Note that the wood has been painted before the brick has been applied.

Figure 10-32. Here the gable siding is added and painted before brick siding is added.

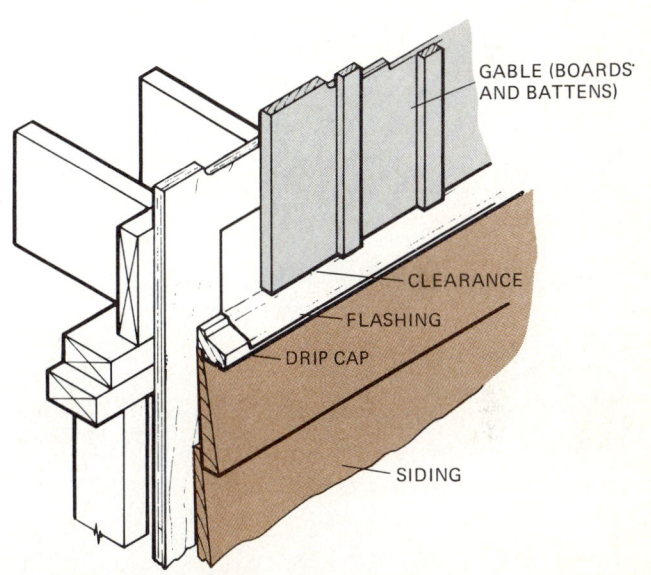

Figure 10-34. Different types of siding can be used for the gable and the wall. *(Courtesy Forest Products Laboratory)*

often covered with different siding. Brick exterior walls topped by wooden gable walls are common. Different types of wood and fiber sidings may also be combined. See Figures 10-34 and 10-35. Different textures, colors, and directions are used to make pleasing contrasts.

It is important to apply the gable siding in such a way that water is shed properly from gable to wall. For brick or stone, the gable must be framed to overhang the rest of the wall. See Figure 10-36. This overhang provides an allowance for the thickness of the stone or brick.

For wood, plywood, or fiber siding, special drainage joints are used. Also metal strips and separate wooden moldings are used.

gables and painted, then a brick siding is laid. See Figures 10-32 and 10-33.

Gable walls may be sided in several ways. Gable siding may be the same as that on the rest of the walls. In this case, the gable and the walls are treated in the same way. However, gable walls are

Finishing the Exterior Walls 273

Figure 10-35. Smooth gable panels contrast with the brick siding.

Figure 10-36. This gable is framed to overhang the wall.

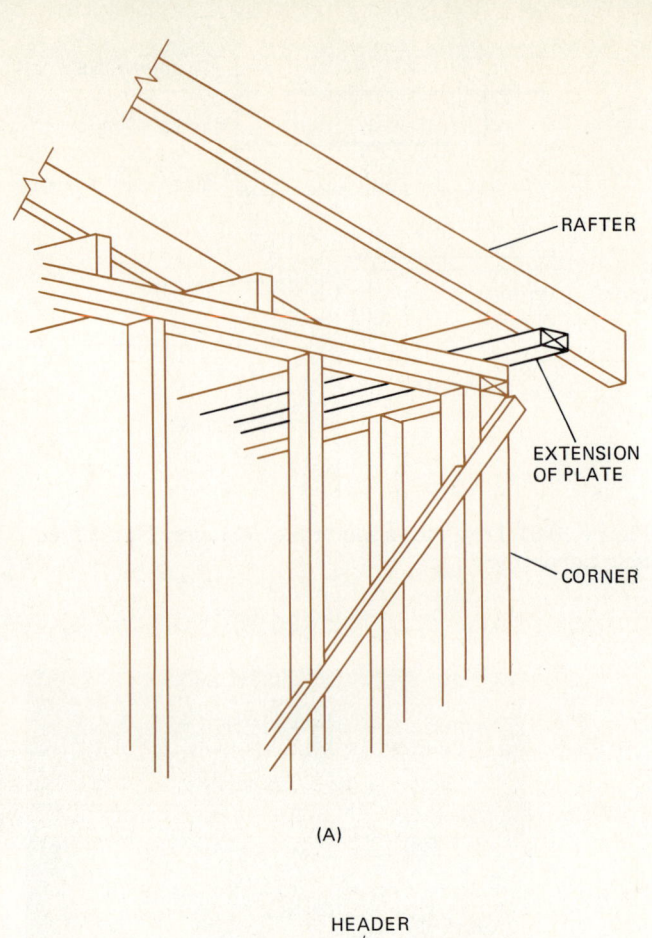

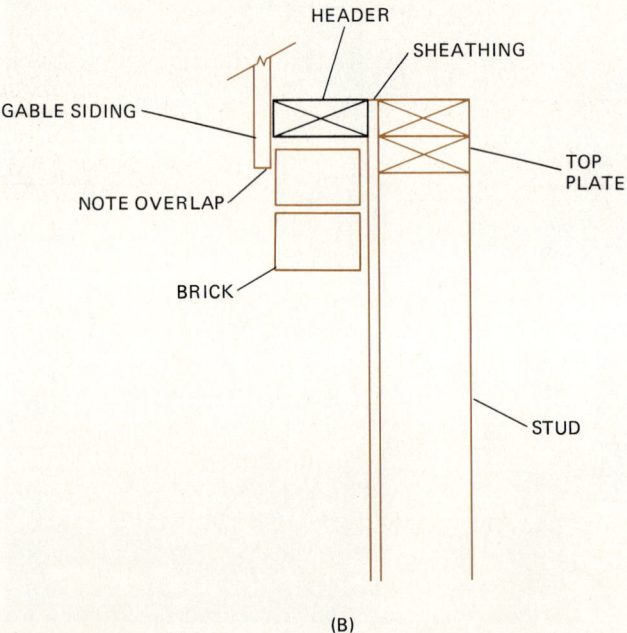

Figure 10-37. (A) Extending the top plate so that gable siding overhangs a brick or masonry wall. (B) With a solid header added on the bottom.

FRAMED OVERHANGS • Gable ends can be framed to overhang a wall. See Figure 10-36. Remember that this must be done to allow for the thickness of a brick wall.

Several methods are used. One of the most common is to extend the top plate as in Figure 10-37A. This puts the regular end rafter just over the wall for the overhang. A short (2-foot) piece is nailed to the top plate. The rafter birdsmouth is cut 1½ inch deeper than the others. A solid header is added on the bottom as shown in Figure 10-37B. The gable is framed in the usual manner. Gable siding is done before wall siding (Figures 10-38 and 10-39).

DRAINAGE JOINTS • Drainage joints are made in several ways. As mentioned before, they are important because water must be shed from gable to wall properly. Otherwise, water would run inside the siding and damage the wall.

274 Carpentry Fundamentals

Figure 10-38. Gable siding is done before wall siding. *(Fox and Jacobs)*

Figure 10-39. Installing gable siding. *(Fox and Jacobs)*

Drainage joints are used wherever two or more pieces are used, one above the other. Figure 10-40 shows the most common drainage joints. In one method, molded wooden strips are used. These strips are called drip caps.

Some siding, particularly plywood panels, has a rabbet joint cut on the ends. These are overlapped to stop water from running to the inside.

Metal flashing is also used. This flashing may be used alone or with drip caps.

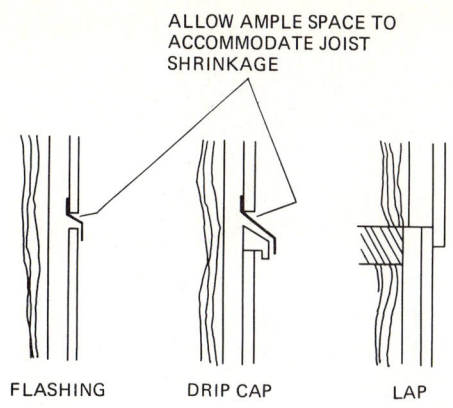

Figure 10-40. Drainage joints between gable and siding. *(Boise-Cascade)*

At the bottom of panel siding, a special type of drip cap is used. It is called a water table. It does the same thing, but has a slightly different shape. The water table may also be used with or without metal flashing.

• INSTALLING THE SIDING •

Several shapes of siding are used. Boards, panels, shakes, and shingles are the most common. These may be made of many different materials. However, many of the techniques for installing different types of siding are similar. Shape determines how the siding is installed. For example, wood or asbestos shingles are installed in about the same way. Boards made of any material are installed alike. Panels made of plywood, hardboard, or fiber are installed in a similar manner. Special methods are used for siding made from vinyl or metal.

Putting Up Board Siding

There are three major types of board siding. These are plain boards, drop siding, and beveled siding. Each type is put up in a different manner.

Board siding may be of wood, plywood, or composition material. This material is a type of fiberboard made from wood fibers. Generally, plain boards and drop siding are made from real wood. Composition siding is usually made in the plain beveled shape.

PLAIN BOARDS • Plain boards are applied vertically. This means that the length runs up and down. There are no grooves or special edges to make the boards fit together. There are three major board patterns used. The *board and batten* pattern is shown in Figure 10-41. In the board and batten style, the board is next to the wall. The board is usually not nailed in place except sometimes at the top end. It is sometimes nailed at the top to hold it in place. A narrow opening is left between the boards. This

Finishing the Exterior Walls **275**

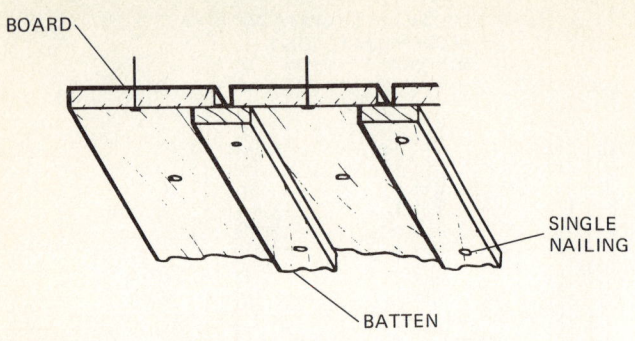

Figure 10-41. Board and batten vertical siding. *(Courtesy Forest Products Laboratory)*

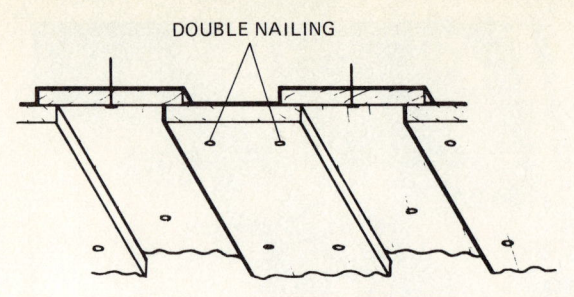

Figure 10-43. The Santa Rosa or board and board style of vertical siding. *(Courtesy Forest Products Laboratory)*

NOTE: Nail for first board should be 8d or 9d. Nail for second board should be 12d.

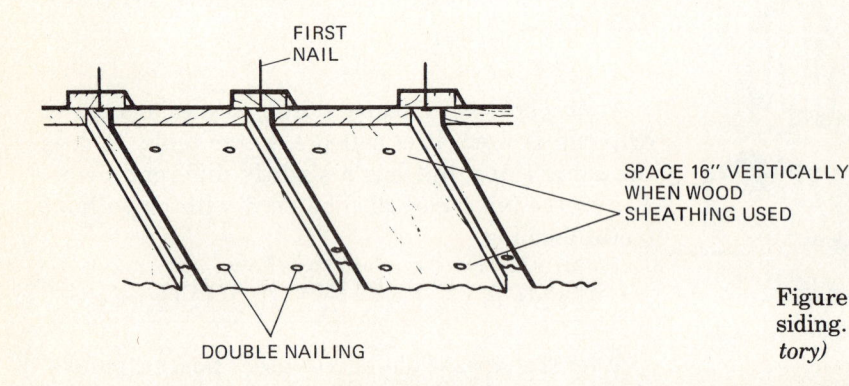

Figure 10-42. Batten and board vertical siding. *(Courtesy Forest Products Laboratory)*

opening is covered with a board called a *batten*. The batten serves as the weather seal. The nails are driven through the batten but not the boards, as shown in Figure 10-41.

The *batten and board* style is the second pattern. The batten is nailed next to the wall. A typical nailing pattern is shown in Figure 10-42. The wide board is then fastened to the outside as shown.

Another style is the Santa Rosa style. It is shown in Figure 10-43. All the boards are about the same width in this style. A typical nail pattern is shown in the figure. The inner board can be thinner than the outer board.

Plain board siding can be used only vertically. This is because it will not shed water well if it is used flat or horizontally. The surface appearance of the boards may vary. The boards may be rough or smooth. For the rough effect, the boards are taken directly from the saw mill. In order to make a smooth board finish, the rough-sawn boards must be surfaced.

DROP SIDING • Drop siding differs from plain boards. Drop siding has a special groove or edge cut into it. This edge lets each board fit into the next board. This makes the boards fit together and resist moisture and weather. Figure 10-44 shows some types of drop siding that are used. Nailing patterns for each type are also shown.

BEVELED SIDING • *Beveled* siding is made with boards that are thicker at the bottom. Figure 10-45 shows the major types of beveled siding. Common beveled siding is also called *lap* siding. The nailing pattern for lap siding is shown in Figure 10-46.

The minimum amount of lap for lap siding is about 1 inch. For 10-inch widths, which are common, about 1½ inches of lap is suggested. Most lap siding today is made of wood fibers. However, in many areas wood siding is still used. When wood siding is used, the standard nailing pattern shown in Figure 10-46 should be used. Some authorities suggest nailing the nails through both boards. However, this is not recommended. All wood tends to expand and contract, and few pieces do so evenly. Nailing the two boards together causes them to bend and bow. When this occurs, air and moisture can easily enter.

Another type of beveled siding is called *rabbeted beveled siding*. This is also called *Dolly Varden siding*. The Dolly Varden siding has a groove cut into the lower end of the board. This groove is called a rabbet. When installed from the bottom up, each successive board should be rested firmly on the one beneath it. The nails are then driven into the top of the boards.

Siding Layout

For all types of board siding, two or three factors are important to remember. Boards should be spaced to lay even with windows and doors. One board should fit against the bottom of a window. Another should rest on the top of a window. This way there is no cutting for an opening. Cutting should also

276 Carpentry Fundamentals

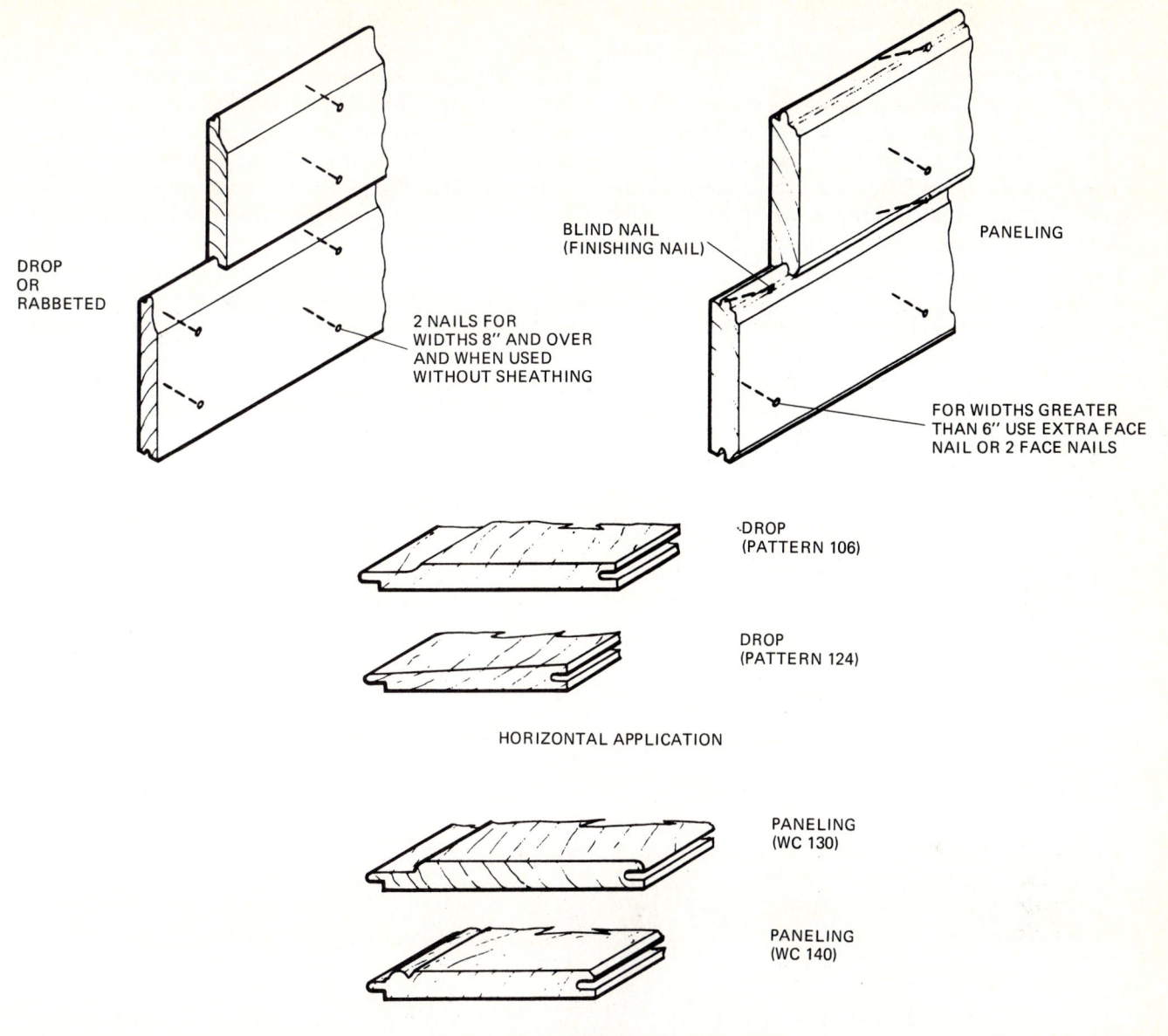

Figure 10-44. Shapes and nailing for drop siding. *(Courtesy Forest Products Laboratory)*

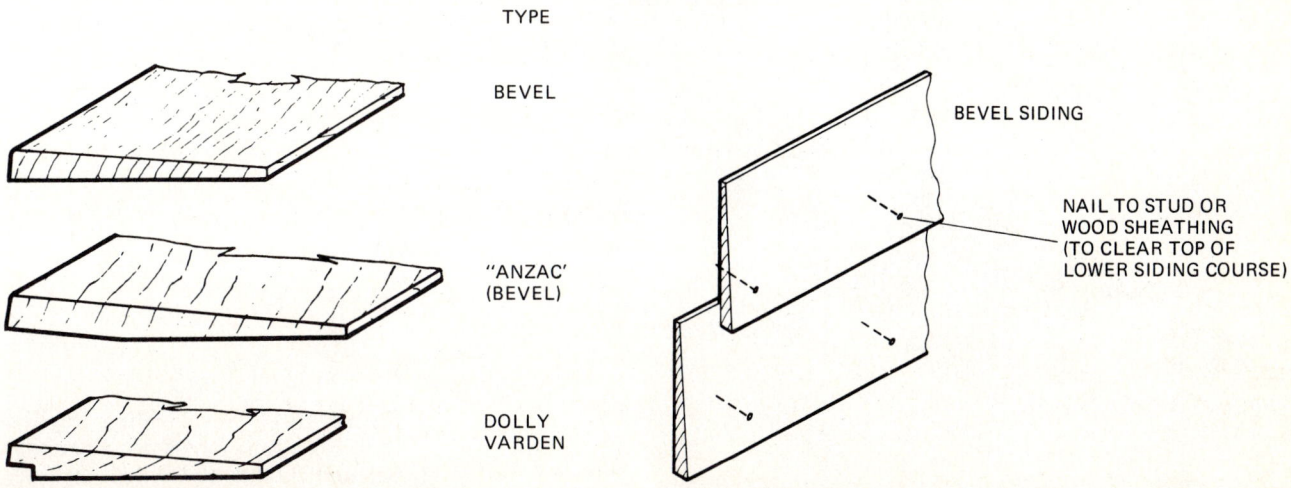

Figure 10-45. Major types of bevel siding. *(Courtesy Forest Products Laboratory)*

Figure 10-46. Nailing pattern for lap siding. *(Courtesy Forest Products Laboratory)*

Finishing the Exterior Walls   **277**

be avoided for vertical siding. The spacing should be adjusted so that a board rests against the window on either side.

When siding is installed, guide marks are laid out on the wall. These are used to align the boards properly. To do this, a pattern board, called a story pole, is used just as in wall framing. The procedure for laying out horizontal lap siding will be given. The procedures for other types of siding are similar. This procedure may be adapted accordingly.

PROCEDURE • Choose a straight 1-inch board. Cut it to be exactly the height of the area to be sided. Many siding styles are allowed to overlap the foundation 1 to 2 inches. Find the width of the siding and the overlap to be used. Subtract the overlap from the siding width. This determines the spacing between the bottom ends of the siding. The bottom end of lap siding is called the *butt* of the board. For example, siding 10 inches wide is used. The overlap is 1½ inches. Thus, the distance between the board butts is 8½ inches. On the story pole, lay out spaces 8½ inches apart.

Place the story pole beside a window. Check to see if the lines indicating the spacing between the boards line up even with the top and bottom of the windows. The amount of overlap over the foundation wall may be varied slightly. The amount of lap on the siding may also be changed. Small changes will make the butts even with the tops of the windows. They will also make the tops even with the window bottoms. See Figure 10-47. Story pole markings should be changed to show adjustments. New marks are then made on the pole.

Use the story pole to make marks on the foundation. Also make marks around the walls at appropriate places. Siding marks should be made at edges of windows, corners, and doors. See Figure 10-48. If the wall is long, marks may also be made at intervals along the wall. In some cases, a chalk line may be used to snap guide lines. A chalk line is often used to snap a line on the foundation. This shows where the bottom board is nailed.

Nailing

Normally siding is installed from the bottom to the top. The first board normally overlaps the foundation at least 1 inch. The first board is tacked in place and checked for level. After leveling, the first

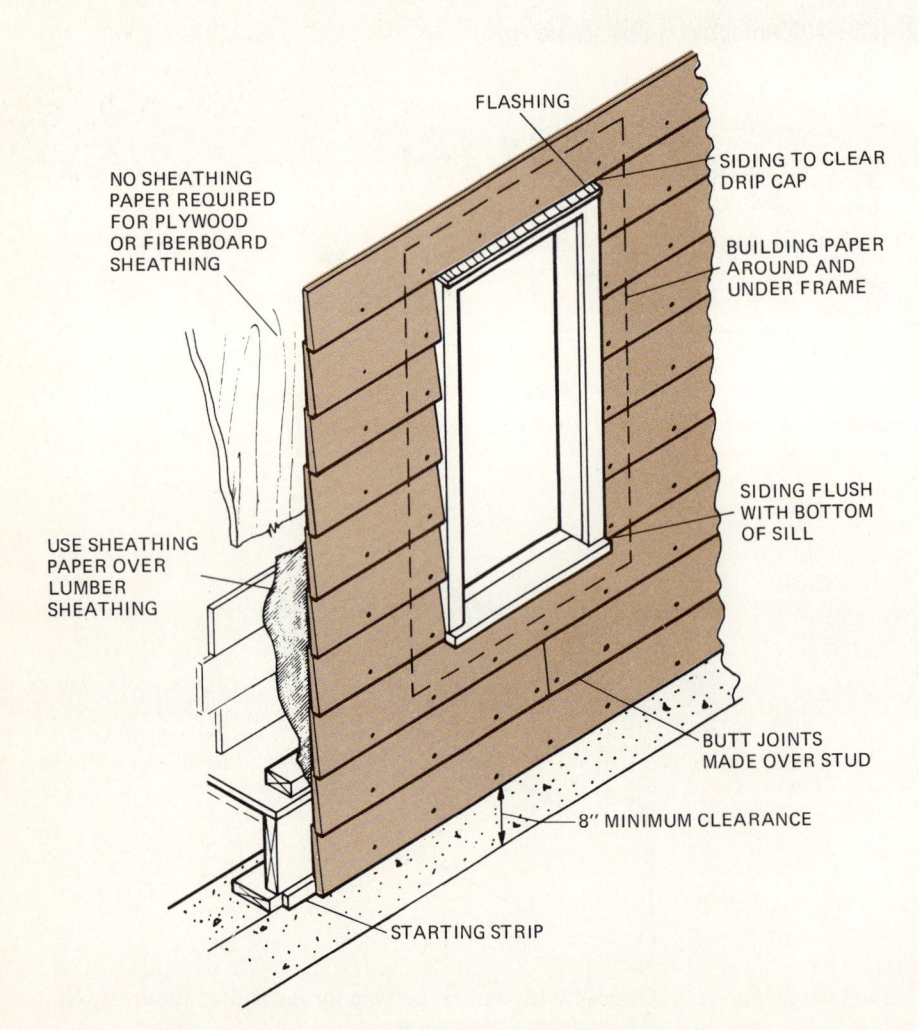

Figure 10-47. Bevel siding application. Note that pieces are even with top and bottom of window opening. *(Courtesy Forest Products Laboratory)*

278 Carpentry Fundamentals

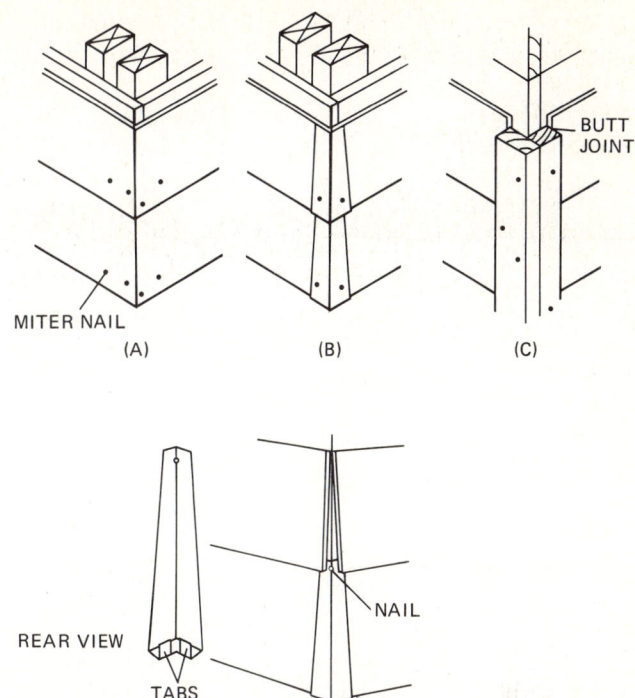

Figure 10-49. Methods of finishing corners. *(Courtesy Forest Products Laboratory)*

at the top. Two sets of nails are used on the top board. The first set of nails is nailed near the tops of the board. These may be nailed firm. Then, the bottom nail is driven 1½ inches from the butt of the board. This nail is not nailed down firm. About ¾ inch should be left sticking out. The butt of the board is pried up and away from the wall. A nail bar or the claw of a hammer is used. The board is left this way. Then, the next board is pushed against the nails of the first board. The second board is checked for level. Hold the level on the bottom of the board and check the second level. When the board is level, it is nailed at a place 1½ inches from the butt. Again, the nail in the second board is left out. About ¾ inch should be left sticking out. The board is nailed down at the butt for its length. Then the first board is nailed down firm. The second board is then pried up for the third board. This process is repeated until the siding has been applied.

Corner Finishing

There are three ways of finishing corners for siding. The most common ways are shown in Figure 10-49. Corner boards can be used for all types of siding. In one style of corner board, the siding is butted next to the boards. In this way, the ends of the siding fit snugly and the corner is the same thickness as the siding. See Figure 10-49C.

In Figure 10-49B special metal corner strips are used. These are separate pieces for each width of

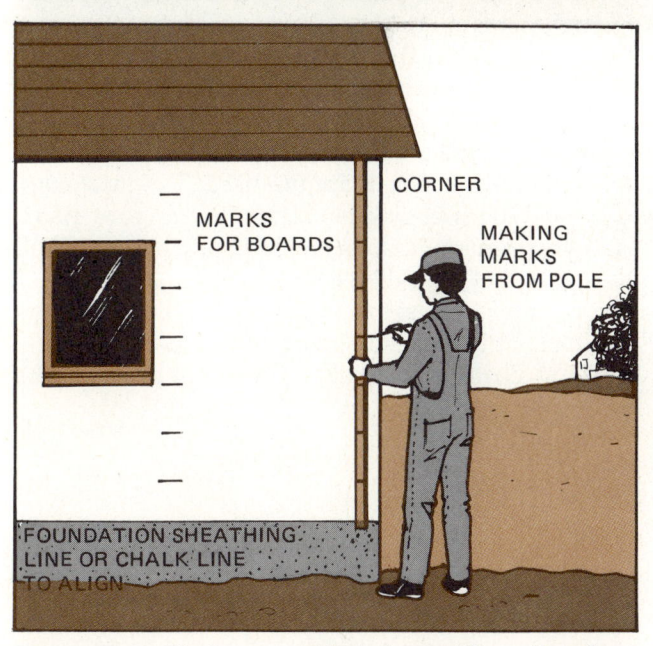

Figure 10-48. Marks are made to keep siding aligned on long walls.

board is then nailed firm. Usually bottom boards will be placed for the whole wall first. Then the other boards are put up, bottom to top. The level of the siding is checked after each few boards.

Siding can also be installed from the top down. This is done when scaffolding is used. The layout is the same. The lines should be used to guide the butts of the boards. However, the first board is nailed

Finishing the Exterior Walls **279**

board. The carpenter must be careful to select and use the right size corner piece for the board siding used.

In other methods, the siding is butted at the corners. The corner boards are then nailed over the siding. See Figure 10-50. This type of corner board is best used for plywood or panel siding. However, it is also common for board siding.

Another common method uses metal corners for lap siding. Figure 10-49B shows metal corner installation. These corners have small tabs at the bottom. They fit around the butt of each board. At the top is a small tab for a nail. The corners are installed after the siding is up. The bottom tab is put on first. Tabs are then put on other boards from bottom to top. The corners can be put on last. This is because the boards will easily spread apart at the bottoms. A slight spread will expose the tab for nailing.

Another method of finishing corners is to miter the boards. Generally, this method is used on more expensive homes. It provides a very neat and finished appearance. However, it is not as weatherproof. In addition, it requires more time and is thus more expensive. To miter the corners, a miter box should be used. Lap siding fits on the wall at an angle. It must be held at this angle when cut. A small strip of wood as wide as the siding is used for a brace. It is put at the base of the miter box. See Figure 10-51. This positions the siding at the proper angle for cutting. If this is not handy, an estimate may be made. It is generally accurate enough. This method is shown in Figure 10-52. A distance equal to the thickness of the siding is laid off at the top. The cut is then made on the line as shown.

Inside corners are treated in two ways. Both metal flashing and wooden strips are used. The most common method uses wooden strips. See Figure 10-53. A ¾-inch-square strip is nailed in the corner. Then each board of the siding is butted against the corner strip. After the siding is applied, the corner is caulked. This makes the corner weather tight.

Metal flashing for inside corners is similar. The

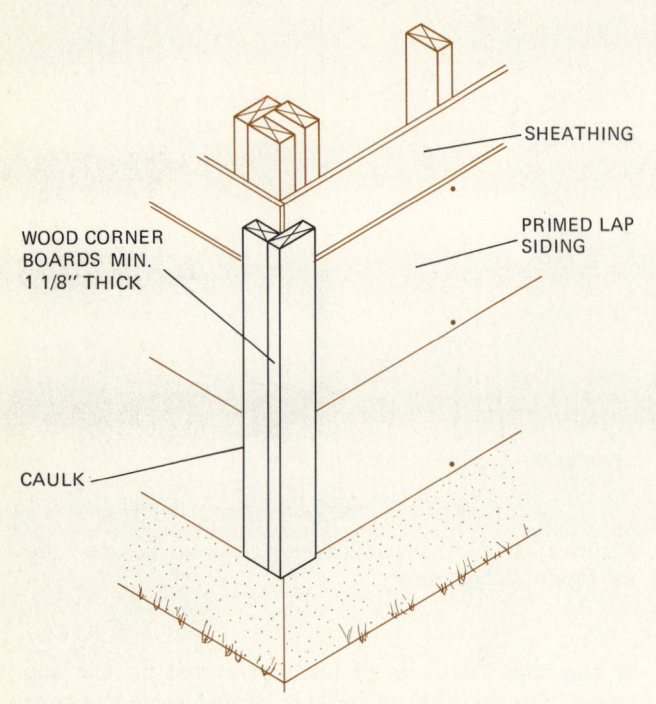

Figure 10-50. Wood outside corner detail. *(Boise-Cascade)*

Figure 10-51. A strip must be used to cant the siding for mitering.

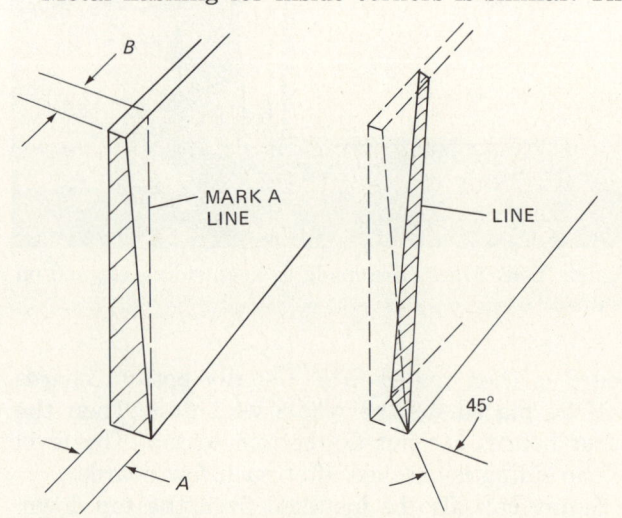

Figure 10-52. To cut miters without a miter box, make the top distance *B* equal to the thickness of *A*. Then cut back at about 45°.

280 Carpentry Fundamentals

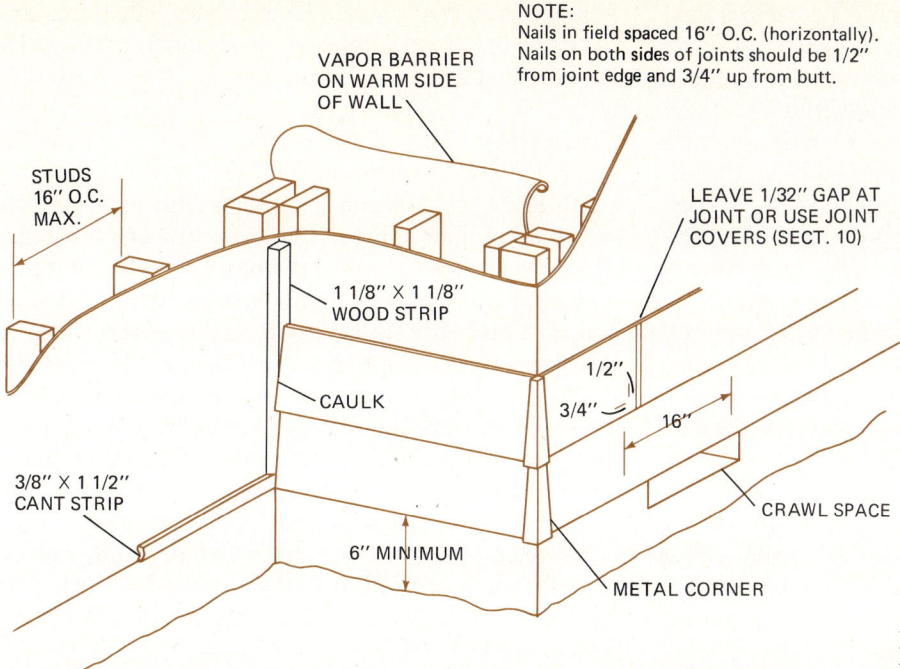

Figure 10-53. Inside corners may be finished with square wooden strips. *(Boise Cascade)*

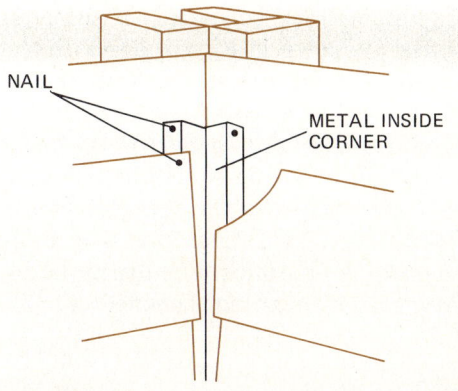

Figure 10-54. Inside corners may be finished with metal strips.

metal has been bent to resemble a wooden square. The strips are nailed in the corner as shown in Figure 10-54. As before, each board is butted against the metal strip.

• PANEL SIDING •

Panel siding is now widely used. It has several advantages for the builder. It provides a wider range of appearance. Panels may look like flat panels, boards, or battens and boards. Panels can also be grooved, and they can look like shingles. The surface textures range from rough lumber to smooth, flat board panels. Panels can show diagonal lines, vertical lines, or horizontal lines. Also, panels may be used to give the appearance of stucco or other textures.

Figure 10-55. Panel siding is often stapled in place. *(Fox and Jacobs)*

Panel siding is easier to lay out. There is little waste and the installation is generally much faster. It is often stapled in place, rather than nailed. See Figure 10-55. As a rule, corners are made faster and easier with panel siding.

Panel siding can be made from a variety of materi-

Finishing the Exterior Walls **281**

als. Plywood is commonly used. Hardboard and various other fiber composition materials are also used. A variety of finishes is also available on panel siding. These are prefinished panels. Panels may be prefinished with paint, chemicals, metal, or vinyl plastics.

The standard panel size is 4 feet wide and 8 feet long. The most common panel thickness is 5/8 inch. However, other panel sizes are also used. Lengths up to 14 feet are also available. Thickness ranges from 5/16 to 3/4 inch thick in 1/16-inch intervals.

The edges of panel siding may be flat. They may also be grooved in a variety of ways. Grooved edges form tight seams between plywood panels. Seams may be covered by battens. They may also be covered by special edge pieces. Outside corners may be treated in much the same manner as corners of regular board siding. Corners may be lapped and covered with corner boards. They may be covered with metal. They may also be mitered. Inside corners may be butted together. The edge of one panel is butted against the solid face of the first.

As a rule, panels are applied from the bottom up. However, they may be applied from top to bottom.

• SHINGLE AND SHAKE SIDING •

Shingles and shakes and often used for exterior siding. They are very similar in appearance. However, shakes have been split from a log. They have a rougher surface texture. Shingles have been sawn and are smoother in appearance. The procedure is the same for either shingles or shakes.

Shingles are made from many different materials. Shakes are made only from wood. The standard lengths for wood shakes and shingles are 16, 18, and 24 inches.

Shingles

Shingles may be made of wood, flat composition, or mineral fiber composition. The last is a combination of asbestos fiber and portland cement. Shingles for roofs are lapped about two-thirds. About one-third of the shingle is exposed. When the shingles are laid, or coursed, there are three layers on the roof. However, when shingles are used for siding, the length exposed is greater. Slightly more than one-half is exposed. This makes a two-layer thickness instead of three as on roofs. For asbestos mineral sidings, sometimes less lapping is used.

Wooden shingles are made in random widths. Shingles will vary from 3 to 14 inches in width. The better grades will have more wide shingles than narrow shingles.

Nailing

All types of shingles may be nailed in either of two ways. The first is shown in Figure 10-56. In the first method nailing strips are used. Note that a moisture barrier is almost always used directly under shingles. Builder's felt (tar paper) is the most common moisture barrier. Shingles should be spaced like lap siding. The butt line of the shingles should be even with the tops of wall openings. Top lines should be even with the bottoms. For utility buildings such as garages, no sheathing is needed. A mois-

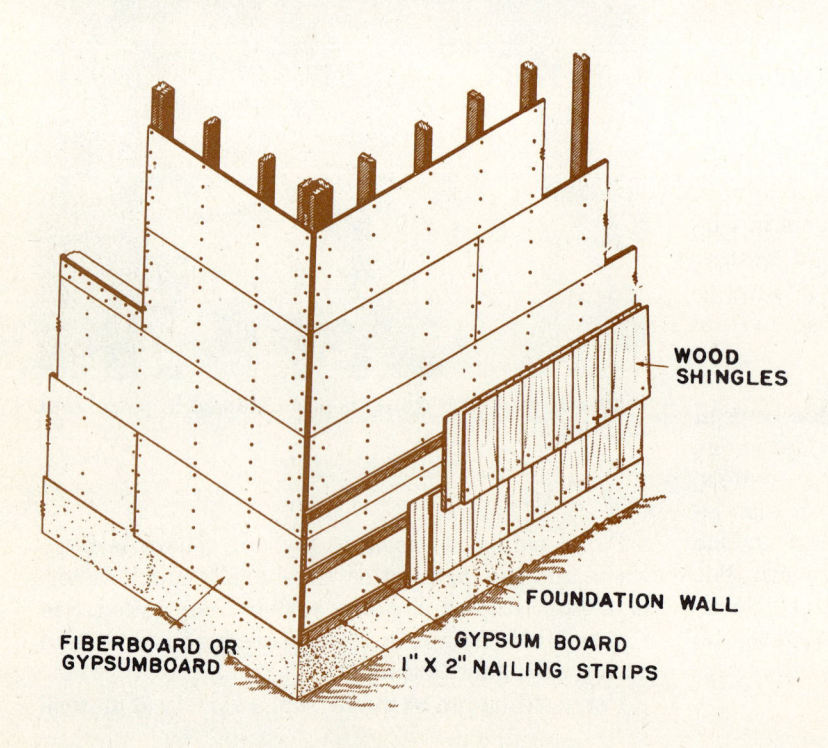

Figure 10-56. Nail strips are used over studs and sheathing. Shingles are then nailed to the strips. At least two nails are used for each shingle.

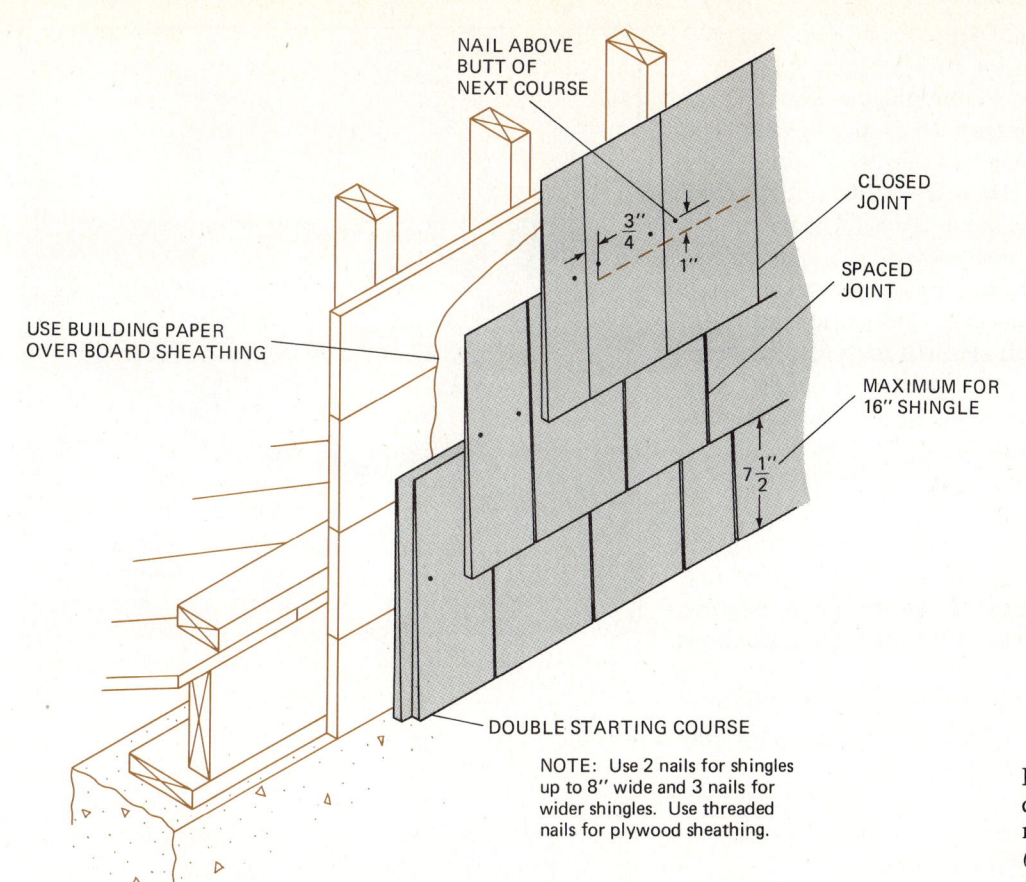

Figure 10-57. Single-course shingles nailed directly to solid sheathing. *(Courtesy Forest Products Laboratory)*

ture barrier can be put over the studs. Nailing strips are then nailed to the studs. However, for most residential buildings, a separate sheathing is suggested. Strips are often nailed over the sheathing. This is done over sheathing that is not a good nail base. Shingles may be nailed directly to board or plywood sheathing. This is shown in Figure 10-57.

The bottom shingle course is always nailed in place first. Part of the bottom course can be laid for large wall sections. The course may reach for only a part of the wall. Then that part of the wall can be shingled. The shingled part will be a triangle from the bottom corner upward. As a rule, two layers of shingles are used on the bottom course. The first layer is nailed in place, and a second layer is nailed over it. The edges of the second shingle should not line up with the edges of the first. This makes the siding much more weather-resistant. This layout is shown in Figure 10-57.

Another technique for shingles is called *double coursing*. This means that two thicknesses are applied. The first layer is often done with a cheaper grade of shingle. Again, each course is completed before the one above is begun. The edges should alternate for best weather-resistive qualities. Double coursing is shown in Figure 10-58. The outside shingle covers the bottom of the inside shingle. This gives a dramatic and contrasting effect.

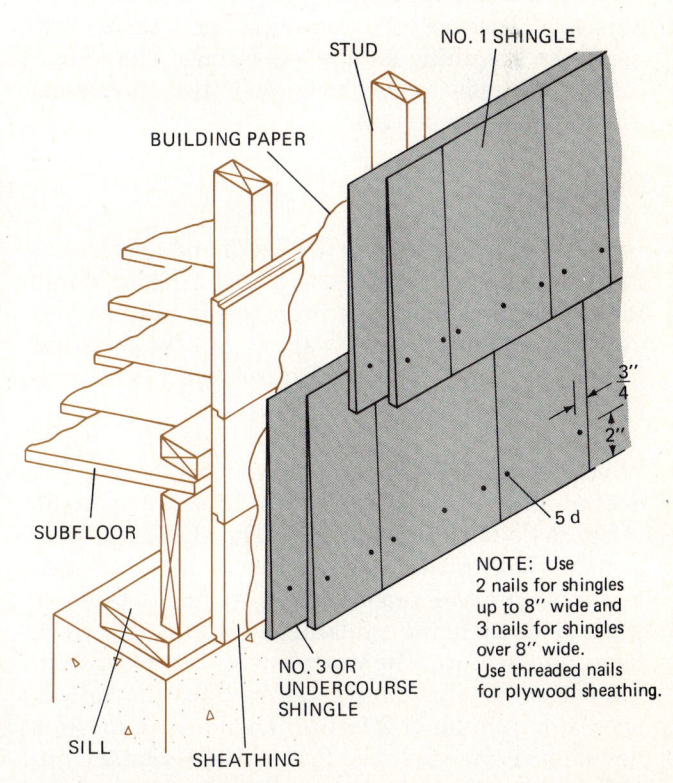

Figure 10-58. Double-coursed shingle siding. *(Courtesy Forest Products Laboratory)*

Finishing the Exterior Walls **283**

Shakes

The shake is similar to a shingle, but shakes may be made into panels. Figure 10-59 shows this type of siding. Shake panels are real wooden shakes glued to a plywood base. The strips are easier and quicker to nail than individual shakes. Also, shakes can be spaced more easily. A more consistent spacing is also attained. The individual panels may be finished at the factory. Color, spacing, appearance, and texture can be factory-made. Panels may also be combined with insulation and weatherproofing. The panels are applied in the same manner as shingles.

Corners

Corners are finished the same as for other types of siding. Three corner types are used. These are corner boards, metal corners, and mitered corners. Again, mitering is generally used for more expensive buildings.

Corners may also be woven. Shingles may be woven as in Figure 10-60. This is done on corners and edges of door and window openings.

· PREPARATION FOR OTHER WALL FINISHES ·

There are other common methods for finishing frame walls. These include stucco, brick, and stone. As a rule, the carpenter does not put up these wall finishes. However, the carpenter sometimes prepares the sheathing for these coverings. The preparation depends upon how well the carpenter understands the process.

Stucco Finish

Stucco is widely used in the South and Southwest. It is durable and less expensive than brick or stone. Like brick or stone, it is fireproof. It may be put over almost any type of wall. It can be prepared in several colors. Several textures can also be applied to give different appearances.

WALL PREPARATION • A vapor barrier is needed for the wall. The vapor barrier can be made of builder's felt or plastic film. The sheathing may also be used for the vapor barrier. It is a good idea to apply an extra vapor barrier over most types of walls. This includes insulating sheathing, plywood, foam, and gypsum. The vapor barrier is applied from the bottom up. Each top layer overlaps the bottom layer a minimum of 2 inches. Then a wire mesh is nailed over the wall. See Figure 10-61. Staples are generally used rather than nails. A solid nail base is essential. The mesh should be stapled at 18- to

Figure 10-59. Shakes are often made into long panels. (Shakertown)

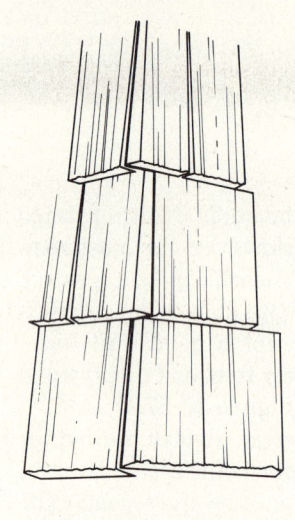

Figure 10-60. Shingle and shake corners may be woven or lapped.

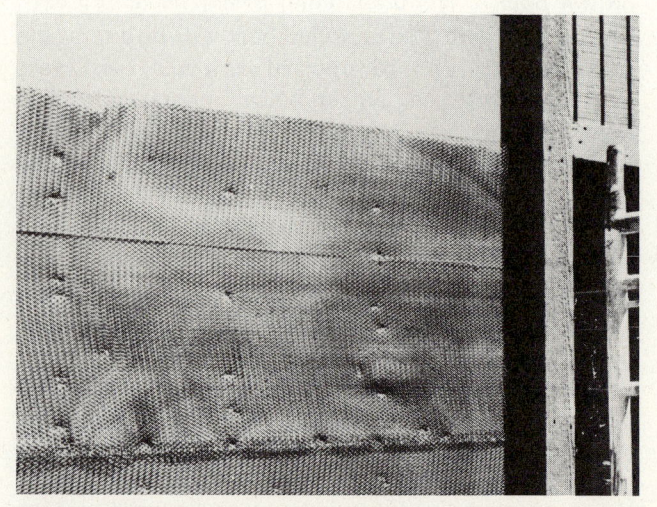

Figure 10-61. A wire mesh is nailed to the wall. Stucco will then be applied over the mesh.

284 Carpentry Fundamentals

Figure 10-62. Stucco is spread over mesh with a trowel.

Figure 10-63. Ties help hold brick or stone to the wall. They are nailed to studs and embedded in the mortar.

24-inch intervals. The intervals should be in all directions—across, up, and down. The wire mesh may be of lightweight "chicken wire." However, the mesh should be of heavier material for large walls.

The mesh actually supports the weight of the stucco. The staples only hold the stucco upright, close to the wall.

APPLY THE STUCCO • Usually two or three coats are applied. The first coat is not the finish color. The first coat is applied in a very rough manner. It is applied carefully but it may not be even in thickness or appearance. It is called the "scratch" coat. Its purpose is to provide a layer that sticks to the wire. The coat is spread over the mesh with a trowel. See Figure 10-62.

This scratch coat should be rough. It may be scratched or marked to provide a rough surface. The rough surface is needed so that the next coat will stick. A trowel with ridges may be used. The ridges make grooves in the coat. These grooves may run in any direction. However, most should run horizontally. Another way to get a rough surface is to use a pointed tool. After the stucco sets, grooves are scratched in it with the tool.

One or more coats may be put up before the finish coat is applied. These coats are called "brown" coats. They are scratched so that the next coats stick better. The last coat is called the finish coat. It is usually white or tan in color. However, dyes may be added to give other colors. The colors are usually light in shade.

Brick and Stone Coverings

Brick and stone are used to cover frame walls. The brick or stone is not part of the load-bearing wall. This means that they do not hold up any roof weight. The covering is called a brick or stone veneer. It adds beauty and weatherproofing. Such walls also increase the resistance of the building to fires.

The preparation for either brick or stone is similar. First, a moisture barrier is used. If a standard wall sheathing has been used, no additional barrier is needed. However, it is not bad practice to add a separate moisture barrier.

The carpenter may then be asked to nail *ties* to the wall. These are small metal pieces, shown in Figure 10-63. Ties are bent down and embedded in the mortar. After the mortar has set, these ties form a solid connection. They hold the brick wall to the frame wall. They also help keep the space between the walls even. Small holes are usually left at the

Finishing the Exterior Walls 285

bottom. These are called *weep* holes. Moisture can soak through brick and stone. Moisture also collects at the bottom from condensation. The small weep holes allow the moisture to drain out. By draining the moisture, damage to the wood members is avoided. The joints are trimmed after the brick is laid (Figure 10-64).

• ALUMINUM SIDING •

Aluminum siding is widely used on houses. Aluminum is used for new siding or can be applied over an old wall.

A variety of vertical and horizontal styles are used. Probably the most common type looks like lap siding. However, even this type may come as individual "boards" or as panels two or three "boards" wide. See Figure 10-65. These types of siding may be hollow-backed or insulated. See Figure 10-66.

Aluminum siding may also look like shingles or

Figure 10-64. Joints are trimmed after brick is laid. *(Fox and Jacobs)*

(A)

(B)

(C)

(D)

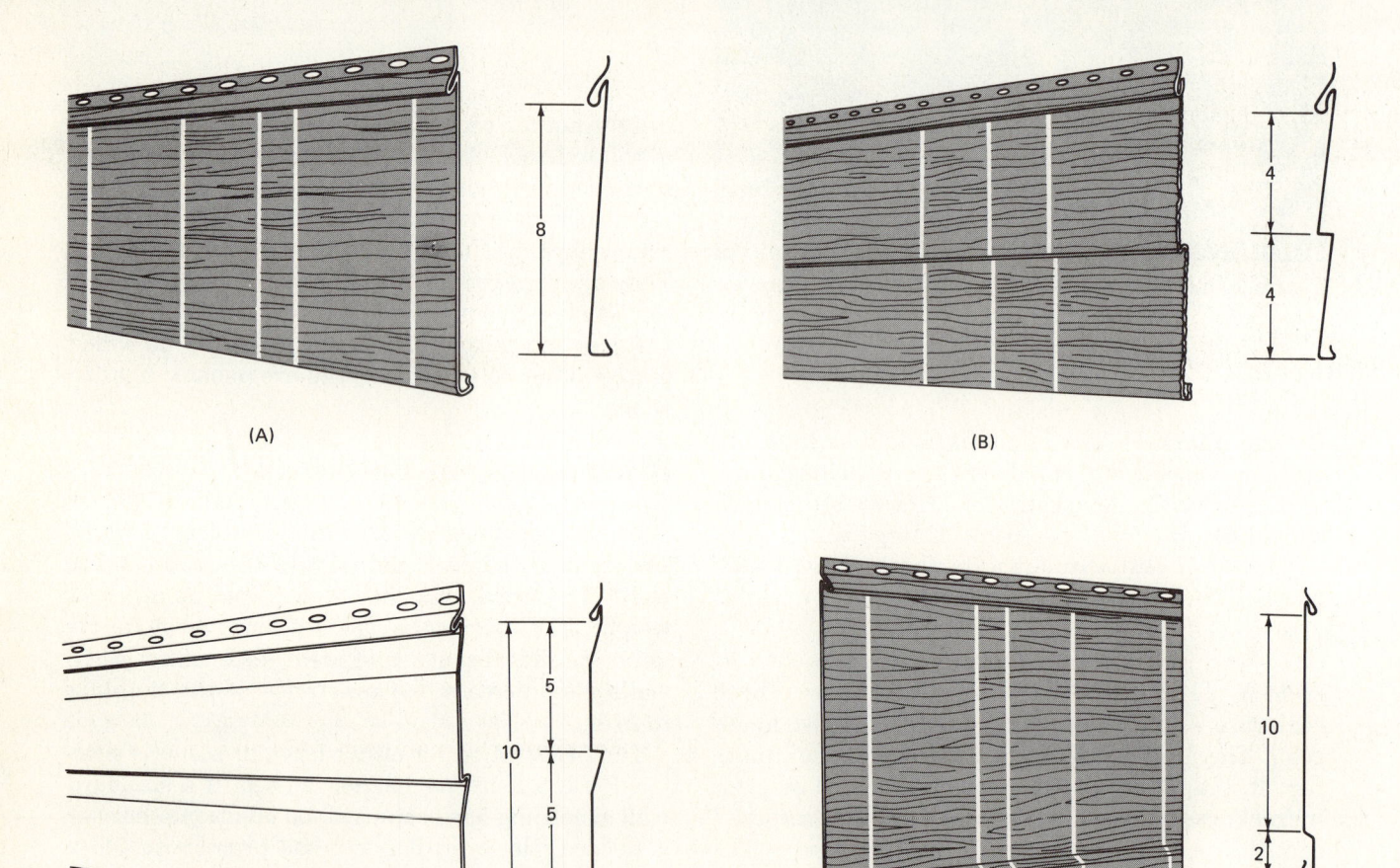

Figure 10-65. The most common type of aluminum siding looks like lap siding. (A) Single-"board" lap siding. (B) Two-"board" panel lap siding. (C) Double-"board" drop siding. (D) Vertical "board" siding.

286 Carpentry Fundamentals

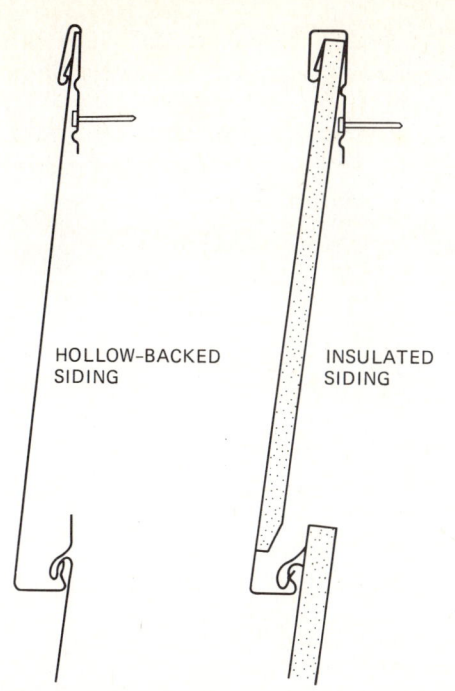

Figure 10-66. Aluminum siding may be hollow-backed or insulated. (ALCOA)

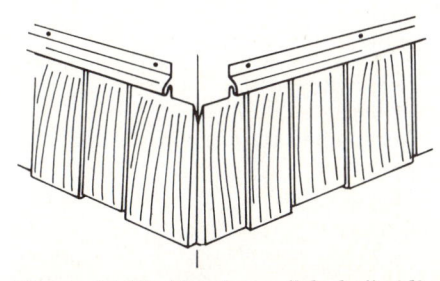

Figure 10-67. Aluminum "shake" siding (ALCOA)

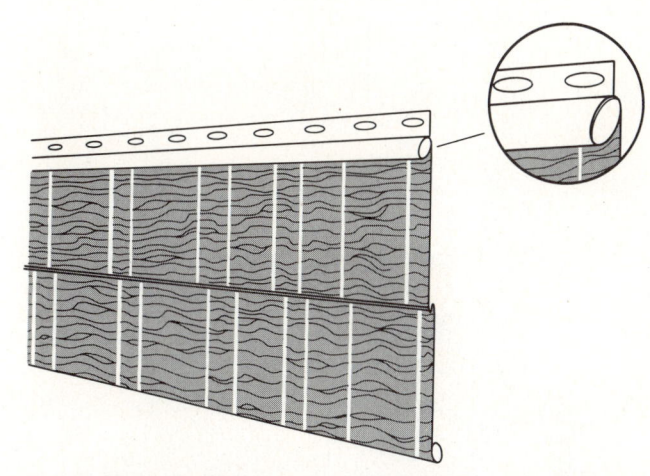

Figure 10-68. Panels are nailed at the top and interlock at the bottom. (ALCOA)

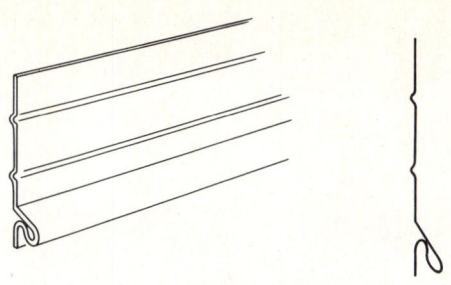

Figure 10-69. A starter strip is nailed at the bottom of the wall. The lower edge of the first board can interlock with it for support. (ALCOA)

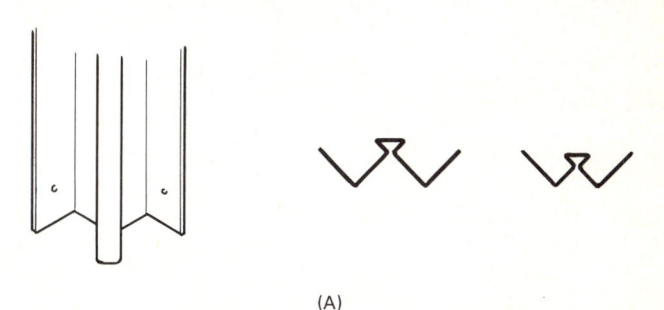

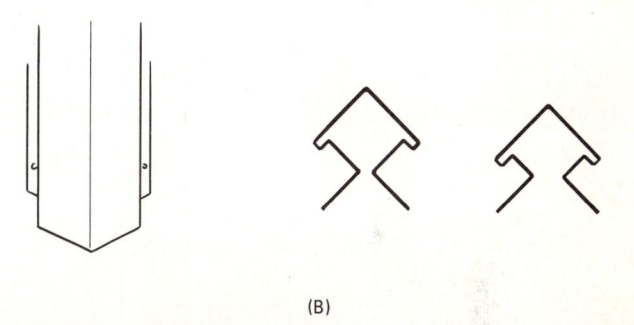

Figure 10-70. Corner strips for aluminum siding. (ALCOA) (A) Inside. (B) Outside.

shakes, as in Figure 10-67. For all types of aluminum siding, a variety of surface texture and colors are available.

Aluminum siding is put up using a special system. Note in Figure 10-68 that the top edge has holes in it. All nails are driven in these holes. The bottoms or edges of the pieces interlock. In this way, the tops are nailed and the bottoms interlock. Each edge is then attached to a nailed portion for solid support.

To start, a special starter strip is nailed at the bottom. See Figure 10-69. Then corner strips, as in Figure 10-70, are added. Special shapes are placed around windows and other features as in Figure 10-71. For these, the manufacturer's directions should be carefully followed.

Next, the first "board" is nailed in place. Note that it is placed at the bottom of the wall. The bottom of the first piece is interlocked with the starter strip and nailed. It is best to start at the rear of the house and work toward the front. This way, overlaps do not show as much. For the same reason, factory-cut ends should overlap ends cut on the site. See Figure 10-72. Backer strips, as in Figure 10-73, should be used at each overlap. These provide strength at the overlaps. Also, overlaps should be

Finishing the Exterior Walls **287**

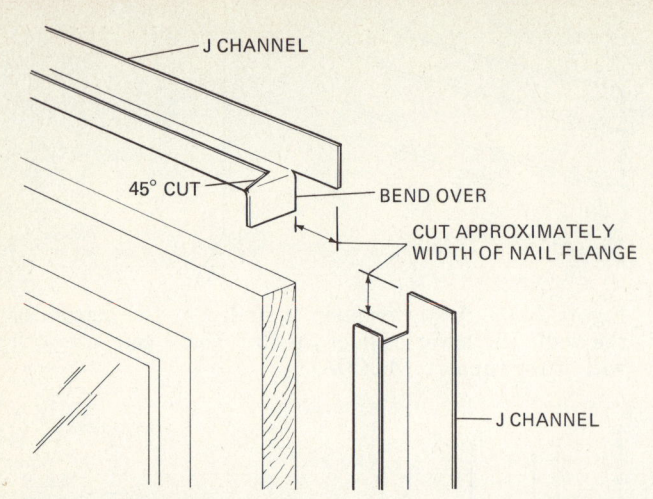

Figure 10-71. Special pieces are used around windows and gables. *(ALCOA)*

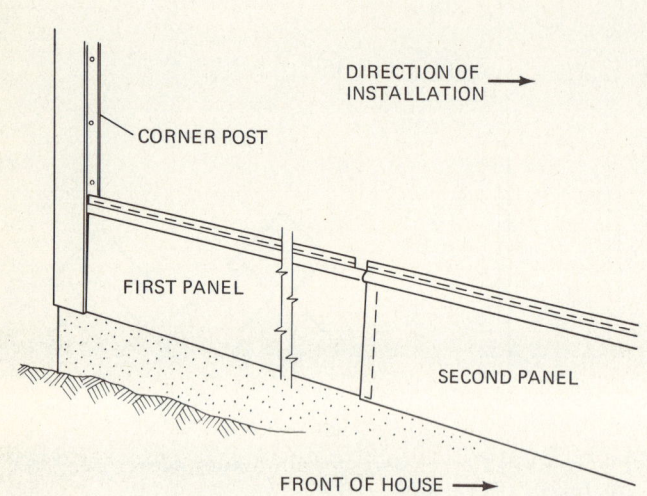

Figure 10-72. Siding is started at the back. *(ALCOA)*

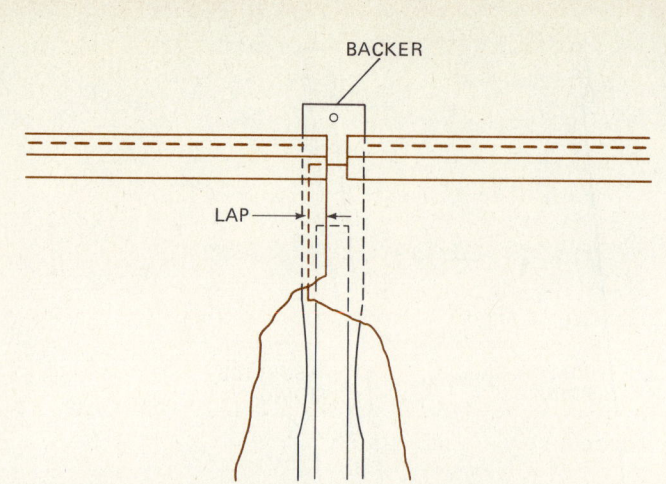

Figure 10-73. Backer strips support the ends at overlaps. *(ALCOA)*

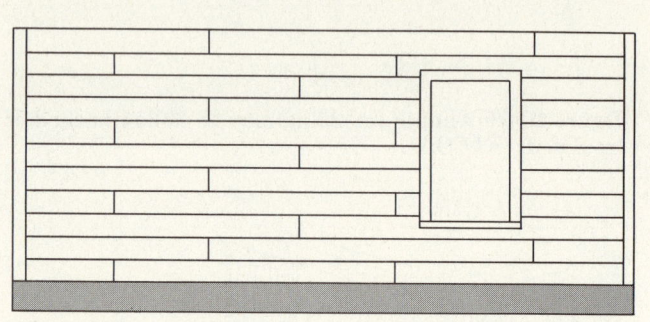

Figure 10-74. Overlaps should be equally spaced for best appearance. *(ALCOA)*

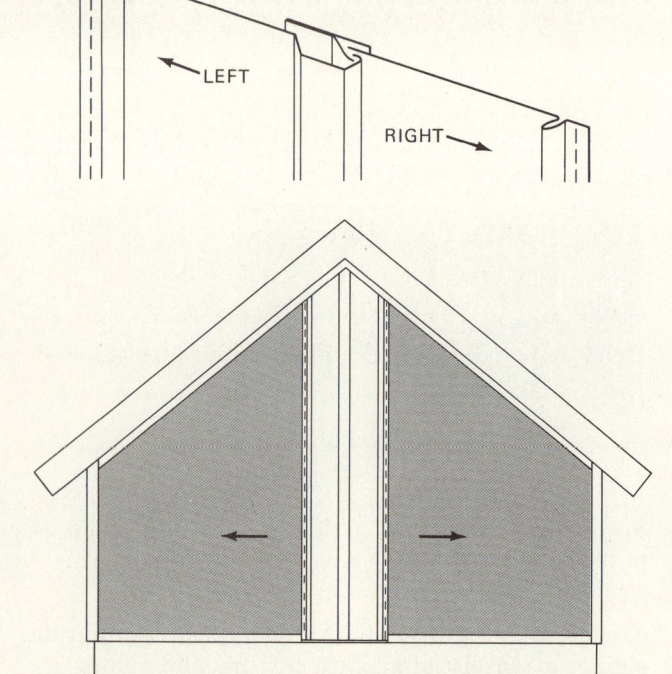

Figure 10-75. Vertical aluminum siding is started at the center. *(ALCOA)*

spaced evenly, as in Figure 10-74. Grouping the overlaps together gives a poor appearance.

Allowance must be made for heat expansion. Changes in temperature can cause the pieces to move. To allow movement, the nails are not driven up tightly. It is best to check the instructions that come with the siding.

Vertical Aluminum Siding

Most procedures for vertical siding are the same. Strips are put up at corners, windows, and eaves.

However, the starter strip is put near the center. A plumb line is dropped from the gable peak. A line is then located half the width of a panel to one side. The starter strip is nailed to this line. Panels are then installed from the center to each side. See Figure 10-75.

288 Carpentry Fundamentals

SOLID VINYL SIDING

Solid vinyl siding is also used widely. It is available in both vertical and horizontal applications. A wide variety of colors and textures are also available.

As with aluminum siding, vinyl siding is nailed on one edge. The other edge interlocks with a nailed edge for support. See Figure 10-76. Special pieces are again needed for joining windows, doors, gables, and so forth.

As with aluminum siding, the finish is a permanent part of the siding. No painting will ever be needed. It may be cleaned with a garden hose.

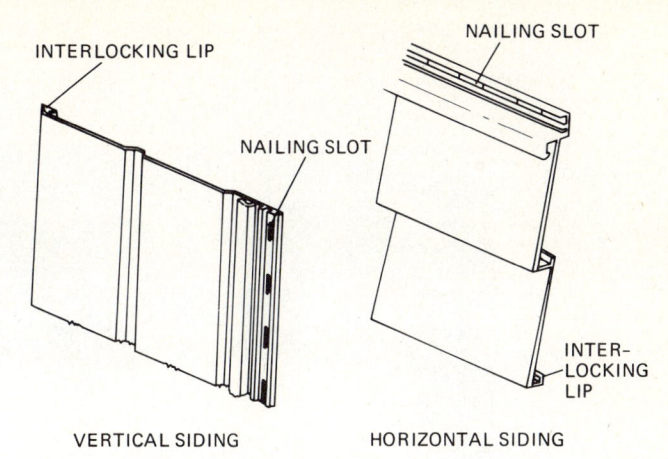

Figure 10-76. Solid vinyl siding.

WORDS CARPENTERS USE

cornice	shakes	bundle	fascia	batten
rake	scaffold	square	soffit	stucco
wall bracket	bevel	trestle jack	vent	veneer
siding	drop siding	ladder jack	flashing	
panels	lap	eaves	water tables	

STUDY QUESTIONS

10–1. What are the methods for enclosing eaves?

10–2. What sequence is used for finishing an exterior wall?

10–3. What can happen to painted wooden walls that have no vapor barrier?

10–4. How far should siding nails penetrate into the nail base?

10–5. How much siding will be needed for:
   Wall size = 8 feet high × 30 feet long
   Openings = one, 3 × 4 feet
   Overlap = 1½ inches
   Siding width = 10 inches

10–6. Why is some siding started from the top?

10–7. What is the usual starting point for siding?

10–8. When are guard rails required on scaffolds?

10–9. What is a wall bracket?

10–10. What are the advantages of steel double-pole scaffolds?

10–11. How thick should platform planks be?

10–12. List six rules for ladder safety.

10–13. List six rules for scaffold safety.

10–14. Why are soffit vents used?

10–15. What types of board siding are used?

10–16. Why are boards not nailed together on lap siding?

10–17. Why is siding spaced even with openings?

10–18. What is the difference between shingles and shakes?

10–19. Why are shingles double-coursed on walls?

10–20. What is a "square" in shingling?

10–21. How are gable overhangs framed?

10–22. How is wooden gable and wall siding joined?

10–23. What holds brick veneer to a wall?

10–24. How are corners treated for wood panel siding?

10–25. What should be done to prepare a wall for stucco?

Finishing the Exterior Walls

11
INSULATING FOR THERMAL EFFICIENCY

Insulation makes a difference in the amount of energy used to heat or cool a house. Previously, houses were not insulated because of the low cost of energy. Today, things have changed. It is no longer economical to build without insulating.

Seventy percent of the energy consumed in a house is used for heating and cooling. This is according to a study conducted by the U.S. Department of Energy. Another 20 percent is used to heat water. Only 10 percent is used for lighting, cooking, and appliances. See Figure 11-1.

Insulation can help conserve as much as 30 percent of the energy lost in a home. Note the air leakage test results shown in Figure 11-2. Wall outlets are a source of 20 percent of this leakage. This can be stopped by using socket sealers. See Figure 11-3. These are foam insulation pads. They are placed between the faceplate and the socket. Windows and doors have a leakage problem, too. They can be treated in a number of ways to cut down on leakage. The sole plate is an important place to check for proper insulation. To conserve energy, new codes call for leakage seals over corners and sole plates (Figure 11-4). The ductwork in the heating or cooling system can be insulated with tape. Special tape is made for the job. See Figure 11-5.

There are a number of skills to be learned in this chapter. They are

- How to check for insulation requirements
- How to apply insulation
- How to caulk around windows and doors
- How to apply various insulation materials.

• TYPES OF INSULATION •

Batts and blankets are typically made of Fiberglas. They may also be made of cellulose fibers. Batts are best for use by carpenters and do-it-yourselfers. They can be easily applied in attics and between exposed ceiling joists.

Loose fill comes in bags. It can be poured or blown

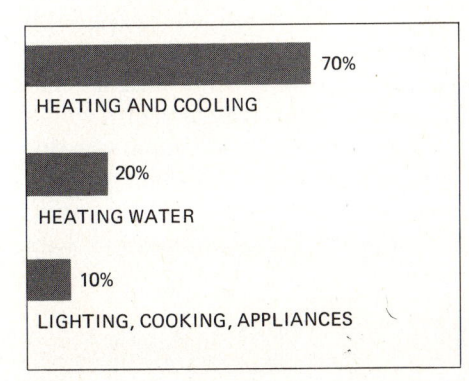

Figure 11-1. Consumption of energy in the home. *(US Dept. Energy)*

Figure 11-3. Socket sealers. *(Manco Tape)*

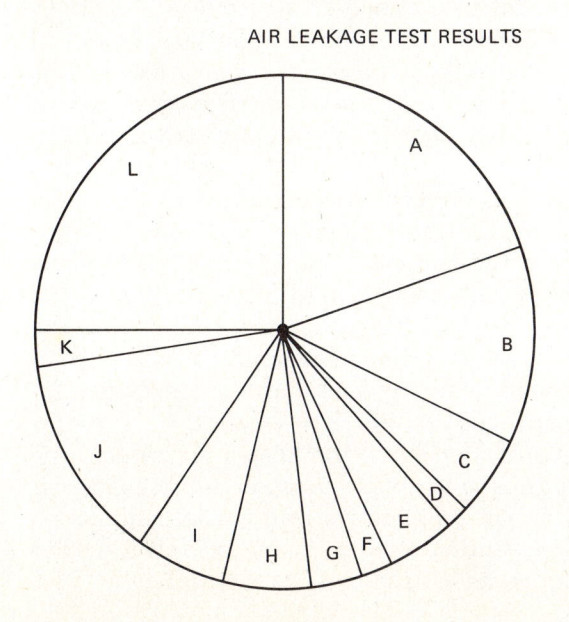

AIR LEAKAGE TEST RESULTS

A — 20% WALL OUTLETS
B — 12% EXTERIOR WINDOWS
C — 5% RECESSED SPOT LIGHTS
D — 1% BATH VENT
E — 5% EXTERIOR DOORS
F — 2% SLIDING GLASS DOOR
G — 3% DRYER VENT
H — 5% FIREPLACE
I — 5% RANGE VENT
J — 14% DUCT SYSTEM
K — 3% OTHER
L — 25% SOLEPLATE

Figure 11-2. Air leakage test results. *(Manco Tape)*

Insulating for Thermal Efficiency 291

Figure 11-4. New codes call for leakage seals over corners and sole plates. *(Fox and Jacobs)*

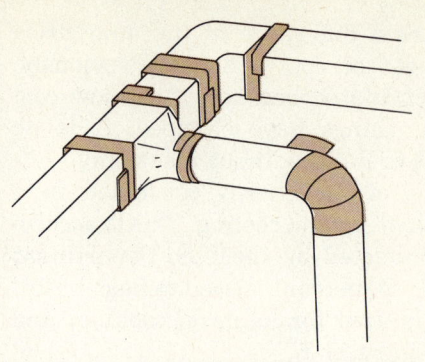

Figure 11-5. Insulating tape for ductwork. *(Manco Tape)*

into the joist space in open attics. It can be blown by special machines into closed spaces such as finished walls.

Rigid types of plastic insulation come in boards. Foams can be applied with a spray applicator. Plastic has high insulation quality. Thinner layers of plastic may provide the same protection as thicker amounts of other materials. Thinner sections may provide higher *R values*. R values indicate the relative insulating qualities of a material.

All insulation material should have a vapor barrier. This barrier provides resistance to the passage of water vapor.

Another type of Fiberglas insulation is the tongue-and-groove sheathing that is available for exterior walls.

Batts and blankets (except friction-fit) have an asphalt-impregnated paper vapor barrier. Or, in some cases, they have a foil backing. All other types should have a 2-mil (0.002-inch) plastic sheet installed as a vapor barrier.

• HOW MUCH IS ENOUGH •

For years the recommended amount of insulation was R-19 for ceilings. R-11 was suggested for walls almost anywhere in the country.

Recently there have been changes in these recommendations. The map in Figure 11-6 shows how much is needed in various locations. The first number is for the ceiling. The second number is for the wall. Insulate the floor if the basement is not heated. These are the values recommended by Owens-Corning. Owens-Corning is a manufacturer of insulation.

In some parts of the country, utilities and cities have separate standards. Be sure to check the local requirements for your area. In most instances local codes follow Figure 11-7. At present it appears that *more is better*. This has not been proven to everyone's satisfaction. More research needs to be done in this area. Different types of insulation are being developed. Foams appear to be the best bet for insulating an older house—that is, if it has no previous wall insulation.

• WHERE TO INSULATE •

Batts are made in specific thicknesses. Choose the proper R value for the space. Allow the material to expand. Fit it closely at the edges. Don't overlook the space between closely fitted timbers.

Vapor barrier protection is important. It should always be installed toward the warm side of the area being insulated. Vapor barriers must not be torn or broken. Repair any breaks with a durable tape.

Provide proper ventilation. Natural venting is important. It controls moisture and relieves heat buildup in the summer. See Figure 11-8. In some cases there isn't enough natural ventilation. A fan is usually needed to exhaust the attic during the summer. This is when heat builds up and causes the air conditioner to work harder. See Figure 11-9. The air space of the attic is often 60°F hotter than outside. Removing the heated air reduces the heat transfer to the home. Keep in mind that power venting is not an effective solution in conjunction with ridge venting. Power venting isn't always effective with vents located close to the unit.

292 Carpentry Fundamentals

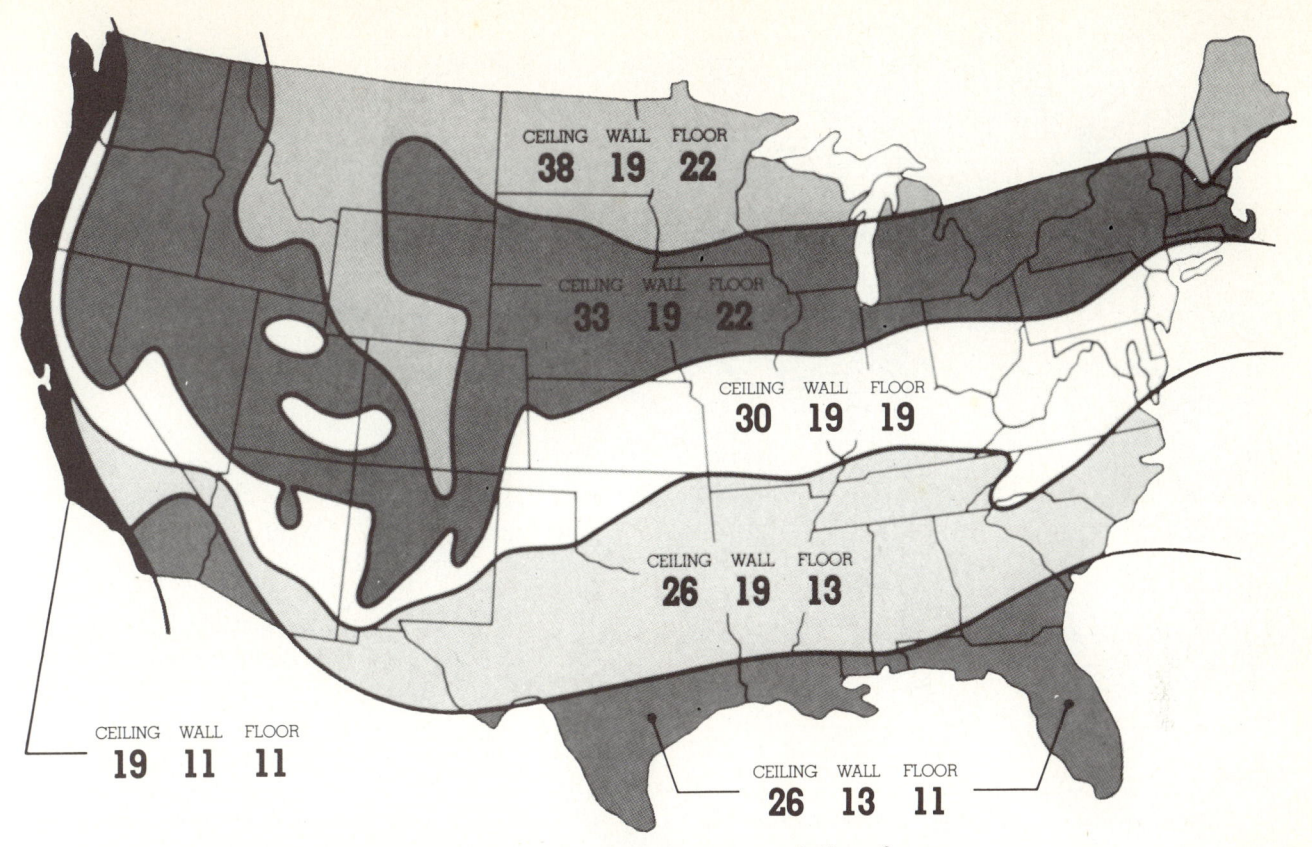

Figure 11-6. Owens-Corning Fiberglas minimum insulation recommendations for energy-efficient homes.

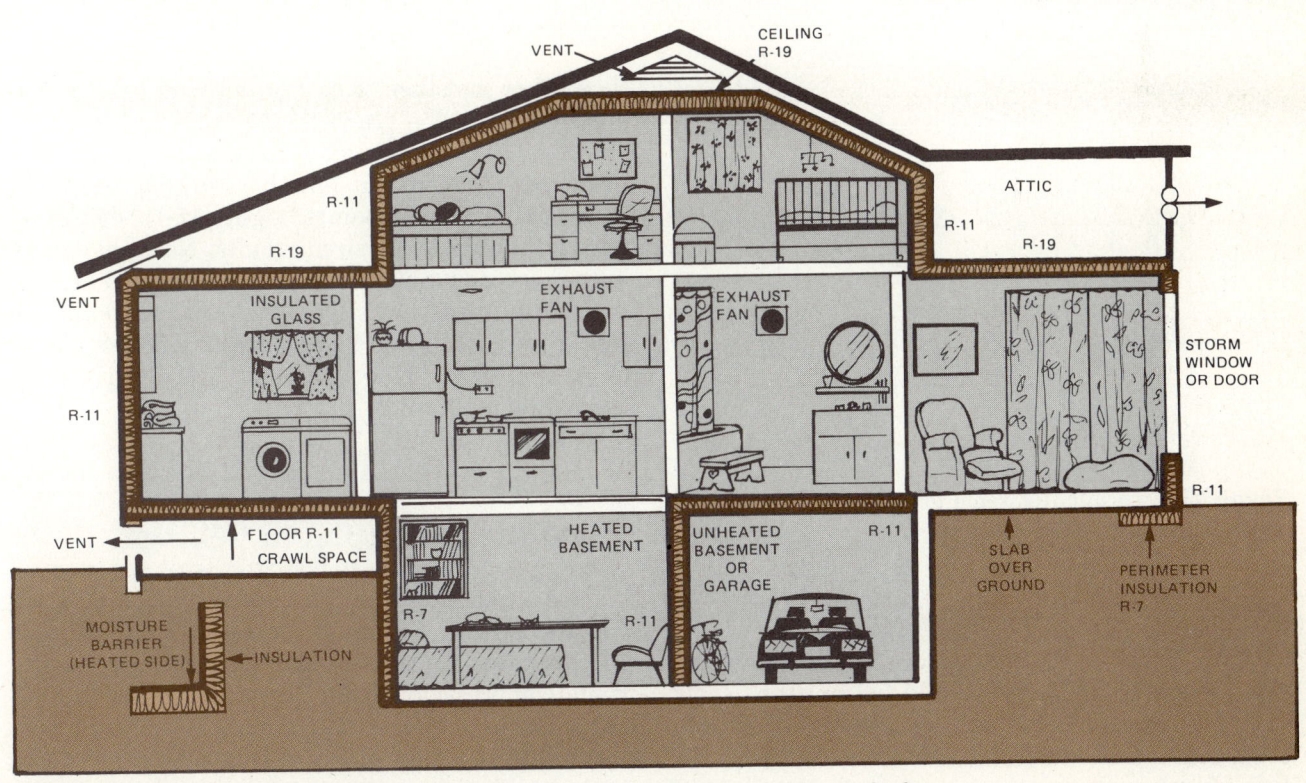

Figure 11-7. Recommendations for insulating your home. Note the R values and where they are assigned. *(New York State Electric and Gas)*

Insulating for Thermal Efficiency 293

EAVE VENT

GABLE VENT

ROOF VENT

CUPOLA VENT

RIDGE VENT

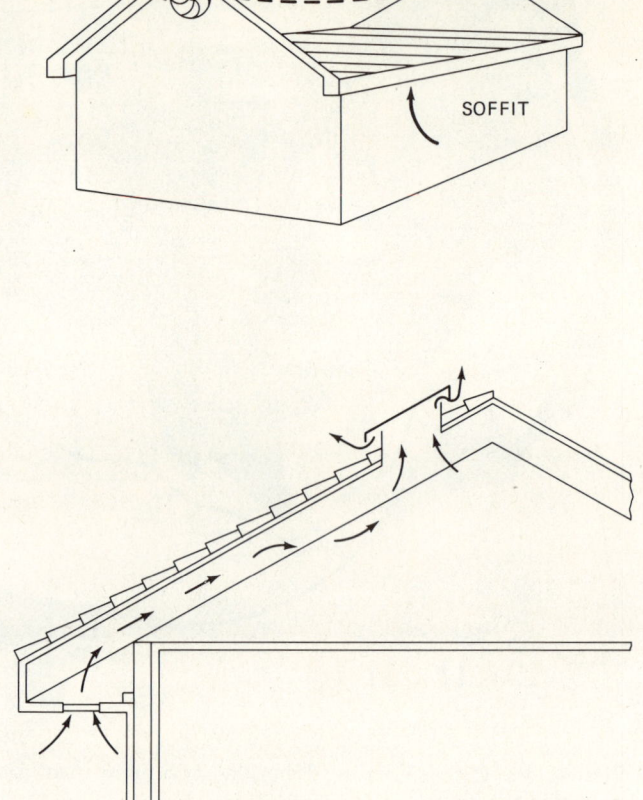

FOUNDATION VENT

SOFFIT

Figure 11-9. Attic ventilation—forced air removal. *(New York State Electric and Gas)*

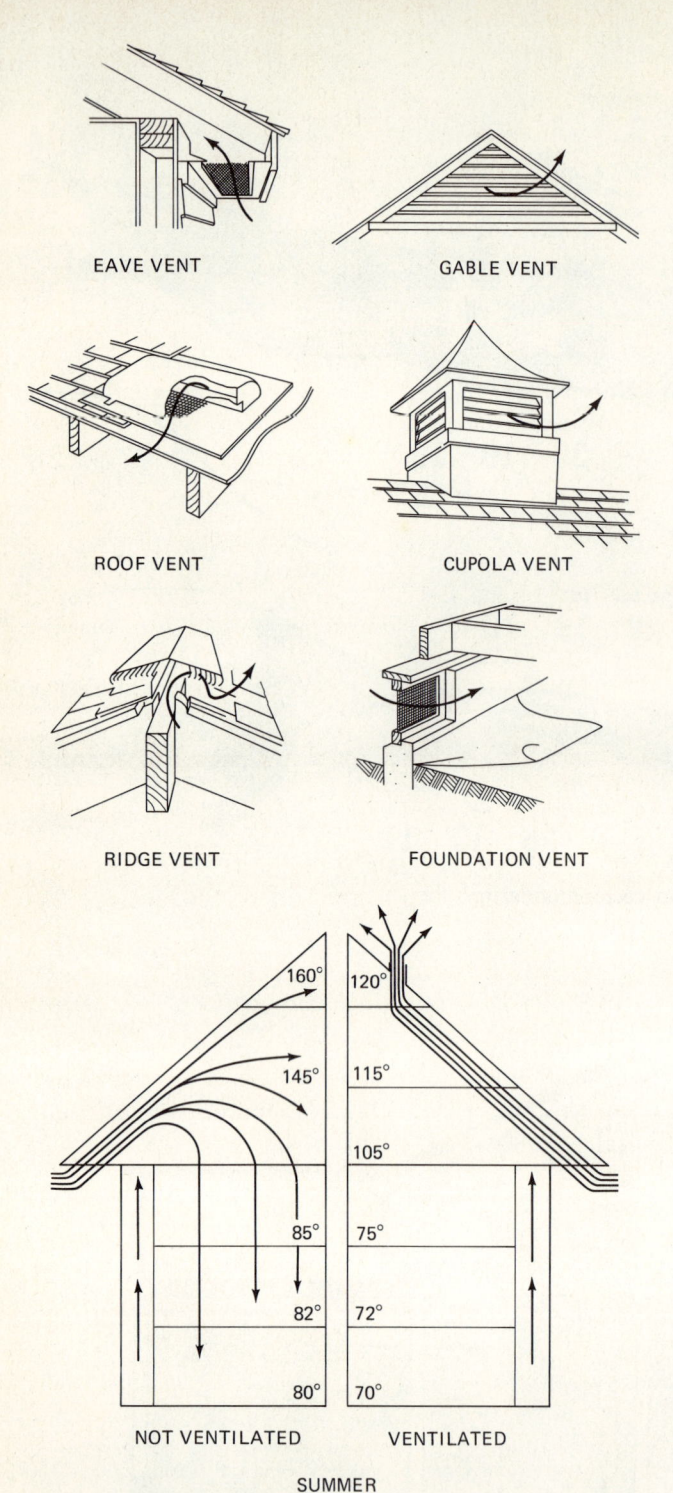

NOT VENTILATED — VENTILATED
SUMMER

Figure 11-8. Location of vents for adequate ventilation. *(New York State Electric and Gas)*

ture. This moisture is generated through the exposed earth in the crawl space.

Figure 11-11 shows some of the points which may need special attention. Note how the vapor barrier is installed at these points. Figure 11-12 shows the best insulation for desired results. Note that some houses are now built with 2×6-inch instead of 2×4-inch studs in the walls. This allows for more insulation. Note how 1-inch foam sheathing board is used on the outside. This gives the added insulation needed for 2×4 studs.

• INSTALLING INSULATION •

Installing Insulation in Ceilings

Push batts up through joist spaces. Pull them back down even with bottom of the joists. See Figure 11-13. This ensures full R-value performance. It prevents compression of the insulation. When high-R batts are used, it allows the batts to expand over the top of the joist. Compressing the batt reduces its insulating value.

CRAWL SPACE • Crawl spaces need ventilation. See Figure 11-10. There should be moisture seals in crawl spaces. Sealed and unsealed crawl spaces are treated differently. Vents in the basement wall help circulate the outside air. They draw out mois-

294 Carpentry Fundamentals

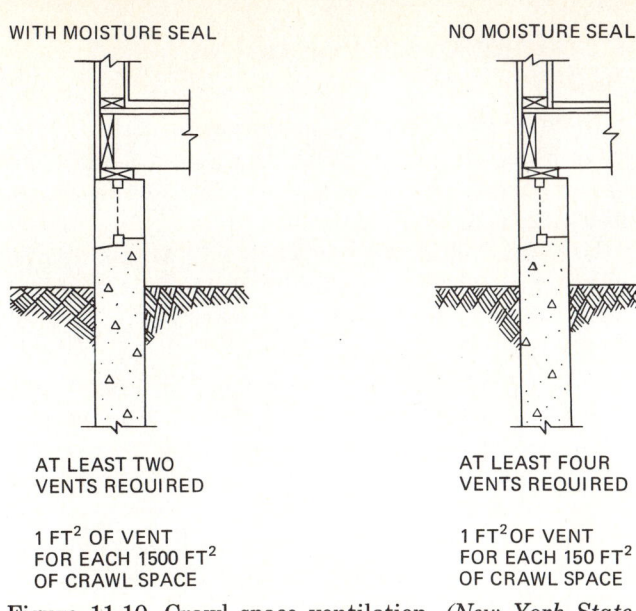

Figure 11-10. Crawl space ventilation. *(New York State Electric and Gas)*

Overlap the top plate. Do not block eave vents. Make sure the ends butt tightly against each other.

Obstructions (such as electrical boxes for light fixtures) should have insulation installed over them.

CAUTION: Insulation must be kept at least 3 inches away from recessed light fixtures.

Installation Safety

You should keep in mind some of the health hazards involved with installing Fiberglas. It has tiny fibers of resin (glass) that can be breathed in and cause trouble in your lungs. Just be careful and take the proper precautions. If you work with Fiberglas all day, make sure you wear a mask over your nose and mouth.

Notice that the people in the illustrations on page 297–298 are wearing long sleeves and gloves. This will prevent the itching that this type of material can lead to. You should wear goggles and not just glasses. These will keep the fibers out of your eyes and prevent damage to your vision.

A hat and proper long pant legs are also called for in this job. Proper shoes are needed, since you are working around a construction area. The nails are still exposed in some types of work, so make sure the shoes can protect your feet from the nails.

The usual safety procedures should be followed in using the ladder and climbing around an attic or any place that may be elevated from floor level. A hard hat helps to keep you from getting roofing nails in your scalp when you are installing insulation near the underside of the roof.

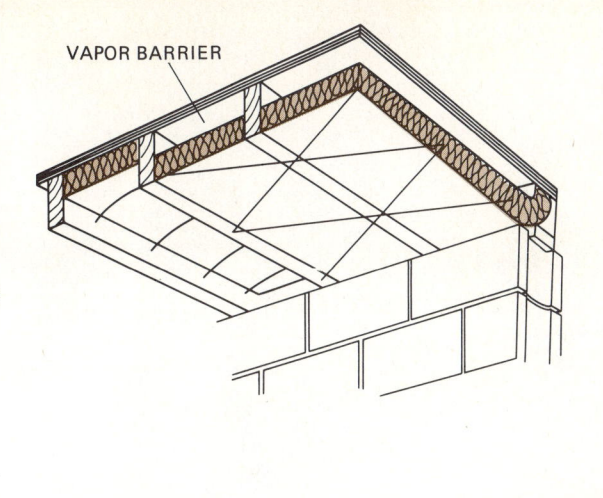

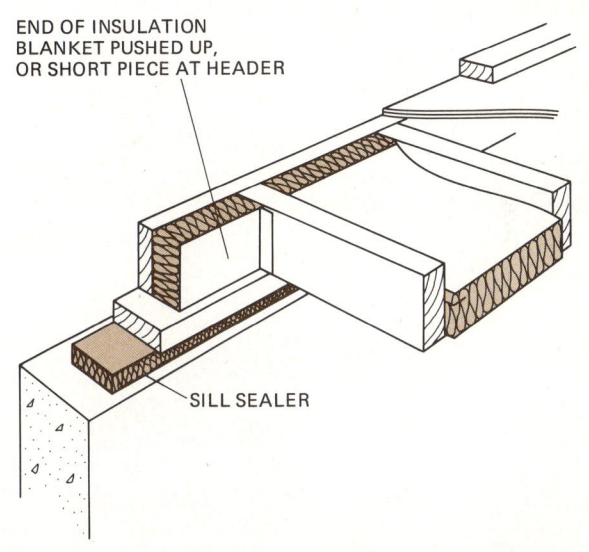

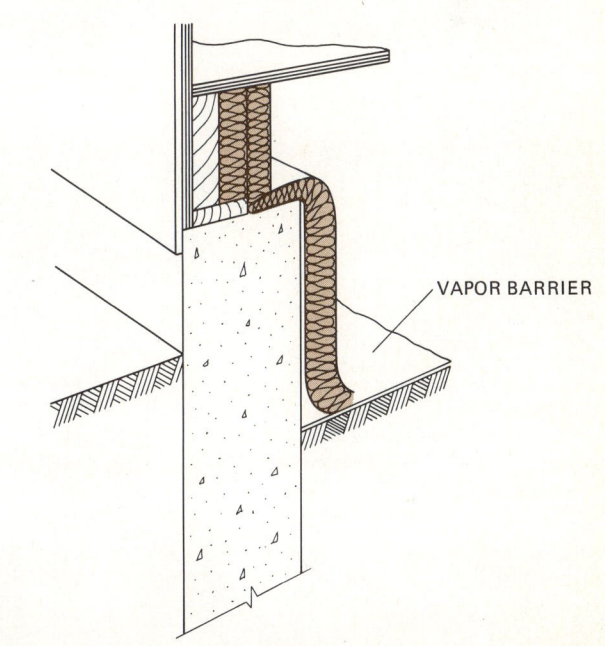

Figure 11-11. Note placement of the vapor barrier in the installation of insulation. *(New York State Electric and Gas)*

Insulating for Thermal Efficiency   **295**

R-44
WITH 2 × 6
OR 2 × 8
CEILING JOISTS

Use
6½" thick R-22 Fiberglas insulation between joists and 6½" thick R-22 Fiberglas insulation across the joists.

R-38
WITH 2 × 6
CEILING JOISTS

Use
6" thick R-19 Fiberglas insulation between ceiling joists and 6" thick R-19 Fiberglas insulation across the joists.

R-30
WITH 2 × 6
CEILING JOISTS

Use
6" thick R-19 Fiberglas insulation between ceiling joists and 3½" thick R-11 Fiberglas insulation across the joists.

R-30
WITH 2 × 6
OR 2 × 8
CEILING JOISTS

Use
10" thick R-30 Fiberglas insulation. It compresses between ceiling joists and expands over them to form a continuous insulation blanket.

R-30
WITH 2 × 4
TRUSS
CONSTRUCTION

Use
3½" thick R-11 Fiberglas insulation between ceiling joists and 6" thick R-19 Fiberglas insulation across the joists.

(A)

Installing Insulation in Unfloored Attics

Install faced building insulation in attics which have no existing insulation. See Figure 11-14. The vapor barrier should face down toward the warm-in-winter side of the structure.

Use unfaced insulation when you are adding to existing insulation. See Figure 11-15. Batts or blankets can be laid either at right angles or parallel to existing insulation.

Start from the outside and work toward the center of the attic. See Figure 11-16. Lay insulation in long runs first and use leftovers for shorter spaces. Be sure to butt insulation tightly at joints for a complete barrier.

Insulation should extend far enough to cover the top plate. But it should not block the flow of air from the eave vents. See Figure 11-17.

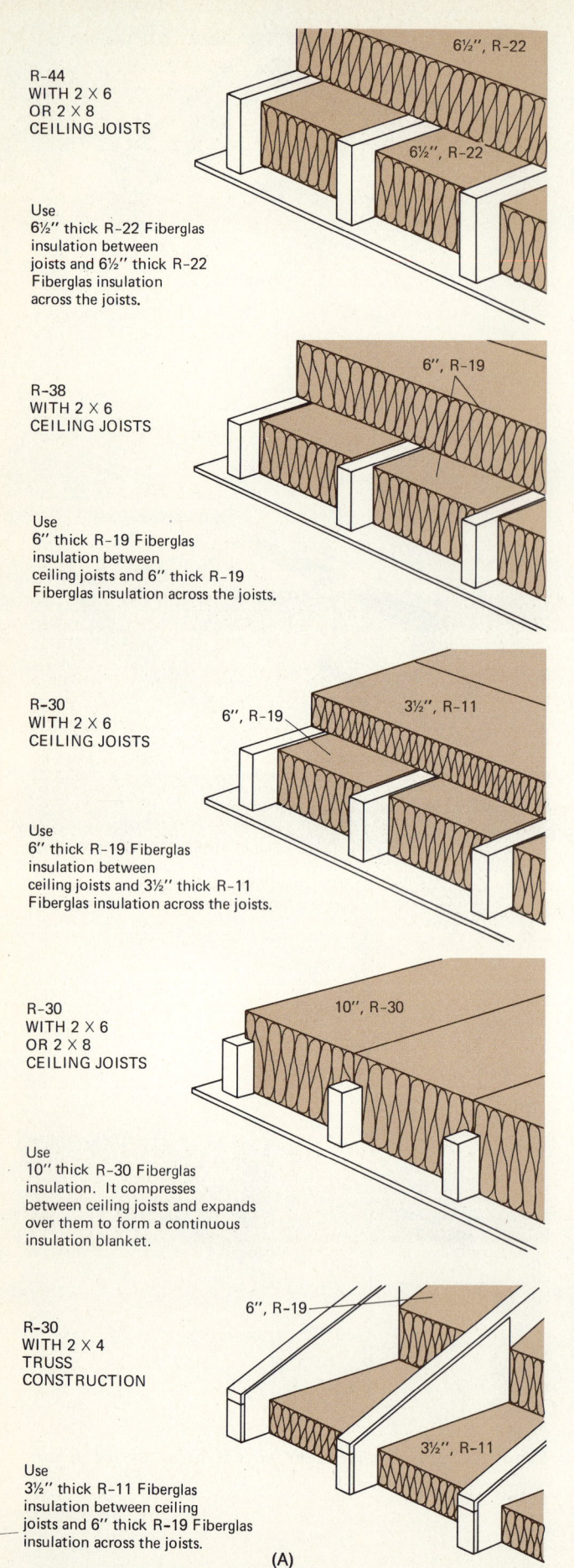

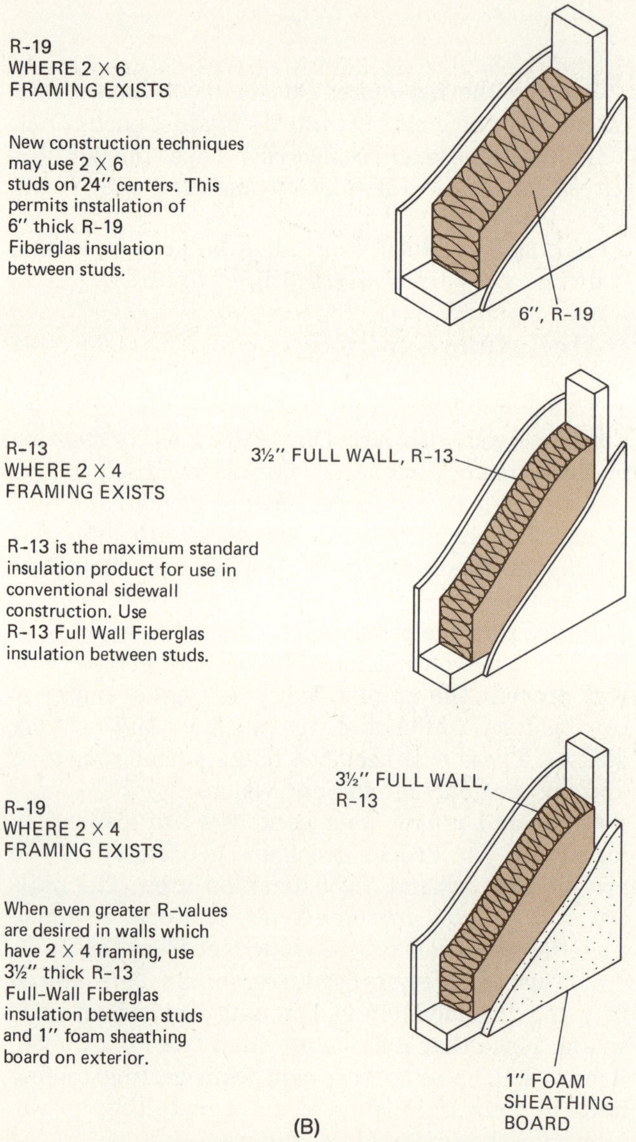

R-19
WHERE 2 × 6
FRAMING EXISTS

New construction techniques may use 2 × 6 studs on 24" centers. This permits installation of 6" thick R-19 Fiberglas insulation between studs.

R-13
WHERE 2 × 4
FRAMING EXISTS

R-13 is the maximum standard insulation product for use in conventional sidewall construction. Use R-13 Full Wall Fiberglas insulation between studs.

R-19
WHERE 2 × 4
FRAMING EXISTS

When even greater R-values are desired in walls which have 2 × 4 framing, use 3½" thick R-13 Full-Wall Fiberglas insulation between studs and 1" foam sheathing board on exterior.

(B)

Figure 11-12. (A) Optimum insulation for ceilings and attics. *(Certain-Teed)* (B) Optimum insulation for sidewalls. *(Certain-Teed)*

Figure 11-13. Installing insulation in ceilings. Note the long-sleeved shirt, gloves, hat, respirator, and goggles. (Owens-Corning)

Figure 11-14. Install insulation with a vapor barrier facing in attics with no existing insulation. The vapor barrier should face the warm-in-winter side of the structure. (Owens-Corning)

Figure 11-15. Use insulation without a vapor barrier facing when adding to existing insulation. Batts or blankets can be laid at right angles or parallel to existing insulation. (Owens-Corning)

Insulation should be pushed under wiring unless the wiring will compress the insulation. The ends of batts or blankets should be cut to fit snugly around cross-bracing and wiring. Fill spaces between the chimney and the wood framing with unfaced Fiberglas insulation.

Scuttle holes, pulldown stairways, and other attic accesses should also be insulated. Insulation can be glued directly to scuttle holes. Pulldown stairways may need a built-up framework to lay batts on and around.

Installing Insulation in Floored Attics

If the attic is not used for storage, unfaced batts or blankets may be laid directly on top of the floor. See Figure 11-18.

Floorboards must be removed before insulation is installed if the attic is to be used for storage. Since the flooring will be replaced, the R value will be

Figure 11-16. Work toward the center of the attic, laying long runs of insulation first. Fill in with short pieces. Be sure to butt insulation tightly at all joists. Insulation should cover top plate but must not block eave vents. Insulation must be kept at least 3″ away from recessed light fixtures. (Owens-Corning)

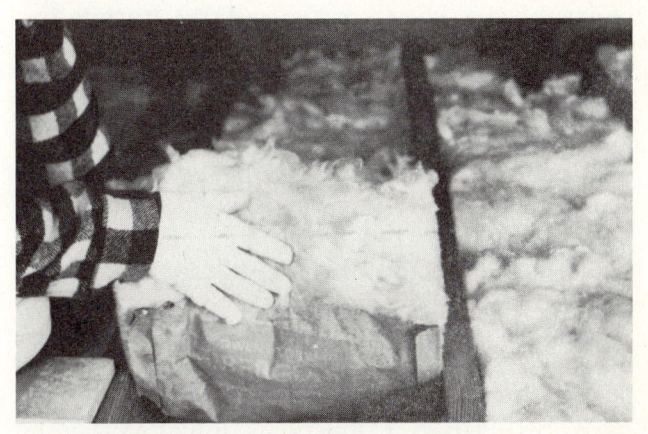

Figure 11-17. If the attic is used for storage, the floorboards must be removed. The flooring will limit the R value to the amount of uncompressed insulation which can fit between the joists. (Owens-Corning)

Insulating for Thermal Efficiency   **297**

Figure 11-18. If the attic is not used for storage, unfaced batts or blankets may be laid directly on top of the floor. *(Owens-Corning)*

Figure 11-20. Floors over heated areas should be insulated to the R values recommended. *(Owens-Corning)*

Figure 11-19. Once the flooring has been removed, the attic should be insulated in this manner. *(Owens-Corning)*

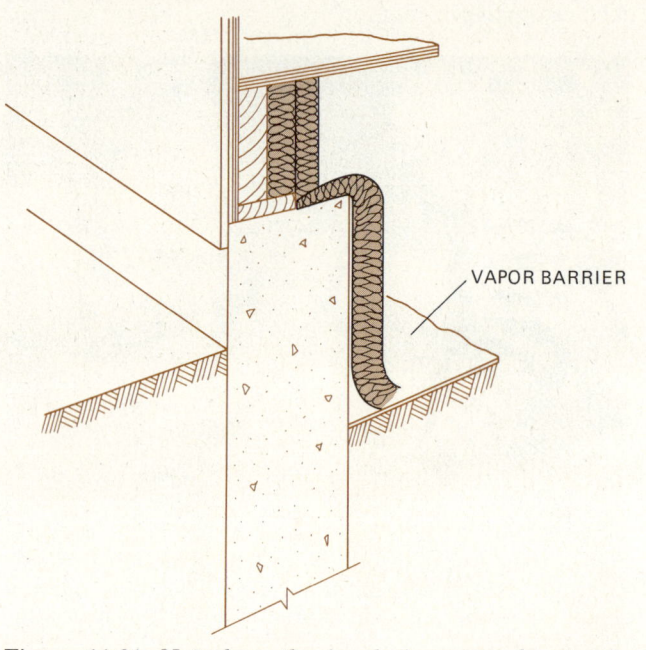

limited. It is limited to that of the amount of uncompressed insulation that can be fitted between the joists.

Remove the flooring in the attic. It should then be insulated as shown in Figure 11-19. Replace the floorboards and stairsteps.

Installing Insulation in Floors

Floors over unheated areas should be insulated. The recommended insulation R values should be used. See Figure 11-20. With faced insulation, the vapor barrier should be installed face up. This is toward the warm-in-winter side of the house. Insulation should overlap the bottom plate. It should also overlap the band joist. See Figure 11-21.

BOW SUPPORTS • Fiberglas insulation can be supported by heavy-gage wires. The wire may be bowed or wedged into place under the insulation.

Figure 11-21. Note how the insulation vapor barrier is installed in the crawl space.

CRISSCROSS WIRE SUPPORT • Insulation may also be supported by lacing wires around nails located at intervals along the joists.

CHICKEN-WIRE SUPPORT • Chicken wire nailed to the bottom of floor joists will also support the insulation.

Insulating Basement Walls

Masonry walls in conditioned areas may be insulated. Install a furring strip or 2×4 framework. Nail the top and bottom strips in place. Then, install ver-

298 Carpentry Fundamentals

tical strips 16 or 24 inches O.C. Next cut small pieces of insulation. Push the pieces in place around the band joist. Get insulation between each floor joist and against the band joist.

Install insulation either faced or unfaced. Unfaced insulation will require a separate vapor barrier. Do not leave faced insulation exposed. The facing is flammable. It should be covered with a wall covering. Cover the insulation as soon as it has been installed.

Insulating Crawl Spaces

Measure and cut small pieces of insulation to fit snugly against the band joist. This is done so that there is no loss of heat through this area. Push pieces into place at the end of each joist space.

CAUTION: Do not use faced insulation; the facing is flammable.

Use long furring strips. Nail longer lengths of insulation to the sill. Extend it 2 feet along the ground into the crawl space. Trim the insulation to fit snugly around the joists.

On walls that run parallel to the joists, it is not necessary to cut separate header strips. Simply use longer lengths of insulation and nail (with furring strips) directly to the band joist. Install the insulation. Lay the polyethylene film under the insulation and the entire floor area. This prevents ground moisture from migrating to the crawl space. Use 2×4-inch studs or rocks to help hold the insulation in place.

Installing Insulation in Walls

Insulating walls can pay dividends. Figure 11-22 shows that the sidewalls lose 15.2 percent of the heat lost in a ranch house. In the two-story house, where there is more wall area, the loss is 30.8 percent. This means it is very important to properly insulate walls.

Figure 11-23 shows how insulation fits into the wall area. It is placed between the studs. The outside wood is also an insulation material.

Insulation for walls can be R-19. Use a combination of R-11 insulation with 1-inch (R-8) *high-R sheathing* or R-13 insulation with ¾-inch (R-8) *high-R sheathing* on the interior or exterior. See Figure 11-24. R-13 insulation may also be used with ⅝-inch *high-R sheathing* only on the exterior.

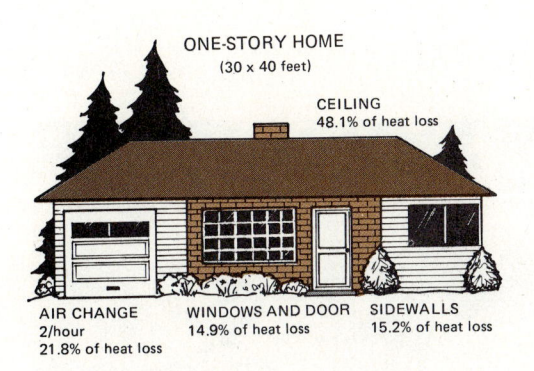

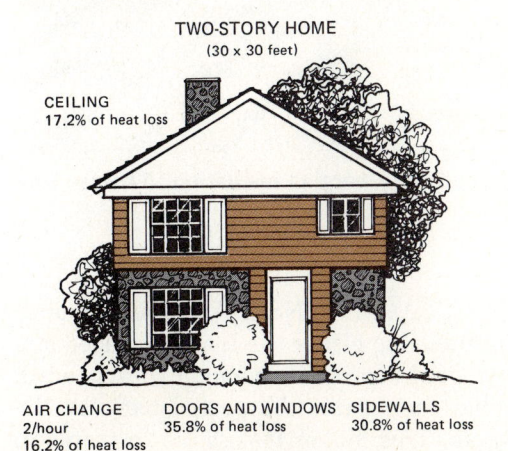

Figure 11-22. Heat loss in a one-story and a two-story house. *(New York State Electric and Gas)*

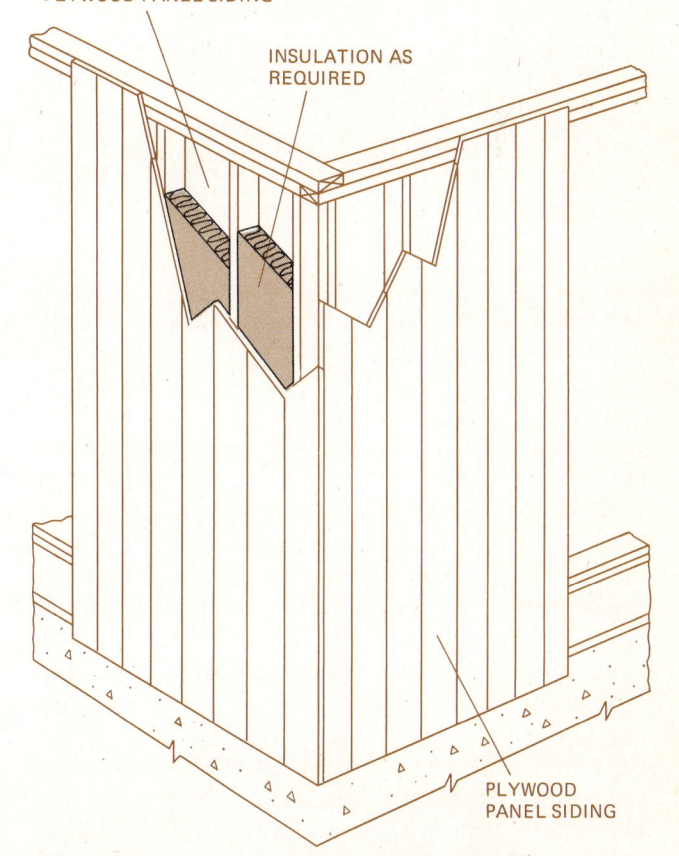

Figure 11-23. Placement of insulation inside a wall between studs. *(American Plywood Association)*

Insulating for Thermal Efficiency

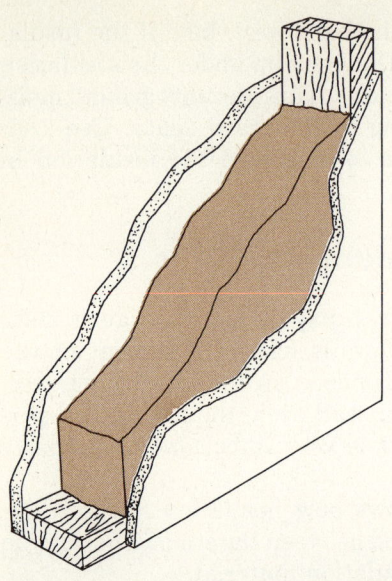

Figure 11-24. R-19 insulation can be obtained by combining R-11 insulation with 1" (R-8) high-R sheathing or combining R-13 insulation with ¾" (R-8) sheathing on the exterior. R-13 insulation may also be used with ⅝" high-R sheathing only on the exterior. *(Owens-Corning)*

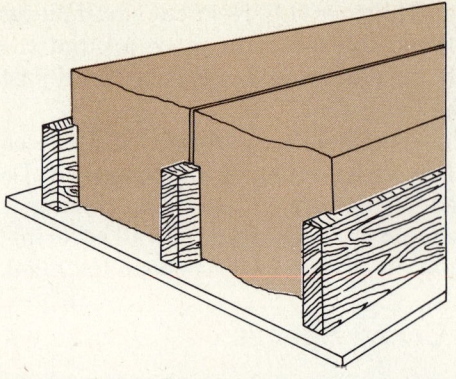

Figure 11-26. Single-layer Fiberglas high-R batts (R-26, R-30, R-38) are made in full 16" and 24" widths so that they expand over the top of the joists for a continuous barrier to heat flow. *(Owens-Corning)*

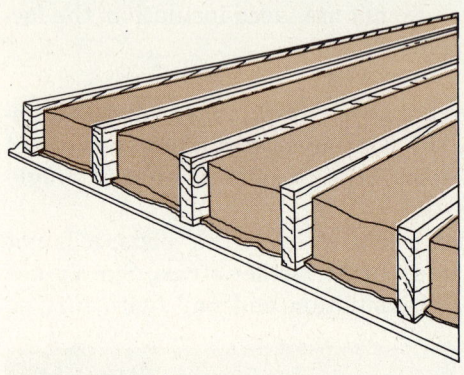

Figure 11-27. R values for floors can be achieved with a single layer of Fiberglas blanket insulation.

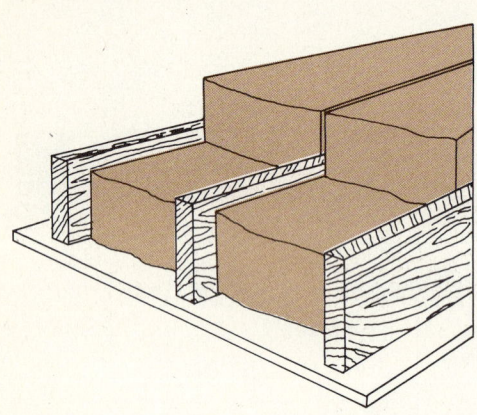

Figure 11-25. R-30 can be obtained by using a single R-30 blanket or R-19 and R-11 Fiberglas blankets in combination. *(Owens-Corning)*

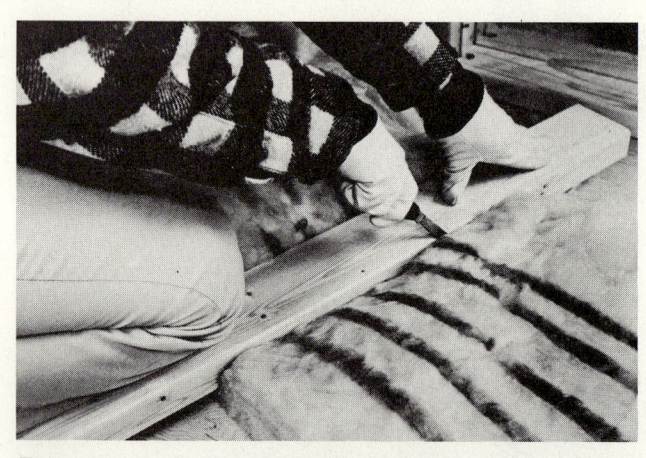

Figure 11-28. Cutting Fiberglas insulation. *(Owens-Corning)*

In placing R-30, use a single-layer blanket. Or you can use R-19 and R-11 blankets. This combination is shown in Figure 11-25.

High-R batts (R-26, R-30, and R-38) are made in full 16- and 24-inch widths. They expand over the top of the joists. This presents a continuous barrier to heat flow. See Figure 11-26. Floors can have R-values. Choose the right single layer of blanket insulation. See Figure 11-27.

CUTTING FIBERGLAS INSULATION • Measure the length of Fiberglas insulation needed. Place the insulation on a piece of scrap plywood or wallboard. Compress the material with one hand. Cut the material with a sharp knife. See Figure 11-28.

INSTALLING FACED INSULATION • Install faced insulation using either the insert or faced stapling method. Cut lengths 1 inch longer than needed. This is so that facing may be peeled back at the top and bottom. This makes it possible to staple to plates. Make sure the insulation fills the entire cavity. See Figure 11-29.

Figure 11-29. Installing faced insulation.

Figure 11-30. Insert stapling. *(Owens-Corning)*

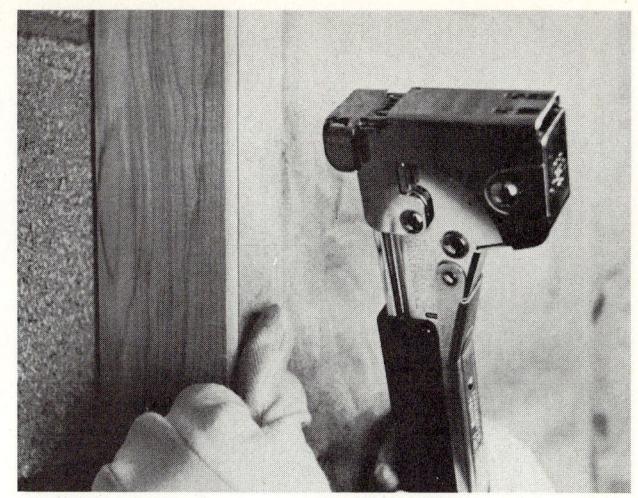

Figure 11-31. Faced stapling. *(Owens-Corning)*

Figure 11-32. Installing unfaced insulation. *(Owens-Corning)*

INSERT STAPLING • Insulation is pushed into the stud or joist cavity. Vapor barrier flanges are stapled to the sides of the studs. Make sure that flanges do not gap. This will allow vapor to penetrate. See Figure 11-30.

FACED STAPLING • Vapor barrier flanges are overlapped. They are then stapled to the edge of the stud. Flanges must be kept smooth. This is necessary for proper application of the wall finish. See Figure 11-31.

INSTALLING UNFACED INSULATION • Wedge unfaced batts into the cavity. Make sure they fit snugly against the studs. Also fit the top and bottom plates snugly. See Figure 11-32.

FILLING SMALL GAPS • Hand-stuff small gaps around window and door framing. This can be done with scrap pieces. Cover the stuffed areas with a vapor-resistant material. See Figure 11-33. Insulation must fit snugly against both the top and bottom plates. It should also fit against the sides of the stud

Insulating for Thermal Efficiency 301

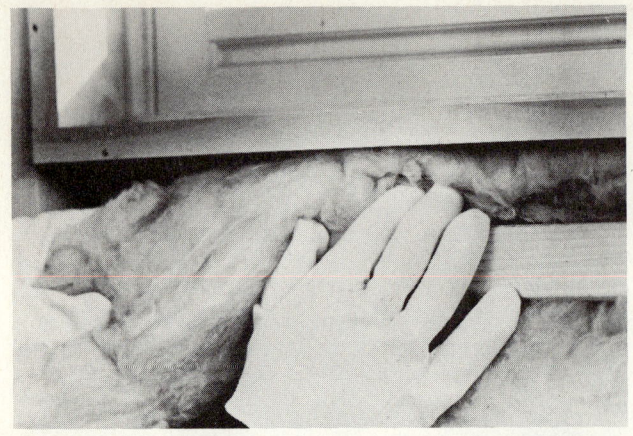

Figure 11-33. Filling small gaps. *(Owens-Corning)*

Figure 11-34. Close doesn't count. *(Owens-Corning)*

Figure 11-35. Insulating behind wiring and pipes. *(Owens-Corning)*

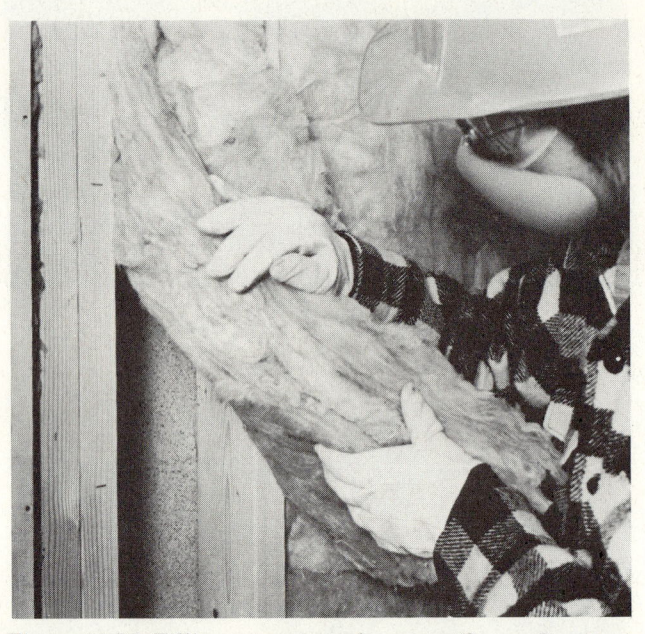

Figure 11-36. Filling narrow stud spaces. *(Owens-Corning)*

opening. See Figure 11-34. Improperly installed, it will permit heat to escape.

INSULATING BEHIND WIRING AND PIPES • Insulation should be fitted behind or around heat ducts. It should also fit closely around pipes and electrical boxes. This will help keep pipes from freezing. It also prevents unnecessary heat loss. Repair any tears in the vapor barrier of faced insulation. This prevents condensation problems. See Figure 11-35.

FILLING NARROW STUD SPACES • With unfaced insulation you can cut to stud width. Wedge the insulation into place. This area must be covered with a vapor barrier. With faced insulation, make sure the vapor barrier is properly stapled to the studs. See Figure 11-36.

INSULATING AROUND WIRING • Split unfaced insulation and install it on both sides of the wiring. Install faced insulation behind the wiring. Pull it through to permit stapling. See Figure 11-37.

• VAPOR BARRIERS AND MOISTURE CONTROL •

Excess moisture may be a concern in a well-insulated house. Excess air changes have been eliminated. Therefore, some consideration must be given to the tight house. The insulation has been properly located. It keeps the inside air inside. This means it doesn't require too much energy to reheat it. It doesn't have to be recooled too much during the summer.

302   Carpentry Fundamentals

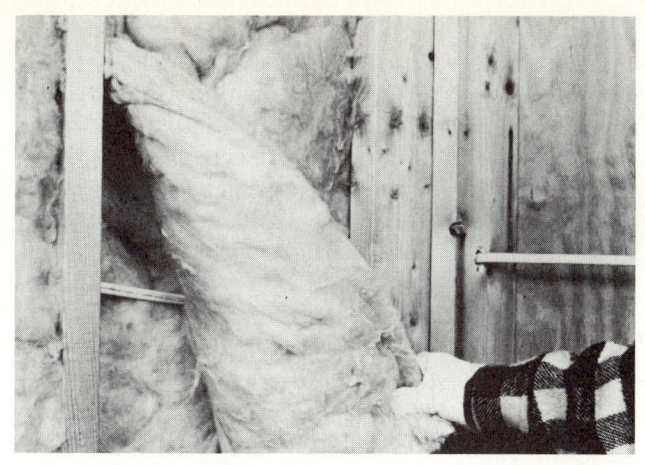

Figure 11-37. Insulation around wiring. *(Owens-Corning)*

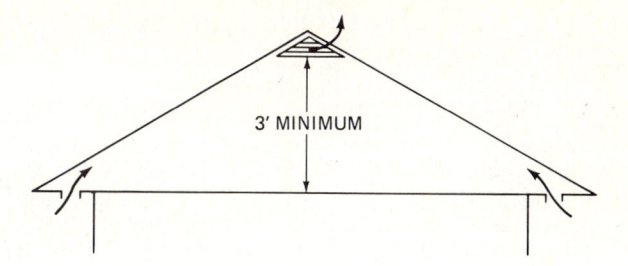

NOTE: There is 1 ft² inlet and 1 ft² outlet for each 600 ft² of ceiling area, with at least half of the vent area at the top of the gables and the balance at the eave.

(A)

(B)

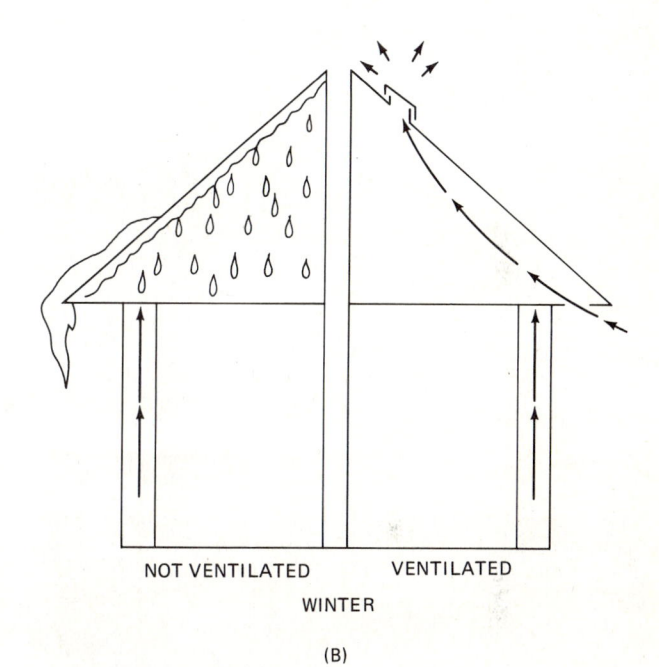

Figure 11-38. (A) Spacing of attic vents. (B) Ventilation can control condensation.

Condensation

A house must *breathe*. Condensation which is not caused by insulation can be controlled. Condensation is simply water vapor. Water in the form of a fog comes from a variety of sources in a home. It is produced in baths, in the kitchen, and in the laundry. It will distribute itself throughout the house. Unless it is controlled, it can condense. Just as frost forms on the outside of an iced glass on a warm day, vapor condenses on house surfaces. Slowly, but surely, it can cause damage.

MOISTURE CONTROL • There are two ways to control moisture. They are elimination and ventilation. Elimination means to get rid of the source of moisture. This can be leaks, wet ground in crawl spaces, or small leaks in the structure.

Ventilation means simply getting rid of moisture. It carries excess humidity outside. This is done with vent fans or hoods in the kitchen. The bathroom may be vented to the outside. Automatic clothes dryers must be vented to the outside. They are a major source of moisture.

Vent fans at high-moisture-producing locations are helpful. There could be the fan over the kitchen range. The bathroom may have a fan to remove moisture. Any of these fans can be equipped with a humidistat that will turn it on when the humidity gets to a preset level.

ATTIC VENTILATION • Attic ventilation is very important in many ways. It helps eliminate moisture in the attic. That moisture can leak into the living quarters, ruining the ceiling as it does so. Or, it can rot studs or other parts of the house. One way to provide ventilation is to place vents in the end of the roof structure. See Figure 11-38. They can be various shapes. Some use cupolas to allow air to move upward and out. Some use a vented ridge section to allow hot air to escape in the summer and a constant flow of air all during the year.

These openings must be covered to prevent insects, birds, snow, and rain from entering. There are a number of materials used to cover the openings. Table 11-1 shows some of them.

Keep in mind that it takes 1 square foot of inlet area and 1 square foot of outlet area for each 600 square feet of ceiling area. At least half of the vent should be at the top of the gables and the balance should be at the eave.

• THERMAL CEILINGS •

Thermal ceiling panels have insulating values up to R-12. They are the only types of insulation which can be used for some types of homes. Take for in-

Insulating for Thermal Efficiency **303**

TABLE 11-1 ATTIC VENTILATION COVERINGS.

Type of Covering	Size of Covering
¼-inch hardware cloth and rain louvers	2 × net vent area
8-mesh screen	1¼ × net vent area
¼-inch hardware cloth	1 × net vent area
8-mesh screen and rain louvers	2¼ × net vent area
16-mesh screen	2 × net vent area
16-mesh screen and rain louvers	3 × net vent area

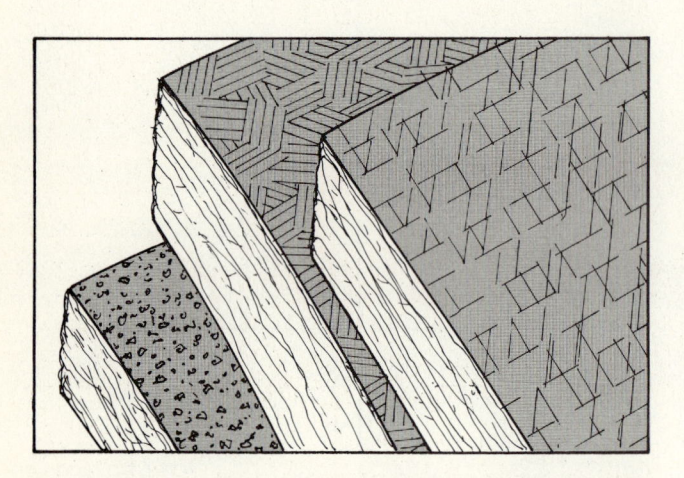

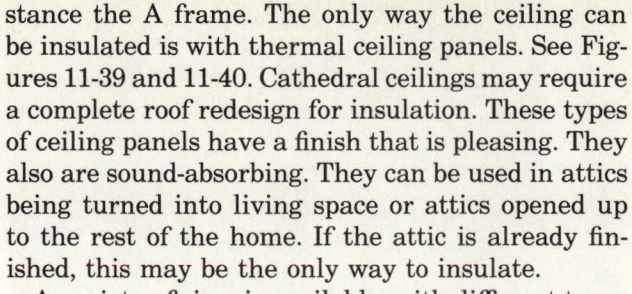

stance the A frame. The only way the ceiling can be insulated is with thermal ceiling panels. See Figures 11-39 and 11-40. Cathedral ceilings may require a complete roof redesign for insulation. These types of ceiling panels have a finish that is pleasing. They also are sound-absorbing. They can be used in attics being turned into living space or attics opened up to the rest of the home. If the attic is already finished, this may be the only way to insulate.

A variety of sizes is available, with different types of finishes on the face. There are sizes and faces to suit almost every installation requirement. They come in sizes from 2×4 feet up to 4×16 feet. Figure 11-41 shows some of the applications for these panels.

Installing Thermal Ceiling Panels

There are a number of ways to install thermal ceiling panels.

GRID SYSTEM • This uses the suspended interlocking metal grid system. It can be installed on the 2×4-foot or 4×4-foot system. A larger-faced grid system is needed for 3-inch R-12 panels. See Figure 11-42A.

Figure 11-39. Thermal ceiling panels are easy to install. (Owens-Corning)

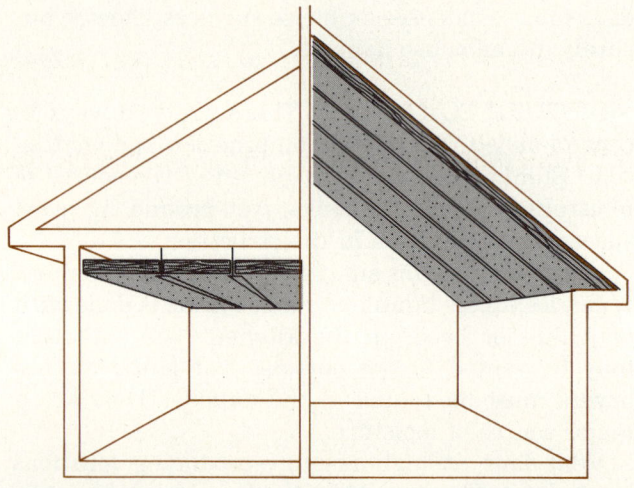

Figure 11-41. Thermal ceiling panels for cathedral ceilings and regular ceilings. (Owens-Corning)

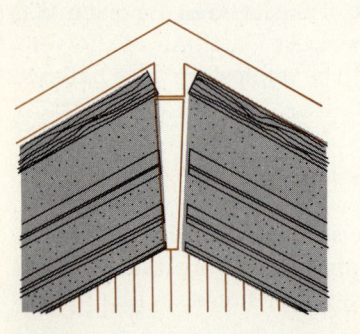

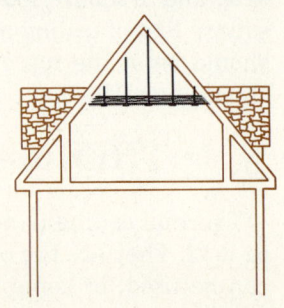

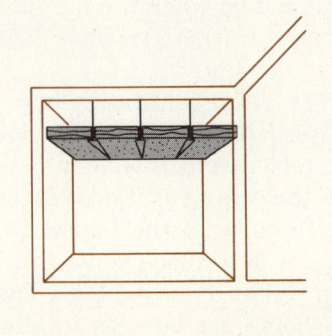

Figure 11-40. Typical installations of thermal ceiling panels. (Owens-Corning)

(A)

(B)

(C)

(D)

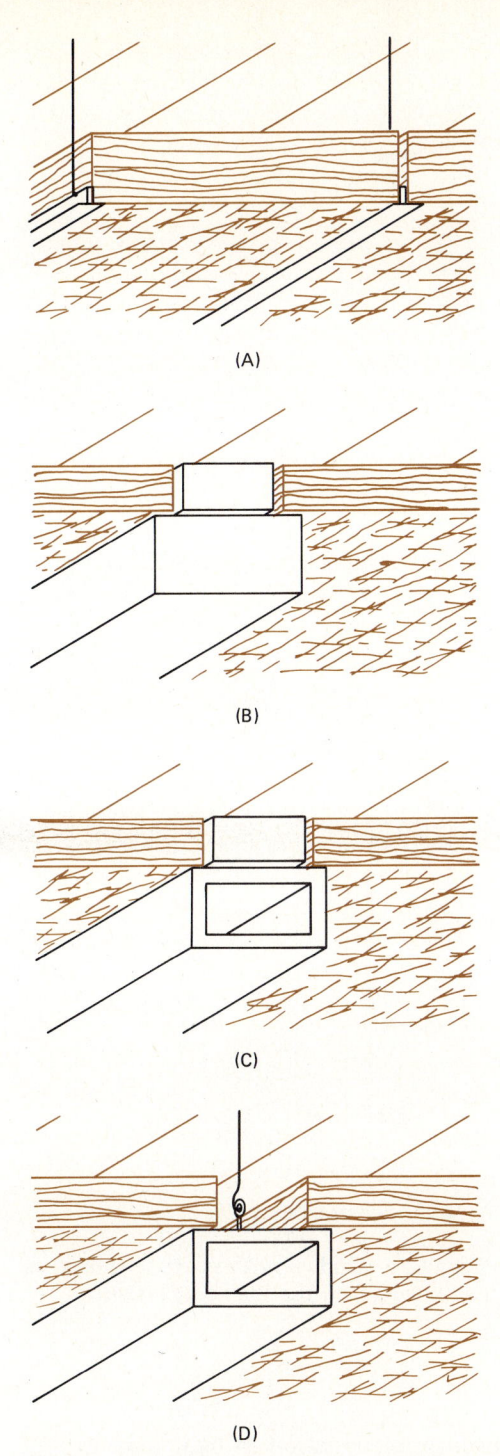

Figure 11-42. Methods of supporting thermal ceiling panels. *(Owens-Corning)* (A) Grid system. (B) Solid wood beams. (C) False-bottom beams. (D) Suspended beams.

SOLID WOOD BEAMS • Solid beams as in Figure 11-42B are used in many homes. The beam becomes a support system for the insulation panels. The panels can be removed and replaced if they are damaged.

Figure 11-43. Cutting a plastic beam.

FALSE-BOTTOM BEAMS • A system of 1×6-inch false-bottom or box beams provides the beauty of beams without the expense. When wood beam support is used, wood block spacers must be installed between the existing ceiling or joists and the beam. This is so that panels are not compressed when the beams are nailed up.

SUSPENDED BEAMS • False beams may also be used to lower a ceiling like a grid system. Beams are suspended. They use screw eyes attached to the existing ceiling joists.

Decorative Beams

The suspended-beams method of putting up ceiling insulation is easy.

In Figure 11-43 note how the plastic beams are cut to length with a knife.

Make a sketch on paper showing the pattern you want for your beams. Now it is simple to cut the beams to size. Use a knife or sharp tool with a razor blade.

Measure the exact length you need. Cut the beams to size. See Figure 11-44. Use a ruler or chalk line to mark where you want to install the beam. Beams can be glued to other material. They can be glued to the ceiling with an adhesive. You may want to glue a piece of furring strip down the groove (Figure 11-45) of the beam. This will allow you to attach screw hooks. They will be wired to the old ceiling later. See Figure 11-42D. Angles can be cut in the beam with a miter box.

Insulating for Thermal Efficiency **305**

Figure 11-44. A ceiling using plastic foam beam.

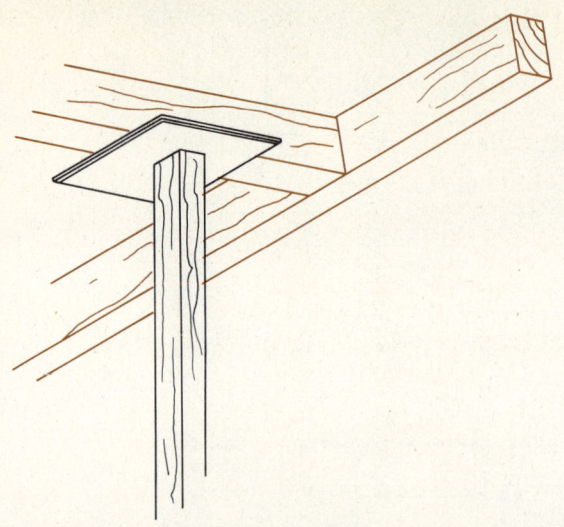

Figure 11-46. Method of supporting beam while the adhesive sets.

Figure 11-45. Note the groove in the beam. You may want to glue a furring strip in it.

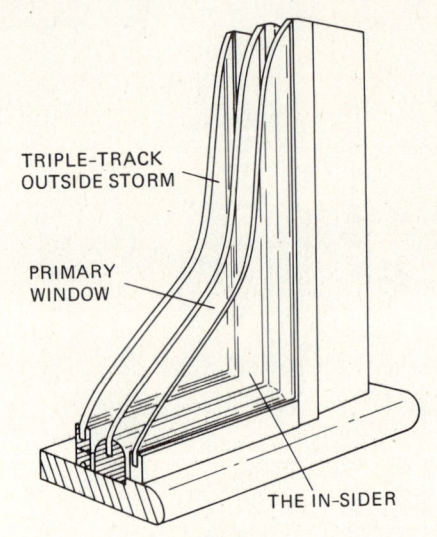

Figure 11-47. Standard window with storm window on exterior and the "insider" on the inside. *(Plaskolite)*

If you want to glue the beam up, use a brace such as that shown in Figure 11-46. This can be done if you want to make a ceiling resemble the one in Figure 11-42B. Nicks and scratches in this type of plastic foam beam can be touched up with shoe polish. Since the beam is made of foam, it also is very good insulation.

• STORM WINDOWS •

A single pane of glass is a very efficient heat transmitter. Window and door areas approach 20 percent of the sidewall area of the average home. It is important that heat loss from this source be minimized.

Properly fitted storm windows are a must for the well-insulated home. They come in styles suitable to the design of any home.

Older windows have storms fitted on the outside. This can cause some reduction in heat loss. In fact an old window with a triple-deck outside storm window reduces heat loss by 67 percent. Adding another layer of insulating glass or plastic, such as in Figure 11-47, makes it possible to reduce heat loss by up to 93 percent. The inside storm window is easily added to any window. Figures 11-48 through 11-54 show how the inside storm window is made and placed into position. Also check Table 11-2.

WEATHERSTRIPPING A WINDOW • There are three basic types of weatherstripping for double-hung windows. The foam-rubber type is probably

306   Carpentry Fundamentals

Figure 11-48. Measuring for the inside storm window. *(Plaskolite)*

Figure 11-49. Cutting the inside window. *(Plaskolite)*

Figure 11-50. Cutting trim for the window. *(Plaskolite)*

Figure 11-51. Placing the trim around the plastic. *(Plaskolite)*

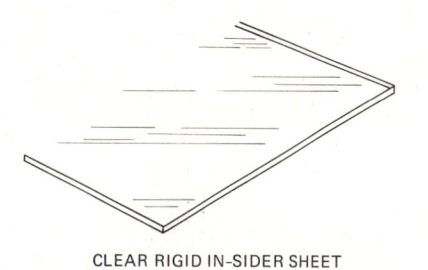

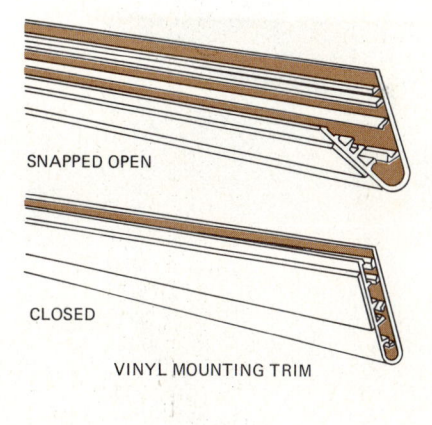

CLEAR RIGID IN-SIDER SHEET

SNAPPED OPEN

CLOSED

VINYL MOUNTING TRIM

WINDOWSILL TRIM

JOINER STRIP

Figure 11-52. Pieces that go into the making of a storm window for the inside. *(Plaskolite)*

Insulating for Thermal Efficiency **307**

Figure 11-53. Installing the window. *(Plaskolite)*

Figure 11-54. Placing the plastic inside the window. *(Plaskolite)*

TABLE 11-2 STORM WINDOWS

Window (Glass Area Only)	Approximate R of Unit*
Single glazing	0.88
Double glazing with ¼-inch air space	1.64
Double glazing with ½-inch air space	1.73
Single glazing plus storm window	1.89
Double glazing plus storm window	2.67

Courtesy New York State Electric and Gas Corp.
* R value for vertical air space is for air space from ¾ to 4 inches thick. R values for glass are actual for type of window listed. R (resistance) indicates the amount of heat a material will prevent from passing through it in a given time. The higher the R value, the more heat the material will hold back, and hence, the better the insulation of that material.

the best. It is the easiest to work with. See Figure 11-55. Each of the weatherstripping materials comes in coils. It often comes in complete cut-to-size kits. Metal-frame windows (such as casemate, awning, or jalousie styles) use invisible vinyl tape, joint-sealing self-adhesive foam rubber, or casemater aluminum strip.

INSTALLING WEATHERSTRIPPING ON A WINDOW • Metal strip insulation should be nailed at three positions. The sash channels inside the frame (making sure you leave the pulleys near the top uncovered) are one location. Underneath the upper sash on the inside is another location. Check Figure 11-56. It shows how the spring metal type of window is insulated with weatherstripping.

Vinyl and adhesive types of weatherstripping can be installed as shown in Figure 11-57. Vinyl strips should be nailed in place at three locations (see Figure 11-58):

1. On the outer part of the bottom rail of the upper sash.
2. On the same location on the lower sash.
3. On the outer surface of the parting strips.

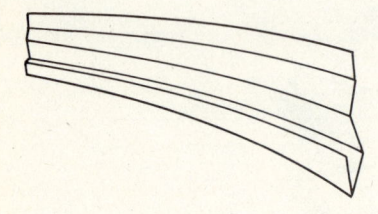

THIN SPRING-METAL INSULATION (A)

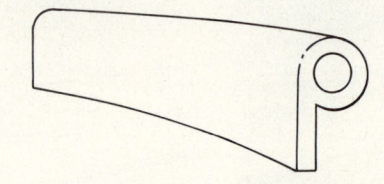

VINYL INSULATION (B)

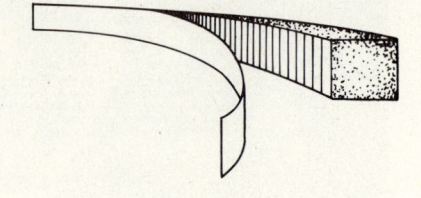

FOAM-RUBBER INSULATION WITH ADHESIVE BACKING (C)

Figure 11-55. Basic types of strip or roll weather stripping.

308 Carpentry Fundamentals

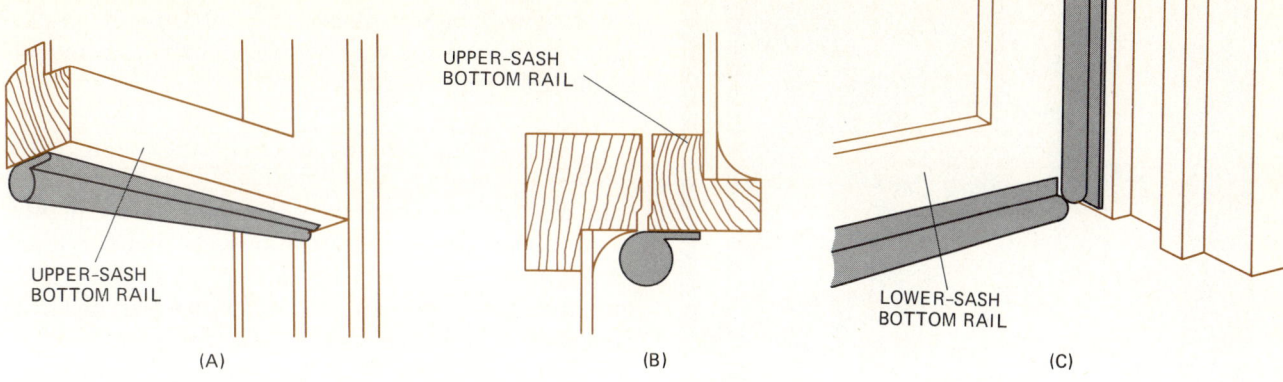

Figure 11-56. Vinyl tape and self-adhesive foam-rubber strip installation.

Figure 11-57. Applying clear tape around windows. *(Manco Tape)*

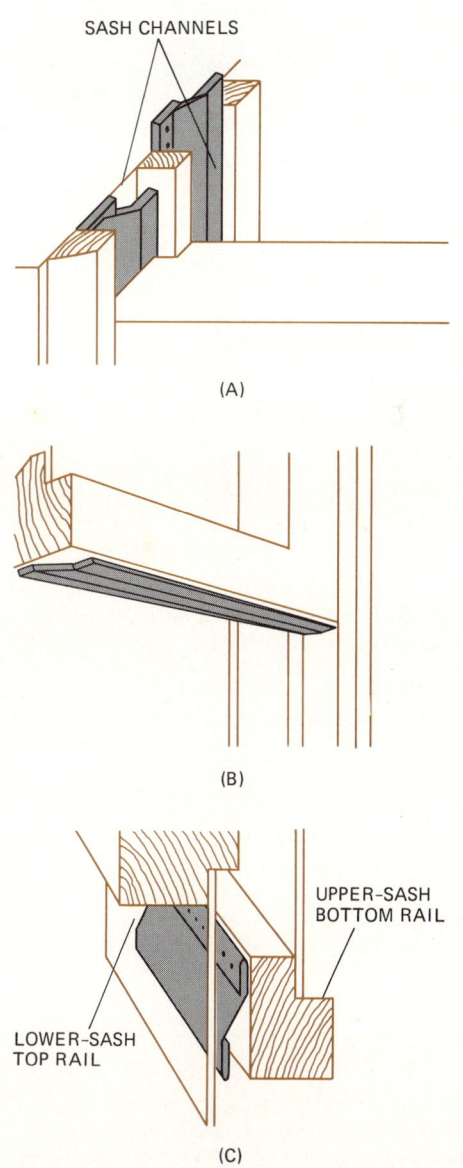

Figure 11-58. Metal strip insulation nailed at three positions: (A) sash channels inside the frame, (B) underneath the upper sash, and (C) on the lower sash bottom rail.

Insulating for Thermal Efficiency **309**

Figure 11-59. Metal storm door with full glass panel.

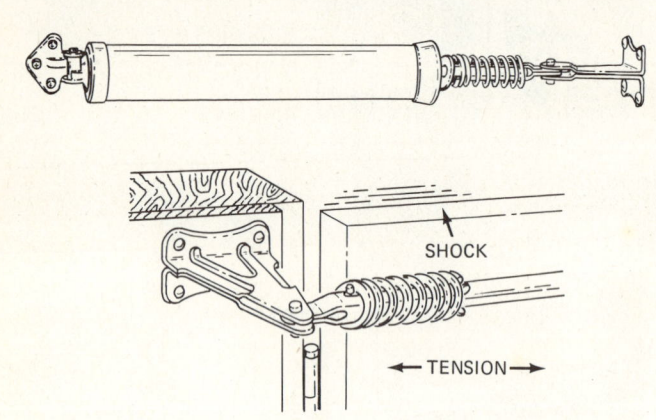

Figure 11-60. Storm door closer (top) and spring for keeping the door from being blown off its hinges. *(Stanley Tools)*

TABLE 11-3 STORM DOORS

Door Type	Approximate R of Unit*
Solid wood, 1-inch	1.56
Solid wood, 2-inch	2.33
Solid wood, 1-inch plus metal/glass storm door	2.56
Solid wood, 2-inch plus metal/glass storm door	3.44
Solid wood, in-inch plus wood/glass (50%) storm door	3.33
Solid wood, 2-inch plus wood/glass (50%) storm door	4.17
Doors with rigid insulation core	up to 7

Courtesy of New York State Electric and Gas Corp.

* R (resistance) indicates the amount of heat a material will prevent from passing through it in a given time. The higher the R value, the more heat the material will hold back, hence the better the insulation. To find the R value of a building material, multiply above R value by actual thickness of the material.

There are other methods which can help. Invisible polyethylene tape replaces the old type of caulk. This fits around the windows to seal them from the inside. See Figure 11-57.

• STORM DOORS •

In most of the country it is necessary to install storm doors. These fit on the outside and serve as screen doors during the summer. The glass and screen can be taken out and exchanged as needed. See Figure 11-59. Of course, the doors aren't too effective unless they have door closers to make sure they close. See Figure 11-60. Storm doors usually come in a prehung package. They are easily installed by simply placing screws through the holes and checking for plumb. They are designed for the do-it-yourselfer. In most cases homes do not come with storm doors. The first owner has to install them. This is also true of the mailbox. The first owner has to purchase and attach the mailbox.

Take a look at Table 11-3. It will show you just how important the storm door really is. It can make quite a difference in heat loss.

• SEALANTS •

There are a number of materials on the market for sealing air leaks in a house. Leaks appear around doors, windows, and chimneys. There are a number of caulking compounds made to fill these gaps. Some caulks will last longer than others. However, one of the best appears to be a paintable silicone sealant made by Dow-Corning. See Figure 11-61. These sealants are usually packaged in 11-ounce tubes suitable for use in a caulking gun. Some, of course, are in tubes for use inside the house.

Acrylic Latex Sealant This sealant cures to a rubbery seal. Used inside and outside, it can seal any joint up to ½×½ inch in size. It is also used between common construction materials where movement may be expected. It can be cleaned from tools or hands with water. Do not use under standing water or when rain is expected. It has an expected life of 8 to 10 years. It is paintable in 30 to 60 minutes after application.

Vinyl Latex Caulk This is a middle-priced sealant. It has the same advantages and restrictions as the acrylic latex sealant. Life expectancy is 3 to 5 years. See Figure 11-62.

Butyloid Rubber Caulk Especially useful in narrow openings up to ⅜ inch, this is used where movement is expected. It can be used in glass-to-metal joints. It is also used between overlapping metal. It can be used to seal gutter leaks or under

310 Carpentry Fundamentals

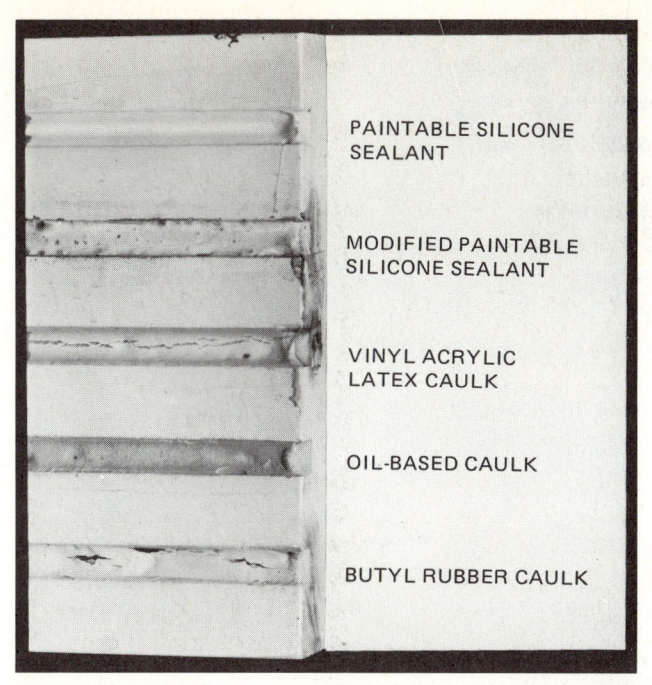

Figure 11-61. Sealant samples ¼" wide by ¼" deep after 600 hours in an accelerated weathering machine. This is the equivalent of more than one year weathering outdoors in Florida. Only the paintable silicone sealant seems to have held up. *(Dow-Corning)*

metal or wooden thresholds. This type of caulk is useful under shower door frames. Life expectancy of the caulk is 7 to 10 years.

Roof Cement Roof cement may be used on wet or dry surfaces. Cold weather application is possible. It stops leaks around vent pipes, spouts, valleys, skylights, gutters, and chimneys. This cement can be used to tack down shingles. It is also useful for sealing cracks in chimneys or foundations. Roof cement comes in black. It is not paintable.

Concrete Patch This is a ready-made concrete and mortar patching compound. It contains portland cement. This patching compound is quick and easy. It is used to repair cracks in concrete sidewalks, steps, sills, or culverts. It is also recommended for cracked foundations, tuck pointing, and stucco repair.

• WINTERIZING A HOME •

There are a number of ways to winterize a home. They all deal with insulation and stopping air leaks. Check the 33 ways listed in Figure 11-63. See how effective they are at saving energy.

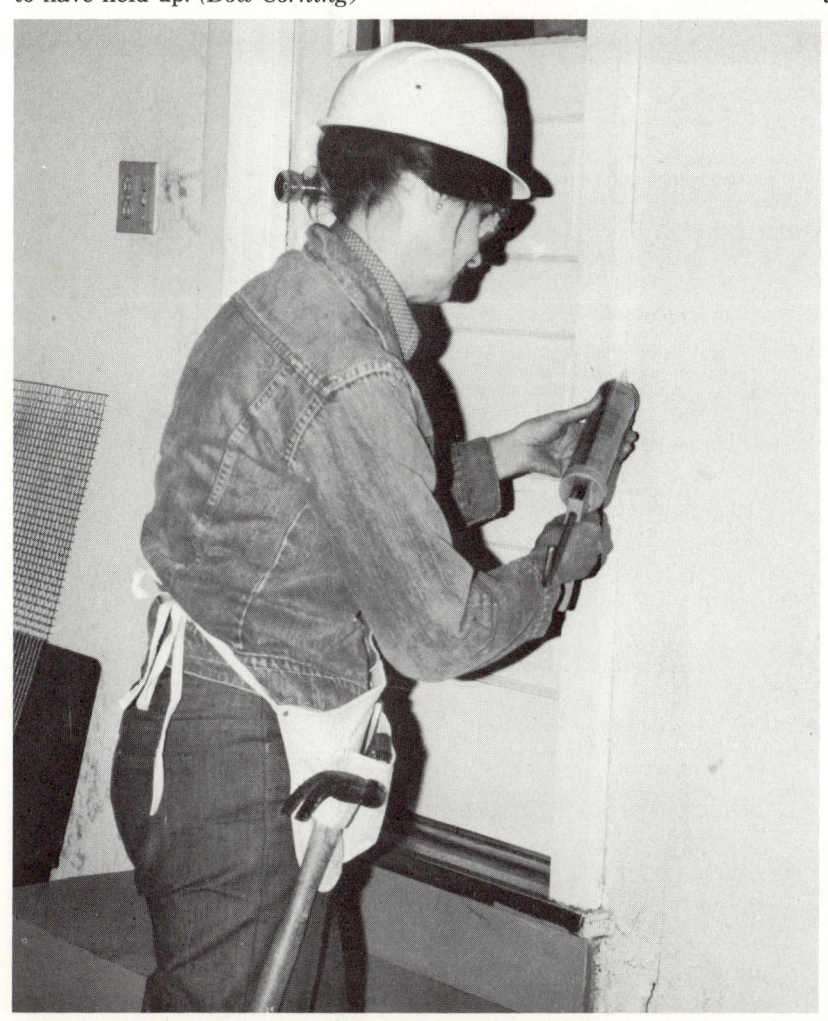

Figure 11-62. Caulking around doors.

Insulating for Thermal Efficiency **311**

33 Ways To Winterize Your Home

KEEPING THE COLD AIR OUT
1. Seal all outside doors, including basement doors, with weather-stripping material. In some cases, you can use old carpet strips.
2. Put masking tape around moving parts of windows. Caulk around window and door frames, including those in the basement. You can stop drafts under doors by placing rugs at the bottom.
3. Check pipes entering your home. You can keep the cold air out by packing rags around them.
4. Check light bulb fixtures for air leaks.
5. Put tape over unused keyholes.
6. Make sure unused flue or chimney covers fit tightly.
7. Keep fireplace dampers closed tightly when not in use.
8. Seal your foundation and sill plate with caulking material, insulation or rags.

KEEPING THE WARM AIR IN
9. Install storm windows and storm doors. If you don't have storm windows, you can substitute plastic sheeting but make sure it's tacked tightly all around the edges.
10. Install insulation between warm and cold areas. Begin by insulating the attic floor.
11. Wherever possible, carpet floors. If your attic floor can't be insulated, lay down a carpet.
12. Close off rooms you don't use, particularly those with the biggest windows and the largest outside walls.

MAINTAINING YOUR HEATING SYSTEM
13. Keep your heating system clean.
14. Make sure there is nothing blocking your registers, radiators or baseboard heaters.
15. Keep cold air returns clear.
16. Change or clean furnace filters monthly.
17. Have a qualified person adjust your heating system.

ADJUSTING YOUR LIVING HABITS
18. Keep the thermostat set at 68 degrees during the day. If this seems too cold for you, try wearing a sweater.
19. Set the temperature back at least 3 degrees at night.
20. If you're going away for a few days, set the thermostat at 60 degrees before you leave.
21. Try lowering the temperature in those rooms you don't spend much time in by adjusting registers, radiators or thermostats.
22. Keep humidifiers at the 30 percent mark or place pans of water on warm air registers or radiators. You'll feel more comfortable at relatively lower temperatures simply by maintaining the right humidity in your home.
23. Cover windows with drapes or curtains.
24. Open your drapes during the day to let the sunshine in and close them at night to keep the cold air out.
25. Try locating your furniture away from cold outside walls and windows.

SOME DO'S
26. Do fix leaky faucets, especially hot water taps.
27. Do use cold water for clothes washing.
28. Do turn off the lights, TV, radio or record player when not needed.

SOME DON'TS
29. Don't permanently fasten windows and doors shut—they may be needed for an emergency.
30. Don't use kitchen appliances to heat your home.
31. Don't use portable heaters as the main source of heat—be particularly cautious with oil or gas space heaters not vented or vented to your chimney.
32. Don't seal off attic ventilation.
33. Don't put insulation over recessed light fixtures.

Figure 11-63. Thirty-three ways to winterize your home.

• WORDS CARPENTERS USE •

insulation	vapor barrier	crisscross wire support	thermal ceilings	vinyl
energy	crawl space	support	grid system	caulk
ductwork	moisture seal	furring strips	false bottom	storm door
batts	faced insulation	insert stapling	suspended beams	sealant
Fiberglas	unfaced insulation	faced stapling	decorative beams	roof cement
cellulose fiber		moisture control	storm window	
R values	floorboards	ventilation	weatherstripping	

312 Carpentry Fundamentals

· STUDY QUESTIONS ·

11–1. Why weren't older homes insulated?

11–2. What does insulation do?

11–3. What is a vapor barrier?

11–4. What is meant by moisture control?

11–5. What does R value mean?

11–6. What is an insulation batt?

11–7. Why do crawl spaces have to be insulated?

11–8. What is the purpose of ventilation?

11–9. Which way should the vapor barrier face in an insulated home?

11–10. What are floorboards?

11–11. What does faced insulation mean?

11–12. What is the meaning of unfaced insulation?

11–13. Why can't you place faced insulation close to a chimney or source of heat?

11–14. Why should floors over unheated basements be insulated?

11–15. How do you cut Fiberglas insulation?

11–16. What is insert stapling?

11–17. How do you fill narrow stud spaces with insulation?

11–18. What is condensation?

11–19. How do you prevent condensation?

11–20. If you have a house with 1200 square feet of ceiling space and need to ventilate it, how much attic ventilation space would be needed if you used 8-mesh hardware cloth to cover the vent area?

Insulating for Thermal Efficiency

12
PREPARING INTERIOR WALLS AND CEILINGS

The inside walls and ceilings of a building are finished after the outside walls are finished and after doors and windows have been installed. This makes the building weathertight and protects the interior from damage. This is important because most interior materials are not weatherproof. Then the interior can be finished on bad days, when rain or snow makes work on the exterior impractical.

As a rule, ceilings are finished first. Then the walls are finished. Finally floors are completed. It does no damage to drop tools on unfinished floors. Workers will drop tools and materials as they work. Doing floors last prevents damage.

Both walls and ceilings can be finished the same way. This is because the same materials can be used for both walls and ceilings. Several steps can be involved. It is common for carpenters to apply vapor barriers. Other items to be installed include insulation, plumbing, and electrical wiring. The carpenter usually does not do the plumbing and wiring. However, the carpenter does install the insulation.

The most common interior material is gypsum board. Gypsum board is made from a chalk-like paste. This chalk is the gypsum. It is spread evenly between two layers of heavy paper. When the gypsum hardens, it forms a rigid board. The paper cover adds to its strength and durability. Finish may be applied directly to the paper. Wood panels or other materials may be applied over the gypsum board. Gypsum board is also called by other names. These names include *sheetrock* and *drywall*. Other interior wall panels are made of wood, plaster, and hardboard. Wooden boards are also used for interior walls.

Carpenters use special skills in covering walls and ceilings. These skills are related to working the various materials. Plaster is held up by strips called *lath*. Lath can be made of several things. Wood boards, special drywall, or metal mesh may be used. Carpenters often nail the wood edge guides and lath. However, the plasterer, not the carpenter, puts up the plaster. Skills carpenters need to prepare internal walls and ceilings include:
- Measuring and cutting wall materials
- Installing insulation
- Installing gypsum board
- Putting up paneling
- Preparing walls for plaster
- Putting up board panels
- Installing ceiling materials

• SEQUENCE •

As a rule, ceilings are done before either walls or floors. The carpenter usually does not select the interior materials. However, the carpenter must plan the way in which the work is done. To do this, the carpenter must know the various processes. The carpenter must also use the right process. The sequence for the carpenter to use is

Make sure the material and the fasteners are available.
Plan the correct work sequence.
Put insulation in the walls.
Put in the wall material.
Put in trim and molding.

• PUTTING INSULATION IN WALLS •

Today nearly all exterior walls are insulated. Insulation is also commonly used in interior walls. Interior wall insulation reduces sound transmission between rooms or apartments. Insulation must be put in before the walls are covered.

Most insulation today is made of loose woven fibers. The material comes in standard-width rolls or strips. These are made to fit between the wall studs. There are two commonly used standard widths. One is for walls with studs that are 16 inches O.C. The other is for walls with studs that are 24 inches O.C. The insulation is usually made from glass or mineral fibers. The fibers are woven into a thick mat. The mat is then glued to a heavy paper backing. The backing is a little wider than the mat. This way, the edges can be used to nail the insulation to the studs. The insulation may come in long rolls, or it may come in precut lengths. The precut lengths are called batts.

The paper side of the material is placed toward the living space. See Figure 12-1. The paper backing

Figure 12-1. The paper back on the insulation should face the living space. It also acts as a vapor barrier.

Preparing Interior Walls and Ceilings 315

may also be coated with metal foil. The foil helps reduce the energy loss. See Figure 12-2.

To put up insulation, the material must first be cut. Strips are cut to the proper length for the walls. The strips are called batts. The batts are then placed between the studs. The top of the batts should firmly touch the top plate in the walls. The batts should reach fully from top to the bottom. They should reach without being stretched. Either nails or staples may be used as fasteners. The fasteners may be driven on the inside of the studs. The edges of the paper are sometimes lapped on the edge of the studs. Fasteners are started at the top of the material. The nails should be placed about 12 inches apart. Fasteners are alternated from side to side. See Figure 12-3.

• INSTALLING A MOISTURE BARRIER •

After the insulation has been installed, the carpenter should check the plans carefully. Vapor barriers are often put up on the inside. Plastic film from 2 to 4 mils thick may be used. It is fastened across the interior studs as in Figure 12-4. Moisture barriers are essential in cold climates. They keep

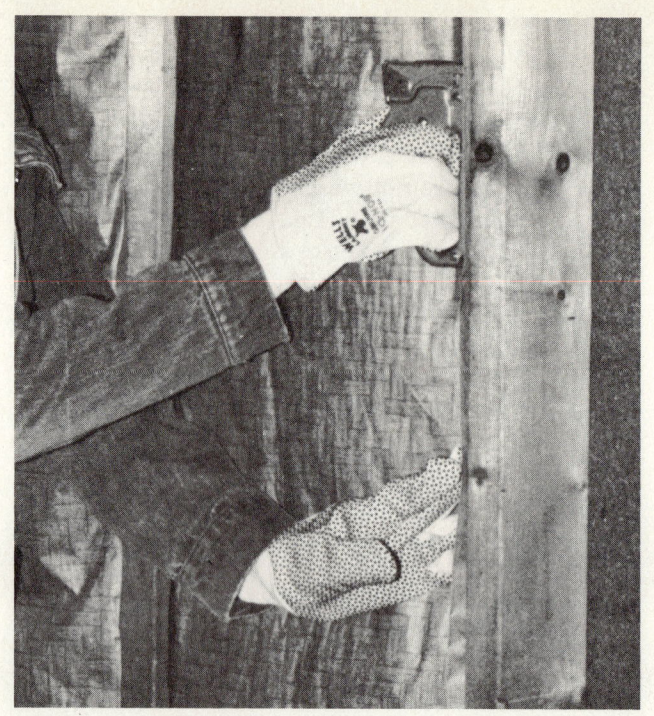

Figure 12-3. Insulation is stapled between the studs.

Figure 12-2. Installing foil-backed insulation between the studs. Note that the foil is toward the living space. It is also the vapor barrier.

316   Carpentry Fundamentals

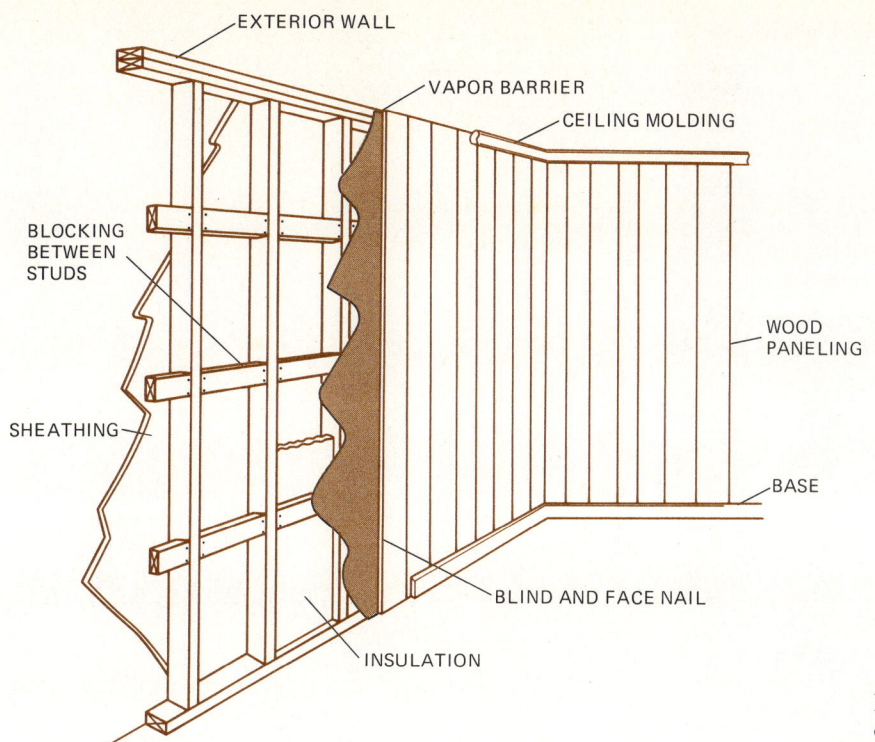

Figure 12-4. Plastic film may be applied over the inside of the studs and insulation for a vapor barrier. *(Courtesy Forest Products Laboratory)*

cold air and moisture from entering the building. Some sheathing can act as a vapor barrier. Also, the paper or foil back on the insulation acts as a vapor barrier. However, vapor barriers are needed on all outside walls and may also be used on interior walls. Vapor barriers may be made of plastic film, insulation paper, sheathing, or builder's felt.

• PUTTING UP GYPSUM BOARD •

Gypsum board is also called sheetrock or drywall. It is probably the most common interior wall material today. There are several advantages of using drywall. It is quick and easy to apply. This reduces labor costs. It is not an expensive material. And, there is no need to wait for it to dry.

Drywall comes in several sizes and thicknesses. Widths of 16, 24, and 48 inches are common. Lengths of 48 and 96 inches are used. One of the most often-used sizes is the standard 4×8-foot sheet. This large size means there are fewer seams to finish.

The edges of the sheets are finished or shaped in various ways. For most houses, the tapered edge is used. See Figure 12-5. The edge is tapered to make a sunken bed for the seam. This way, the edges may be covered and hidden. The edges are first coated with plaster and tape. Later coats are added to make a flat, smooth surface. See Figure 12-6. This flat surface can then be covered, papered, or painted without seams showing through.

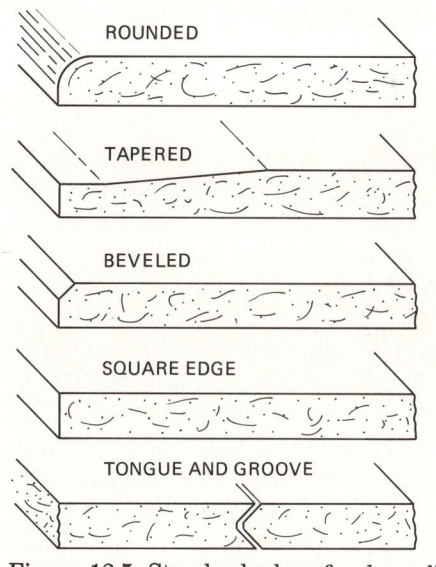

Figure 12.5. Standard edges for drywall. *(Gypsum Assn.)*

Drywall is held up by nails, screws, or adhesives. A special hole is made for nails and screws. This hole is made by the hammer on the last stroke. It is called a *dimple*. The hole, or dimple, holds the plaster. The plaster helps cover the nail and screw

Preparing Interior Walls and Ceilings **317**

Figure 12-6. After the sheetrock is nailed in place, the seams and nail dimples are covered.

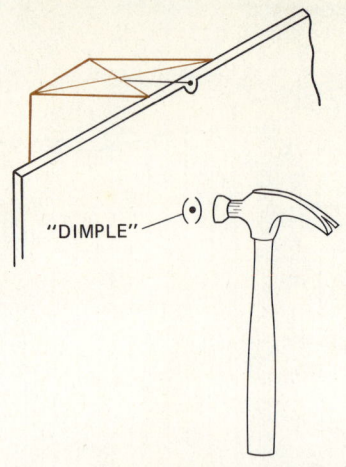

Figure 12-7. A dimple, or depression, is made around the nail head. It is later covered with plaster. *(Courtesy Forest Products Laboratory)*

Figure 12-8. Drywall sheets are often used for plaster lath.

to give a smooth surface. Figure 12-7 shows the dimple. On ceilings, nails or screws are also used with adhesives. However, not as many fasteners are used with an adhesive. Adhesives are not used alone on ceilings.

It is quicker to use adhesives rather than nails. However, some building codes prohibit the use of adhesives without nails. But fewer fasteners are used, and so the dual adhesive-nail combination is still fast. Panels can be put up very quickly with the adhesives. Then the nails may be added. No special holding is needed for nailing.

The thickness of drywall sheets can vary. Drywall is very fire-resistant. Layers can be used to increase the resistance to fire. Several layers can increase the strength. They will also reduce the transmission of noise between rooms. However, most construction is done with one thickness (single ply). There are three common thicknesses of drywall sheets. These are ⅜, ½, and ⅝ inch. In some locations, ceilings are made of ⅜-inch thickness. This reduces the weight being held. However, local building codes often detail what thickness can be used.

Drywall sheets can be used as a base for other types of walls. Drywall sheets are used for plaster lath. The sheets may be solid, or they may have holes (perforations) to help hold the plaster. See Figure 12-8. Drywall sheets can also be used for backing thin finished panels. Drywall itself is also available with several types of surface finish.

Drywall sheets are applied in two ways. These are called horizontal (or parallel) and perpendicular applications. However, this refers to the direction of the long side with respect to the studs or joists. It does not refer to the direction with respect to the floor. For example, in perpendicular applications the long side is perpendicular, or at right angles, to the stud or joist. Horizontal application means that the long side runs in the same direction as the stud or joist. See Figure 12-9.

318   Carpentry Fundamentals

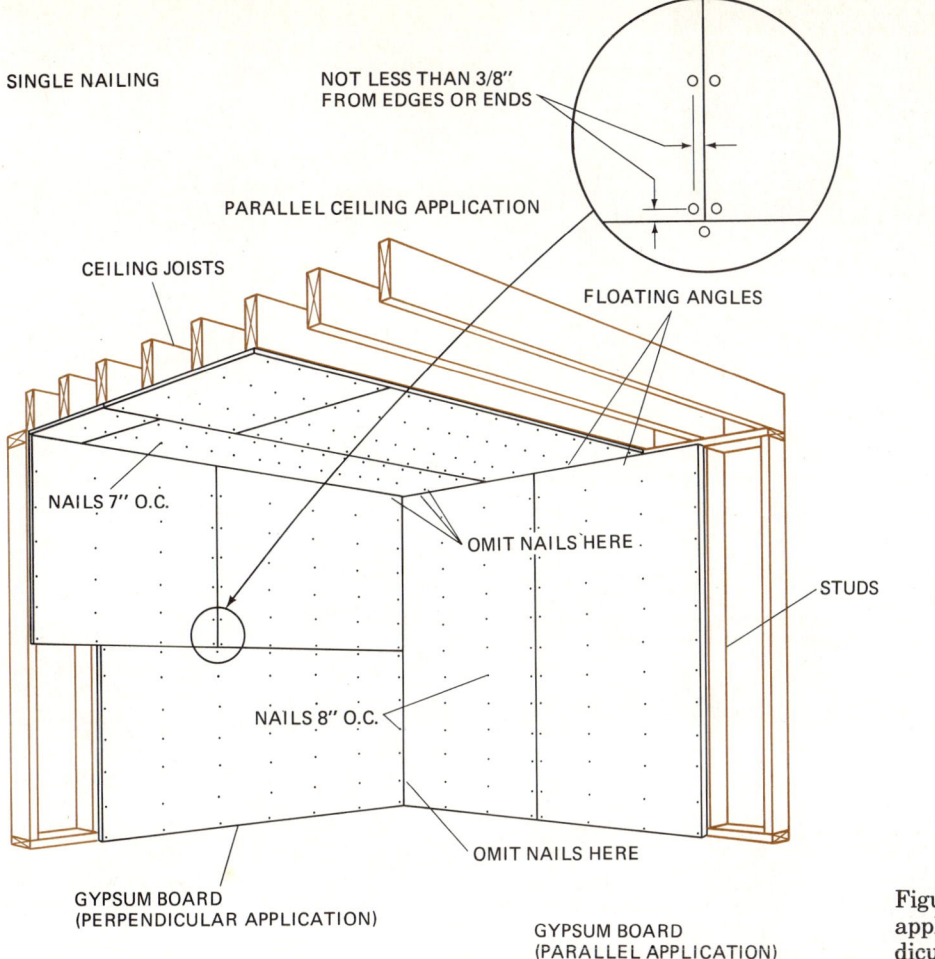

Figure 12-9. Drywall sheets can be applied either parallel or perpendicular. *(Gypsum Assn.)*

Putting Up the Ceiling

Drywall is usually applied to ceilings first. Then the walls are covered. Drywall should be put on a ceiling in a horizontal application. Most manufacturers recommend that the first piece be placed in the middle of the ceiling. Then additional sheets are nailed up. A circular pattern from the center toward the walls is used. This procedure makes sure that the joists are properly spaced. The drywall acts as a brace to hold the joists in place. Joist spacing is important. Edges and ends of the drywall sheets can be joined only on well-spaced joists. Nonsupported joints will move. Movement can ruin the taped seams. This will give a bad appearance.

MEASURE AND LOCATE FIRST PIECE • Before the first sheet is applied, two base walls are selected. One wall should be parallel to the joists. See Figure 12-10. The distance to the approximate center of the room is measured. Intervals of 4 feet are marked from this base wall. Next, the distance from the second wall to the center is found. Intervals of 4 feet are marked off. Intervals of 4 feet are used because the standard sheet size is 4×8 feet. The first sheet is located so that whole sheets can be applied from the center to that one wall. Cutting is done only on one side of the room. This saves time in trimming and piecing.

Before the first panel is applied, measure the distances between the joists at the center. The distance should be the proper 16 inches O.C. or 24 inches O.C. If the joists are spaced correctly, the first panel may be applied. The center nails should be driven first. Then joists should be spaced properly by slight pressure. Edges of the panels should rest on the centers of the joists. This provides a nail base for both panels.

STRONGBACKS • Strongbacks are braces used across ceiling joints. They are made for two reasons. The first reason is to help space the joists properly. To space the joists, a flat board is nailed across the tops of them. The second reason is to help even up the bottom edges of the ceiling joists. One of the disadvantages of drywall is that studs and joists

Preparing Interior Walls and Ceilings

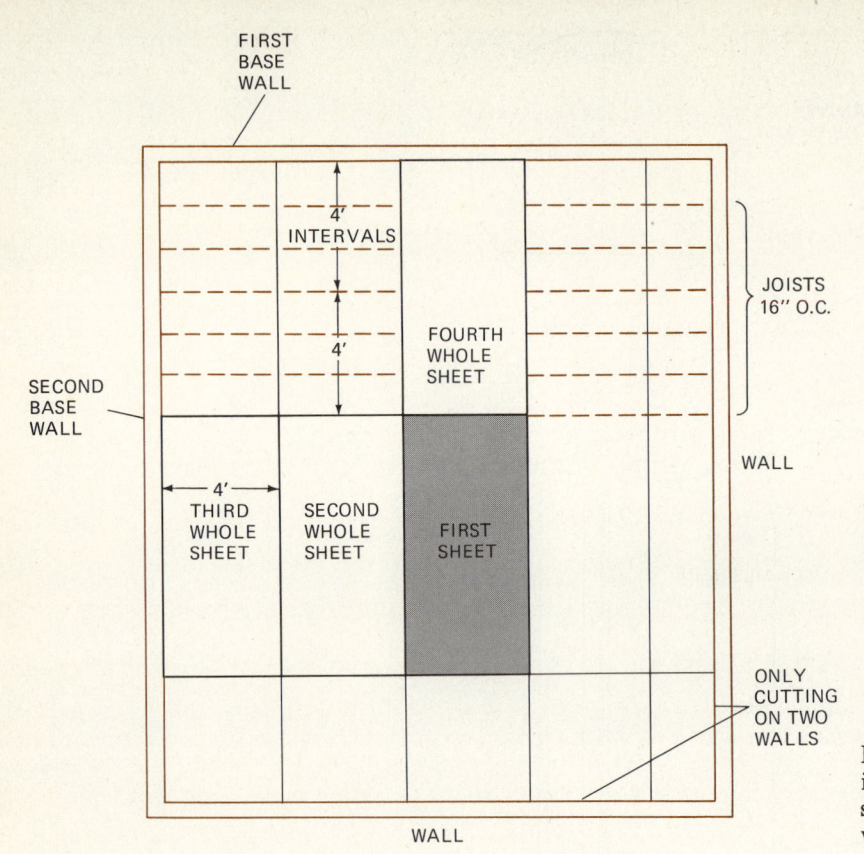

Figure 12-10. For ceilings, the first sheet is near the center. Measure so that whole sheets can be used toward the two base walls.

must be very even. Drywall cracks easily when it is nailed over uneven joists or studs.

To make a strongback, two boards are used. The bottom board is laid flat across the tops of the joists. A 2×4 should be used. The end of the board is nailed to the joist nearest the wall. Two 16d common nails should be used. The proper distance (16- or 24-inch marks) is measured. Pressure is applied to move the joist to the proper spacing. Then the joist is nailed at the proper spacing. This is done to each joist across the distance involved. This braces the joists into the proper spacing on centers.

The strongback is completed with the second brace. See Figure 12-11. The second piece is made from a 2×4- or 2×6-inch board turned on edge. The edge provides a brace to even the joists to the same height. One end of the second brace is nailed to the end of the first brace. The carpenter should then stand on the flat piece over each joist. The carpenter's weight is used to force the joists to an even line. When the joists are even, the piece is nailed in place.

Applying Ceiling Sheets

The ceiling sheets must be held up for nailing. Special braces are used in the nailing process. These temporary braces are called *cradles*. Figure 12-12 shows a cradle being used. Cradle braces are shaped like a large T. The leg of the cradle is about ⅛ to ¼ inch longer than the ceiling height. This way, the cradle can be wedged in place to hold the panel. Once the sheet has been nailed, the brace can be removed and used again.

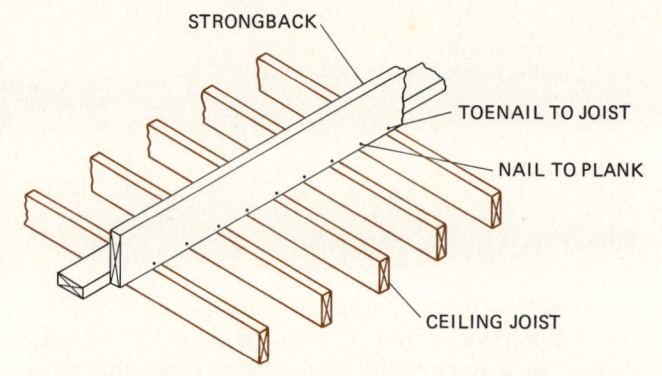

Figure 12-11. Strongbacks help even up and space joists. The flat piece is nailed first. *(Courtesy Forest Products Laboratory)*

Sometimes, several workers are involved. Then, some workers will hold materials in place and others will nail. Scaffolds can also be used to let the carpenters stand at a good working height. Stilts, as in Figure 12-13, are also used. However, workers on stilts should not try to lift materials from the floor. A worker on stilts cannot bend or lean over. That person can only brace, nail, or plaster seams and joints.

320 Carpentry Fundamentals

Figure 12-12. A T-shaped "cradle" can be used to hold up pieces for nailing. *(Gold Bond)*

FASTENERS • Sheets are fastened at the center of the panel first. Then fasteners are placed toward the edges. Fasteners are driven at intervals on ceilings. The spacing depends on the type of sheetrock used. Nails, screws, and staples are all used. Special nails are used for sheetrock. Some are smooth and some have ridges on the shank of the nail. Special screws are also used. A special attachment is used to drive them. It releases the screw when it is driven in place. Nails and screws for sheetrock are shown in Figure 12-14. As a rule, the nail or screw should penetrate into solid wood about 1 inch. Thus, for ⅜-inch drywall, a nail 1⅜ inch long would be recommended. When staples are used, the crown of the staple should be perpendicular to the joist or stud.

EDGE SPACING • The seams of the sheets should be staggered. The edges of one junction should not align with the edges of another junction. This is part of a process called *floating*. Figure 12-15 shows how edges are floated or staggered. This allows each sheet to reinforce the next sheet. It makes a stronger, more rigid wall or ceiling. Floating reduces expansion and contraction of the walls and ceilings. With this controlled, the finished seams are less likely to crack. By eliminating the nails, the sheets are allowed to move or "float."

Cutting Gypsum Board

Gypsum board is easily cut. Careful measurements should be taken before cutting. It is a good idea to take all measurements at least twice to be sure. Sometimes, measurements are checked by measuring from two different places.

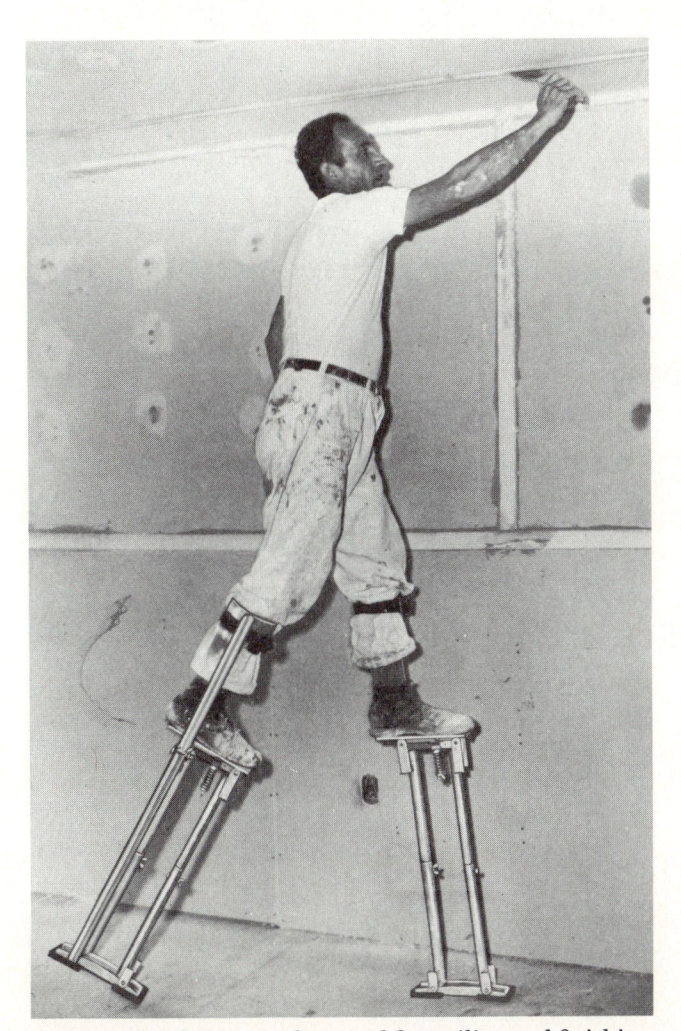

Figure 12-13. Stilts are often used for nailing and finishing ceilings. *(Goldblatt Tools)*

Preparing Interior Walls and Ceilings 321

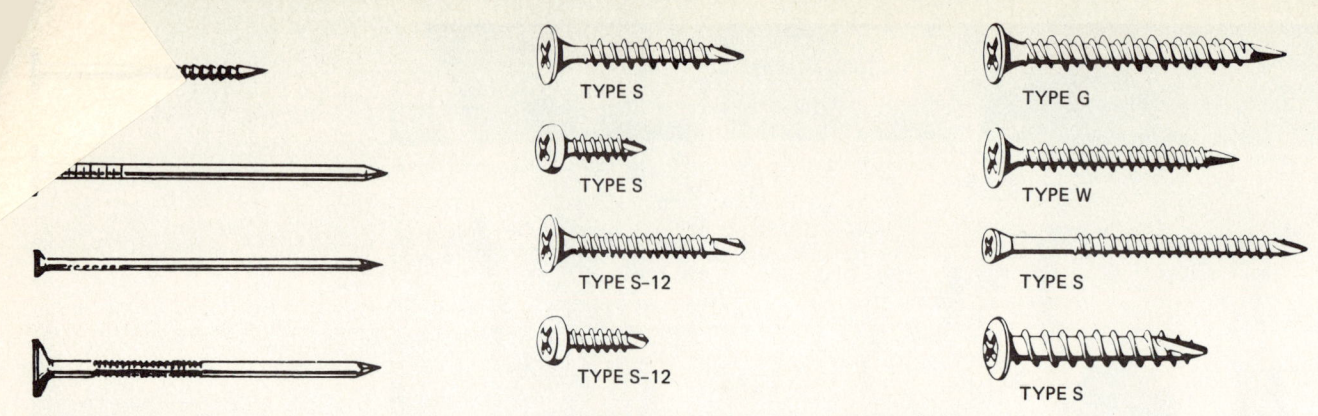

Figure 12-14. Typical nails and screws for sheetrock. Each is used for a certain application. *(Gypsum Assn.)*

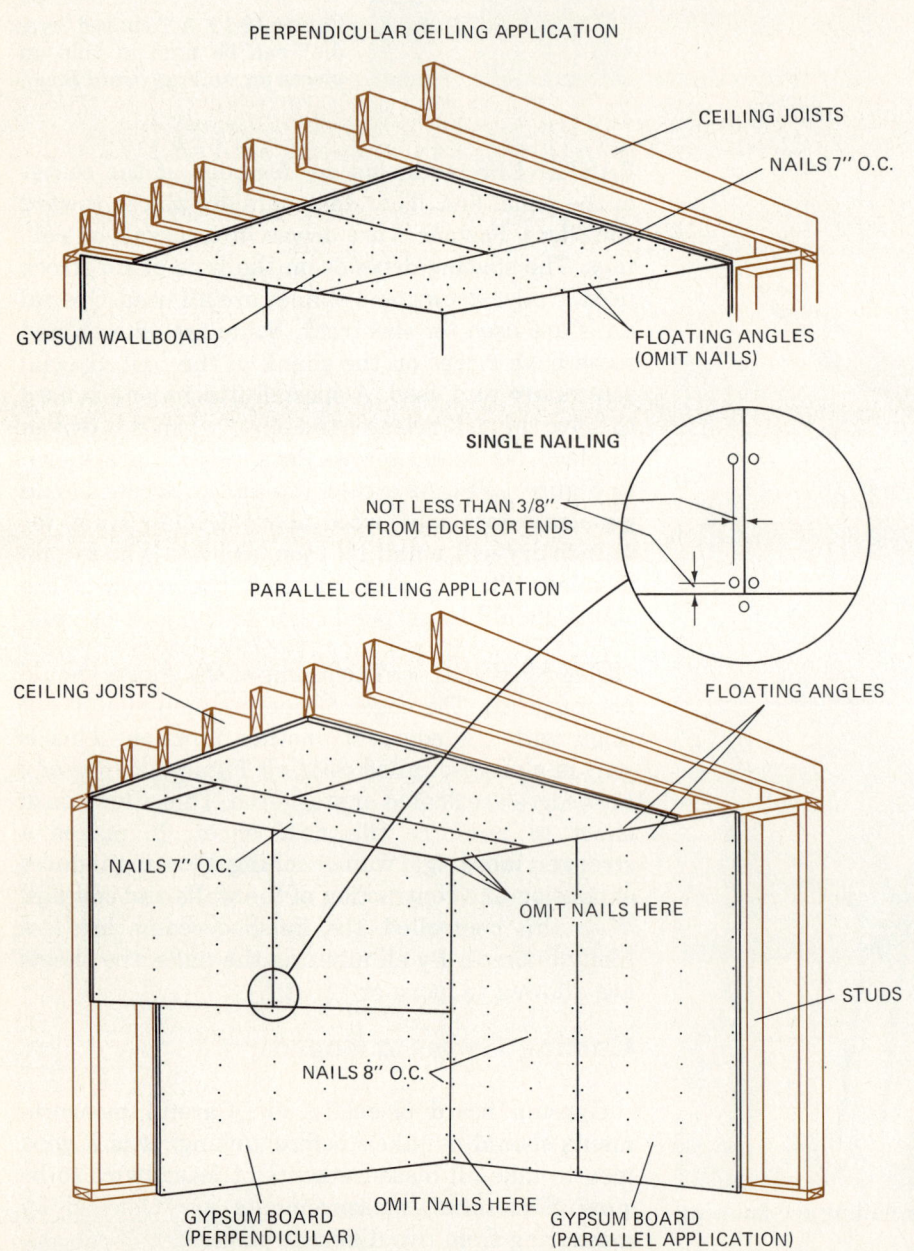

Figure 12-15. Floating angle construction helps eliminate nail popping and corner cracking. Fasteners at the intersection of walls or ceiling are omitted. Note that seams are not matched. *(Gypsum Assn.)*

322   Carpentry Fundamentals

Figure 12-16. Sheetrock is marked and scored with a knife for cutting. *(Gold Bond)*

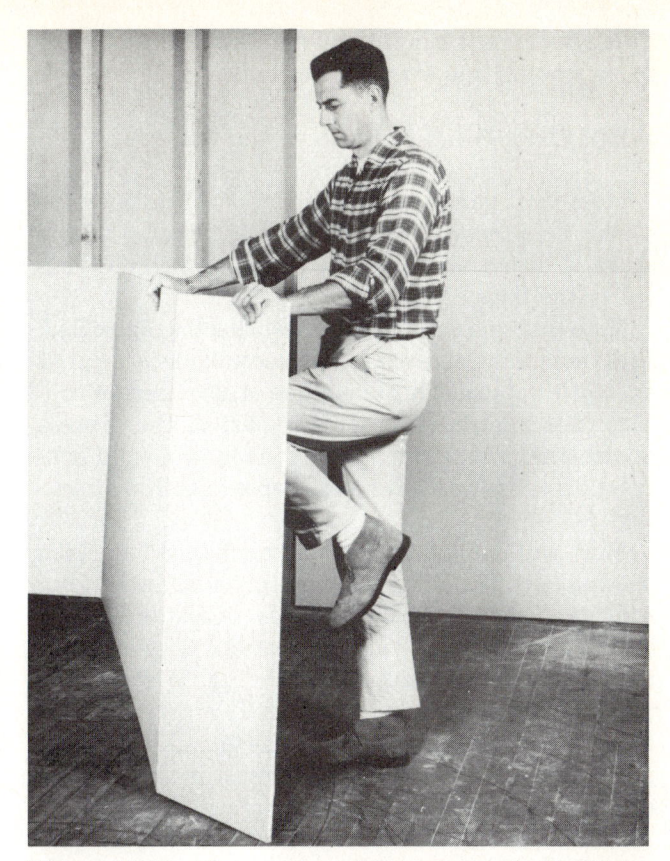

Figure 12-17. After scoring, sheetrock can be "cut" by breaking it over the knee or a piece of lumber. *(Gold Bond)*

CUTTING STRAIGHT LINES • Straight lines are cut after being carefully marked. A pencil line is made on the good, finished side of the sheet. See Figure 12-16. This line is then scored with a sharp knife. The cut should go through the paper and slightly into the gypsum core.

The board or sheet is then placed over a board or a rigid back. The carpenter can sometimes use a knee for a back. See Figure 12-17. The board is then snapped or broken on the scored line. The paper on the back side is then cut with a knife. This finishes the cut. See Figure 12-18.

CUTTING OPENINGS • Openings must be carefully measured and marked. Then, the lines are heavily scored. Next the piece may be strongly tapped with a hammer. If a sheetrock hammer is used, the hatchet end may be used. The cutting edge is placed into the score. The blade is then pushed through with even pressure. The back side is cut with a knife and the section falls free.

Another method is to use a special device to cut the holes. This device resembles a box with teeth on the edges. The teeth are placed in the outlet box in the wall. The approximate center is marked on the panel. The panel is put in place. The handle of the special tool is forced into the marked center.

Figure 12-18. After scoring and breaking the paper back may be cut. This separates the pieces.

Prongs are engaged into a slot in the teeth. The teeth quickly cut through the gypsum core. The tooth plate and core are simply pulled out. Thin saws, knives, and punches are also commonly used.

Preparing Interior Walls and Ceilings **323**

Care should be taken with any method. Edges should be as smooth and even as possible.

Applying Wall Sheets

Most wall panels are applied horizontally to the studs. This means the long edge is parallel to the studs. It also means that the long edge will be vertical to the floor.

Note that one corner of the first sheet is not nailed. It is held in place when the second piece at that corner is applied. This technique is also used on ceilings. See Figure 12-19. This is part of the process of floating. Floating allows for expansion and contraction of the walls. It helps keep seams from cracking.

Most ceilings are slightly more than 8 feet from the subfloor. This slight distance gives some working space and clearance. With this extra distance, standard 8-foot lengths will not catch. The drywall should be butted against the ceiling. It should be raised off the floor when applied. To start, one edge of the panel is laid on the floor. The top is laid against the wall studs. The sheet of drywall is then raised off the floor. Then it is nailed in place. A kick lever as in Figure 12-20 is used. Kick levers may be purchased or may be made from lumber. A carpenter can step on the pedal to raise the panel. By using a kick lever, the carpenter keeps both hands free for nailing and holding (Figure 12-21).

The first piece is usually put at a corner of the room. The work is done from corner to corner. The carpenter does not start in the middle of a wall. Nails, staples, or screws are driven at 8-inch intervals on the studs.

CORNER TREATMENTS • Outside edges and corners are reinforced with metal strips. Outside edges and corners are easily damaged and broken. The metal strips help prevent damage to these exposed edges. The metal strips are called *beads*. The most commonly used bead is the corner bead as shown in Figures 12-22 and 12-23. Corners are butted together. The bead is then used to cover the corner. The reinforcement bead protects the corner or edge. The bead helps to prevent ugly damage to the edge or corner.

As a rule, wood is not covered with plaster. Plaster and wood expand and contract at different rates. The difference in movement of these two materials

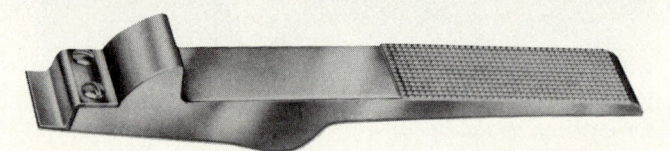

Figure 12-20. A kick lever. *(Courtesy Golblatt Tools)*

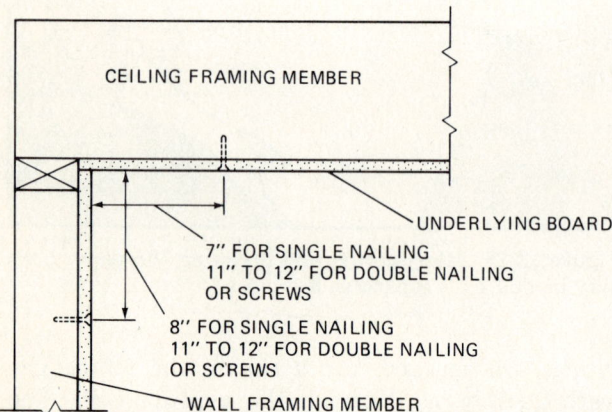

Figure 12-21. A kick lever being used to raise the panel in place. *(Goldblatt Tools)*

Figure 12-19. Techniques for floating corners. *(Gypsum Assn.)*

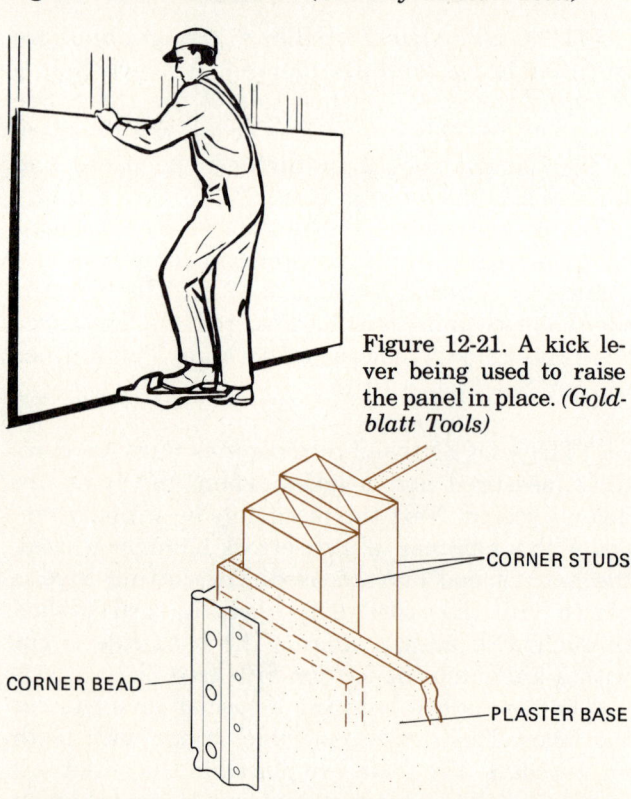

Figure 12-22. Corners are butted together. A metal bead is then applied to protect the corner. *(Courtesy Forest Products Laboratory)*

324 Carpentry Fundamentals

would cause cracking. The wood is covered with drywall. This reduces cracking.

Inside corners are normally not reinforced with beads. They are taped just as are other seams. The taping and beading process will be explained later.

Double-Ply Construction

Two layers of drywall are often used. The second layer increases the ability of the wall or ceiling to resist fire. It also helps reduce the amount of noise transmitted from one room to another.

Double-ply ceilings are glued and then nailed. However, the nails are driven from 16 to 24 inches apart. The length of the nail is also increased. A longer nail will penetrate through the second thickness of the drywall.

The joints of each layer should overlap. The joint of the top layer should occur over a solid sheet. Figure 12-24 shows this. Adhesives are applied, as in

CORNERBEAD (Numbers indicate width of flanges, i.e.—118 is 1 1/8-inches wide flange)
CB—100 X 100
CB—118 X 118
CB—114 X 114
CB—100 X 114
CB—PF (Paper flange, steel corner, combination bead)

"L" BEAD (Numbers indicate thickness of board to be used)
L-38
L-12
L-58
L-34

"LK" BEAD (For use with Kerfed jamb)
LK

"U" BEAD (Numbers indicate thickness of board to be used, i.e.—38 is 3/8 inches)
U-38
U-12
U-58
U-34

"LC" BEAD (Numbers indicate thickness of board to be used)
LC-38
LC-12
LC-58
LC-34

Figure 12-23. Metal trim and casing. *(Gypsum Assn.)*

FINISH LAYER 3/8" or 1/2" TAPERED EDGE GYPSUM BOARD
CEILING JOISTS
BASE LAYER 3/8" OR 1/2" BACKER BOARD OR GYPSUM BOARD
LAMINATING ADHESIVE (APPLY WITH NOTCHED TROWEL OR MECHANICAL SPREADER)
WOOD FRAMING MEMBERS
FASTENERS (SINGLE NAILS SHOWN)
JOINT TREATMENT (ALL JOINTS AND CORNERS)
BASEBOARD

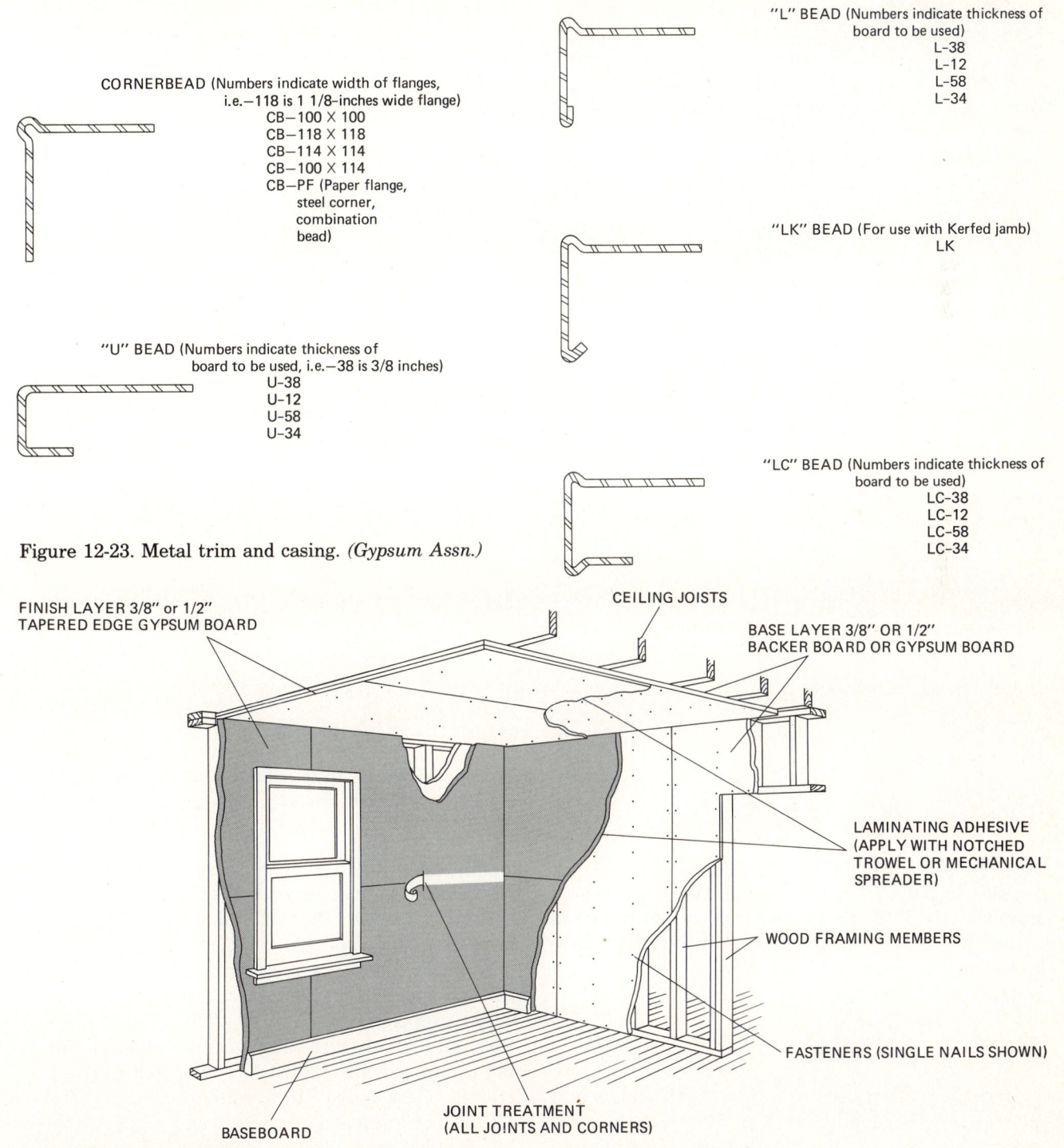

Figure 12-24. Double-ply application. Note that joints are not in the same place for each layer. *(Gypsum Assn.)*

Preparing Interior Walls and Ceilings   **325**

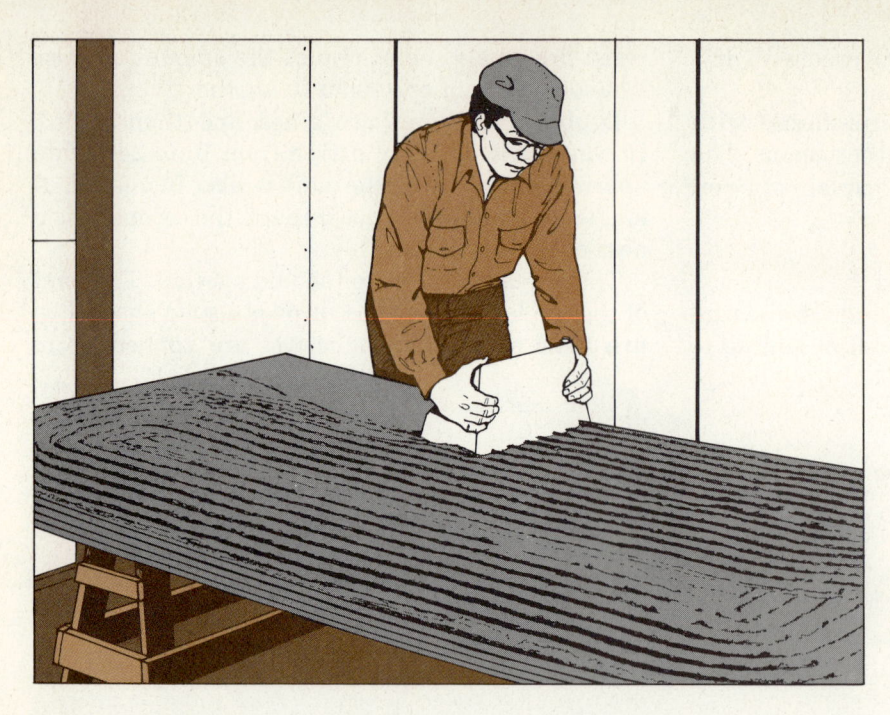

Figure 12-25. Notched spreaders are used to apply adhesive for ceiling sheet lamination.

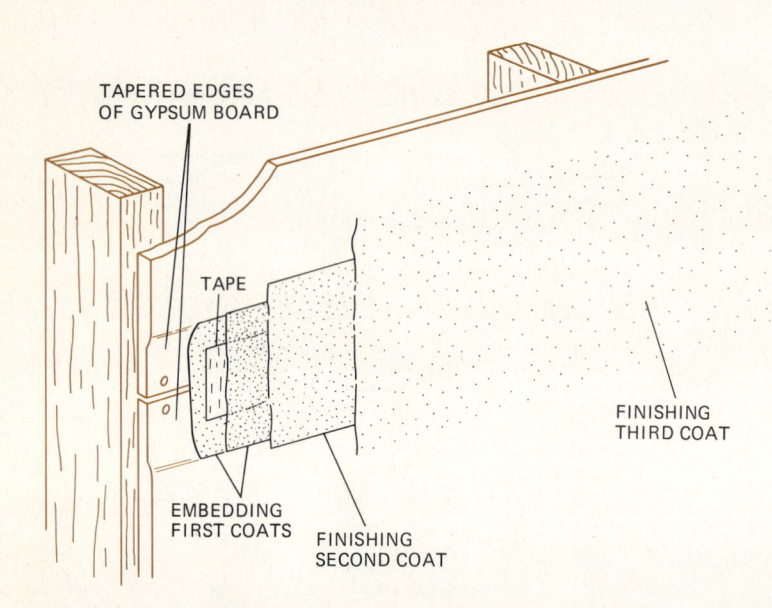

Figure 12-26. Seams are finished in layers. *(Gypsum Assn.)*

Figure 12-25. The sheets of drywall should be firmly pressed in place. It is best to nail the glued sheets immediately. If they must be left, a temporary brace should be used. A brace will hold the panels firmly until the glue dries. Nails are driven when the brace is removed.

Finishing Joints and Seams

Drywall is often used as a base. Both panels and plaster may be applied over it. However, in most cases, it is finished directly. It may be painted or papered. Before gypsum board is finished, the seams should be smoothly covered. Some building codes require seams to be covered even when the drywall is merely a base. The covered seams increase the fire-resistive properties.

Seams are covered with successive coats of plaster and special tape. The tape may be either paper or Fiberglas mesh (Figure 12-26). Figure 12-27A through F shows how the layers are applied. Covering seams are also called *finishing joints*. The process is often called "taping and bedding" (Figure 12-28). The seams on all corners and edges are taped and bedded with plaster. However, when paneling or plaster is applied, it is not always necessary.

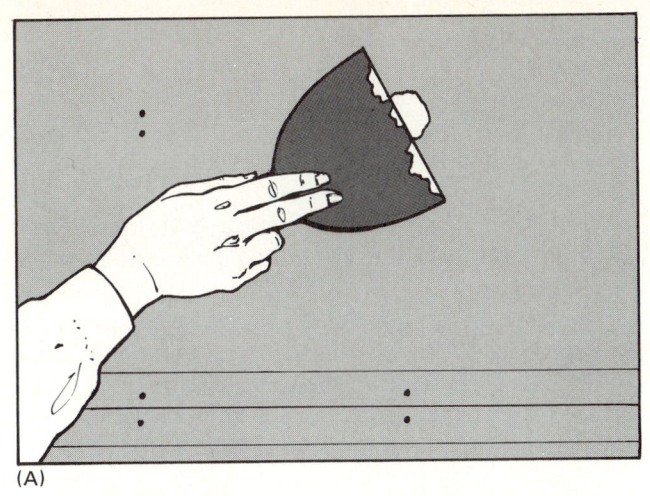

Figure 12-27. (A) Spot nail heads with a first coat of compound either as a separate operation before joint treatment or after applying tape. *(Gold Bond)*

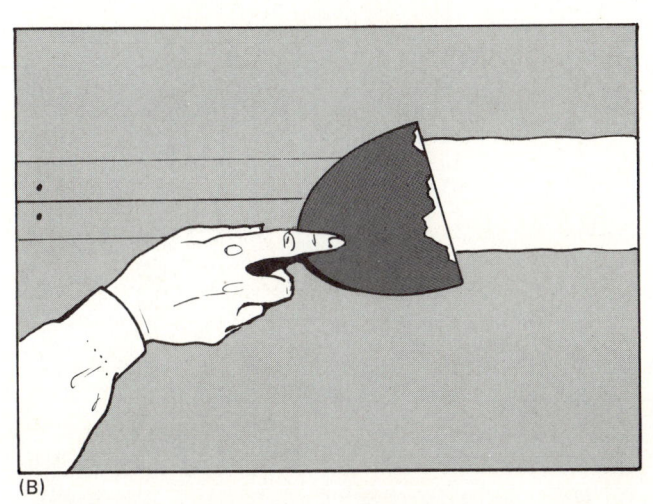

(B) The first coat of joint compound fills the channel formed by the tapered edges of the wallboard. *(Gold Bond)*

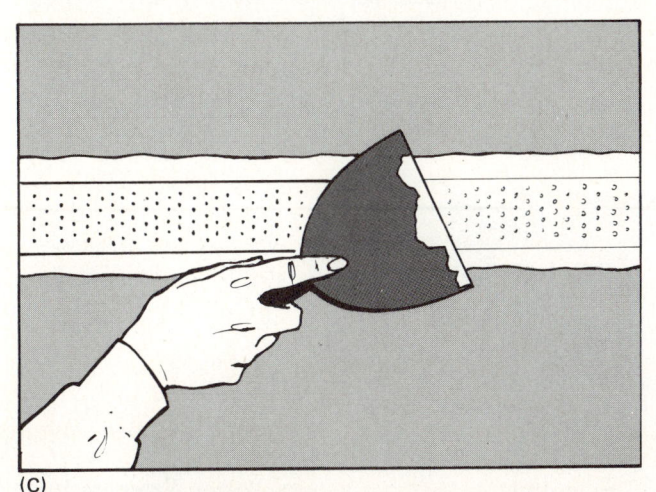

(C) Tape is embedded directly over the joint for the full length of the wall. Smooth joint compound around and over the tape to level the surface. *(Gold Bond)*

(D) The first finishing coat is applied after the first coat has dried. Apply it thinly and feather out 3 to 4 inches on each side of the joint. Apply a second coat to the nailheads if needed. *(Gold Bond)*

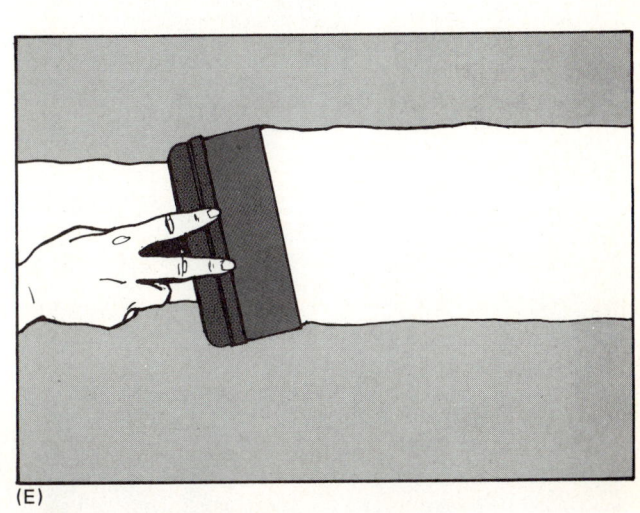

(E) The second finishing coat is applied when the last coat has dried. Spread it thinly, feathering out 6 to 7 inches on each side of the joint. Finish nail spotting may be done at this time. *(Gold Bond)*

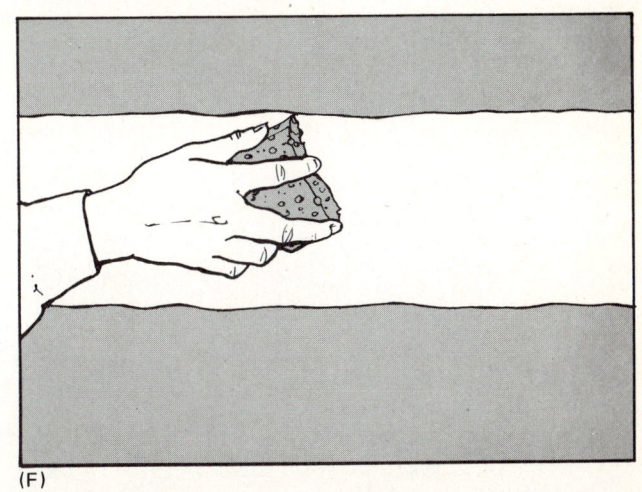

(F) After 24 hours, smooth the finished joints with a damp sponge. If light sanding is required, use a respirator to avoid inhaling the dust. *(Gold Bond)*

Preparing Interior Walls and Ceilings

Usually, the carpenter does not finish drywall seams. This is normally done by workers in the trowel trades. Plasterers or sheetrock workers do most of this.

• PREPARING A WALL FOR TUBS AND SHOWERS •

Walls around tubs and showers must be carefully prepared. This prevents harmful effects from water and water vapor. Walls are finished with tile, panels, and other coverings. Water-resistant gypsum board can be used as a wall base. Special water-resistive adhesives should also be used. Edges and openings around pipes and fixtures should be caulked. A waterproof nonhardening caulking compound should be used. The caulking should be flush with the gypsum. The wall finish should be applied at least 6 inches above tubs. It should be at least 6 feet above shower bases. See Figure 12-29. Figure 12-30 shows a tub support board. It is made from 2×4-inch lumber, and is nailed over the gypsum wallboard.

Another technique for preparing walls is shown in Figure 12-31. In this situation, gypsum wallboard is not used. Instead, a vapor barrier is applied directly over the studs. It can be made of water-resistant sheathing paper or plastic film. A metal spacer

Figure 12-28. Taping and bedding sheetrock seams. (Courtesy Fox & Jacobs)

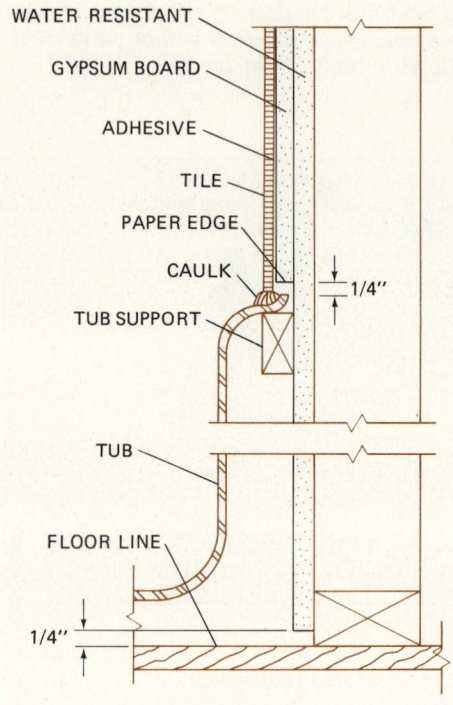

Figure 12-30. Tubs are supported on the walls. Note the two layers of gypsum board and the gap used. (Gypsum Assn.)

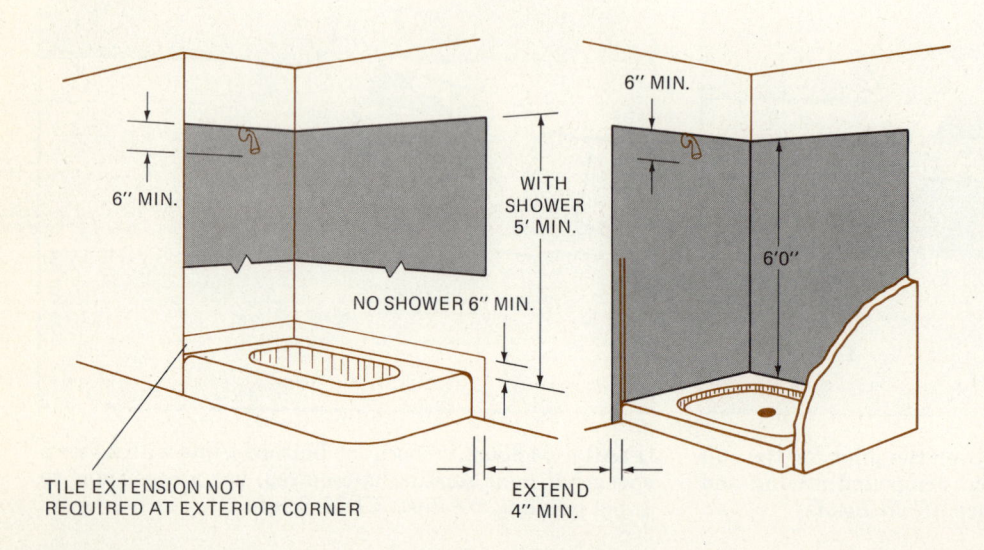

Figure 12-29. Water resistive wall finishes should be applied around bathtubs and showers. (Gypsum Assn.)

328   Carpentry Fundamentals

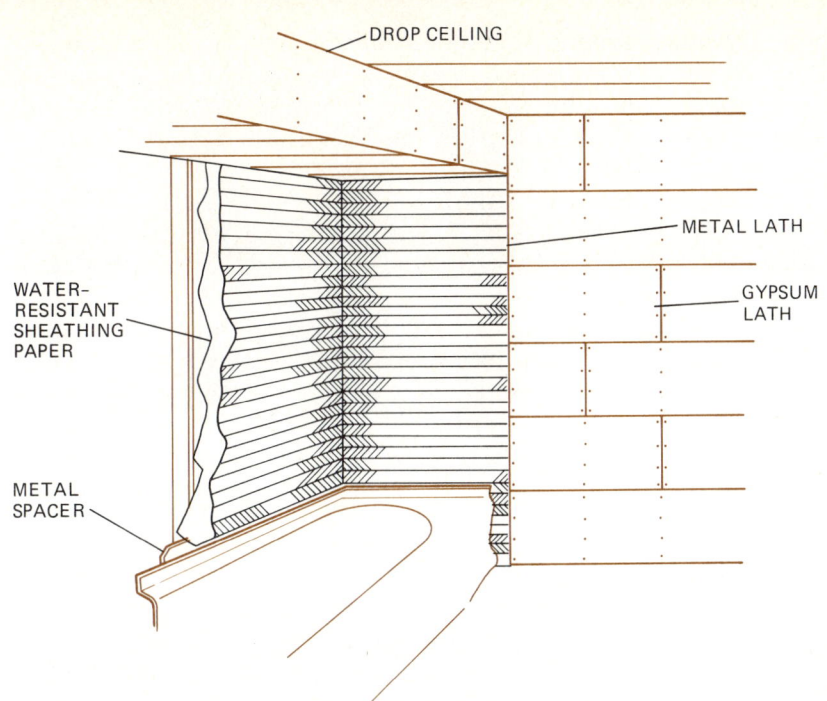

Figure 12-31. Vapor barrier and metal lath directly over studs. *(Courtesy Forest Products Laboratory)*

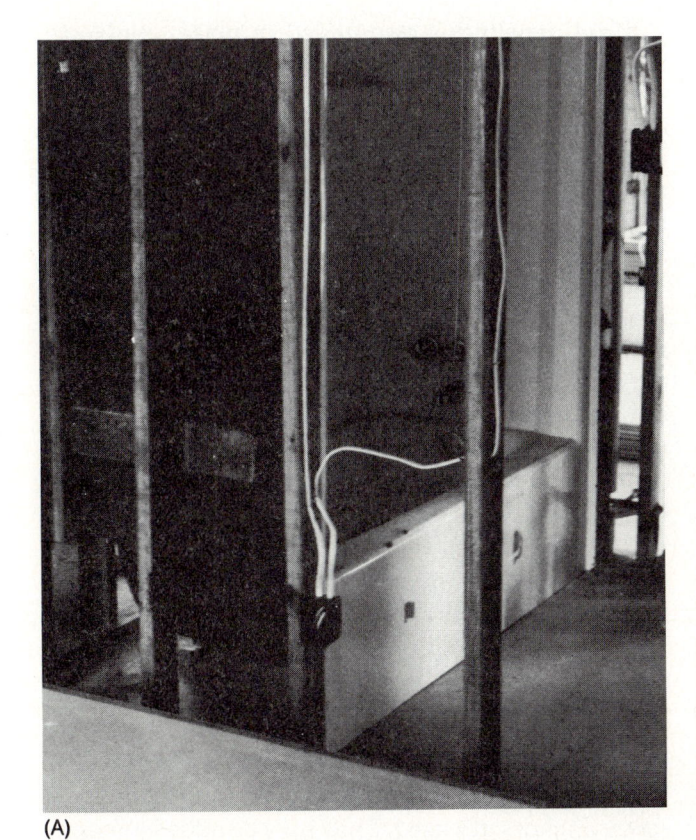

Figure 12-32. (A) Special tub and shower units are framed and braced by carpenters. (B) This shower stall is made of Fiberglas. No base wall is needed. *(Corl Industries)*

is used around the edge of the tub. See the figure. Next, metal lath is applied over the vapor barrier. The wall is then finished by applying plaster. A water-resistive plaster is used for the tub or shower area.

Special tub and shower enclosures are also used. These may be made of metal or Fiberglas. See Figure 12-32. These are framed and braced to manufacturer's recommendations. No base wall or vapor barrier is used.

Preparing Interior Walls and Ceilings **329**

• PANELING WALLS •

Walls are often finished with standard size panels. They are made of wood, fiberboard, hardboard, or gypsum board. Panels made from hardboard, fiberboard, or gypsum are usually prefinished. The surfaces of these panels can be made to resemble wood or tile. Paint, stain, or varnish is not needed. This gives a fast finish with little labor. They can also be covered with wallpaper or plastic laminates. Paint and other materials lend an extremely wide range of appearances. Wood panels are also commonly used. They provide a wide range of wood grains and finishes. Panels may be prefinished. However, unfinished panels are also used. Stains may be applied to customize the appearance of the interior.

All panels come in standard 4×8-foot sizes. They are put up with nails, screws, or glue. Special nails and screws are available. These are colored to match the surface finish. Plain nails are also used. Casing or finish-head nails are recommended. These should be set below the surface and filled. This filling should match the color and texture of the surface.

Adhesives are applied in patterns, as shown in Figure 12-25. On studs where panels join, two lines of adhesives are used.

Panels are available in various materials and range from 1/8 to 1 inch in thickness. As a rule, wood paneling 1/4 inch thick needs no base wall. However, the wall is more substantial if a base wall is used, especially with thinner panels. The wall is stronger and more fire-resistant. Many building codes require gypsum base walls. Gypsum drywall is the most common base wall for panels. Some codes require it to have the joints covered. The 3/8-inch thickness is commonly used.

EDGES AND CORNERS • Panel edges and corners are finished in several ways. Figure 12-33 shows several methods. Divider clips are used between panels. To install the panel, a special trick is often used. The edge trim is nailed down first. Then one edge of the panel is inserted into one clip. The second edge is moved away from the wall. Pressure is applied against the edge. This causes the panel to bow slightly. The second edge of the panel can then be slipped into the second clip. When the pressure is released, the edge will spring into the clip. The panel will be held flat against the wall. The panel can then be pressed against the adhesives or nailed in place.

SPACING AND NAILING • Panels are often fastened directly to the base walls. Then, panels should be carefully spaced. A line may be snapped at 4-foot intervals to show where panels should join.

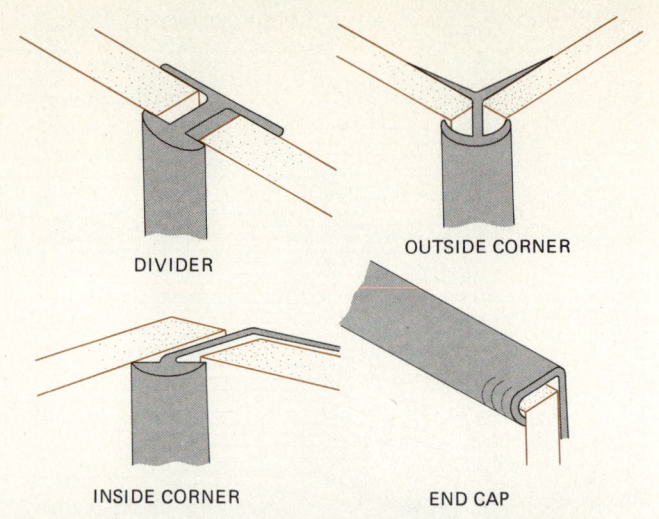

Figure 12-33. Joining edges and corners of paneling.

These lines may be used for guides. They show where the adhesives are applied. They also show where the panels are applied. Chalk lines may be snapped at 4-foot intervals on vertical or horizontal references.

Wood paneling is also nailed in place. Nails are placed approximately 4 inches O.C. around the edges. Nails should be placed at 8-inch intervals on the inside studs. Panels made of laminates, hardboards, and other materials may require special methods. The carpenter should always read the manufacturer's suggestions. These usually describe special methods and materials that should be used.

PANELS OVER STUDS • First, crooked or uneven studs should be straightened. They can be planed or sawed to make a flat base. Sometimes panels cannot be joined over studs. Then special boards are added between the studs. This is often true for vertical applications of gypsum board. Fire stops and special studs may be added.

PANELS OVER MASONRY • Panels are often used over concrete walls. Basements are prime examples. They are paneled in both new and remodeling jobs.

Special nail bases must be provided. They are made from wood boards called *furring strips*. Furring strips are fastened to the walls first. Top and bottom plates are also needed. Masonry nails or screw anchors are used. They may be added horizontally or vertically.

A vapor barrier should also be used. It may be applied over the masonry or over the furring strips. In either case, insulation should also be added. It is placed between the furring strips. See Figures 12-34 through 12-37. After insulation and vapor barriers have been installed, the wall cover can be applied. Either adhesives or nails can be used.

330 Carpentry Fundamentals

Figure 12-34. A vapor barrier is applied over a concrete wall. This is a must for best paneling. *(Masonite)*

Figure 12-35. Furring strips are attached to the concrete wall. They are aligned with the level. They form a nail base for the panels. *(Masonite)*

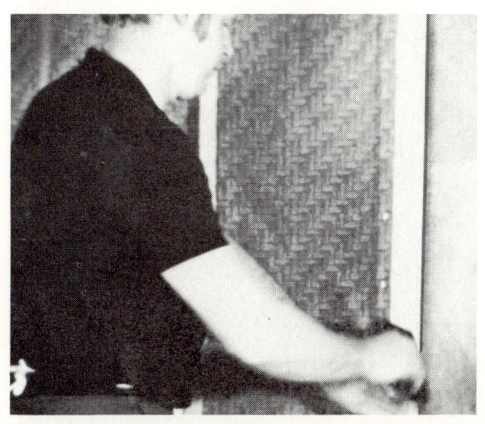

Figure 12-36. Insulation is added between the furring strips. *(Masonite)*

Figure 12-37. Finally, the panels are nailed in place. A level may be used to align the panels. *(Masonite)*

Board Walls

Walls can also be finished with boards. Figure 12-38 shows a boarded interior. The boards may be wood, plywood, or composition. The boards may be prefinished or unfinished. Most are shaped on the edges for joining. Tongue-and-groove joints and rabbeted joints are common.

Boards laid horizontally (parallel to the floor) are braced in several places by the studs. For vertical-board walls, the boards are nailed to the top and bottom plates. Sometimes more nail base is needed. Then special headers or fire stops can be used. See Figure 12-39.

The tongue-and-groove joints are strong. Also, nails can be hidden in the tongues of the boards. This way, the nails are covered by the next board. Boards are tapped into place with a block and ham-

Figure 12-38. Boards may be used for attractive interior walls. *(Weyerhauser)*

Preparing Interior Walls and Ceilings 331

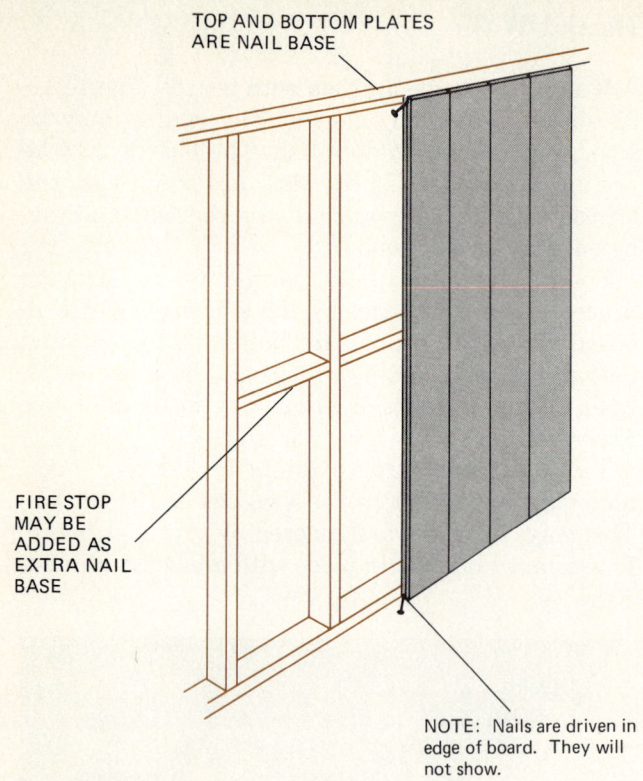

Figure 12-39. Nailing vertical board walls.

mer. See Figure 12-40. Boards may be applied horizontally, vertically, or diagonally. See Figures 12-41, 12-42, and 12-43. However, as in Figure 12-39, special bracing may be needed for vertical or diagonal boards.

• PREPARING A WALL FOR PLASTER •

The carpenter does not apply plaster to walls as a rule. However, the carpenter sometimes prepares the wall for the plaster. This includes nailing edge strips around the walls, doors, and windows. These strips are called *grounds*. The grounds may be made of wood or metal. They are used to help judge the thickness of the plaster. They also guide the application of the plaster.

The carpenter will often apply the lath that holds up the plaster. The most common lath is made from drywall. Both solid and perforated drywall is used. Figure 12-44 shows perforated drywall used for lath. Metal mesh, as in Figure 12-45, is also used. Wooden lath is no longer used much because it costs more and takes longer to install.

Nailing Plaster Grounds

Plaster grounds are strips of wood or metal. They are nailed around the edges of a wall. Grounds are

Figure 12-40. Boards may be forced into place using a scrap block and hammer. The scrap block protects the edges.

332 Carpentry Fundamentals

Figure 12-41. Vertical board walls accent height. (California Redwood Assoc.)

Figure 12-42. Horizontal wood boards provide a natural look that goes well with modern designs. (Potlatch)

Figure 12-43. Diagonal boards can give dramatic wall effects. (California Redwood Assoc.)

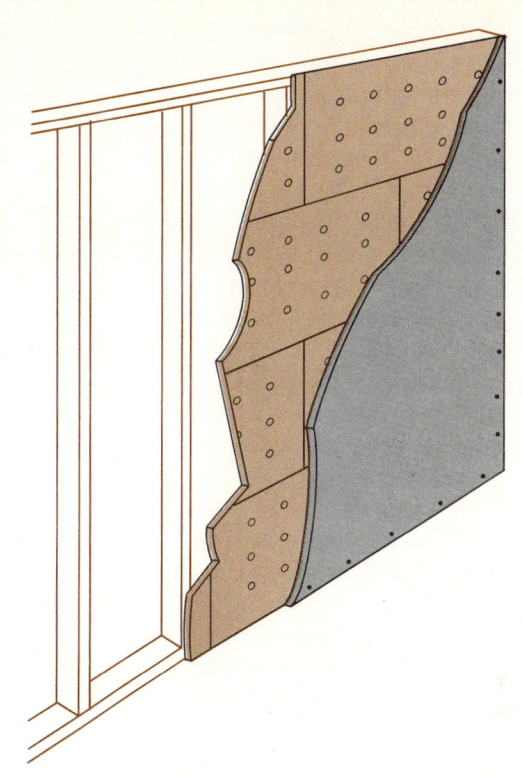

Figure 12-44. Most plaster lath for houses is made of perforated gypsum board.

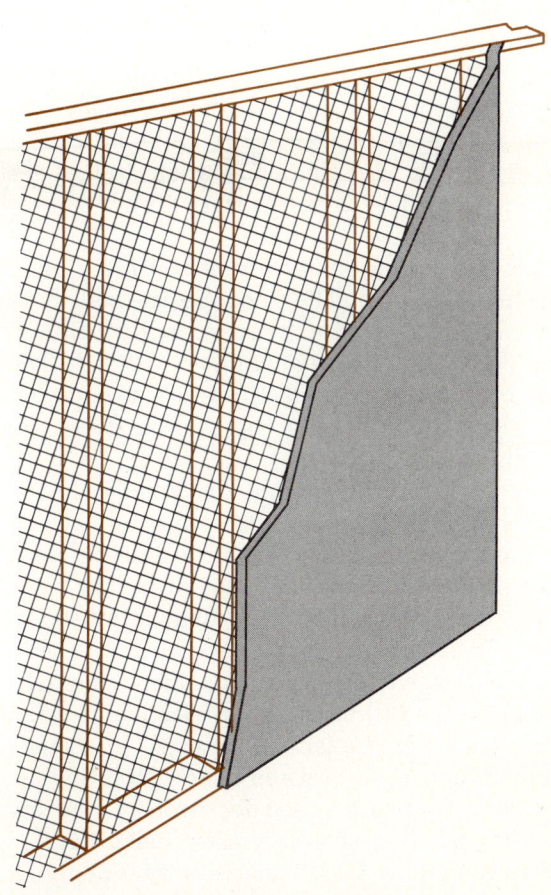

Figure 12-45. Wire lath for plaster.

Preparing Interior Walls and Ceilings   333

also nailed around openings, corners, and floors. The frame of the door or window is usually put up first. Grounds are nailed next to the frames. Sometimes the edges of the window or door casing serve as the grounds.

GROUNDS ON INTERIOR DOORS AND OPENINGS • Plaster is usually applied before the doors are installed. A temporary ground is nailed in place. Figure 12-46 shows grounds around an interior opening. There are two methods that are used. The standard width of the plastered wall is 5¼ inches. A piece of lumber 5¼ inches wide may be used. Note that it is centered on the stud and nailed in place. Thus the outside edges of the board form the grounds. These help guide the application of the plaster. The other method is shown in Figure 12-46B. This method uses less material. Two strips are nailed in place as shown. Here, the carpenter must be careful with the measurements.

Plaster is normally applied to a certain thickness for both the lath and the plaster together. Two thicknesses are often used, either ¾ or ⅞ inch. Local building codes and practices determine which is used. If ⅜-inch drywall lath is used for a ⅞-inch wall, then ½ (4/8) inch of plaster must be applied.

The plaster is applied in two or three coats. The first coat is called the "scratch" coat. The scratch coat is the thickest coat. It is put on and allowed to harden slightly. It is then scratched to make the next coat hold better. The next coat may be either a "brown" coat or the finish coat. The finish coat is a thin coat. In most residential building it is ⅛ inch thick or less. For this application, the guides become like screeds in concrete work. The final coat is put on as a flat, even surface.

The plaster ground strips are left in place after the plaster is applied. They become the nail base for the molding and trim. Trim is applied around ceilings, floors, and door and window openings. The grounds on interior door openings are usually removed. Then, the jambs for the interior doors are built or installed.

Plastering is normally done by people in the trowel trades. Two types of finish coats may be put on by these workers. The *sand float* finish gives a textured finish. This can then be painted. The *putty* finish is a smoother finish. It is commonly used in kitchens and bathrooms. In these locations, the plaster is often painted with a gloss enamel. This hard finish makes the wall more water-resistive. In addition, an insulating plaster can be used. It is made of vermiculite, perlite, or some other insulating aggregate. It may also be used for wall and ceiling finishes. Figure 12-47 shows plaster being applied with a trowel. Figure 12-48 shows a worker applying plaster with a machine.

• FINISHING MASONRY WALLS •

Masonry walls (brick, stone, or concrete block) are not comfortable walls. They are common in basements. However, they are cold and they sometimes sweat. This can give a room a cold, clammy feeling. To make them comfortable, they are finished with

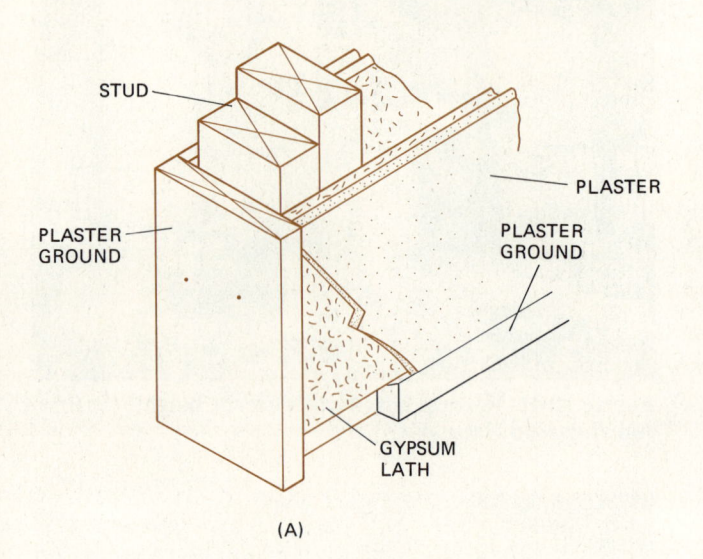

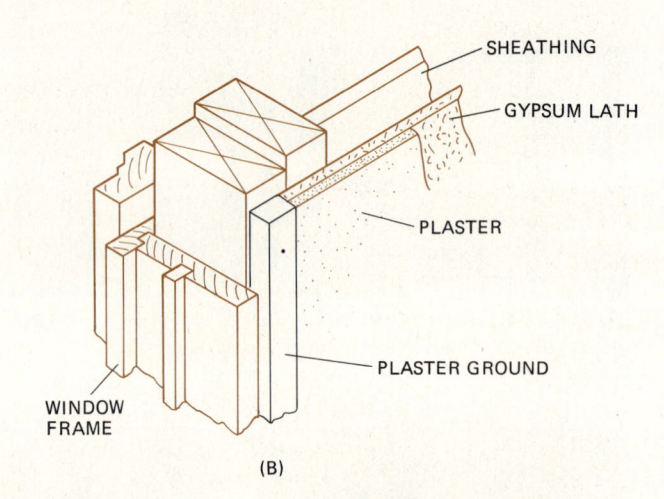

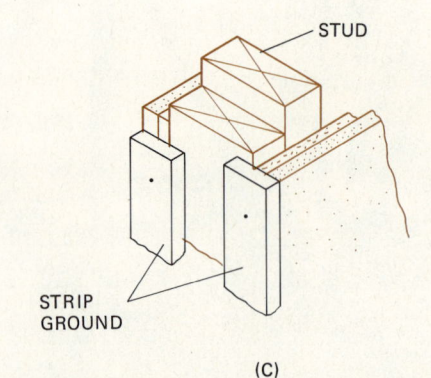

Figure 12-46. Plaster grounds around openings. (A) Recommended for doors. (B) Recommended for windows. (C) Temporary. *(Courtesy Forest Products Laboratory)*

334 Carpentry Fundamentals

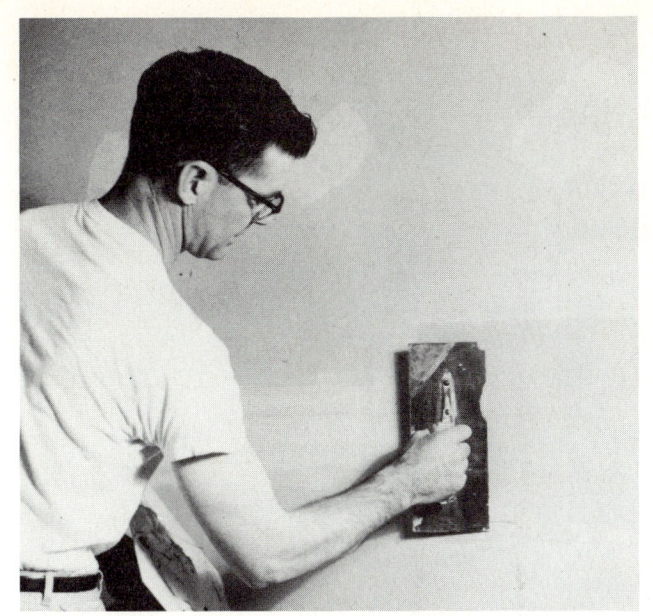

Figure 12-47. Plaster being applied with a trowel. (Gold Bond)

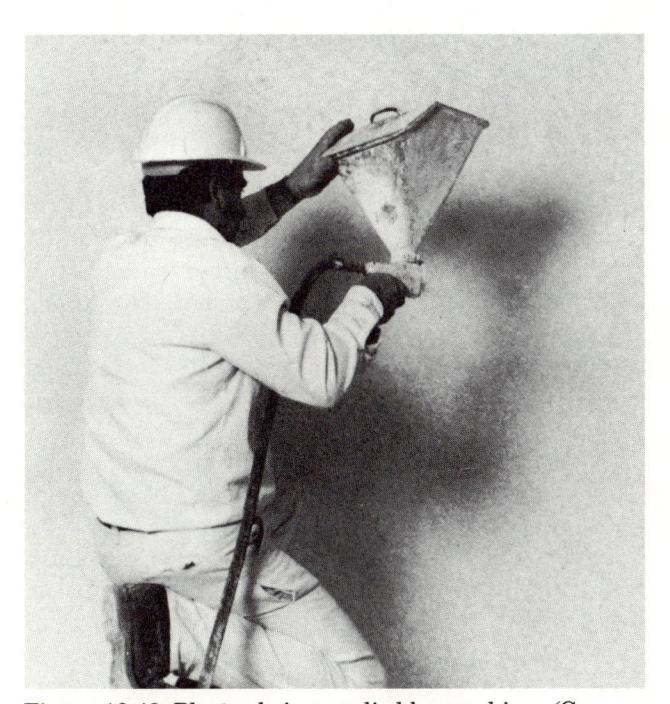

Figure 12-48. Plaster being applied by machine. (Gypsum Assn.)

Figure 12-49. Special decorative trim can be used with ceiling tile. (Armstrong Cork)

care. A vapor barrier and insulation should be installed. These are then covered with paneling or drywall. This was explained earlier.

• INSTALLING CEILING TILE •

Ceiling tiles can be installed over gypsum board, plaster, joists, or furring strips. Ceiling tile is available in a wide variety of surface textures and appearance. It is also available in a variety of sizes. A standard tile is 12×12 inches. However, sizes such as 24×24, 24×48, and 16×32 inches are also available. Appearance, texture, light reflection, and insulation effect should be considered. Also, tiles can help absorb and deaden sounds. This ability is called the *acoustical quality*. Other factors in choosing tile include fire resistance, cost, and ease of installation.

Ceiling tile can be used for new or old ceilings. It can add new decoration or improve an old appearance. Special decorative trim, as in Figure 12-49, can be added. These artificial beams and corner supports are made of rigid foam. They greatly enhance the appearance. Figure 12-50 shows the overall appearance.

Putting Tiles over Flat Ceilings

This method is good for both new and old ceilings. Tiles are usually applied with a special cement. Interlocking tiles may also be used in this situation. To install the ceiling tiles, first brush the loose dirt and grime from the ceiling. Then locate the center of the ceiling. Snap chalk lines on the surface of the ceiling to provide guides. Then check the square-

Figure 12-50. A basement area using ceiling tile and artificial beams and posts. *(Armstrong Cork)*

ness of the lines with the walls. Make sure that the tiles on opposite walls are equally spaced. If the space is not even, tiles on each wall should be equally trimmed. Tiles should not be trimmed on only one side of the ceiling. The tile on both sides must be trimmed an equal amount. This makes the center symmetrical and pleasing in appearance.

Apply cement to the back of the tile. See Figure 12-51. The cement may be spread evenly or spotted. Use the chalk lines as guides. Then press each tile firmly in place. The work should progress from the center out in a spiral.

Using Furring Strips to Install Ceiling Tile

Small tiles may be attached to ceiling joists. However, furring strips are also needed. They provide a nailing base. Figure 12-52 shows how the furring strips provide the nailing base. Small strips of wood 1×2 or 1×3 inches are used. These are nailed with one nail at each joist. They are perpendicular to the joists as shown. Insulation may be added to conserve energy or deaden sound. The furring strips are spaced to the length of the tile. Each edge of the tile should rest in the center of a strip.

The tile may be fastened with staples, adhesives, or nails. Nails or staples are applied to the inside of the tongue. This way, the next tile covers the nail so that it cannot be seen.

Installing Suspended Ceilings

Most suspended ceilings use larger panels. The 24×24 or 24×48-inch panels are common. The panels are suspended into a metal grid. This grid is composed of special T-shaped braces. These are called *cross T's* and *runners*. The grid system is suspended on wires from the ceiling or the joists.

Suspended ceilings are popular for remodeling. They also give a pleasing appearance in new situations. These ceilings are economical, quick, and easy to install. They provide good sound-deadening qualities. They also give access to pipes and utilities. Lighting fixtures may be built into the grid system. This way no special brackets are used.

A suspended ceiling is installed in five steps. First, nail the molding to the wall at the proper ceiling height. This supports the panels around the edge of the room. Second, attach hanger wires to the joists. These are usually nailed in place at 4-foot intervals. Third, fasten the main runners of the metal grid frame to the hanger wires. Check to make sure that the main runners are suspended at the proper distance. Or the main runners may be nailed directly to the joists. See Figures 12-53 and 12-54. Fourth, snap the cross T into place between the runners. The cross T and runners form a grid. See Figure 12-55. Fifth, lay the ceiling panels into the grid. See Figure 12-56.

Figure 12-51. Applying five spots of adhesive to the back of ceiling tiles. *(Armstrong Cork)*

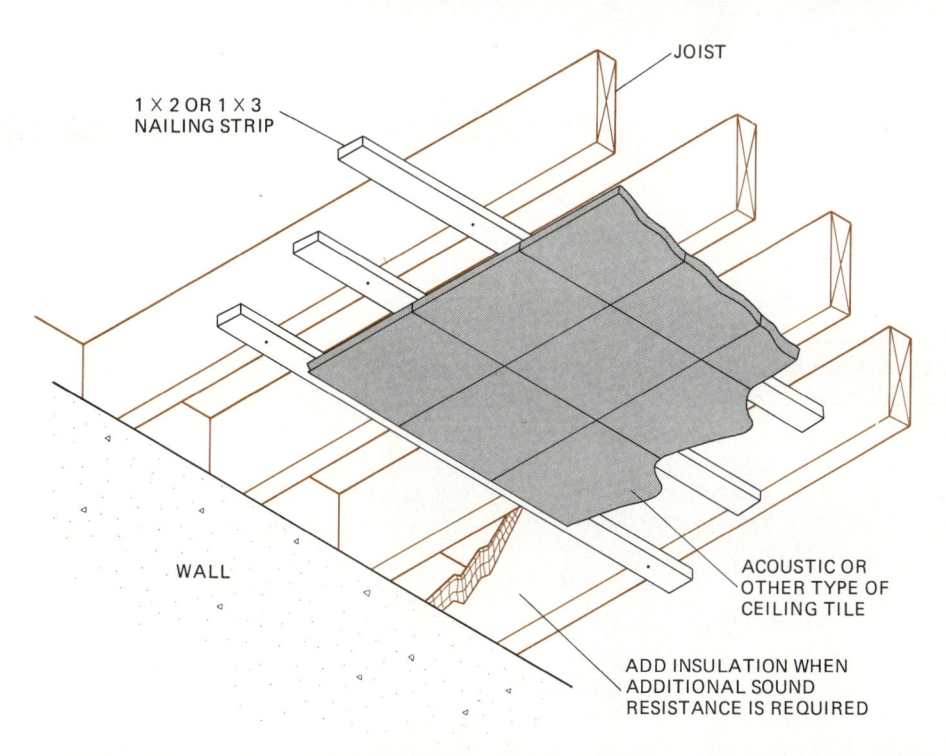

Figure 12-52. Furring, or nailing, strips are used as a nailing base for ceiling tile. Nails are driven in edges so that they will not show. *(Courtesy Forest Products Laboratory)*

Figure 12-53. Main runners can be suspended from the joists. *(Armstrong Cork)*

Preparing Interior Walls and Ceilings **337**

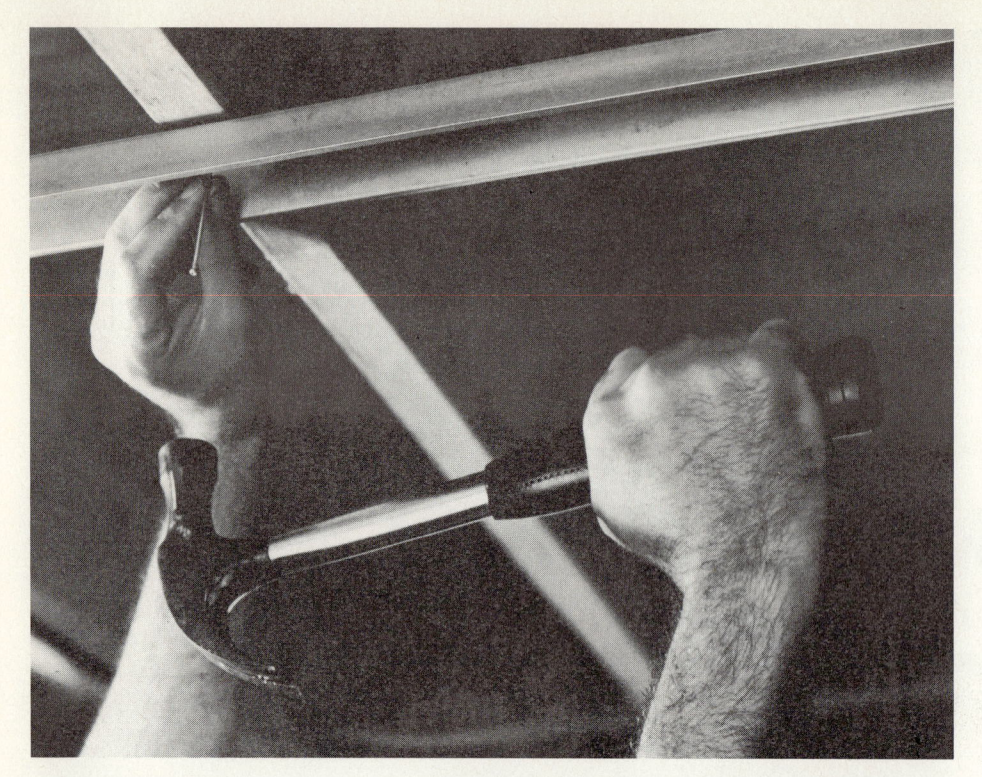

Figure 12-54. Main runners can also be nailed directly to joists. *(Armstrong Cork)*

Figure 12-55. Cross T pieces are then attached to the runner to form a grid. *(Armstrong Cork)*

Figure 12-56. Last, the panels are laid in place. The pattern of the tile helps to disguise the grid. *(Armstrong Cork)*

Figure 12-57. Panels with grooved edges may be used for a system where no grid shows. *(Armstrong Cork)*

Figure 12-58. No metal grid shows when grooved-edge tile is used. *(Armstrong Cork)*

Concealed Suspended Ceilings

A variation of the suspended ceiling is also used. The metal grid is visible. There is a special technique to hide this. Panels with a grooved edge are used. See Figures 12-57 and 12-58. The metal then holds the panels from the groove. In this system, panels are installed in much the same manner. However, the metal grids fit into the grooves. As shown, once the tiles are in place, no grid system is visible.

• WORDS CARPENTERS USE •

gypsum	lath	floating	furring strips	acoustical
sheetrock	insulation batts	corner beads	plaster grounds	
drywall	strongback	taping and bedding	suspended ceilings	
plaster	cradle brace			

Preparing Interior Walls and Ceilings 339

• STUDY QUESTIONS •

12–1. What materials are used for interior walls?

12–2. What are the advantages of gypsum wallboard?

12–3. Why is gypsum wallboard used under other panels?

12–4. How are vapor barriers installed?

12–5. What is the sequence for covering a wall?

12–6. What are the three methods of applying wall panels?

12–7. Why should a carpenter know how plaster is applied?

12–8. What are some other names for gypsum board?

12–9. Why are strongbacks used?

12–10. Why are ceiling panels started near the center of the room?

12–11. What is floating?

12–12. How is gypsum board cut?

12–13. How are ceiling panels held for nailing?

12–14. How are wall panels held for nailing?

12–15. What are the nailing intervals for walls and ceilings?

12–16. What is done for bath or kitchen walls?

12–17. What is done to edges and corners?

12–18. What ways are used to hang ceiling tile?

12–19. How are boards for walls positioned in place?

12–20. Why should ceiling tiles be started in the exact center?

13
FINISHING THE INTERIOR

The interior of a building is the last part finished. The frame has been built and covered. Exterior walls and roof are complete. Exterior doors and windows are done. The interior walls and ceilings have also been built. However, they have not been finished. Also, windows and doors on interior sections have not been installed.

Finishing the interior consists of several things. Interior doors are installed, and the molding and trim are applied to the inside of the windows and around the interior doors. Cabinets are then built or installed. Cabinets are needed in bathrooms and kitchens. Other special cabinets or shelves (called built-ins) are also built.

Then plumbing is installed. The woodwork is finished with paint or stain, and the walls and ceilings are finished. Then wiring is connected to the electrical outlets and to built-in units. Appliances such as dishwashers, ovens, lights, and other things are installed and connected. Finally, the final floor layer is finished to complete the interior.

Skills needed to finish the interior include:

- Installing cabinets in kitchens and baths
- Building shelves and cabinets
- Applying interior trim and molding
- Applying finishes
- Painting or papering walls
- Installing floor materials

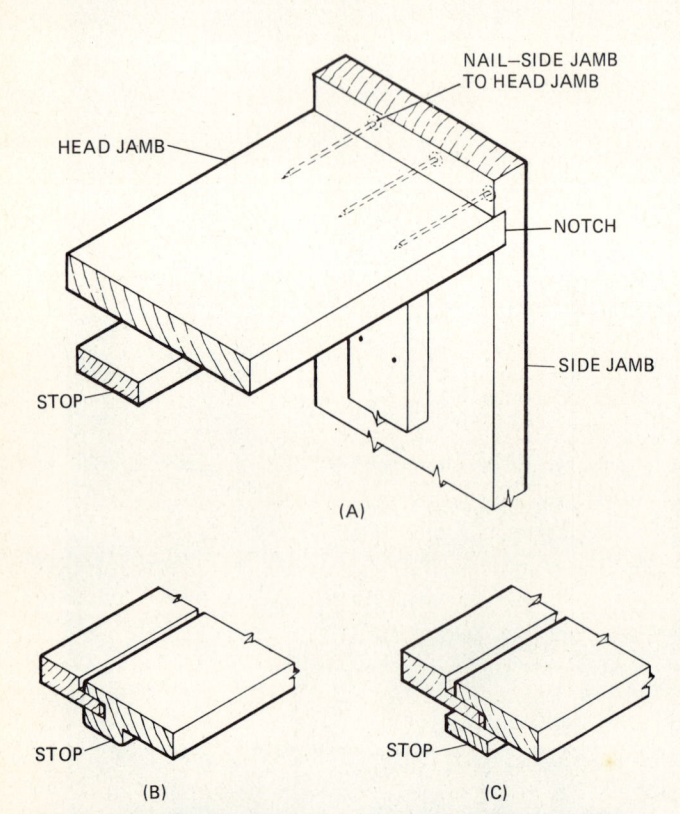

Figure 13-1. Types of interior door jambs. *(Courtesy Forest Products Laboratory)*

• SEQUENCE FOR FINISHING THE INTERIOR •

As a rule, the sequence can be varied. However, a few factors must be considered in planning the sequence. The first factor is the type of floor involved. Some buildings are constructed on a slab. Here, the interior may be completed before the finish flooring is applied. However, with a wooden frame, two layers of flooring are laid. Here the second layer of flooring may be applied before the interior is finished. The second layer, however, is not finished until later. For a frame building on a wooden frame floor, the general sequence should be:

Install cabinets in kitchens and baths. } In any sequence
Install interior doors.
Trim out interior doors and windows.
Paint or stain wood trim and cabinets.
Finish walls with paint, texture paint, or paper.
Install electrical appliances
Lay the finish floor.
Finish floors by sanding, staining, varnishing, or laying linoleum or carpet.

• INTERIOR DOORS AND WINDOW FRAMES •

Interior door units are installed to finish the separation of rooms and areas. Then the insides of the windows must be cased or trimmed. Trim is also applied to all interior door units. The trim is needed to cover framing members. It also seals these areas from drafts and airflow.

Installing Interior Doors

Openings for interior doors are larger than the door. They are higher and wider than the actual door width. Frames are usually 3 inches higher than the door height. They are usually 2½ inches wider than the door width. This provides space for the members of the door. Interior door frames are made up of two side pieces and a headpiece. These cover the rough opening frame. The pieces that cover these areas are called *jambs*. They consist of side jambs and head jambs. Other strips are installed on the jamb. These are called *stops*. The stops, also called door stops, form a seat for the door. It can latch securely against the stops. Most jambs are made in one piece as in Figure 13-1A. However, two- and three-piece adjustable jambs are also made. Adjustable jambs are used because they can be adapted to a variety of wall thicknesses. See Figure 13-1B.

Today, interior door frames can be purchased with the door prehung. These are ready for installing.

342 Carpentry Fundamentals

Sometimes, however, the carpenter must cut off the door at the bottom. This is done to give the proper clearance over the finished floor material.

DOOR FRAMES • The door unit is assembled first. Then it is adjusted to be vertical and square. This is called being *plumb*. The adjustment is made with small wedges, usually made from shingles, as shown in Figure 13-2. The jamb is plumbed vertical and square. Then nails are driven through the wedges into the studs as shown. Any necessary hinging and adjusting of the door height is made at this point. For instructions on installing door hinges, see the chapter on exterior doors and windows.

Trim is attached after the door frame is in place. The door is first properly hung. Then trim, called *casing*, may be applied. Casing is the trim around the edge of the door opening. Casing is also applied over interior door and window frames. Casing is nailed to the jamb on one side. On the other side it is nailed to the plaster ground or the framing stud. It should be installed to run from the bottom of the finished floor. Note, in Figure 13-3, that two procedures may be used for the top casing. The top casing piece can be elaborately shaped. Other pieces of molding are used to enhance the line. This is often done to fit historical styles. A slight edge of the jamb is exposed, as shown in Figure 13-3C.

Next, the door stop may be nailed in place. The door is closed. The stop should be butted close to the hinged door. However, a clearance of perhaps $1/16$ inch should be allowed. The door stop is usually $1\frac{1}{2}$ or 3 inches wide. It is wide enough for two nails to be used. The finishing nails may be driven as shown. When a stop is spliced, a beveled cut in either direction is made. See Figure 13-4. Casing may be shaped in several ways, as shown in Figure 13-5.

DOOR DETAILS • Two interior door styles are the flush and the panel door. Other types of doors are also commonly used. They include folding doors and louvered doors. These are known as novelty doors.

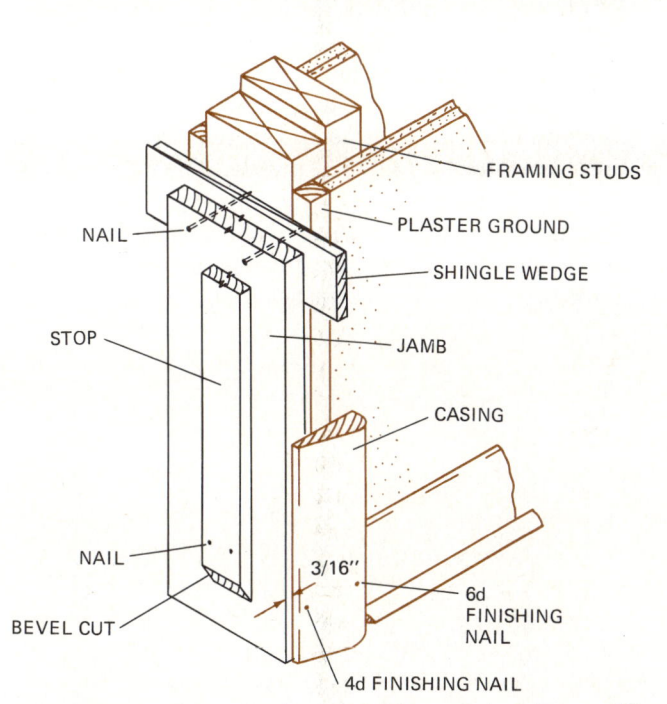

Figure 13-2. Door jambs are shimmed in place with wedges, usually made from shingles. *(Courtesy Forest Products Laboratory)*

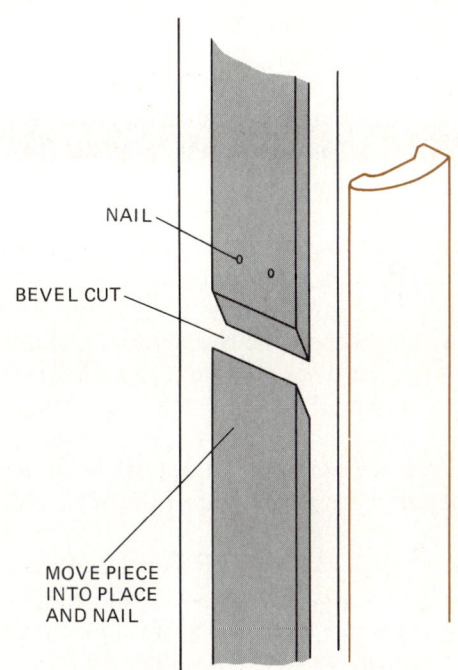

Figure 13-4. Miter joints should be used to splice stop and casing pieces where necessary.

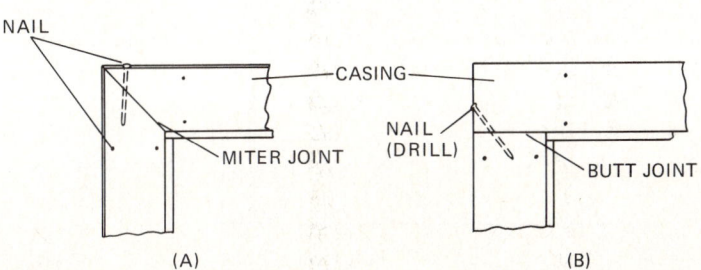

Figure 13-3. Two types of joints for door casing. Note the decoration in (C). This does not change the joint used.

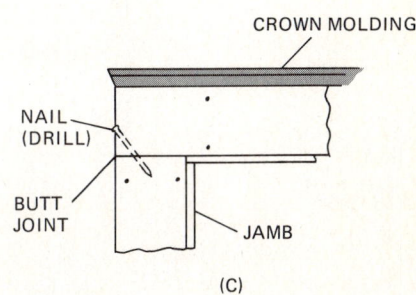

Finishing the Interior 343

Most standard interior doors are 1⅜ inches thick. They are used in common widths. Doors for bedrooms and other living areas are 2'6". Bathrooms usually have doors with a minimum width of 2'4". Small closets and linen closets have doors with a minimum width of 2'0". Novelty doors come in varied widths for special closets and wardrobes. They may be 6 feet or more in width. In most cases, regardless of the door style, the jamb, stop, and casing are finished in the same manner.

The standard height for interior and exterior doors is 6'8". However, for upper stories, 6'6" doors are sometimes used.

The flush interior door is usually a hollow-core door. This means that a framework of some type is covered with a thin outside layer. The inside is hollow, but may be braced and stiffened with cores. These cores are usually made of cardboard or some similar material. They are laid in a zigzag or circular pattern. This gives strength and stiffness but little weight. Cover layers for hollow-core doors are most often hardwood veneers. The most commonly used wood veneers are birch, mahogany, gum, and oak. Other woods are occasionally used as well. Doors may also be faced with hardboard. The hardboard may be natural, or it may be finished in a variety of patterns. These patterns include a variety of printed wood-grain patterns.

HINGING DOORS • Doors should be hinged to provide the easiest and most natural use. Doors should open or swing in the direction of natural entry. The door should open and rest against a blank wall if possible. It should not obstruct or cover furniture or cabinets. It should not bar access to or from other doorways. Doors should never be hinged to swing into hallways. Doors should not strike light fixtures or other fixtures as they are opened.

INSTALLING DOOR HARDWARE • Hardware for doors includes hinges, handles or locksets, and strike plates. Hinges are sold separately. Door *sets* include locks, handles, and strike plates. They are available in a variety of shapes and classes. Special locks are used for exterior doors. Bathroom door sets have inside locks with safety slots. The safety slots allow the door to be opened from the outside in an emergency. Bedroom locks and passage locks are other classes. Bedroom locks are sometimes keyed. Often, they have a system similar to the bathroom lock. They are also available without keyed or emergency opening access. Passage "locks" cannot be locked.

To install door sets, two or more holes must be drilled. Most locksets today feature a bored set. This is perhaps the simplest to install. A large hole is drilled first. It allows the passage of the handle assembly. The hole is bored in the face of the door at the proper spacing. The second hole is bored into the edge of the door. The rectangular area for the face plate is routed or chiseled to size. Variations occur from manufacturer to manufacturer. The carpenter should refer to specifications with the locks before drilling holes. The door handle should be 36 to 38 inches from the floor. Other dimensions and clearances are shown in Figure 13-6. A machine used to cut mortises and bores for locks is shown in Figure 13-7.

A second type of lockset is called a mortise lock. Its installation is shown in Figure 13-8. For the mortise lock, two holes must be drilled in the face of the door. One is for the spindle. The other is for the key. Also, the area inlet into the edge of the door is larger. This requires more work. As shown in the figure, setting the face plate also requires more work than the other style.

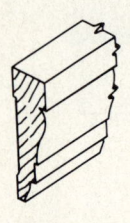

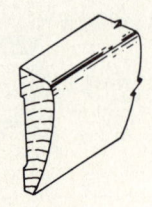

Figure 13-5. Two common casing shapes. (A) Colonial (B) Ranch *(Courtesy Forest Products Laboratory)*

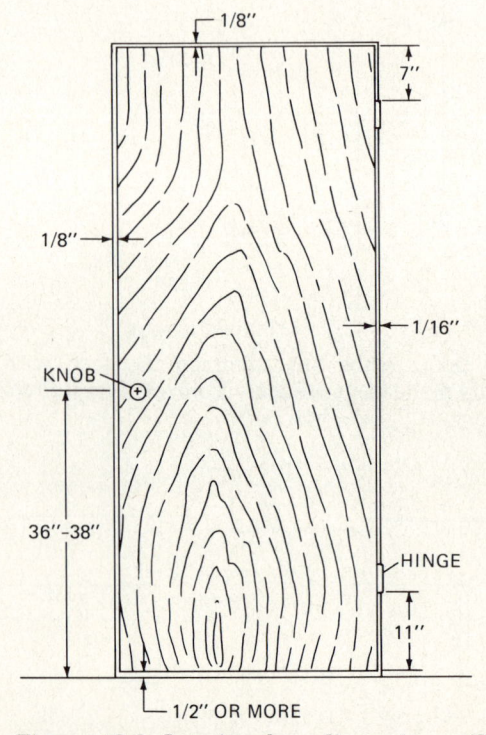

Figure 13-6. Interior door dimensions. *(Courtesy Forest Products Laboratory)*

344 Carpentry Fundamentals

INSTALLING HINGES • Interior doors need two hinges, but exterior doors need three. The third hinge offsets the tendency of doors to warp. They tend to do this because the weather on the outside is very different from the "weather" on the inside.

Special door hinges should be used in all cases. Loose-pin *butt* hinges should be used. For 1¾-inch-thick exterior doors, 4x4-inch butt hinges should be used. For 1⅜-inch-thick interior doors, 3½×3½-inch butt hinges may be used.

First, the door is fitted to the framed opening. The proper clearances shown in Figure 13-6 are used. Hinge halves are then laid on the edge of the door. A pencil is used to mark the location. The portion of the edge of the door is inlet or removed. This is called cutting a "gain." A chisel is often used. But, for many doors, gains are cut using a router with a special attachment.

The hinge should be placed square on the door. Both top and bottom hinges should be placed on the door. The door is then blocked in place. The hinge location on the jamb is marked. The door is removed, and the hinges are taken apart. The hinge halves are screwed in place. Next, the gain for the remaining hinge half is cut on the door jamb. The door is then positioned in the opening. The hinge pins are inserted into the hinge guides. The door can then be opened and swung. If the hinges are installed properly, the door will swing freely. Also, the door will close and touch the door stops gently. If the door tends to bind, it should be removed and planed or trimmed for proper fit. Also, note that a slight bevel is recommended on the lock side of the door. This is shown in Figure 13-9.

INSTALLING THE STRIKE PLATE • The strike plate holds the door in place. It makes contact with the latch and holds it securely. This holds the door in a closed position. See Figure 13-10. To install the strike plate, close the door with the lock and latch in place. Mark the location of the latch on the door jamb. This locates the proper position of the strike plate. Outline the portion to be removed with a pencil. Use a router or a chisel to remove the portion required. Note that two depths are required for the strike plate area. The deeper portion receives the lock and latch. It may be drilled the same diameter used for the lockset.

FINISHING DOOR STOPS • Stops are first nailed in place with nail heads out. This way their positions can be changed after the door is hung. Once the door has been hung, the stops are positioned. Then they are securely nailed in place. Use 6d finish or casing nails. The stops at the lock side should be nailed first. The stop is pressed gently against the face of the door. The pressure is applied to the entire height. Space the nails 16 to 18 inches apart in pairs as in Figure 13-10.

Next, nail the stop behind the hinge side. However, a very slight gap should be left. This gap should be approximately the thickness of a piece of cardboard (1/32 inch). A matchbook cover makes a good

Figure 13-7. A machine used for cutting mortises and bores for locks. *(Courtesy Rockwell International, Power Tool Division)*

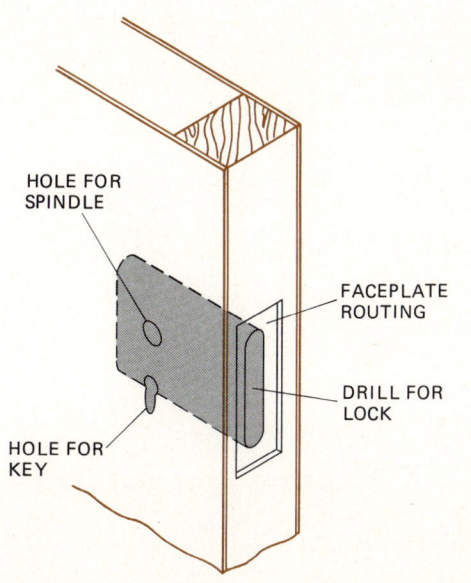

Figure 13-8. Mortise lock preparation. *(Courtesy Forest Products Laboratory)*

Finishing the Interior 345

Figure 13-9. (A) Door details. *(Courtesy Forest Products Laboratory)* (B) Using templates and fixtures to cut hinge gains *(Courtesy Rockwell International, Rockwell Power Tools Division)*

gage. After the lock and the hinge door stops have been nailed in place properly, the door is checked. When the clearances are good, the head jamb stop is nailed.

• WINDOW TRIM •

There are two major ways of finishing window interiors. The window interior should be finished to seal against drafts and air currents. Trim is applied to block off the openings. The trim also gives the window a better appearance. The trim completely covers the rough opening.

Finishing Wooden Frame Windows

The traditional frame building has wooden double-hung windows. The frame of the window consists of a jamb around the sides and top. The bottom piece is a sill. The bottom sill is sloped. Water on the outside will drain to the outside. The frame covers the inside of the rough opening. However, there is still a gap between the window frame and the interior wall. This must be covered to seal off the window from air currents. Two ways are commonly used to trim, or case, around the window.

MAKING A WINDOW STOOL • A separate ledge can be made at the bottom on the inside. See Figure 13-11. The bottom piece extends like a ledge into the room. This bottom piece is called a *stool*. However, a finishing piece is applied directly beneath it. This is called the *apron*. See Figure 13-11. The sides and top of the window are finished with casing.

CASING A WINDOW • The second method of finishing windows also uses trim. However, trim or

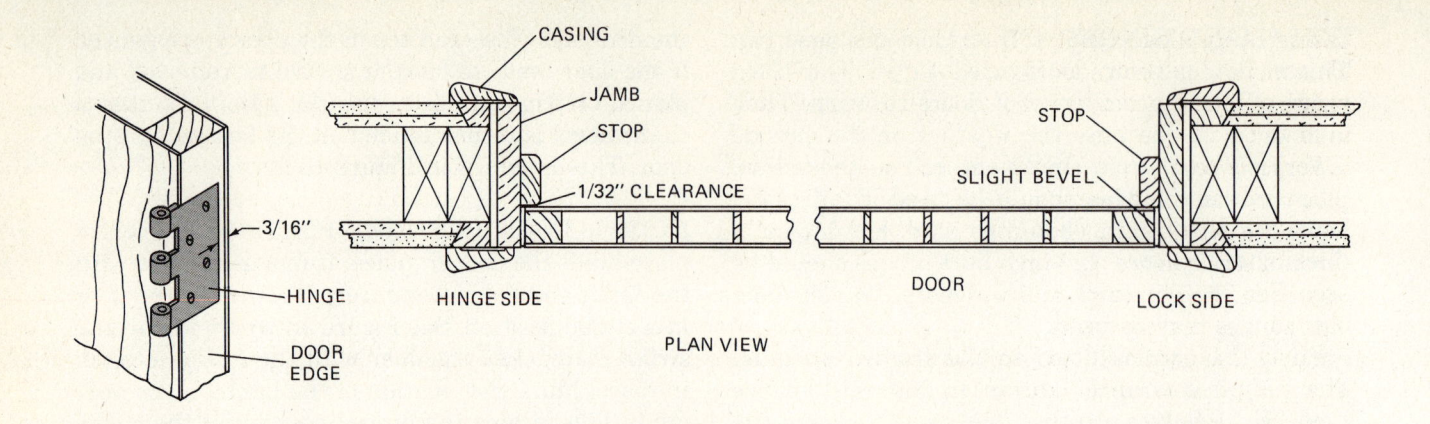

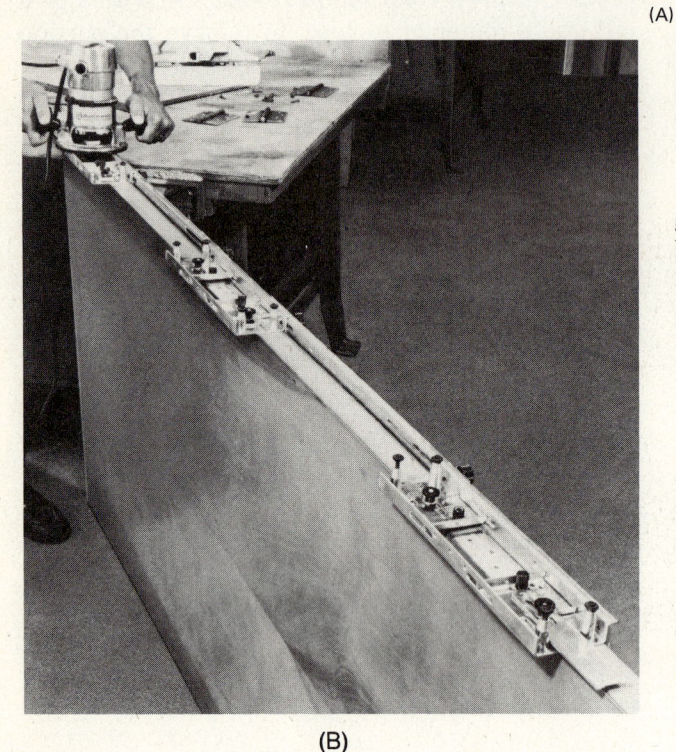

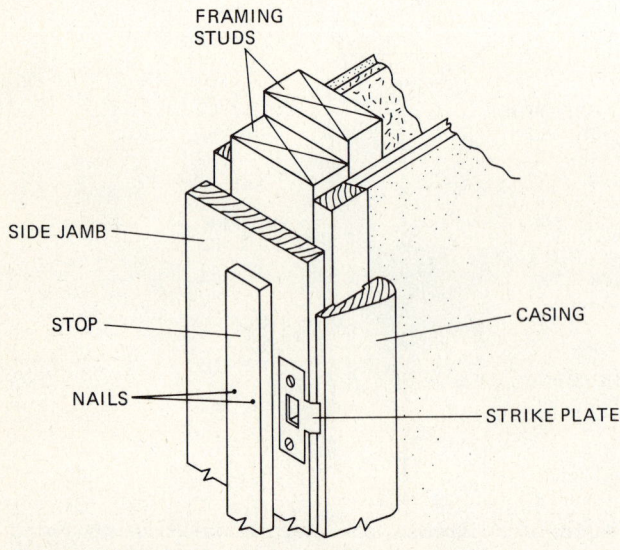

Figure 13-10. Strike plate details.

346 Carpentry Fundamentals

casing is applied completely around the opening. This method is shown in Figure 13-12. This method requires fewer pieces, takes less time to install, and is faster.

Sometimes, the carpenter also nails the stops in place around the window jamb. The stop in the window provides a guide for moving the window. It also forms a seal against air currents. To install the window stop, the window is fully closed. Next, the bottom stop is placed against the window sash. If the window is unpainted, 1/32 inch (about the thickness of a stiff piece of cardboard) is left between the stop and the sash. The stop is nailed in place with paired nails. This was also done for the doors. Next, the side stops are placed on the side jamb. The same amount of clearance is allowed. The stops on each side are nailed in place. The nail heads are left protruding. Then, the window is opened and shut. If the window slides evenly and smoothly, the stops are nailed down in place. If the window binds, not enough clearance was used. If the window wobbles, too much clearance was used. In these cases, pull the nails and reposition them.

Finishing Metal Window Frames

The use of windows with metal frames is increasing. See Figure 13-13. The metal frame is nailed to the frame members of the rough opening. This is done after the wall has been erected. Later, the window is sealed with insulation, caulk, or plastic foam. The window is finished in two steps. The inside of the rough opening is framed with plain boards. These boards are trimmed flush with the finish wall. Then, casing is applied to cover the gap between

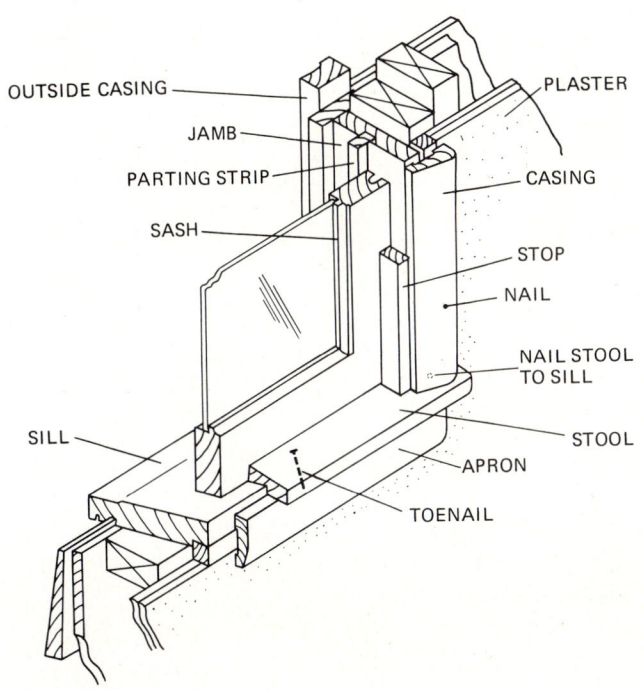

Figure 13-11. Stool and apron window trim. *(Courtesy Forest Products Laboratory)*

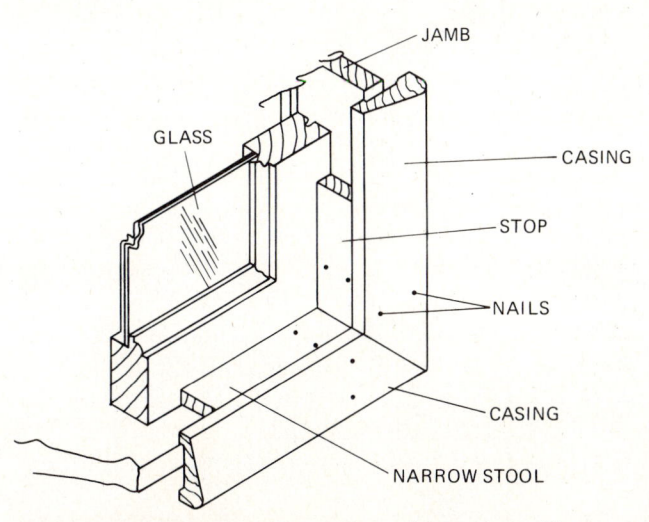

Figure 13-12. Finishing a window with casing. *(Courtesy Forest Products Laboratory)*

Figure 13-13. Metal window frames have different details.

Finishing the Interior **347**

the wall and the window framing. See Figure 13-14.

For a different appearance, the window may be cased with a stool and apron. See Figure 13-15. A flush-width board is used on the sides and tops. However, for the bottom, a wider board is used. The ends are notched to allow the side projections as in the figure. For a more finished appearance, the edges can be trimmed with a router. The apron is then added to finish out the window.

• CABINETS AND MILLWORK •

Millwork is a term used for materials made at a special factory, or mill. It includes both single pieces of trim and big assemblies. Interior trim, doors, kitchen cabinets, fireplace mantels, china cabinets, and other units are all millwork. Most of these items are sent to the building ready to install. So the carpenter must know how to install units.

However, not all cabinets and trim items are made at a mill. The carpenter must know how to both construct and install special units. This is called *custom* work. Custom units are usually made with a combination of standard dimension lumber and molding or trim pieces. Also, many items that are considered millwork do not require a highly finished appearance. For example, shelves in closets are considered millwork. But they generally do not require a high degree of finish. On the other hand, cabinets in kitchens or bathrooms do require better work.

Various types of wood are used for making trim and millwork. If the millwork is to be painted, pine or other soft woods are used most often. However, if the natural wood finish is applied, hardwood species are generally preferred. The most common woods include birch, mahogany, and ash. Other woods such as poplar and boxwood are also used. These woods need little filling and can be stained for a variety of finishes.

Installing Ready-Built Cabinets

Ready-built cabinets are used most in the kitchens and bathrooms. They may be made of metal, wood, or wood products. The carpenter should remember that overhead cabinets may be used to hold heavy dishes and appliances. Therefore, they must be solidly attached. Counters and lower cabinets must be strong enough to support heavy weights. However, they need not be fastened to the wall as rigidly as the upper units. Ready-builts are obtained in widths from 12 to 48 inches. The increments are 3 inches.

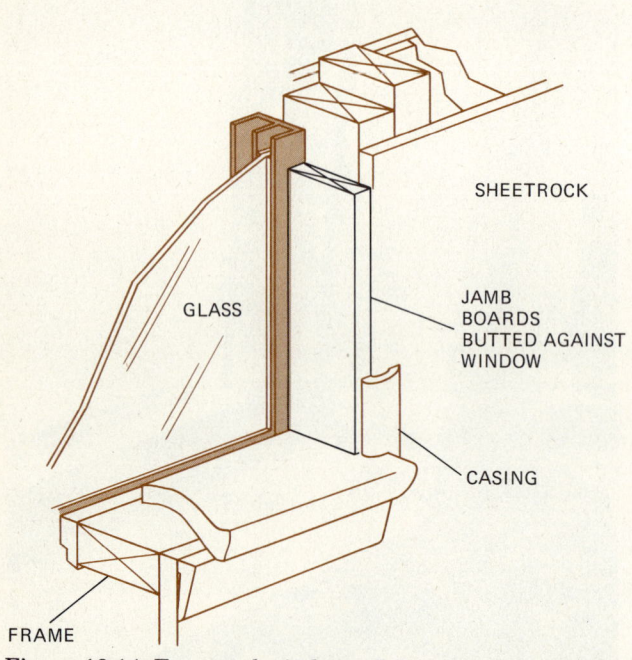

Figure 13-14. For metal windows, the jamb is butted next to the window unit. The jamb is then nailed to the rough opening frame and cased.

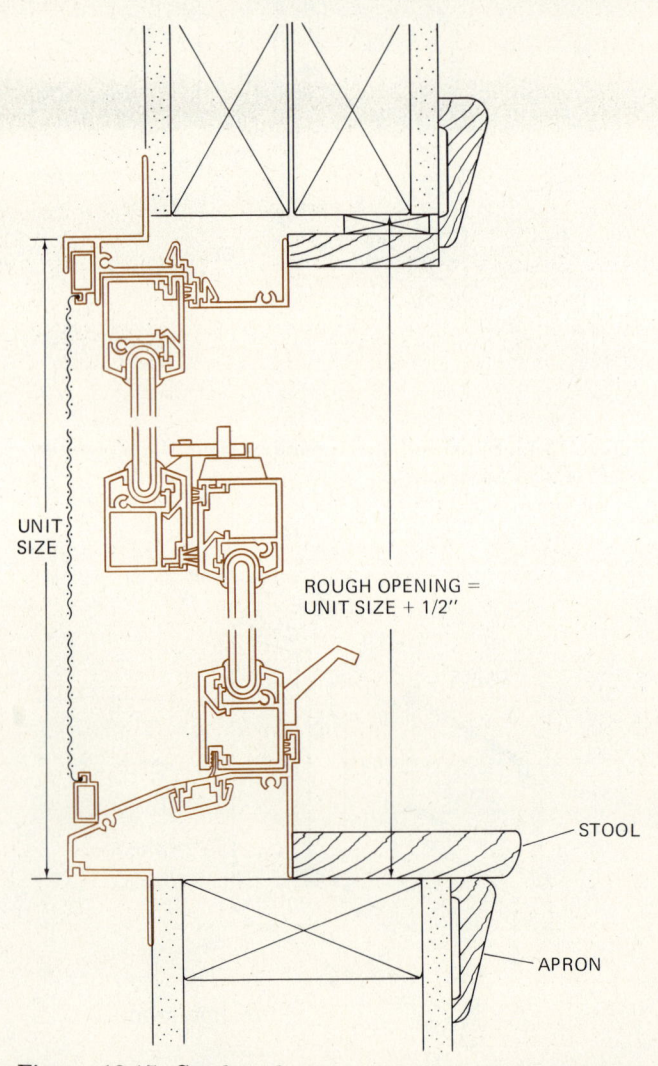

Figure 13-15. Stool and apron detail for an aluminum window. *(ALCOA)*

348 Carpentry Fundamentals

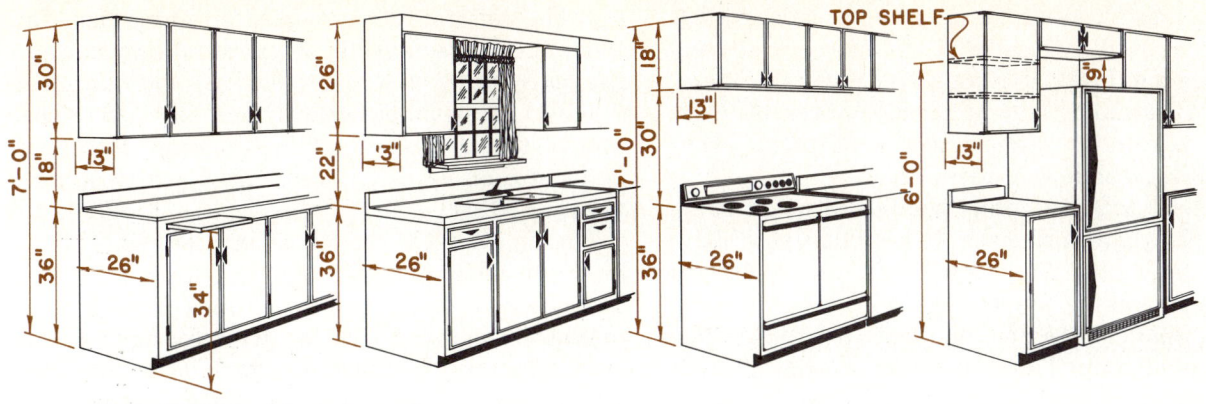

Figure 13-16. Typical cabinet dimensions.

Figure 13-17. (A) Appliance is built-in to extend.

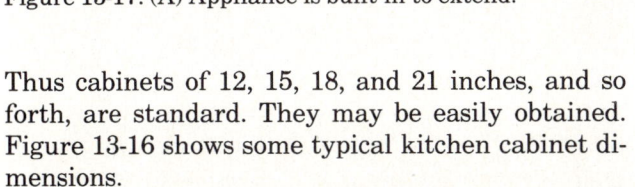

(B) Appliance is built-in flush.

Thus cabinets of 12, 15, 18, and 21 inches, and so forth, are standard. They may be easily obtained. Figure 13-16 shows some typical kitchen cabinet dimensions.

Ordering cabinets takes careful planning. There are many factors for the builder or carpenter to consider. The finish, the size and shape of the kitchen, and the dimensions of built-in appliances must be considered. For example, special dishwashing units and ovens are often built-in. The cabinets ordered should be wide enough for these to be installed. Also, appliances may be installed in many ways. Manufacturer's data for both the cabinet and the appliance should be checked. Special framing may be built around the appliance. The frame can then be covered with plastic laminate materials. This provides surfaces that are resistant to heat, moisture, and scratching or marring. See Figures 13-17A and B.

Finishing the Interior **349**

Sinks, dishwashers, ranges, and refrigerators should be carefully located in a kitchen. These locations are important in planning the installation of cabinets. Plumbing and electrical connections must also be considered. Also, natural and artificial lighting can be combined in these areas.

Cabinets and wall units should have the same standard height and depth. It would be poor planning to have wall cabinets with different widths in the same area.

Five basic layouts are commonly used in the design of a kitchen. These include the sidewall type as shown in Figure 13-18A. This type is recommended for small kitchens. All elements are located along one wall.

The next type is the parallel or pullman kitchen. See Figure 13-18B. This is used for narrow kitchens and can be quite efficient. Arrangement of sink, refrigerator, and range is critical for efficiency. This type of kitchen is not recommended for large homes or families. Movement in the kitchen is restricted, but it is efficient.

The third type is the L shape. See Figure 13-18C. Usually the sink and the range are on the short leg. The refrigerator is located on the other. This type of arrangement allows for an eating space on the open end.

The U arrangement usually has a sink at the bottom of the U. The range and refrigerator should be located on opposite sides for best efficiency. See Figure 13-18D.

The final type is the island kitchen. This type is becoming more and more popular. It promotes better utilization and has better appearance. This arrangement makes a wide kitchen more efficient. The island is the central work area. From it, the appliances and other work areas are within easy reach. See Figure 13-18E.

Screws should be used to hang cabinet units. The screws should be at least No. 10 three-inch screws. The screws reach through the hanging strips of the cabinet. They should penetrate into the studs of the wall frame. Toggle bolts could be used when studs are inaccessible. However, the walls must be made of rigid materials rather than plasterboard.

To install wall units, one corner of the cabinet is fastened. The mounting screw is driven firm. The other end is then plumbed level. While someone

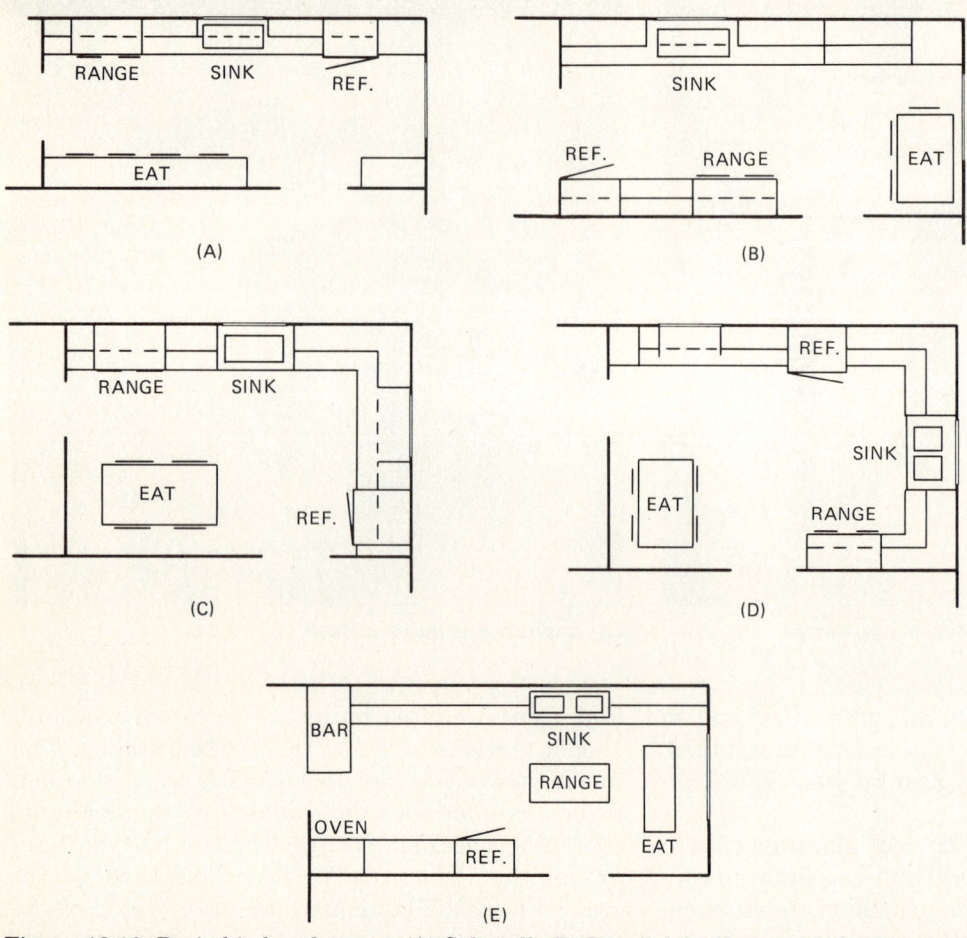

Figure 13-18. Basic kitchen layouts. (A) Sidewall. (B) Parallel (pullman). (C) L shape. (D) U shape. (E) Island. *(Courtesy Forest Products Laboratory)*

350 Carpentry Fundamentals

holds the cabinet, a screw is driven through at the second end. Next, screws are driven at each stud interval.

When installing counter units, care must be taken in leveling. Floors and walls are not often exactly square or plumb. Therefore, care must be taken to install the unit plumb and level. What happens when a unit is not installed plumb and level? The doors will not open properly and the shelves will stick and bind.

Shims and blocking should be used to level the cabinets. Shingles or planed blocks are inserted beneath the cabinet bases.

The base unit and the wall unit are installed first. Then the countertop is placed. Countertops may be supplied as prefabricated units. These include dashboard and laminated tops. After the countertops are applied, they should be protected. Cardboard or packing should be put over the top. It can be taped in place with masking tape. It is removed after the building is completed. Sometimes the cabinets are hung but not finished. The countertop is installed after the cabinets are properly finished.

Ready-built units may be purchased three ways. First, they may be purchased assembled and prefinished. Counters are usually not attached. The carpenter must install them and protect the finished surfaces.

Second, cabinets may be purchased assembled but unfinished. These are sometimes called *in-the-white* and are very common. They allow all the woodwork and interior to be finished in the same style. Such cabinets may be made of a variety of woods. Birch and ash are among the more common hardwoods used. The top of the counter is usually provided but not attached. Also, the plastic laminate for the countertop is not provided. A contractor or carpenter must purchase and apply this separately.

In the third way, the cabinets are purchased unassembled. The parts are precut and sanded. However, the unit is shipped in pieces. These are put together and finished on the job.

Cabinets are sometimes a combination of special and ready-built units. Many builders use special crews that do only this type of finish work. The combination of counter types gives a specially built look with the least cost.

Making Custom Cabinets

Custom cabinets are special cabinets that are made on the job. Many carpenters refer to these and any type of millwork as *built-ins*. See Figure 13-19. These jobs include building cabinets, shelves, bookcases, china closets, special counters, and other items.

A general sequence can be used for building cabinets. The base is constructed first. Then a frame is made. Drawers are built and fitted next. Finally the top is built and laminated.

PATTERN LAYOUT • Before beginning the cabinet, a layout of the cabinet should be made. This may be done on plywood or cardboard. However, the layout should be done to full size if possible. The layout should show sizes and construction methods involved. Figure 13-20 shows a typical layout.

Custom cabinets can be made in either of two ways. The first involves cutting the parts and assembling them *in place*. Each piece is cut and attached to the next piece. When the last piece is done, the cabinet is in place. The cabinet is not moved or positioned.

For the second procedure, all parts are cut first. The cabinet unit is assembled in a convenient place. Then, it is moved into position. The cabinet is leveled and plumbed and then attached.

Several steps are common to cabinet making. A bottom frame is covered with end and bottom panels. Then partitions are built and the back top strip is added. Facing strips are added to brace the front. Drawer guides and drawers are next. Shelves and doors complete the base.

MAKING THE BASE • The base for the cabinets is made first. Either 2×4- or 1×4-inch boards may be used for this. No special joints are used. See Figure 13-21. Then the end panel is cut and nailed to the base. The toeboard, or front, of the base covers the ends. The end panel covers the end of the toeboard. A temporary brace is nailed across the tops

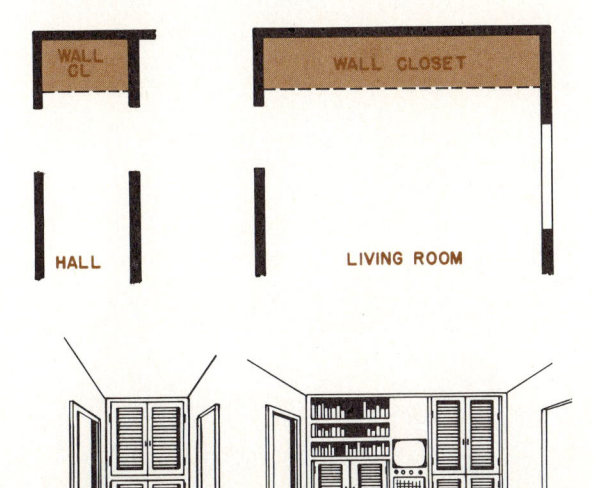

Figure 13-19. Typical built-ins.

Finishing the Interior 351

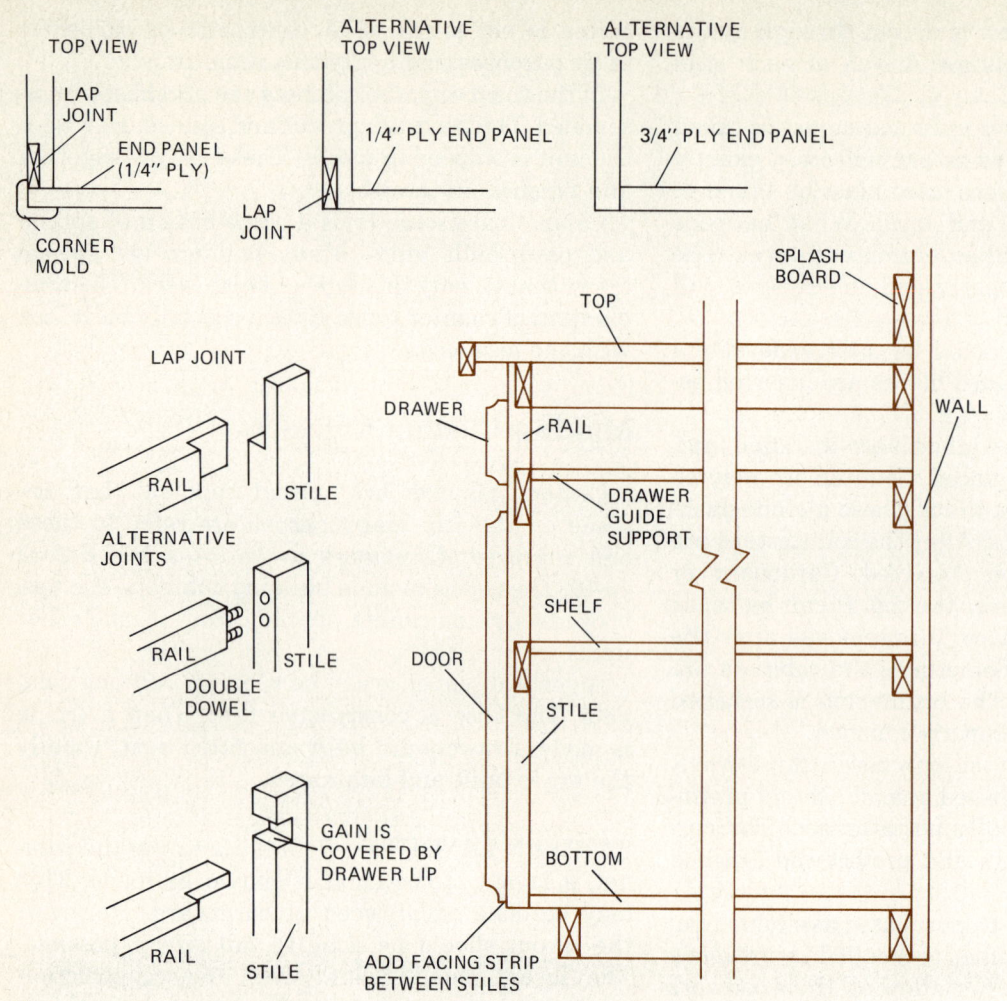

Figure 13-20. A cabinet layout. Dimensions and joints would also be added.

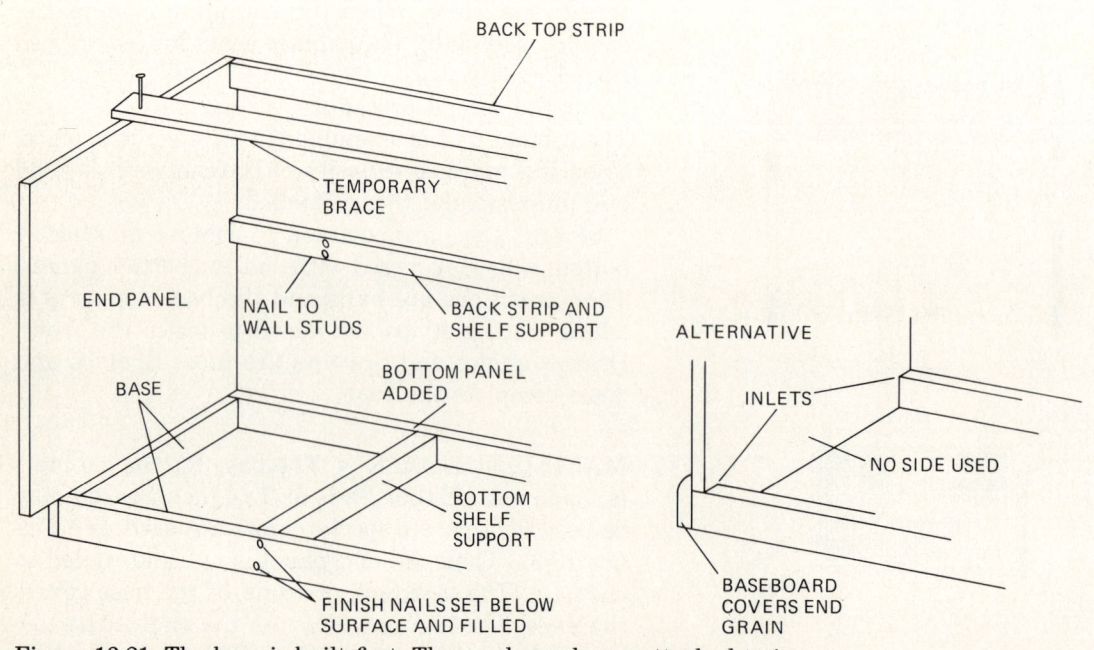

Figure 13-21. The base is built first. Then end panels are attached to it.

352 Carpentry Fundamentals

of the end panels. This braces the end panels at the correct spacing and angles.

Bottom panels are cut next. The bottom panels serve as a floor over the base. The partition panels are placed next. They should be notched on the back top. This allows them to be positioned with the back top strip.

The back top strip is nailed between the end panels as shown. The temporary brace may then be removed. The partition panels are placed into position and nailed to the top strip. For a cabinet built in position, the partitions are toenailed to the back strip. The process is different if the cabinet is not built in place. The nails are driven from the back into the edges of the partitions. The temporary braces may be removed.

CUTTING FACING STRIPS • Facing strips give the unit a finished appearance. They cover the edges of the panels, which are usually made of plywood. This gives a better and more pleasing appearance. The facing strips also brace and support the cabinet. And, the facing strips support the drawers and doors. Because they are supports, special notches and grooves are used to join them.

The vertical facing strip is called a *stile*. The horizontal piece is called the *rail*. They are notched and joined as shown in Figure 13-22. Note that two types of rail joints are used. The flat or horizontal type uses a notched joint. The vertical rail type uses a notched lap joint in both the rail and the stile.

As a rule, stiles are nailed to the end and partition panels first. The rails are then inserted from the rear. Glue may be used, but nails are preferred. Nail holes are drilled for nails near the board ends. The hole is drilled slightly smaller than the nail diameter. Finishing nails should be driven in place. Then the head is set below the surface. The hole is later filled.

DRAWER GUIDES • After the stiles and rails, the drawer assemblies are made. The first element of the drawer assembly is the drawer guide. This is the portion into which the drawer is inserted. The drawer guides act as support for the drawers. They also guide the drawers as they move in and out. The drawer guide is usually made from two pieces of wood. It provides a groove for the side of the drawer. The side of the guide is made from the top piece of wood. It prevents the drawer from slipping sideways.

A strip of wood is nailed to the wall at the back of the cabinet. It becomes the back support for the drawer guide. The drawer guide is then made by cutting a bottom strip. This fits between the rail and the wall. The top strip of the guide is added next. See Figure 13-22. It becomes the support for the drawer guide at the front of the cabinet. Glue and nails are used to assemble this unit. The glue is applied in a weaving strip. Finish nails are then driven on alternate sides about 6 inches apart to complete the assembly. Each drawer should have two guides. One is on each side of the drawer.

There are three common types of drawer guides that are made by carpenters. These are side guides, corner guides, and center guides. Special guide rails may also be purchased and installed. These include special wheels, end stops, and other types of hardware. A typical set of purchased drawer guides is shown in Figure 13-23.

The corner guide has two boards which form a corner. The bottom edge of the drawer rests in this corner. See Figure 13-24A. The side guide is a single piece of wood nailed to a cabinet frame. As in Figure 13-24B, the single piece of wood fits into a groove cut on the side of the drawer. The guides serve both as a support and as guides.

For the center guide, the weight of the drawer is supported by the end rails. However, the drawer is kept in alignment by a runner and guide. See Figure 13-24C. The carpenter may make the guides by nailing two strips of wood on the bottom of the drawer. The runner is a single piece of wood nailed to the rail.

In addition to the guides, a piece should be installed near the top. This keeps the drawer from falling as it is opened. This piece is called a *kicker*. See Figure 13-25. Kickers may be installed over the guides. Or, a single kicker may be installed.

Drawer guides and drawer rails should be sanded smooth. Then they should be coated with sanding

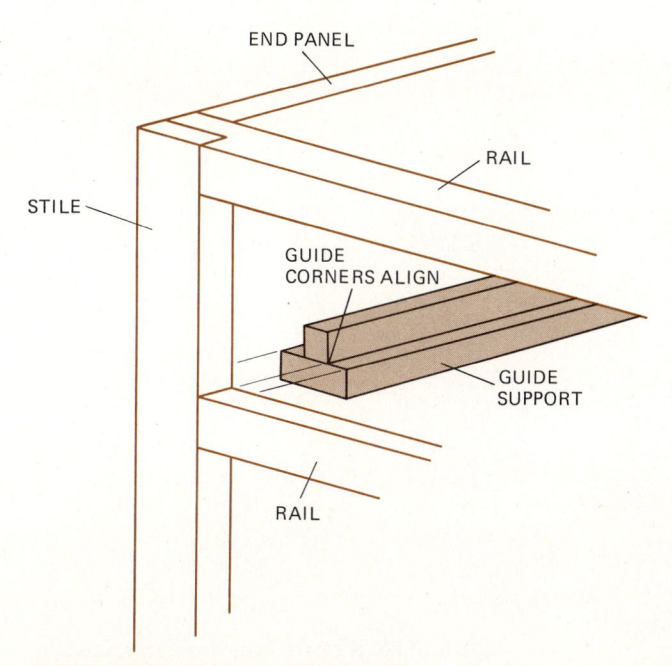

Figure 13-22. Drawer guide detail.

Finishing the Interior 353

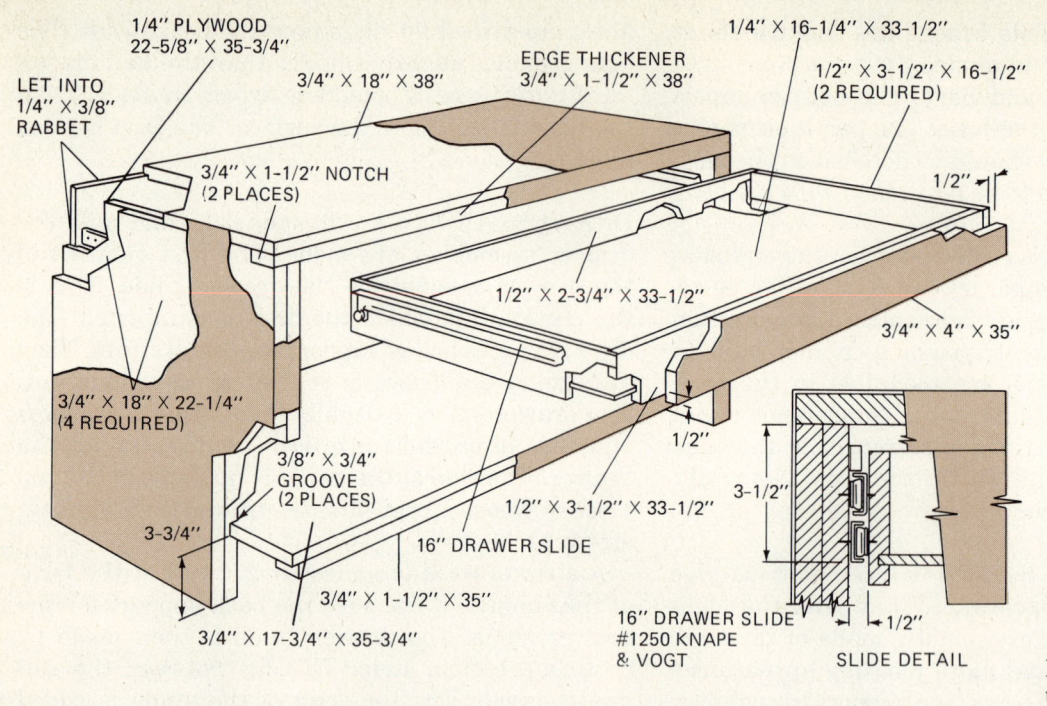

Figure 13-23. A built-in unit with factory-made drawer guides. *(Formica)*

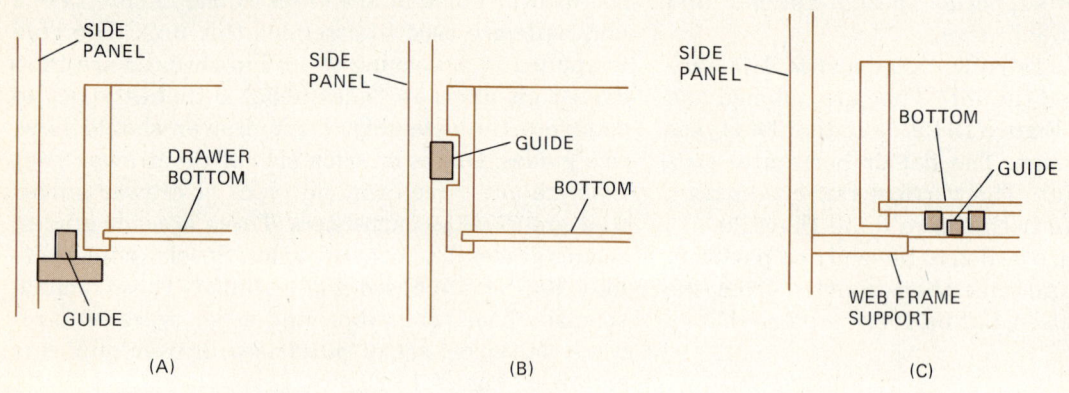

Figure 13-24. Carpenter-made drawer guides. (A) Corner guide. (B) Side guide. (C) Center guide.

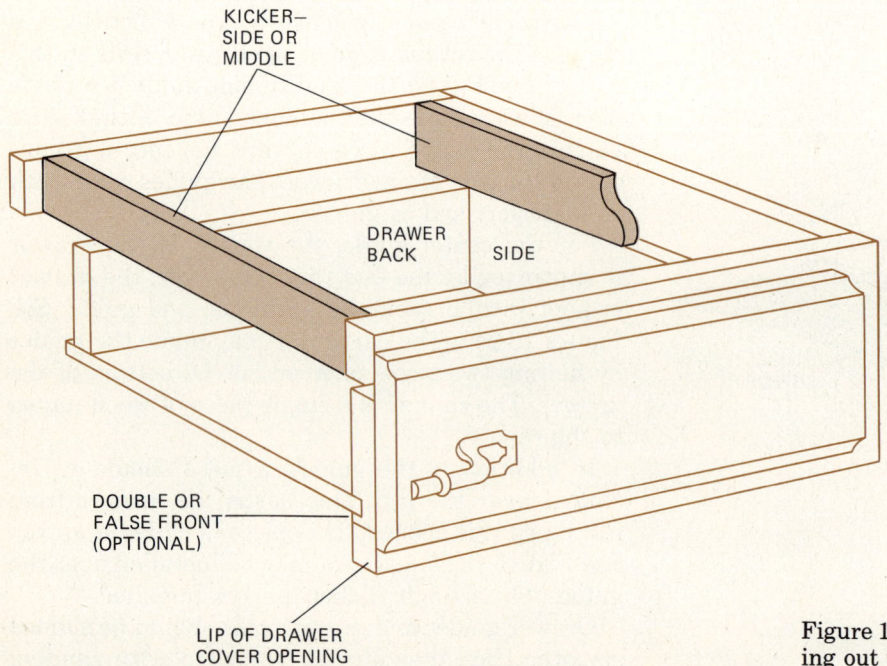

Figure 13-25. Kickers keep drawers from tilting out when opened.

354 Carpentry Fundamentals

sealer. The sealer should be sanded lightly and a coat of wax applied. In some cases, the wood may be sanded and wax applied directly to the wood. However, in either case, the wax will make the drawer slide better.

MAKING A DRAWER • Most drawers made by carpenters are called *lip drawers*. This type of drawer has a lip around the front. See Figure 13-26. The lip fits over the drawer opening and hides it. This gives a better appearance. It also lets the cabinet and opening be less accurate.

The other type of drawer is called a *flush drawer*. Flush drawers fit into the opening. For this reason, they must be made very carefully. If the drawer front is not accurate, it will not fit into the opening. Binding or large cracks will be the result. It is far easier to make a lip drawer.

Drawers may be made in several ways. As a rule, the procedure is to cut the right and left sides as in Figure 13-27. The back ends of these have dado joints cut into them. Note that the back of the drawer rests on the bottom piece. The bottom is grooved into the sides and front piece as shown.

The front of the drawer should be made carefully. It takes the greatest strain from opening and closing. The front should be fitted to the sides with a special joint. Several types of joint may be used. These are shown in Figure 13-28.

The highest quality work can feature a special joint called the dovetail joint. However, this joint is expensive to make and cut. As a rule, other types of dovetail joints are used. The dovetailed dado joint is often used.

Drawers are made after the cabinet frame has been assembled. Fronts of drawers may be of several shapes. The front may be paneled as in Figure 13-29A. Or it can be molded, as in Figure 13-29B. Both styles can be made thicker by adding extra boards.

The wood used for the drawer fronts is selected and cut to size. Joints are marked and cut for assembly. Next, the groove for the bottom is cut.

Then the stock is selected and cut for the sides. As a rule, the sides and backs are made from different wood than the front. This is to reduce cost. The drawer front may be made from expensive woods finished as required. However, the interiors are

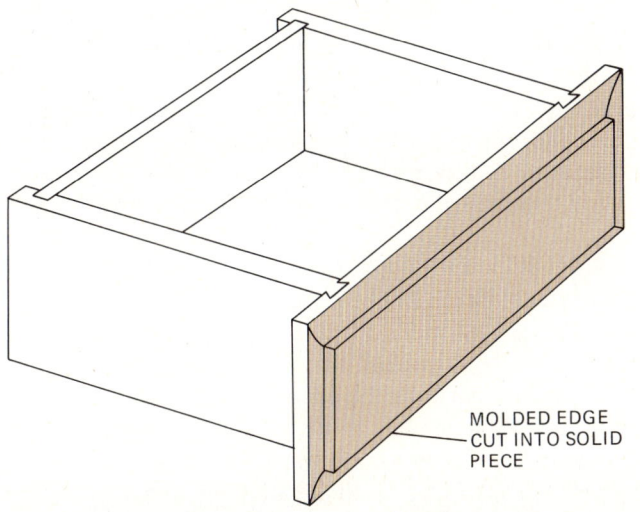

Figure 13-26. Lip drawers have a lip that covers the opening in the frame. Drawer fronts can be molded for appearance.

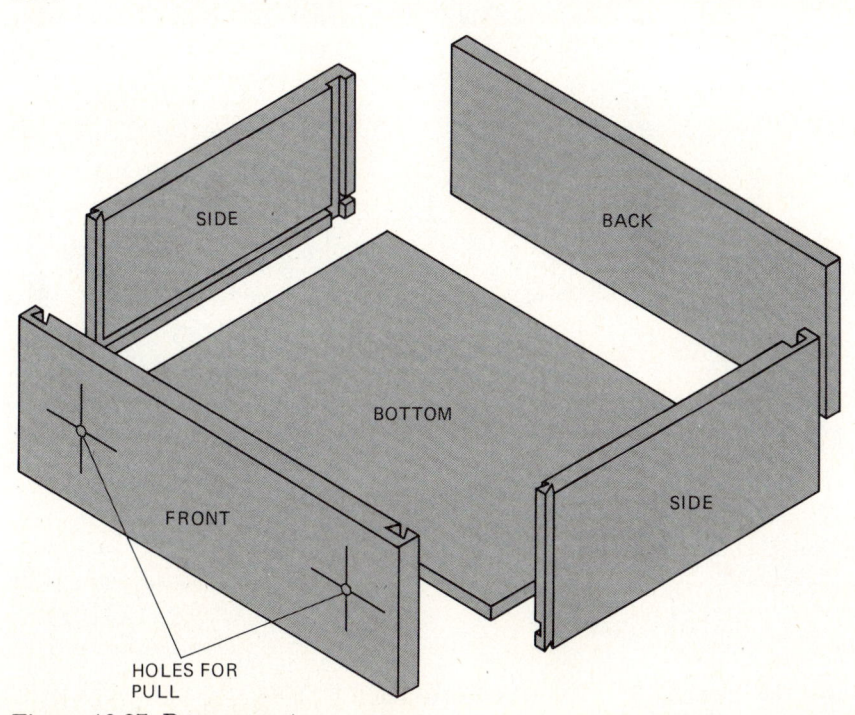

Figure 13-27. Drawer parts.

Finishing the Interior 355

made from less expensive materials. They are selected for straightness and sturdiness. Plywood is not satisfactory for drawer sides and backs.

The joints for the back are cut into the sides. The joints should be cut on the correct sides. Next, the grooves for the drawers are cut in each side. Be sure that they align properly with the front groove.

Next, the back piece is cut to the correct size. Again, note that the back rests on the bottom. No groove is cut. All pieces are then sanded smooth.

A piece of hardboard or plywood is chosen for the bottom. The front and sides of the drawer are assembled. The final measurements for the bottom are made. Then, the bottom piece is cut to the correct size. It is lightly sanded around the edges. Then the bottom is inserted into the grooves of the bottom and front. This is a trial assembly to check the parts for fit. Next, the drawer is taken apart again. Grooves are cut in the sides for the drawer guides. Also, any final adjustments for assembly are made.

The final assembly is made when everything is ready. Several processes may be used. However, it is best to use glue on the front and back pieces. Sides should not be glued. This allows for expansion and contraction of the materials.

The drawer is checked for squareness. Then one or two small finish nails are driven through the bottom into the back. One nail should be driven into each side as well. This will hold the drawer square. The bottom of the drawer is numbered to show its location. A like number is marked in the cabinet. This matches the drawer with its opening.

CABINET DOOR CONSTRUCTION • Cabinet doors are made in three basic patterns. These are shown in Figure 13-30. A common style is the rabbeted style. A rabbet about half the thickness of the wood is cut into the edge of the door. This allows the door to fit neatly into the opening with a small clearance. It also keeps the cabinet door from appearing bulky and thick. The outside edges are then rounded slightly. The door is attached to the frame with a special offset hinge. See Figures 13-31 and 13-32.

The molded door is much like the lip door. See Figure 13-30B. As can be seen in the figure, the back

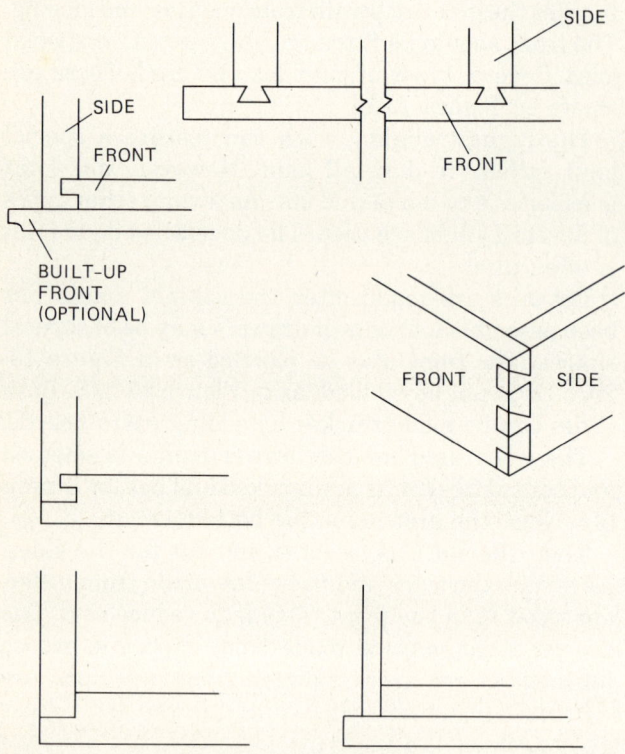

Figure 13-28. Joints for drawer fronts.

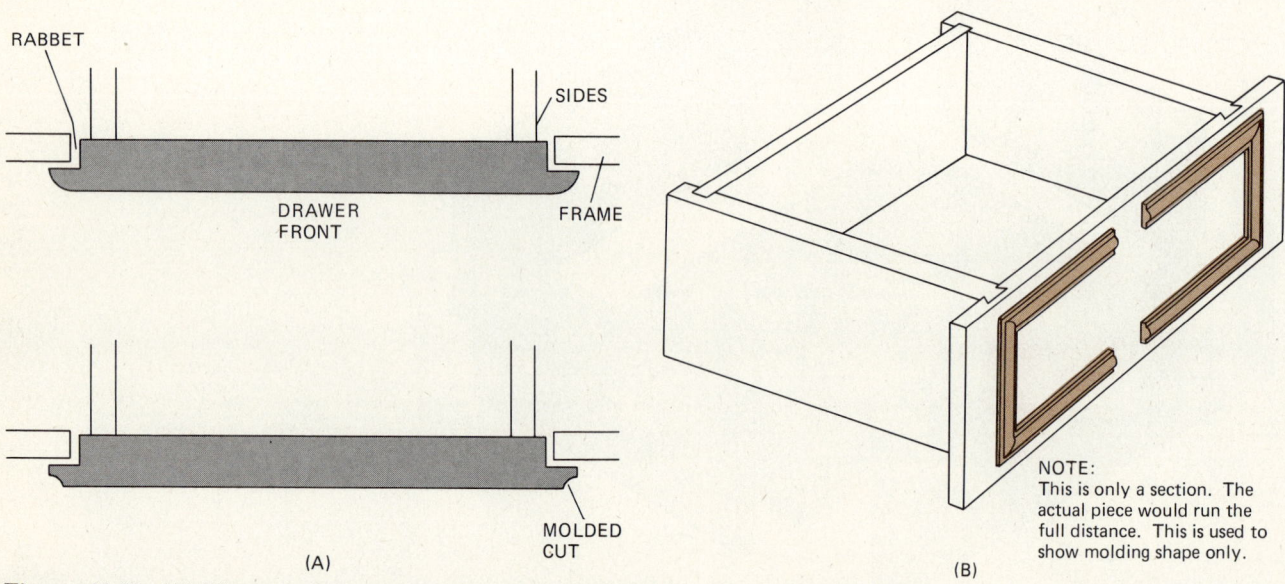

Figure 13-29. (A) These drawer types cover the opening. The accuracy needed for fitting is less. (B) Molding can be used to give a paneled effect.

356 Carpentry Fundamentals

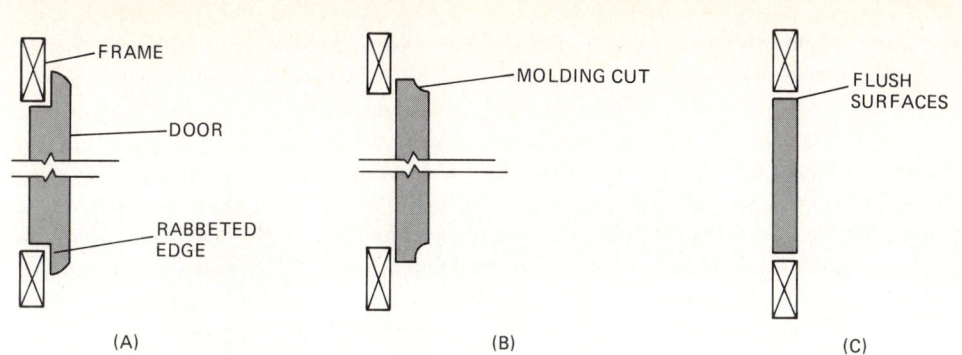

Figure 13-30. Cabinet door styles. (A) Rabbeted (B) Molded (C) Flush

Figure 13-31. A colorful kitchen with rabbeted lip drawers and cabinets. Note the lack of pulls. (Armstrong Cork)

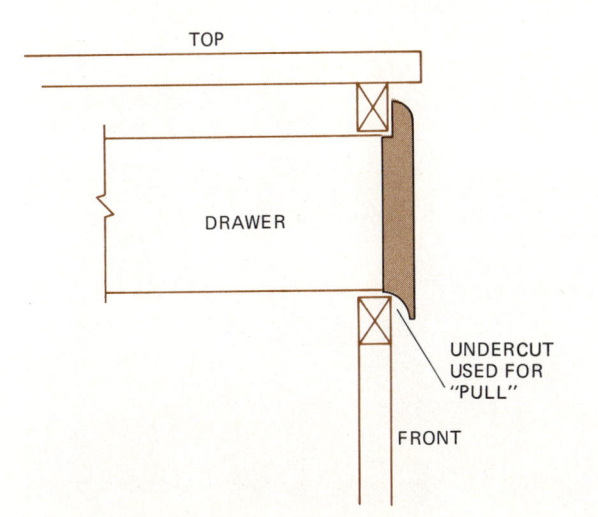

Figure 13-32. How drawers and doors are undercut so that pulls need not be used.

side of the door fits over the opening. However, no rabbet is cut into the edge. This way, no part of the door fits inside the opening. The edge is molded to reduce the apparent thickness. This can be done with a router. Or, it may be done at a factory where the parts are made. This method is becoming more widely used by carpenters. Its advantages are that it does not fit in the opening. That means no special fitting is needed. Also, special hinges are not required. The appearance gives extra depth and molding effects. These are not available for other types of cabinets.

Both styles of door are commonly paneled. Paneled doors give the appearance of depth and contour. Doors may be made from solid wood. However, when cabinets are purchased, the panel door is very common. The panel door has a frame much like that of a screen door. Grooves are cut in the edges of

Finishing the Interior **357**

the frame pieces. The panel and door edges are then assembled and glued solidly together. Cross sections of solid and built-up panels are shown in Figure 13-33.

Flush doors fit inside the cabinet. These appear to be the easiest to make. However, they must be cut very carefully. They must be cut in the same shape as the opening. If the cut is not carefully made, the door will not fit properly. Wide or uneven spaces around the door will detract from its appearance.

SLIDING DOORS • Sliding doors are also widely used. These doors are made of wood, hardboard, or glass. They fit into grooves or guides in the cabinet. See Figure 13-34. The top groove is cut deeper than the bottom groove. The door is installed by pushing it all the way to the top. Then the bottom of the door is moved into the bottom groove. When the door is allowed to rest at the bottom of the lower groove, the lip at the top will still provide a guide. Also, special devices may be purchased for sliding doors.

MAKING THE COUNTERTOP • Today, most cabinet tops are made from laminated plastics. There are many trademarks for these. These materials are usually $\frac{1}{16}$ to $\frac{1}{8}$ inch thick. The material is not hurt by hot objects, and does not stain or peel. The laminate material is very hard and durable. However, in a thin sheet it is not strong. Most cabinets today have a base top made from plywood or chipboard. Usually, a ½-inch thickness, or more, is used for counter tops made on the site.

Also, specially formed counters made from wood products may be purchased. These tops have the plastic laminates and the mold board permanently formed into a one-piece top. See Figure 13-35. This material may be purchased in any desired length. It is then cut to shape and installed on the job.

The top pieces are cut to the desired length. They

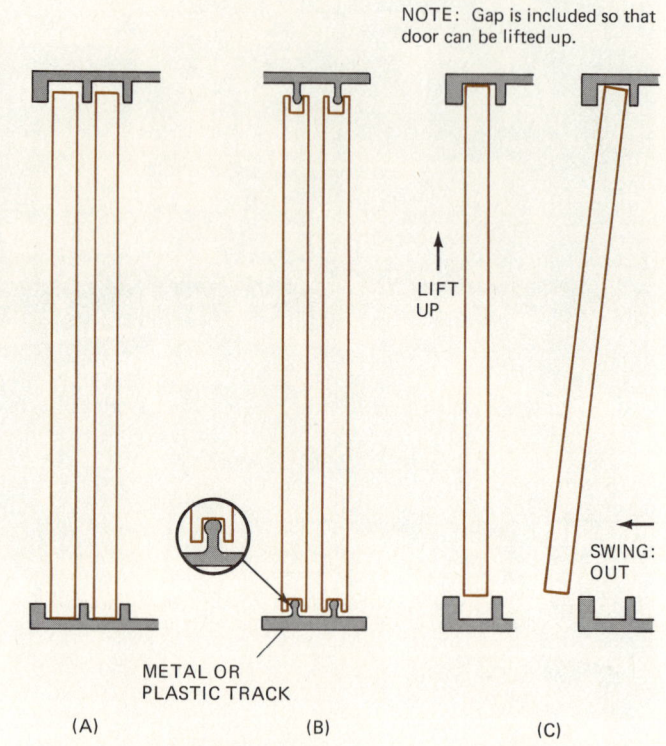

Figure 13-34. Sliding doors. (A) Doors may slide in grooves. (B) Doors may slide on a metal or plastic track. (C) To remove, lift up, and swing out.

Figure 13-33. (A) Cross section of a solid door panel. (B) Cross section of a built-up door panel.

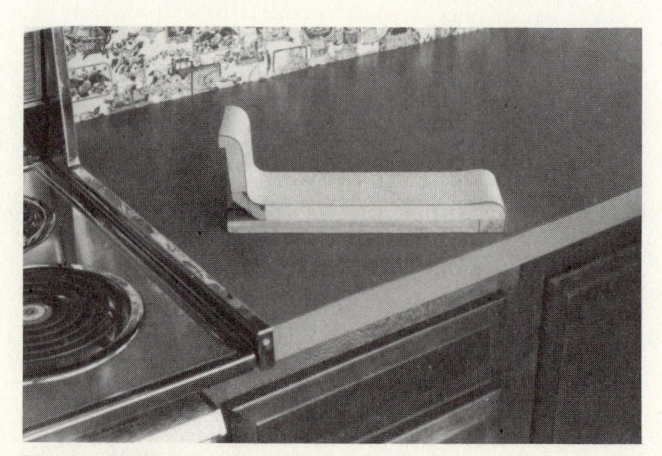

Figure 13-35. Counter tops may be flat or may include splash boards.

358   Carpentry Fundamentals

may be nailed to the partitions or to the drawer kickers on the counter. The top should extend over the counter approximately ⅜ inch. Next, sides or rails are put in place around the top. These pieces can be butted or rabbeted as shown in Figure 13-36. These are nailed to the plywood top. They may also be nailed to the frame of the counter. Next, they should be sanded smooth. Uneven spots or low spots are filled.

Particle board is commonly used for a base for the countertop. It is inexpensive, and it does not have any grain. Grain patterns can show through on the finished surface. Also, the grain structure of plywood may form pockets. Glue in pockets does not bond to the laminate. Particle board provides a smooth, even surface that bonds easily.

Once the counter has been built, the top is checked. Also, any openings should be cut. Openings can be cut for sinks or appliances. Next, the plastic laminate is cut to rough size. Rough size should be ⅛ to ¼ inch larger in each dimension. A saw is used to cut the laminate as in Figure 13-37.

Next, contact cement is applied to both the laminate and the top. Contact cement should be applied with a brush or a notched spreader. Allow both surfaces to dry completely. If a brush is used, solvent should be kept handy. Some types of contact cement are water-soluble. This means that soap and water can be used to wash the brush and to clean up.

When the glue has dried, the surface is shiny. Dull spots mean that the glue was too thin. Apply more glue over these areas.

It usually takes about 15 or 20 minutes for contact cement to dry. As a rule, the pieces should be joined within a few minutes. If they are not, a thin coat of contact cement is put on each of the surfaces again.

To glue the laminate in place, two procedures may be used. First, if the piece is small, a guide edge is put in place. The straight piece is held over the area and the guide edge lowered until it contacts. The entire piece is lowered into place. Pressure is applied from the center of the piece to the outside edges. The hands may be used, but a roller is better.

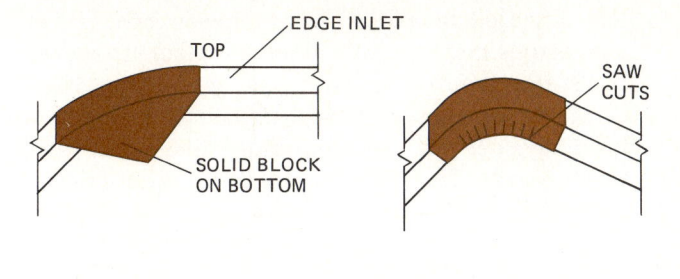

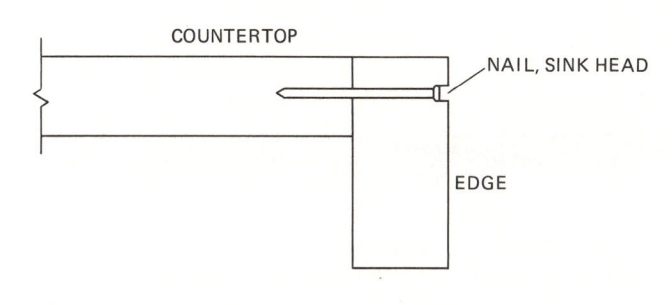

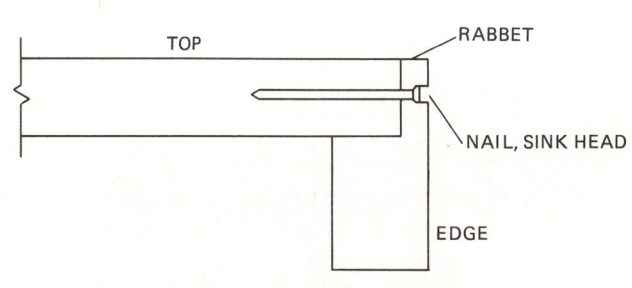

Figure 13-36. Blocking up counter edges.

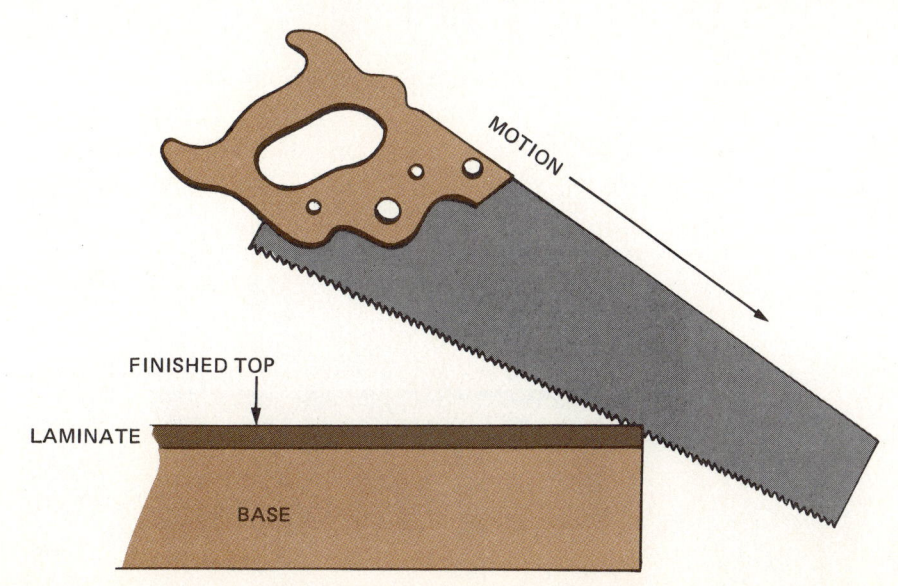

Figure 13-37. Cut into the laminate to avoid chipping.

Finishing the Interior **359**

For larger pieces, a sheet of paper is used. Wax paper may be used, but almost any type of paper is acceptable. The glue is allowed to dry first. Then the paper is placed on the top. The laminate is placed over the paper. Then, the laminate is positioned carefully. The paper is gently pulled about 1 inch from beneath the laminate. The position of the laminate is checked. If it is in place, pressure may be applied to the exposed edge. If it is not in place, the laminate is moved until it is in place. Then, the exposed edge is pressed until a bond is made. Then the paper is removed from the entire surface. Pressure is applied from the middle toward the edges. See Figure 13-38A to M.

TRIM FOR LAMINATED SURFACES • The edges should be trimmed. The pieces were cut slightly oversize to allow for trimming. The tops should extend over the sides slightly. The tops and corners should be trimmed so that a slight bevel is exposed. This may be done with a special router bit as in Figure 13-39A. It may also be done with a sharp and smooth file as shown in Figure 13-39B.

The back of most countertops has a raised portion. This is called a *splash board*. Splash boards and countertops may be molded as one piece. However, splash boards are also made as two pieces. Then, metal cove and cap strips are applied at the corners. Building codes may set a minimum height for these splash boards. The FHA requires a minimum height of 4 inches for kitchen counters.

INSTALLING HARDWARE • Hardware for doors and drawers means the knobs and handles. These

Figure 13-38. Applying plastic laminate to countertops.

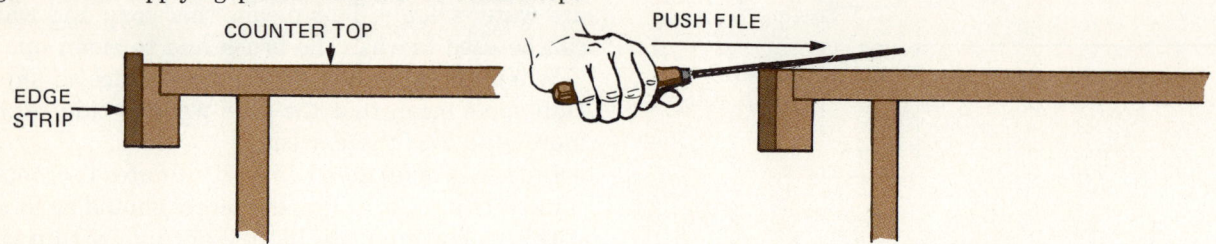

(A) Apply the edge strip.

(B) Trim the edge strip flush with the top.

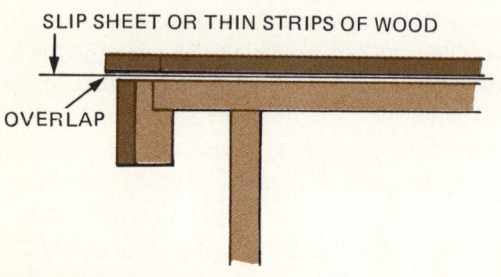

(C) Apply glue. Lay a slip sheet or sticks in place when the glue is dry.

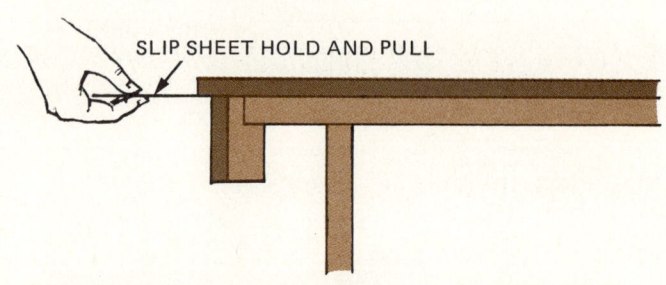

(D) Position the top. Remove the slip sheet or sticks.

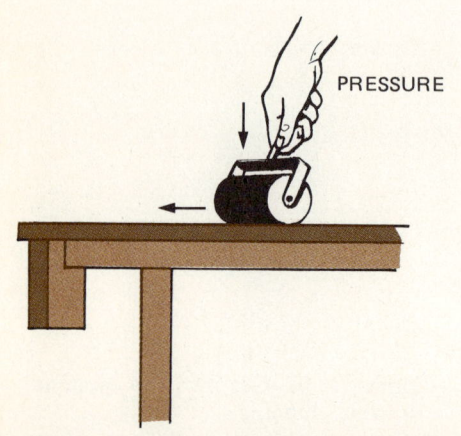

(E) Apply pressure from center to edge.

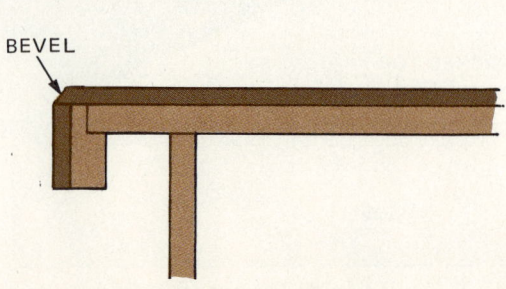

(F) Trim the edge at slight bevel.

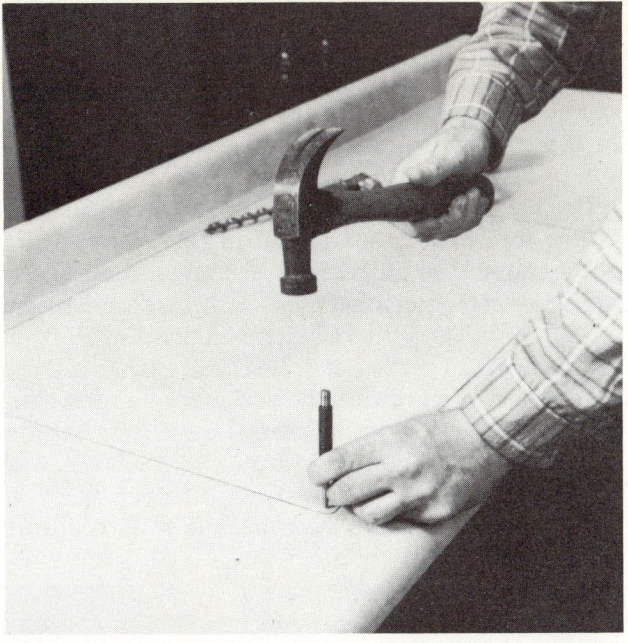

(G) To cut holes for sinks, first center punch for drilling. *(Formica)*

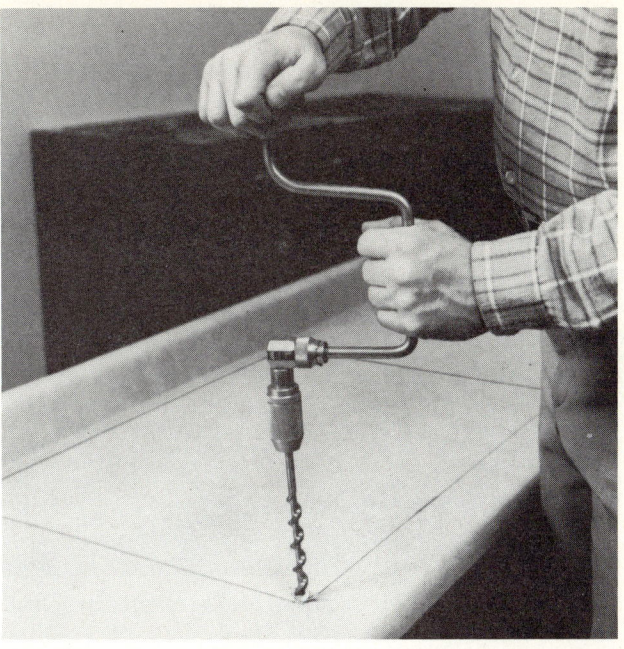

(H) Next drill holes at corners. *(Formica)*

(I) Then cut out opening. *(Courtesy Rockwell International, Power Tool Division)*

(J) Lay sink in opening. *(Formica)*

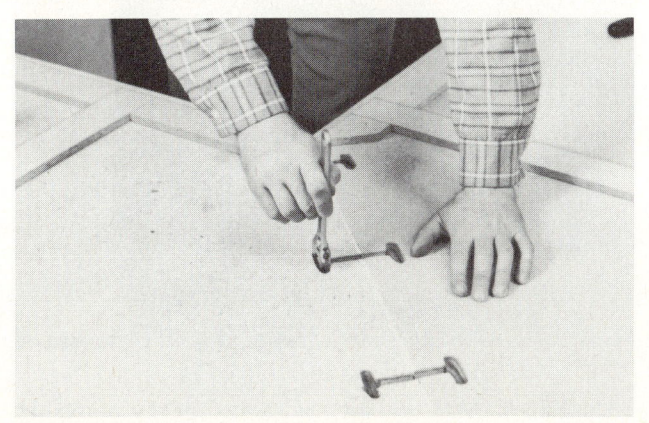

(K) Mitered corners can be held with special clamps placed in cut-outs on the bottoms. *(Formica)*

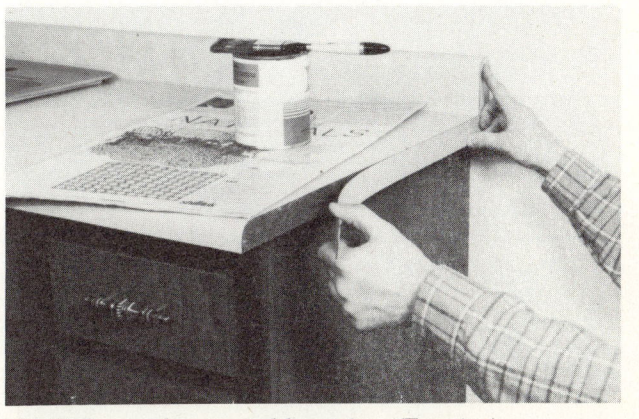

(L) Ends may be covered by strips. *(Formica)*

Finishing the Interior **361**

are frequently called *pulls*. A variety of styles are available. As a rule, drawers and cabinets are put into place for finishing. However, pulls and handles are not installed. Hinges are applied in many cases. In others, the doors are finished separately. However, pulls are left off until the finish is completed.

Drawer pulls are placed slightly above center. Wall cabinet pulls are placed in the bottom third of the doors. Door pulls are best put near the opening edge. For cabinet doors in bottom units, the position is different. The pull is located in the top third of the door. It is best to put it near the swinging edge. Some types of hardware, however, may be installed in other places for special effects. See Figure 13-40A and B. These pulls are installed in the center of the door panel.

To install pulls, the location is first determined.

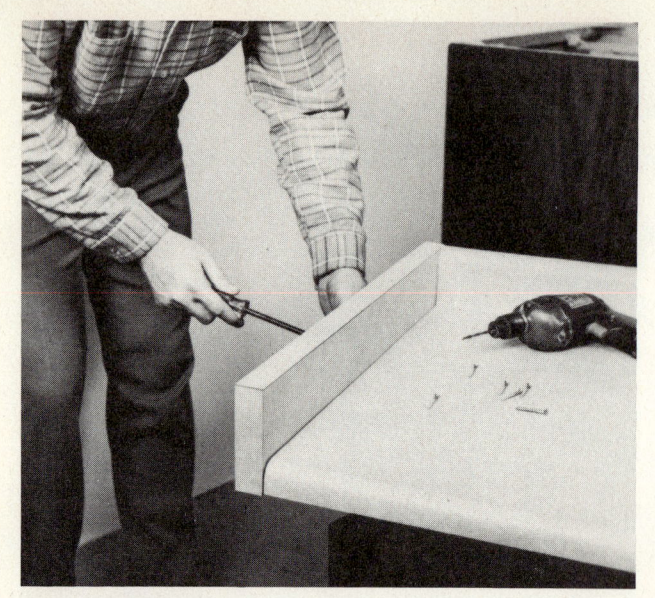

Figure 13-38 *(cont.)*(M) Ends may also be covered by splash boards. *(Formica)*

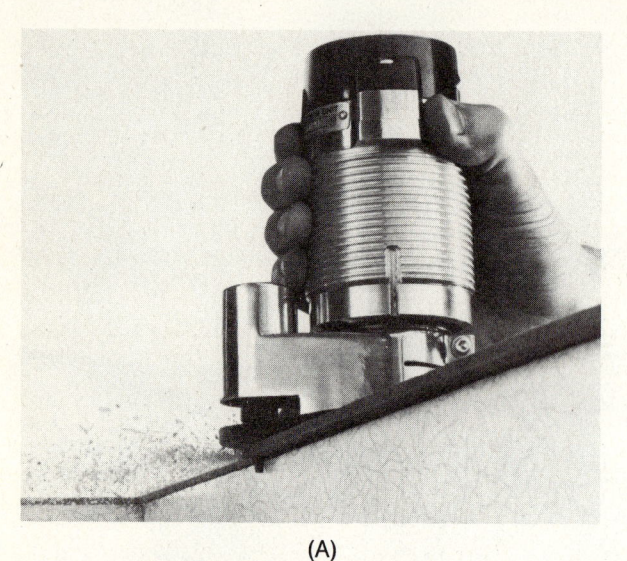

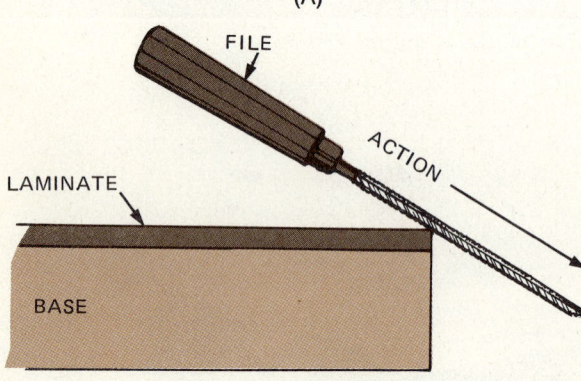

Figure 13-39. (A) Edges may be trimmed with edge trimmers or routers. *(Courtesy Rockwell International, Power Tool Division)* (B) Edges may also be filed. Note direction of force.

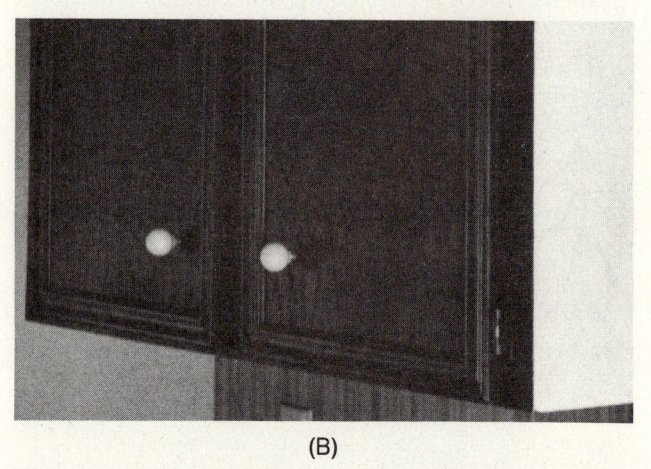

Figure 13-40. (A) Pulls may be located in the center for dramatic effect. (B) Regular pull location.

362   Carpentry Fundamentals

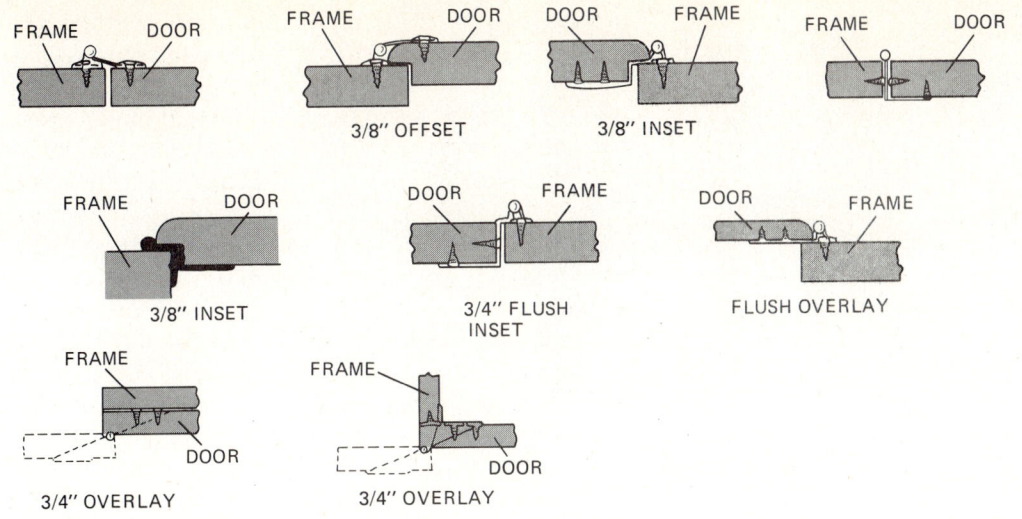

Figure 13-41. Hinge types.

Sometimes a template can be made and used. Whatever method is used, the locations of the holes are found. They are marked with the point of a sharp pencil. Next, the holes are drilled from front to back. It is a good idea to hold a block of wood behind the area. This reduces splintering.

There are other types of hardware. These include door catches, locks, and hinges. The carpenter should always check the manufacturer's instructions on each.

As a rule, it is easier to attach hinges to the cabinet first. Types of hinges are shown in Figure 13-41. Types of door catches are shown in Figure 13-42.

Shelves

Most kitchen and bathroom cabinets have shelves. Also, shelves are widely used in room dividers, bookshelves, and closets. There are several methods of shelf construction that may be used. Figure 13-43 shows some types of shelf construction. Note that each of these allows the shelf location to be changed.

For work that is not seen, shelves are held up by ledgers. Figure 13-44 shows the ledger method of shelf construction.

Special joints may also be cut in the sides of solid pieces. These are types of dados and rabbet joints. Figure 13-45 shows this type of construction.

As a rule, a ledger-type shelf is used for shelves in closets, lower cabinets, and such. However, for exposed shelves a different type of shelf arrangement is used. Adjustable or jointed shelves look better on bookcases and such.

Facing pieces are used to hide dado joints in shelves. See Figure 13-46. The facing may also be used to make the cabinet flush with the wall. The facing can be inlet into the shelf surface.

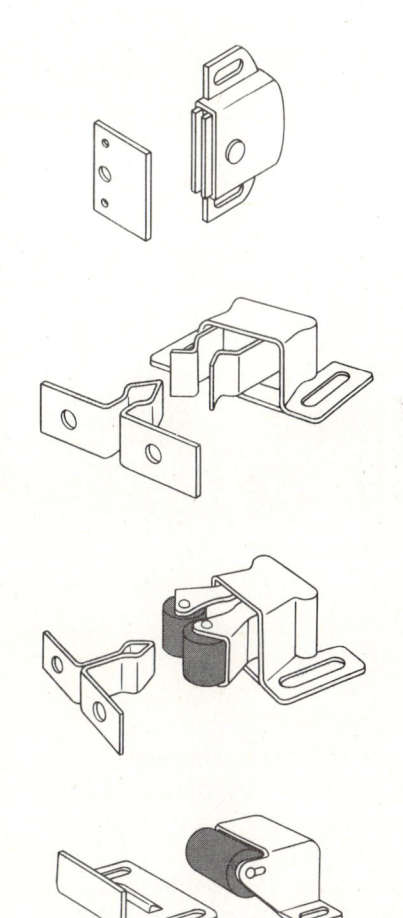

Figure 13-42. Types of cabinet door catches.

Applying Finish Trim

Finish trim pieces are used on the base of walls. They cover floor seams and edges where carpet has been laid. Also, trim is used around ceilings, windows, and other areas. As a rule, a certain procedure

Finishing the Interior **363**

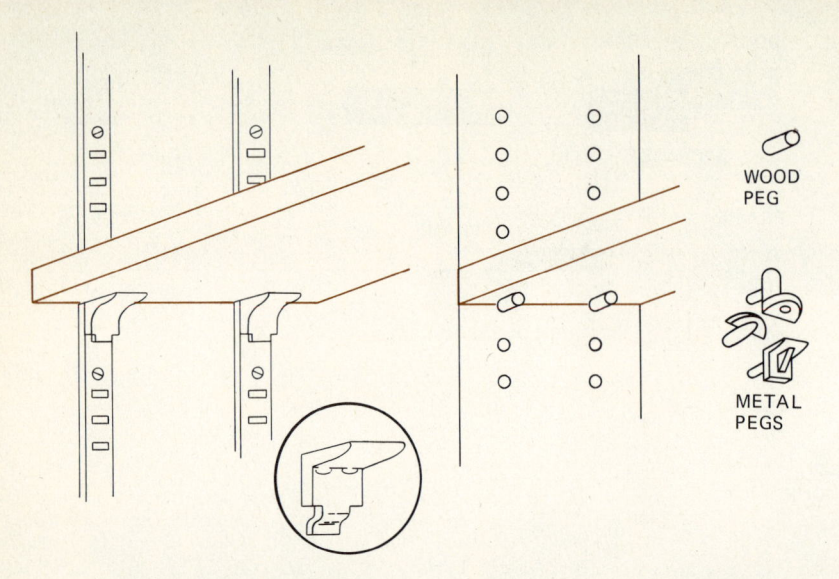

Slotted bookshelf standards and clips are ideal if you want to adjust bookcase shelves, but they add to total cost.

If you use wood or metal pegs set into holes, you must be sure to drill holes at the same level and 3/8 inch deep.

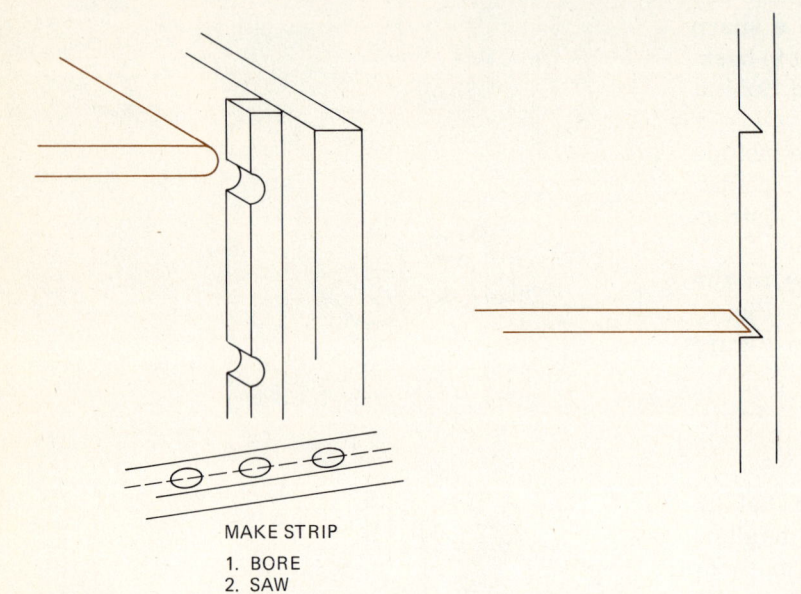

MAKE STRIP
1. BORE
2. SAW

Figure 13-43. Methods of making adjustable shelves.

is followed for cutting and fitting trim pieces. Outside corners, as in Figure 13-47, are cut and fit with miter joints. These are cut with a miter box. See Figure 13-48.

However, trim for inside corners is cut with a different joint. This is done because most corners are not square. Miter joints do not fit well into corners that are not square. Unsightly gaps and cracks will be the result of a poor fit. Instead, a coped joint is used. See Figure 13-47.

For a coped joint, the first piece is butted against the corner. Then the outline is traced on the second piece. A scrap piece is used for a guide. The outline is then cut using a coping saw. See Figure 13-49. The coped joint may be effectively used on any size or shape of molding.

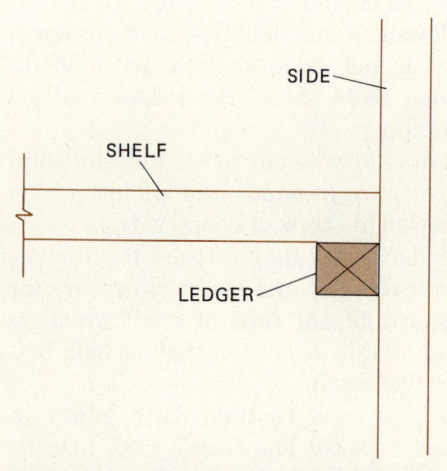

Figure 13-44. Ledgers or cleats are used for shelves and steps where appearance is not important.

364 Carpentry Fundamentals

DADO JOINT—
WITH FASCIA STRIP

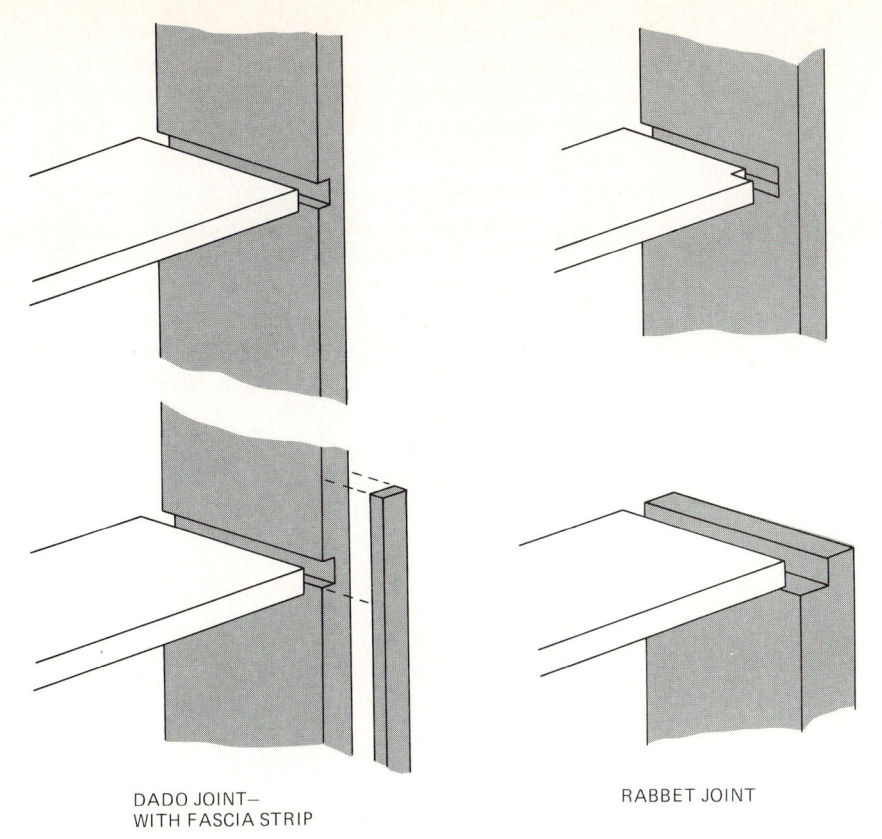

RABBET JOINT

Figure 13-45. Types of dado joints.

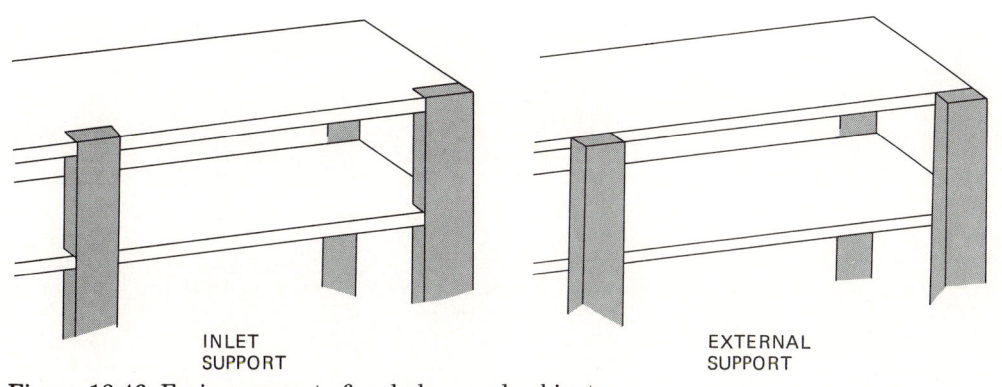

INLET SUPPORT

EXTERNAL SUPPORT

Figure 13-46. Facing supports for shelves and cabinets.

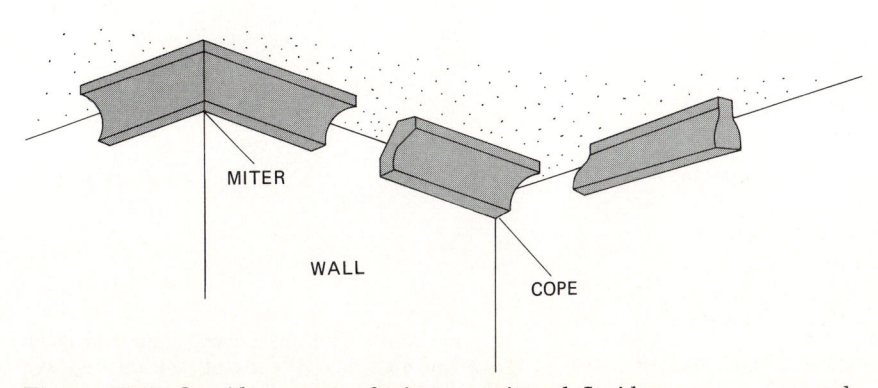

MITER

WALL

COPE

Figure 13-47. Outside corners of trim are mitered. Inside corners are coped.

Finishing the Interior **365**

• APPLYING FINISH MATERIALS •

As a rule, the millwork is finished before the wall surfaces. Many wood surfaces are stained rather than painted. This enhances natural wood effects. Woods such as birch are commonly stained to resemble darker woods. Darker woods are walnut, dark oak, pecan, and so on. These stains and varnishes are easily absorbed by the wall. They are put on first so that they do not ruin the wall finish.

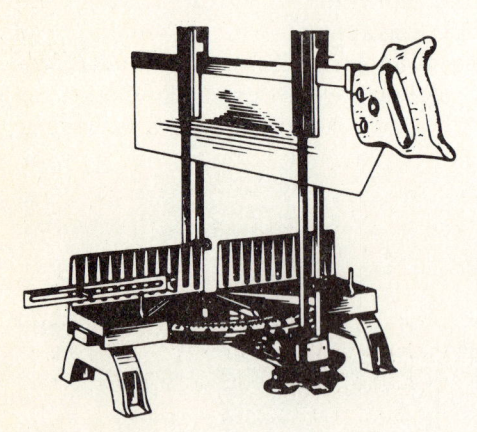

Paint for wood trim is usually a gloss or semigloss paint. These paints are more washable, durable, and costly. Flat or nonglare paints are widely used on walls.

Paints, stains, and varnishes are often applied by spraying. This is much faster than rolling. When spray equipment is used, there is always an overspray. This overspray would badly mar a wall finish. However, the wall finish can be put on easily over the overspray. See Figure 13-50.

To prepare wood for stain or paint, first sand it smooth. Mill and other marks should be sanded until they cannot be seen. Hand or power sanding equipment may be used.

Applying Stain

Stain is much like a dye. It is clear and lets the wood grain show through. It simply colors the surface of the wood to the desired shade. It is a good idea to make a test piece. The stain can be tested

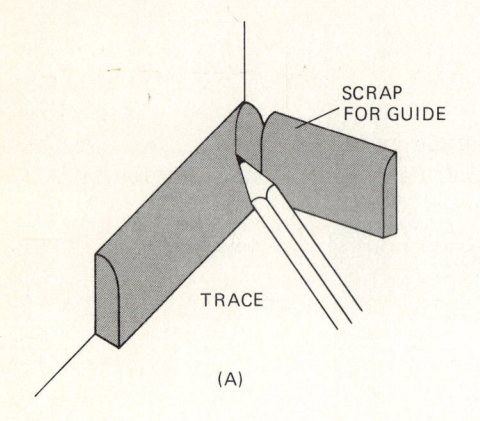

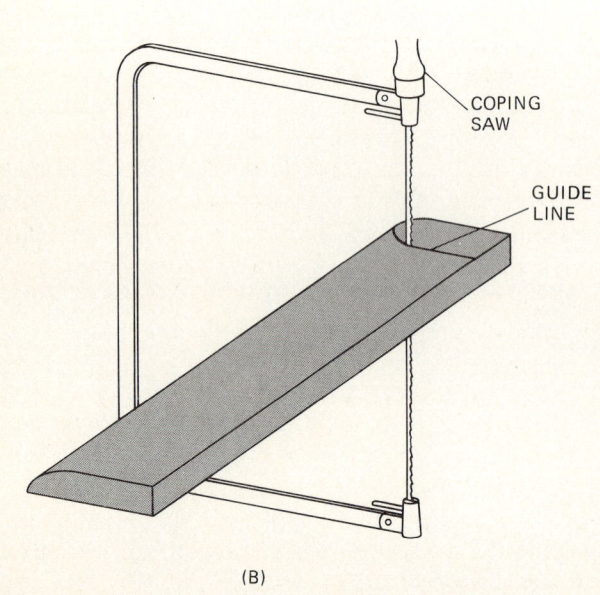

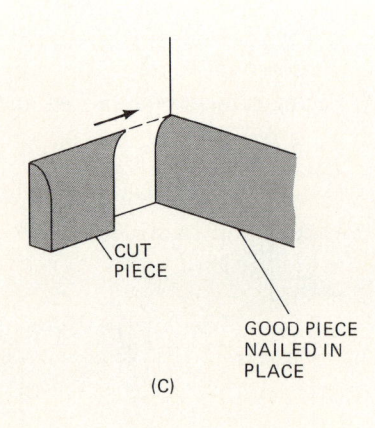

Figure 13-48. A miter box used for cutting miters.

Figure 13-49. Making a coped joint. (A) Tracing the outline. (B) Cutting the outline. (C) Nailing the molding in place.

366 Carpentry Fundamentals

Figure 13-50. Cabinets are stained before walls are painted or papered. Note the overspray around the edges.

on it first. A scrap piece of the same wood to be finished is used. It is sanded smooth and the stain is applied. This lets a worker check to see if the stain will give the desired appearance.

Stain is applied with a brush, a spray unit, or a soft cloth. It is applied evenly and with long, firm strokes. The stain is allowed to sit and penetrate for a few minutes. Then it is wiped with a soft cloth. It is also a good idea to check the manufacturer's application instructions. There are many different types of stains. Thus there are several ways of applying stains.

The stain should dry for a recommended period of time. Then it should be smoothed lightly with steel wool. Very little pressure is applied. Otherwise, the color will be rubbed off. If this happens, the place should be retouched with stain.

The stain is evenly smoothed. Then, varnish or lacquer may be applied with a brush or with a spray gun. Today, most interior finishes are sprayed. See Figure 13-50.

Applying the Wall Finish

Two types of wall finishes are commonly used. The first is wall paint. The other is wallpaper.

PAINTING A WALL • The wood trim for doors and so forth is finished. Then, the walls may be painted. Most wall paint is called *flat* paint. This means that it does not shine. It makes a soft, nonglare surface. Most wall paints are light-colored to reflect light. Before painting, all nail holes or other marks are filled. Patching paste should be used for this.

There are several methods of applying paint. It may be brushed, sprayed, or rolled. Also, special types of paint may be applied. Plain paint is a liquid or gel that gives a smooth, flat finish. However, special "texture" paints are also used. Sand-textured paints leave a slightly roughened surface. This is caused by particles of sand in the paint. Also, thick mixtures of paints are used. These may be rolled on to give a heavy-textured surface. Thicker paints can be used for a shadowed effect.

PUTTING WALLPAPER ON WALLS • Wallpaper is widely used in buildings today. However, the "paper" may not be paper. It may be various types of vinyl plastic films. Also, a mixture of plastic and paper is common. In either case, the wall surface should be prepared. It should be smooth and free of holes or dents. As a rule, a single roll of wallpaper will cover about 30 square feet of wall area. A special wall sealer coat called *size* or *sizing* is used. This may be purchased premixed and ready to use. Powder types may be purchased and mixed by the worker. In either case, the sizing is painted on the walls and allowed to dry. As it dries, it seals the pores in the wall surface. This lets the paste or glue adhere properly to the paper.

A corner is chosen for a starting point. It should be close to a window or door. The width of the wallpaper is measured out from the corner. One inch is taken from the width of the wallpaper. Make a small mark at this distance. The mark should be made near the top of the wall. For a 27-inch roll, 26 inches is measured from the corner. A nail is driven near the ceiling for a chalk line. The chalk line is tied to this nail at the mark. The end of the chalk line is weighted near the floor. The line is allowed to hang free until it comes to a stop. When still, the line is held against the wall. The line is snapped against the wall. This will leave a vertical mark on the wall. See Figure 13-51A.

Next, several strips of paper are cut for use. The distance from the floor to the ceiling is measured. The wallpaper is unrolled on the floor or a table. The pattern side is left showing. The distance from the floor to the ceiling is laid off and 4 inches is added. This strip is cut, and several more are cut.

The first precut strip is laid on a flat surface. The pattern should be face up. The strip is checked for appearance and cuts or damage. Next, the strip is

Finishing the Interior 367

Figure 13-51. Putting wallpaper on walls.

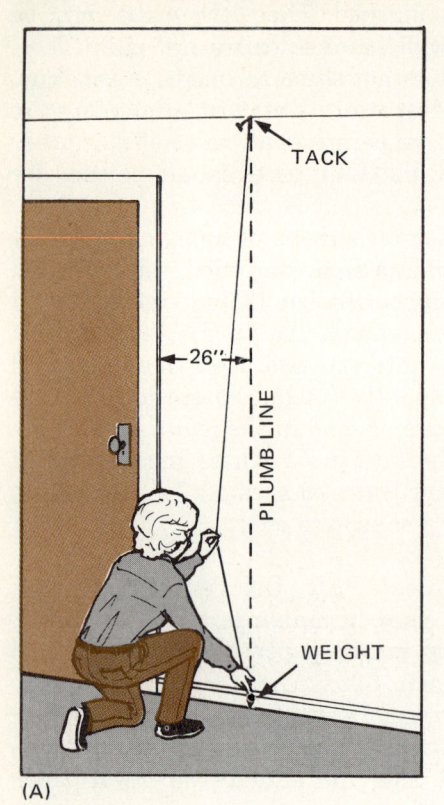

(A) A chalk line and plumb are used to mark the starting point.

(B) After the paper has been measured and cut, paste is brushed on the back.

(C) The bottom and top edges are used to carry the pasted wall covering.

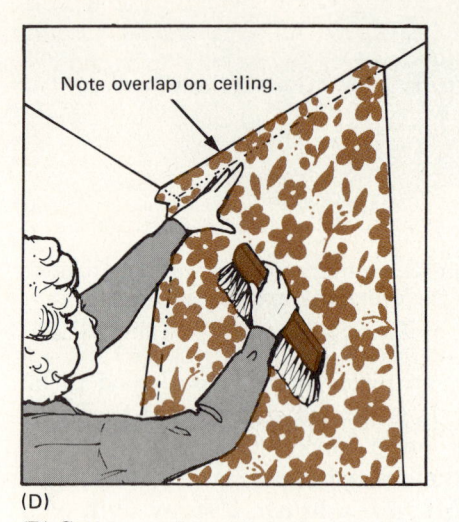

(D) Strips are lapped at the ceiling and brushed down.

(E) Use a putty knife or a straightedge as a guide to trim overlaps.

368 Carpentry Fundamentals

turned over so that the pattern is face down. The paste is applied with the brush. See Figure 13-51B. The entire surface of the paper is covered. Next, the strip is folded in the middle. The pattern surface is on the inside of the fold. This allows the worker to hold both the top and the bottom edges. See Figure 13-51C.

The plumb line is used as a starting point. The first strip is applied at the ceiling. About 1 inch overlaps the ceiling. This will be cut off later. A stiff bristle brush is moved down the strip. See Figure 13-51D. Take care that the edges are aligned with the chalk line. About 1 inch will extend around the corner. This will be trimmed away later. This is an allowance for an uneven corner.

The entire strip is then brushed to remove the air bubbles. The brush is moved from the center toward the edge.

The second strip is prepared in the same manner. However, care is taken to be sure that the pattern is matched. The paper is moved up or down to match the pattern. The edge of the second strip should exactly touch the edge of the first piece.

Next, the second piece is folded down. The edge is exactly matched with the edge of the first piece. The edge should exactly touch the edge of the previous piece. Then, any bubbles are smoothed out. The process is repeated for each strip until the wall is finished. Then the wall is trimmed.

The extra overlap at the ceiling and at the base are cut. A razor blade or knife is used with a straightedge as a guide. See Figure 13-51E.

A new line is plumbed at each new corner. The first strip on each corner is started even with the plumb line. This way each strip is properly aligned. The corners will not have unsightly gaps or spaces.

A diagonal cut is made at each corner of a window. About ½ inch is allowed around each opening. The diagonal cut forms a flap over the molding. This allows the opening to be cut for an exact fit. See Figure 13-51F.

• FLOOR PREPARATION AND FINISH •

The floor is finished last. This really makes sense if you think about it. After all, people will be working in the building. They will be using paint, varnish, stain, plaster, and many other things. All of these could damage a finished floor if they fell on it.

Also, workers will be often entering and leaving the building. Mud, dirt, dust, and trash will be tracked into the building. It is very difficult to paint or plaster without spilling. A freshly varnished floor or a newly carpeted one could be easily ruined. So the floor is finished last to avoid damage to the finish floor.

Special methods are sometimes needed to lay flooring over concrete. Figure 13-52 shows some steps in this. Also, this was discussed in an earlier chapter.

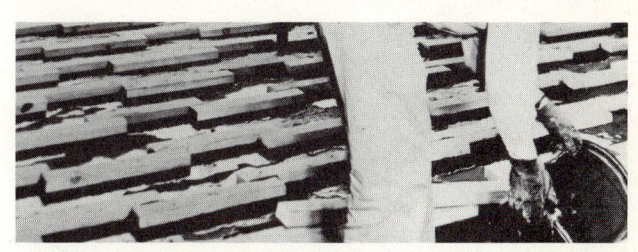

(A) Tar or asphalt is poured on. It acts as a vapor barrier.

(B) Wood strips are positioned on the asphalt.

Figure 13-51 *(cont.)* (F) Cut corners diagonally at windows and doors.

Figure 13-52. Laying wood floors on concrete. *(National Oak Flooring Manufacturers Assn.)*

Finishing the Interior **369**

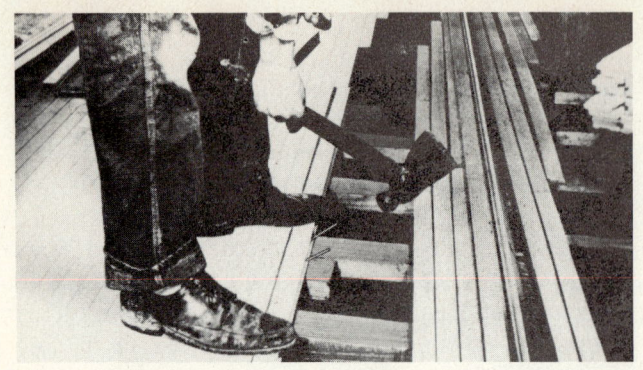

Figure 13-52 *(cont.)* (C) Flooring is nailed to the strips. *(National Oak Flooring Manufacturers Assn.)*

Figure 13-53. Subflooring must be scraped and cleaned. *(National Oak Flooring Manufacturers Assn.)*

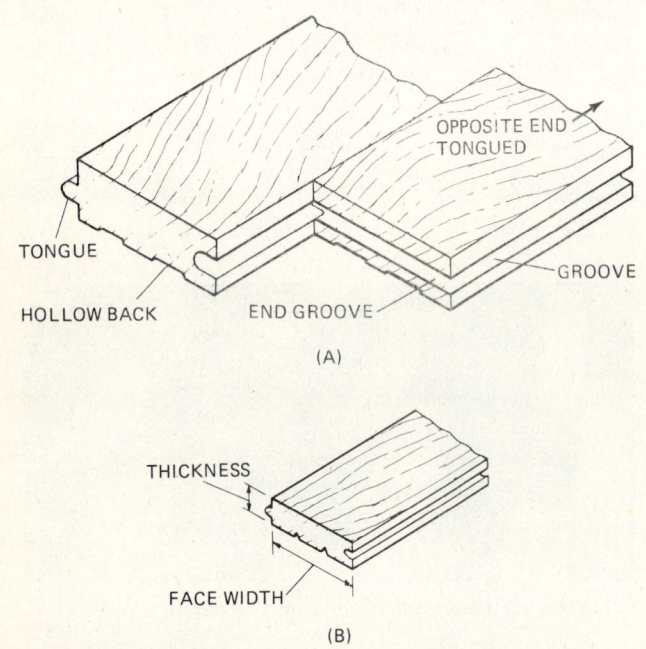

Figure 13-54. Strip flooring. *(Courtesy Forest Products Laboratory)*

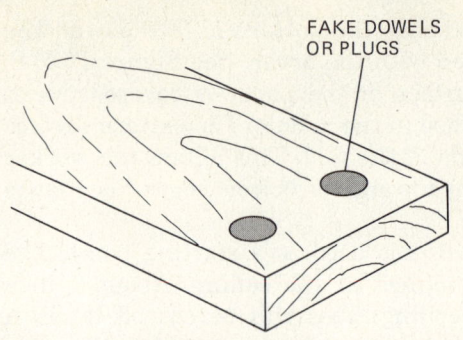

Figure 13-55. Plank flooring.

For all types of floors, the first step is to clean them. See Figure 13-53. The surface is scraped to remove all plaster, mud, and other lumps. Then the floor is swept with a broom.

Laying Wooden Flooring

Wooden flooring most often comes in three shapes. The first is called *strip* flooring (Figure 13-54). The second is called *plank* flooring as in (Figure 13-55). The third type is *block* or *parquet* flooring (Figure 13-56). The blocks may be solid as in Figure 13-57A. Here the grain runs in one direction and the blocks are cut with tongues and grooves. The type shown in Figure 13-57B may be straight-sided. They are made of strips with the grain running in the same direction. Such blocks may have a spline at the bottom or a bottom layer.

The type of block shown in Figure 13-57C is made of several smaller pieces glued to other layers. The bottom layer is usually waterproof.

Most flooring has cut tongue-and-groove joints. Hidden nailing methods are used so that the nails do not show when the floor is done. Strip and plank flooring will also have an undercut area on the bottom. This undercut is shown in Figure 13-54. The undercut or *hollow* helps provide a stable surface for the flooring. Small bumps will not make the piece shift. Strip flooring has narrow, even widths with tongue-and-groove joints on the ends. The strips may be random end matched. Plank flooring comes in both random widths and lengths. It may be drilled and pegged at the ends. However, today most plank flooring has fake pegs that are applied at the factory. The planks are then nailed in much the same manner as strip flooring. Block flooring may be either nailed or glued, but glue is used most often.

Most flooring is made from oak. Several grades and sizes are available. However, the width or size is largely determined by the type of flooring.

Carpeting can cost much more than a finished wood floor. Also, carpeting may last only a few years. As a rule, wood floors last for the life of the building.

Figure 13-56. A parquet floor of treated wood blocks. *(Permagrain Products)*

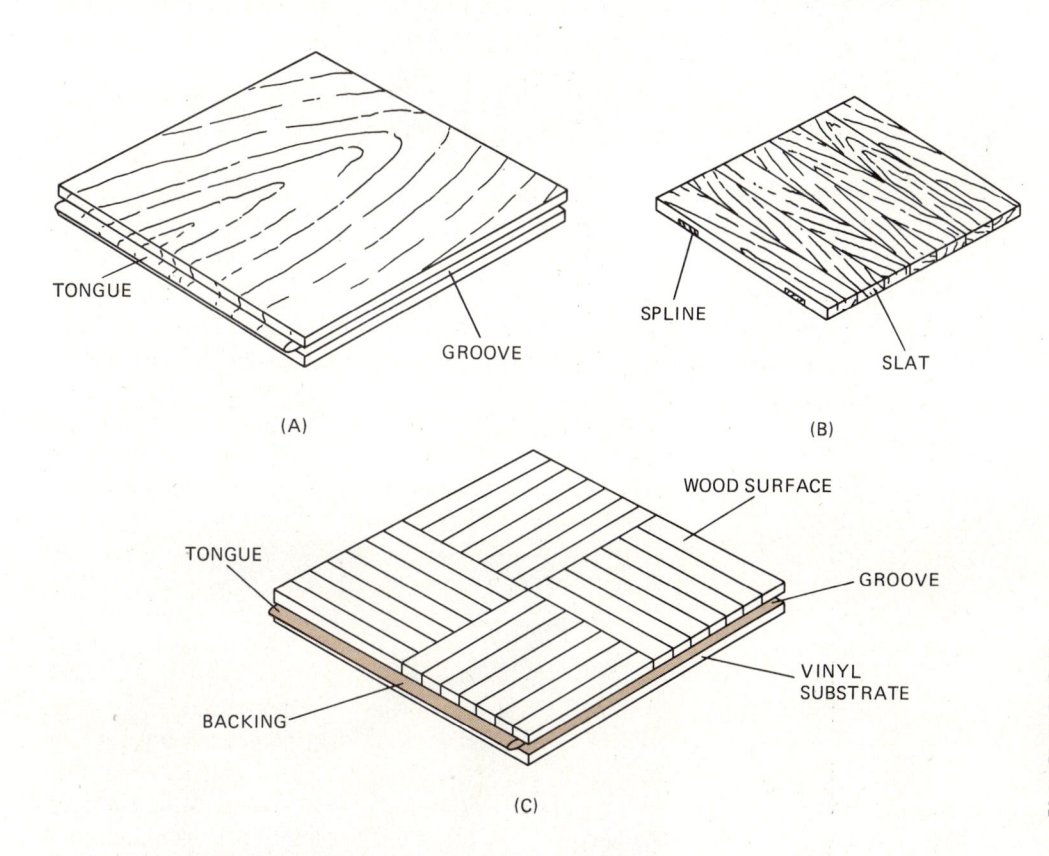

Figure 13-57. Wood block (parquet) flooring. (A) Solid. *(Courtesy Forest Products Laboratory)* (B) Splined. *(Courtesy Forest Products Laboratory)* (C) Substrated.

It is a good idea to have hardwood floors underneath carpet. Then the carpet can be removed without greatly reducing the resale value of the building.

Preparation for Laying Flooring

Manufacturers recommend that flooring be placed inside for several days before it is laid. The bundles should be opened and the pieces scattered around the room. This lets the wood reach a moisture content similar to that of the room. This will help stabilize the flooring. If the flooring is stabilized, the expansion and contraction will be even.

Check the subfloor for appearance and evenness. The subfloor should be cleaned and scraped of all deposits and swept clean. Nails or nail heads should be removed. All uneven features should be planed or sanded smooth. See Figure 13-66.

A vapor barrier should be laid over the subfloor. It can be made of either builder's felt or plastic film. Seams should be overlapped 2 to 4 inches. Then chalk lines should be snapped to show the centers of the floor joists. See Figure 13-58.

INSTALLING WOODEN STRIP FLOORING • Wooden strip flooring should be applied perpendicular to the floor joists. See Figure 13-59. The first strip is laid with the grooved edge next to the wall.

Finishing the Interior **371**

Figure 13-58. Vapor barriers are needed over board subfloors. *(National Oak Flooring Manufacturers Assn.)*

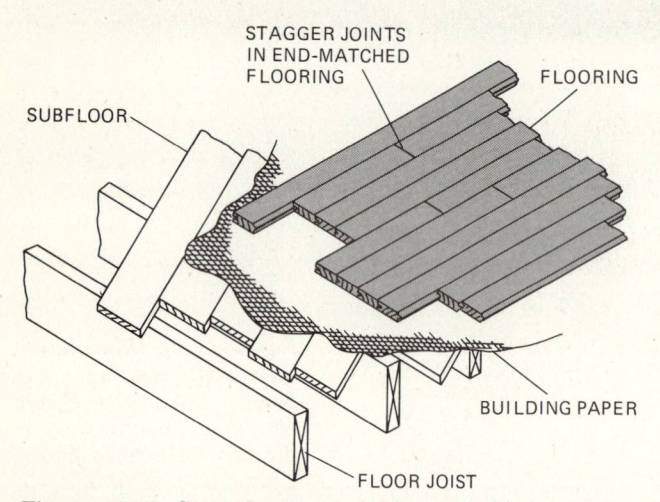

Figure 13-59. Strip flooring is laid perpendicular to the joists. *(Courtesy Forest Products Laboratory)*

At least ½ inch is left between the wall and the flooring. This space controls expansion. Wooden flooring will expand and contract. The space next to the wall keeps the floor from buckling or warping. Warps and buckles can cause air gaps beneath the floor. They also ruin the looks of the floor. The space next to the wall will be covered later by molding and trim.

The first row of strips is nailed using one method. The following rows are nailed differently. In the first row, nails are driven into the face. See Figures 13-60 and 13-61. Later, the nail is set into the wood and covered.

Figure 13-60. Nailing the first strip. *(National Oak Flooring Manufacturers Assn.)*

372 Carpentry Fundamentals

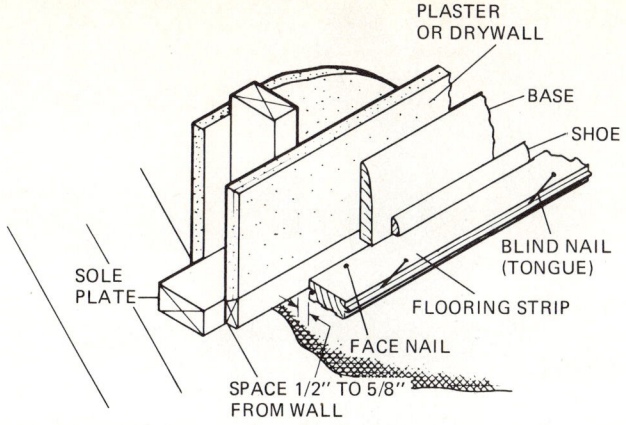

Figure 13-61. In the first strip, nails are driven into the face. *(Courtesy Forest Products Laboratory)*

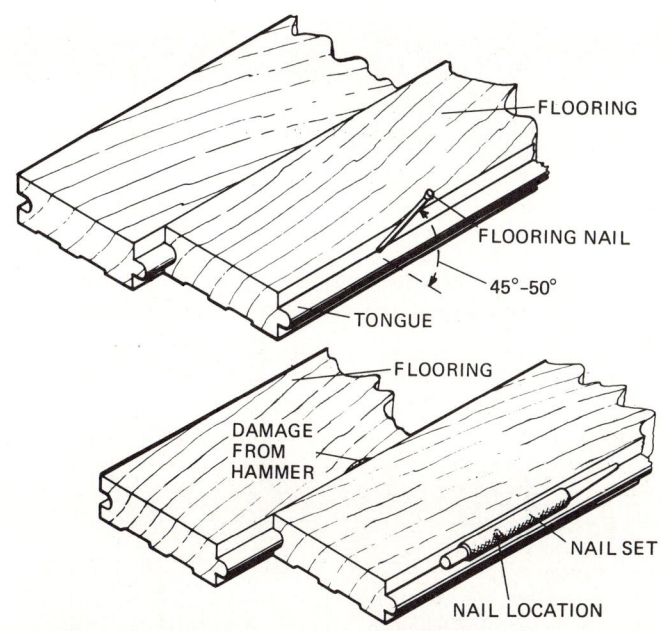

Figure 13-62. Nailing strips after the first. *(Courtesy Forest Products Laboratory)*

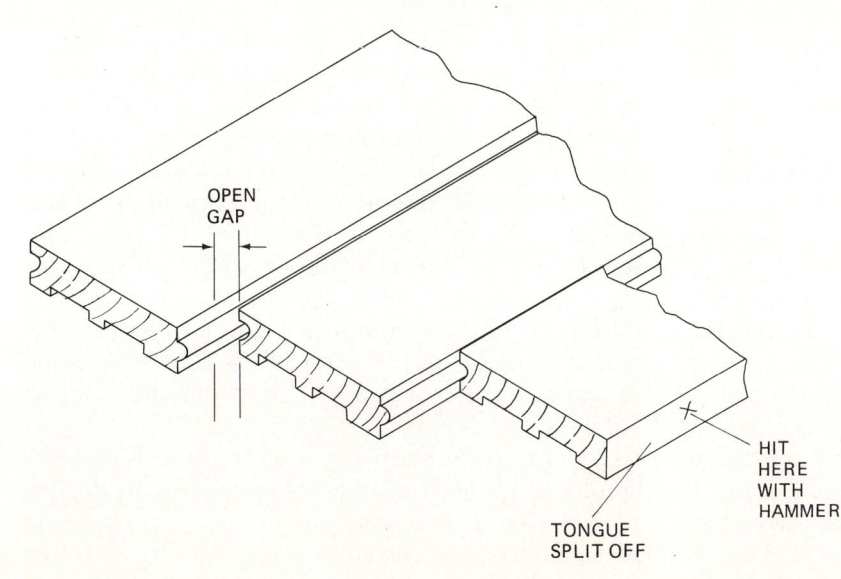

Figure 13-63. A scrap block is used to make flooring fit firmly. This prevents damage to either tongues or grooves.

Hardwood flooring will split easily. Most splits occur when a nail is driven close to the end of a board. To prevent this, drill the nail hole first. It should be slightly smaller than the nail.

The next strip is laid in place as shown in Figure 13-62. The nail is driven blind at a 45° or 50° angle as shown. Note that the nail is driven into the top corner of the tongue. The second strip should fit firmly against the first layer. Sometimes, force must be used to make it fit firmly. A scrap piece of wood is placed over the tongue as shown in Figure 13-63. The ends of the second layer should not match the ends of the first layer. The ends should be staggered for better strength and appearance. The end joints of one layer should be at least 6 inches from the ends on the previous layer. A nail set should be used to set the nail in place. Either the vertical position or the position shown in Figure 13-63 may be used. Be careful not to damage the edges of the boards with the hammer.

The same amount of space (½ inch) is left at the ends of the rows. End pieces may be driven into place with a wedge. Pieces cut from the ends can become the first piece on the next row. This saves material and helps stagger the joints. Figures 13-64 and 13-65 show other steps.

WOODEN PLANK FLOORING • Wooden plank flooring is installed in much the same manner. Today, both types are generally made with tongue-and-groove joints. The joints are on both the edges and ends. The same allowance of ½ inch is made between plank flooring and the walls. The same general nailing procedures are used.

WOOD BLOCK FLOORS • There are two types of wood block floors. The first type is like a wide piece of board. The second type is called parquet. Parquet flooring is made from small strips arranged in pat-

Finishing the Interior 373

Figure 13-64. Boards are laid in position and nailed in place. *(National Oak Flooring Manufacturers Assn.)*

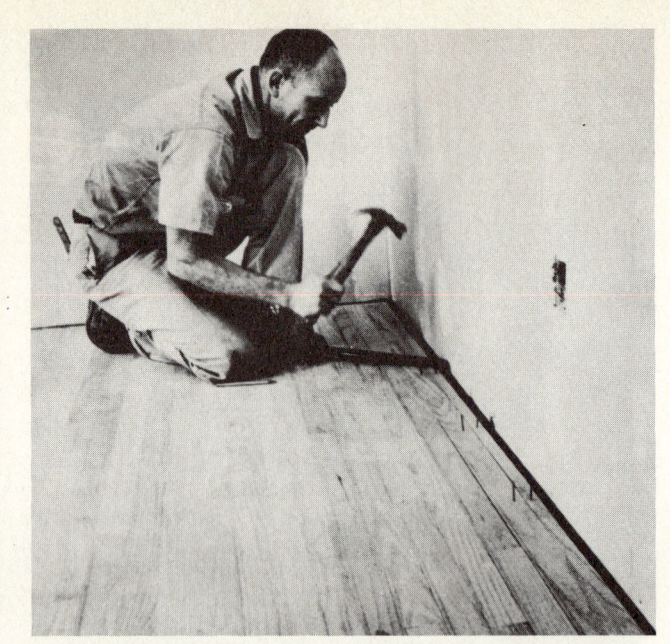

Figure 13-65. The last piece is positioned with a pry bar. Nails are driven in the face. *(National Oak Flooring Manufacturers Assn.)*

terns. Parquet must be laminated to a base piece. Today, both types of block floors are often laminated. When laminated, they are plywood squares with a thick veneer of flooring on the top. Usually, three layers of material make up each block.

Most blocks have tongue-and-groove edges. Common sizes are 3-, 5- and 7-inch squares. With tongue-and-groove joints, the blocks can be nailed. However, today most blocks are glued. When they are glued, the preparation is different than for plank or strip flooring. No layer of builder's felt or tar paper is used. The blocks are glued directly to the subfloor or to a base floor. As a rule, plywood or chipboard is used for a base under the block flooring.

Blocks may be laid from a wall or from a center point. Some patterns are centered in a room. Then, blocks should be laid from the center toward the edges. To lay blocks from the center, first the center point is found. Then a block is centered over this point. The outline of the block is drawn on the floor. Then a chalk line is snapped for each course of blocks. Care is taken that the lines are parallel to the walls.

Sometimes blocks are laid on the floor, proceeding from a wall. The chalk lines should be snapped for each course from the base wall.

Blocks are often glued directly to concrete floors. These floors must be properly made. Moisture barriers and proper drainage are essential. When there is any doubt, a layer process is used. First, a layer of mastic cement is applied. Then a vapor barrier of plastic or felt is applied over the mastic. Then a second layer of mastic is applied over the moisture barrier. The blocks are then laid over the mastic surface.

Wooden blocks may also be glued over diagonal wood subflooring. The same layer process as above is used.

• FINISHING FLOORS •

In most cases, floors are finished after walls and trim. When resilient flooring is applied, no finish step is needed. However, floors of this type should be cleaned carefully.

Finishing Wood Floors

The first step in finishing wood floors is sanding. A special sanding machine is used, such as the one in Figure 13-66. When floors are rough, they are sanded twice. For oak flooring, it is a good idea to use a sealer coat next. The sealer coat has small particles in it. These help fill the open pores in the oak. One of two types of sealer or filler should be used. The first type of sealer is a mixture of small particles and oil. This is rubbed into the floor and allowed to sit a few minutes. Then, a rotary sander or a polisher is used to wipe off the filler. The filler on the surface is wiped off. The filler material in the wood pores is left.

Floors may be stained a darker color. The stain should be applied before any type of varnish or lacquer is used. It is a good idea to stain the molding at the same time. The stain is applied directly after

374 Carpentry Fundamentals

Figure 13-66. After the floor is laid, it is sanded smooth using a special sanding machine. *(National Oak Flooring Manufacturers Assn.)*

the sanding. If a particle-oil type filler is used, the stain should be applied after the filler.

The second filling method uses a special filler varnish or lacquer. This also contains small particles which help to fill the pores. It is brushed or sprayed onto the floor and allowed to dry. The floor is then buffed lightly with an abrasive pad.

After filling, the floor should be varnished. A hard, durable varnish is best. Other coatings are generally not satisfactory. Floors need durable finishes to avoid showing early signs of wear.

Molding can be applied after the floors are completely finished. The base shoe and baseboard are installed as in Figure 13-67. Many workers also finish the molding when the floor is finished.

Base Flooring for Carpet

Often two floor layers are used but neither is the "finished" floor surface. The first layer is the subfloor and the next layer is a base floor. Both base and subflooring may be made from underlayment. Underlayment may be a special grade of plywood or chipboard. In many cases, nailing patterns are printed on the top side of the underlayment.

It is a good idea to bring underlayment into the room to be floored and allow it to sit for several days exposed to the air. This allows it to reach the same moisture content as the rest of the building components.

Base flooring is used when added strength and thickness are required. It is also used to separate resilient flooring or other flooring materials from concrete or other types of floors. It provides a smooth, even base for carpet. It is much cheaper to use a base floor than a hardwood floor under carpets. The costs of nailing small strips and of sanding are saved.

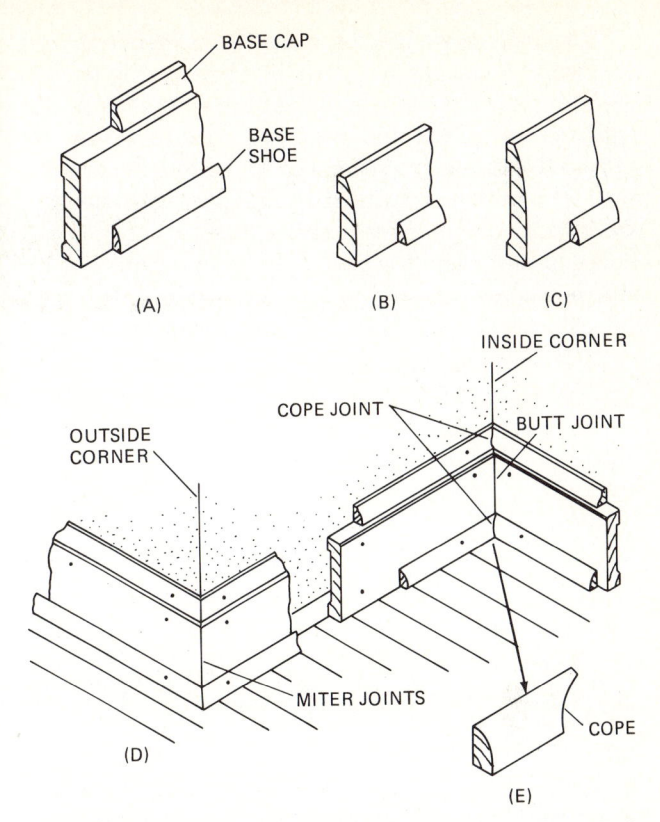

Figure 13-67. Base molding. *(Courtesy Forest Products Laboratory)* (A) Square-edged base. (B) Narrow ranch base. (C) Wide ranch base. (D) Installation. (E) A cope joint. *(Courtesy Forest Products Laboratory)*

• INSTALLING CARPET •

Carpet is becoming more popular for many reasons. It makes the floor a more resilient and softer place for people to stand. Also, carpeted floors are warmer in winter. Carpeting also helps reduce noise, particularly in multistory buildings.

To install carpet, several factors must be considered. First, most carpet is installed with a pad beneath it. When carpet is installed over a concrete floor, a plastic film should also be laid. This acts as an additional moisture seal. The carpet padding may then be laid over the film.

The first step in laying carpet is to attach special carpet strips. These are nailed to the floor. They are laid around the walls of a room to be carpeted. These carpet strips are narrow, thin pieces of wood with long tacks driven through them. They are nailed to the floor approximately ¼ inch from the wall. See Figure 13-68. The carpet padding is then unrolled and cut. The padding extends only to the strips.

Next, the carpet is unrolled. If the carpet is large enough, it may be cut to exactly fit. However, carpet should be cut about 1 inch smaller than the room size.

Finishing the Interior 375

The carpet is wedged between the carpet strip and one wall. A carpet wedge is used as in Figure 13-69. The carpet is then smoothed toward the opposite wall. All wrinkles and gaps are smoothed and removed. Next, the carpet is stretched to the opposite wall. The person installing the carpet will generally walk around the edges pressing the carpet into the tacks of the carpet strips. This is done after each side is wedged. After the ends are wedged, the first side is attached. The same process is repeated. After the first side is wedged the opposite side is attached.

To seam a carpet, special tape is used. This tape is a wide strip of durable cloth. It has an adhesive on its upper surface. It is rolled out over the area or edge to be joined. Half of the tape is placed underneath the carpet already in place. The carpet is then firmly pressed onto the adhesive. Some adhesive tapes use special heating tools for best adhesion. The second piece of carpet is carefully butted next to the first. Be sure that no great pressure is used to force the two edges together. No gaps should be wider than $\frac{1}{16}$ inch. The edges should not be jammed forcefully together, either. If edges are jammed together, lumps will occur. If wide gaps are left, holes will occur. However, the nap or shag of the carpet will cover most small irregularities.

To cut the carpet for a joint, first unroll the carpet. Carefully size and trim the carpet edges as straight as possible. The joint will not be even if the edges are ragged. Various tools may be used. A heavy knife or a pair of snips may be used effectively.

Metal end strips are used where carpet ends over linoleum or tile. The open end strips are nailed to the floor. The carpet is stretched into place over the points on the strip. Then the metal strip is closed. A board is laid over the strip. See Figure 13-70. The board is struck sharply with a hammer to close the strip. Do not strike the metal strip directly with the hammer. Doing so will leave unsightly hammer marks on the metal.

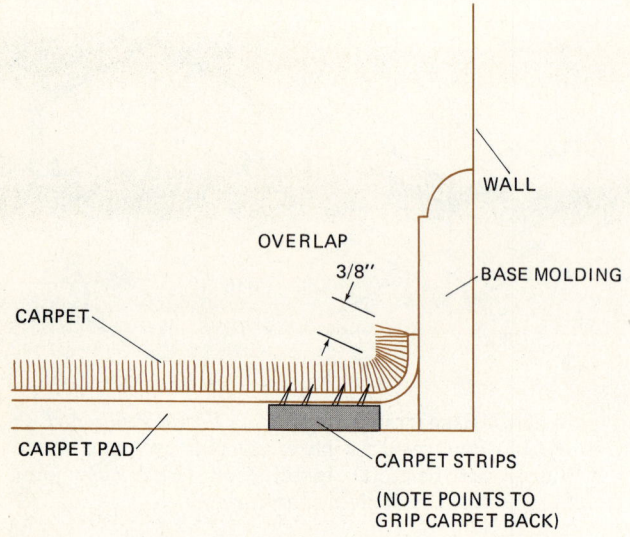

Figure 13-68. Carpet strips hold the carpet in place.

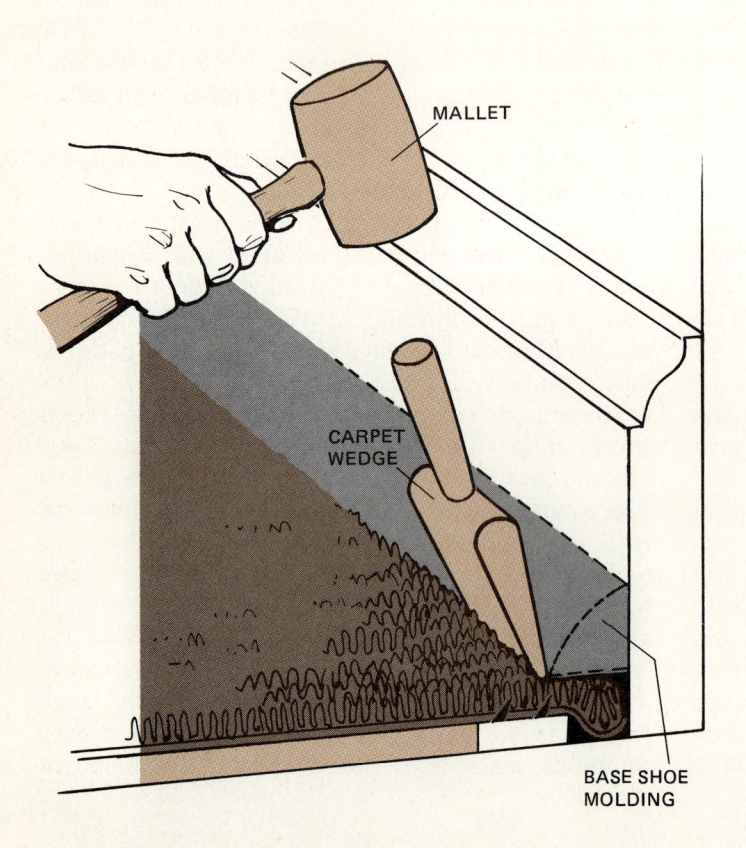

Figure 13-69. Carpet edges are wedged into place. Base shoe molding may then be added.

376 Carpentry Fundamentals

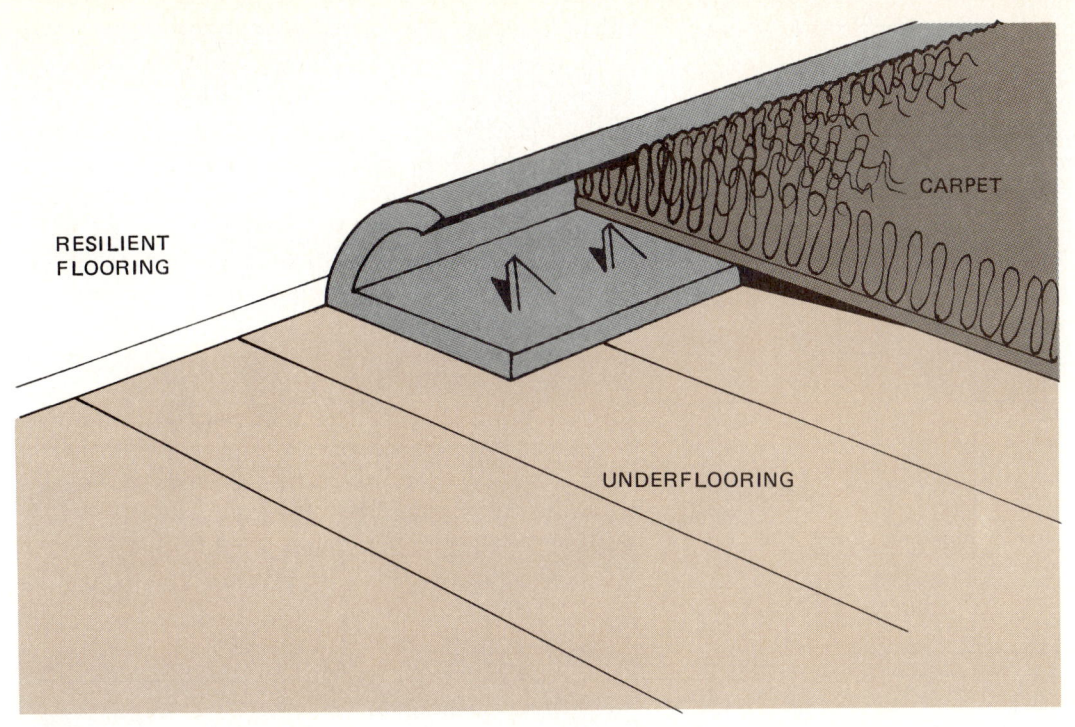

Figure 13-70. A metal binder bar protects and hides carpet edges where carpet ends over linoleum or tile.

• RESILIENT FLOORING •

Resilient flooring is made from chemicals rather than from wood products. Resilient flooring includes compositions such as linoleum and asphalt tile. It may be laid directly over concrete or over base flooring. Resilient flooring comes in both sheets and square tiles. Frequently, resilient flooring sheets will be called linoleum "carpets."

Installing Resilient Flooring Sheets

The first step is to determine the size of the floor to be covered. It is a good idea to sketch the shape on a piece of paper. Careful measurements are made on the floor to be covered. Corners, cabinet bases, and other features of the floor are included. It is a good idea to take a series of measurements. They are made on each wall every 2 or 3 feet. This is because most rooms are not square. Thus, measurements will vary slightly from place to place. The measurements should be marked on the paper.

Next, the floor is cleaned. Loose debris is removed. A scraper is used to remove plaster, paint, or other materials. See Figure 13-53. Then, the area is swept. If necessary, a damp mop is used to clean the area. Neither flooring nor cement will stick to areas that are dirty. Next, the surface is checked for holes, pits, nail heads, or obstructions. Holes larger than the diameter of a nail are patched or filled. Nails or obstructions are removed.

Most rooms will be wider then the roll of linoleum carpet. If not, the outline of the floor may be transferred directly to the flooring sheet. The sheet may then be cut to shape. The shaped flooring sheet is then brought into the room. It is positioned and unrolled. A check is made for the proper fit and shape. Any adjustments or corrections are made. Then, the sheet is rolled up approximately halfway. The mastic cement is spread evenly (about $3/32$ inch thick) with a toothed trowel. The unrolled portion is rolled back into place over the cemented area. Then, the other end is rolled up to expose the bare floors. Next, the mastic is spread over the remaining part of the floor. The flooring is then rolled back into place over the cement. The sheet is smoothed from the center toward the edge.

However, most rooms are wider than the sheets of flooring. This means that two or more pieces must be joined. It is best to use a factory edge for the joint line. Select a line along the longest dimension of the room as shown in Figure 13-71. On the base floor, measure equal distances from a reference wall as shown. These are the same width as the sheet. Then, snap a chalk line for this line. Often more than one joint or seam must be used.

The same measuring process is used. Measurements are taken from the edge of the first sheet. However, for smaller pieces, a different process is used. The center of the last line is found. A line is snapped at right angles. This shows the pattern for the pieces of flooring that must be cut. The second

Finishing the Interior 377

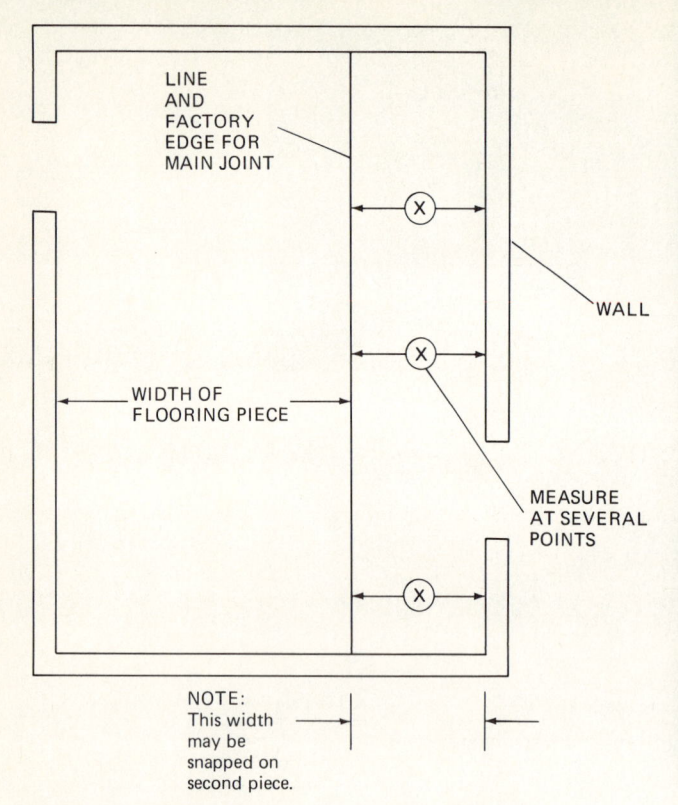

Figure 13-71. Floor layout for resilient flooring.

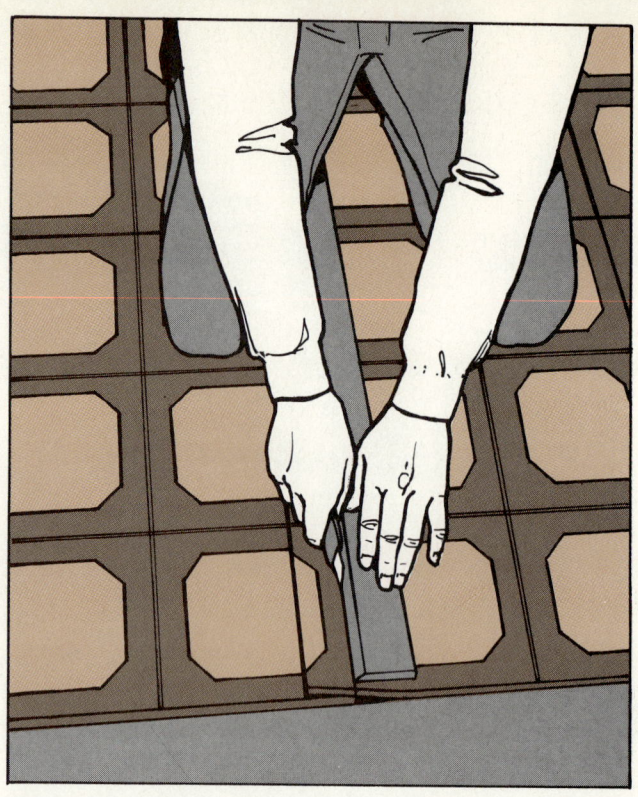

Figure 13-72. Edges are overlapped during cutting. This way seams will match even if the cut is not perfectly straight. *(Armstrong Cork)*

line should run the entire width of the room. A carpenter's square may be used to check the squareness of the lines.

The two chalk lines are now the reference lines. These are used for measuring the flooring and the room area. Measurements to the cabinets or other features are made from these lines. Take several measurements from the paper layout. Walls are seldom square or straight. Frequent measurements will help catch these irregularities. The fit will be more accurate and better.

Next, in a different area unroll the first sheet of flooring to be used. Find the corresponding wall and reference line. The factory edge is aligned on the first line. The dimensions are marked on the resilient flooring. The necessary marks show the floor outline on the flooring sheet. A straight blade or linoleum knife is used to make these cuts. For best results, a guide is used. A heavy metal straightedge is used for a guide when cutting. A check is made to be sure that there is nothing underneath the flooring. Anything beneath it could be damaged by the knife used to cut the flooring.

Next, the cut flooring piece is carried into the area. It is unrolled over the area, and the fit is checked. Any adjustments necessary are made at this point. Next, the material is rolled toward the center of the room. The area is spread with mastic. The flooring is rolled back into place over the cement. The sheet is smoothed as before. Do not force the flooring material under offsets or cabinets. Make sure that the proper cuts are made.

The adhesive should not be allowed to dry more than a few minutes. No more than 10 or 15 minutes should pass before the material is placed. A heavy roller is recommended to smooth the flooring. It should be smoothed from the center toward the edges. This removes air pockets and bubbles.

Where seams are made, a special procedure is recommended. Unroll the two pieces of flooring in the same preparation area. The two edges to be joined are slightly overlapped. See Figure 13-72. Next, the heavy metal guide is laid over the doubled layer. Then both layers are cut with a single motion. The straightedge is used as a guide to make the straightest possible cut. Edges cut this way will match, even if they are not perfectly straight or square. Figure 13-73 shows how edges are trimmed.

To Install Resilient Block Flooring

Resilient block flooring is often called *tile*. This term includes floor tile, asphalt tile, linoleum tile, and others. As a rule, the procedure for all of these is the same. The sizes of these tiles range from 6 to 12 inches square. To lay tile, the center of each of the end walls is found. See Figure 13-74. A chalk

378 Carpentry Fundamentals

Figure 13-73. Trimming the edges. A metal straightedge or carpenter's square is used to guide a utility knife. *(Armstrong Cork)*

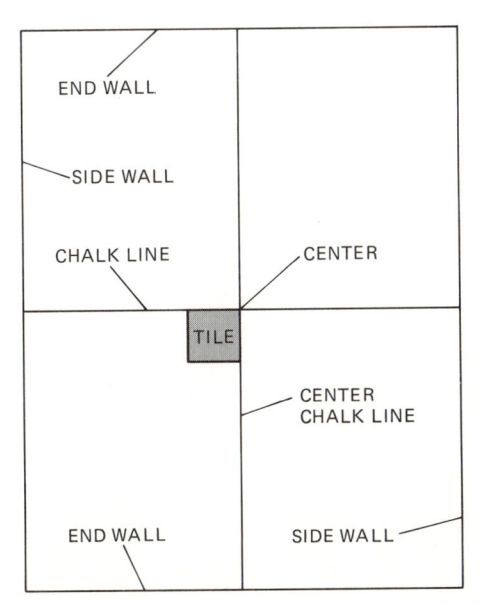

Figure 13-74. Find the intersecting midpoints. *(Armstrong Cork)*

Figure 13-75. Lay tiles in place without cement to check the spacing. *(Armstrong Cork)*

line is tied to each of these points. Lines are then snapped down the middle of the floor. Next, the center of the first line is located. A square or another tile is used as a square guide. A second line is snapped square to the first line.

Next, a row of tiles is laid along the perpendicular chalk line as shown. They are not cemented. Then the distance between the wall and the last tile is measured. If the space is less than half the width of a tile, a new line is snapped. It is placed half the width of a tile away from the center line. See Figure 13-75. A second line is snapped half the width of a tile from the perpendicular line. The first tile is then aligned on the second snapped line. The tile can now be cemented. The first tile becomes the center tile of the room.

Another method is used if the distance at the side walls is greater than one-half the width of a tile. Then the tile is laid along the first line. This way, no single tile is the center.

The first two courses are laid as a guide. Then the mastic is spread over one-quarter of the room area. See Figure 13-76. Lay the tiles at the center first. The first tiles should be laid to follow the snapped chalk lines. Tiles should not be slid into place. Instead, they are pressed firmly into position

Finishing the Interior **379**

as they are installed. It is best to hold them in the air slightly. Then the edges are touched together and pressed down.

The first quarter of the room is covered. Only the area around the wall is open. Here, tiles must be cut to fit.

CUTTING THE TILES • A loose tile is placed exactly on top of the last tile in the row. See Figure 13-77A. Then a third tile is laid on top of this stack. It is moved over until it touches the wall. The edge of the top of the tile becomes the guide. See Figure 13-77B. Then, the middle tile is cut along the pencil line as shown. This tile will then have the proper spacing.

A pattern is made to fit tile around pipes or other shapes. The pattern is made in the proper shape from paper. This shape is traced on the tile. The tile is then cut to shape.

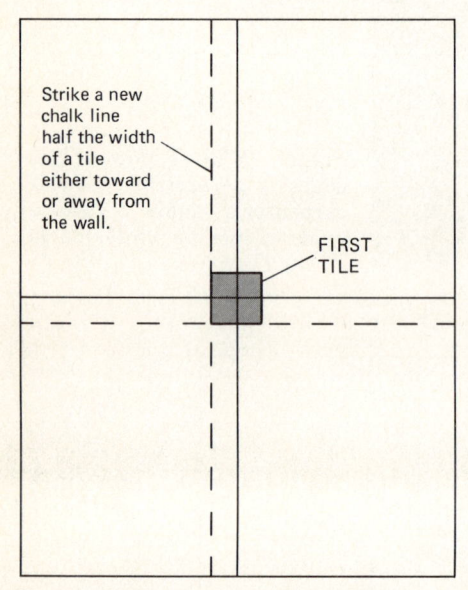

Figure 13-76. The next chalk line is used as a guide for laying tile. This method places one tile in the center.

• LAYING CERAMIC TILE •

Several types of ceramic tile are often used. Ceramic tile, quarry tile, and brick are used. They are desirable where water will be present. Bathrooms and wash areas often have tile. However, ceramic tiles also make pleasing entry areas and lobbies. Two methods are used for installing ceramic floor tile.

The first method uses a cement-plaster combination. This process is more difficult. A special concrete is used as the bed for the tile. This bed should be carefully mixed, poured, leveled and troweled smooth. It should be allowed to sit for a few minutes. Just before it hardens, it is still slightly plastic. Then the tiles are embedded in place. The tiles should be thoroughly soaked in water if this method is used. They should be taken out of the water one at a time. They are allowed to drain slightly before they are used. The tiles are then pressed into place in the

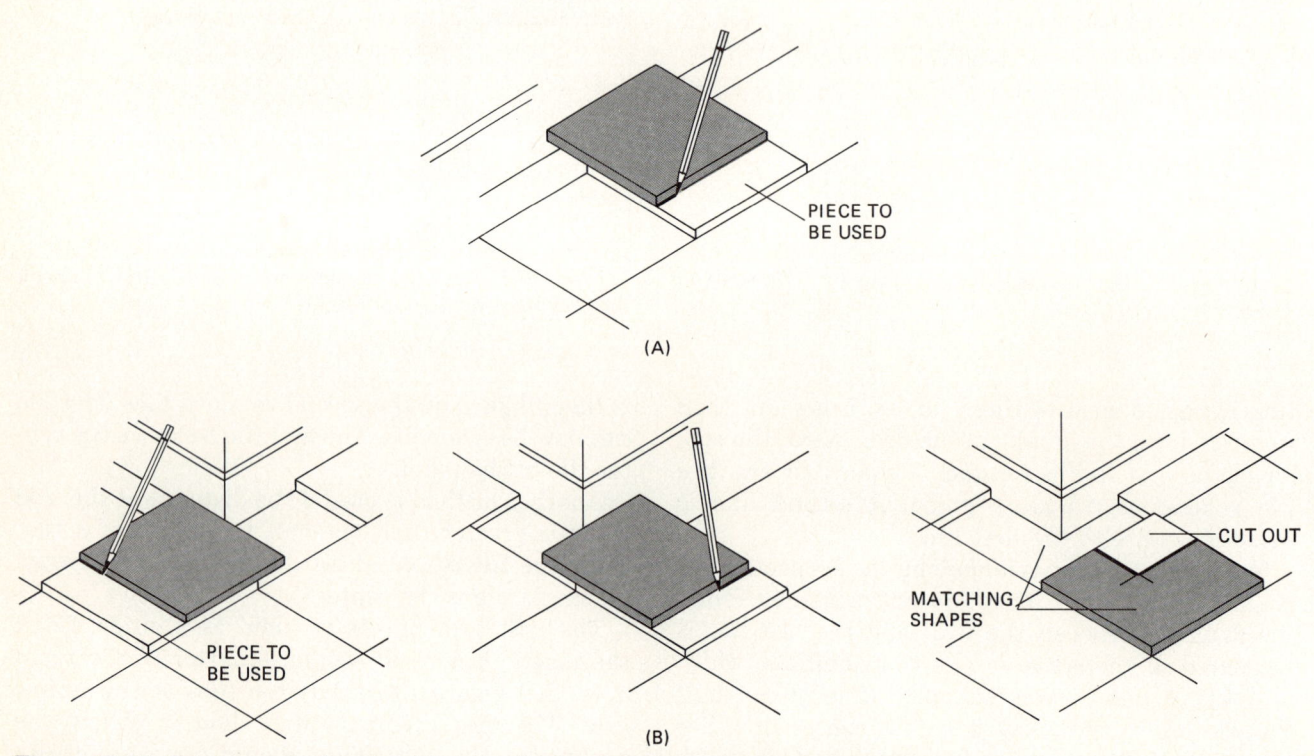

Figure 13-77. Cutting and trimming tile to fit. (A) Marking tile for edges. (B) Marking tile for corners.

380   Carpentry Fundamentals

cement base. All the tiles are installed. Then special grout is pressed into the cracks between the tiles. This completely fills the joints between the tiles. The grouted joints should be cleaned and tooled within a very few hours after installation.

Adhesives are most widely used today. Adhesives are much easier to use. The adhesives are spread evenly on the tiles. This is much like laying resilient tile. The adhesive should be one that is recommended for use with ceramic tiles. After the adhesive has been spread, the tiles are placed. Then a common grout is forced between the tiles. It is wiped and allowed to dry.

Tiles are available as large individual pieces. They are also available as assortments. Small tiles may be preattached to a cloth mesh. This mesh keeps the tiles arranged in the pattern. This also makes the tile easy to space evenly.

• WORDS CARPENTERS USE •

stop	hang a door	rail	size	carpet tape
jamb	window stool	kicker	strip flooring	resilient flooring
trim	window apron	plastic laminates	planks	factory edge
casing	in-the-white	pulls	blocks	grout
panel doors	built-ins	coped joint	parquet	
hollow-core doors	facing	stain	mastic	
hardware	drawer guide	texture paint	underlayment	
strike plate	stile	sealer coat	carpet strips	

• STUDY QUESTIONS •

13-1. How is expansion controlled on strip flooring?

13-2. When is a vapor barrier used under finish flooring?

13-3. How are board ends kept from splitting?

13-4. How are floor tiles laid out?

13-5. Why are floor tiles started in the center?

13-6. When is strip flooring laid? When is it finished?

13-7. How are seams in linoleum cut?

13-8. When is door trim applied?

13-9. What is the easiest way to lay ceramic tile?

13-10. How is interior trim used on metal window frames?

13-11. What are built-ins?

13-12. What is the sequence for hanging doors?

13-13. What are three ways of making shelves?

13-14. List the steps in building a cabinet counter.

13-15. How are ready-built cabinets hung?

13-16. How is plastic laminate applied?

13-17. Why are coped joints used?

13-18. Why are coatings smoothed from the center toward the edges?

13-19. When should floors be stained?

13-20. What are the advantages of lip drawers?

Finishing the Interior

14
SPECIAL CONSTRUCTION METHODS

In building, carpenters have many jobs to do. They build the frames and cover them. However, the carpenter must often do other jobs. The regular methods are not always used for the frames. Special jobs include fireplaces, chimneys, and stairs. The post-and-beam method of building is also included.

Carpenters need special skills to do the special jobs. Almost every building will involve one or more of these special jobs. Specific skills that can be learned in this unit are:

- Lay out, cut, and install stair parts
- Design, cut, and make fireplace frames according to the manufacturer's or builder's specifications
- Plan and build basic post-and-beam frames

• STAIRS •

Carpenters should know how to make several types of stairs. Main stairs should be pleasing and attractive. This is because they are visible in the living or working areas. They should also be easy to climb, safe, and sturdy. It is a good idea to make them as wide as possible. Service stairs are not as visible. They are mainly used to give occasional access or entry. They are used in basements, attics, and other such areas. Service stairs are not as wide or attractive as main stairs. Service stairs can be steeper and harder to climb. However, all stairs should be sturdy and safe to use.

Stair Parts

The carpenter will most often build stairs around a notched frame. See Figure 14-1. This frame is called a *carriage*. The carriage is also called a *stringer* by some carpenters. The carriage should be made of 2×10-inch or 2×12-inch boards. This way one solid board extends from the top to the bottom. The other stair parts are then attached to the carriage.

Figure 14-1. A notched frame called a carriage is the first step in building stairs.

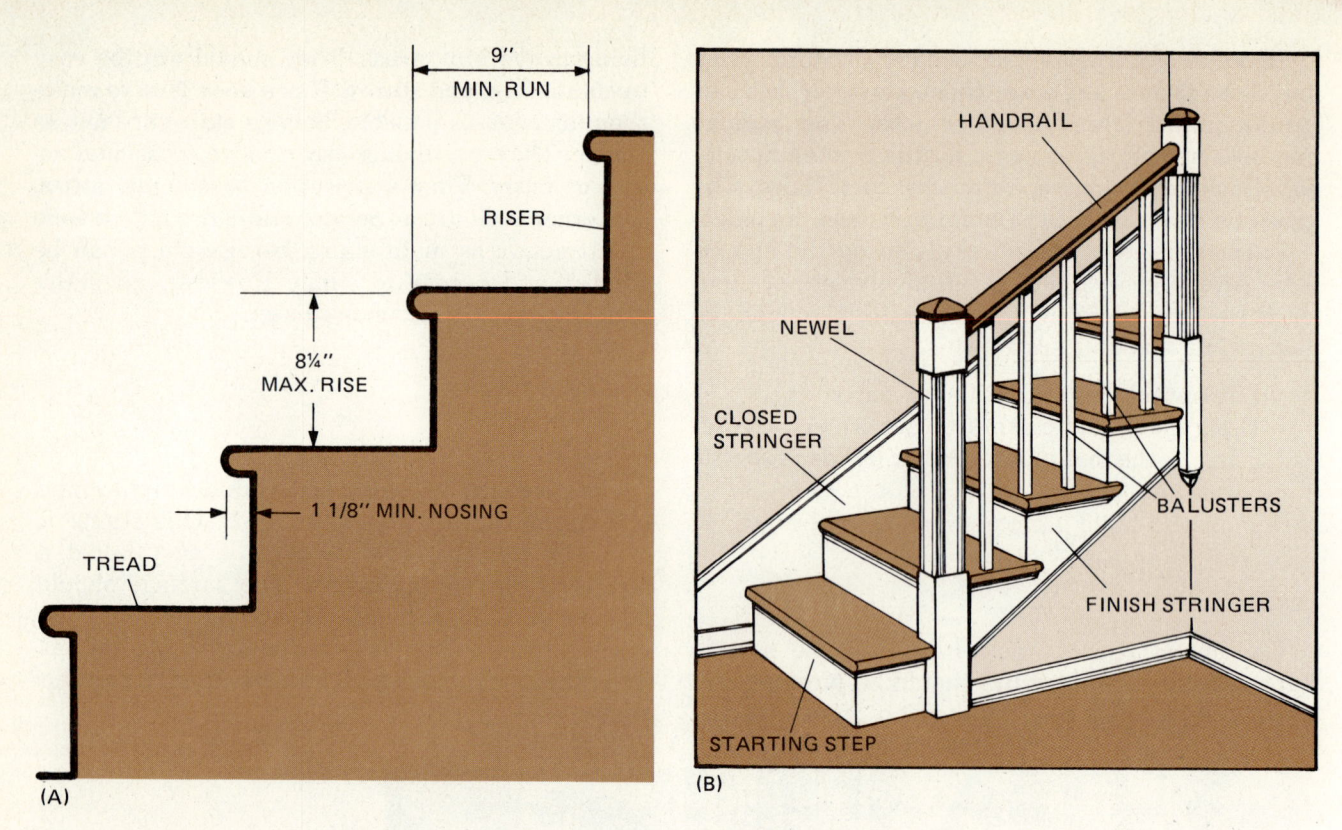

Figure 14-2. Stair parts. (A) Risers and treads. *(Courtesy Forest Products Laboratory)*
(B) Railing. The newels, balusters, and rails together are called a balustrade.

Figure 14-3. A landing is used to change direction.

The part on which people step is called the *tread*. The vertical part at the edge is called *riser*. A stair unit is made of one tread and one riser. The unit *run* is the width of the tread. The unit *rise* is the height of the riser. The tread is usually rounded on the front edge. The rounded edge also extends over the riser. The part that hangs over is called the *nose* or *nosing*. See Figure 14-2A.

Stairs must also have a handrail. Where a stair is open, a special railing is used. The fence-like supports are called *balusters*. The hand rail or bannister (which children slide down in stories) rests on the balusters. The end posts are called *newels*. See Figure 14-2B.

Some stairs rise continuously from one level to the next. Other stairs rise only part way to a platform. The platform is called a *landing*. Often landings are used so that the stairs can change direction. See Figure 14-3.

Stair Shapes

Stairs are made in several shapes. See Figure 14-4A through E. Stairs may be in open areas or have walls on one or both sides. Stairs with no wall or one wall on the side are called *open* stairs. Stairs that have walls on both sides are called *closed* stairs.

STRAIGHT-RUN STAIRS • The straight run is the simplest stair shape. This type of stair rises in a straight line from one level to the next. The stairs may be open or closed. See Figure 14-4A and B.

L-SHAPED STAIRS • Figure 14-4C shows L-shaped stairs. These stairs rise in two sections around a corner. To make these, the carpenter builds two sets of straight runs. The first is to the platform or landing. The second is from the landing to the next level.

U-SHAPED STAIRS • The stairs in Figure 14-4D are U-shaped stairs. They are much like L-shaped stairs. However, two corners are turned instead of one. As a rule, a platform or landing is used at each corner. Again the carpenter makes straight-run stairs from one platform to the next.

WINDERS • Some stairs have steps that "fan" or turn as in Figure 14-4E. No landing is used to turn a corner. These turning steps are called *winders* or winding stairs. These are the hardest to make and they take special framing. As a rule, the stairs are pie-shaped only on corners. However, some special

Figure 14-4. (A) Straight-run stairs do not change direction. This stair is part open.

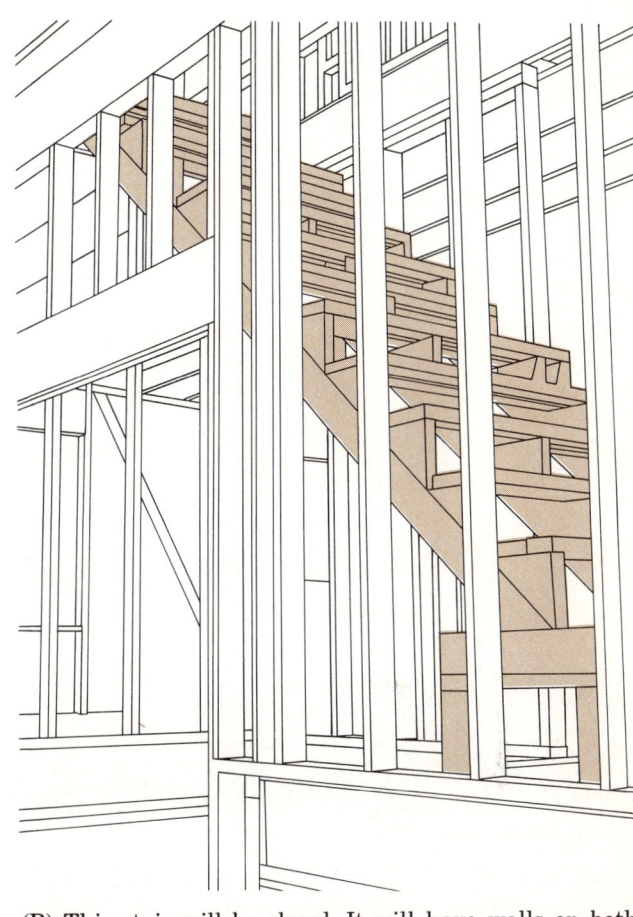

(B) This stair will be closed. It will have walls on both sides.

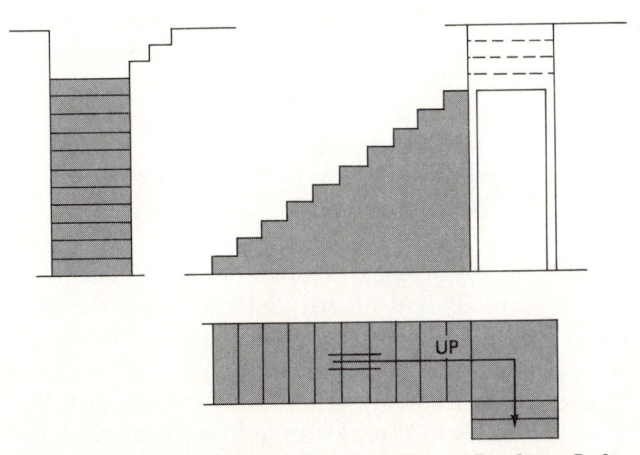

(C) A long L-shaped stair. *(Courtesy Forest Products Laboratory)*

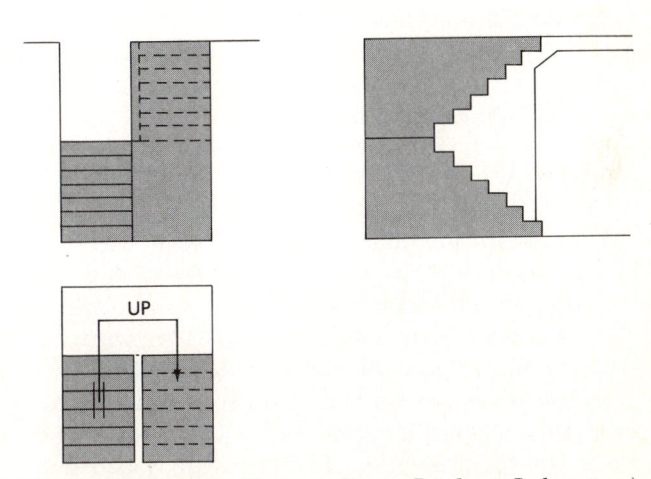

(D) U-shaped stairs. *(Courtesy Forest Products Laboratory)*

Special Construction Methods **385**

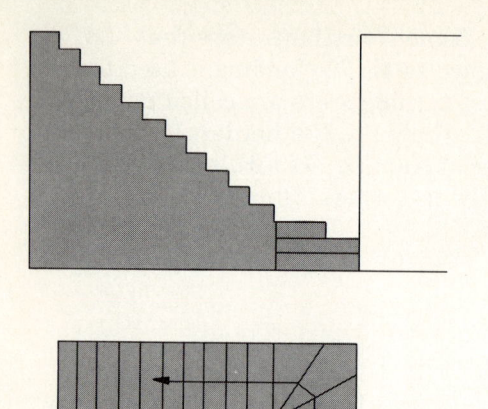

Figure 14-4 *(cont.)* **(E)** A winder saves space. It does not use a landing to change direction. *(Courtesy Forest Products Laboratory)*

THREE RULES FOR STAIR DESIGN
1. UNIT RISE + UNIT RUN = 17" to 18"
   $(7" + 10\frac{3}{8}" = 17\frac{3}{8}")$
2. 2 UNIT RISES + ONE UNIT RUN = 24" to 25"
   $[2(7) + 10\frac{3}{8} = 24\frac{3}{8}"]$
3. UNIT RISE × UNIT RUN = 72" to 75"
   $(7 \times 10\frac{3}{8} = 72\frac{5}{8})$

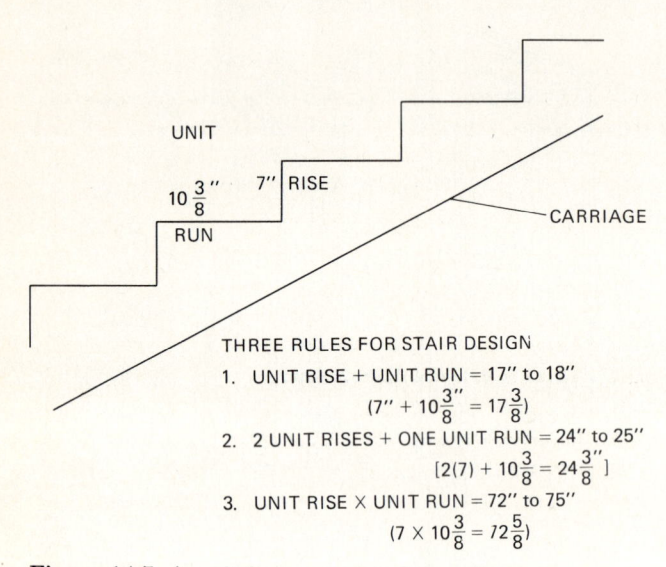

Figure 14-5. A unit is one run and one rise.

jobs may feature the true winding staircase.

Winders are considered attractive by many people. However, carpenters make winders only when there is not enough room for a landing. The winder is not as safe as a regular landing.

Stair Design

Most stairs are made with 2×12-inch carriages rising through a framed opening. As a rule, the framed opening is made when the floors are framed. The carpenter may check for proper support and framing before making the carriage attachments.

Of course, carpenters build stairs as the designer or architect specifies. However, carpenters should know the fundamentals of stair design. Sometimes, architects or designers will not specify all the details. In these cases, carpenters must use good practices to complete the job.

Stairs should rise at an angle of 30° to 35°. The minimum width of a tread is about 9 inches. There is no real maximum limit for tread width. However, a long step results when the tread is very wide. As a rule, treads 9 inches wide are used only for basements and service stairs. For main stairs, the tread width is normally 10 to 12 inches.

The riser for most main stairs should not be very high. If the riser is too high, the stairs become steep and hard to climb. As a rule, 8¼ inches is considered the maximum riser height. Less than that is desirable. For main stairs, 7 to 7½ inches is both more comfortable and safer.

FIND THE NUMBER OF STEPS • The amount of rise per step determines the number of steps in stairs. If a 7-inch rise is desirable, then the total rise is divided by 7 inches. The total rise is the total distance from the floor level of one landing to the next. A two-story building with an 8-foot ceiling is an example. The width of the floor joists and second-story finished flooring is added. The second-floor joists are 2×10 inches. The total rise would be about 8'10". For 7-inch risers, this distance (8'10") is divided by 7:

8'10" = 106"
Height/rise = number of steps

$$\frac{106}{7} = 15.14$$

15 would be the number of steps

In practice, most stairs in houses have 13, 14, or 15 steps. Fourteen steps (a riser of about 7½ inches) is probably the most common.

USING RATIOS • To find the step width, a riser-to-tread ratio is used. There are three rules for this ratio. Remember, one tread and one riser is considered one unit. See Figure 14-5.

1. Unit rise + unit run = 17 to 18.
2. Two unit rises + one unit run = 24 to 25.
3. Unit rise × unit runs = 72 to 75.

Now, the total length of run is divided by the number of steps. If this length is 13'0", then the step width would be 156 inches divided by 15. This would be approximately 10⅜ inches.

These two numbers (7 inches for the riser and 10⅜ inches for the tread) are used to check the stair shape. To apply the rules:

1. Allowable unit rise + unit run = 17 to 18. Actual sum is

    7 inches + 10 3/8 inches = 17 3/8

    This figure is OK.
2. Allowable length of two unit rises + one unit run = 24 to 25. Actual length is

    2 × 7 =   14 inches
    1 × 10 3/8 = +10 3/8 inches
    ────────────
                24 3/8

    This is OK
3. Allowable value of unit rise × unit run = 72 to 75. Actual value is

    7 × 10 3/8 = 72 5/8.

    This is also OK.

Sometimes, two or three combinations have to be tried to get a good design. Experience has shown that these ratios give a safe, comfortable, and well-designed stair.

HEADROOM • Headroom is the distance between the stairs and the lowest point over the opening. The minimum distance should be above 6'4". It is better to have 6'8". The headroom can be quickly checked. The carpenter can lay a board at the desired angle. Or a line can be strung. The distance between the board or line and the lowest point can be measured.

STAIR WIDTH • The width of the stair is the distance between the rail and the wall or baluster. This is the *walking area*. This area between the rail and the wall is called the *clear* distance. Narrow stairs are not as safe as wide stairs. Also, it is hard to move furniture and appliances in and out through narrow stairs. The width of the stair is a major factor.

The minimum clear distance allowed by the FHA is 2'8" on main stairs. Sometimes, 2'6" is used on service stairs. However, added width is always desirable.

Tread width on winders should be equal to the regular width. Winder treads are checked at the midpoint. Or, the tread width can be checked at the regular walking distance from the handrail. For example, tread width on a stair is 10 3/8 inches. Then, the tread width on the middle of the winder should also be 10 3/8 inches. See Figure 14-6.

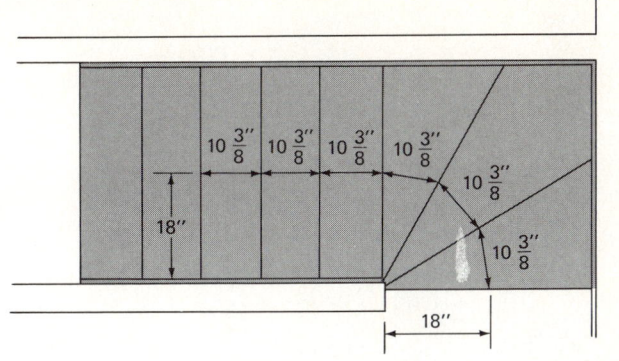

Figure 14-6. In a winder, tread widths should be equal on the main path. *(Courtesy Forest Products Laboratory)*

Sequence in Stair Construction

Carpenters should follow a given sequence in building the stairs. The stair is laid out on the carriage piece. The step (riser and tread) notches are cut. Two carriage pieces are used for stairs up to 3 feet in width. Three or more carriage boards are used for stairs that are wider than 3'0". One is on each side, and one is in the center.

Carriage boards are often positioned as walls and floors are framed. However, the stairs are not finished until the walls have been finished. Often, 2-inch lumber is used for temporary steps during the construction. See Figure 14-7. Stairs that rest on concrete floors are not built until the concrete floor is finished.

Three methods of making stairs are used. The first two methods require the carpenter to make the carriage. The treads are finished later. In the first method, regular lumber is used for a tread. These treads are later covered with finished hardwood flooring.

In the second method, hardwood treads are nailed directly to the carriage. These boards are usually oak 1 1/8 or 1 1/16 inches thick.

The third method is called a *housed carriage*. The carriage board has grooves to receive the treads and risers. As a rule, these pieces are precut at a factory or mill. The treads are inserted from the top to the bottom. Glue and wedges are used to wedge the stairs tightly in place. See Figure 14-8. This type of stair is very sturdy and normally does not squeak.

Carriage Layout

Riser height and tread width are found from the ratios. The example given used a 7-inch riser with a 10 3/8-inch tread. A 2×12-inch board of the proper

Figure 14-7. Two-inch lumber is used for temporary steps during construction. It can be left as a base for finish treads.

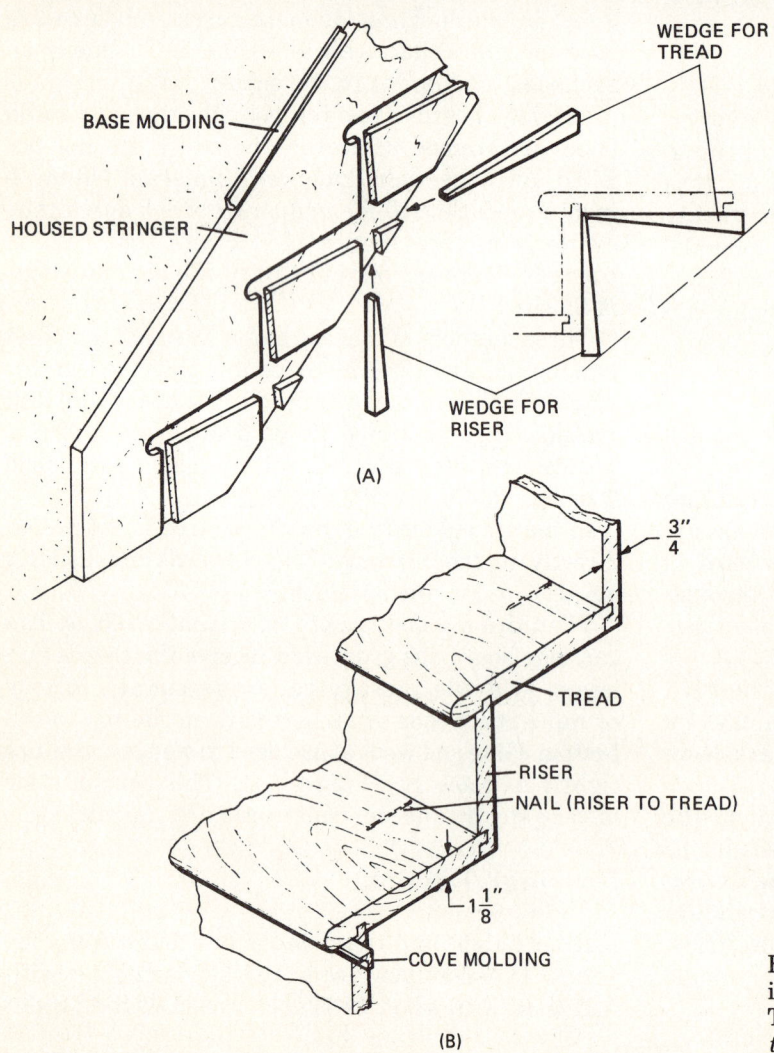

Figure 14-8. Housed carriage stairs. (A) Housing. *(Courtesy Forest Products Laboratory)* (B) Tread detail. *(Courtesy Forest Products Laboratory)*

388 Carpentry Fundamentals

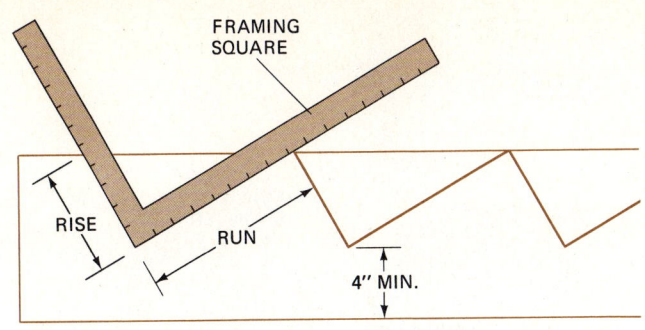

Figure 14-9. Lay out carriage with a framing square.

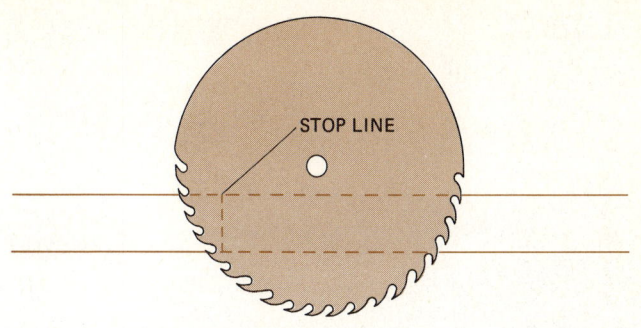

Figure 14-11. A circular saw should not be used, since it will cut past the line.

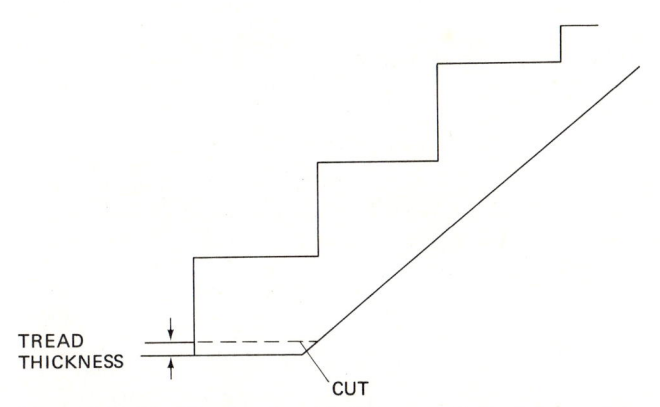

Figure 14-10. The carriage bottom must be trimmed to make the first rise correct.

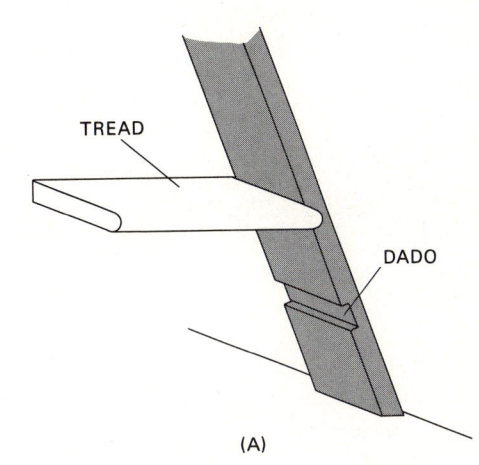

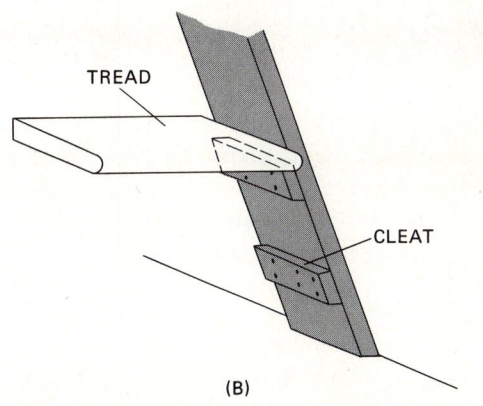

Figure 14-12. Exterior stair construction. (A) Dadoed carriage. (B) Cleated carriage.

length is selected. The board is laid across the stair area to see whether the board is long enough. If so, the board is placed on a sawhorse or work table.

A framing square is used as in Figure 14-9. The square is placed at the top end of the carriage as shown. The short blade of the square is held at the corner. The outside edge is used for these measurements. The height of the riser is measured down from the corner on the outside tongue. The blade is swung until the tread width is shown on the scale. This is shown when the tread width (10⅜ inches) meets the edge of the board. Lines are drawn along the edges of the squares.

This process is repeated for all the steps. On the last step, two things are done. First, a riser is marked for the first step height. See Figure 14-10. Then the height is adjusted to allow for the tread thickness. This must be done. When the tread is nailed to the carriage, it will raise the riser distance. For example, suppose a carriage is cut so that the full first step rises 7 inches. Then a board 1⅛ inches thick is added for the tread. This would make the total riser height 8⅛ inches. So the thickness of the tread is subtracted as in Figure 14-10.

Notches are cut. A bayonet saw or a handsaw should be used. It is not a good idea to use a circular saw. The blade does not cut vertically through the wood. See Figure 14-11.

For some outside stairs, the carriage is cut differently. The layout is the same. However, the steps are dadoed into place as in Figure 14-12. To do this, the dado is cut on each side of the line. Cut the dado one-third to one-half the thickness of the car-

Special Construction Methods **389**

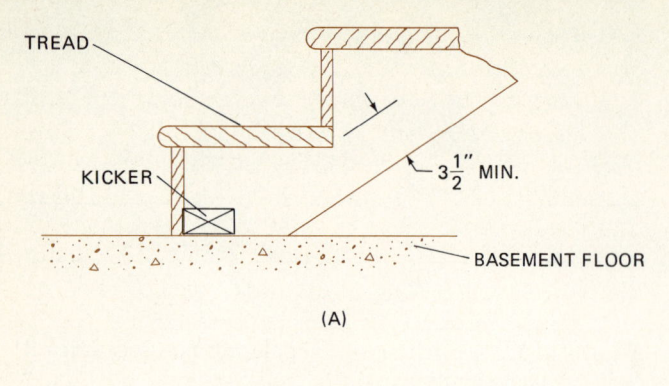

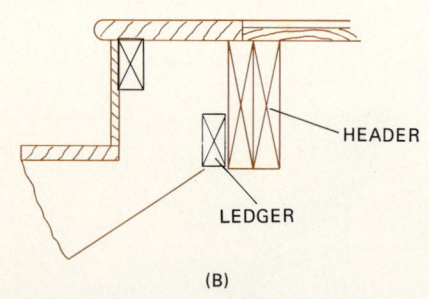

Figure 14-13. (A) A kicker anchors the bottom of the carriage. (B) The ledger anchors the top. *(Courtesy Forest Products Laboratory)*

riage. Standard 2-inch lumber is only 1½ inches thick. The dado should be cut ½ inch (one-third the thickness) to ¾ inch (one-half the thickness) deep.

Frame the Stairs

First, the ledger for the top is cut. Then it is nailed in place at the top of the landing or second story. The ledger is made from 2×4-inch lumber. It is nailed with 16d common nails. See Figure 14-13. Next, a kicker is cut from 2×4-inch lumber. It is nailed (16d common nails) at the bottom as in Figure 14-13. Then, notches are cut in the carriage for the kicker and ledger. Next, a backing stringer is nailed on the walls. This should be a clear piece of 1-inch lumber. It gives a finished appearance to the stairs. Use 8d common nails. See Figure 14-14. Next, the carriage is nailed to the ledgers, kickers, wall, and backing strips. Heavy spikes are recommended. Next, the outside carriage is nailed to the ledger and kicker.

The outside finish board is nailed to the outside of the carriage. If the finish board is 2 inches thick, 16d finish nails should be used. If the outside board is 1-inch lumber, an 8d finish nail should be used. After the wall is complete, the newel is nailed in place. A finish stringer can be used as in Figure 14-15. The finish stringer is cut from a 1×12-inch board. A finish stringer is blocked and nailed in place as shown. Finish nails should be used.

Temporary steps are now removed from the carriage. As a rule, cutting is avoided on stair pieces.

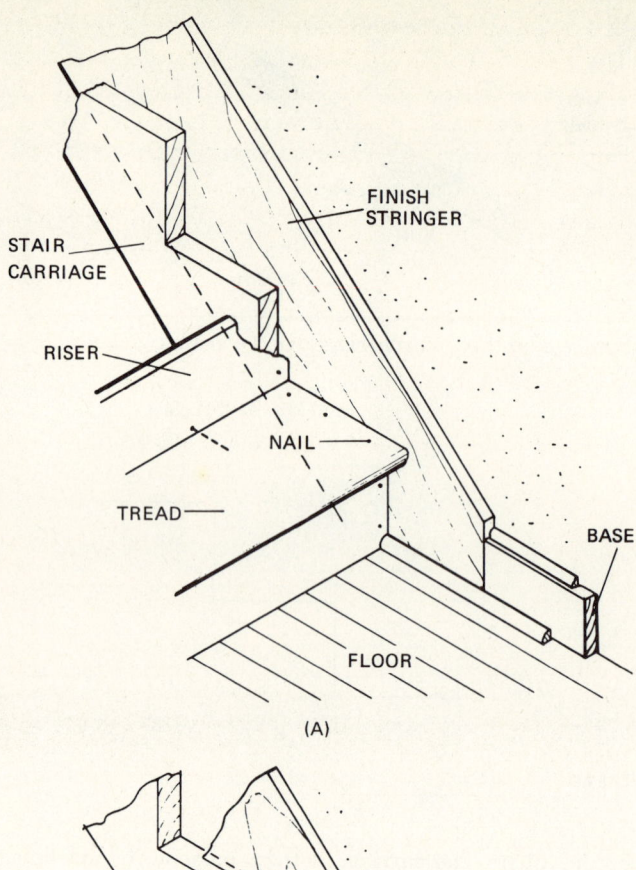

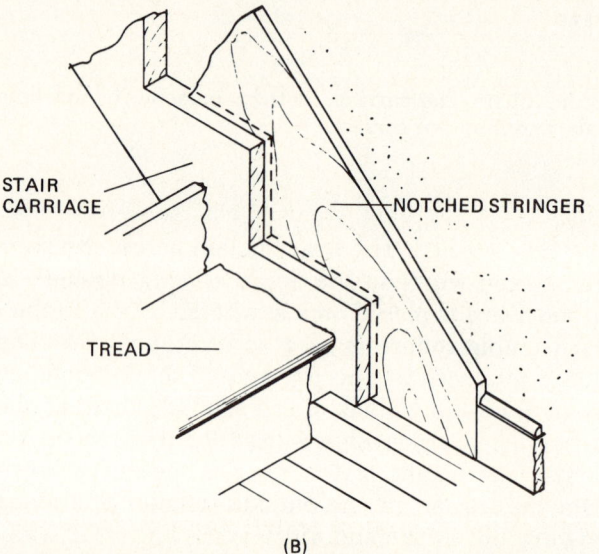

Figure 14-14. Backing or finish stringers on a wall. *(Courtesy Forest Products Laboratory)*

Tread overhang less than 1½ inches is ignored. The overhang becomes the nosing. However, excess width is cut from the back side.

In hardwoods it's a good idea to drill holes for nails. Doing this keeps the nails from splitting the board. The drilled hole should be slightly smaller than the nail. The nails should be set and filled after they are driven. Of course, only finish nails are used.

First, the bottom riser is nailed in place. It must be flush with the top of the carriage riser. It is normal to have a small space at the bottom of the riser.

The space at the bottom can be covered by molding. The riser is always laid first. A convenient number of steps is worked at one time. This includes all the risers that can be easily reached. Probably two or three steps can be worked at the same time. Next, the bottom tread is placed in position. It is moved firmly against the riser. The tread is drilled and nailed as needed. Treads are laid until one is on each placed riser. Then, the worker moves up the steps. Again, a convenient number of risers and treads are laid. This process is repeated until the stair is finished.

The baluster is started when all the treads are installed. The ends of balusters are round like dowels. Remember, holes are located by the center of the hole. The centers of the holes for the balusters are located. They are spaced and located carefully. The holes are drilled to receive the baluster ends. The top newel post is attached as specified by the manufacturer. Next, the bottom newel post is fastened in place. Large spikes should be used. It is best to attach the newel post to the subflooring. Sometimes, carpenters must first chisel an area in the finished floor. This receives the bottom of the newel post. Next, glue is put on the round bottoms of the balusters. These are inserted in the holes drilled in the treads. Glue is never put in the hole. All the bottom pieces are glued and placed. Glue is then put on the top stems of the balusters. Then the top rail, or bannister, is laid over the tops of the balusters. It is tapped firmly into position and allowed to dry. Then, finish caps, molding, handrails, and trim are installed.

Install Housed Carriages

To install housed carriages, it is best to work from the back. Here, the carpenter should start at the top and work toward the bottom. The first riser is applied and lightly wedged. Then the tread is attached. The wedge is coated with glue and tapped firmly into place. Then the riser wedge is coated with glue. It is then wedged firmly into place. Refer to Figure 14-8A.

Wedged and glued stairs squeak less than other types. Stair squeaks are usually caused by loose nails. Since the closed carriage has fewer nails, it squeaks less.

Although this type of stair is commonly precut, carpenters sometimes cut these on the site. Special template guides and router bits should be used. See Figure 14-16.

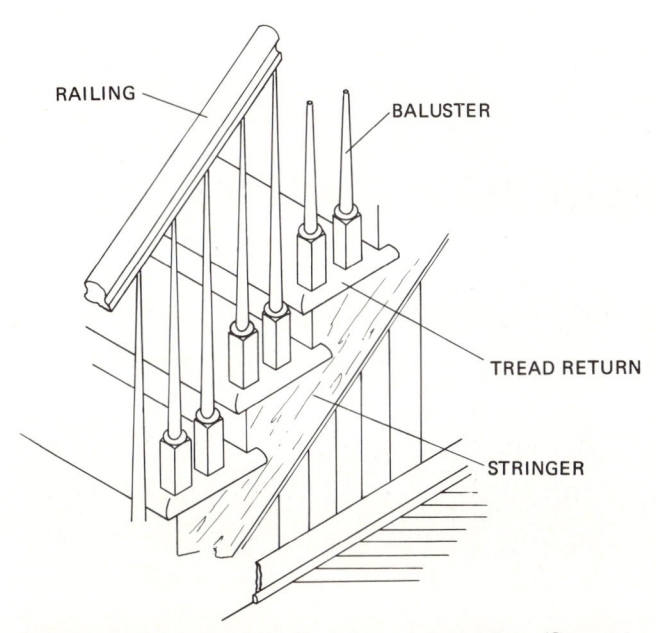

Figure 14-15. A finish stringer on open stairs. (Courtesy Forest Products Laboratory)

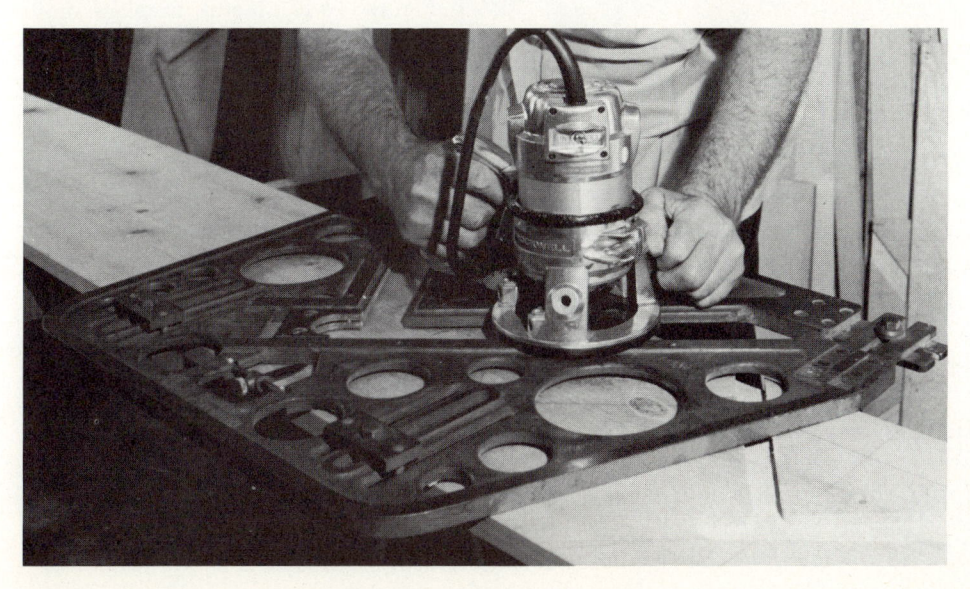

Figure 14-16. Templates are used to rout the housed stringer on the site. (Courtesy Rockwell International, Power Tool Division)

Special Construction Methods 391

Figure 14-17. Metal fireplace units must be enclosed by frames. *(Martin Industries)*

• FIREPLACE FRAMES •

As a rule, the carpenter does not make a fireplace. Fireplaces are usually made of brick, stone, or concrete block. These are constructed by people in the trowel trades. However, the carpenter must make framed openings for the fireplace and flues (or chimney).

Also, today many builders use metal fireplace units. The chimneys for these units are made of metal too. These fireplaces and chimneys are installed in framed openings. See Figure 14-17.

Fireplaces make attractive areas. They give the appearance and feeling of comfort and security. Fireplaces are low in heating efficiency. However, they are popular devices to help offset fuel bills. Fireplaces can be made with various types of ventilators and reflector devices. These allow the heat to be directed and radiated into the living area. This gives greater efficiency in heating. Chimneys are also used to vent furnaces, water heaters, and other fuel-burning appliances. Often these vents are brought through the roof inside the chimney unit. This makes the roof line more attractive. Fewer unsightly vents pierce the roof structure.

Rough Openings in Ceilings and Roofs

The rough openings in ceilings and roofs are framed in by the carpenter. The carpenter makes these openings from the plans. These should be made as the roof and walls are constructed. They should be built before the fireplace is started. Special care must be taken to maintain proper clearances. A minimum distance of 2 inches is required between all frame members and the chimney. This allows for uneven settling of the chimney foundation. It also gives additional safety from fire.

The sequence used for constructing stair openings is also used here. First, an inner joist is nailed in place. This becomes the side of the opening. Then a first header is nailed in place. This boxes in the sides of the rough opening. Then tail joists or rafters are nailed in place. After this step, second headers are nailed on the insides. Next, the outer joist or rafter is added to complete the rough opening. After the rough opening is framed, the deck for the roof is applied.

At this point the fireplace, flue, and chimney should be built. Proper flashing techniques should be used where the chimney goes through the roof. After the flashing is installed, the roof is applied. See Figure 14-18A through C.

Figure 14-18. (A) Openings in roofs must be framed. Metal units may be added after a house is finished. *(Martin Industries)*

(B) A metal unit being set in place on a new home.

(C) The same unit after framing and finishing.

Special Construction Methods   393

Fireplace Types

Several types and shapes of fireplaces are common. Others, such as the one in Figure 14-19, are used occasionally. These special units add beauty, warmth, and utility to a living area.

A shape that is commonly used is the standard rectangular shape. Another shape is the double opening, with open areas on the face and one side. "See-through" fireplaces have openings on both faces, as in Figure 14-19. This fireplace serves two rooms as shown.

Free-standing fireplaces are also popular. These are usually metal units located over a hearth area on the floor. See Figure 14-20. Another version of the free-standing fireplace is shown in Figure 14-21. This type of fireplace features an open fire without walls. A large hood is placed over the unit for the chimney.

Fireplaces are made of many materials. Brick (Figure 14-22) and natural stone are the most common. However, these are metal units faced with brick and stone. The insides must be lined with firebrick. Most bricks will absorb moisture. Fire changes the moisture to steam. This expands and cracks the brick. Firebrick is a special brick. It has been glazed so that it will not absorb moisture. Using firebrick, which does not have moisture, reduces cracking.

Many fireplaces are made from masonry materials. They are constructed piece by piece in a proper shape. However, getting the correct shape is very critical. A poorly designed fireplace will not draw the smoke. This will make the smoke enter the house or building and cause discomfort.

To make the design of the fireplace easier, builders often use metal units. There are two types of these units. One is a plain unit made of sheet metal. The sheet-metal unit is placed on a reinforced concrete

Figure 14-19. "See through" fireplaces serve two rooms. *(Potlatch)*

394 Carpentry Fundamentals

Figure 14-20. A free-standing metal fireplace. *(Majestic)*

Figure 14-22. This metal fireplace is faced with brick and trimmed with a formal wooden mantel. *(Martin Industries)*

Figure 14-21. A free-standing fireplace that is open on all sides. *(Weyerhauser)*

Special Construction Methods **395**

hearth or bottom. A properly designed frame is built around it. The outside may be faced with brick or stone as desired.

Another type of prefabricated metal fireplace has hollow chambers or tubes. These tubes pass through the heated portion. They allow air to be drawn in and heated. The heated air is circulated into the living area. These openings or tubes are vented to the front of the fireplace. Since heated air rises naturally, natural ventilation occurs. See Figure 14-23. These tubes can be connected to the regular furnace air system. Special blowers are also used to increase their heating effect. These systems make the fireplace more efficient.

General Design Factors

The design of a fireplace and chimney is extremely important. Certain practices must be followed. The heat from the burning fire will cause the smoke to rise. It must pass through the chimney to the outside. As a rule, the chimney and fireplace are built as a single unit. A special foundation is required for this combined unit. The chimney is a vertical shaft. The smoke from the burning fire passes through it to the outside. The chimney may also be used for vents or flues from furnaces and water heaters. Footings for the fireplace should extend beneath the frostline. Also they should extend 6 inches or more beyond the sides of the fireplace. Chimney walls should be at least 4 inches thick. Manufacturers' specifications should always be followed closely.

The lower part of the fireplace is called the *hearth*. See Figure 14-24. The hearth has two sections. The first is the outer hearth. This is the floor area directly in front of the fireplace. The second part is the bottom of the chamber where the fire is built. The front hearth should be floored with fire-resistive material. Tile, brick, and stone are all used. These contrast with the regular floor. The contrast makes the hearth decorative as well as practical. Some fire-

A. Fan Kit. Increases the flow of heated room air.
B. Duct Kits. Divert heated air into adjacent rooms, even upstairs, with the flick of a lever.
C. Glass Enclosure Kits. To save warmed room air from going up the chimney. Tempered glass, framed in matte black or antique brass trim.
D. Outside Air Kits. Feed the fire with outside air, rather than warm room air.

■ Heated air
☐ Room air
■ Outside air

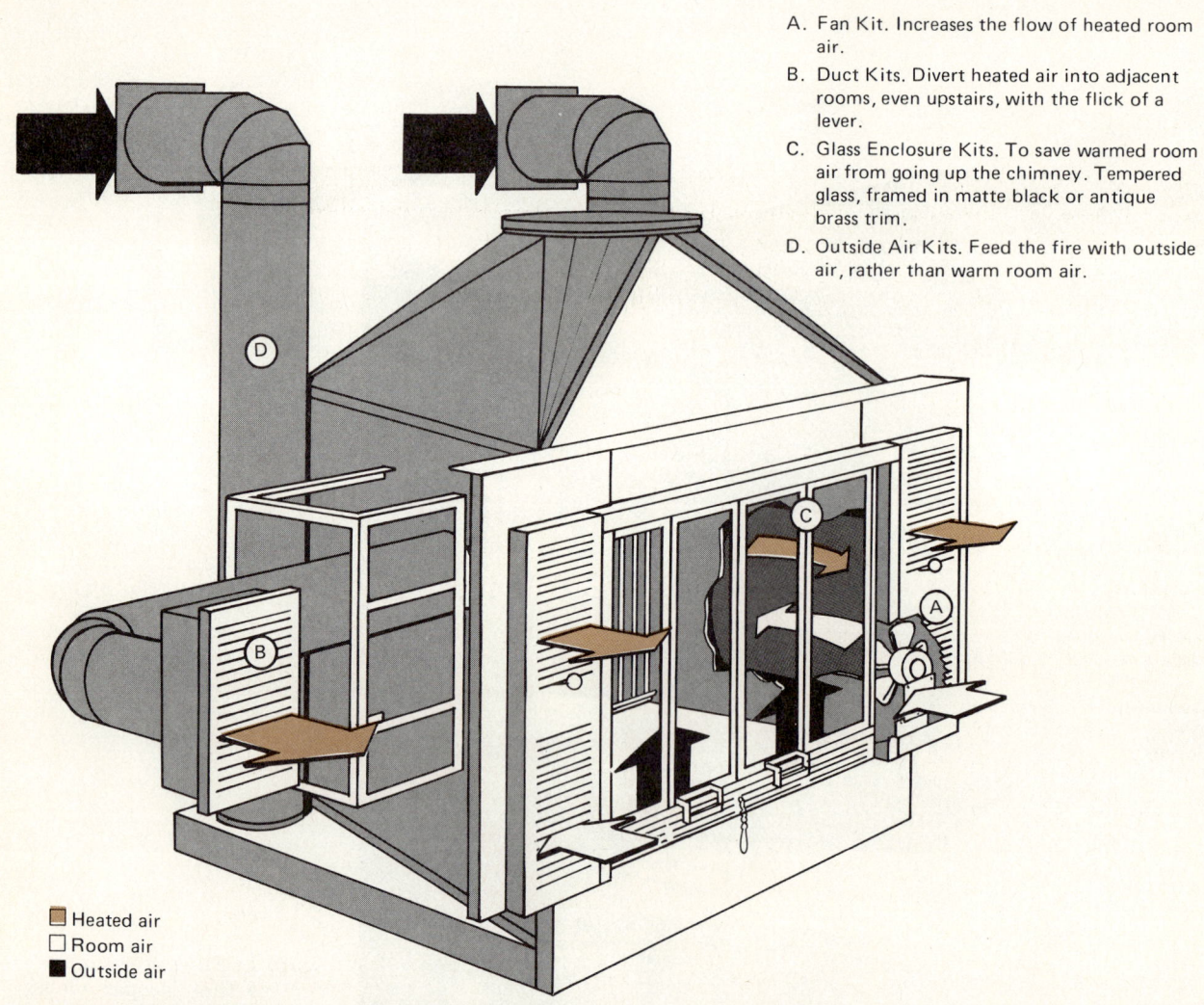

Figure 14-23. A metal fireplace unit with an air circulation system. *(Majestic)*

396 Carpentry Fundamentals

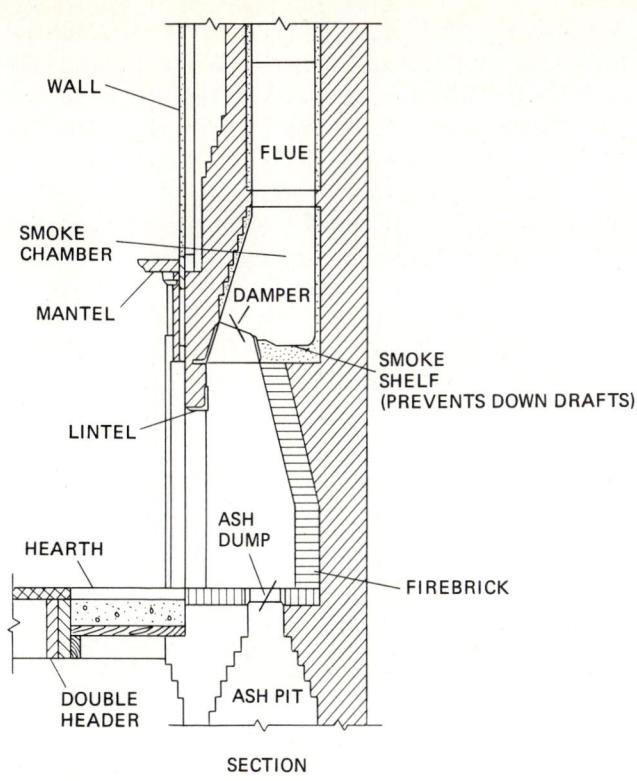

Figure 14-24. Fireplace parts.

places have a raised hearth. See Figure 14-25. The raised hearth is also an advantage in some cases. As in Figure 14-26, a special ash pit may be built at floor level. This is an advantage when houses are built on concrete slabs. The raised hearth allows the clean-out unit to be installed.

The back hearth, the back wall, and the sides of the fireplace are lined with firebrick. The back wall should be approximately 14 inches high. The depth from the front lintel to the back wall should be 16 to 18 inches. See Figure 14-24. Sidewalls and the top portion of the back are built at an angle. This angle helps the brick to radiate and reflect heat into the room. The back wall should slope forward. The amount of slope is almost one-half the height of the opening. This slope reflects heat to the room. It also directs the smoke into the throat area of the fireplace.

FIREPLACE THROAT • The throat is the narrow opening in the upper part of the fireplace. See Figure 14-24. The throat makes a ledge called a *smoke shelf*. The smoke shelf does three things for the fireplace. First, it provides a barrier for downdrafts. The smoke and heated air rise rapidly in the chimney.

Figure 14-25. A fireplace may have a raised hearth. *(Martin Industries)*

Special Construction Methods **397**

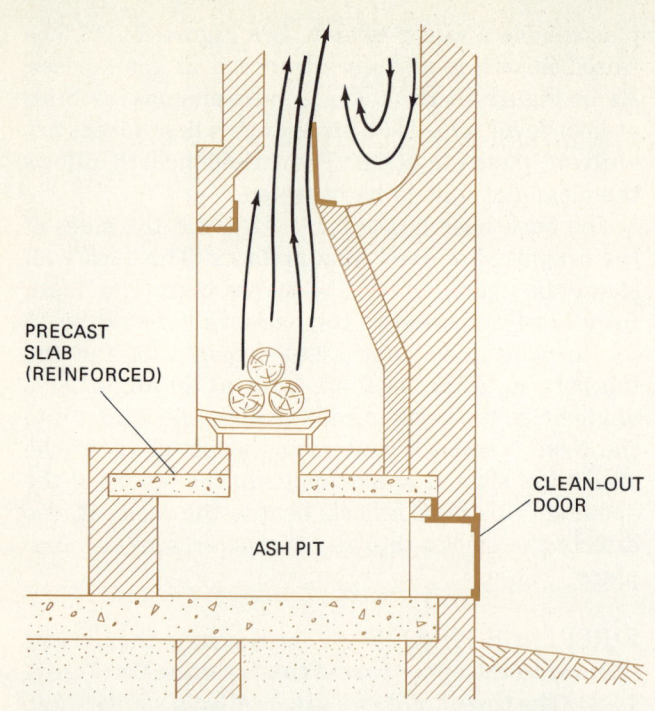

Figure 14-26. Proper construction blocks downdrafts that can push smoke into a room.

This movement causes an opposite movement of cold air from the outside. This is called a *downdraft* (Figure 14-26). The smoke shelf blocks the downdraft as shown. It also keeps the downdraft from disrupting the movement of the smoke. The throat narrows the area in the chimney. This increases the speed of the smoke. The faster speed helps create a good draft. Without proper throat design, the chimney will not draw well. The smoke will reenter the living area.

Second, the smoke shelf provides a rain ledge. The ledge shields the burning fire from direct rainfall. Third, the throat and smoke shelf provide a fastening ledge for the damper.

The area of the throat should not be less than that of the flue. For best results, the throat area should be slightly bigger than the flue. For this reason, the throat opening should be long. Its length should be the full width of the fireplace. The throat width is approximately one-half the width of the flue. To have the same area, the throat must be longer than the flue.

FIREPLACE DAMPER • The narrowest part of the throat has a door. The door is called a *damper*. Special mechanisms keep the damper open in normal use. The damper should open as in Figure 14-24. This way, the door helps form a wall at the bottom of the smoke shelf. When the fireplace is not being used, the damper may be closed. This keeps heated or cooled air from being lost through the flue.

SMOKE CHAMBER • The *smoke chamber* is the area above the smoke shelf. The throat and the smoke shelf are the full width of the fireplace. The smoke chamber narrows from the width of the fireplace to that of the flue. This also helps to reduce the velocity of downdrafts. The effects of wind gusts are lessened here. A smoke chamber is necessary for proper draft. Smoke chambers should be plastered smooth with fire clay.

CHIMNEY FLUE • The chimney flue is the passage through the chimney. Its size is determined by the area of the fireplace opening. The flue area should be one-tenth the opening area. Lower chimney heights are often used in modern buildings. Here, flues should be about 10 percent larger than this ratio.

A rule of thumb concerning flue areas is as follows: Fourteen square inches of flue are allowed per square foot of the opening of the main fireplace. For example, a fireplace opening has a height of 30 inches (2½ feet). The width is 36 inches (3 feet). The area of the fireplace opening would be 7.5 square feet. The chimney flue area should be 14 square inches for every square foot. For the opening of 7.5 square feet, the flue size should be 7.5 × 14. This would be a total of 105 square inches. A good flue size would be 10½ × 10 inches.

Flues should be as smooth as possible on the inside. This prevents buildup of soot, oily smoke, or tar. Fire clay or mortar is most often used to line the flues. Industrial chimneys are exposed to chemical action. Special porcelain liners may be needed for these. For most buildings today, tile flue liners are used.

The drawing capability of a chimney is affected by the air differential. The air rising from the chimney is hot. The outside air is cold. The difference in the temperature of these two is the air differential. The greater the difference, the better the chimney will draw. Locating chimneys on outside walls does not help this. Locating them on inside sections conserves the heat of the air rising in the chimney. Thus, chimneys located on the inside are usually more efficient.

FLUES AND CHIMNEYS • Chimneys are made of several different materials. For standard masonry chimneys, brick, stone, and concrete block are all used. Special metal chimneys are also used. In many cases, the metal chimneys are encased with special coverings. See Figure 14-27. These covers must be built according to manufacturer's specifications. However, in most cases metal standoffs must be used to hold the metal chimney away from the wood members. The chimney cover is framed from wood and siding materials. Some chimney covers have a fake brick or masonry appearance.

398 Carpentry Fundamentals

(A)

(B)

Figure 14-27. The sequence for installing special chimney covers. *(Martin Industries)*

Special Construction Methods **399**

(C)

(D)

Figure 14-27 *(cont.)*.

400 Carpentry Fundamentals

(E)

Figure 14-27 *(cont.)*

Many chimneys have special covers over them. These keep out rain or snow. Several methods are used to fasten these covers to the chimney. Some types slip inside the flue opening. Other types are fastened around the outside areas of the chimney by pressure bands. These are simply clamps that are tightened against the masonry of the chimney. Other types must be fastened by special expanding screw anchors. For these, holes are drilled into the brick or stone of the chimney. The special screw anchors are inserted and expanded firmly against the brick. This provides a firm anchor for regular bolts. These screw anchors should always be placed in the solid brick or stone. Placing them into the mortar will cause the joint to crack. This will loosen the masonry from its mortar bed. This weakens the chimney.

• POST-AND-BEAM CONSTRUCTION •

Post-and-beam construction has long been used. Warehouses, barns, and other large buildings are all examples of post-and-beam construction. Designers are now using post-and-beam construction for schools, homes, and office buildings as well. Post-and-beam construction can have wider, longer rooms and higher ceilings. Exposed beams, as in Figure 14-28, add beauty, line, and depth to the building. Beams can be laminated or boxed in several shapes. This makes unusually shaped buildings possible.

Wall supports are widely spaced. Wide spacing makes possible many very large windows. Using many windows makes a building interior light and cheerful. Wooden beams give a natural appearance. These and the natural wood finishes can add greatly to a feel of natural beauty. The actual construction methods are simple. Floors and roofs are constructed in a similar manner. Several different materials and techniques may be used with pleasing effect.

Both roof decking and floors are done in much the same manner. Wooden planks, plywood sheets, and special panels may all be used on either floors or roofs. In many construction processes, the roof and ceiling are the same. This cuts down the complexity and cost. Only one layer is used instead of two or three. The walls may combine a variety of glass, stone, brick, wood, and metal materials. There are other names for the post-and-beam type of construction. It is sometimes called *plank and beam* when used in floors and roofs. For ceilings, the term *cathedral ceiling* is often used.

Special Construction Methods **401**

Figure 14-28. Post-and-beam construction allows wider, longer open areas. *(Weyerhauser)*

Figure 14-29. Widely spaced floor or roof joists are decked with long planks. *(Weyerhauser)*

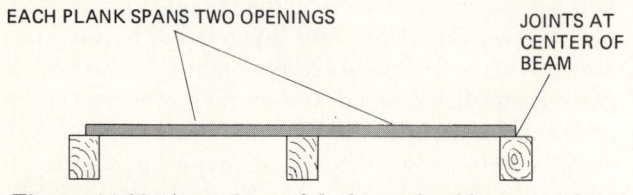

EACH PLANK SPANS TWO OPENINGS　　JOINTS AT CENTER OF BEAM

Figure 14-30. Any piece of decking should span at least two openings.

General Procedures

Regular footing and foundation methods are used. The floor beams or joists are laid as in Figure 14-29. Decking is applied directly over the floor beams as shown. Grooved-plank decking or plywood sheets are common. As a rule, any piece of decking should span at least two openings. Figure 14-30 shows spans. End-joined boards may be used so that joints need not occur over beams. However, if end-joined pieces are not used, joints should be made over floor beams.

WALL STUD POSTS • In regular framing, wall pieces are nailed together while they are flat on the subfloor. This is called the conventional system. Studs are spaced 12 to 24 inches apart.

Post-and-beam stud posts may also be nailed to sole and top plates. See Figure 14-31. However, these posts should not be less than 4 inches square. They may be spaced 4 to 12 feet apart. This gives a much wider area between supports. This allows larger glass areas without special headers to help support the roof.

Stud posts may also be applied directly to floors. See Figure 14-32. In either method, various metal straps and cleats are used. They anchor the stud posts to the floor or sole plate. The posts should be attached directly over the floor joists or beam. Where sole plates are used, the stud posts may be attached at any point.

Stud posts should not be attached directly to concrete floors exposed to the outside weather. A special metal bracket should be used. This bracket raises

402　Carpentry Fundamentals

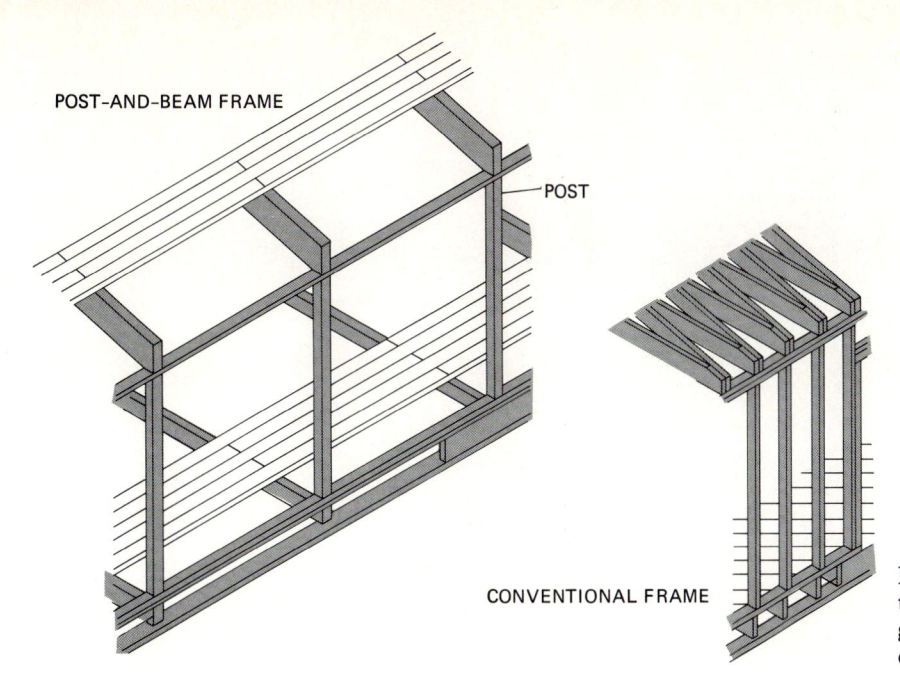

Figure 14-31. Posts may be nailed to top and sole plates. However, note the greater space between posts than for conventional frames.

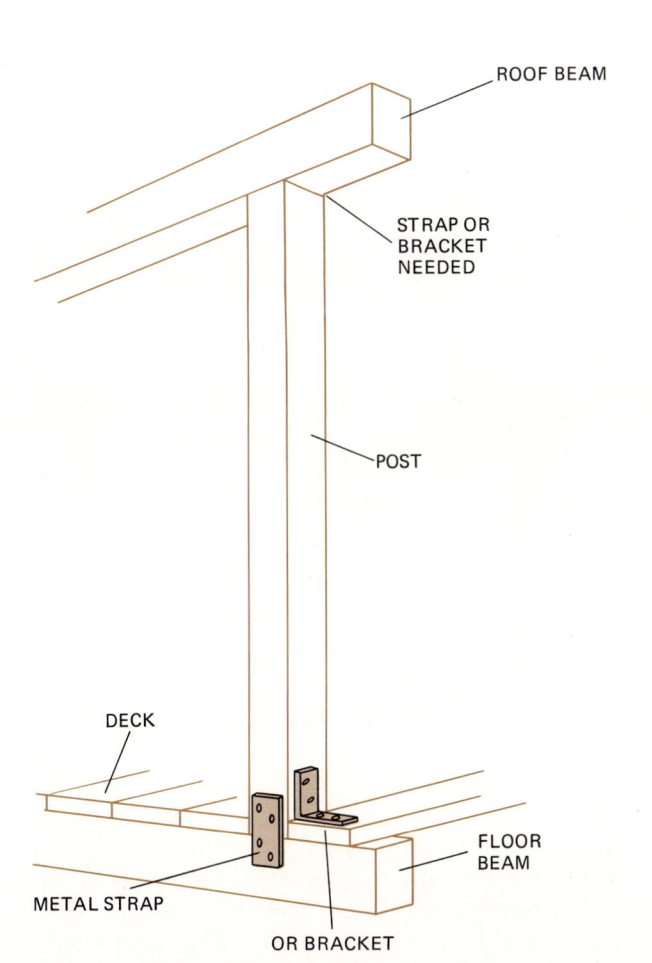

Figure 14-32. Posts may also be anchored directly to floors. They must be anchored solidly.

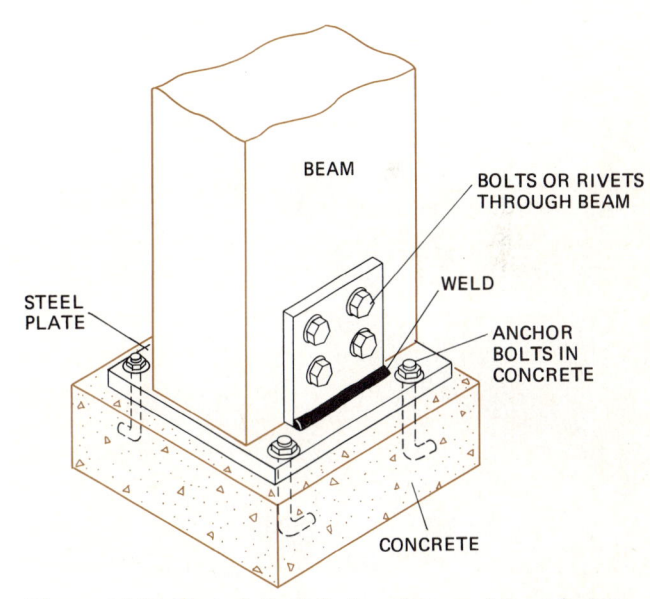

Figure 14-33. Foot plates help to reduce moisture damage.

the post off the concrete as in Figure 14-33. These brackets must be used to help prevent rot and deterioration of the post caused by moisture. Outside concrete retains moisture. This moisture leads to wood decay. Any wood piece that directly touches the concrete can decay. The special metal brackets make an air space so that the wood does not touch the

Special Construction Methods **403**

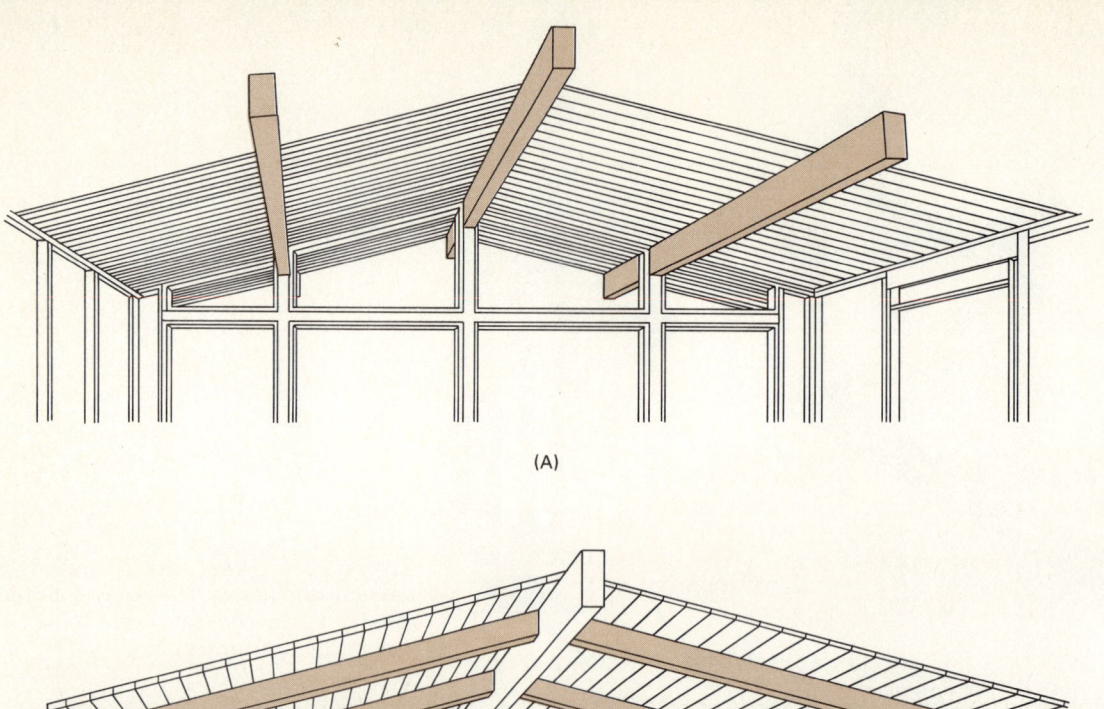

Figure 14-34. (A) Longitudinal beams run the length of a building. (B) Cross or transverse beams run across the width.

Figure 14-35. Special beam shapes may combine wall and roof framing. *(Weyerhauser)*

concrete. The circulating air keeps the wood dry and reduces decay.

BEAMS • Beams are supports for roof decking. Two systems are used. Beams may run the length of the building as in Figure 14-34A. They may also run across the building (Figure 14-34B). Both straight gable and shed roof shapes are common. Other beams may be the frame of a roof and wall combined. Beams may combine the wall studs and roof beams. The special beam shapes shown in Figure 14-35 are typical. These beams combine interior beauty with structural integrity. They are common in churches, schools, and similar buildings.

Beams are made in four basic types. These are solid beams, laminated beams, bent laminated beams, and plywood box beams. Solid beams are made of solid pieces of wood. Obviously, for larger beams, this becomes impractical. To make larger beams, pieces of wood are laminated and glued together. Laminated beams may be used horizontally or vertically. To provide graceful curved shapes, beams are both laminated and curved to shape.

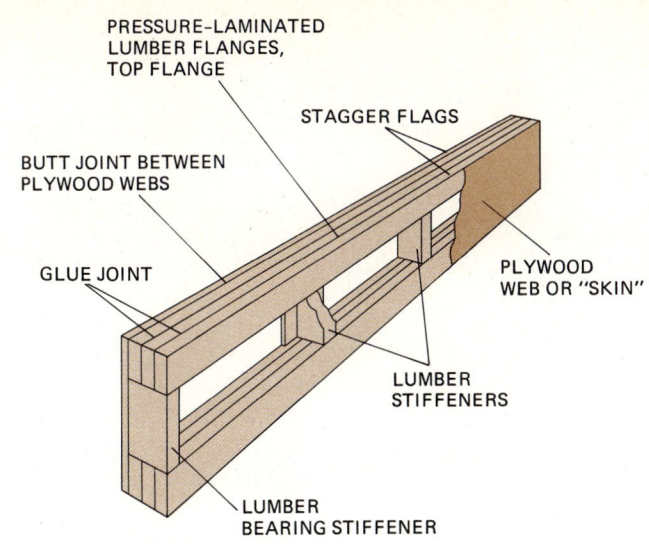

Figure 14-36. Box beams look massive, but they are made of plywood over board frames.

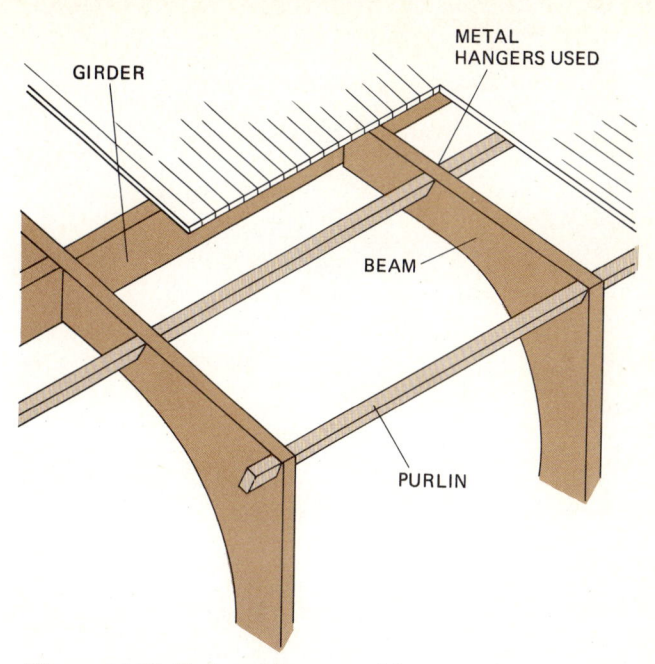

Figure 14-37. Post-and-beam roof frame.

These are built up of many layers of wood. Each layer is bent to the shape desired. They are extremely strong and fire-resistant and provide graceful beauty and open space. Laminated beams have a more uniform moisture content. This means that they expand and contract less. Plywood box beams provide a heavy appearance, great strength, and comparatively light weight. They are made as in Figure 14-36. They have one or more wooden beams at top and bottom. These beams are stiffened and spaced by lumber spacers as shown. However, the interiors are hollow. The entire framework is then covered with a plywood skin as shown. These beams appear to be solid beams. However, they are much lighter than solid wood.

Beam systems may also be used. These use small beams between larger beams. This allows large beams to be spaced wider apart. The secondary beams are called *purlins*. See Figure 14-37.

Beams must be anchored to either posts or top plates. There are two ways to anchor beams. The first is to use special wooden joints. These are cut into the beams, posts, and other members. To prevent side slippage and other shear forces, special pegs, dowels, and splines are used. Many of the old buildings in Pennsylvania and other areas were built in this manner. However, today, this method is time-consuming and costly. Today, the second method is used more often. In this method, special metal or wooden brackets are used. See Figure 14-38. There are many of these bracket devices. They hold two or more pieces together and prevent slipping and twisting. They should be bolted rather than nailed. Nails will not hold the heavy pieces. Although lag bolts may be used, through bolts are both stronger and more permanent.

Figure 14-38. Special brackets are often used to hold beams and other parts together. *(Weyerhauser)*

Special Construction Methods 405

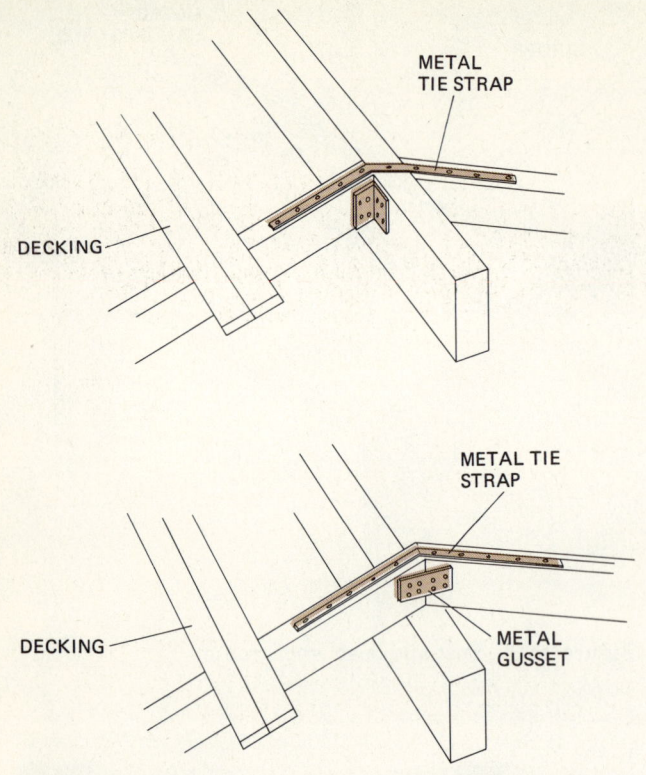

Figure 14-39. Beams are tied to girders with gussets and straps.

Ridge beams are used for gable construction using post-and-beam methods. Where beams are joined to the ridge beams, special straps are used. See Figure 14-39. These straps may resemble gussets and are applied to the sides. However, an open appearance without brackets can be made. The straps are inlet and applied across the top as shown.

Purlins are held by hangers and brackets like joist hangers. They are most often made of heavy metal. Most metal fasteners are made specifically for each job to the designer's recommendations.

BEARING PLATES • Where several beams join, special allowances must be made for expansion and contraction. The surface of this joining area should also be enlarged. All beams may then have enough area to support them. Special steel plates are used for these types of situations. These are called *bearing plates*. Bearing plates are used at tops of posts, bottoms of angular braces, and other similar situations.

DECKING • Common decking materials include plywood boards and 2-inch-thick lumber. Decking is often exposed on the interior for a natural look. Both lumber and plywood should be grooved for best effect. The grooves should be on both the edges and ends. Great strength is needed because of the longer spans between supports.

Other materials used for roof coverings include concrete and special insulated gypsum materials. Common post-and-beam construction has insulation

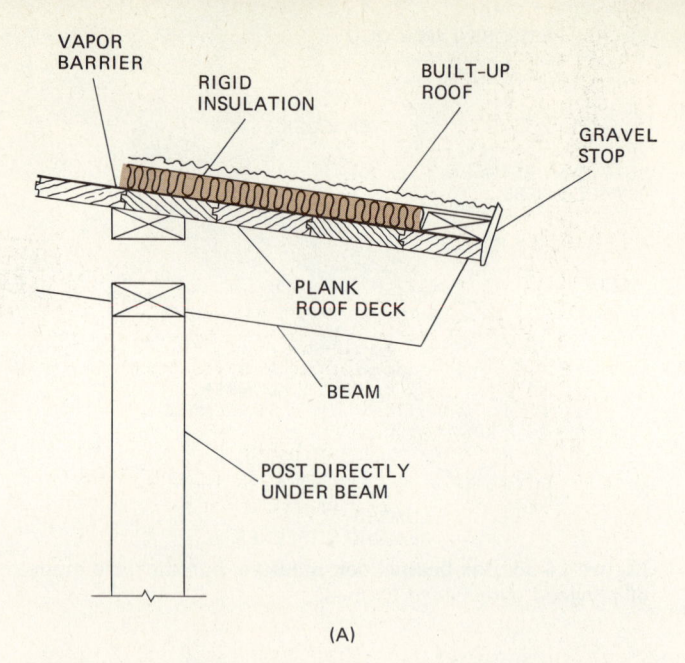

Figure 14-40. (A) Insulation can be applied on top of either beams or roof decking. (B) A composition roof is then laid. *(Gold Bond)*

applied on top of roof decking. See Figure 14-40. A moisture barrier should be used on top of the decking and beneath the insulation. A rigid insulation should be used on top of the decking. It should be strong enough to support the weight of workers without damage. As a rule, a built-up roof is applied over this type of construction. Roofs of this type generally are shed roofs with very little slope. When gables are used, a low pitch is deemed better. Built-up roofs may not be used on steep-pitched roofs.

However, on special shapes, several different roofing techniques may be used. Decking or plywood may be used to sheath the beams or bents. Metal or wooden roofing can be applied over the decking. Commonly sheet copper, sheet steel, and wood shingles are all used.

STRESSED-SKIN PANELS • Stressed-skin panels are large prebuilt panels. They are used for walls,

406 Carpentry Fundamentals

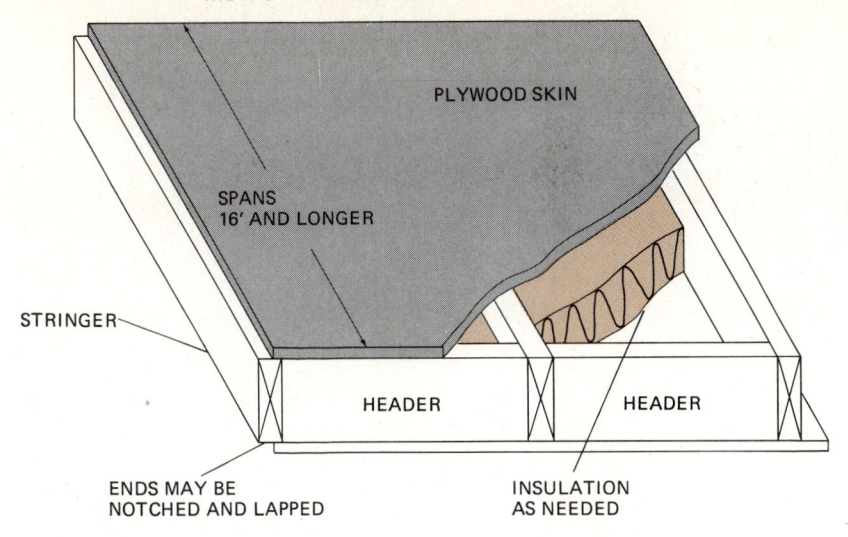

Figure 14-41. Stressed-skin panels are used on floors and roof decks.

floors, and roof decks. They can be made to span long distances and still give strong support. They are built in a factory and hauled to the building site. They cost more to make, but they save construction time.

The panels are formed around a rigid frame. Interior and exterior surfaces are included. The panels can also include ducts, pipes, insulation, and wiring. One surface can be roof decking. The interior side can be insulating sound board to make the finished ceiling. The exterior would be heavy plywood for the roof decking. A standard composition roof could be applied over the panels. The same techniques can be used for wall and floor sections. See Figure 14-41. The panels are usually assembled with glue and fasteners such as nails or staples.

• WORDS CARPENTERS USE •

carriage	balusters	winder	damper	bearing plate
stringer	bannister	flue	post and beam	stressed-skin
tread	newels	hearth	purlin	panel
riser	landing	firebrick	hangers	

• STUDY QUESTIONS •

14-1. Why is firebrick used in fireplaces?

14-2. How many steps would be cut for stairs with:
 Rise of 8′2″
 Run of 14′0″
 Riser: 7 inches

14-3. For the steps above, what would the tread width be?

14-4. Why should fireplaces be carefully designed?

14-5. What is the opening-to-flue ratio?

14-6. Where is the damper located?

14-7. What would be a good flue size for a fireplace with an opening 30 inches high and 48 inches long?

14-8. What are the advantages of metal fireplaces?

14-9. What are the advantages of post-and-beam construction?

14-10. How are posts and beams connected?

14-11. Where are bearing plates used?

14-12. Would you embed an exterior beam into concrete? Why?

14-13. List the types of wood beams used.

14-14. Why are laminated beams better than solid wooden beams?

14-15. How can a stressed-skin panel be used for both roof decking and finished ceilings?

14-16. What kind of insulation is used over lumber roof decking?

14-17. What type of roof is not used on steep-pitched roofs?

14-18. What is a "housed carriage"?

14-19. What causes stairs to squeak?

14-20. Can stud posts be fastened at any point on a subfloor?

Special Construction Methods

15 MAINTENANCE AND REMODELING

Planning is an important part of any job. The job of remodeling is no exception. It takes a plan to get the job done correctly. The plan must have all the details worked out. This will save money, time, and effort. The work will go smoothly if the bugs have been worked out before the job is started.

• PLANNING THE JOB •

Maintenance means keeping something operating properly. It means taking time to make sure a piece of equipment will operate tomorrow. It means doing certain things to keep a house in good repair. Many types of jobs present themselves when it comes to maintenance. The carpenter is the person most commonly called upon to do maintenance. This may range from the replacement of a lock to complete replacement of a window or door.

Remodeling means just what the word says. It means changing the looks and the function of a house. It may mean you have to put in new kitchen cabinets. The windows may need a different type of opening. The floor may be old and need a new covering. The basement may be in need of remodeling. It may require new paneling or tiled floors.

Working on a house when it is new calls for a carpenter who can saw, measure, and nail things in the proper place. Working on a house after it is built calls for many types of operations. The carpenter may be called upon to do a number of different things related to the trade.

Diagnosing Problems

A person who works in maintenance or remodeling needs to know what the problem actually is. That person must be able to find out what causes a problem. The next step is, of course, to decide what to do to correct the problem.

If you know how a house is built, you should be able to repair it. This means you know what has to be done to properly construct a wall or repair it if damaged. If the roof leaks, you need to know where and how to fix it.

In other words, you need to be able to diagnose problems in any building. Since we are concerned with the residential types here, it is important to know how things go wrong in a well-built home. It is important too to be able to repair them.

Before you can remodel a house, or add on to it, you need to know how the original was built. You need to know what type of foundation was used. Can it support another story, or can the soil support what you have in mind? Are you prepared for the electrical loads? What about the sewage? How does all this fit into the addition plans? What type of consideration have you given the plumbing, drainage, and other problems?

Identifying Needed Operations

If peeling paint needs to be removed and the wall repainted, can you identify what caused the problem? This will be important later when you choose another paint. Figure 15-1 shows what can happen with paint.

Paint problems by and large are caused by the presence of moisture. The moisture may be in the wood when it is painted. Or it may have seeped in later. Take a look at Figure 15-1 to see just what causes cracking and alligatoring. Also notice the causes of peeling, flaking, nailhead stains, and blistering.

Normal daily activity in the home of a family of four can put as much as 50 pints of water vapor in the air in one day. Since moisture vapor always seeks an area of lesser pressure, the moisture inside the house tries to become the same as the outside pressure. This equalization process results in moisture passing through walls and ceilings, sheathing, door and window casings, and the roof. The eventual result is paint damage. This occurs as the moisture passes through the exterior paint film.

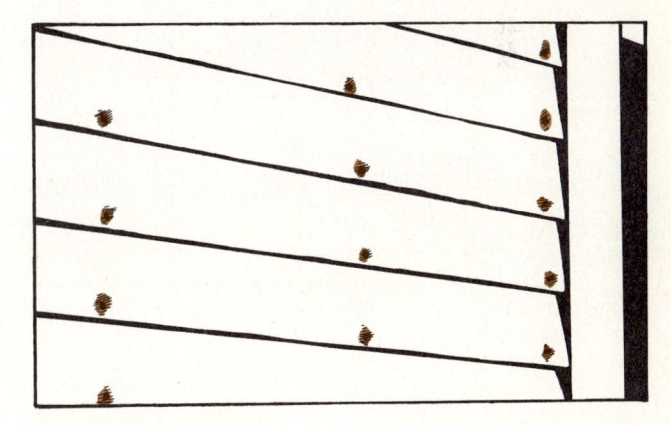

NAILHEAD STAINS
CONDITION:
  Nailhead stains are caused by moisture rusting old or uncoated nails.
CORRECTION:
- All moisture problems must be corrected before repainting.
- Sand or wire brush stained paint and remove rust down to bright metal of nailhead.
- Countersink nail if necessary.
- Apply primer to nailheads. Allow to dry.
- Caulk nail holes. Allow to dry. Sand smooth.
- Apply primer to surface, following label directions.
- Apply topcoat according to label directions.

Figure 15-1. Causes of paint problems on houses. (*Grossman Lumber*)

Special Construction Methods

PEELING

CONDITION:
  Peeling is caused by moisture being pulled through the paint (by the sun's heat), lifting paint away from the surface.

CORRECTION:
- All moisture problems must be corrected before surface is repainted.
- Remove all peeling and flaking paint.
- Scrape and sand all peeling paint to bare wood including several inches around damaged areas. Feather edges.
- Apply primer according to directions on label.
- Apply topcoat according to directions.

BLISTERING

CONDITION:
  Blistering is actually the first stage of the peeling process. It is caused by moisture attempting to escape through the existing paint film, lifting the paint away from the surface.

CORRECTION:
- All moisture problems must be corrected before repainting.
- Scrape and sand all blistering paint to bare wood several inches around blistered area.
- Feather or smooth the rough edges of the old paint by sanding.
- Apply primer according to label directions.
- Apply topcoat according to label directions.

FLAKING

CONDITION:
  Siding alternatively swells and shrinks as moisture behind it is absorbed and then dries out. Paint film cracks from swelling and shrinking and flakes away from surface.

CORRECTION:
- All moisture problems must be corrected before surface is repainted.
- Scrape and sand all peeling paint to bare wood including several inches around damaged areas. Feather edges.
- Apply primer according to label directions.
- Apply topcoat according to label directions.

CRACKING AND ALLIGATORING

CONDITION:
  Cracking and "alligatoring" are caused by (a) paint that is applied too thick, (b) too many coats of paint, (c) paint applied over a paint coat which is not completely dry, or (d) an improper primer.

CORRECTION:
- Removal of entire checked or alligatored surface may be necessary.
- Scrape and sand down the surface until smooth. Feather edges.
- Apply primer. Follow label directions.
- Apply topcoat. Follow label directions.

Figure 15-1 *(cont.)*. Causes of paint problems on houses. *(Grossman Lumber)*

Sequencing Work to be Done

In making repairs or doing remodeling, it is necessary to schedule the work properly. It is necessary to make sure things are done in order. For example, it is difficult to paint if there is no wall to paint. This may sound ridiculous, but it is no more so than some other problems associated with getting a job done.

It is hard to nail boards onto a house if there are no nails. Somewhere in the planning you need to make sure nails are available when you need

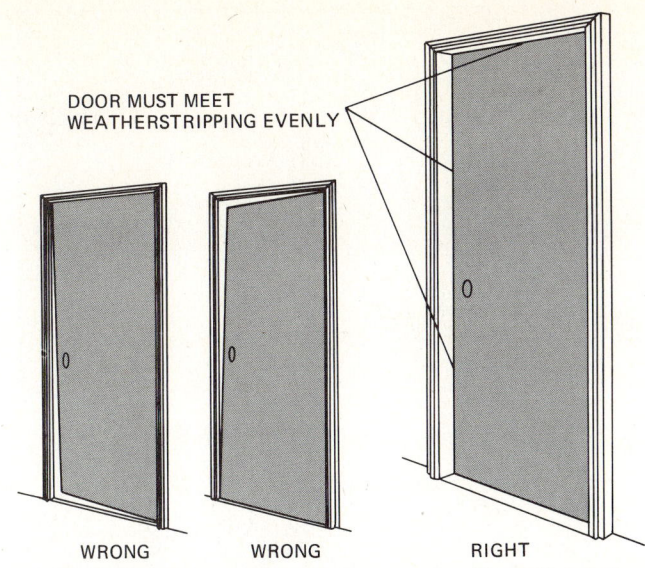

Figure 15-2. Right and wrong ways of nailing the strike jamb. (General Products)

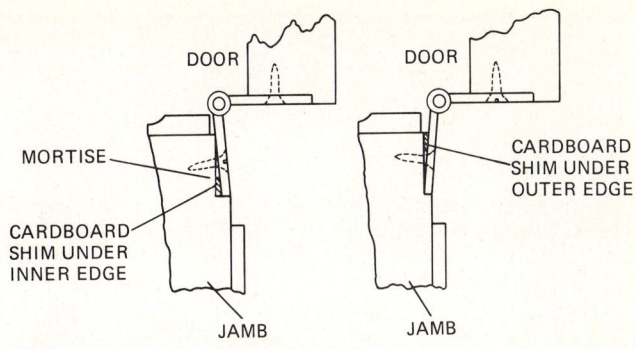

Figure 15-3. Hinge adjustments for incorrectly fitting doors. Note the shim placement.

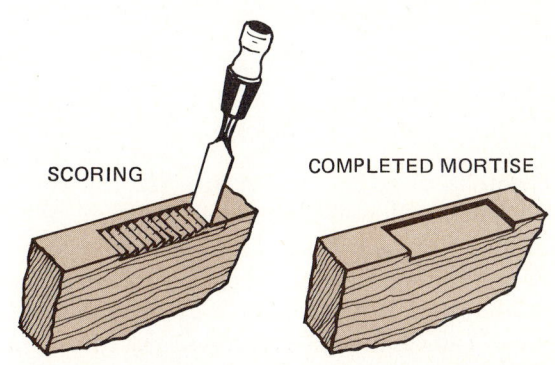

Figure 15-4. Removing extra wood from a door. Be careful not to remove too much. (Grossman Lumber)

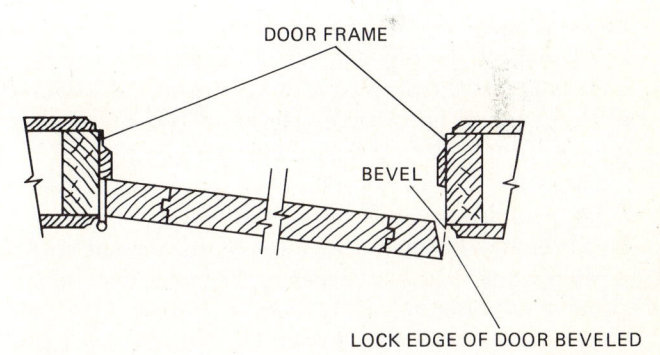

Figure 15-5. Beveling the lock stile of a door. (Grossman Lumber)

them. If things are not properly planned—in sequence—you could be ready to place a roof on the house and not have the proper size nails to do the job. You wouldn't try to place a carpet on bare joists. There must be a floor or subflooring first. This means there is some sequence that must be followed before you do a job or even get started with it.

Make a checklist to be sure you have all the materials you need to do a job *before* you get started. If there is the possibility that something won't arrive when needed, try to schedule something else so the operation can go on. Then you can pick up the missing part later when it becomes available. This means that sequencing has to take into consideration the problems of supply and delivery of materials. The person who coordinates this is very important. This person can make the difference between the job being a profitable one or a money loser.

• MINOR REPAIRS AND REMODELING •

When doors are installed, they should fit properly. That means the closed door should fit tightly against the door stop. Figure 15-2 shows how the door should fit. If it doesn't fit properly, adjust the strike jamb side of the frame in or out. Do this until the door meets the weatherstripping evenly from top to bottom.

Adjusting Doors

The strike jamb can be shimmed like the hinge jamb. Place one set of shims behind the strike plate mounting location. Renail the jamb so that it fits properly. If it is an interior door, there will be no weatherstripping. Figure 15-3 shows how to shim the hinges to make sure the door fits snugly. In some cases it may be necessary to remove some of the wood on the door where the hinge is attached. See Figure 15-4. This shows how the wood is removed with a chisel. You must be careful not to remove the back part of the door along the outside of the door.

If the door binds or does not fit properly, you may have to remove some of the lock edge of the door. Bevel it as shown in Figure 15-5 so that it fits. An

Maintenance and Remodeling 411

Figure 15-6. Trimming the width of a door. (Grossman Lumber)

old carpenter's trick is to make sure the thickness of an 8d nail is allowed all around the door. This usually allows enough space for the door to swell some in humid weather and not too much space when winter heat in the house dries out the wood in the door and causes it to shrink slightly.

Occasionally it is necessary to remove the door from its hinges by removing the hinge pins. Take the door to a vise or workbench so that the edge of the door can be planed down to fit the opening. If the amount of wood to be removed from the door is more than ¼ inch, it will be necessary to trim both edges of the door equally to one-half the width of the wood to be removed. Trim off the wood with a smooth or jack plane as shown in Figure 15-6.

Sometimes it is necessary to remove the doors from their hinges and cut off the bottom. This may be necessary because the doors were not cut to fit a room where there is carpeting. If this is necessary, remove the door carefully from its hinges. Mark off the amount of wood that must be removed from the bottom of the door. Place a piece of masking tape over the area to be cut. Redraw your line on the masking tape. It should be in the middle of the tape. Tape the other side of the door at the same distance from the bottom. Set the saw to cut the thickness of the door. Cut the door with a power saw or handsaw. Cutting through the tape will hold the finish on the wood. You will not have a door with splinters all along the cut edge. If the door is cut without tape, it may have splinters. This can look very bad if the door has been prefinished.

If the top of the door binds, it should be beveled slightly toward the stop. This can let it open and close more easily.

In some cases the outside door does not meet the threshold properly. It may be necessary to obtain

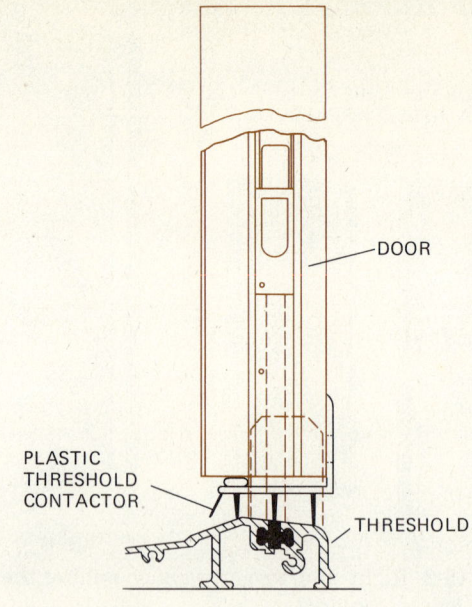

Figure 15-7. Adding a piece of plastic to a threshold to make sure the bottom of the door fits snugly. This keeps rain from entering the room as well as keeping out cold air in the winter. (General Products)

a thicker threshold or a piece of plastic to fit on the bottom of the door as in Figure 15-7.

Adjusting Locks

Installing a lock can be as easy as following instructions. Each manufacturer furnishes instructions with each new lock. However, in some cases you have to replace one that is around the house and no instructions can be found. Take a look at Figure 15-8 for a step-by-step method of placing a lock into a door. Note that the lockset is typical. It can be replaced with just a screwdriver. There are 18 other brands that can be replaced by this particular lockset.

Various locksets are available to fit the holes that already exist in a door. Strikes come in a variety of shapes too. You should choose one that fits the already-grooved door jamb.

A deadbolt type of lock is called for in some areas. This is where the crime rate is such that a more secure door is needed.

Figure 15-9 shows a lockset that is added to the existing lock in the door. This one is key-operated. This means you must have two keys to enter the door. It is very difficult for a burglar to cause this type of lock bolt to retract.

In some cases it is desired that once the door is closed, it is locked. This can be both an advantage and a disadvantage. If you go out without your key, you are in trouble. This is especially true if no one else is home. However, it is nice for those who are

412 Carpentry Fundamentals

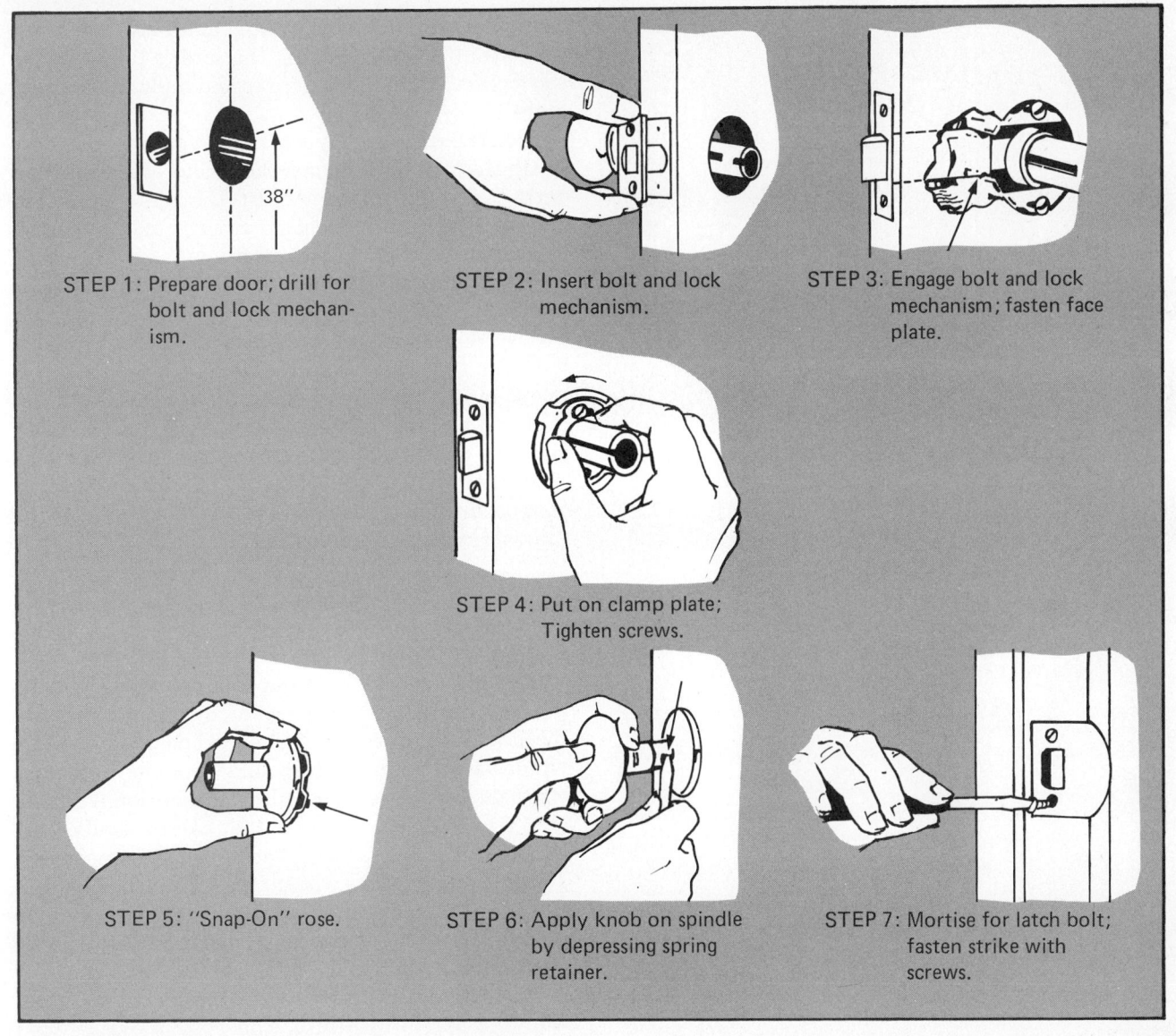

STEP 1: Prepare door; drill for bolt and lock mechanism.

STEP 2: Insert bolt and lock mechanism.

STEP 3: Engage bolt and lock mechanism; fasten face plate.

STEP 4: Put on clamp plate; Tighten screws.

STEP 5: "Snap-On" rose.

STEP 6: Apply knob on spindle by depressing spring retainer.

STEP 7: Mortise for latch bolt; fasten strike with screws.

Figure 15-8. Fitting a lockset into a door. (*National Lock*)

Figure 15-9. Key-operated auxiliary lock. (*Weiser Lock*)

a bit absent-minded and forget to lock the door once it is closed. The door locks automatically once the door is pushed closed.

Figure 15-10 shows how to insert a different type of lock. It is designed to fit into the existing drilled holes. All you need is a screwdriver to install it. Figure 15-11 shows how secure the lock can be with double cylinders to fit through the holes on the door jamb.

About the only trouble with a lockset is the loosening of the doorknob. It can be tightened with a screwdriver in most cases. If the lockset has two screws near the knob on the inside of the door, simply align the lockset and tighten the screws. If it is another type—with no screws visible—just release the small tab that sticks up inside the lockset; it can be seen through the brass plate around the knob close to the wooden part of the door. This will allow the

Maintenance and Remodeling   413

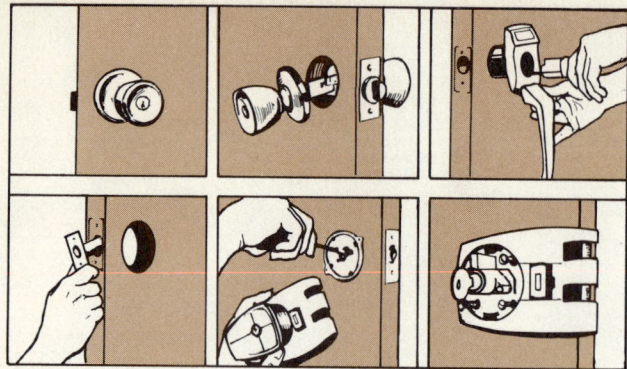

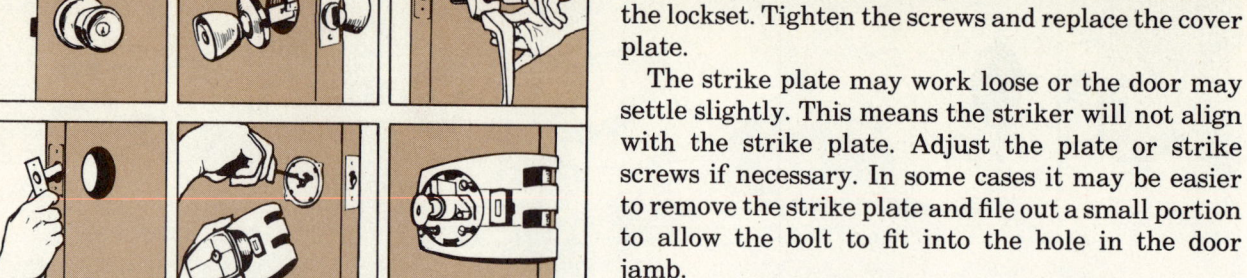

Figure 15-10. Putting in a lockset. *(Weiser Lock)*

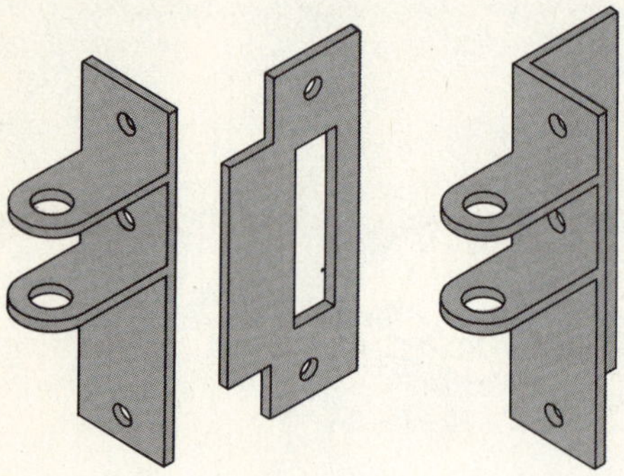

Figure 15-11. Strikes for single- and double-cylinder locks. *(Weiser Lock)*

brass plate or ring to be rotated and removed. Then it is a matter of tightening the screws found inside the lockset. Tighten the screws and replace the cover plate.

The strike plate may work loose or the door may settle slightly. This means the striker will not align with the strike plate. Adjust the plate or strike screws if necessary. In some cases it may be easier to remove the strike plate and file out a small portion to allow the bolt to fit into the hole in the door jamb.

If the doorknob becomes green or off color, remove it. Polish the brass and recoat the knob with a covering of lacquer and replace. In some cases the brass is plated and will be removed with the buffing. Here you may want to add a favorite shade of good metal enamel to the knob and replace it. This discoloring does occur in bathrooms where the moisture attacks the doorknob lacquer coating.

Installing Drapery Hardware

One often-overlooked area of house building is the drapery hardware. People who buy a new house are faced with the question, "What do I do to make these windows attractive?" Installing window hardware can be a job in itself. Installing conventional adjustable traverse rods is a task that can be easily done by the carpenter, in some cases, or the homeowner. There are specialists who do these things. However, it is usually an extra service that the carpenter gets paid for after the house is finished and turned over to the homeowner.

Decorative traverse rods are preferred by those who like period furniture. This type of drapery rod is shown in Figure 15-12. Installation of this type

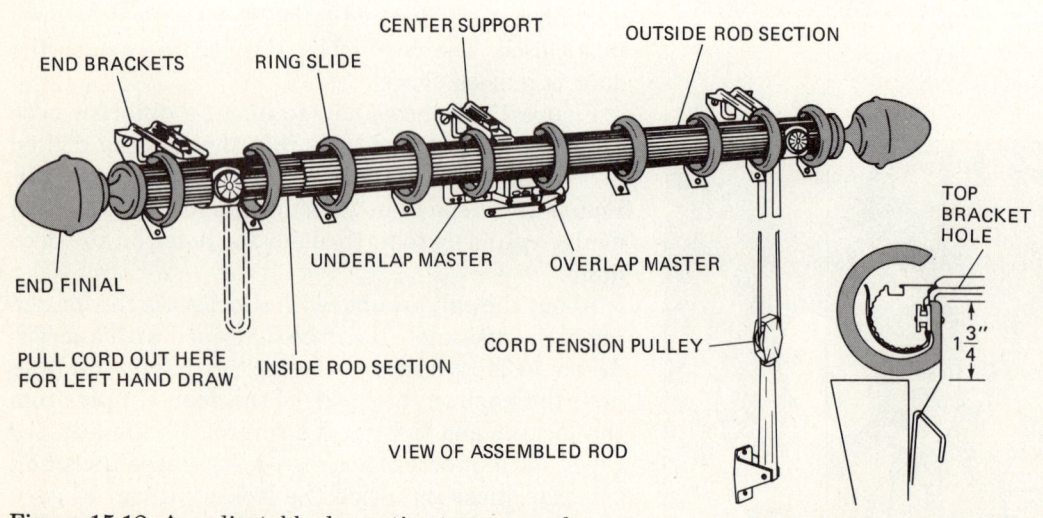

Figure 15-12. An adjustable decorative traverse rod.

414 Carpentry Fundamentals

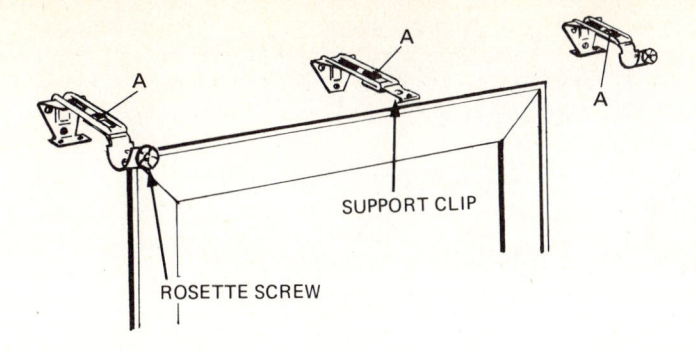

Install end brackets above and 6" to 18" to the sides of casing. When installing on plaster or wallboard and screws do not anchor to studding, plastic anchors or other installation aids may be needed to hold brackets securely. Place center supports (provided with longer size rods only) equidistant between end brackets. Adjust bracket and support projection at screws. "A." *NOTE: Attach rosette screws to end brackets if not already assembled.*

Figure 15-13. Install end brackets and center support.

Place end finials on rod. Finial with smaller diameter shaft fits in "inside" rod section, larger one fits in "outside" section. Place last ring between bracket and end finial. (See Figure 15-17A.)

Extend rod to fit brackets. Place rod in brackets so tongue on bottom of bracket socket fits into hole near end of rod. Tighten rosette screw.

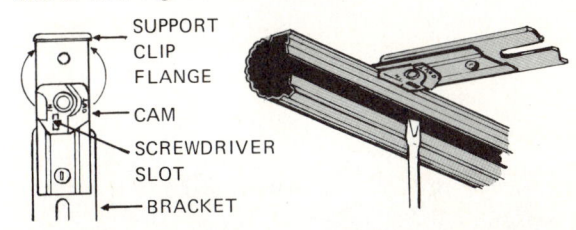

When center supports are used, insert top of rod in support clip flange. If outside rod section, turn cam with screwdriver counterclockwise to lock in rod. If inside rod section, turn cam clockwise.

Figure 15-14. Place rod in bracket.

Right hand draw is standard. If left hand draw is desired, simply pull cord loop down from pulley wheels on left side of rod.

Fasten tension pulley to sill, wall or floor at point where cord loop falls from traverse rod. Lift pulley stem and slip nail through hole in stem. Pull cord up through bottom of pulley cover. Take up excess slack by pulling out knotted cord from back of overlap master slide. See Figure 15-16. Re-knot and cut off extra cord. Remove nail. Pulley head may be rotated to eliminate twisted cords.

Figure 15-15. Install cord tension pulleys.

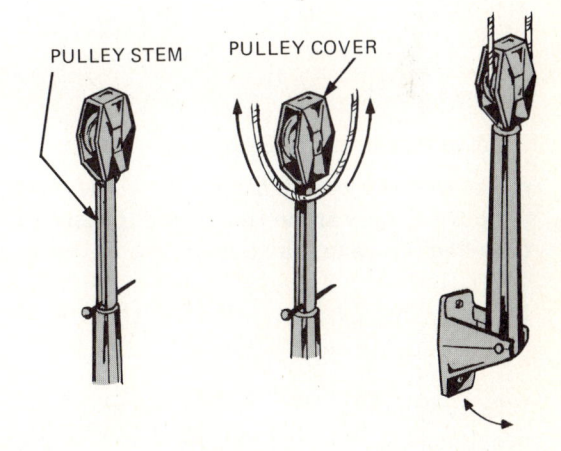

of rod is shown step by step in Figures 15-13 through 15-17.

Now that you have been studying some of the different types of windows in Chapter 9, have you wondered how the draperies would be fitted?

Figure 15-18 shows some of the regular-duty types of corner and bay window drapery rods. Note the various shapes of bay windows that are made. The rods are made to fit the bay windows.

Figure 15-19 shows more variations of bay windows. These call for some interesting rods. Of course, in some cases the person looks at the cost of the

Maintenance and Remodeling **415**

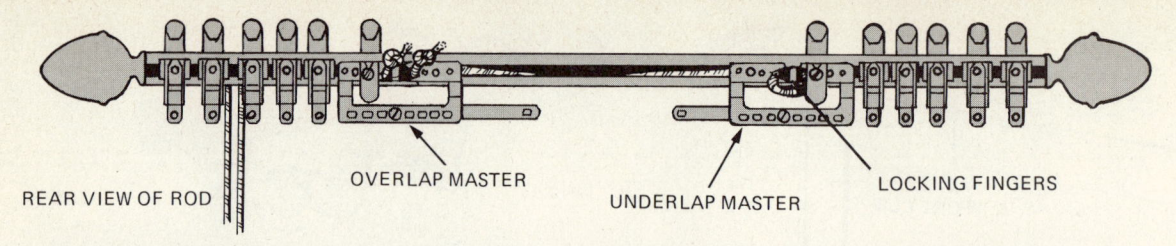

REAR VIEW OF ROD — OVERLAP MASTER — UNDERLAP MASTER — LOCKING FINGERS

Pull draw cord to move overlap master to end of rod. Holding cords taut, slide underlap master to opposite end of rod. Then fasten cord around locking fingers on underlap master to stop underlap master from slipping. Then center both masters.

Figure 15-16. Adjusting master slides.

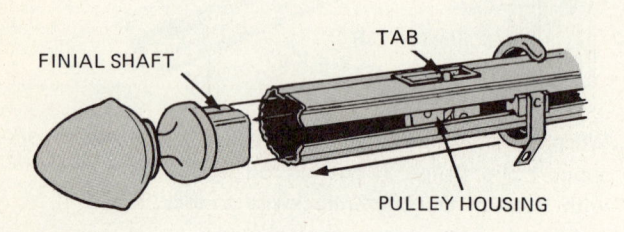

FINIAL SHAFT — TAB — PULLEY HOUSING

Slip off end finial. Remove extra ring slides by pulling back on tab and sliding rings off end of rod. Be sure to leave the last ring between pulley housing and end finial.

Figure 15-17. (A) Remove unused ring slides.

To train draperies, "break" fabric between pleats by folding the material toward the window. Fold pleats together and smooth out folds down the entire length of the draperies. Tie draperies with light cord or cloth. Leave tied two or three days before operating.

(B) Training draperies.

rods and hardware and decides to put one straight rod across the bay and not follow the shape of the windows. This can be the least expensive, but loses the effect of having a bay window in the first place.

Repairing Damaged Sheetrock Walls (Drywall)

In drywall construction the first areas to show problems are over joints or fastener heads. Improper application of either the board or the joint treatment may be at fault. Other conditions existing on the job can also be responsible for reducing the quality of the finished gypsum board surface. A discussion of some of these conditions follows.

PANELS IMPROPERLY FITTED

Cause Forcibly wedging an oversize panel into place. This bows the panel and builds in stresses. The stress keeps it from contacting the framing. See Figure 15-20.

Result After nailing, a high percentage of the nails on the central studs probably will puncture the paper. This may also cause joint deformation.

Remedy Remove the panel. Cut it to fit properly. Replace it. Fasten from the center of the panel toward the ends and edges. Apply pressure to hold the panel tightly against the framing while driving the fasteners.

PANELS WITH DAMAGED EDGES

Cause Paper-bound edges have been damaged or abused. This may result in ply separation along the edge. Or it may loosen the paper from the gypsum core. Or, it may fracture or powder the core itself. Damaged edges are more susceptible to ridging after joint treatment.

Remedy Cut back any severely damaged edges to the sound board before application.

416 Carpentry Fundamentals

REGULAR DUTY CUSTOM CEILING SETS

ONE-WAY TRAVERSE SETS

| 1787R | RIGHT TO LEFT | 1787L | LEFT TO RIGHT | | | |

TWO CORD TRAVERSE SETS
WITH 1 PAIR MASTER CARRIERS CENTER CLOSING

| 1788 | | | |

TWO CORD TRAVERSE SETS
WITH 2 PAIRS OF MASTER CARRIERS CENTER CLOSING

| 1789 | | | |

CRATING CHARGE: 10% OF CURVED ROD ORDER

REGULAR DUTY CORNER, CURVED AND BAY WINDOW CUSTOM SETS

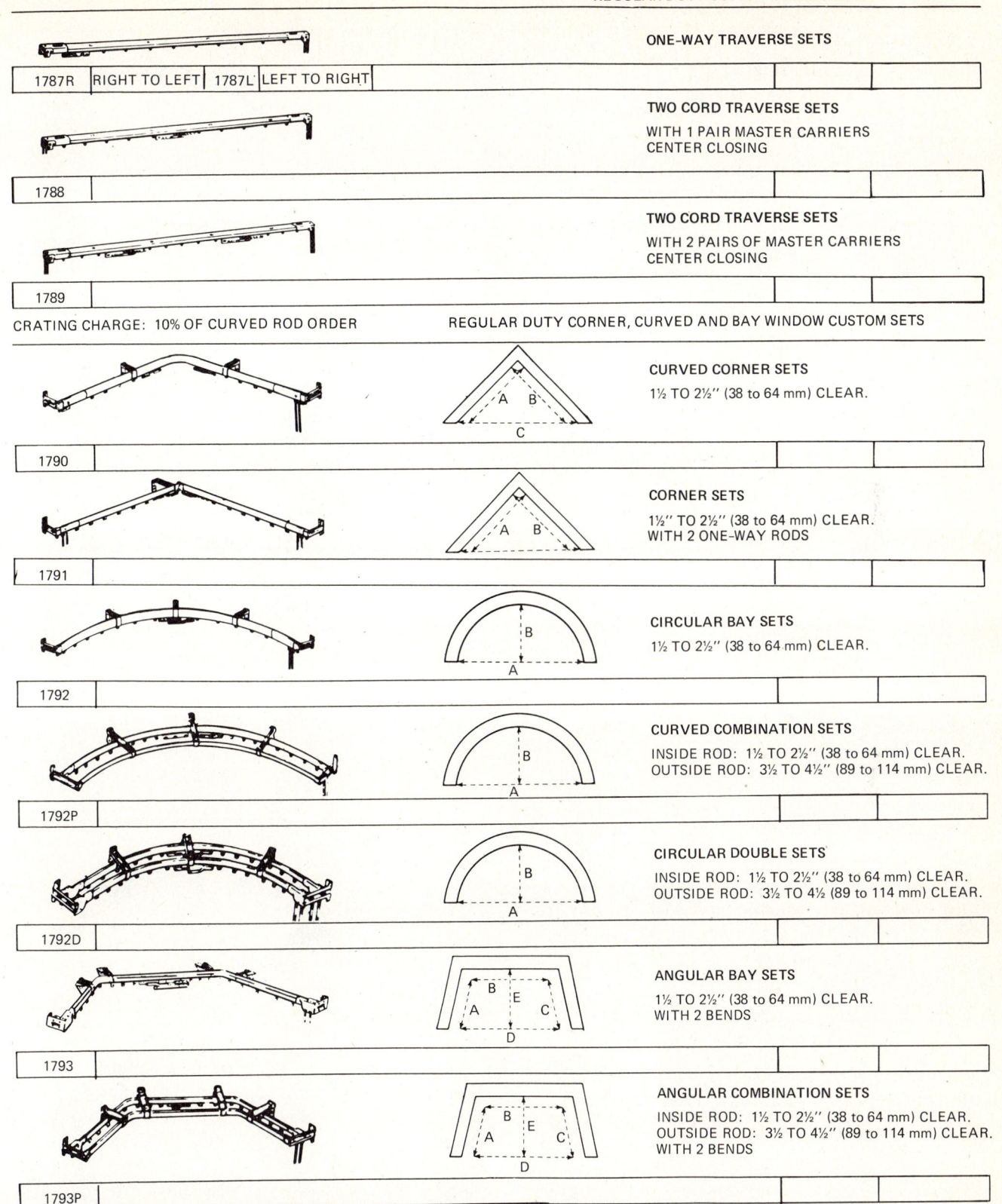

CURVED CORNER SETS
1½ TO 2½" (38 to 64 mm) CLEAR.

| 1790 | | | |

CORNER SETS
1½" TO 2½" (38 to 64 mm) CLEAR.
WITH 2 ONE-WAY RODS

| 1791 | | | |

CIRCULAR BAY SETS
1½ TO 2½" (38 to 64 mm) CLEAR.

| 1792 | | | |

CURVED COMBINATION SETS
INSIDE ROD: 1½ TO 2½" (38 to 64 mm) CLEAR.
OUTSIDE ROD: 3½ TO 4½" (89 to 114 mm) CLEAR.

| 1792P | | | |

CIRCULAR DOUBLE SETS
INSIDE ROD: 1½ TO 2½" (38 to 64 mm) CLEAR.
OUTSIDE ROD: 3½ TO 4½ (89 to 114 mm) CLEAR.

| 1792D | | | |

ANGULAR BAY SETS
1½ TO 2½" (38 to 64 mm) CLEAR.
WITH 2 BENDS

| 1793 | | | |

ANGULAR COMBINATION SETS
INSIDE ROD: 1½ TO 2½" (38 to 64 mm) CLEAR.
OUTSIDE ROD: 3½ TO 4½" (89 to 114 mm) CLEAR.
WITH 2 BENDS

| 1793P | | | |

Figure 15-18. Regular-duty custom ceiling sets of traverse rods; regular-duty corner, curved, and bay window custom sets. *(Kenney)*

Maintenance and Remodeling

REGULAR DUTY CORNER, CURVED AND BAY WINDOW CUSTOM SETS

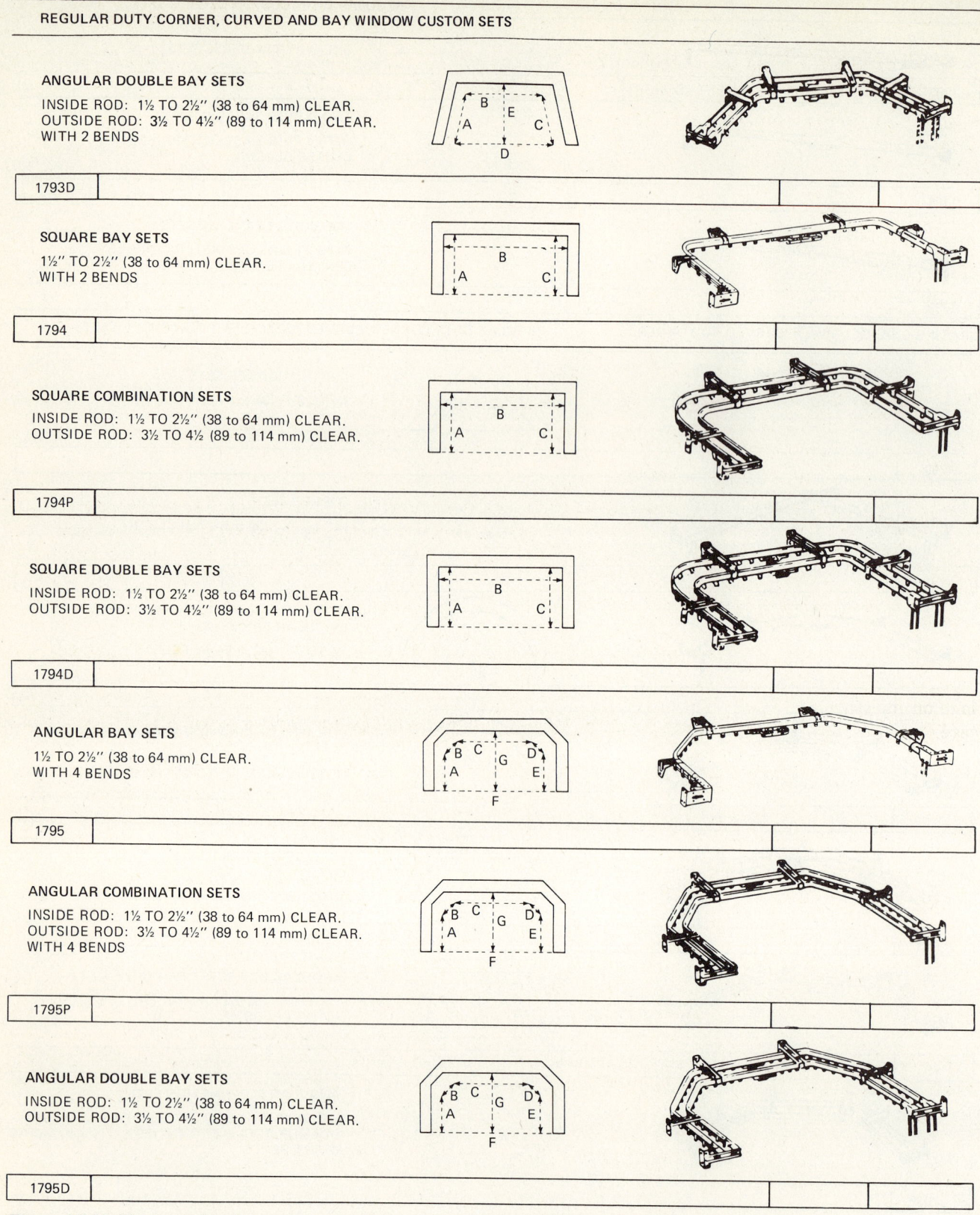

ANGULAR DOUBLE BAY SETS

INSIDE ROD: 1½ TO 2½" (38 to 64 mm) CLEAR.
OUTSIDE ROD: 3½ TO 4½" (89 to 114 mm) CLEAR.
WITH 2 BENDS

1793D

SQUARE BAY SETS

1½" TO 2½" (38 to 64 mm) CLEAR.
WITH 2 BENDS

1794

SQUARE COMBINATION SETS

INSIDE ROD: 1½ TO 2½" (38 to 64 mm) CLEAR.
OUTSIDE ROD: 3½ TO 4½ (89 to 114 mm) CLEAR.

1794P

SQUARE DOUBLE BAY SETS

INSIDE ROD: 1½ TO 2½" (38 to 64 mm) CLEAR.
OUTSIDE ROD: 3½ TO 4½" (89 to 114 mm) CLEAR.

1794D

ANGULAR BAY SETS

1½ TO 2½" (38 to 64 mm) CLEAR.
WITH 4 BENDS

1795

ANGULAR COMBINATION SETS

INSIDE ROD: 1½ TO 2½" (38 to 64 mm) CLEAR.
OUTSIDE ROD: 3½ TO 4½" (89 to 114 mm) CLEAR.
WITH 4 BENDS

1795P

ANGULAR DOUBLE BAY SETS

INSIDE ROD: 1½ TO 2½" (38 to 64 mm) CLEAR.
OUTSIDE ROD: 3½ TO 4½" (89 to 114 mm) CLEAR.

1795D

Figure 15-19. More regular-duty corner, curved, and bay window custom rods. *(Kenney)*

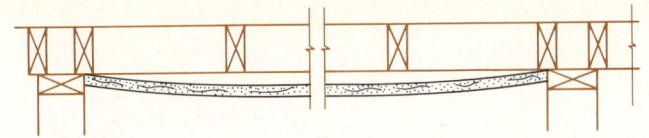

Figure 15-20. Forceably fitted piece of gypsum board. (*US Gypsum*)

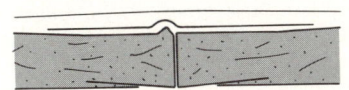

Figure 15-21. Damaged edges and their effect on a joint. (*US Gypsum*)

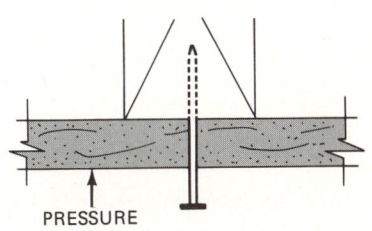

Figure 15-22. Apply hand pressure (see arrow) while nailing the board to the stud. (*US Gypsum*)

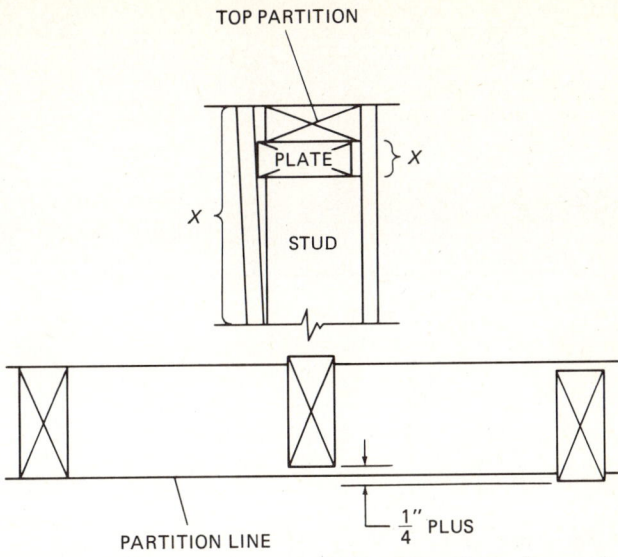

Figure 15-23. If framing members are bowed or misaligned, shims are needed if the wall board is to fit properly. (*US Gypsum*)

sponding plug from a sound piece of gypsum. Sand the edges to an exact fit. If necessary, cement an extra slat of gypsum panel to the back of the face layer to serve as a brace. Butter the edges and finish as a butt joint with joint compound.

Prevention Avoid using board with damaged edges that may easily compress. Damaged edges can take on moisture and swell. Handle sheetrock with care.

PANELS LOOSELY FASTENED

Cause Framing members are uneven because of misalignment or warping. If there is lack of hand pressure on the panel during fastening, loosely fitting panels can result. See Figure 15-21.

Remedy When panels are fastened with nails, during final blows of the hammer use your hand to apply additional pressure to the panel adjacent to the nail. See Figure 15-22.

Prevention Correct framing imperfections before applying the panels. Use screws or adhesive method instead of nails.

SURFACE FRACTURED AFTER APPLICATION

Cause Heavy blows or other abuse have fractured finished wall surface. If the break is too large to repair with joint compound, do the following.

Remedy In the shape of an equilateral triangle around the damaged area, remove a plug of gypsum. Use a keyhole saw. Slope the edges 45°. Cut a corre-

FRAMING MEMBERS OUT OF ALIGNMENT

Cause Because of misaligned top plate and stud, hammering at points X in Figure 15-23 as panels are applied on both sides of the partition will probably result in nail heads puncturing the paper or cracking the board. If framing members are more than ¼ inch out of alignment with adjacent members, it is difficult to bring panels into firm contact with all nailing surfaces.

Remedy Remove or drive in problem fasteners and drive new fasteners only into members in solid contact with the board.

Prevention Check the alignment of studs, joists, headers, blocking, and plates before applying panels. Correct before proceeding. Straighten badly bowed or crowned members. Shim out flush with adjoining surfaces. Use adhesive attachment.

MEMBERS TWISTED

Cause Framing members have not been properly squared with the plates. This gives an angular nailing surface. See Figure 15-24. When panels are applied, there is a danger of fastener heads puncturing the paper or of reverse twisting of a member as it dries out. This loosens the board and can cause fastener pops.

Maintenance and Remodeling

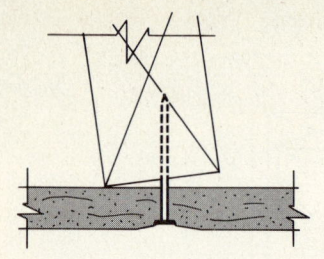

Figure 15-24. Framing members improperly squared. *(US Gypsum)*

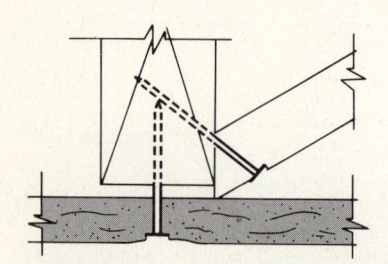

Figure 15-25. Nail head goes through but doesn't pull the board up tightly against the stud. The stud is prevented from meeting the nailing surface by a piece of bridging out of place. *(US Gypsum)*

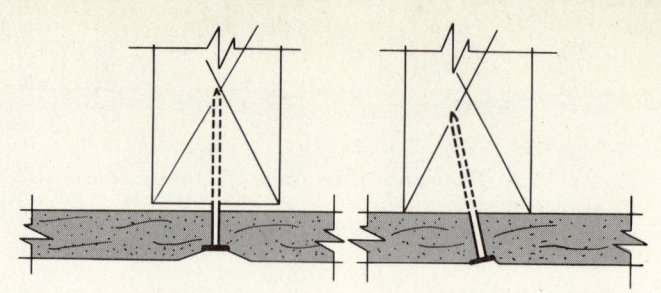

Figure 15-26. Puncturing of the face paper. *(US Gypsum)*

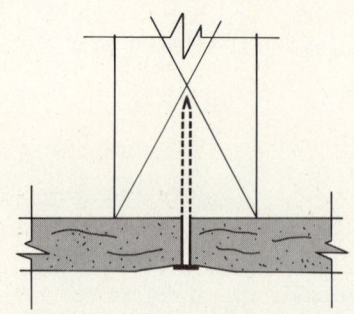

Figure 15-27. Proper dimple made with the nail head into the drywall board. *(US Gypsum)*

Remedy Allow the moisture content in the framing to stabilize. Remove the problem fasteners. Refasten with carefully driven screws.

Prevention Align all the twisted framing members before you apply the board.

FRAMING PROTRUSIONS

Cause Bridging, headers, fire stops, or mechanical lines have been installed improperly. See Figure 15-25. They may project out past the face of the framing member. This prevents the board or drywall surface from meeting the nailing surface. The result can be a loose board. Fasteners driven in this area of protrusion will probably puncture the face paper.

Remedy Allow the moisture content in the framing to stabilize. Remove the problem fasteners. Realign the bridging or whatever is out of alignment. Refasten with carefully driven screws.

PUNCTURING OF FACE PAPER

Cause Poorly formed nail heads, careless nailing, excessively dry face paper, or a soft core can cause the face paper to puncture. Nail heads which puncture the paper and shatter the core of the panel are shown in Figure 15-26. They have little grip on the board.

Remedy Remove the improperly driven fastener. Properly drive a new fastener.

Prevention Correcting faulty framing and driving nails properly produce a tight attachment. There should be a slight uniform dimple. See Figure 15-27 for the proper installation of the fastener. A nail head bears on the paper. It holds the panel securely against the framing member. If the face paper becomes dry and brittle, its low moisture content may aggravate the nail cutting. Raise the moisture content of the board and the humidity in the work area.

NAIL POPS FROM LUMBER SHRINKAGE

Cause Improper application, lumber shrinkage, or a combination of the two. With panels held reasonably tight against the framing member and with proper-length nails, normally only severe shrinkage of the lumber will cause nail pops. But if panels are nailed loosely, any inward pressure on the panel will push the nail head through its thin covering pad of compound. Pops resulting from "nail creep" occur when shrinkage of the wood framing exposes nail shanks and consequently loosens the panel. See Figure 15-28.

Remedy Repairs usually are necessary only for pops which protrude 0.005 inch or more from the face of the board. See Figure 15-28. Smaller protrusions may need to be repaired if they occur in a smooth gloss surface or flat-painted surface under extreme lighting conditions. Those which appear before or during decoration should be repaired immedi-

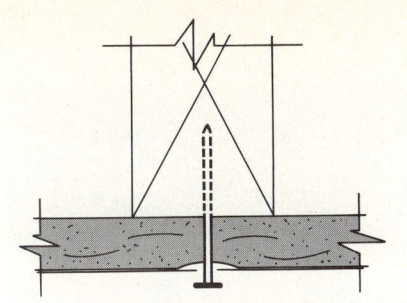

Figure 15-28. Popped nail head. *(US Gypsum)*

Figure 15-29. Cutting plastic laminate with a saber saw. *(Grossman Lumber)*

Figure 15-30. Cutting plastic laminate with a handsaw. *(Grossman Lumber)*

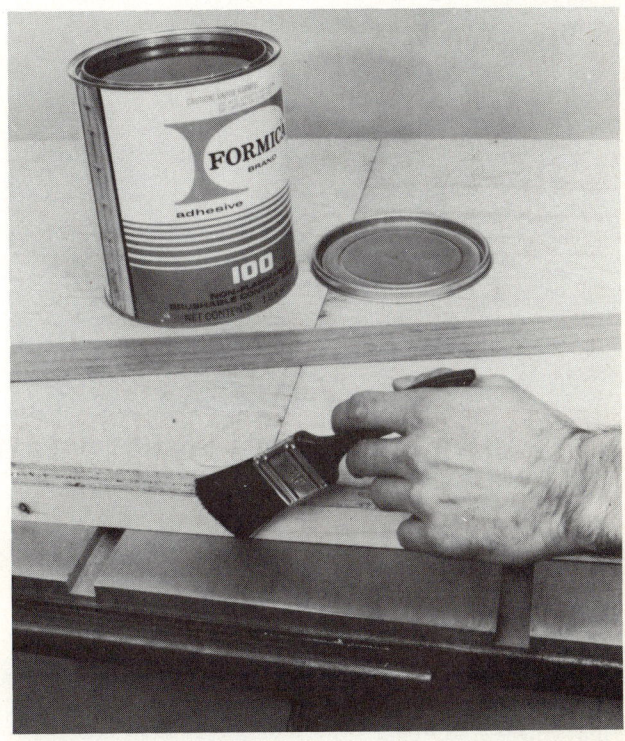

Figure 15-31. Applying the strip of laminate to the edge. (A) Flop the top right side up. Apply contact cement to laminate and core self-edge. *(Formica)*

ately. Pops which occur after one month's heating or more are usually caused wholly or partly by wood shrinkage. They should not be repaired until near the end of the heating season. Drive the proper nail or screw about 1½ inches from the popped nail while applying sufficient pressure adjacent to the nail head to bring the panel in firm contact with the framing. Strike the popped nail lightly to seat it below the surface of the board. Remove the loose compound. Apply finish coats of compound and paint.

These are but a few of the possible problems with gypsum drywall or sheetrock. Others are cracking, surface defects, water damage, and discoloration. All can be repaired with the proper tools and equipment. A little skill can be developed over a period of time. However, in most instances it is necessary to redecorate a wall or ceiling. This can become a problem of greater proportions. It is best to make sure the job is done correctly the first time. This can be done by looking at some of the suggestions given under *Prevention*.

Installing New Countertops

Countertops are usually covered by a plastic laminate. Formica is usually applied to protect the wooden surface and make it easier to clean. (However, Formica is only one of the trademarks for plastic laminates.) Plastic laminates are easy to work with since they can be cut with either a power saw or a handsaw. Just be sure to cut the plastic laminate face down when using a portable electric saw. This minimizes chipping. See Figure 15-29.

If you use a handsaw as in Figure 15-30, make sure you use a low angle and cut only on the downward stroke.

Before applying the laminate to the surface of the counter, you have to coat both the laminate and the counter with adhesive. See Figure 15-31A. This is usually a contact cement. That means both sur-

Maintenance and Remodeling 421

Figure 15-31 *(cont.)*. Applying the strip of laminate to the edge.

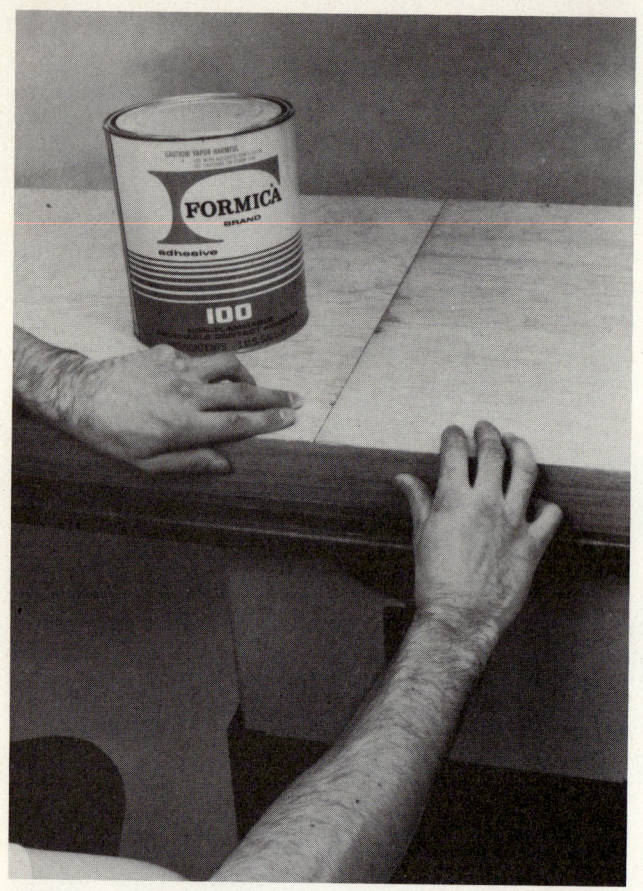

(B) Bond the strip of laminate to the edge using your fingertips to keep the surfaces apart as you go. *(Formica)*

(C) Apply pressure immediately after bonding by using a hammer and a clean block of hardwood. *(Formica)*

faces must be dry to the touch before they are placed in contact with one another. See Figure 15-31B and C. Make sure the cement is given at least 15 minutes and no more than 1 hour for drying time. See Figure 15-32. If the cement sinks into the work surface, it may be best to apply a second coat. See Figure 15-33A and B. Make sure it is dry before you apply the laminate to the work surface of the counter.

In order to get a perfect fit, you may have to place a piece of brown paper on the work surface. Then slide the plastic over the paper. Slip the paper from under the plastic laminate slowly. As the paper is removed, the two adhesive surfaces will contact and stick. Don't let the cemented surfaces touch until they are in the proper location. Figure 15-33C and D shows another method of applying the laminate.

Once the paper has been removed and the surfaces are aligned, apply pressure over the entire area. Carefully pound a block of wood with a hammer on the laminated top. This bonds the laminate to the surface below. See Figure 15-33E for another method.

Figure 15-32. Placing the glue-coated plastic laminate on the bottom over a piece of paper. *(Grossman Lumber)*

422   Carpentry Fundamentals

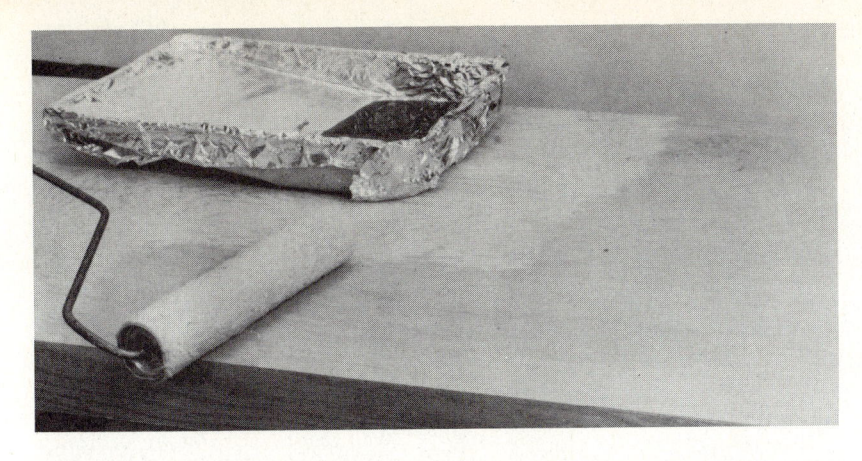

Figure 15-33. (A) Use a paint roller and work from a tray to apply the contact cement. Lining the tray with aluminum foil speeds cleanup later. *(Formica)*

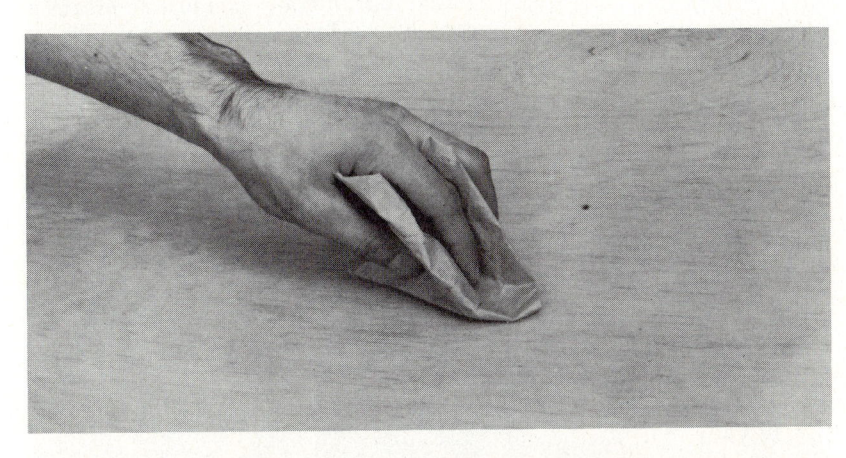

(B) Surfaces are ready for bonding when the glue does not adhere to clean kraft paper. *(Formica)*

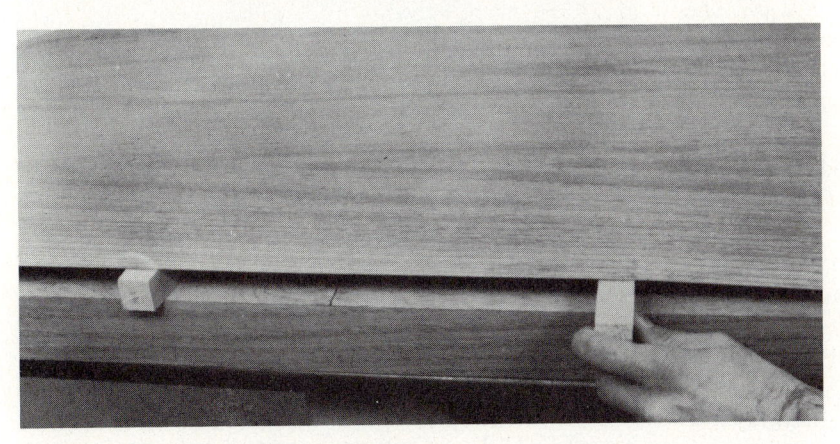

(C) Align the laminate over the core. As you remove the ¾" sticks, the laminate is bonded to the wood. *(Formica)*

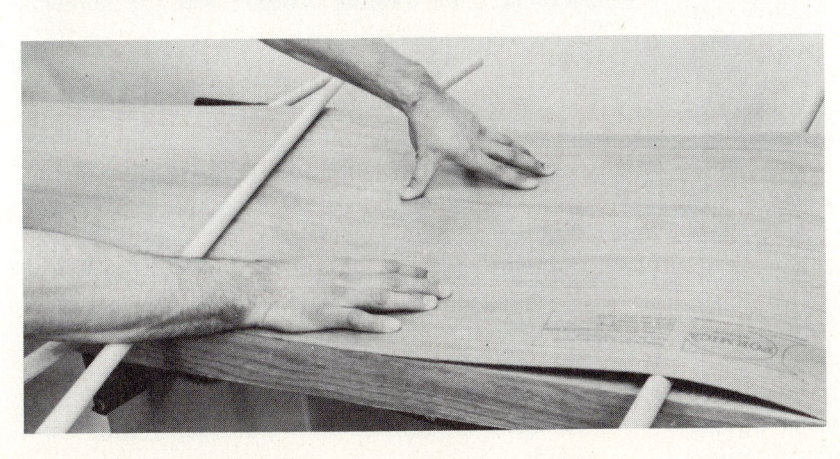

(D) Dowel rods can also be used to prevent bonding until they are removed. *(Formica)*

Maintenance and Remodeling **423**

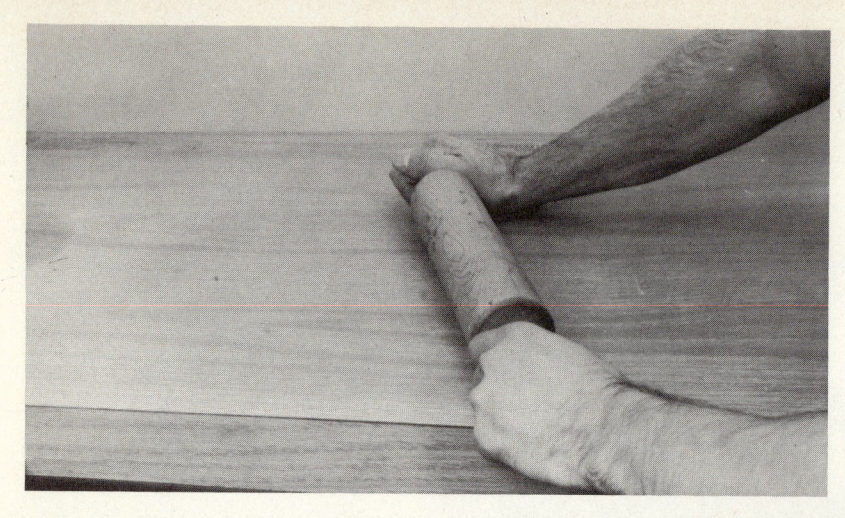

Figure 15-33 *(cont.)*. (E) Use an ordinary rolling pin to apply pressure to create a good bond. *(Formica)*

TRIMMING EDGES • Use a router as shown in Figure 15-34 to finish up the edges. A saber saw or hand plane can be used if you do not have a router. The edge of the plywood that shows can be covered with plastic laminate, wood molding, or metal strips made for the job. Figure 15-35 shows the cleanup operation after the laminate has been applied.

Fitting the sink in the countertop is another step in making a finished kitchen. You will need a clamp-type sink rim installation kit. Apply the kit to your sink to be sure it is the correct size. Then, cut the countertop hole about ⅛ inch larger all around than the rim. Make the cutout with a keyhole saw, a router, or a saber saw. See Figure 15-36.

Start the cutout by drilling a row of holes and leaving a ¹⁄₁₆-inch margin to accommodate the leg of the rim. Install the sink and rim. Or, you can go on to the *backsplash*.

The backsplash is a board made from ¾-inch plywood which is nailed to the wall. It is perpendicular to the countertop. Finish off the backsplash with the same plastic laminate before mounting it to the wall permanently. In some cases it may be easier to mount the plywood to the wall with an adhesive, since the drywall is already in place at this step in the construction. Use a router to trim the edges of the backsplash. It should fit flush against the countertop. You may want to finish it off with the end grain of the plywood covered with the same plastic laminate or with metal trim.

Figure 15-37 shows how the sink is installed in the countertop. Also note the repair method used to remove a bubble in the plastic laminate.

Repairing a Leaking Roof

In most areas when a reroofing job is under consideration, a choice must be made between removing the old roofing or permitting it to remain. It is generally not necessary to remove old wood shingles, old asphalt shingles, or old roll roofing before applying a new asphalt roof—that is, if a competent inspection indicates that:

1. The existing deck framing is strong enough to support the weight of workers and the additional new roofing. This means it should also be able to support the usual snow and wind loads.

Figure 15-34. Using a router to edge the plastic laminate. *(Formica)*

424 Carpentry Fundamentals

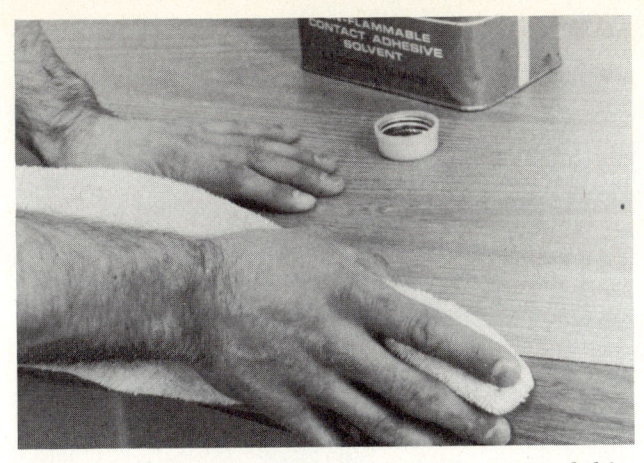

Figure 15-35. Wipe the surfaces clean with a rag and thinner. Use the thinner very sparingly, since it can cause delamination at the edges. (*Formica*)

Figure 15-36. Marking around the sink rim for the cut out in the counter top. (*Grossman Lumber*)

2. The existing deck is sound and will provide good anchorage for the nails used in applying the new roofing.

OLD ROOFING TO STAY IN PLACE • If the inspection indicates that the old wood shingles may remain, the surface of the roof should be carefully prepared to receive the new roofing.

This may be done as follows:

1. Remove all loose or protruding nails, and renail the shingles in a new location.
2. Nail down all loose shingles.
3. Split all badly curled or warped old shingles and nail down the segments.
4. Replace missing shingles with new ones.
5. When shingles and trim at the eaves and rakes are badly weathered, and when the work is being done in a location subject to the impact of unusually high winds, the shingles at the eaves and rakes should be cut back far enough to allow for the application, at these points, of 4- to 6-inch wood strips, nominally 1 inch thick. Nail the strips firmly in place, with their outside edges projecting beyond the edges of the deck the same distance as did the wood shingles. See Figure 15-38.
6. To provide a smooth deck to receive the asphalt roofing, it is recommended that beveled wood "feathering" strips be used along the butts of each course of old shingles.

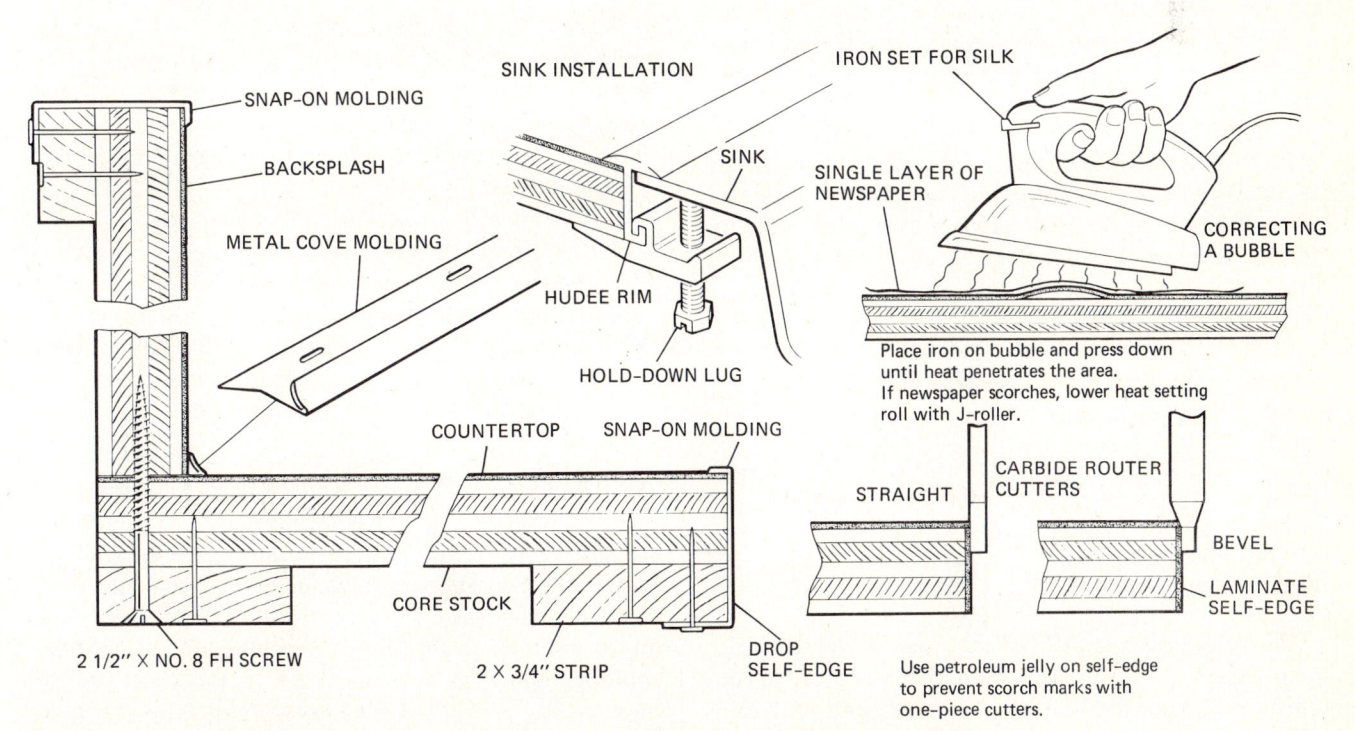

Figure 15-37. Details of installing a sink in a laminated counter top. Note the method used to remove a bubble caught under the laminate. (*Formica*)

Maintenance and Remodeling **425**

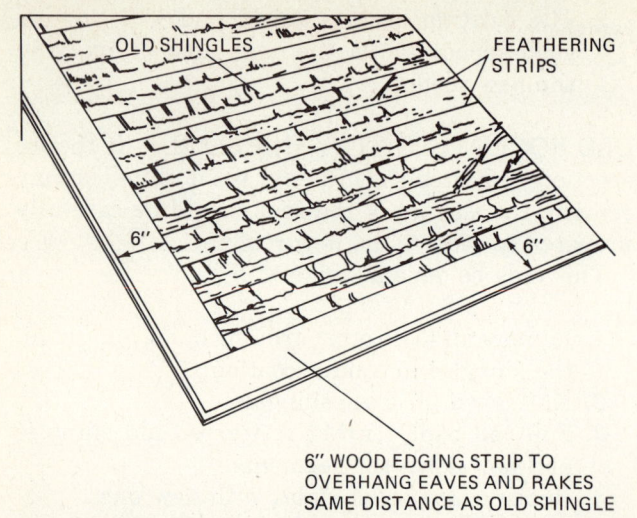

Figure 15-38. Treatment of rakes and eaves when reroofing in windy location. *(Bird and Son)*

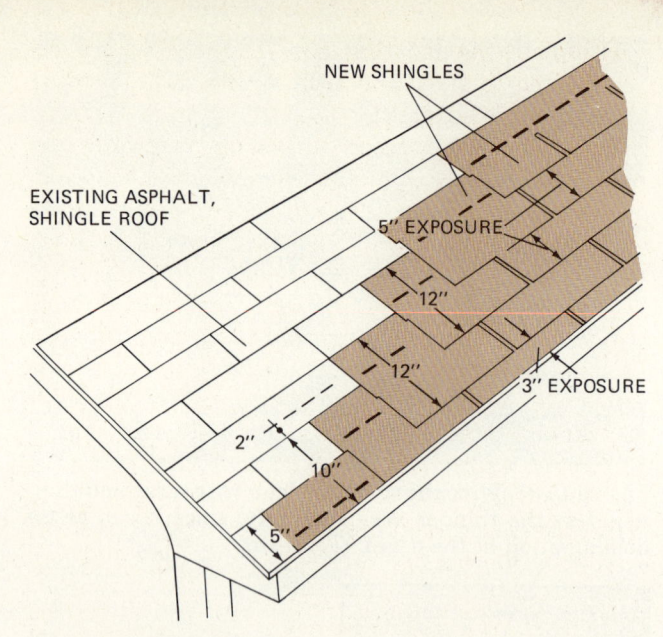

Figure 15-39. Exposure of new shingles when reroofing. *(Bird and Son)*

OLD ROOFING (ASPHALT) SHINGLES TO REMAIN IN PLACE • If the old asphalt shingles are to remain in place, nail down or cut away all loose, curled, or lifted shingles. Remove all loose and protruding nails. Remove all badly worn edging strips and replace with new. Just before applying the new roofing, sweep the surface clean of all loose debris.

SQUARE-BUTT STRIP SHINGLES TO BE RECOVERED WITH SELF-SEALING SQUARE-BUTT STRIP SHINGLES • The following application procedure is suggested to minimize uneven appearance of the new roof. All dimensions are given assuming that the existing roof has been installed with the customary 5-inch shingle exposure.

Starter Course Cut off the tabs of the new shingle using the head portion equal in width to the exposure of the old shingles. This is normally 5 inches for the starter shingle. See Figure 15-39.

First Course Cut 2 inches from the top edge of a full-width new shingle. Align this cut edge with the butt edge of the old shingle.

Second Course Use a full-width shingle. Align the top edge with the butt edge of the old shingle in the next course. Although this will reduce the exposure of the first course, the appearance should not be objectionable, as this area is usually concealed by the gutter.

Third Course and All Others Use full-width shingles. Align the top edges with the butts of the old shingles. Exposure will be automatic and will coincide with that of the old roof.

OLD LOCK-DOWN OR STAPLE-DOWN SHINGLES • These shingles should be removed before reroofing. They have an uneven surface, and the new shingles will tend to conform to it. If a smoother-surface base is desired, the deck should be prepared as described in *Old Roofing to Be Removed*, below.

NEW SHINGLES OVER OLD ROLL ROOFING • When new asphalt roofing is to be laid over old roll roofing without removing the latter, proceed as follows to prepare the deck:

1. Slit all buckles, and nail segments down smoothly.
2. Remove all loose and protruding nails.
3. If some of the old roofing has been torn away, leaving pitchy knots and excessively resinous areas exposed, cover these defects with sheet-metal patches made from galvanized iron, painted tin, zinc, or copper having a thickness approximately equal to 26 gauge.

OLD ROOFING TO BE REMOVED • When the framing supporting the existing deck is not strong enough to support the additional weight of roofing and workers during application, or when the decking material is so far gone that it will not furnish adequate anchorage for the new roofing nails, the old roofing, regardless of type, must be removed before new roofing is applied. The deck should then be prepared for the new roofing as follows:

1. Repair the existing roof framing where required to level and true it up and to provide adequate strength.
2. Remove all rotted or warped old sheathing and replace it with new sheathing of the same kind.
3. Fill in all spaces between boards with securely nailed wood strips of the same thickness as the old deck. Or, move existing sheathing together and sheath the remainder of the deck.
4. Pull out all protruding nails and renail sheathing firmly at new nail locations.
5. Cover all large cracks, slivers, knot holes, loose knots, pitchy knots, and excessively resinous areas with sheet metal securely nailed to the sheathing.
6. Just before applying the new roofing, sweep the deck thoroughly to clean off all loose debris.

OLD BUILT-UP ROOFS

If the Deck Has Adequate Support for Nails When the pitch of the deck is below 4 inches per foot but not less than 2 inches per foot, and if the deck material is sound and can be expected to provide good nail-holding power, any old slag, gravel, or other coarse surfacing materials should first be removed. This should leave the surface of the underlying felts smooth and clean. Apply the new asphalt shingles directly over the felts according to the manufacturer's recommendations for low-slope application.

If the Deck Material Is Defective and Cannot Provide Adequate Security All old material down to the upper surface of the deck should be removed. The existing deck material should be repaired. Make it secure to the underlying supporting members. Sweep it clean before applying the new roofing.

PATCHING A ROOF • In some cases it is not necessary to replace the entire roof to plug a leak. In most instances you can visually locate the place where the leak is occurring. There are a number of roof cements which can be used to plug the hole or cement the shingles down. In some cases it is merely a case of backed-up water. To keep this from happening, heating cables may have to be placed on the roof to melt the ice. See Chapter 8 for more details.

Replacing Guttering

Guttering comes in both 4- and 5-inch widths. The newer aluminum gutter is usually available in 5-inch widths. The aluminum has a white baked-on finish. It does not require soldering, painting, or priming. It is easily handled by one person. The light weight can have some disadvantages. Its ability to hold ice or icicles in colder climates is limited. It can be damaged by the weight of ice buildup. However, its advantages usually outweigh its disadvantages. Not only is it lighter, prefinished on its exterior, and less expensive, but it is also very easy to put up.

The aluminum type of guttering is used primarily as a replacement gutter. The old galvanized type requires a primer before it is painted. In most cases, it does not hold paint even when primed. It becomes an unsightly mess easily with extremes in weather. It is necessary to solder the galvanized guttering. These soldered points do not hold if the weather is such that it heats up and cools down quickly. The solder joints are subject to breaking or developing hairline cracks which leak. These leaks form large icicles in northern climates and put an excessive load on the nails which support the gutter. Once the nails have been worked loose by the expansion and contraction, it is only a matter of time before the guttering begins to sag. If water gets behind the gutter and against the fascia board, it can cause the board to rot. This further weakens the drainage system. The fascia board is used to support the whole system.

Figure 15-40A shows that the drainage system should be lowered at one end so that the water will run down the gutter to the downspout. About ¼ inch for every 10 feet is sufficient for proper drainage. Figure 15-40B shows a hidden bracket hanger. It is used to support the gutter.

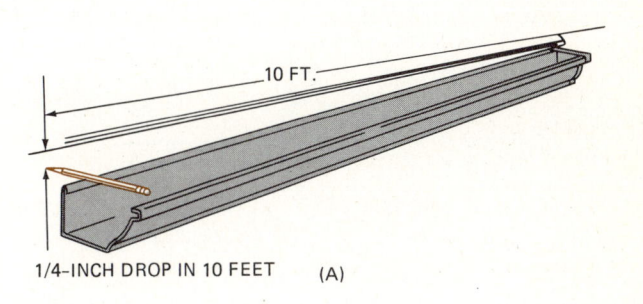

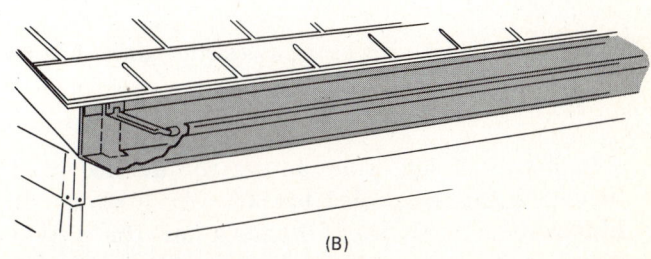

Figure 15-40. Replacing an existing gutter and downspout system. *(Sears, Roebuck & Co.)*

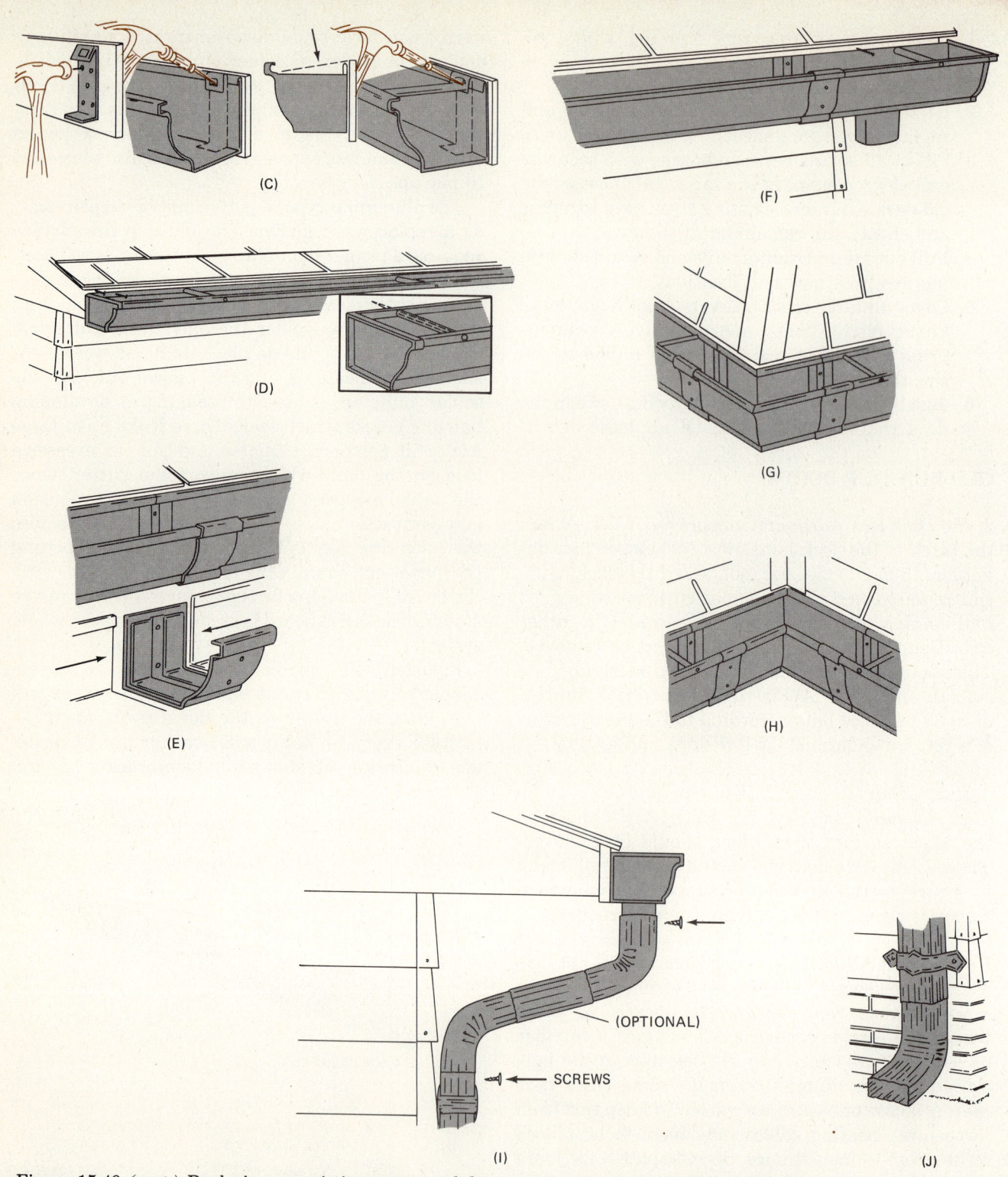

Figure 15-40 (cont.). Replacing an existing gutter and downspout system. (Sears, Roebuck & Co.)

In Figure 15-40C you can see the method used to mount this concealed bracket. The rest of the illustrations in Figure 15-40 show how the system is put together to drain water from the roof completely. Each of these pieces has a name. The names are shown in Figure 15-41. If you are planning a system, its cost is easily estimated by using the form provided in Figure 15-41. Note the tools needed for installing the system. The caulking gun comes in handy to caulk places that were left unprotected by removal of the previous system. The holes for the support of the previous system should be

428　Carpentry Fundamentals

TOOLS NEEDED

The only tools required in the installation of guttering are those that are commonly found in the home, such as the tools shown at the left. The instructions indicate what job each tool performs.

Figure 15-41. How to select the right fittings and the right quantity. *(Sears, Roebuck & Co.)*

Figure 15-42. Unroll the flooring and allow it to sit face up overnight before installing it. *(Armstrong Cork)*

caulked. Newer types of cements can be used to make sure each connection of the inside and outside corners and the end caps are watertight.

Once the water leaves the downspout, it is spread onto the lawn or it is conducted through plastic pipes to the storm sewer at the curb. Some locations in the country will not allow the water to be emptied onto the lawn. The drains to the storm sewer are placed in operation when the house is built. This helps control the seepage of water back into the basement once it has been pumped out with a sump pump. The sump pump also empties into the drainage system and dumps water into the storm sewer in the street.

Replacing a Floor

Many types of floor coverings are available. You may want to check with a local dealer before deciding just which type of flooring you want. There are continuous rolls of linoleum or there are 9×9-inch squares of tile. There are 12×12-inch squares of carpet that can be placed down with their own adhesive. The type of flooring that is chosen determines the type of installation method to use.

STAPLE-DOWN FLOOR • This type of floor is rather new. It can be placed over an old floor. A staple gun is used to fasten down the edges.

Figures 15-42 through 15-44 show the procedure required to install this type of floor. This type of flooring is so flexible it can be folded and placed in the trunk of even the smallest car. Unroll the flooring in the room and move it into position (Figure 15-42). In the 12-foot width (it also comes in 6-foot width), it covers most rooms without a seam. In the illustration it is being laid over an existing vinyl floor. It can also be installed over plywood, particle board, concrete, and most other subfloor materials.

To cut away excess material (see Figure 15-43),

Maintenance and Remodeling 429

Figure 15-43. Use a utility knife and carpenter's square to make sure the cut is straight. *(Armstrong Cork)*

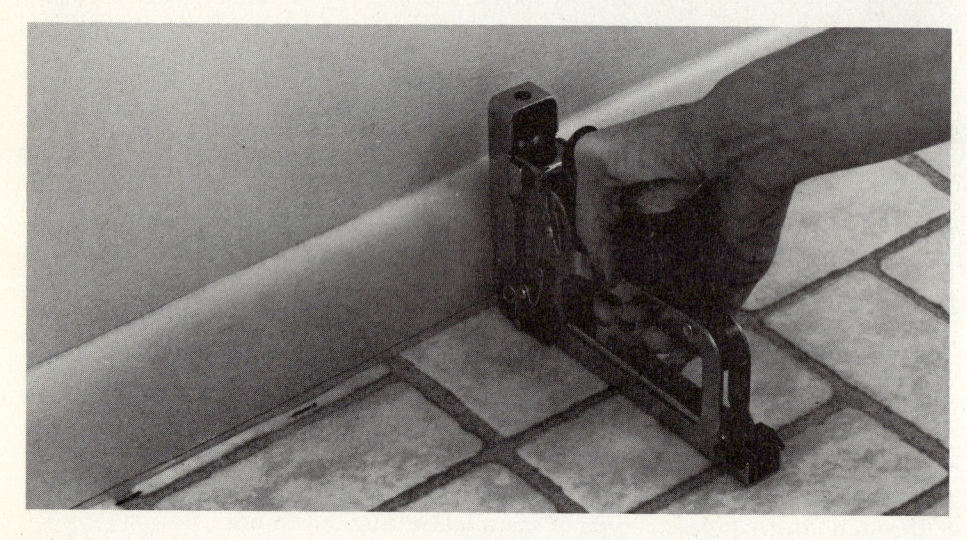

Figure 15-44. Place a staple every 3" along the kickboard. These staples will be covered by molding. *(Armstrong Cork)*

use a metal straightedge or carpenter's square to guide the utility knife. Install the flooring with a staple every 3 inches close to the trim. (See Figure 15-44.) This way the quarter-round trim can be installed over the staples and they will be out of sight. Cement is used in places where a staple can't penetrate. For a concrete floor, use a special adhesive around the edges. Any dealer who sells this flooring has the adhesive.

The finished job looks professional even when it is done by a do-it-yourselfer. This flooring has a built-in memory. When it was rolled face-side-out at the factory for shipment, the outer circumference of the roll was stretched. After it is installed in the home, the floor gently contracts, trying to return to the dimensions it had before it was rolled up. This causes any slack or wrinkles that might have been left in the flooring to gradually be taken up by the memory action.

NO-WAX FLOOR • One of the first rooms that comes in for improvement is the kitchen. Remodeling may be a major project, but renewing a floor is fairly simple.

Figure 15-45. Note the tools needed to install no-wax flooring. *(Armstrong Cork)*

Figure 15-46. Room measurements are essential to a good job. *(Armstrong Cork)*

No-wax flooring comes in the standard widths—6 and 12 feet. In some instances it doesn't need to be tacked or glued down. However, in most cases it should be cemented down. It fits directly over most floors, provided they are clean, smooth, and well bonded. Make sure any holes in the existing flooring are filled and smoothed over. In the case of concrete basement floors, just vacuum them or wash them thoroughly and allow them to dry.

Figure 15-45 shows a couple getting ready to install a new floor. The tools needed are a carpenter's square, chalk line, adhesive, trowel, and knife or scissors.

The key to a perfect fit is taking accurate room measurements. See Figure 15-46. Diagram the floor plan on a chart, noting the positions of cabinets, closets, and doorways.

After transferring the measurements from the

Maintenance and Remodeling 431

Figure 15-47. Cut along the chalk lines using a sharp knife and straightedge. *(Armstrong Cork)*

Figure 15-48. Spread the adhesive on half of the floor and place the flooring material down. Then do the other half. *(Armstrong Cork)*

chart to the flooring material, cut along the chalk lines. Use a sharp knife and a straightedge. See Figure 15-47. Transfer the measurements and cut the material in a room where the material can lie flat. Cardboard under the cut lines will protect the knife blade.

Return the material to the room where it is to be installed. Put it in place. Roll back one-half of the material. Spread the adhesive. Unroll the material onto the adhesive while it is still wet. Repeat the same steps with the rest of the material to finish the job. See Figure 15-48.

The finished job makes any kitchen look new. All that is needed in the way of maintenance is a sponge mop with detergent.

FLOOR TILES • Three of the most popular kinds of self-adhering tiles are vinyl-asbestos, no-wax, and vinyl tiles that contain no asbestos filler.

The benefit of no-wax tiles is obvious from the name. They have a tough, shiny, no-wax wear surface. You pay a premium price for no-wax tiles.

Vinyl tiles are not no-wax, but they are easier to clean than vinyl-asbestos tiles. Their maintenance benefits are the result of a nonporous vinyl wear surface that resists dirt, grease, and stains better than vinyl-asbestos.

Any old tile or linoleum floor can be covered with self-adhering tiles provided the old material is smooth and well bonded to the subfloor. Just make sure the surface is clean and old wax is removed.

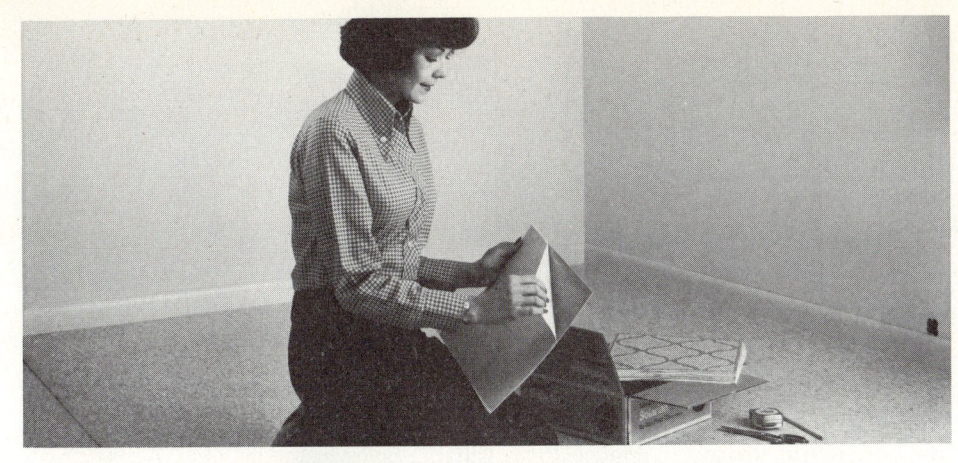

Figure 15-49. Peel the paper off the back of a floor tile and place the tile in a predetermined spot. *(Armstrong Cork)*

Figure 15-50. Do one section of the room at a time. The tiles can be maneuvered into position and pressed into place. *(Armstrong Cork)*

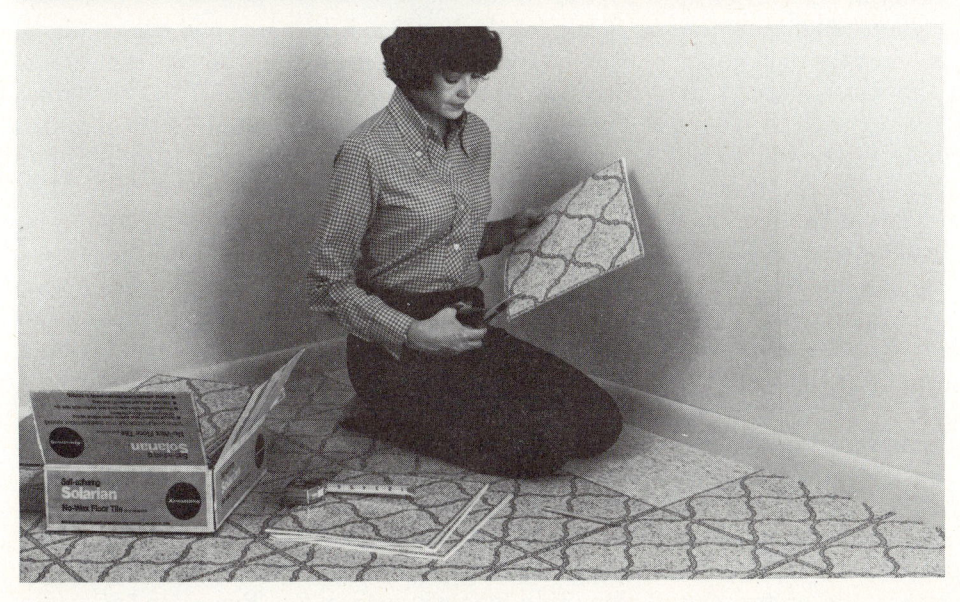

Figure 15-51. Border tiles for the edges of the room can be easily cut to size with a pair of household shears. *(Armstrong Cork)*

Putting Down the Tiles Square off the room with a chalk line. Open the carton, and peel the protective paper off the back of the tile. See Figure 15-49.

Do one section of the room at a time. The tiles are simply maneuvered into position and pressed into place. See Figure 15-50.

Border tiles for the edges of the room can be easily cut to size with a pair of ordinary household shears. See Figure 15-51.

It doesn't take long for the room to take shape.

Maintenance and Remodeling **433**

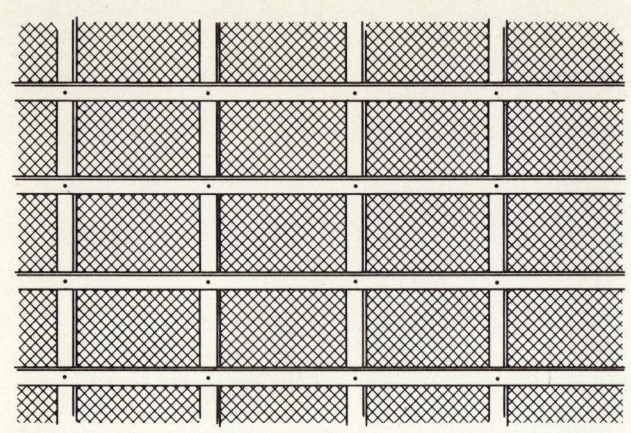

Figure 15-52. Applying vertical and horizontal furring strips on an existing wall. *(Valu)*

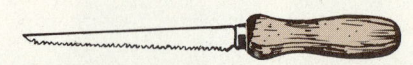

Figure 15-53. Cutting holes in the paneling for electrical switches and outlets. *(Valu)*

There is no smell from the adhesive. The floor can be used as soon as it is finished. This type of floor replacement or repair is commonplace today.

Paneling a Room

The room to be paneled may have cracked walls. This means you'll need furring strips to cover the old walls and provide good nailing and shimming for a smooth wall. See Figure 15-52. Apply furring strips (1×2s or 1×3s) vertically at 16-inch intervals for full-size 4×8-foot panels. Apply the furring strips horizontally at 16-inch intervals for random-width paneling. Start at the end of the wall farthest from the main entrance to the room. The first panel should be plumbed from the corner by striking a line 48 inches out. Trim the panel in the corner (a rasp works well) so that the panel aligns on the plumb line. Turn the corner, and butt the next panel to the first panel. Plumb in the same manner as the first panel.

To make holes in the paneling for switch boxes and outlets, trace the box's outline in the desired place on the panel. Then drill holes at the four corners of the area marked. Next, cut between the holes with a keyhole saw. Then rasp the edges smooth. See Figure 15-53.

Before you apply the adhesive, make sure the panel is going to fit. If the panel is going to be stuck right onto the existing wall, apply the adhesive at 16-inch intervals, horizontally and vertically. See Figure 15-54A.

SPACING • Avoid a tight fit. Above grade, leave a space approximately the thickness of a matchbook cover at the sides and a ⅛-inch space at top and bottom. Below grade, allow ¼ inch top and bottom. Allow not less than 1/16 inch space (the thickness of a dime) between panels in high humidity areas. See Figure 15-54B.

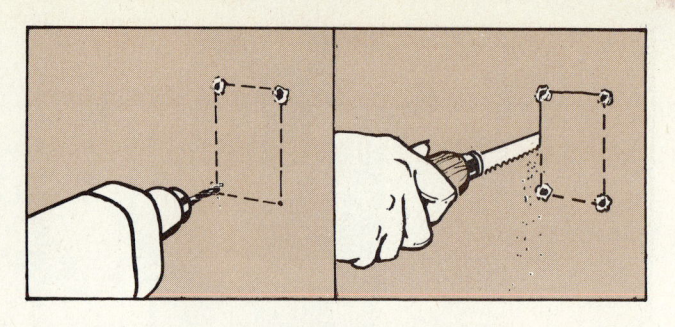

VAPOR BARRIER • A vapor barrier is needed if the panels are installed over masonry walls. It doesn't matter if the wall is above or below grade. See Figure 15-54C. If the plaster is wet or the masonry is new, wait until it is thoroughly dry. Then condition the panels to the room.

ADHESIVES • In Figure 15-55 a caulking gun is being used to apply the adhesive to the furring strips. In some cases it is best to apply the adhesive to the back of the panel. Follow the directions on the tube. Each manufacturer has different instructions.

Hammer a small nail through the top of the panel. It should act as a hinge. Press the panel against the wall or frame to apply glue to both sides. See Figure 15-56A.

Pull out the bottom of the panel and keep it about 8 to 10 inches away from the wall or framework. Use a wood block. See Figure 15-56B. Let the adhesive become tacky. Leave it for about 8 to 10 minutes.

CAUTION: Methods of application for some types of panel adhesives may differ. Always check the instructions on the tube.

Let the panel fall back into position. Make sure that it is correctly aligned. The first panel should butt one edge against the adjacent wall. The panel should be completely plumb. Trim the inner edge as needed. This is so that the outer edge falls on a stud.

SCRIBING • Use a compass to mark panels so that they fit perfectly. It can be used to mark the variations in a butting surface. See Figure 15-56C.

NAILING • Use finishing nails (3d) or brads (1¼-inch) or annular hardboard nails (1-inch). Begin at the edge. Work toward the opposite side. Never nail

434 Carpentry Fundamentals

Figure 15-54. (A) Make sure the panel fits before you apply adhesive. (*Valu*)

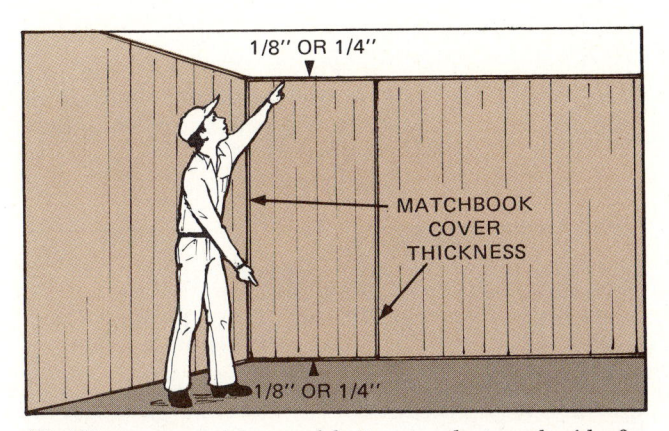

(B) Allow space at top and bottom and at each side for the panel to expand. (*Abitibi*)

(C) Apply a vapor barrier to the walls before you place the furring strips onto the wall. (*Abitibi*)

opposite ends first, then the middle. With a hammer and a cloth-covered block, hammer gently to spread the adhesive evenly. See Figure 15-56D. If you are using adhesive alone to hold the panels in place, don't remove the nails until the adhesive is thoroughly dry. Then, after they're out, fill the nail holes with a matching putty stick.

Moldings can be glued into place or nailed.

Figure 15-55. Use a caulking gun to apply the adhesive. (*Valu*)

Figure 15-56. (A) Hammer a nail through the top of the panel. This will act as a hinge so that the panel can be pressed against the wall to apply the glue to both surfaces. (*Valu*) (B) Pull out the bottom of the panel and keep it about 8 to 10 inches away from the wall with a wood block until the adhesive becomes tacky. (*Valu*) (C) Use a compass to mark panels so that they fit perfectly against irregular surfaces. (*Abitibi*) (D) Use a hammer and a block of wood covered with a cloth to spread the adhesive evenly. (*Valu*)

Installing a Ceiling

There are a number of methods used to install a new ceiling in a basement or a recreation room. In fact, some very interesting tiles are available for living rooms as well.

REPLACING AN OLD CEILING • The first thing to do in replacing an old, cracked ceiling is to lay out the room. See Figure 15-57. It should be laid out accurately to scale. Use ½" = 1'-0" as a scale. Do the layout on graph paper.

Then, on tracing paper, draw ½-inch squares representing the 12-inch ceiling tile. Lay the tracing

Maintenance and Remodeling 435

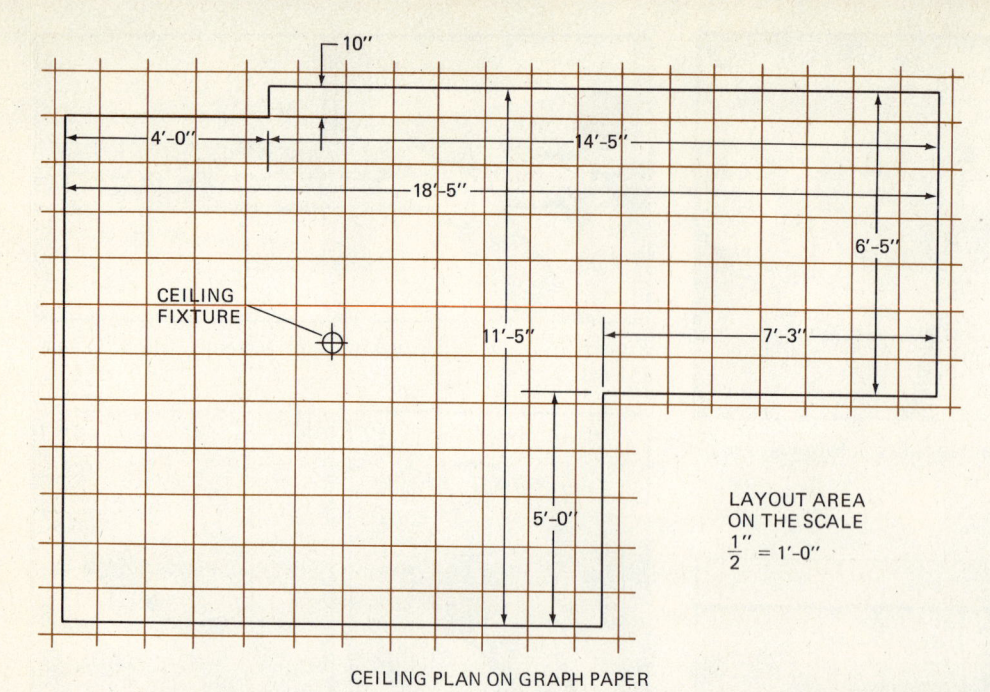

Figure 15-57. Lay out your room accurately to scale on graph paper. *(Grossman Lumber)*

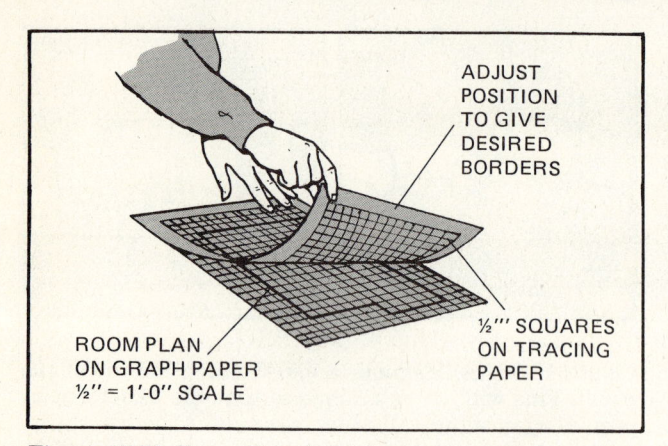

Figure 15-58. Draw ½-inch squares on tracing paper. Lay the tracing paper over the ceiling plan. Adjust the paper until the borders are even. Border pieces should never be less than half a tile wide. *(Grossman Lumber)*

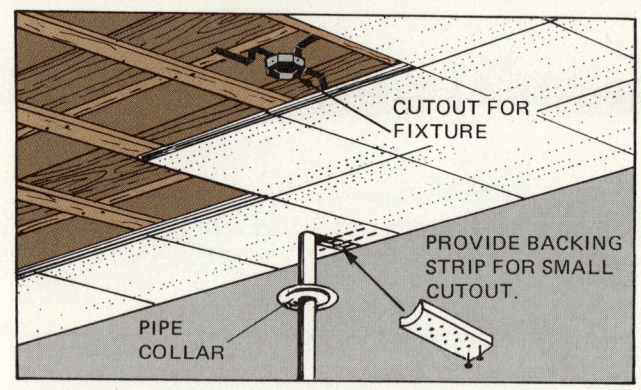

Figure 15-59. Fiber tile is stapled to furring strips applied to the joists. *(Grossman Lumber)*

paper over the ceiling plan. Adjust the paper until the borders are even. Border pieces should never be less than half a tile wide. Use cove molding at the walls to cover the trimmed edges of the tile. See Figure 15-58. Tile can be stapled or applied with mastic. This is especially true of fiber tile. It has flanges which will hold staples. See Figure 15-59. You can staple into wallboard or into furring strips nailed to joists. When applying tile to furring strips, follow your original layout but start in one corner and work toward the opposite corner.

If you apply tile with mastic, snap a chalk line as shown in Figure 15-60. This will find the exact center of your ceiling. In applying the mastic to the ceiling tile, put a golf-ball-sized blob of mastic on each corner of the tile. See Figures 15-61 and 15-62. It is best to apply a blob of mastic in the middle of the tile too, as shown in Figure 15-62.

Apply the first tile where the lines cross. Then, work toward the edges. Set the tile just out of position, then slide it into place, pressing firmly to ensure a good bond and a level ceiling. See Figure 15-63.

Precut holes for lighting fixtures and pipes. Use a sharp utility knife. Always be sure to make the cutouts with the tile face up. See Figure 15-64.

THE DROP CEILING • If an old ceiling is too far damaged to be repaired inexpensively, it may be best to install a drop ceiling. This means the ceiling will be completely new and will not rely upon the old ceiling for support. This way the old ceiling does not have to be repaired.

Figure 15-65 shows the first operation of a drop ceiling. Nail the molding to all four walls at the desired ceiling height. Either metal or wood molding

436 Carpentry Fundamentals

Figure 15-60. If you are going to glue the tiles to the ceiling, snap a chalk line to find the exact center of the ceiling. *(Grossman Lumber)*

Figure 15-61. Place a blob of mastic at each corner and in the middle of the tile before sliding it into place on the ceiling surface. *(Grossman Lumber)*

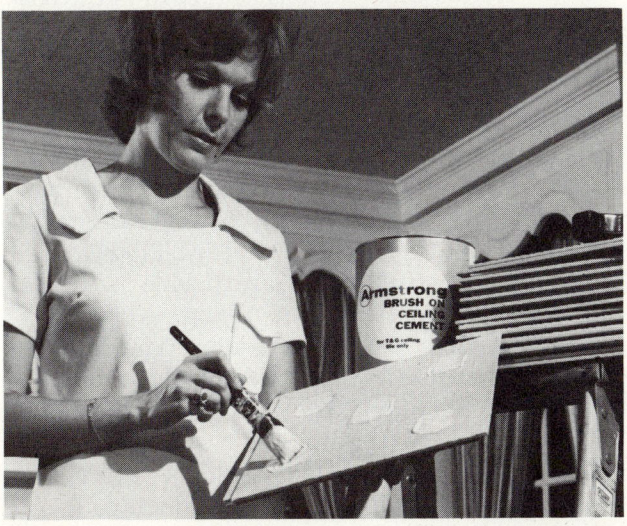

Figure 15-62. If the old ceiling is smooth and structurally sound, new ceiling tiles can be installed with five spots of adhesive. *(Armstrong Cork)*

Figure 15-63. Apply the first tile where the lines cross on the ceiling. Work toward the edges. *(Grossman Lumber)*

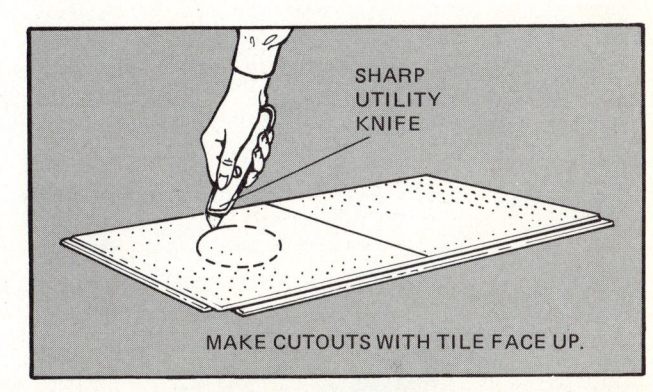

Figure 15-64. Precut holes for lighting fixtures and pipes. Use a sharp utility knife. Always be sure to cut the tiles with the face up. *(Grossman Lumber)*

Figure 15-65. For a new suspended ceiling, nail molding to all four walls. *(Armstrong Cork)*

Maintenance and Remodeling 437

Figure 15-66. A suspended ceiling in a basement is begun the same way. Nail molding on all four walls. *(Armstrong Cork)*

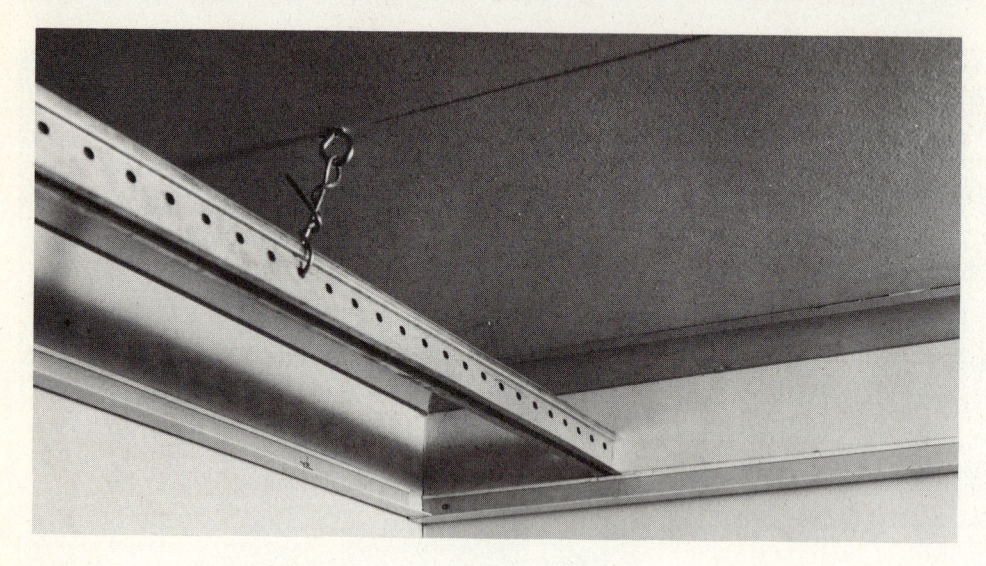

Figure 15-67. Main runners are installed on hanger wires. *(Armstrong Cork)*

may be used. Figure 15-66 shows how a basement dropped ceiling is begun the same way, by nailing molding at the desired height.

The next step is to install the main runners on hanger wires as shown in Figure 15-67. The first runner is always located 26 inches out from the sidewall. The remaining units are placed 48 inches O.C., perpendicular to the direction of the joists. Unlike conventional suspended ceilings, this type requires no complicated measuring or room layout. Figure 15-68 shows how the main runner is installed by fastening hanger wires at 4-foot intervals—the conventional method for a basement dropped ceiling.

After all main runners are in place, begin installing ceiling tile in a corner of the room. Simply lay the first 4 feet of tile on the molding, snap a 4-foot cross T onto the main runner, and slide the T into a special concealed slot on the leading edge of the tile. See Figure 15-69. This will provide a ceiling without the metal supports showing.

In the conventional method, as shown in the basement installation in Figure 15-70, cross T's are installed between main runners. The T's have tabs that engage slots in the main runner to lock firmly in place. After this is done, the tiles are slid into place from above the metal framework and allowed to drop into the squares provided for them.

In Figure 15-71 tile setting is continuing. The tiles and cross T's are inserted as needed. Note how all metal suspension members are hidden from view in the finished portion of the ceiling.

The result of this type of ceiling is an uninterrupted surface. There is no beveled edge to produce a line across the ceiling. All supporting ceiling metal is concealed.

One of the advantages of dropping a ceiling is the conservation of energy. Heat rises to the ceiling. If the ceiling is high, the heat is lost to the room. The lower the ceiling, the less the volume to be heated and the less energy needed to heat the room.

438   Carpentry Fundamentals

Figure 15-68. In the basement installation, install the main runners by fastening them to hanger wires at 4-foot intervals. *(Armstrong Cork)*

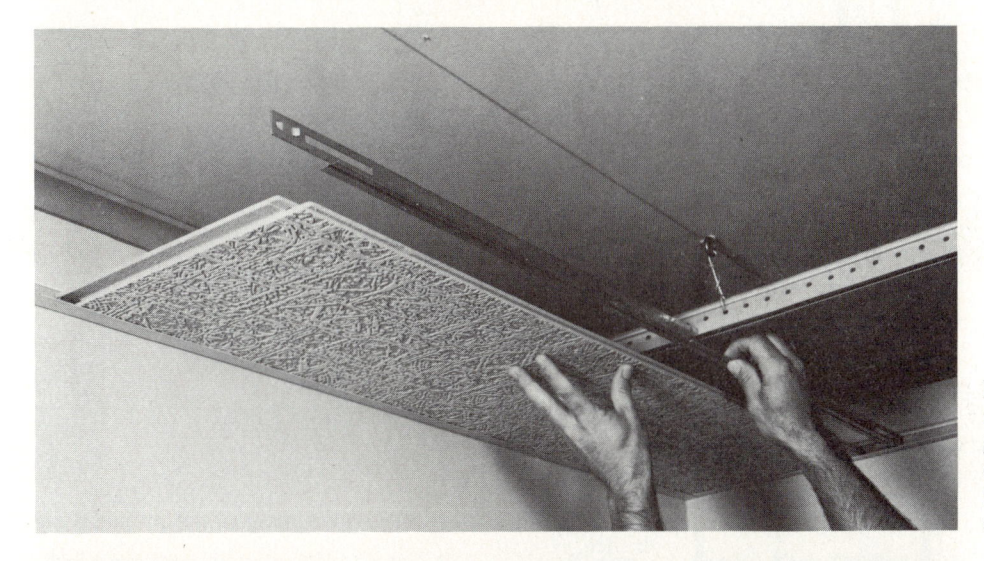

Figure 15-69. After all the main runners are in place, begin installing ceiling tile in a corner of the room. Note how the T slides into the tile and disappears. *(Armstrong Cork)*

Figure 15-70. When the main runners are in place, install cross T's between them. The T's have tabs that engage slots in the main runner to lock firmly in place. This type of suspended ceiling will have the metal showing when the job is finished. *(Armstrong Cork)*

Maintenance and Remodeling 439

Figure 15-71. Continue across the room in this manner. Insert the tiles and cross T's. (Armstrong Cork)

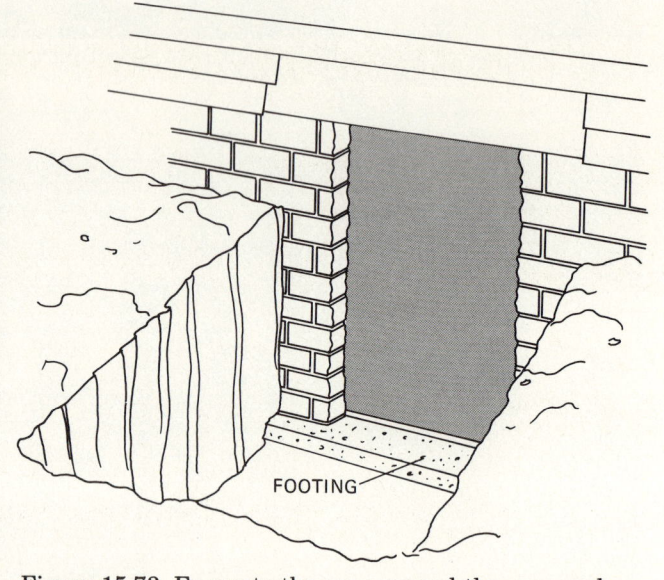

Figure 15-73. Excavate the area around the proposed entrance. (Bilco)

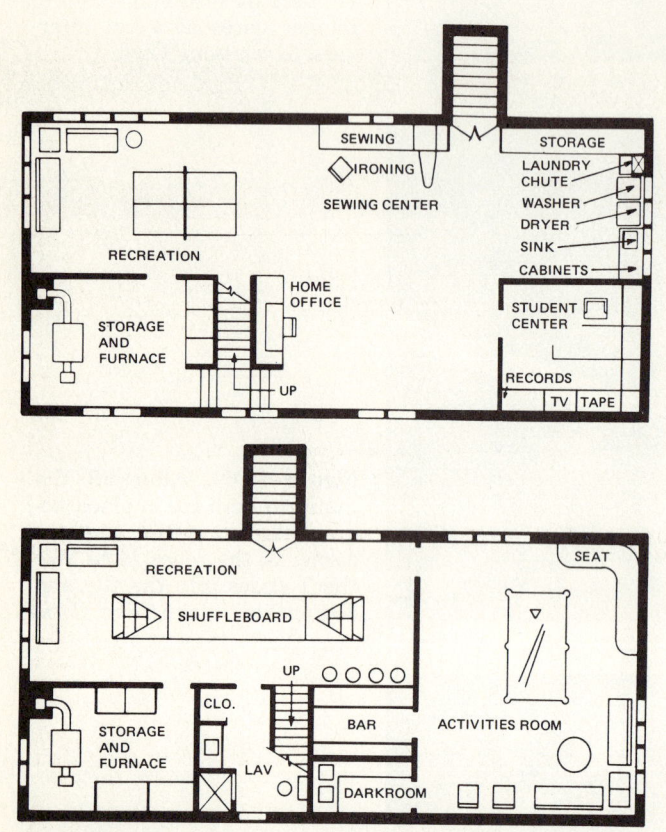

Figure 15-72. Note the location of the outside basement door or entry on these drawings. (Bilco)

Replacing an Outside Basement Door

Many houses have basements that can be made into playrooms or activity areas. The workshop that is needed but can't be put anywhere else will probably wind up in the basement. Basements can be very difficult to get out of if a fire starts around the furnace or hot-water-heater area. One of the safety measures that can be taken is placing a doorway directly to the outside so you don't have to try to get up a flight of stairs and then through the house to escape a fire.

Figure 15-72 shows how an outside stairway can be helpful for any basement. This type of entrance or exit from the basement can be installed in any house. It takes some work, but it can be done.

Figure 15-73 shows how digging a hole and breaking through the foundation in stages will permit using tools to get the job done. A basement wall of concrete blocks is easier to break through than a poured wall.

Start by building the areaway. This is done by laying a 12- to 16-inch concrete footing. This footing forms a level base for the first course of concrete block. See Figure 15-74. Allow the footing to set for 2 to 3 days before laying blocks for the walls.

The areaway for a size C door is shown in Figure 15-75. Figure 15-75A shows the starting course. Fig-

440 Carpentry Fundamentals

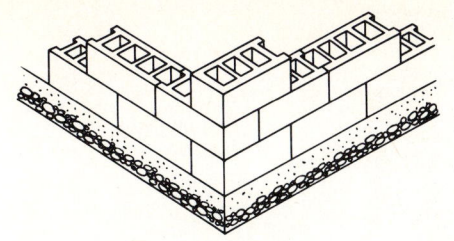

Figure 15-74. Start the blocks on top of the footings. (Bilco)

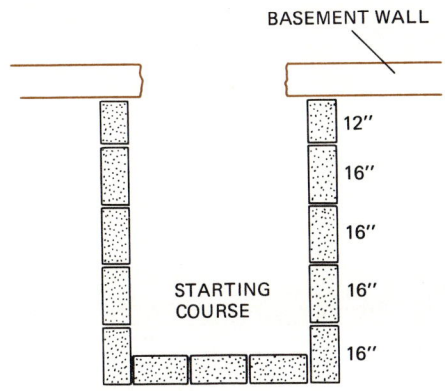

Figure 15-75. (A) The starting course, looking down into the hole. (Bilco)

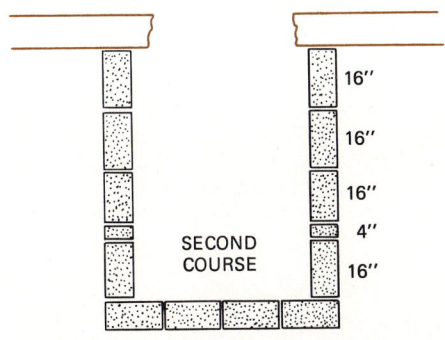

(B) The second course, on top of the first, looks like this. (Bilco)

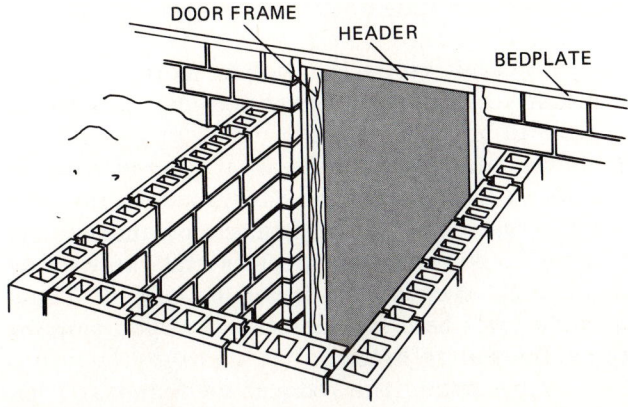

Figure 15-76. The finished block work ready for the cap of concrete. (Bilco)

Figure 15-77. Forms for the concrete cap are built around the concrete block. (Bilco)

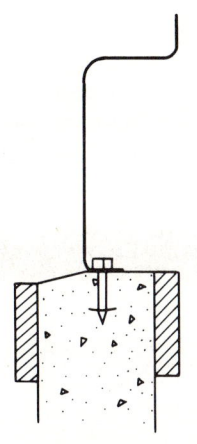

Figure 15-78. Anchor bolts are embedded into the concrete cap to allow attachment of the metal doors. (Bilco)

ure 15-75B shows how the next course of blocks is laid. The top course should come slightly above ground level. See Figure 15-76. It should be about 3 inches from the required areaway height as given in the construction guide provided with the door.

Build up the areaway to the right height in the manner shown. You may want to waterproof the outside of the new foundation. Use the material recommended by your local lumber dealer or mason yard.

Now build a form for capping the wall. See Figure 15-77. Fill the cores of the top blocks halfway up with crushed balls of newspaper or insulation. The cap to be poured on this course will bring the areaway up to the required height.

When the cap is an inch or so below the desired height, set the door back in position with the header flange between the siding and the sheathing underneath. Make sure the frame is square. Insert the mounting screws with the spring-steel nuts in the side pieces and the sill. Embed them in the wet concrete to hold the screws tightly. See Figure 15-78. Continue pouring the cap. Bring the concrete flush

Maintenance and Remodeling 441

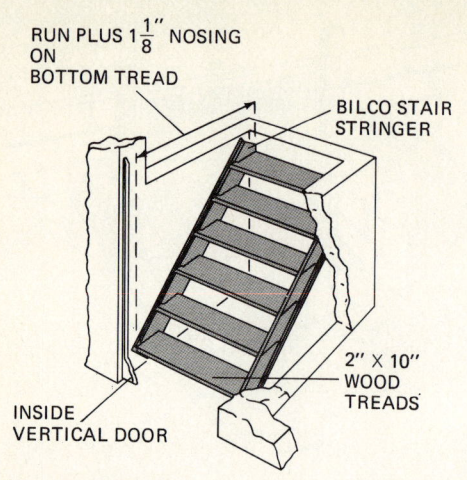

Figure 15-79. The stairs are installed inside the blocked-in hole. *(Bilco)*

with the bottom of the sill and side-piece flanges. Do not bring the capping below the bottom of the door. The door should rest on top of the foundation. With a little extra work, the cap outside the door can be chamfered downward as shown in Figure 15-78. This ensures good drainage. Trowel the concrete smooth and level.

Install a prehung door in the wall of the basement. These come in standard sizes. The door should be selected to fit the hole made in the wall. Use the widest standard unit that fits the entryway you have built.

Figure 15-79 shows how the steps are installed in the opening. The stringers for the steps are attached to the walls of the opening. See Figure 15-80.

The outside doors will resemble those shown in Figure 15-81. There are a number of designs available for almost any use. Lumber for the steps is 2×10s cut to length and slipped into the steel stair stringers.

Seal around the door and foundation with caulk. Seal around the door in the basement and the wall. Allow the stairwell to *air out* during good weather by keeping the outside doors open. This will allow the moisture from the masonry to escape. After it has dried out, the whole unit will be dry.

• CONVERTING EXISTING BUILDING SPACES TO OTHER USES •

There is never enough room in any house. All it takes is a few days after you unpack all your belongings to find out that there isn't enough room for everything. The next thing new homeowners do is look around at the existing building to see what

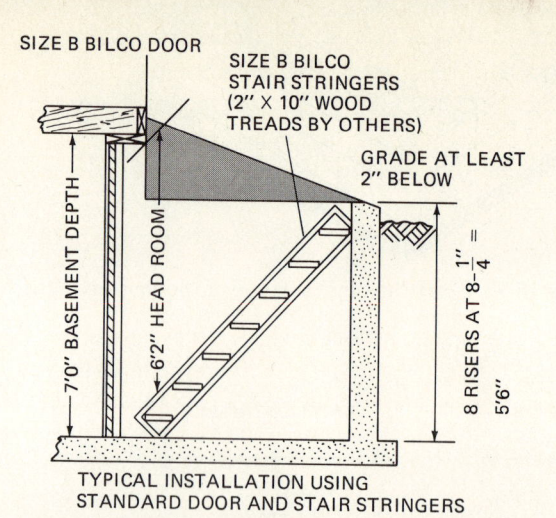

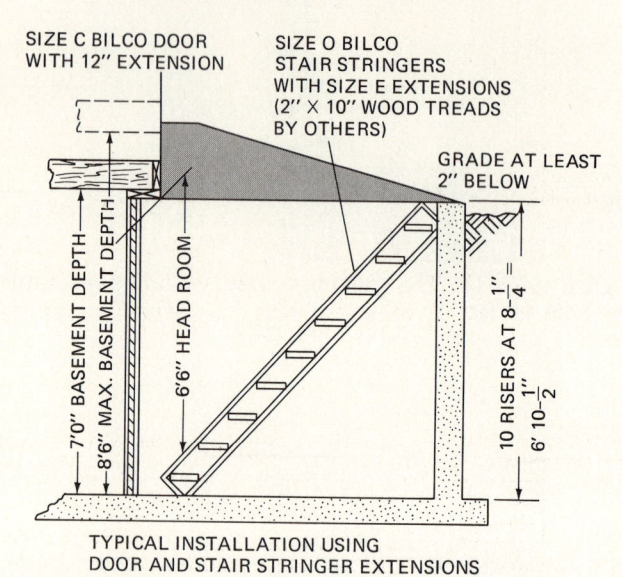

Figure 15-80. Typical door installations. *(Bilco)*

space can be converted to other uses. It is usually too expensive to add on immediately, but it is possible to remodel the kitchen, porch, or garage.

Adding a Bathroom

As the family grows, the need for more bathrooms becomes very apparent. First you look around for a place to put the bathroom. Then you locate the plumbing and check to see what kind of a job it will be to hook up the new bathroom to the cold and hot water and to the drains. How much effort will be needed to hook up pipes and run drains? How much electrical work will be needed? These questions will have to be answered as you plan for the additional bathroom.

Start by getting the room measurements. Then make a plan of where you would place the various necessary items. Figure 15-82 shows some placement

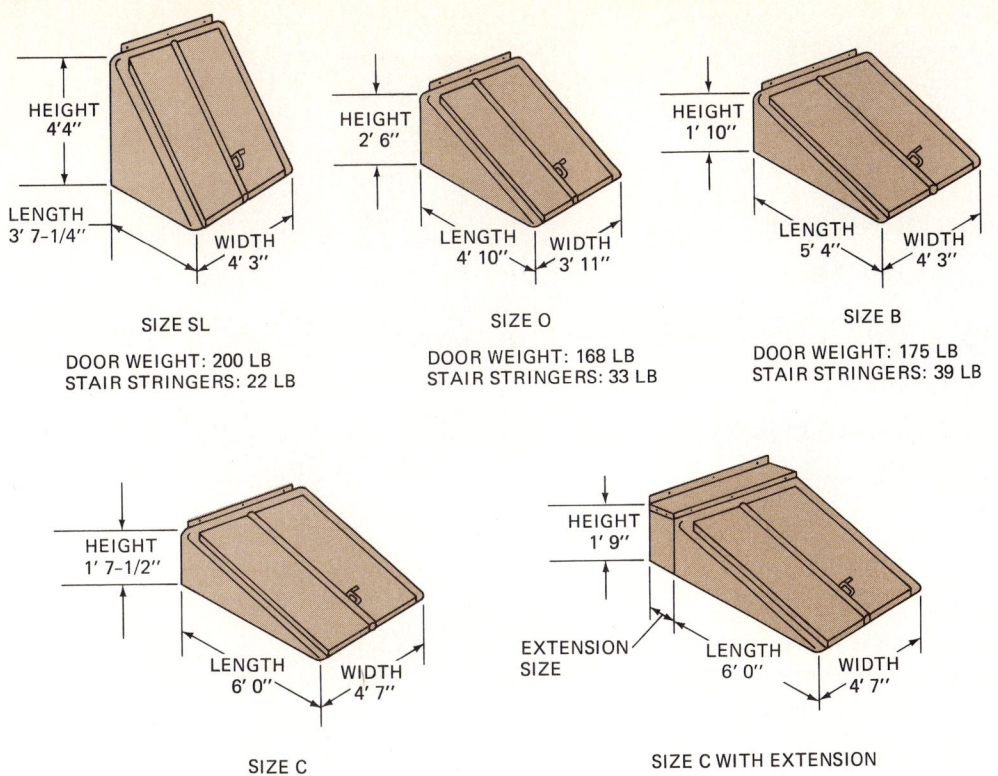

SIZE SL
DOOR WEIGHT: 200 LB
STAIR STRINGERS: 22 LB

SIZE O
DOOR WEIGHT: 168 LB
STAIR STRINGERS: 33 LB

SIZE B
DOOR WEIGHT: 175 LB
STAIR STRINGERS: 39 LB

SIZE C
DOOR WEIGHT: 196 LB
STAIR STRINGERS: 44 LB

SIZE C WITH EXTENSION
EXTENSION WEIGHTS: 6" 20 LB,
12" 52 LB, 18" 70 LB, 24" 95 LB
SIZE E STRINGER EXTENSIONS: 18 LB

Figure 15-81. Various shapes and sizes of outside basement doors. *(Bilco)*

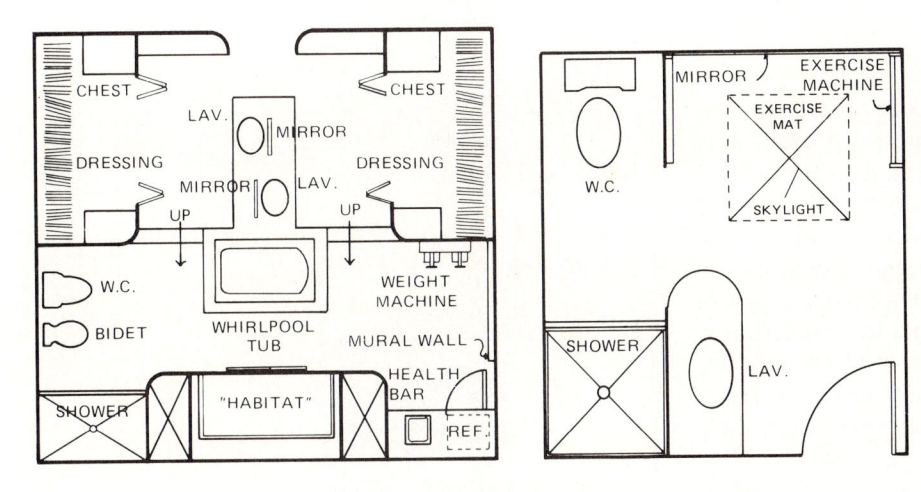

Figure 15-82. Floor plans for bathrooms. *(Kohler)*

possibilities. Your plan can be as simple as a lavatory and water closet, or you can expand with a shower, a whirlpool bath, and a sauna. The amount of money available will usually determine the choice of fixtures.

Look around at books and magazines as well as literature of manufacturers of bathroom fixtures. Get some ideas as to how you would rearrange your own bathroom or make a new one.

If you have a larger room to remodel and turn into a bathroom, you might consider what was done in Figure 15-83. This is a Japanese bathroom dedicated to the art of bathing. Note how the tub is fitted with a shower to allow you to soap and rinse on the bathing platform before soaking as the Japanese do.

You may not want to become too elaborate with your new bathroom. All you have to do then is decide how and what you need. Then draw your ideas for arrangements. Check the plumbing to see if your idea will be feasible. Order the materials and fixtures and then get started.

Maintenance and Remodeling 443

Figure 15-83. Japanese bathroom design. The platform gives the illusion of a sunken bath. Many of these features can be incorporated into a remodeled room which can serve as a bathroom. *(Kohler)*

Providing Additional Storage

CEDAR-LINED CLOSET • Aromatic red cedar closet lining is packed in a convenient, no-waste package which contains 20 square feet. It will cover $16\frac{1}{3}$ square feet of wall space for lifetime protection from moths. In order to install the cedar, follow these steps. See Figure 15-84.

1. Measure the wall, ceiling, floor, and door area of the closet. Figure the square footage. Use the length times width to produce square feet.
2. Cedar closet lining may be applied to the wall either vertically or horizontally. See Figure 15-85. When applied to the rough studs, pieces of cedar must be applied horizontally.
3. When applying the lining to the wall, place the first piece and flush against the floor with grooved edge down. Use small finishing nails to apply the lining to the wall. See Figure 15-86.
4. When cutting a piece of cedar to finish out a course or row of boards, always saw off the tongued end so that the square sawed-off end will fit snugly into the opposite corner to start the new course. See Figure 15-87.
5. Finish one wall before starting another. Line the ceiling and floor in a similar manner. Each piece of cedar is tongued and grooved for easy fit and application. See Figure 15-88.

The cedar-lined closet will protect your woolens for years. However, in some instances you may not have a closet to line. You may need extra storage space. In this instance take a look at the next section.

Figure 15-84. Cedar-lined closet. The aromatic red cedar serves as a moth deterrent. *(Grossman Lumber)*

Figure 15-85. Placing a piece of cut-to-fit red cedar lining vertically in a closet. *(Grossman Lumber)*

Figure 15-86. Work from the bottom up when applying the cedar boards. *(Grossman Lumber)*

BUILDING EXTRA STORAGE SPACE • Any empty corner can provide the back and one side of a storage unit. This simple design requires a minimum of materials for a maximum of storage. Start with a 4-foot unit now and add 2 feet later. See Figure 15-89.

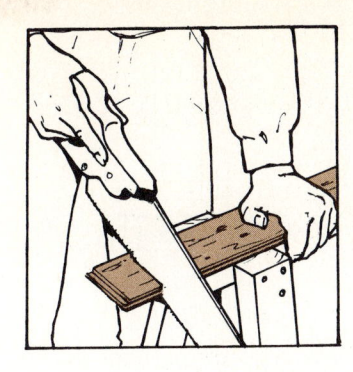

Figure 15-87. To start a new course, saw off the tongued end. The square sawed end will fit snugly into the opposite corner. (Grossman Lumber)

Figure 15-88. Finish one wall before starting another. (Grossman Lumber)

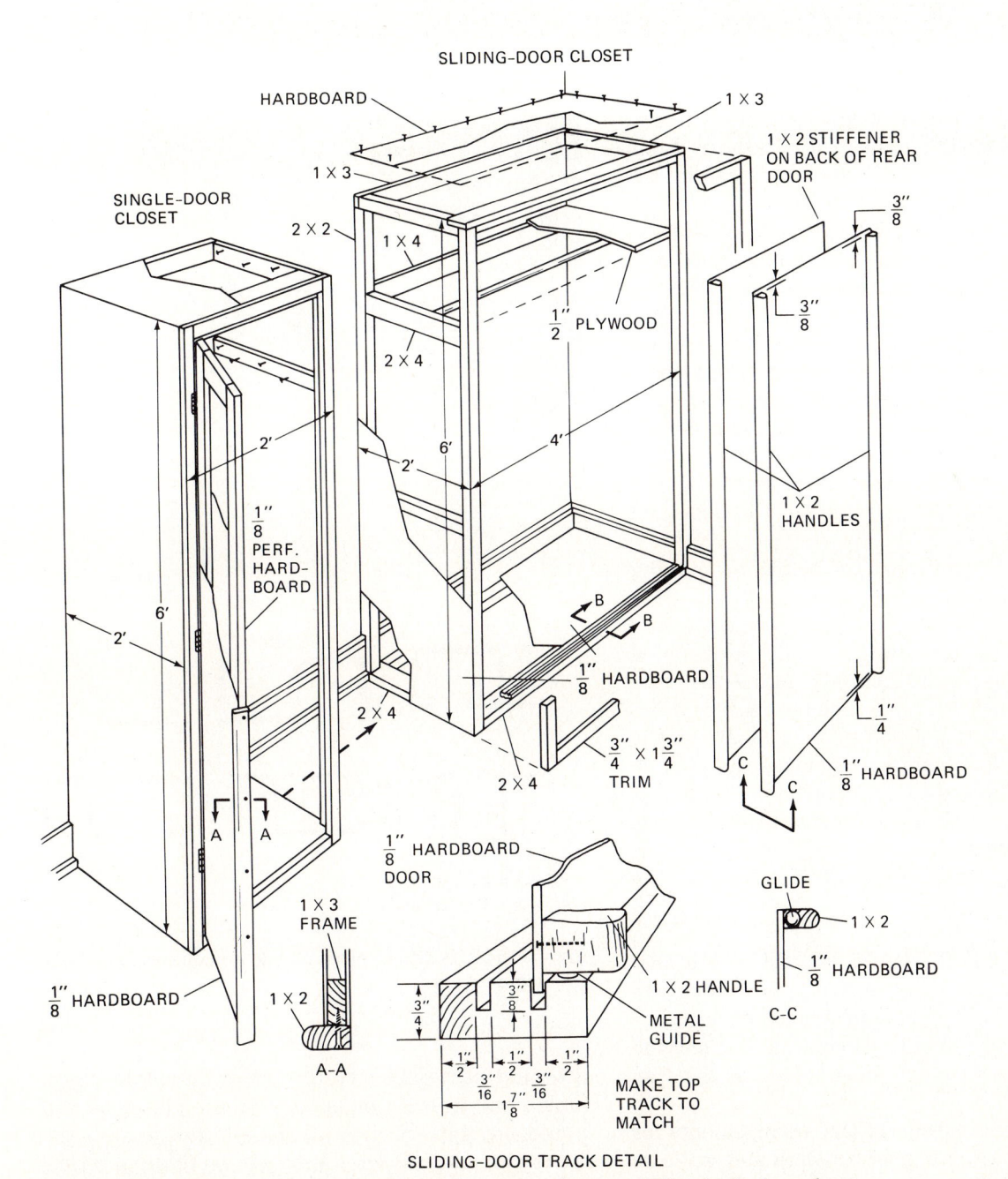

Figure 15-89. Extra storage space. The sliding door closet and the single-door closet. (Grossman Lumber)

Maintenance and Remodeling 445

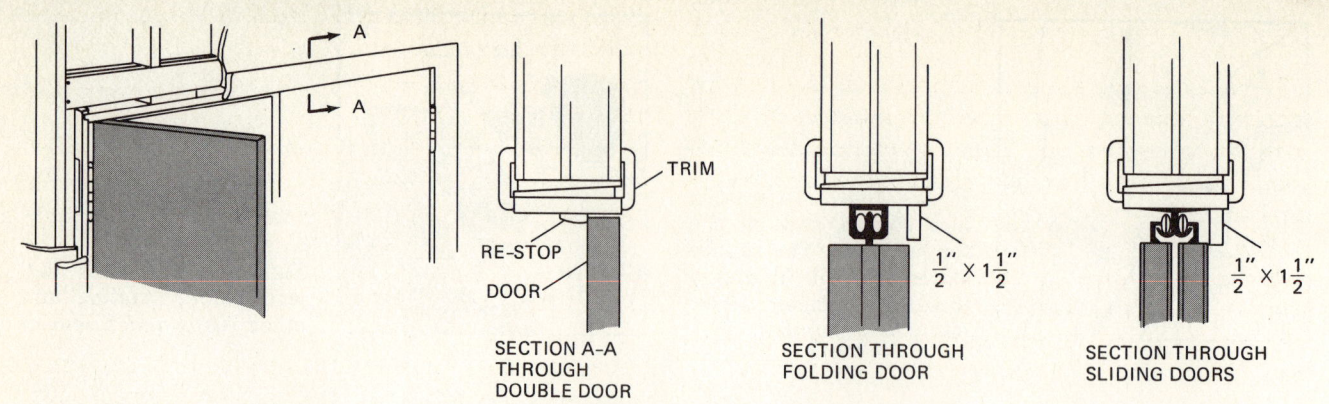

Figure 15-90. Details of the door installation for the closets in Figure 15-89. (Grossman Lumber)

Use a carpenter's level and square to check a room's corners for any misalignment. Note where the irregularities occur. The basic frame is built from 1-inch and 2-inch stock lumber. It can be adjusted to fit any irregularities in the walls or floor. Nail the top and shelf cleats into the studs. Cut the shelf from ½-inch plywood. Slip it into place and nail it to the cleat. Attach the clothes rod with the conventional brackets. Fit, glue, and nail the floor, side, and top panels. Make them from ⅛-inch hardboard. You can buy metal or hardwood sliding doors and tracks. Or you can make the doors and purchase only the tracks. Doors are made of single sheets of ⅛-inch hardboard stiffened with full-length handles of 1×2 trim set on the edge. Metal glides on the ends of handles carry the weight of the door and prevent binding.

Figure 15-90 shows some of the details for making the door operational.

A swinging door unit is built almost like the double door unit. The framework should be fastened to the wall studs where possible. Make the door frame of 1×3s laid flat with ⅛-inch hardboard (plain) on the face and ⅛-inch perforated hardboard on the inside. Again, a full length 1×2 makes the handle.

Attach three hinges to the door. Make sure that the pins line up so that the door swings properly. Place the door in the opening and raise it slightly. Mark the frame and chisel out notches for the hinges.

OTHER TYPES OF STORAGE SPACE • For adequate, well-arranged storage space, plan your closets first. Minimum depth of closets should be 24 inches. The width can vary to suit your needs. But provide closets with large doors for adequate access. See Figure 15-91.

Figure 15-92 shows some of the arrangements for closets. Use 2×4 framing for dividers and walls so that shelves may be attached later. Carefully measure and fit ⅜-inch gypsum board panels in place. Finish the interior or outside of the closet as you would any type of drywall installation. Install the doors you planned for. These can be folding, bifold, or a regular prehung type.

Figure 15-91. A variation of cabinet storage designs. (Grossman Lumber)

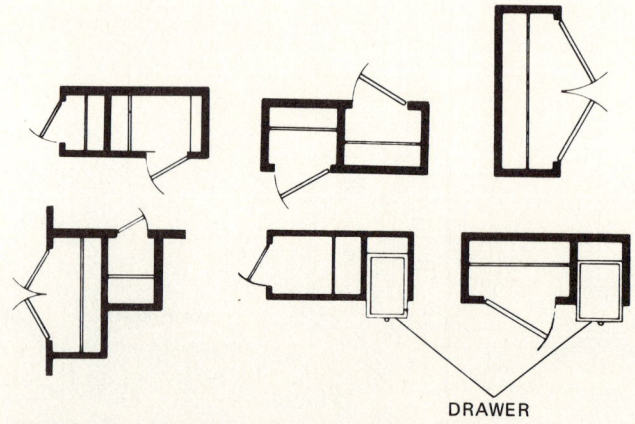

Figure 15-92. Different plans and door openings for closets. (Grossman Lumber)

446 Carpentry Fundamentals

Remodeling a Kitchen

In remodeling a kitchen the major problem is the kitchen cabinets. There are any number of these available already made. They can be purchased and installed to make any type of kitchen arrangement desired.

The countertops have already been covered. The color of the countertop laminate must be chosen so that it will blend with the flooring and the walls. This is the job of the person who will spend a great deal of time in the kitchen.

PLANNING THE KITCHEN • The kitchen begins with a set of new cabinets for storage and work areas. The manufacturer of the cabinets will supply complete installation instructions.

Check the drawing and mark on the walls where each unit is to be installed. Mark the center of the stud lines. Mark the top and the bottom of the cabinet so that you can locate them easily at the time of installation.

As soon as the installation is completed, wipe the cabinets with a soft cloth dampened with water. Dry the cabinets immediately with another clean soft cloth. Follow this cleaning with a very light coat of high-quality liquid or paste wax. The wax helps keep out moisture and causes the cabinets to wear longer.

FINISHING UP THE KITCHEN • After the cabinets have been installed, it is time to do the plumbing. Have the sink installed and choose the proper faucet for the sink.

Kitchen Floor Now it's time to put down the kitchen floor. In most cases the flooring preference today is carpet, although linoleum and tile are also used. It is easier to clean carpet—just vacuuming is sufficient. It is quieter and can be wiped clean easily if something spills. If a total remodeling job has been done, it may be a good idea to paint or wallpaper the walls before installing the flooring. A new range and oven are usually in order, too. The exhaust hood should be properly installed electrically and physically for exhausting cooking odors and steam. This should complete the kitchen remodeling. Other accents and touches here and there are left up to the user of the kitchen.

Enclosing a Porch

One of the first things to do is to establish the actual size you want the finished porch to be. In the example shown here, a patio (16×20 feet) is being enclosed. A quick sketch will show some of the possibilities (Figure 15-93). This becomes the foundation plan. It is drawn ¼ inch to equal 1 foot. The plan can then be used to obtain a building permit from the local authorities.

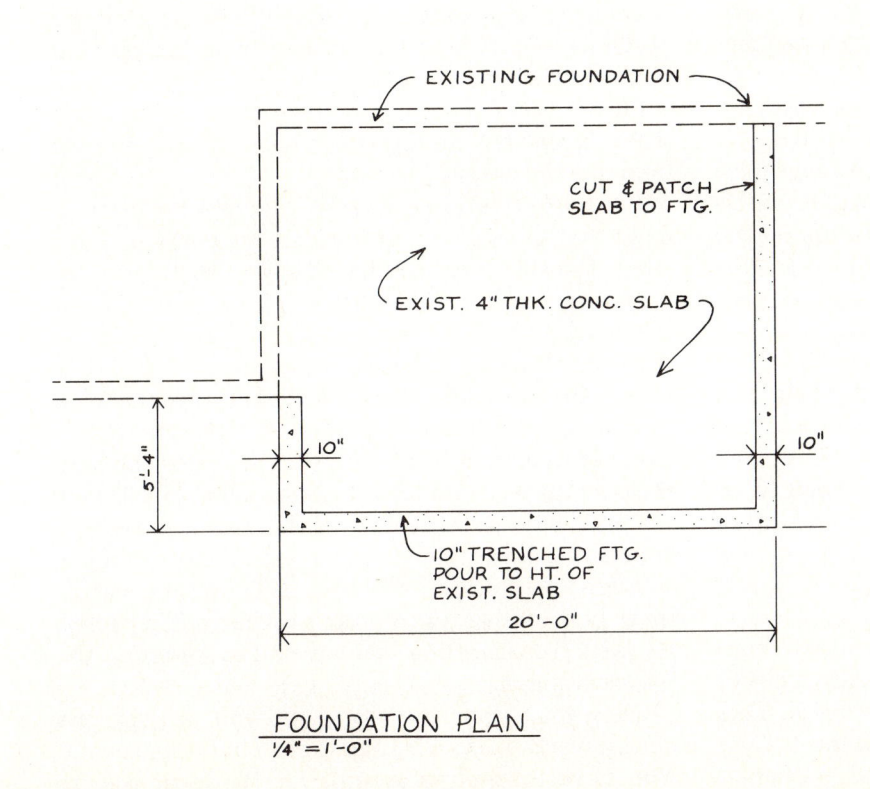

Figure 15-93. Foundation plan for an addition to an already existing building: enclosing a patio.

Maintenance and Remodeling

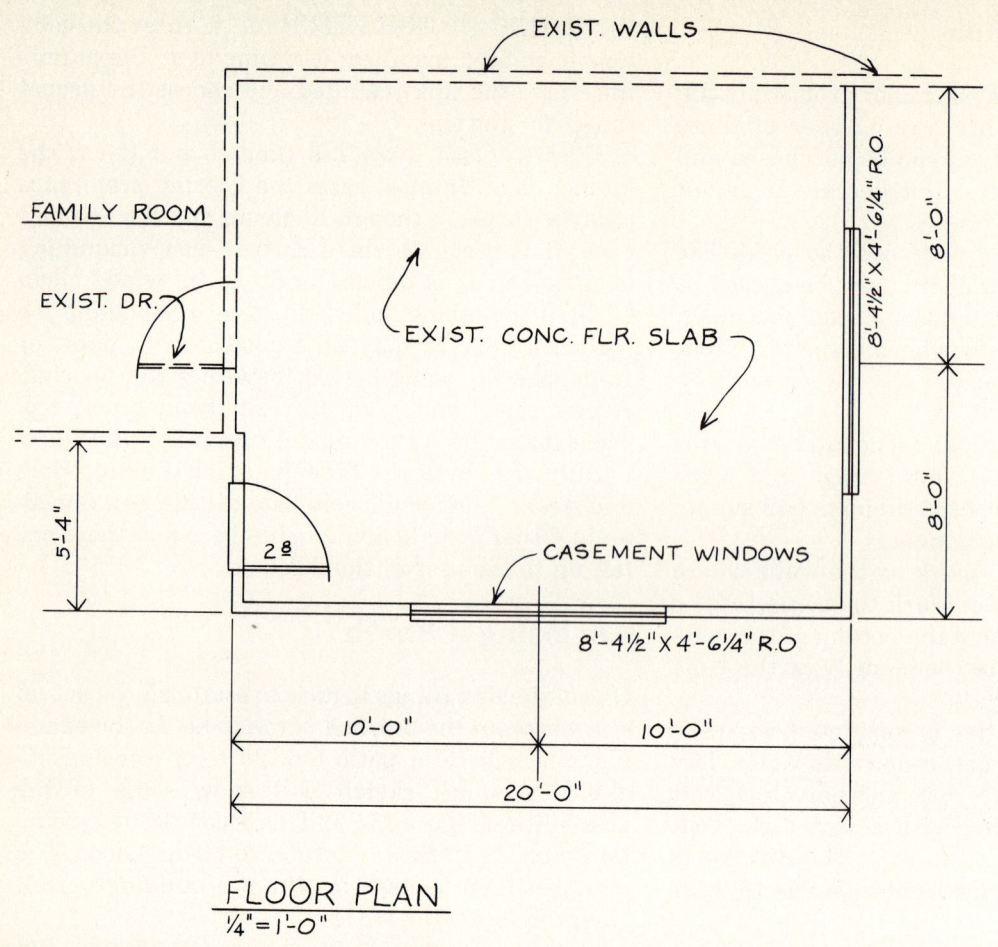

Figure 15-94. Floor plan for an addition.

The floor plan is next. See Figure 15-94. It shows the location of the doors and windows and specifies their size.

A cross section of the addition or enclosure is next. See Figure 15-95. Note the details given here. It deals with the actual construction. The scale here is $\frac{1}{2}'' = 1'\text{-}0''$. Note that the roof pitch is to be determined with reference to the window location on the second floor of the existing building. Figure 15-98A shows what actually happened to the pitch as determined by the window on the second floor. As you can see from the picture, the pitch is not 5:12 as called for on the drawing. Because of the long run of 16 feet, the 2×8 ceiling joists, 16 inches O.C., had to be changed to 2×10s. This was required by the local code. It was a good requirement, since with the low slope on the completed roof, the pile-up of snow would have caused the roof to cave in. See Figure 15-98A.

ELEVATIONS • Once the floor plan and the foundation plan have been completed, you can begin to think about how the porch will look enclosed. This is where the elevation plans come in handy. They show you what the building will look like when it is finished. Figure 15-96A shows how part of the porch will look when it is extended past the existing house in the side elevation.

Figure 15-96B shows how the side elevation looks when finished. Note the storm door and the outside light for the steps.

The rear elevation is simple. It shows the five windows that allow a breeze through the porch on days when the windows can be opened. See Figure 15-97 for a view of the enclosed porch viewed from the rear.

A side elevation is necessary to see how the other side of the enclosure will look. This shows the location of the five windows needed on this side to provide ventilation. Figure 15-98A illustrates the way the enclosure should look. Figure 15-98B shows how the finished product looks with landscaping and the actual roof line created by the second-floor window location. With this low-slope roof, heating cables must be installed on the roof overhang. This keeps ice jams from forming and causing leaks inside the enclosed porch.

Once your plans are ready and you have all the details worked out, it is time to get a building permit. You have to apply and wait for the local board's

448 Carpentry Fundamentals

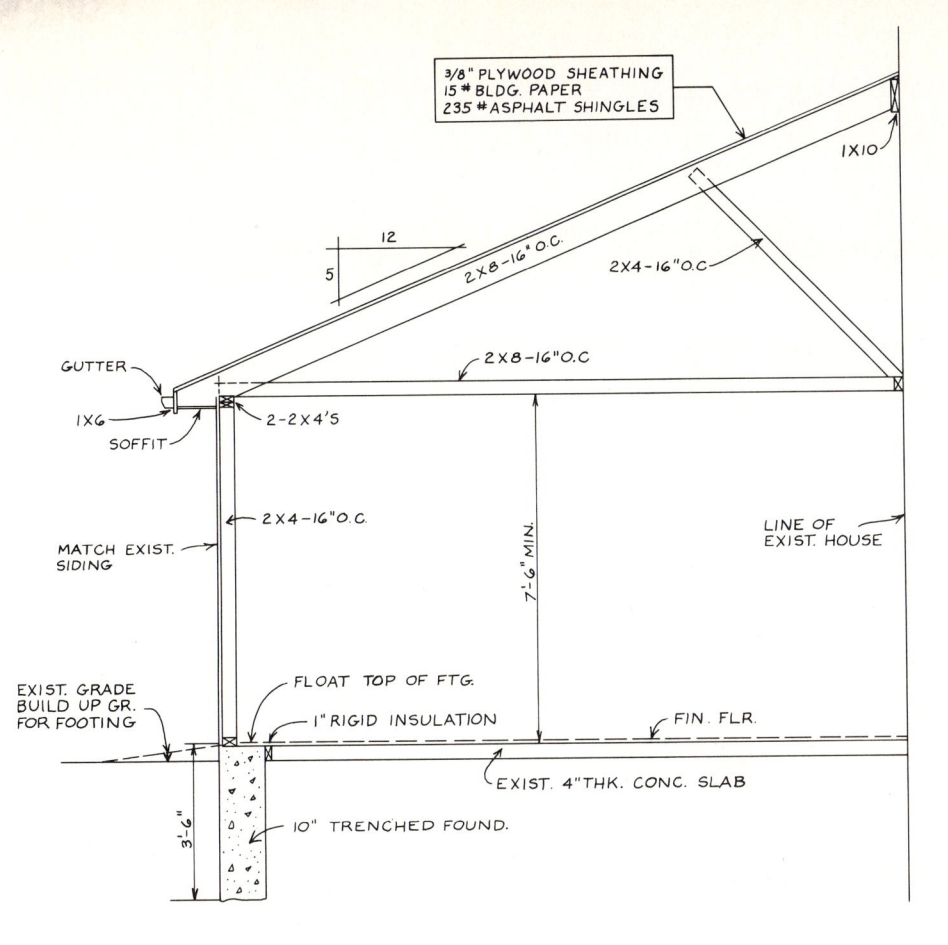

Figure 15-95. Cross section view of the additon.

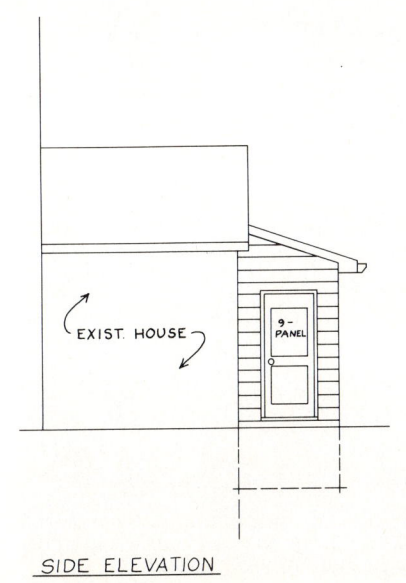

Figure 15-96. (A) Side elevation of the addition.

(B) What the side will look like when finished.

Maintenance and Remodeling 449

REAR ELEVATION
1/4" = 1'-0"

Figure 15-97. Rear elevation of the addition.

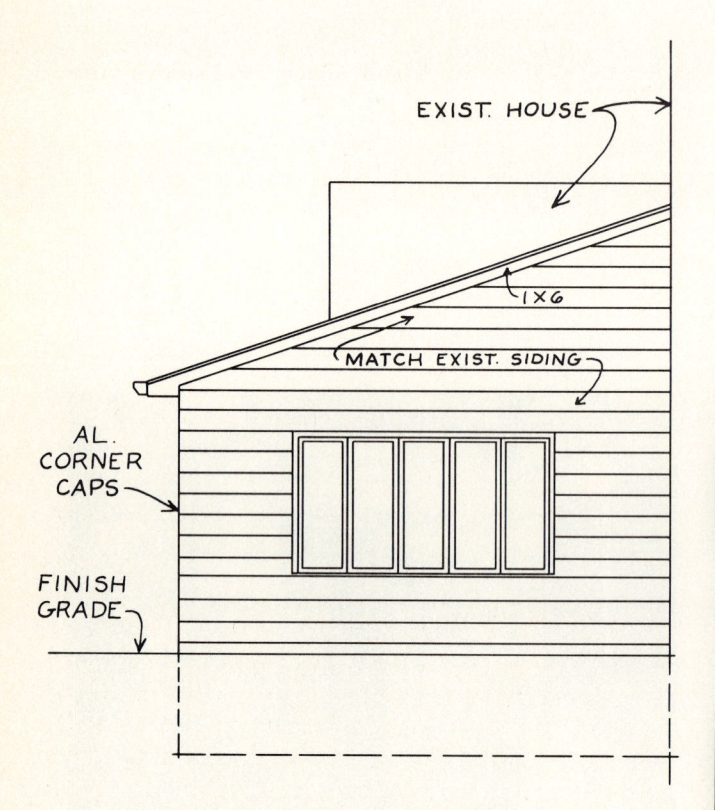

SIDE ELEVATION

Figure 15-98. (A) Side elevation of the addition.

(B) What the side elevation will look like when finished.

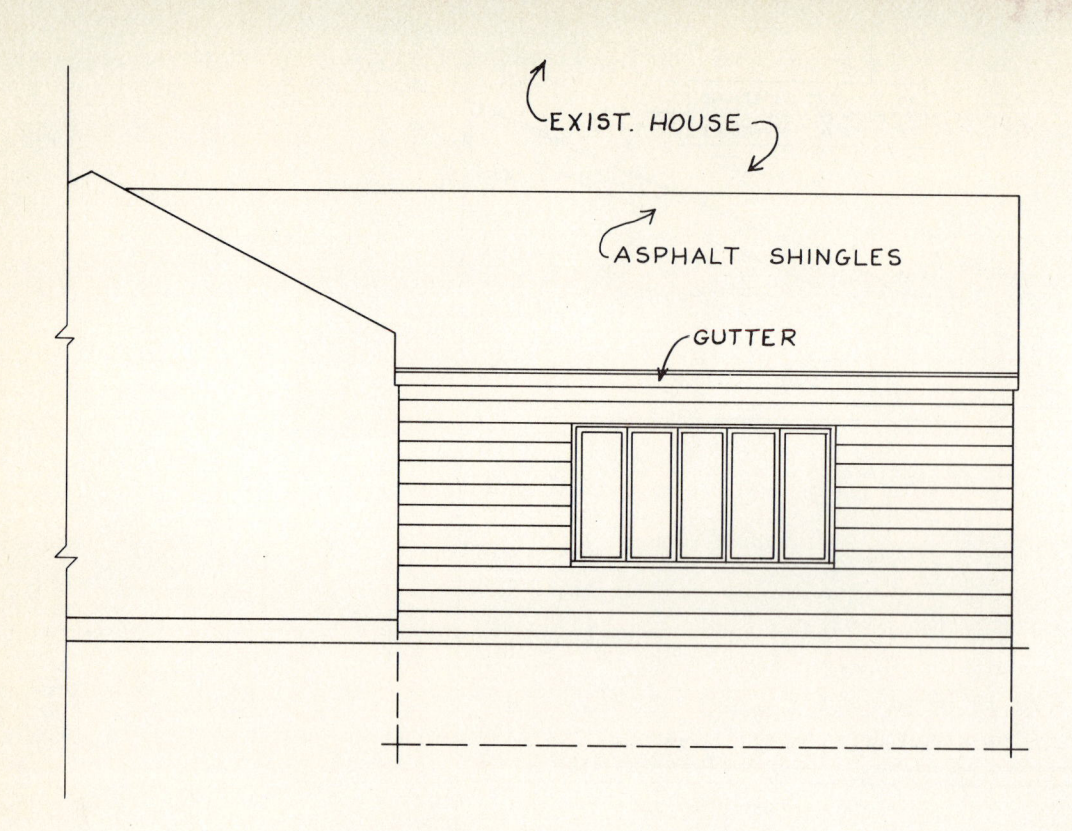

450 Carpentry Fundamentals

Figure 15-99. Application for a building permit.

decision. If you comply with all the building codes, you can go ahead. This can become involved in some communities. The building permit shown in Figure 15-99 shows some of the details and some of the people involved in issuing a building permit. This building permit is for the porch enclosure shown in the previous series of drawings and pictures.

STARTING THE PROJECT • Once you have the building permit, you can get started. The first step is to dig the hole for the concrete footings or for the trench-poured foundation shown here. Once the foundation concrete has set up, you can add concrete blocks to bring the foundation up to the existing level. After the existing grade has been established with the addition, you can proceed as usual for any type of building. Put down the insulation strip and the sole plate. Attach the supports to the existing wall and put in the flooring joists. Once the flooring is in, you can build the wall framing on it and push

Maintenance and Remodeling 451

the framing up and into place. Put on the rafters and the roof. Get the whole structure enclosed with a nail base or plywood on the outside walls. Place the windows and doors in. Put on the roof and then the siding. Trim the outside around the door and windows. Check the overhang and place the trim where it belongs. Put in the gutters and connect them to the existing system. If there is to be electrical wiring, put that in next. In most porches there is no plumbing, so you can go on to the drywall. Finish the ceiling and walls.

After the interior is properly trimmed, it should be painted. The carpeting or tile can be placed on the floor and the electrical outlets covered with the proper plates. The room is ready for use.

FINISHING THE PROJECT • Throughout the building process, the local building inspector will be making calls to see that the code is followed. This is for the benefit of both the builder and the owner.

There is another *benefit* (to the local government) of the inspections and the building permit. Once the addition is inhabited, the structure can go on the tax roles and the property tax can be adjusted accordingly.

• Adding Space to Existing Buildings •

Adding space to an already-standing building requires some special considerations. You want the addition to look as if it "belongs." That means you should have the siding the same as the original or as close as possible. In some cases you may want it to look added-on so you can contrast the new with the old. However, in most instances, the intention is to match up the addition as closely as possible.

First, the additional space must have a function. You may need it for a den, an office, or a bedroom. In any of these instances you don't need water or plumbing if you already have sufficient bathroom facilities on that floor. You will need electrical facilities. Plan the maximum possible use of the building and then put in the number of electrical outlets, switches, and lights that you think you can use. Remember, it is much cheaper to do it now than after the wallboard has been put up.

Planning an Addition

In the example used here, we will add a 12×22-foot room onto an existing, recently built house. The addition is to be used as an office or den. It is located off the dining room, so it is out of the way of through-the-house foot traffic. The outside wall will also serve to deaden the sound. It was insulated when the house was built. This means that only three walls will have to be constructed. Make a rough sketch of what you think will be needed. See Figure 15-100 for an example. Note that the light switch has been added and that two lights over the bookcases will be added later. Note the outlets for electricity and heat. The windows are of two different sizes. The local code calls for a certain amount of square footage in the windows for ventilation. One window has been taken from the existing upstairs bedroom. It is too large and will interfere with the roof line of this addition. A smaller window is put in up there, and the one used in the bedroom is moved down to the addition. Only one window needs to be purchased. Only one door is needed. These should be matched up with the existing doors and windows so that the house looks complete from the inside, too.

Elevations must be drawn up to show to the building code inspectors. You must get a building permit; therefore the elevation drawings will definitely be needed. See Figures 15-101, 102, and 103 for the front, rear, and side elevations. This information will help you obtain a building permit. If you use Andersen's window numbers, the town board will be able to see that these windows provide the right amount of ventilation.

SPECIFICATIONS • To make the finished product fit your idea of what you wanted, it is best to write a list of specifications. Make sure you list everything you want done and how you want it done. For example, the basement, siding, overhang, flooring, windows, door, walls, roof, and even the rafters should be specified.

SPECIFICATIONS

Addition for 125 Briarhill in Town of Amherst (See detailed sketches for rear, front, and side elevations)

BASEMENT

Crawl space—skim coat of concrete over gravel.
Drainage around footings and blocks to prevent moisture buildup.
Blocks on concrete footings—42 inches deep.
To be level with adjoining structure.
To be waterproofed. Fill to be returned and leveled around the exterior.

SIDING

National Gypsum Woodrock Prefinished (as per existing).
Size to match existing.
Must match at corners and overlap in rear to look as if built with original structure.
To be caulked around windows.

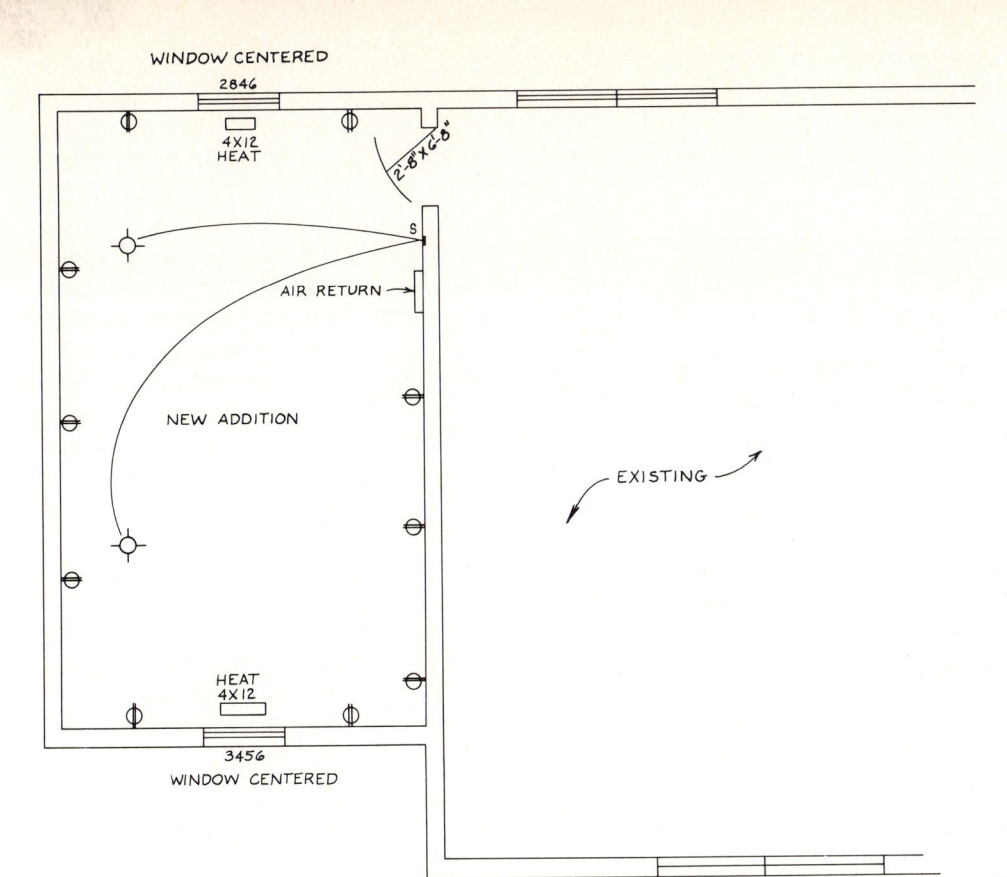

Figure 15-100. Floor plan for an addition.

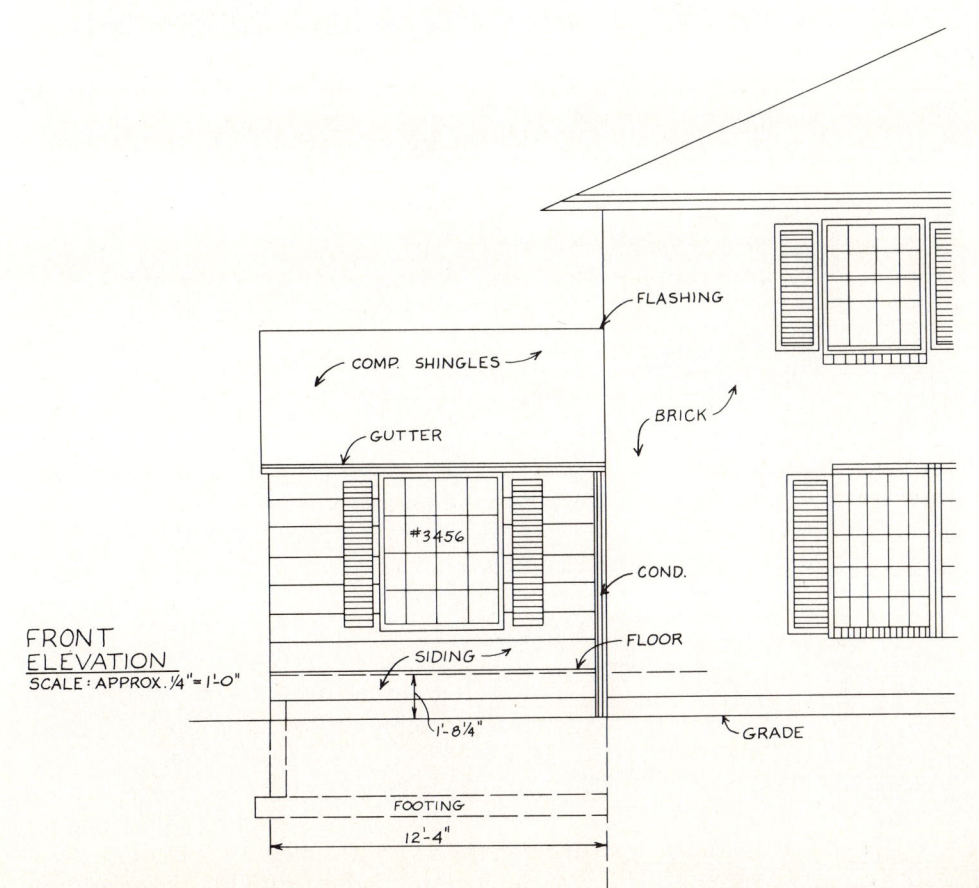

Figure 15-101. Front elevation.

Maintenance and Remodeling 453

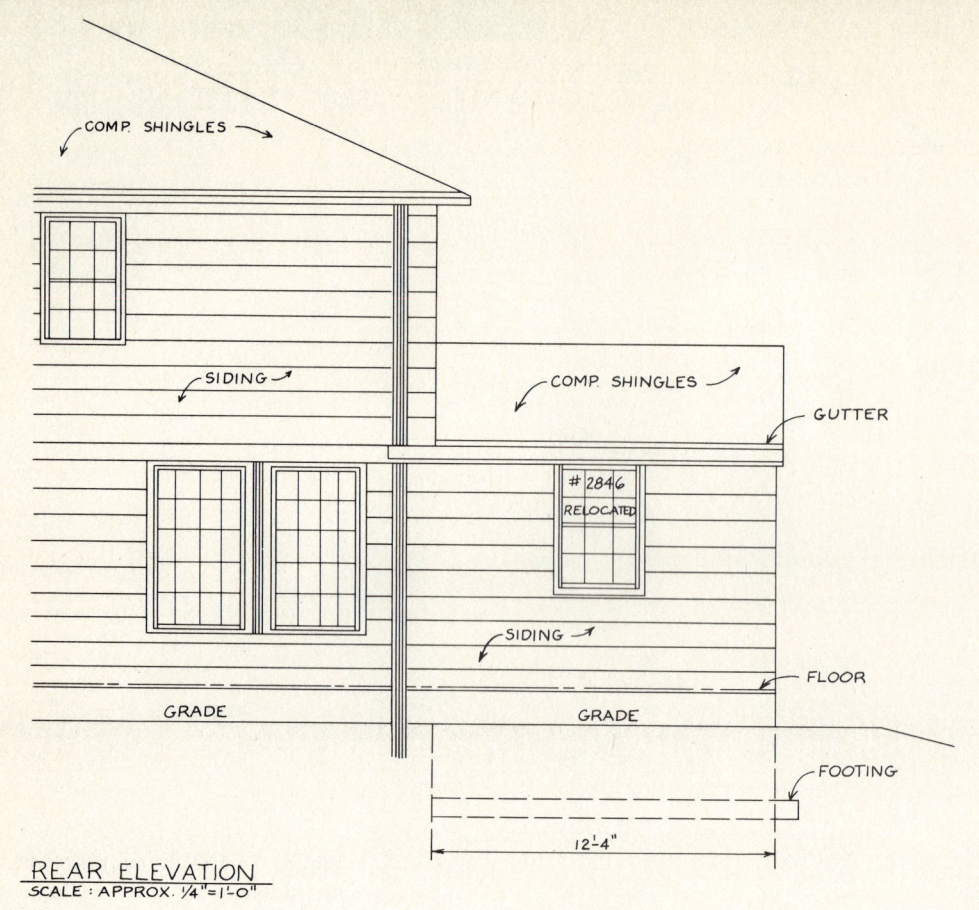

Figure 15-102. Rear elevation.

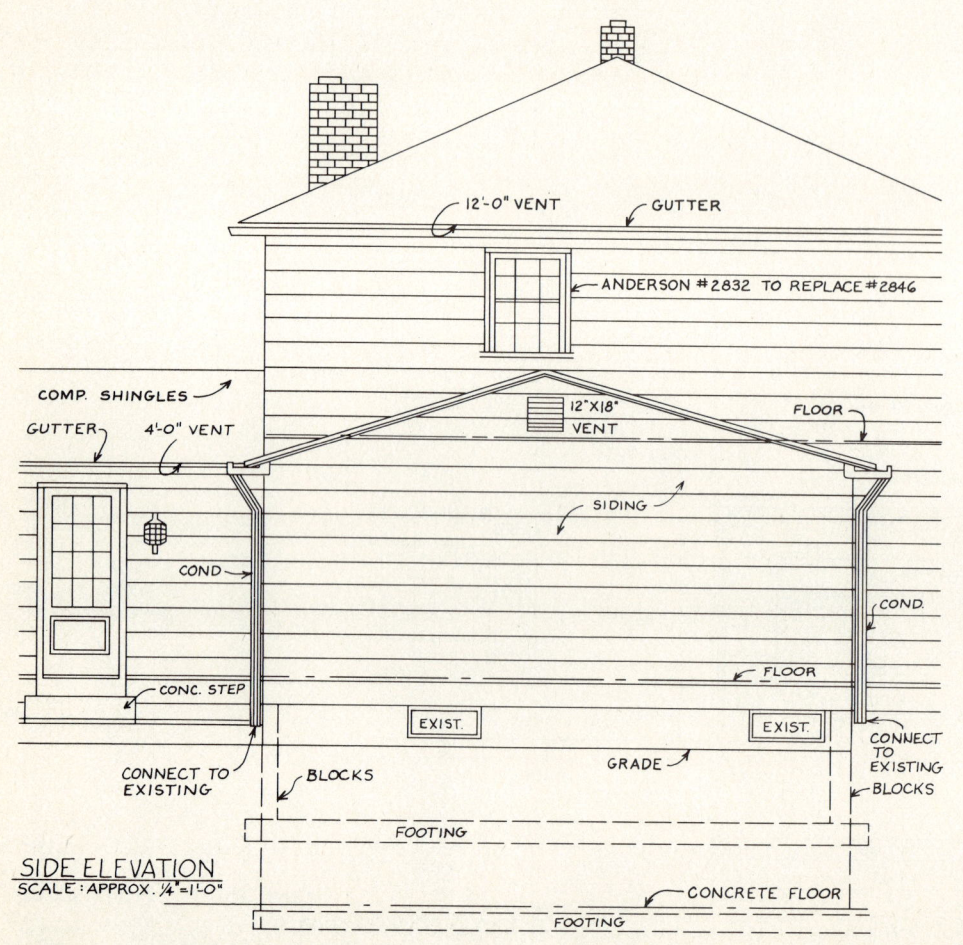

Figure 15-103. Side elevation.

454 Carpentry Fundamentals

OVERHANG

To match existing as per family room extension (see sheet 5).
Gutters front and rear to match existing.
Downspouts (conduits) to match existing and to connect.
Ventilation screen in Upson board overhang.
Exterior fascia and molding to match existing.

FLOORING

Kiln-dried 2×10 (construction or better) on 16-inch centers.
Plywood subfloor, 5/8-inch exterior, A–D.
Plywood subfloor, 1/2-inch A–D, interior.
Fiberglas insulation strip between sole plate and blocks.
Must meet with existing room. Allowance made for carpeting.
Must be level.

WINDOWS

Andersen No. 3456 in front; No. 3456 placed in rear of addition.
Replace existing window with Andersen No. 2832 and finish interior.
Windows to be vinyl-coated and screened as per existing.
Shutters on front window section to match existing.

DOOR

Interior with trim as per existing, mounted and operating.
Size, 32"×6'8"; solid, flush, walnut with brass lock and hardware.
Dining room to be left in excellent condition.

PERMITS

To be obtained by the contractor.

WALLS

Studs, 2×4 kiln-dried (construction or better), 16 inches O.C.
Double or better headers at windows and door and double at corners where necessary.

ROOF

Composition shingles as per existing, black of same weight as existing.
Felt paper under shingles and attached to 5/8-inch exterior plywood (A–D) sheathing.

RAFTERS

Kiln-dried 2×4 truss type as per existing, 16 inches on-centers.
Roof type to be shown in attached drawings. Check with existing garage to determine type of construction if necessary.

In order to add any information you may have left out, you can make drawings illustrating what you need. Figure 15-104 shows a cornice detail that ensures there is no misunderstanding of what the overhang is. Other details are also present in this drawing. The scale of the cornice is 1½" = 1"-0".

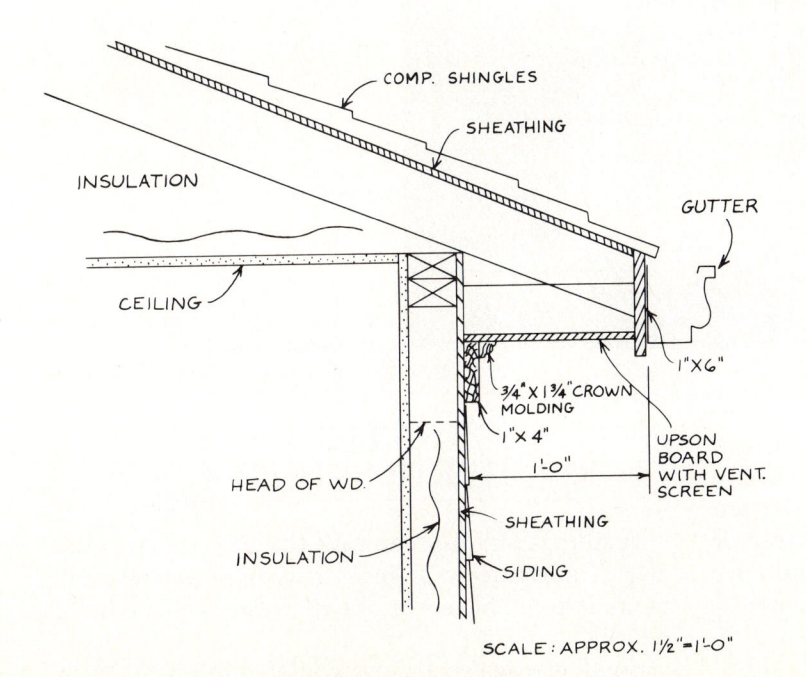

Figure 15-104. Cornice detail.

Maintenance and Remodeling

Figure 15-105. Side elevation in finished form.

Figure 15-106. Rear elevation in finished form.

The contractor is protected when the details of work to be done are spelled out in this way. There is little or no room for argument if things are written out. A properly drawn contract between contractor and owner should also be executed to make sure both parties understand the financial arrangements. Figures 15-105 and 15-106 show the completed addition as it looks with landscaping added.

• CREATING NEW STRUCTURES •

A storage building can take the shape of any number of structures. In most instances you want to make it resemble some of the features of the home nearby.

Small storage sheds are available in precut pack-

RUSTIC

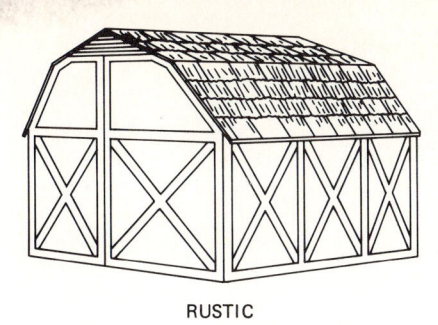

GREENHOUSE
COVER WITH GLASS, PLEXIGLAS OR POLYVINYL

CONTEMPORARY

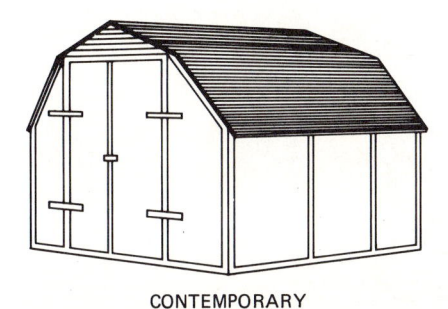

Figure 15-107. Various designs for storage facilities. *(Courtesy of TECO)*

ages from local lumber yards. Figure 15-107 shows three versions of the same package. It can be varied to meet the requirements of the buyer. Lawnmowers, bicycles, or almost anything can be stored in a structure of this type.

All you need is a slab to anchor the building permanently. In some instances you may not want to anchor it, so you just drive stakes in the ground and nail the sole plate to the stakes. Most of the features are simple to alter if you want to change the design.

Before you choose any storage facility, you should know just how you plan to use it. This will determine the type of structure. It will determine the size, and in some cases the shape. The greenhouse in Figure 15-107 is nothing more than the rustic or contemporary shed with the siding left off. The frame is covered with glass, Plexiglas, or polyvinyl according to your taste or pocketbook.

Custom-Built Storage Shed

In some cases it is desirable to design your own. The example used here, shown in Figure 15-108, was designed to hold lawnmowers and yard equipment. It was designed for easy access to the inside. Take a look at the overhead garage door in the rear of the shed. The 36-inch door was used along with the window to make it more appealing to the eye. It also has a hip roof to match the house to which it is a companion. See Figure 15-109.

The design is 10×15 feet. See Figure 15-110. That produces a 2-to-3 ratio which isn't too bad to look at. A 3-to-4 ratio is also very common. The concrete slab was placed over a bed of crushed rocks and

Figure 15-108. View of one end of a storage shed.

Maintenance and Remodeling 457

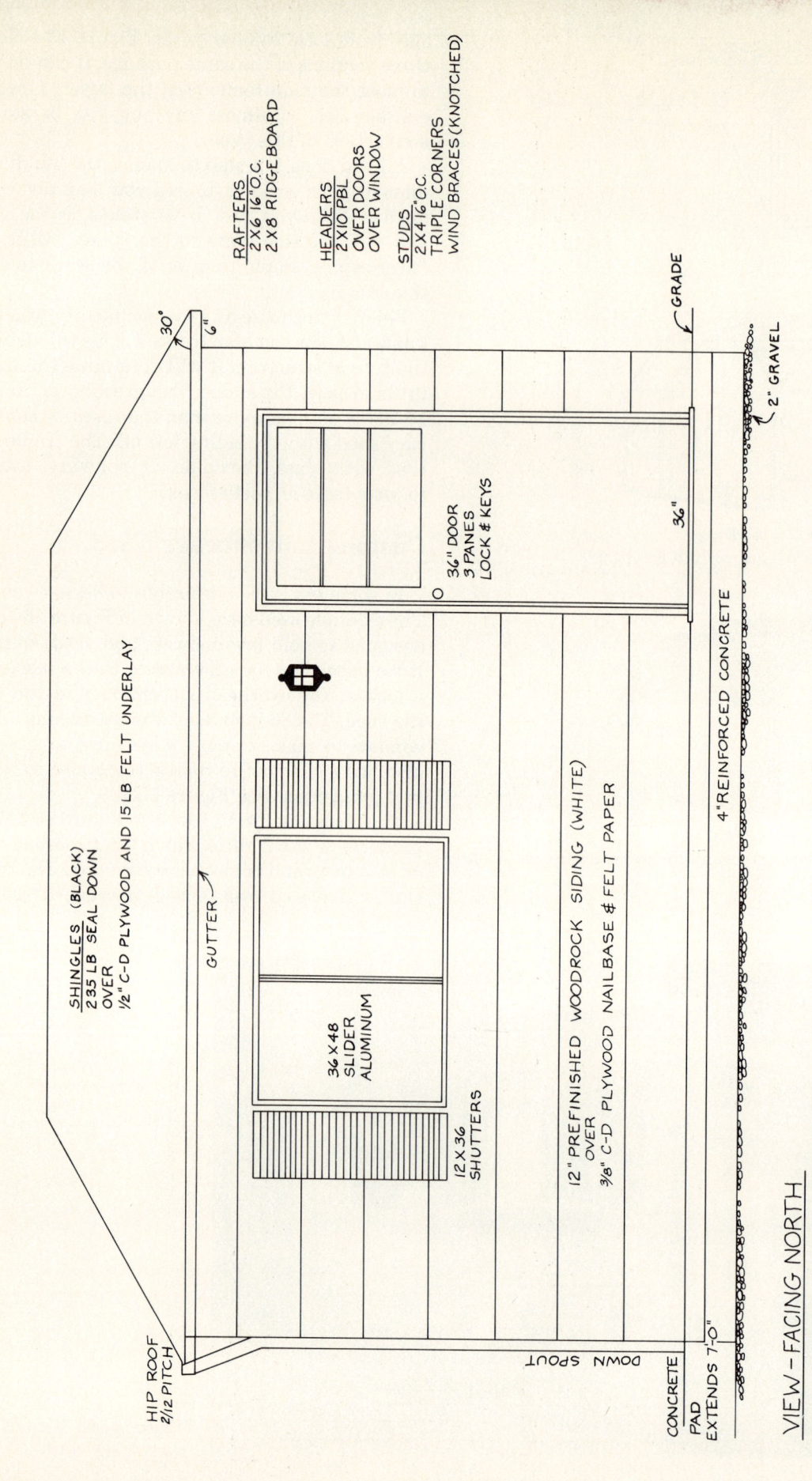

VIEW – FACING NORTH
Figure 15-109. View of the finished storage shed.

RAFTERS
2×6 16" O.C.
2×8 RIDGE BOARD

HEADERS
2×10 PBL
OVER DOORS
OVER WINDOW

STUDS
2×4 16" O.C.
TRIPLE CORNERS (KNOTCHED)
WIND BRACES (KNOTCHED)

SHINGLES (BLACK)
235 LB SEAL DOWN
OVER
1/2" C-D PLYWOOD AND 15 LB FELT UNDERLAY

HIP ROOF
2/12 PITCH

GUTTER

36" DOOR
3 PANES
LOCK & KEYS

36 X 48
SLIDER
ALUMINUM

12 X 36
SHUTTERS

12" PREFINISHED WOODROCK SIDING (WHITE)
OVER
3/8" C-D PLYWOOD NAILBASE & FELT PAPER

DOWN SPOUT

CONCRETE PAD EXTENDS 7'-0"

4" REINFORCED CONCRETE

GRADE

2" GRAVEL

458 Carpentry Fundamentals

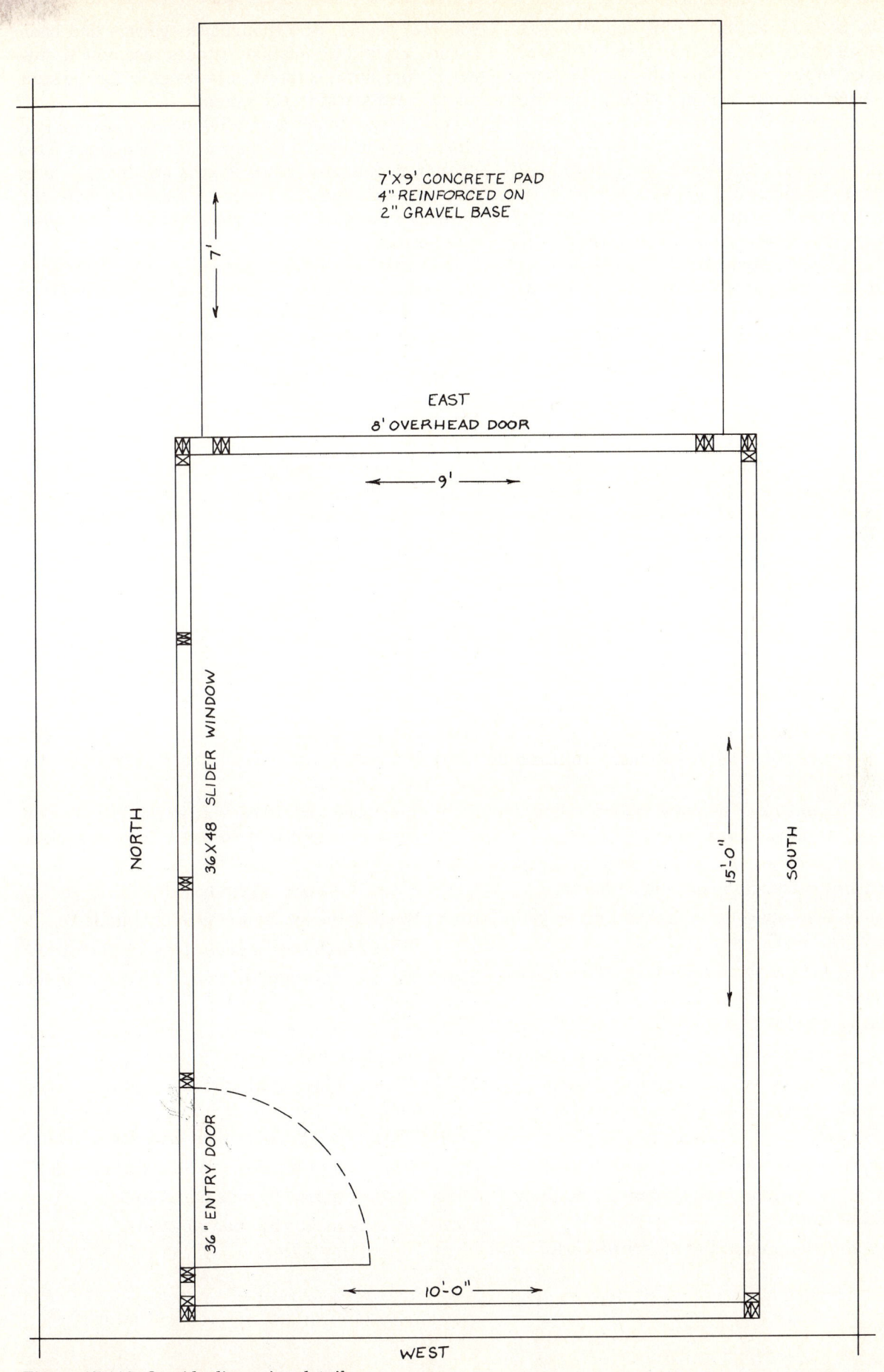

Figure 15-110. Outside dimension details.

Maintenance and Remodeling

anchored by bolts embedded in the concrete slab. An 9-foot-wide and 7-foot-long slab is tapered down from the floor to the yard. This allows rider lawnmowers to be driven into the shed. The outside pad also serves as a service center.

Wires serving the structure are buried underground and brought up through a piece of conduit. They enter the building near the small door. There is only one window, so the wall space can be used to hang yard tools. The downspouts empty into the beds surrounding the structure to water the evergreens. An automatic light switch turns on both lights at sundown and off again at dawn.

The overhead door faces the rear of the property. This produced some interesting comments from the concrete people, who thought the garage had been turned around by mistake. It does resemble a one-car garage, but it is specifically designed for the storage of yard working equipment.

Don't forget to get a building permit. Even a tool shed requires a building permit in some areas. This doesn't necessarily mean it goes on the tax rolls, but does call for a number of inspections by the building inspector, which helps protect both builder and owner.

Buildings for storage take all shapes. They may be garages or barns. The design of a new structure should be carefully chosen to harmonize with the rest of the buildings on the property.

• WORDS CARPENTERS USE •

remodeling	lockset	laminate	asphalt shingles	scribing
alligatoring	deadbolt lock	Formica	square butt strip	mastic
blistering	strike	adhesives	shingle	swinging door
nailhead stains	traverse rod	contact cement	guttering	patio
peeling	right-hand draw	bonds	downspouts	elevations
flaking	sheetrock	saber saw	sump pump	
jamb	drywall	molding	vinyl-asbestos tile	
shim	nail creep	backsplash	vapor barrier	

• STUDY QUESTIONS •

15-1. What is the meaning of the word "maintenance?"

15-2. What do you need to know about a house before you start remodeling it?

15-3. How does moisture passing through the exterior paint cause problems with a house?

15-4. What is the importance of scheduling in getting a job done?

15-5. What is meant by shimming a strike jamb of a door?

15-6. What is a lockset?

15-7. What is a deadbolt lock? Where is it used?

15-8. What is a traverse rod? Where is it used?

15-9. Why do bay windows need special rods for curtains?

15-10. What is another name for gypsum board?

15-11. How do you eliminate popped nails and their unsightliness?

15-12. Why do nails used to install drywall pop?

15-13. What do you use to cut plastic laminates for countertops?

15-14. How does contact cement work? Why do you have to wait for it to dry before you place the two pieces together?

15-15. What is a back splash board?

15-16. What is meant by a course of shingles?

15-17. What are the two standard sizes of floor tiles?

15-18. What is the standard size of a piece of paneling?

15-19. What do you mean when you say the panel should be hung perfectly plumb?

15-20. What is a drop ceiling? Where would you use one?

15-21. What type of cedar is used to line a closet?

15-22. What is a bifold door? Where do you use it?

15-23. What is meant by side elevation?

15-24. What is meant by rear elevation?

16
THE CARPENTER AND THE CONSTRUCTION INDUSTRY

In this unit you will learn how the carpenter fits into the construction industry. You will learn how to become a carpenter. What's in the future for carpenters is also discussed. How you measure up will be checked out as we go along. Things you can learn to do are:

- Check out the future of the carpenter
- Check out building codes
- See how a building permit is obtained
- Take a test to check your skills
- Become a community planner

• INTRODUCTION •

The construction industry is made up of many companies. These companies build or erect structures. The structures may be of many different types. The three major types of construction are

Industrial and commercial
Civil
Residential

Examples of industrial buildings are factories. Commercial buildings include shopping plazas. Office buildings also can be commercial. Refineries are industrial structures. Carpenters are needed for all these buildings. They do different types of work for each type of structure.

Civil construction usually means tax money is paying for the building. This type of structure serves the community. It is built by the state, local, or federal government. It may be a post office or a library. In some cases office buildings are also included here. Schools are civil buildings. Some bridges, dams, and roads are also classed in this group.

Residential construction involves homes for people. This type of construction usually means single-family homes. In some cases it also means apartments. Residential houses are found in cities, in suburbs, and in the country. See Figures 16-1 and 16-2.

• OPPORTUNITIES IN CARPENTRY •

Construction crafts are the largest of the skilled worker groups. They make up the largest number of the country's labor force. More than 3.3 million people were employed in construction in 1976. Almost 3 out of every 10 skilled workers were in construction crafts.

There are more than two dozen skilled crafts. These crafts vary in size. Carpentry is just one of these skilled crafts. Other craft workers are painters, plumbers, and electricians. Carpenters alone number more than 1 million. About a third of all

Figure 16-1. Residential construction usually means single-family homes. Here carpenters get ready to push walls into place. (*Western Wood Products*)

construction people are carpenters.

Carpenters have a bright future. They are needed in most parts of the country. There is always something to be built. Even during a recession buildings are going up. Government building increases when the economy turns down. This means a carpenter can have a job almost anytime. Of course, the carpenter must be willing to move around a bit.

Qualifications

A carpenter must be properly trained. Carpentry is a skilled trade or craft. That means a person has to have experience so that he or she can do the job quickly. Not only must the job be done quickly, but it must also be done right. This calls for experience. Experience can lead to skill. A skilled person has done a job many times. A skilled person is also very good at a particular job. This does not come overnight. You have to work at it. It usually takes 4 years of working with someone before you become a skilled carpenter.

APPRENTICESHIP • Most training people recommend a formal apprenticeship. This means a definite period of on-the-job training, which is usually 4 years. It includes classroom work too. You learn all about tools, materials, and the principles of the trade. An *apprentice* is a student or learner. Some people do not take this route. In some cases, on-the-job training is enough to learn the trade. An employer usually works with a training program to give you a good background in the trade. A good vocational school background helps. You should have math, industrial arts, and science courses in high school. Being a high-school graduate is to your advantage.

Figure 16-2. Residential construction also involves teamwork. Here carpenters work as a team to place the walls into position. *(Western Wood Products)*

The Carpenter and the Construction Industry **463**

Employment growth and replacement needs will create large numbers of job openings in the construction occupations.

SELECTED CONSTRUCTION OCCUPATIONS
AVERAGE ANNUAL OPENINGS, 1976-85   (IN THOUSANDS)

- CARPENTERS
- CONSTRUCTION LABORERS
- OPERATING ENGINEERS (CONSTRUCTION MACHINERY)
- PAINTERS AND PAPERHANGERS
- PLUMBERS AND PIPEFITTERS

SOURCE: Bureau of Labor Statistics

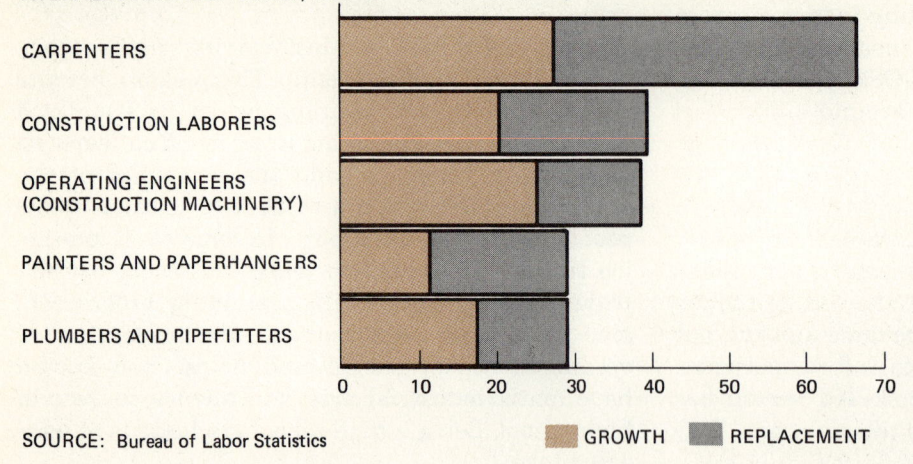

GROWTH   REPLACEMENT

Figure 16-3. Employment growth and replacement needs create job openings. Note the position of carpenters at the top. (*U.S. Bureau of Labor Statistics*)

Long-Range Outlook

Figure 16-3 shows what the future of the carpentry trade looks like at this time. Notice how the carpenter has a very good future. Growth and replacement are very high. Growth means new jobs that demand more carpenters. Replacement means jobs held by those who die or retire and have to be replaced. Of course, some people are promoted or go into other businesses. They, too, have to be replaced.

There is always demand for carpenters in the remodeling field. People remodel when housing gets too expensive. This is the present case. There is also a growing demand for repair work. Repair work on highway systems, dams, bridges, and similar projects needs carpenters.

• ROLE OF THE CARPENTER ON THE JOB •

A carpenter has many roles to play. The carpenter must be able to do many things. Some of them are directly related to working with wood. However, today carpenters also work with metal studs in new buildings. A carpenter is a person who repairs various types of structures. But carpentry is also a trade which demands a great deal of work with wood. Wood has to be laid out, cut, shaped, and fitted at certain points. Power and hand tools are used on the job. A good knowledge of the tools will make your production greater. This, in turn, makes you more valuable on the job.

A carpenter must be a responsible person. Much depends upon safe work habits.

Primary Responsibilities

First, a carpenter must be skilled. A skilled person works well with others. A skilled person will be able to do a job properly the first time. This person must be able to assume responsibility for a job. Time and ability are most important. Time means money to the contractor and to the carpenter. Ability means you don't have to do a job over to get it right. This ability to do the job comes from being alert and always learning new techniques. Knowing how to do a job better the next time is very important.

The construction of a building frame is very important. Its safety and the safety of all those inside it for years is a primary responsibility. Poor workmanship can produce dangerous conditions. It is up to the carpenter to keep up to date with the latest materials. New techniques are developed all the time. The carpenter should be sharp enough to catch on to these techniques quickly. This is a job that demands an open mind. A person must study all the time. There are magazines and booklets which can help keep you updated.

Being on time and doing a good day's work is very much appreciated. It may mean the difference between getting a job the next time and not getting one.

Working as a Member of a Construction Team

It is important to be able to work with a team. Figure 16-2 shows how difficult it is for one person to do some of the jobs a carpenter has to do. Teamwork is very important. The ability to get along with others on the job can make the difference between getting a job done and not being able to do it at all.

A team effort is usually called for in putting together a wall, roof, or floor. Siding and trim are also a team effort. In some instances it is best to specialize in a particular job. You can do your particular job and work as part of the team while someone else is doing another specialty. Put it all together and the team produces a good-quality building.

Participating in Union Activities

Not all carpenters belong to unions. A great many work for themselves. They are called *self-employed*. This type of carpenter is independent. That means you can set your own work hours within limits. You do have the responsibility of getting the job. You must do the job at the price you agreed upon even if you don't make any money doing the work. It takes very little investment to go into business for yourself. However, it does mean you are risking a lot. You may have to pay your own insurance. You will have to take responsibility for those who work for you. This can mean a lot of hours off the job just getting people to work. Then, again, you're responsible for the quality of the job done. Some people like to work this way and others do not. The residential carpenter is the one most likely to be nonunion.

In commercial and industrial building the carpenter is usually a union member. The union gets the job and furnishes the labor or carpenters. This means they do the work for you in the areas of money and quality of the work. You are responsible for doing a good job. You are free to concentrate on doing a good job. You are also paying someone else a fee for their time and effort on your behalf. They get you insurance for sickness. They also get you vacation time. Retirement benefits and other things are part of the union package. This calls for you, as a union member, to be responsible to your group. You should have a say in what the union does for you. A union member should vote on all things presented. Strikes are part of the scene with unions. This is a weapon used to get good working conditions and benefits.

For most government jobs and industrial work, the carpenter must belong to a union. The reason is usually that people in government and industry are also unionized. They can work with this arrangement better.

A union is only as good as its members. In order to have a good union, you have to work within it. You have to see to it that you are represented in all matters that concern you. Ask questions, and don't hesitate to give your opinion when it is asked for in meetings. Good unions have good members. Good members are concerned members. They vote. They are active in the bargaining sessions. Everyone who belongs to a union has certain rights. See to it that yours are protected. The only way to do that is to keep informed on what your union is doing for you and others in your group.

• BROADENING HORIZONS IN CARPENTRY •

Carpentry, like other trades, is constantly changing. You have to keep up or become dated. A dated carpenter is soon too far behind to be of much value to the trade.

New Building Materials

Figure 16-4 shows a house made of plywood. Only a few years ago all floors and walls were made of pieces of wood that measured 1×6 inches. Plywood has some special features. It is stronger, it has a

Figure 16-4. Building materials used in construction are changing. Note the extensive use of plywood in a modern house. *(Western Wood Products)*

Figure 16-5. Carpenters must learn to adapt to new materials like this composition nail base used for siding.

good nail surface, and it is good insulation. It is easily placed in position, and it can be bought cheaply compared with single pieces of siding or flooring.

New materials are used as a nail base for siding. Carpenters have to adapt to these types of material and be able to handle them. See Figure 16-5.

Changing Construction Procedures

Construction procedures are changing. Note that in Figure 16-6 the wall construction is different from the usual. This is a double-wall partition. It is used to separate rooms in apartments so that sound is not easily transmitted between rooms.

Figure 16-7 shows how adhesives can be used to apply panels. These panels are held in place with glue instead of nails. This makes the carpenter's job easier. A carpenter can do more work in less time. This helps the carpenter to demand more money. More productivity means more pay. New materials and new construction methods are being developed. They make the job easier and quicker. This calls for a carpenter who can adapt to new ways of doing things. New clips are available for holding 2×4s and 2×10s. They are also designed to hold plywood sheets. Using these clips cuts installation time. They also make buildings stronger. The carpenter then becomes a valuable person on the job.

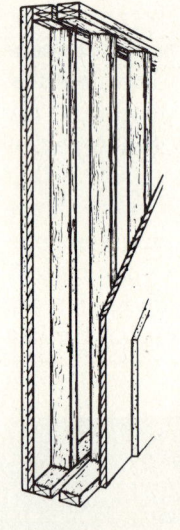

Figure 16-6. Construction procedures are also changing. For example, double walls are used to help keep the noise down in apartments.

Innovations in Building Design

Architects are coming up with new designs. These new designs call for different ways of doing things. The carpenter has to be able to work with new woods and new combinations of materials. Figure 16-8 shows one of the newer designs for a "modern" home. A mixture of brick and wood is used for the outside covering. The inside also calls for some new

466 Carpentry Fundamentals

Figure 16-7. Adhesives are used extensively in the building industry today. *(U.S. Gypsum)*

Figure 16-8. Modern construction using brick and wood. A number of various materials are used in house exteriors.

The Carpenter and the Construction Industry **467**

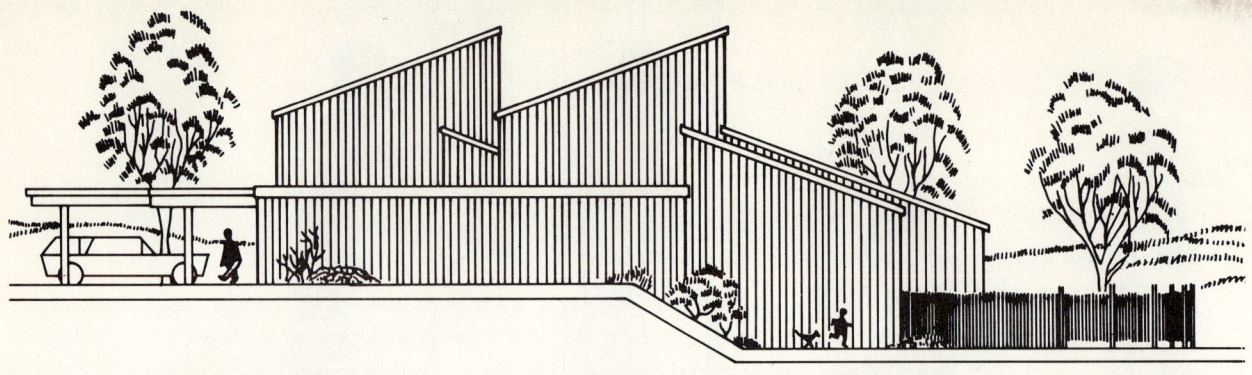

Figure 16-9. Buildings of various shapes and sizes use wood today. The carpenter is needed to apply acquired skill in a building of this nature.

methods, since the open ceiling is used here. Different window sizes call for a carpenter with ability to innovate on the job. Doors are different from the standard types. This calls for an up-to-date carpenter, or one who can adapt to the job to be done.

Figure 16-9 shows a modern design for a condominium building. This uses many carpentry skills. Note the different angles and the wood siding. Even the fence calls for close following of drawings.

Different types of home units can draw upon the carpenter's abilities. The skilled carpenter can adapt to the demands made by newer designs.

New materials and new ways of doing things are being developed. The carpenter today has to be able to adapt to the demands. The carpenter has to keep up to date on new materials and new techniques. New designs will demand an even more adaptable person in this trade in the future.

A person interested in doing something new and different can surely find it in carpentry.

• BUILDING CODES AND ZONING PROVISIONS •

Building Codes

Building codes are laws. They are written to make sure buildings are properly constructed. They are for the benefit of the buyer. They also benefit all the people in a community. If an expensive house is built next to a very inexpensive one, it lowers the property value of the expensive house. Codes are rules which direct people who build homes. They say what can and cannot be done with a particular piece of land. Some land is hard to build on. It may have special surface problems. There may be mines underneath. There are all kinds of things which should be looked into before building.

If a single-family house is to be built, the building codes determine the location, materials, and type of construction. Codes are written for the protection of individuals and the community. In areas where there are no codes, there have been fires, collapsed roofs, and damage from storms.

Figure 16-11 shows a building permit. This is one way the community has to check on building. You have to obtain a permit to add on to a house. It is necessary to get a permit if you build a new house. Figure 16-10 shows the permit for a tool shed. One reason for requiring a building permit is to let the tax assessor know so that the property value can be changed.

Note in Figure 16-11 how the application for a building permit is filled out. Note also the possibilities to be checked off. The town board is required to sign, since they are responsible to the community for what is built and where.

In Figure 16-10, look at the number of inspections required before the building is approved. Each inspection is made by the proper inspector. This way the person who buys the house or property is protected.

The certificate of occupancy is shown in Figure 16-12. This is required before a person can move into the house. The certificate is given with the owner in mind. It means the house has been checked for safety hazards before anyone is allowed to live in it.

Community Planning and Zoning

Zoning laws or codes are designed to regulate areas to be used for building. Different types of buildings can be placed in different areas for the benefit of everyone. Shopping areas are needed near where people live. Working areas or zones are needed to supply work for people. In most cases, we like to keep living and working areas separated. This is because of the nature of the two types of building.

NOTICE
INSTRUCTIONS AND REQUIRED INSPECTIONS

A. Builder's name, phone number, property house number, and building permit number must be displayed on all construction or building projects.

B. A reasonable means of ingress must be provided to each structure and each floor. (Planks or other; and ladders or stairs from floor-to-floor.)

C. The following inspections are mandatory on all construction within the Town of Amherst:
  1. FIRST WALL: When footing is ready to be poured, including trench pour.
  2. PRE-LATH: Before insulating and after ALL electrical, telephone, plumbing, and heating rough work is complete, including metal gas vent and range-hood exhaust duct.
  3. DRAIN TILE: Subfloor and/or footing drain tile, before pouring of concrete floor.
  4. FINAL: Make application for Certificate of Occupancy, file copy of survey with application. Everything must be completed. All work must be finished, including sidewalks and drive aprons.) Each structure must have a house number displayed and visible from the street.

NOTE: An approved set of building plans and plot plans shall be made available to the Building Inspector at the above four inspections.

Call the Building Department 24 hours before you are ready for inspection.
Phone 555-6200, Ext. 42, 43, or 49

Figure 16-10. Notice of the inspections needed when a building permit is required for a house.

Some people don't like to live near a factory with its smoke and noise.

Living near a sewage treatment plant can also be very unpleasant. It is important to make sure certain types of buildings are located near one another and away from living zones.

Some areas are designated as industrial. Others are designated as commercial. Still others are marked for use as residential areas. Residential means homes. Homes may be single houses or apartments. An industrial building cannot be built in a residential area. All over the United States there are regional planning boards. They decide which areas can be used for what. Master plans are made for communities. Master plans designate where various types of buildings are located.

Community plans also include maps and areas outlines as to types of buildings. Parks are also designated in a community plan. Streets are given names, and maps are drawn for developments. A development means land to be developed for housing or other use. Figure 16-13 shows a typical plan for a residential development. Note how the streets are laid out. They are not straight rows. This has a tendency to slow down traffic. The safety of children is important in a residential area. It is best to have residential areas off main traveled roads or streets.

Overbuilding

Where there has been residential, commercial, and industrial overbuilding, the utilities have not been able to keep up. The additional activity puts a strain on the existing facilities. That is another reason for having a community plan.

For example, there can be problems with sewage. Plants may not be big enough to handle the extra water and effluent. Storm sewers may not be available or may not be able to handle the extra water. Water can accumulate quickly if there is a large paved surface. This water has to be drained fast during a rainstorm. If it is not, flooding of streets and houses causes damage. Some areas are just too flat to drain properly. It takes a lot of money to build sewers.

Sewers are of two types. The sanitary sewer takes the fluid and solids from toilets and garbage disposals. This is processed through a plant before the water is returned to a nearby river or creek. The storm sewer is usually much larger in diameter than the sanitary sewer. It has to take large volumes of water. The water is dumped into a river or creek without processing.

Local communities can have a hand in controlling their growth and problems. They can form commu-

The Carpenter and the Construction Industry **469**

APPLICATION FOR BUILDING PERMIT
Town of Amherst, Erie County, N.Y.

Account No. **10-95-130**
Application No. **2**
WEEK OF **10-25-8X**
Permit No. **1356** Date **10-27** 19 **8X**
Applied For **10-18** 19 **8X**

APPLICATION IS HEREBY MADE FOR PERMISSION TO
- [X] Erect
- [ ] Remodel
- [ ] Alter
- [ ] Extend

- [X] Frame
- [ ] Brick
- [ ] " Veneer
- [ ] Stone

- [ ] Concrete Blk.
- [ ] " Reinforced
- [ ] Vinyl or Plastic
- [ ] Steel

STRUCTURE

TO BE USED AS A
- [ ] Single Dwelling
- [ ] Dbl. Dwelling
- [ ] Apartment
- [ ] Add. to S.D.

- [ ] Prvt. Garage
- [ ] Store Bldg.
- [ ] Office Bldg.
- [X] Shed **TOOL**

- [ ] Tank
- [ ] Pub. Garage
- [ ] Service Sta.
- [ ] Swim. Pool

- [ ] Sign
- [ ] Street Sidewalk Conc.
- [ ] Parking Area

Size of Completed
- [X] Building
- [ ] Swimming Pool
- [ ] Sign

15 ft. wide **10** ft. long **8** ft. high _____ diam. if round
1 stories **150'** habitable area _____ ground area _____ sign face area

Building will be located on the (REAR, FRONT) of Lot No. **57** M.C. No. **2259** House No. **425**
NE**S**W side of **BRIANHILL** street, beginning **85** feet from **REAR LOT LINE**
What other buildings, if any, are located on same lot? **S.D.**
The estimated cost of Structure exclusive of land is $ **500.—**
How many families will occupy entire building when completed? _____ SFHA _____ Zoning **R3**
Restrictions _____

Site Plan # _____ Date approved _____ Variances granted _____

Name of building contractor **PERMA-STONE OF BFL.** Address **10157 MAIN ST.**
Name of plumbing contractor _____ Address _____
Name of Elec. Cont. _____ Address _____
Name of Heating Cont. _____ Address _____

I, the undersigned have been advised as to the requirements of the Workmen's Compensation law, and declare that, (check the following).
  A. [X] I have filed the required proof, as affirmed by my Insurance carrier.
  B. [ ] I have no people working directly for me, therefore I require no Workmen's Compensation.
Should there be any change in my status during the exercise of this permit, I will so advise the Building Dept. and immediately comply with all requirements.
  The undersigned has submitted plans, specifications and a plot plan in duplicate which are hereto attached, incorporated into and made a part of this application.
  In consideration of the granting of the permit hereby petitioned for, the undersigned hereby agrees that if such permit is granted he will comply with the terms thereof, the Laws of the State of New York, the Ordinances of the Town of Amherst, and the Regulations of the various departments of the Town, County of Erie, and the State of New York; that he will preserve the established building line; request all necessary inspections & authorize & provide the means of entry to the premises & building to the Building Inspector, and that he will not use or permit to be used the structure or structures covered by the permit until sanitary facilities are completely furnished.
  The undersigned hereby certifies that all of the information in this petition is correct and true.

ITEM	FEE
Trees on Town Hwys. _@_	
San. Sewer Dist. # _ Trib. to _	
Water Line Size _ # BR _	
SWDD # _	
MIN.	$5.—
Cubage	
TOTAL FEE	$5.—

JOHN DOE
Record Owner
425 BRIANHILL
Address
WILLIAMSVILLE, N.Y., 1400

John Doe
Owners or Agents Signature
Subscribed and sworn to before me this **18**
day of **18 OCT**, 19 **8X**
Richard Delaney
Notary Public, Erie County, New York

I do certify that I have examined the foregoing petition and building plans and plot plan and that they conform to Ordinances of Town of Amherst.

S. W. Zenty
Building Commissioner

Receipt is hereby acknowledged of the sum of $ **5.—**, being the permit fee established by the Town Board of Town of Amherst, N.Y.

John Shearer
Town Clerk

This permit shall expire **April 27**, 19 **8X** if building has not commenced.
ORIGINAL

470 Carpentry Fundamentals

TOWN OF AMHERST
CERTIFICATE OF OCCUPANCY

Date: June 25, 19XX   No. 1800

This Certifies that the building located at and known as
125 Briarhill
(NO.)   (STREET)

Sub Lot No. 57   Map Cover No. 2259

and used as a single-family dwelling with private garage
(KIND OF OCCUPANCY)

has been inspected and the use thereof conforms to the Amherst Zoning Ordinance, and other controlling laws and is hereby:

( ) Approved

(x) Approved subject to the planting of the required number of trees not later than Oct. 15, 19XX.

Any alteration of the property or change in the use voids this certificate and a new certificate will be required.

COMMISSIONER OF BUILDING

Figure 16-12. Certificate of Occupancy. *(Town of Amherst, NY)*

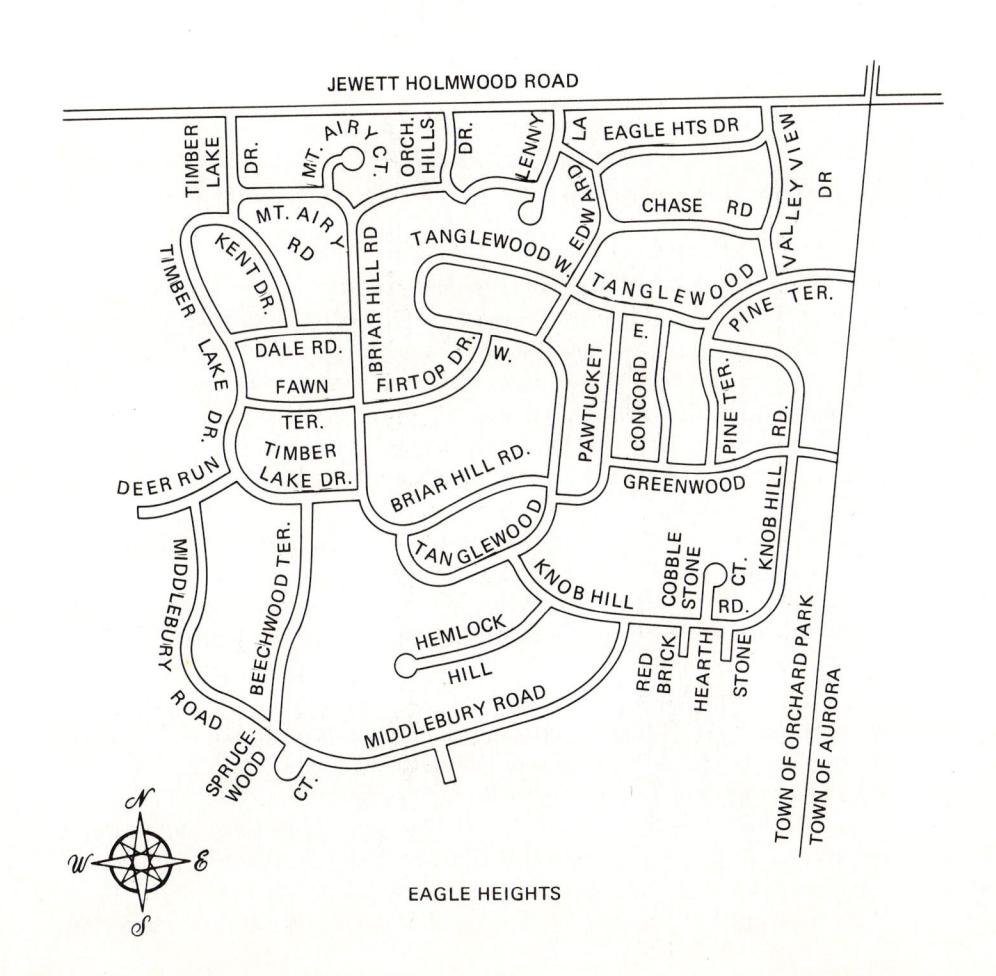

Figure 16-13. Map of a planned subdivision development.

The Carpenter and the Construction Industry   **471**

Figure 16-14. Poured concrete houses. Example of an early type of manufactured house. *(Universal Form Clamp)*

nity planning boards. These boards enforce zoning requirements, and the development of the community can progress smoothly.

• TRENDS AND EFFECTS •

Manufactured Housing

Some of the early attempts at manufacturing housing are shown in Figure 16-14. Here are a number of houses that look alike. They have been made one after the other, as in a factory setup. They are made of poured concrete. The forms were moved from one place to the next and concrete was poured to make a complete house. As you can see, the housing looks rather dull. It would be hard to find your house if you didn't know the house number.

Factory-produced buildings are relatively recent. They are made in a number of sizes and shapes. Some of them are used as office buildings, as in Figure 16-15. Large sections of the building are made in a factory. They are shipped to the construction site. Here they are bolted together. In this case, the parts make a very interesting building.

One of the advantages of this type of building is the minimum of waste and lost time in its manufacture. Construction workers do not have to move from one building to another, but can do the same job day after day. They can work inside. After a while the worker becomes very skilled at the job. Little material is wasted. Such things as plumbing, floors, walls, and electric facilities are included in the package.

Figure 16-16 shows another type of commercial building. This bank was built in sections inside a factory and then assembled on site. Everything is measured closely. This means little time is wasted on the job. The building can be put together in a short time if everything fits. It is very important that the foundation and the water and plumbing lines are already in place in the floor.

Types of Factory-Produced Buildings

There are two types of factory-produced buildings. *Modular* buildings are constructed of modules. The module is completely made at the factory. A module is a part of a building, such as a wall or a room. The modular technique is very efficient. All it takes at the site is a crane to place the module where it belongs. Then the unit is bolted together. This type of construction can be used on hotels and motels. Housing of this type is very practical. It is used as a means of making dormitories for colleges.

In the other type of factory-produced building,

472 Carpentry Fundamentals

Figure 16-15. Office building made in a factory and assembled at the site. *(Butler)*

Figure 16-16. Commercial building made in a factory and assembled on site. *(Butler)*

The Carpenter and the Construction Industry **473**

Figure 16-17. Assembly line for premanufactured apartment house units. *(Western Wood Products)*

only panels are constructed inside. The panels are assembled and erected at the site.

Some companies specialize in factory-produced buildings. Some are specialists in commercial and industrial buildings. Others are specialists in houses.

Some building codes will not allow factory-made housing. It is up to the local community whether or not this type of construction is allowed. In some cases the low-cost construction advantage is lost. The community can have code restrictions that make it expensive to put up such a house or building. In some cases this is good for the community. The type of construction should blend in with the community. If not, the manufactured house can become part of a very big slum in a short time.

A builder who does not adhere to the zoning and building codes in a locality may have to tear down the building. With zoning and building codes, a building permit system is usually used. The permit system enables the local government to monitor construction. This will make sure the type of building fits into the community plan.

Premanufactured Apartments

A combination of plant-built cores and precut woods is a key to volume production. This is especially true with apartment houses. Figure 16-17 shows a well-thought-out utility core production line. Components are fed into the final assembly, where they are integrated into serviceable units. The units get to the job site in a hurry. Production scheduling can then be easily controlled.

Figure 16-18 shows that wood floor joists, precut in the plant, go together quickly. The frame is equipped with wheels. It is then turned over. Sub- and finish flooring is applied. Then the floor is rolled to the main production line.

In Figure 16-19 you see wall panels framed with 2×4s. The bathroom wall uses 2×6s to accommodate the standpipe. For extra strength on the main floor units, special 3×4 studs are used. This size replaces the standard 2×6. This dimension will carry the added weight yet still fit precisely with the upper floor cores.

After the drywall is applied to one side (see Figure

474   Carpentry Fundamentals

Figure 16-18. Lifting a floor section and turning it over for sub- and finish flooring. *(Western Wood Products)*

Figure 16-19. Framing a wall panel. *(Western Wood Products)*

The Carpenter and the Construction Industry **475**

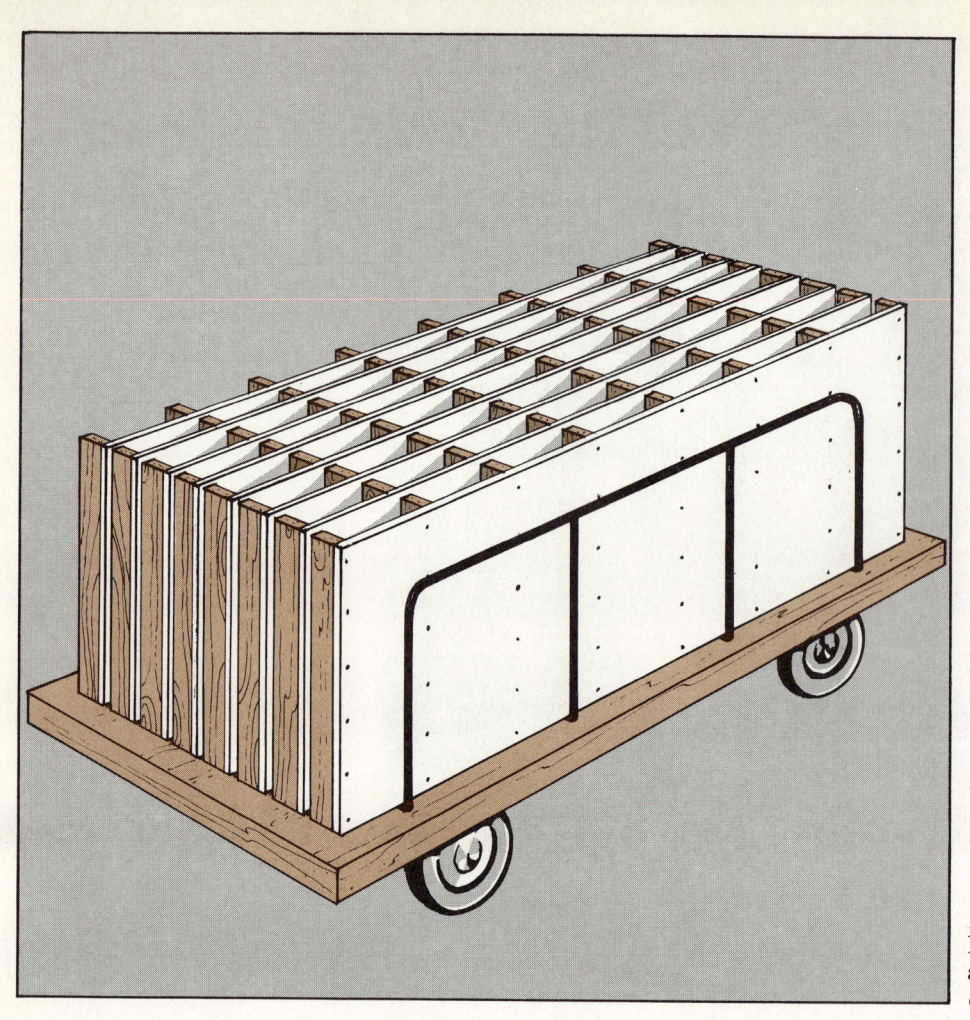

Figure 16-20. Storing panels after drywall installation. *(Western Wood Products)*

16-20), the wall panels are stored. They are stored on wheeled carts. This means they are ready to move to the production line when they are needed. The utility core is beginning to take shape. Carpenters nail the walls in place. Plumbing fixtures are installed before the end wall goes up.

Ceiling panels complete the framing. Next, the plumbing and wiring is done and the furnace and water heater are installed. In the final step, the cabinets and appliances are installed. The interior walls are finished next. See Figure 16-21. Each unit is hooked up and tested before it leaves the plant.

The units are wrapped and shipped to the job site. See Figure 16-22. The cores are stacked on the foundation and hooked up to utilities. Local contractors add the interior and exterior walls. They use either panelized wood framing or conventional construction. This conventional framing is precut to fit the house or apartment unit.

Designs can vary. They can be adapted to meet the floor plan and outside requirements. See Figure 16-23 for an example of manufactured apartments.

Manufactured Homes

One- and two-story homes can be constructed and completely finished in the plant. Interior finishing is completed on outdoor "stations," then the homes are moved onto their concrete foundations. It requires unusually sturdy construction to move a home of this size. A web floor system of 2×4s and plywood is used for strength. See Figure 16-24. Two homes are built simultaneously, with floors built over pits so that workers can install heating ducts, plumbing, and electrical facilities from below.

While the floor system is under construction, carpenters are building wall panels from precut wood. See Figure 16-25. Wood-frame construction is used throughout the house.

At the building site, a giant machine places the foundation. See Figure 16-26. Then, similar equipment moves the finished house from the factory onto the foundation. See Figure 16-27. This is done in just 1 hour and 10 minutes.

Although in-plant production is standardized to

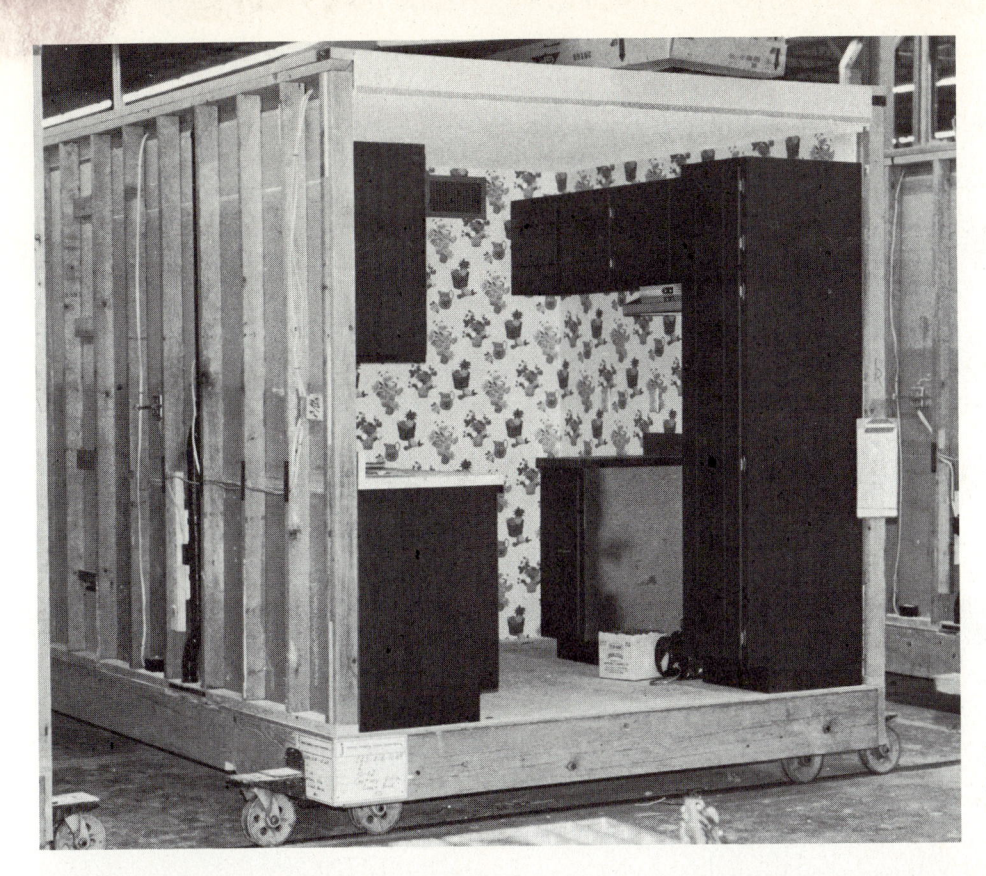

Figure 16-21. Ceiling panels are added to complete the framing. *(Western Wood Products)*

Figure 16-22. Units are wrapped before being shipped to the job site. *(Western Wood Products)*

The Carpenter and the Construction Industry **477**

Figure 16-23. Variations in design are possible in the premanufactured apartment house. *(Western Wood Products)*

Figure 16-24. Floor joists being laid for a manufactured house. *(Western Wood Products)*

478   Carpentry Fundamentals

Figure 16-25. Wall panels attached to the floor system. (*Western Wood Products*)

Figure 16-26. A machine laying the foundation. *(Western Wood Products)*

Figure 16-27. A machine moves the house over the foundation. *(Western Wood Products)*

cut costs, the homes avoid the manufactured look. See Figure 16-28. The wood-frame construction makes it easy to adapt the designs.

The advantages of this type of building are lower cost of materials, less waste, and lower-cost worker skills. It is a great advantage when it comes to maintaining production schedules. The weather does not play too much of a role in production schedules when the house is built indoors.

This type of construction is changing the way homes are being built. It requires a carpenter who can adapt to the changes.

Figure 16-28. With wood frame construction it is easy to change the designs so that each house looks different. *(Western Wood Products)*

• HOW DO I MEASURE UP? •

Most people want to know what it is about them that makes them able to do something. This is also the case with a trade. The trade this time is carpentry. How will I do in this trade? This is the question most often asked. One way to find out is to ask yourself some questions, and don't cheat on the answers.

Skills

1. Can I hit a nail with a hammer without hitting my fingers?
2. Do I know how to hold a hammer properly?
3. Do I know which part of the saw actually cuts wood? Can I use it properly?
4. Can I climb ladders properly?

Interests

1. Am I interested in working with wood?
2. Do I like to handle wood?
3. Do I like to read plans and see them into final form?
4. Am I interested in doing a good job?
5. Am I interested in building things?

Attitudes

1. Do I have a good attitude toward work?
2. Is it easy for me to get up in the morning?
3. Am I pushing myself to go to work?
4. Do I like to work once I get to the job?
5. Do I like the breaks better than doing the job?
6. Do I need to be told to get to work on time?
7. How often must I be reminded to do something?
8. Do I like to work for the money or for the thrill of doing a good job?
9. Can I stand to work at heights?
10. Can I physically work for 8 to 10 hours a day outdoors in summer? In winter?
11. Do I like to take orders?
12. Am I open to suggestions by others?
13. Can I work with others easily?

Knowledge

1. Do I know enough to handle a hammer correctly?
2. Can I use a saw properly?
3. Do I know the standard sizes of lumber?
4. Do I know the standard sizes of plywood?
5. Do I know the standard sizes of windows?
6. Do I know the standard sizes of doors?
7. Do I know how many shingles there are in a square?
8. Do I know how to cut a rafter?
9. Do I know how to make a dormer?
10. Do I know how to frame a house?
11. Do I know what a carpenter's job actually is?
12. Do I like to work and solve problems I know nothing about?

Long-Range Goals

1. What do I want to do 5 years from now?
2. Where do I want to be 10 years from now?
3. Can I become a general contractor?

• WORDS CARPENTERS USE •

residential building	apprenticeship	union	zoning laws	factory-produced houses
commercial building	skilled worker	benefits	community planning	modular houses
construction crafts	on-the-job training	building codes	manufactured housing	
	contractor	building permits		
	team	certificate of occupancy		

The Carpenter and the Construction Industry

• STUDY QUESTIONS •

16–1. What does the future look like for a carpenter?
16–2. What are the opportunities for a carpenter?
16–3. What is an apprentice?
16–4. Are carpenters unionized?
16–5. When are carpenters usually nonunion?
16–6. What is a premanufactured house?
16–7. What is a building code?
16–8. How do you obtain a building permit?
16–9. What does the term *community planning* mean?
16–10. What is meant by *overbuilding*?
16–11. What happens if a builder does not follow local building codes?
16–12. What is the advantage of making a house in a plant?

17 CONSTRUCTION FOR SOLAR HEATING

There is a great deal of interest in solar heating of homes, with the increase in the price of heating oil and natural gas. However, the use of the sun's energy is not inexpensive at this time. It takes some rather expensive devices to make sure the heat is stored and circulated to the proper location when it is needed. The main idea is to change existing architecture to meet the needs of added heating elements on the roof. In some cases the roof line is completely changed to meet the needs of the solar heating. The house is located so as to take advantage of the sun most of the day.

Only a couple of locations in the United States are not suitable for solar heating. There are some spots in the state of Washington and the Buffalo, New York, area. Here the sun doesn't shine long enough to make solar heat a viable alternative.

• TECH HOUSE •

NASA (National Aeronautics and Space Administration) has done some experimental work in the field of solar heating for homes. NASA built the Tech House in Virginia. See Figure 17-1. Preliminary experience indicates that solar energy can provide 70 to 80 percent of the house's annual heating requirements. Eighteen 3×8-foot solar collectors (sixteen for home heating, two for water heating) are mounted on the south-facing roof of the house. This provides the maximum exposure to the winter sun in this location. The most efficient solar collector angle depends upon the latitude where the house is located.

Solar Collectors

The glass of the solar collectors acts as a one-way valve, admitting light and other solar radiation but trapping heat reflected from the interior. This so-called greenhouse effect is commonly experienced when a closed car is left in sunlight.

Water passing through the collectors in Tech House carries the heat to the building heating system. A heat exchanger transfers the heat from the water to ducts that distribute the warm air throughout the house. See Figure 17-2.

If heat is not required, the hot water from the collectors is diverted into a 1900-gallon underground thermal storage tank. Hot water is circulated from the tank to the heat exchanger at night and during cloudy periods to provide heat. The tank stores enough heat for as long as 3½ days of overcast skies.

If a longer stretch of overcast days occurs, the water in the storage tank may drop below 90°F. In this case, the water is used to supply an electric heat pump to heat the house. If skies are overcast for 5 to 7 days and the temperature in the tank drops below 55°F, two deep wells with water at about 55°F are used to supply water to the heat pump. Supplying the heat pump with preheated water reduces the amount of electricity needed to heat the house. A heat pump extracts heat from the outside air or a reservoir of water and pumps the heat into the house.

Air Conditioning

For air conditioning, the heat pump takes heat from the house and ejects the heat into the air or a reservoir of cool water. The house is air-conditioned by the heat pump but uses only about half the electricity for air conditioning that is used in comparably sized conventional homes. The reduction in the air-conditioning load is due to better insulation in the walls and ceiling, shielding windows

Figure 17-1. Home constructed to use solar heating. *(NASA)*

from direct summer sunlight, area temperature control, large attic louvers, the availability of cool water for the heat pump, and night radiators.

The solar collector system used for heating the house is cut off during the summer. The main heat pump transfers heat from the house to the well water. If the heat load exceeds the capacity of the main pump, the auxiliary heat pump, using cool water from the storage tank, helps to cool the house.

At night the water in the storage tank is circulated through night radiators facing north on the garage roof. The radiators radiate the heat from the water to the outside air, reducing the temperature of the water in the tank. See Figure 17-3.

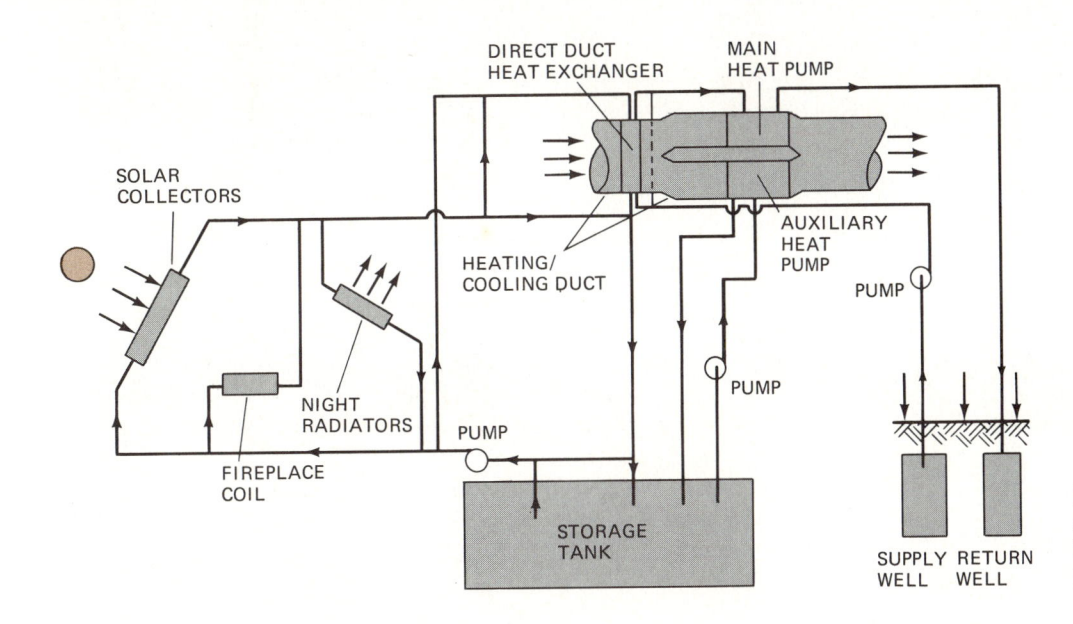

Figure 17-2. Simplified diagram of solar heating and cooling system. *(NASA)*

Figure 17-3. Evaporator-type coils for dissipating heat in the house in hot weather. The water in the house is used to take on heat from what would otherwise be the heating coils. The water is pumped outside to these panels on the garage roof facing north at night when it is cool. The accumulated heat is then dissipated by these panels. *(NASA)*

Construction for Solar Heating **485**

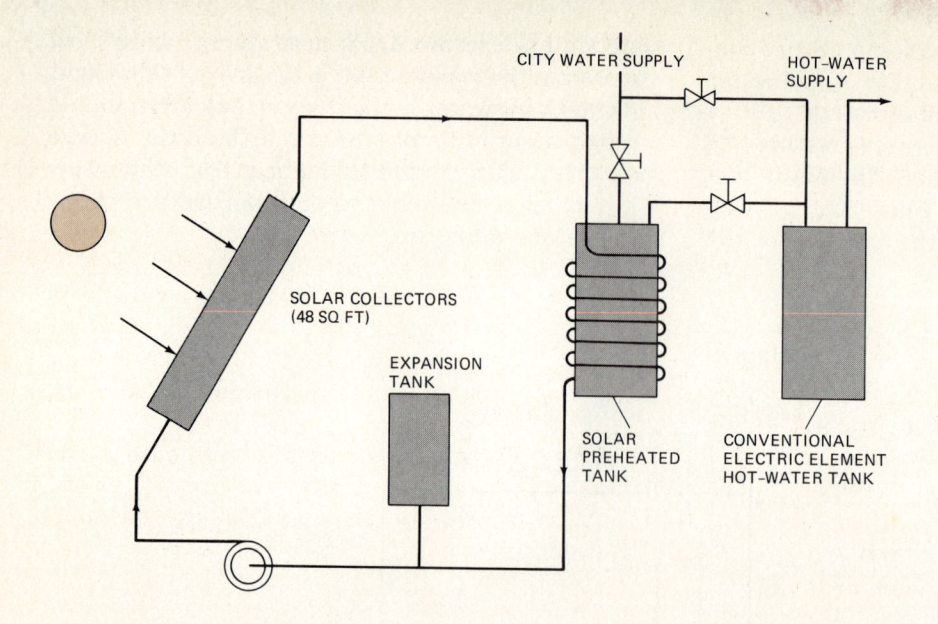

Figure 17-4. Solar hot water system for the house. *(NASA)*

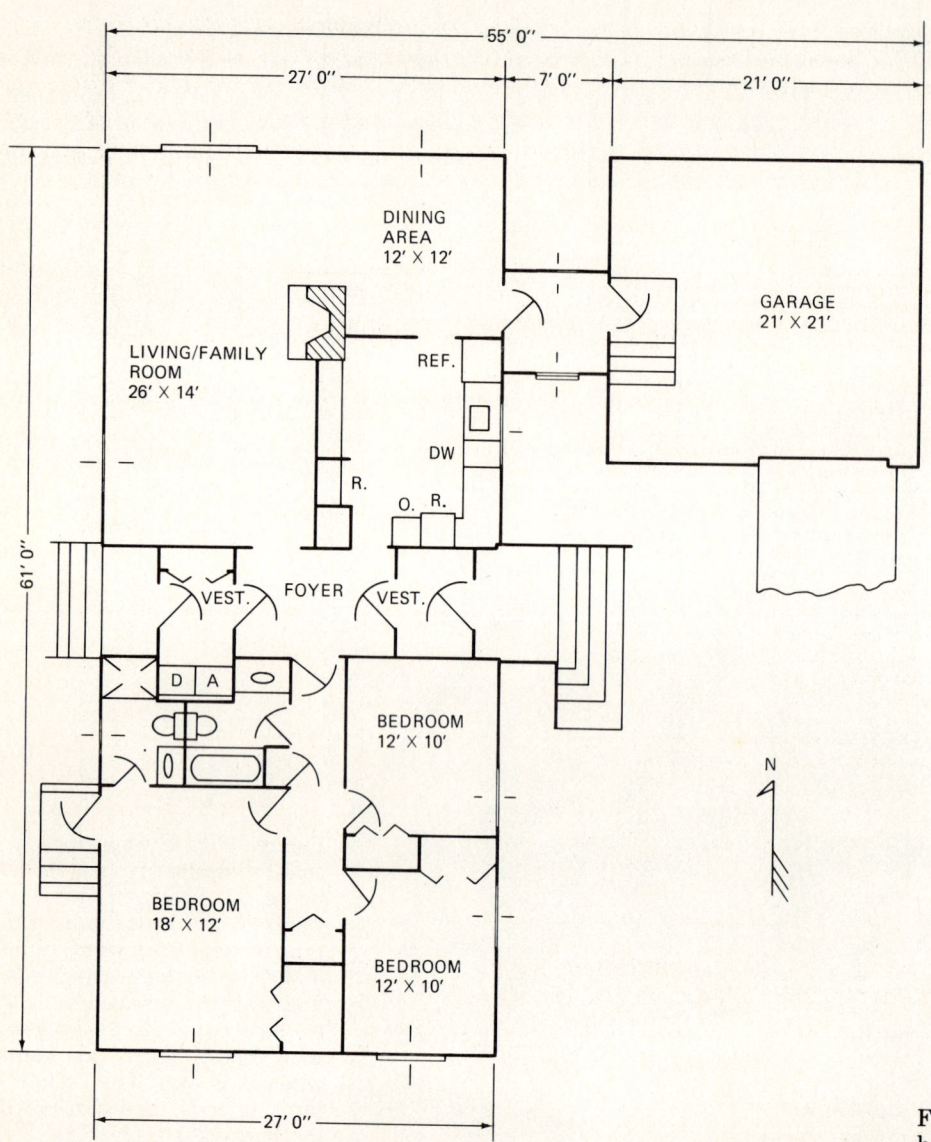

Figure 17-5. Floor plan of the house. *(NASA)*

486 Carpentry Fundamentals

Hot Water

Hot water for the house is also furnished by solar collectors. If the collectors are not sufficient, then an electric element heats the conventional tank to the desired temperature. See Figure 17-4.

Fireplace

The fireplace in Tech House is redesigned, since fireplaces contribute little heat to homes. Note the location of the fireplace in the floor plan. See Figure 17-5. Although people near the fireplace feel radiant heat, most of the heat, which is convective, from an ordinary fireplace sweeps up the chimney. See Figure 17-6.

This house has a redesigned fireplace. It eliminates the waste of room heat and directs much more of the fireplace heat into the home. The location of the fireplace plays a part in this. The carpenter will have to learn the various techniques for conserving energy and build accordingly.

A duct system under the floor of the house supplies outside air to the open hearth for combustion. Low ducts on each side of the fireplace draw room air in. The room air circulates through a double-walled firebox which raises the temperature of the air. The additionally heated room air then returns to the living room through grills above the fireplace.

The major fireplace contributor to house heating, however, is the special fireplace grate which is part of the main heating system. See Figure 17-7. The grate is a coil through which water is circulated. The heated water can be either delivered to the air duct heat exchanger to help heat the house or returned to the underground hot-water storage tank. All these features are expected to increase fireplace efficiency from the usual 10 percent to more than 50 percent.

The water grate system used in this house is available commercially. It can be installed in existing fireplaces, and is adaptable to hot-water, forced air, or electrically heated systems.

Windows

Direct sunlight streaming through the windows of a house heats whatever objects it strikes. These in turn give off heat, which is trapped for a time inside the house. This house uses the so-called greenhouse effect to best advantage. A large window area on the south side permits the maximum amount

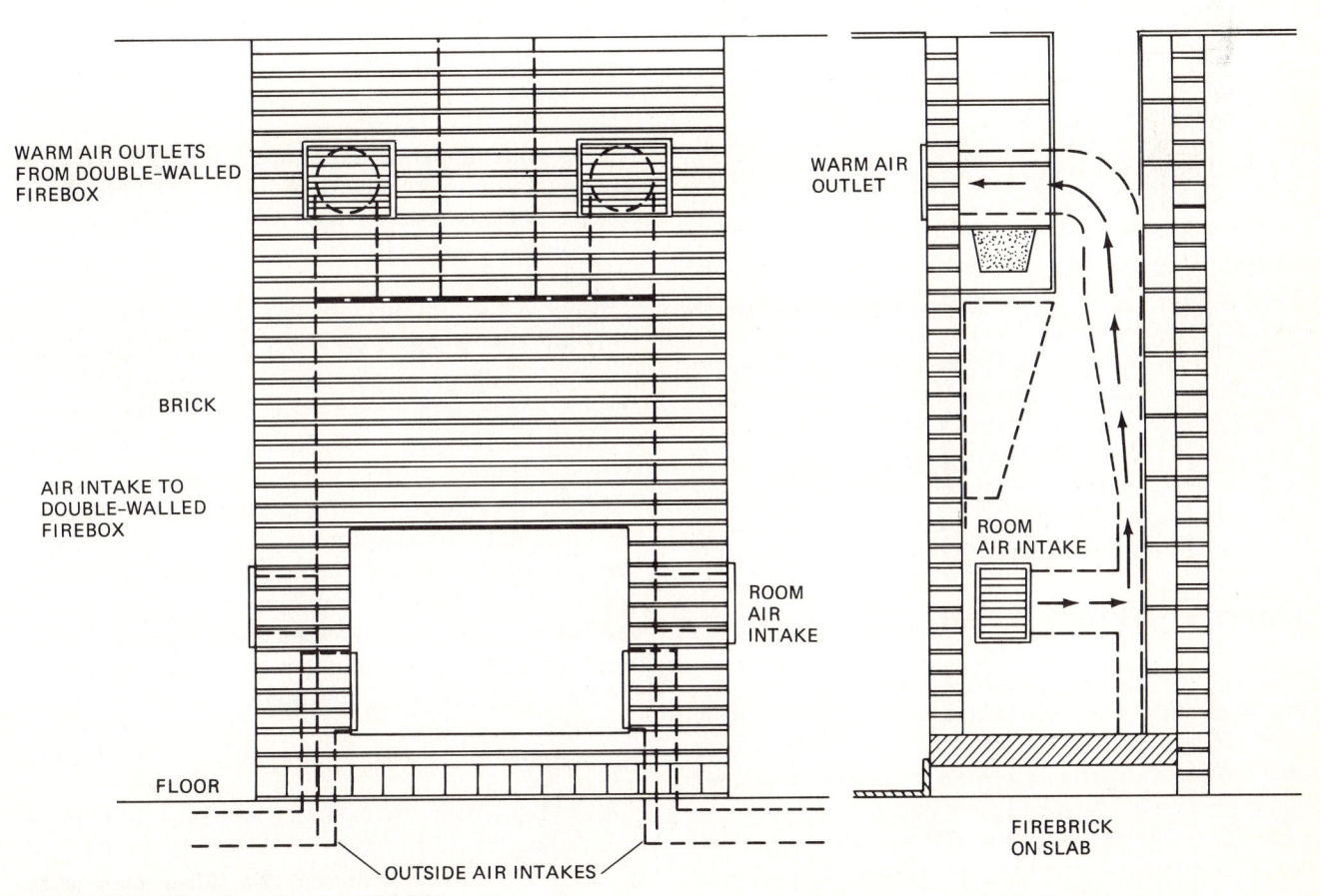

Figure 17-6. Fireplace design. *(NASA)*

Construction for Solar Heating **487**

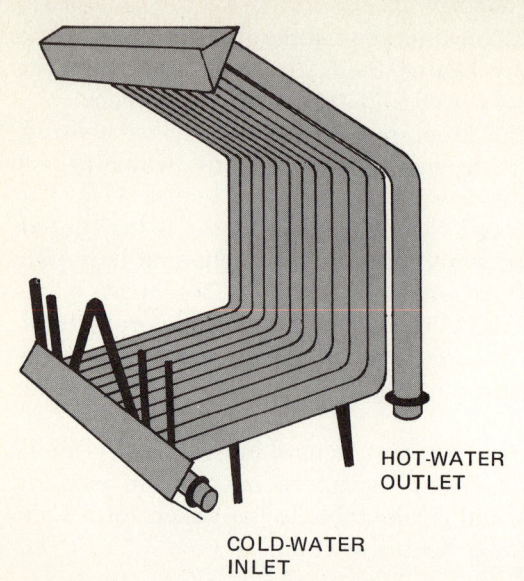

Figure 17-7. Fireplace grate. *(NASA)*

of direct heat from sunlight to enter the house in winter. But a larger-than-normal roof overhang protects the south-facing windows from direct sunlight during the summer. In addition, in the summertime, windows facing west are covered with a reflective plastic sheeting that obstructs the sun's rays. All windows are double-paned for insulation, and a plastic material that has low thermal conductivity separates the framing that holds the panes.

Shutters

Tests have shown that exterior shutters on windows can cut heat loss through windows on cold winter nights by about 65 percent. The shutters can also be rolled down to block direct sunlight during the summer. They can be slightly opened or adjusted to admit light and air.

The shutters have a number of other advantages. Since they cannot be opened from the outside, they contribute to house security. They can protect windows from flying objects whipped up by violent storms. They can reduce the volume of outside noise heard inside. The shutters can be raised or lowered in seconds by a hand crank or electric motor.

Doors and Entry Vestibules

Entry vestibules at the front and rear of the house act as air locks to reduce the quantity of warm or cool air lost when people enter or leave the house. See Figure 17-8. The exterior doors have hot-dipped galvanized steel surfaces for resistance to rust and corrosion and are guaranteed never to warp. Polyurethane foam between the inside and outside metal facings prevents conduction of heat or cold. These doors were illustrated and discussed in detail in Chapter 9 in this book.

Magnetic weatherstripping further reduces heat loss. When the door is closed, an adjustable sill on the door bottom automatically lowers to cut off airflow beneath the door. Heat loss through the house door is about 65 percent less than it would be through a conventional wooden door.

Attic Louvers

Heat developing in attic space can reach 160°F when summer temperatures are 100°F. This adds to the cost of air conditioning. Large ventilation louvers in the attic of the house keep attic temperatures near outside temperatures. Unlike attic fans, louvers use no electricity.

Insulation

Urea-tripolymer foam insulates the house ceiling, exterior walls, and selected interior walls. Approximately 7½ inches of the foam is in the ceiling. About 5½ inches is in exterior walls. See Figure 17-9. In most cases, in standard home construction, 6 inches of insulation is used in the ceiling area. Exterior walls usually get 3½ inches of Fiberglas insulation in standard home construction. In the solar-heated Tech House, 5½ inches are used. In some homes under construction at present, the usual 2×4 studs with 3⅝-inch stud width are being replaced with 2×6 studs, and the 5⅝-inch width is used for additional insulation thickness.

This solar-heated house that NASA built appears to have about 30 percent less heat loss through the ceiling and 45 percent less heat loss through the walls than a house with conventional Fiberglas insulation.

Tripolymer is also nonflammable, forming a charred crust when exposed to flame and extinguishing itself as soon as the flame is removed. Flames cannot advance on tripolymer beyond the point of ignition. Tripolymer is also nontoxic, odor-free, and rodent-resistant.

Tripolymer is cold-setting and nonexpanding, and does not settle or lose its insulating ability. It fills every crack and cranny, blocking noise as well as heat and cold. It flows around pipes, wires, and other obstructions. It can be applied through openings as small as 1½ inches in diameter, making it suitable for insulating existing buildings.

Skylight

A skylight over the hallway serves a number of useful functions. It reduces the need for artifical lighting. Sunlight streaming through the skylight on cold days helps to heat the hallway. The skylight

Figure 17-8. The vestibule has double doors to keep cool air or hot air inside when the door is opened. *(NASA)*

Figure 17-9. Applying urea-tripolymer foam insulation. *(NASA)*

Construction for Solar Heating **489**

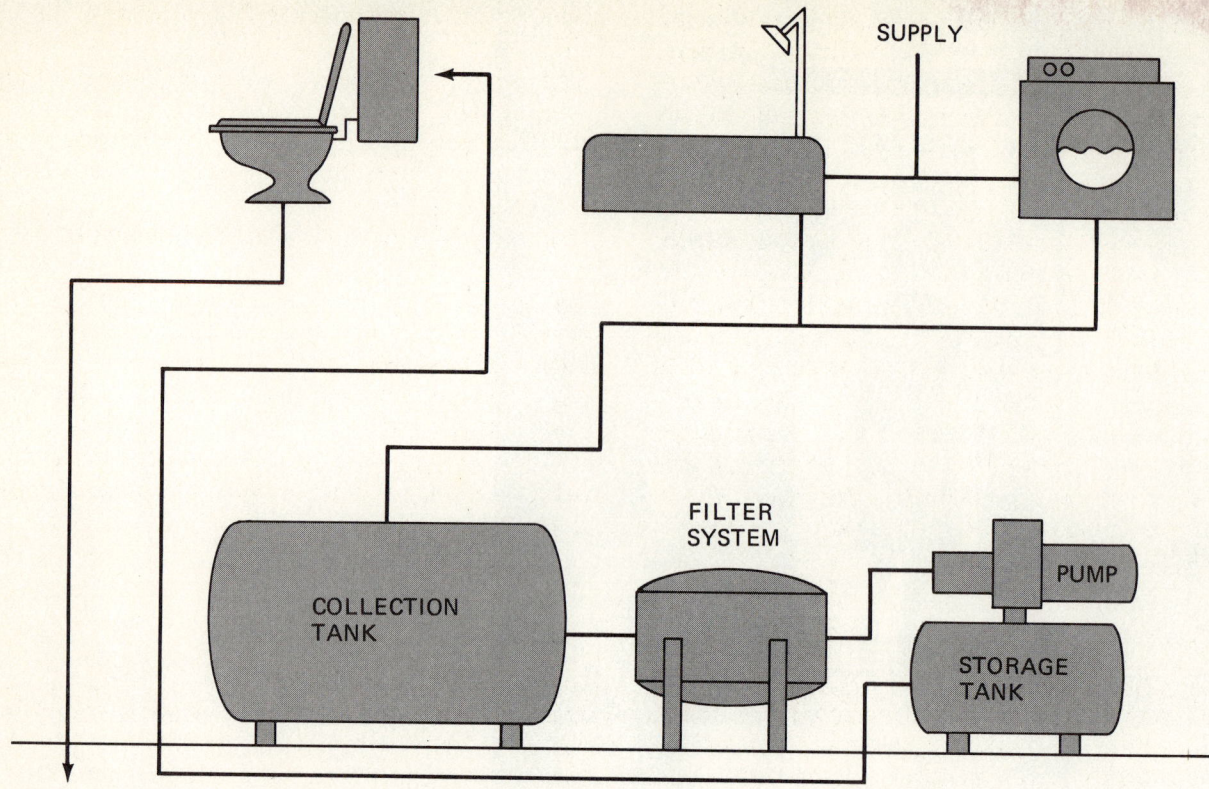

Figure 17-10. Water re-use system. *(NASA)*

can be opened to allow heated air to escape on warm fall or spring days. To block heating from direct sunlight on hot summer days, a reflective plastic sheet is stretched across the bottom of the skylight.

Water Conservation

Another thing the carpenter will observe when constructing this type of home is the water system. Water from the bathtub, washing machine, and shower is collected in a tank. Here it is chlorinated and filtered, and used to supply water for the toilet tanks. This water recycling, together with smaller (low-profile) toilet tanks and water-saving shower heads, reduces water consumption to half of what is normally used. It also trims the sewage load, an advantage for areas plagued by overloaded sewer systems or septic fields. See Figure 17-10.

• NEED FOR SOLAR-HEATED HOUSING •

Builders and manufacturers of homes and housing equipment need new methods and materials to survive in a highly competitive market. Probably the most significant opportunity for change in houses over the next few decades will be in energy management. Our homes consume about 20 percent of the energy used in the United States each year—an amount equal to all imported crude oil.

Tech House was designed with the conservation of energy in mind. It makes some changes in basic design to accommodate the roof panels and the water system. This type of construction is entirely possible today with the equipment available. The carpenter is very much a part of the new building demands. Here is an opportunity to keep up with the developments in the field and to make good use of present skills. The installation of these materials can reduce the homeowner's utility bills for heating, cooling, and water. Over a 20-year period the savings could amount to over $20,000.

This house contains only 1500 square feet of enclosed living space. It is extremely functional, with a living room, three bedrooms, fireplace, dining area, kitchen, two bathrooms, and laundry room plus an attached garage.

While the reduction in energy and water consumption represents a considerable saving in utility costs, it is important to note that additional savings can result from fire-resistant construction and solar energy usage. The use of fire-resistant carpets, drapes, furniture covers, insulation, and other materials can lower fire insurance rates, and many states encourage the use of solar energy by providing tax-break incentives for homeowners.

It should be pointed out that energy-conserving homes are most efficient when they are carefully designed to fit specific sites with their particular characteristics of access, orientation to sun and winds, and history of weather conditions. The house illustrated here was used as one method of doing the job. It has some features which can be used anywhere. These should be studied and incorporated wherever they lend themselves to energy conservation. As the energy problem becomes more acute, more efforts to conserve will become apparent and more efficient housing will be demanded.

• BUILDING MODIFICATIONS •

In order to make housing more efficient, it is necessary to make some modifications in present-day carpentry practices. For instance, the following must be done to make room for better and more efficient insulation:

1. Truss rafters are modified to permit stacking of two 6-inch-thick batts of insulation over the wall plate. The truss is hipped by adding vertical members at each end and directly over the 2×6 studs of the outer wall.
2. All outside walls use thicker (2×6) studs on 2-foot centers to accommodate the thicker insulation batts.
3. Ductwork is framed into living space to reduce heat loss as warmed air passes through the ducts to the rooms.
4. Wiring is rerouted along the sole plate and through notches in the 2×6 studs. This leaves the insulation cavity in the wall free of obstructions.
5. Partitions join outside walls without creating a gap in the insulation. Drywall passes between the sole plates and abutting studs of the interior walls. The cavity is fully insulated.
6. Window area is reduced to 8 percent of the living area.
7. Box headers over the door and window openings receive insulation. In present practice, the space is filled with 2×10s or whatever is needed. Figure 17-9 shows how the window header space can be filled with insulation. This reduces the heat loss through the wood that would normally be located here.

Some of the criticisms voiced so far are:

1. Attics are so full of insulation that they are not usable for storage.
2. The 8 percent window area ignores quality of windows, such as double-paned insulating units.
3. Ductwork creates ugly notches in rooms.

These standards have been adopted by Little Rock, Arkansas. They do not, of course, apply to the nation as a whole. However, they do represent a change from the standards now used. There will have to be many changes in many building codes before the wise use of energy-saving materials and construction methods is adopted nationwide. The next decade should be of much interest to carpenters.

• OTHER METHODS OF SOLAR HEAT USE •

A number of methods are being researched for possible use in solar heating applications. One of the most inexpensive ways of obtaining insulation is building underground.

However, one problem with underground living is psychological. People do not like to live where they can't see the sun or outside.

The idea of building underground is not new. The Chinese have done it for centuries. But the problem comes in selling the public the idea. It will take a number of years of research and development before a move is made in this direciton.

Advantages

There are a number of advantages of going underground. By using a subterranean design, the builder can take advantage of the earth's insulative properties.

The ground is slow to react to climatic temperature changes. It is a perfect year-round insulator. There is a relatively constant soil temperature at 30 feet below the surface. This could be ideal for moderate climates, since the temperature would be a constant 68°F.

By building underground it would be possible to use the constant temperature to reduce heating and cooling costs. A substantial energy reduction or savings could be realized in the initial construction also.

Figure 17-11 shows a roof-suspended earth home. This is one of the designs being researched at Texas A&M University. It uses the earth as a building material, and uses the wind, water, vegetation, and the sun to modify the climate.

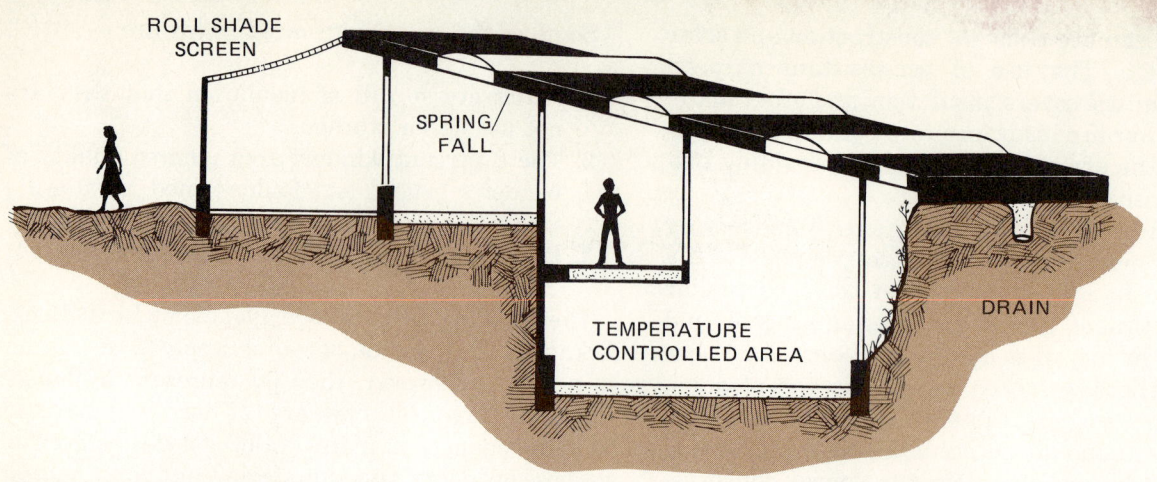

Figure 17-11. Roof-suspended earth home.

Figure 17-12. Another variation of a roof-suspended earth home.

A suitable method of construction uses beams to support the walls. That means structural beams could be stretched across the hole and walls could be suspended from the beams. See Figure 17-12. The inside and outside walls would also be dropped from the beams. That would allow something other than wood to be used for walls, since they would be non-weight-bearing partitions. The roof would be built several feet above the surface of the ground to allow natural lighting through the skylights and provide a view of outside. This would get rid of some of the feeling of living like a mole.

Another design being researched involves tunneling into the side of a hill. See Figure 17-13. This still uses the insulative qualities of the earth and allows a southern exposure wall. This way a conventional front door and windows could be used. Exposing only the southern wall also helps to reduce cooling and heating costs. The steep hills in some areas could be used for this type of housing where it would be impossible to build conventional-type housing.

Many more designs will be forthcoming as the need to conserve energy becomes more apparent. The carpenter will have to work with materials other than wood. The new methods and new materials will demand a carpenter willing to experiment and apply known skills to a rapidly changing field.

492 Carpentry Fundamentals

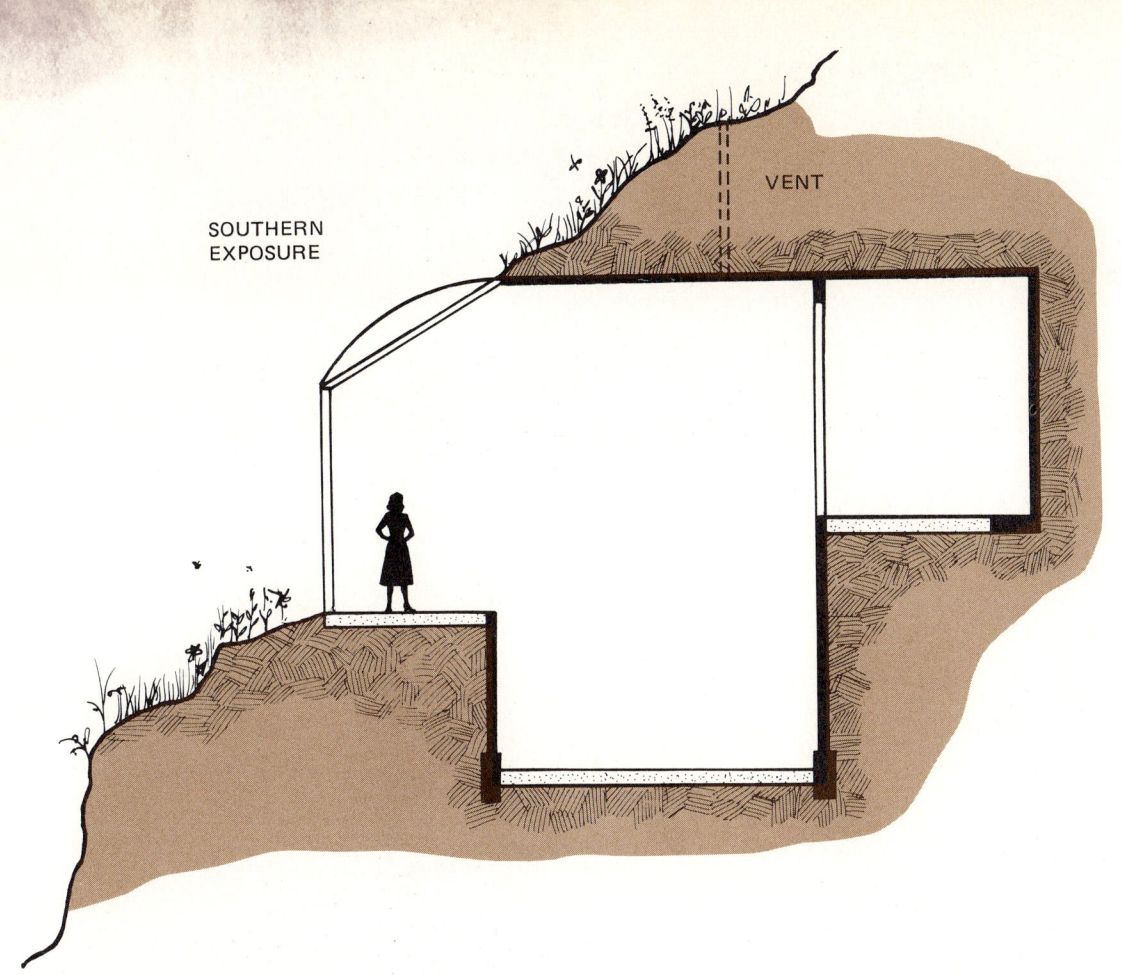

Figure 17-13. Hillside earth home.

• WORDS CARPENTERS USE •

solar collector	grate	skylight	nontoxic	subterranean
convection	shutters	louvres	rodent resistant	suspended roof
duct system	vestibule	urea-tripolymer	energy conservation	

• STUDY QUESTIONS •

17-1 Where are solar-heated houses built?

17-2 How does a solar collector work?

17-3 Explain how the Tech House air-conditioning system works.

17-4 Explain how the fireplace works in Tech House.

17-5 How much can exterior shutters cut heat loss through windows?

17-6 How does the entry vestibule contribute to energy saving in Tech House?

17-7 What temperature does the air in the attic of Tech House reach?

17-8 What is used to insulate Tech House?

17-9 List the functions of a skylight.

17-10 How much energy do our homes consume each year?

17-11 What are the advantages of building a house underground?

17-12 List three types of earth homes.

Construction for Solar Heating   493

GLOSSARY

Access Access refers to the freedom to move to and around a building, or the ease with which a person can obtain admission to a building site.

Acoustical This refers to the ability of tiles on a ceiling to absorb or deaden sound.

Adhesives This is another term for glues.

Aggregate Aggregate refers to sand, gravel, or both in reference to concrete mix.

Alligatoring This occurs when paint cracks and resembles the skin of an alligator.

Anchor bolt An anchor bolt is a steel pin that has a threaded end with a nut and an end with a 90° angle in it. The angled end is pushed into the wet concrete and becomes part of the foundation for anchoring the flooring or sill plates.

Apprenticeship Apprenticeship is the process of working on the job to learn how it is done. An apprenticeship usually lasts four years.

Asphalt shingle This is a composition-type shingle used on roofs. It is made of a saturated felt paper with ground-up pieces of stone embedded and held in place by asphaltum.

Asphalt shingles These are shingles made of asphalt or tar-impregnated paper with a mineral material embedded; they are very fire resistant.

Auxiliary locks Auxiliary locks are placed on exterior doors to prevent burglaries.

Awl An awl is a tool used to mark wood with a scratch mark. It can be used to produce pilot holes for screws.

Awning picture window This type of window has a bottom panel that swings outward; a crank operates the moving window. As the window swings outward, it has a tendency to create an awning effect.

Backsplash A backsplash is the part of a counter top that is vertical and runs along the wall to prevent splashes from marring the wall.

Backsaw This saw is easily recognized since it has a very heavy steel top edge. It has a fine-tooth configuration.

Balloon frame This type of framing is used on two-story buildings. Wall studs rest on the sill. The joists and studs are nailed together, and the end joists are nailed to the wall studs.

Baluster The baluster is that part of the staircase which supports the hand rail or bannister.

Bannister The bannister is that part of the staircase which fits on top of the balusters.

Batten A batten is the narrow piece of wood used to cover a joint.

Batter boards These are boards used to frame in the corners of a proposed building while the layout and excavating work takes place.

Batts Batts are thick pieces of Fiberglas that can be inserted into a wall between the studs to provide insulation.

Bay window Bay windows stick out from the main part of the house. They add to the architectural qualities of a house, and they are used mostly for decoration.

Bearing wall A bearing wall has weight-bearing properties associated with holding up a building's roof or second floor.

Benefits Benefits are the things other than salary which are included in the pay a person receives for doing a job.

Bevel A bevel is a tool that can be adjusted to any angle. It helps make cuts at the number of degrees that is desired. It is a good device for transferring angles from one place to another.

Bifold A bifold is a folding door used to cover a closet. It has two panels that hinge in the middle and fold to allow entrance.

Blistering Blistering refers to the condition that paint presents when air or moisture is trapped underneath and makes bubbles that break into flaky particles and ragged edges.

Blocking Corners and wall intersections are made the same as outside walls. The size and amount of blocking can be reduced. The purpose of blocking is to provide nail surfaces at the corners. These are needed at inside and outside nail surfaces. They are a base for nailing wall covering.

Blockout A form for pouring concrete is blocked out by a frame or other insertion to allow for an opening once the concrete has cured.

Blocks This refers to a type of flooring made of wood. Wide pieces of boards are fastened to the floor, usually in squares and by adhesives.

Bonding This is another word for gluing together wood or plastics and wood.

Bottom or heel cut This refers to the cutout of the rafter end which rests against the plate. The bottom cut is also called the foot or seat cut.

Brace A brace is an inclined piece of lumber applied to a wall or to roof rafters to add strength.

Brace scale A brace scale is a table that is found along the center of the back of the tongue and gives the exact lengths of common braces.

Bridging Bridging is used to keep joists from twisting or bending.

Buck A buck is the same as a blockout.

Builder's level This is a tripod-mounted device that uses optical sighting to make sure that a straight line is sighted and that the reference point is level.

Building codes Building codes are rules and regulations which are formulated in a code by a local housing authority or governing body.

Building permits Most incorporated cities or towns have a series of permits that must be obtained for building. This allows for inspections of the work and for placing the house on the tax rolls.

Built-ins This is a term used to describe the cabinets and other small appliances that are built into the kitchen, bathroom, or family room by a carpenter. They may be custom cabinets or may be made on the site.

Bundle This term refers to the packaging of shingles. A bundle of shingles is usually a handy method for shipment.

Cantilever Overhangs are called cantilevers; they are used for special effects on porches, decks, or balconies.

Carpenter's square This steel tool can be used to check for right angles, to lay out rafters and studs, and to perform any number of measuring jobs.

Carpet strips These are wooden or metal strips with nails or pins sticking out. They are nailed around the perimeter of a room, and the carpet is pulled tight and fastened to the exposed nails.

Carpet tape This is a tape used to seam carpet where it fits together.

Carriage A notched stair frame is called a carriage.

Casement This is a type of window that is hinged to swing outward.

Casing A door casing is that part of the door frame which puts a finish on the opening. It refers to the trim which goes on around the edge of a door opening and also to the trim around a window.

Cathedral ceiling A cathedral ceiling is not flat and parallel with the flooring; it is open and follows the shape of the roof. The open ceiling usually precludes an attic.

Caulk Caulk is any type of material used to close holes or keep out the weather. It is used to seal the walls and windows or door frames. It is usually made of putty or some type of plastic material, and it is flexible and applied with a caulking gun.

Cellulose fiber Insulation material made from cellulose fiber can obtain its material from a number of sources. Cellulose is present in wood and paper, for example.

Cement Cement is a fine powdered limestone that is heated and mixed with other minerals to serve as a binder in concrete mixes.

Ceramic tile Ceramic tile is made of clay and fired to a high temperature; it usually has a glaze on its surface. Small pieces are used to make floors and wall coverings in bathrooms and kitchens.

Certificate of occupancy This certificate is issued when local inspectors have found a house worthy of human habitation. It allows a contractor to sell a house. It is granted when the building code has been complied with and certain inspections have been made.

Chair A chair is a support bracket for steel reinforcing rods that holds the rods in place until the concrete has been poured around them.

Chalk line A chalk line is used to guide a roofer. It is snapped, causing the string to make a chalk mark on the roof so that the roofers can follow it with their shingles.

Chipboard Chipboard is used as an underlayment. It is constructed of wood chips held together with different types of resins.

Chisel A wood chisel is used to cut away wood for making joints. It is sharpened on one end, and the other is hit with the palm of hand or with a hammer to cut away wood for door hinge installation or to fit a joint tightly.

Claw hammer This is the common hammer used by carpenters to drive nails. The claws are used to extract nails that bend or fail to go where they are wanted.

Closed-cut For a closed-cut valley, the first course of shingles is laid along the eaves of one roof area up to and over the valley. It is extended along the adjoining roof section. The distance is at least 12 inches. The same procedure is followed for the next courses of shingles.

Cold chisel This chisel is made with an edge that can cut metal. It has a one-piece configuration, with a head to be hit by a hammer and a cutting edge to be placed against the metal to be cut.

Commercial building A commercial building is designed for business activities.

Common rafter A common rafter is a member which extends diagonally from the plate to the ridge.

Community planning Community planning occurs when a committee or group of people in a community get together and plan what they want the community to look like in terms of housing and buildings as well as parks and recreational areas.

Compass This device has a steel point and a pencil. It can be used to make circles or to mark arcs.

Concrete Concrete is a mixture of sand, gravel, and cement in water.

Contact cement Contact cement is the type of glue used in applying counter-top finishes. Both sides of the materials are coated with the cement, and the cement is allowed to dry. The two surfaces are then placed in contact, and the glue holds immediately.

Contractor A contractor is a person who contracts with a firm, a bank, or another person to do a job for a certain fee and under certain conditions.

Contractor's key This is a key designed to allow the contractor access to a house while it is under construction. The lock is changed to fit a pregrooved key when the house is turned over to the owner.

Coped joint This type of joint is made with a coping saw. It is especially useful for corners that are not square.

Coping saw This saw is designed to cut small thicknesses of wood at any curve or angle desired. The blade is placed in a frame, with the teeth pointing toward the handle.

Corner beads These are metal strips that prevent damage to drywall corners.

Cornice The cornice is the area under the roof overhang. It is usually enclosed or boxed in.

Course This refers to alternate layers of shingles in roofing.

Cradle brace A cradle brace is designed to hold sheetrock or drywall while it is being nailed to the ceiling joists. The cradle brace is shaped like a T.

Crawl space A crawl space is the area under a floor that is not fully excavated. It is only excavated sufficiently to allow one to crawl under it to get at the electrical or plumbing devices.

Cricket This is another term for the saddle.

Cripple jack A cripple jack is a jack rafter with a cut that fits in between a hip and a valley rafter.

Cripple rafter A cripple rafter is not as long as the regular rafter used to span a given area.

Cripple stud This is a short stud that fills out the position where the stud would have been located if a window, door, or some other opening had not been there.

Crisscross wire support This refers to chicken wire that is used to hold insulation in place under the flooring of a house.

Crosscut saw This is a handsaw used to cut wood across the grain. It has a wooden handle and a flexible steel blade.

Curtain wall Inside walls are often called curtain walls. They do not carry loads from roofs or floors above them.

Damper A damper is an opening and closing device that will close off the fireplace area from the outside by shutting off the flue. It can also be used to control draft.

Deadbolt lock A deadbolt lock will respond only to the owner who knows how to operate it. It is designed to keep burglars out.

Deck A deck is that part of a roof which covers the rafters.

Decorative beams Decorative beams are cut to length from wood or plastic and mounted to the tastes of the owner. They do not support the ceiling.

Diagonals Diagonals are lines used to cut across from adjacent corners to check for squareness in the layout of a basement or footings.

Dividers Dividers have two points and resemble a compass. They are used to mark off specific measurements or transfer them from a square or a measuring device to the wood to be cut.

Dormer Dormers are protrusions; they stick out from a roof. They may be added to allow light into an upstairs room.

Double-hung windows Double-hung windows have two sections, one of which slides past the other. They slide up and down in a prearranged slot.

Double plate This usually refers to the practice of using two pieces of dimensional lumber for support over the top section or wall section.

Double trimmer Double joists used on the sides of openings are called double trimmers. Double trimmers are placed without regard to regular joist spacings for openings in the floor for stairs or chimneys.

Downspouts These are pipes connected to the gutter to conduct rainwater to the ground or sewer.

Drain tile A drain tile is usually made of plastic. It generally is 4 inches in diameter, with a number of small holes to allow water to drain into it. It is laid along the foundation footing to drain the seepage into a sump or storm sewer.

Drawer guide The drawers in cabinets have guides to make sure that the drawer glides into its closed position easily without wobbling.

Drop siding Drop siding has a special groove or edge cut into it. The edge lets each board fit into the next board. This makes the boards fit together and resist moisture and weather.

Drywall This is another name for panels made of gypsum.

Ductwork Ductwork is a system of pipes used to pass heated air along to all parts of a house. The same ductwork can be used to distribute cold air for summer air conditioning.

Dutch hip This is a modification of the hip roof.

Eaves Eaves are the overhang of a roof projecting over the walls.

Elevation Elevation refers to the location of a point in reference to the point established with the builder's level or transit. Elevation indicates how high a point is. It may also refer to the front elevation or front view of a building or the rear elevation or what the building looks like from the rear. Side elevations refer to a side view.

Energy Energy refers to the oil, gas, or electricity used to heat or cool a house.

Essex board measure This is a table on the back of the body; it gives the contents of any size lumber. The table is located on the steel square used by carpenters.

Excavation Excavate means to remove. In this case, excavation refers to the removal of dirt to make room for footings, a foundation, or the basement of a building.

Expansion joint This is usually a piece of soft material that is about 1 inch thick and 4 inches wide and is placed between sections of concrete to allow for expansion when the flat surface is heated by the sun.

Exposure Exposure refers to the part of a shingle or roof left to the weather.

Extension form An extension form is built inside the concrete outer form. It forms a stepped appearance so that water will not drain into a building but drain outward from the slab or foundation slab.

Faced insulation This insulation usually has a coating to create a moisture barrier.

Faced stapling Faced stapling refers to the strip along the outer edges of the insulation that is stapled to the outside or 2-inch sides of the 2×4 studs.

Facing Facing strips give cabinets a finished look. They cover the edges where the units meet and where the cabinets meet the ceiling or woodwork.

Factory edge This is the straight edge of linoleum made at the factory. It provides a reference line for the installer.

Factory-produced housing This refers to housing that is made totally in a factory. Complete units are usually trucked to a place where a basement or slab is ready.

Glossary **499**

False bottom This is a system of 1×6-inch false-bottom or box beams that provide the beauty of beams without the expense. The false beams are made of wood or plastic materials and glued or nailed in place. They do not support any weight.

Fascia Fascia refers to a flat board covering the ends of rafters on the cornice or eave. The eave troughs are usually mounted to the fascia board.

FHA The Federal Housing Administration.

Fiberglas Fiberglas is insulation material made from spun resin or glass. It conducts little heat and creates a large dead air space between layers of fibers. It helps conserve energy.

Firebrick This is a special type of brick that is not damaged by fire. It is used to line the firebox in a fireplace.

Fire stops Fire stops are short pieces nailed between joists and studs.

Flaking This refers to paint that falls off a wall or ceiling in flakes.

Flashing Flashing is metal used to cover chimneys or other things projecting through the roofing. It keeps the weather out.

Floating The edges of drywall sheets are staggered, or floated. This gives more bracing to the wall since the whole wall does not meet at any one joint.

Floating Floating refers to concrete work; it lets the smaller pieces of concrete mix float to the top. Floating is usually done with a tool moved over the concrete.

Floorboards This refers to floor decking. Floorboards may be composed of boards, or may be a sheet of plywood used as a subfloor.

Flue The flue is the passage through a chimney.

Flush This term means to be even with.

Folding rule This is a device that folds into a 3×6-inch rectangle. It has the foot broken into 12 inches. Each inch is broken into 16 parts. Snap joints hinge the rule every six inches. It will spread to as much as 6 feet, or 72 inches.

Footings Footings are the lowest part of a building. They are designed to support the weight of the building and distribute it to the earth or rock formation on which it rests.

Form A form is a structure made of metal or wood used as a mold for concrete.

Formica Formica is a laminated plastic covering made for counter tops.

Foundation The foundation is the base on which a house or building rests. It may consist of the footings and walls.

Framing Roof framing is composed of rafters, ridge board, collar beams, and cripple studs.

Framing square This tool allows a carpenter to make square cuts in dimensional lumber. It can be used to lay out rafters and roof framing.

French doors This usually refers to two or more groups of doors arranged to open outward onto a patio or veranda. The doors are usually composed of many small glass panes.

Frost line This is the depth to which the ground freezes in the winter.

Furring strips These are strips of wood attached to concrete or stone. They form a nail base for wood or paneling.

Gable This is the simplest kind of roof. Two large surfaces come together at a common edge, forming an inverted V.

Galvanized iron This material is usually found on roofs as flashing. It is sheet metal coated with zinc.

Gambrel roof This is a barn-shaped roof.

Girder A girder is a support for the joists at one end. It is usually placed halfway between the outside walls and runs the length of the building.

Grade The grade is the variation of levels of ground or the established ground-line limit on a building.

Grid system This is a system of metal strips that support a drop ceiling.

Grout Grout is a white plaster-like material placed into the cracks between ceramic tiles.

Gusset A gusset is a triangular or rectangular piece of wood or metal that is usually fastened to the joint of a truss to strengthen it. It is used primarily in making roof trusses.

Gutter This is a metal or wooden trough set below the eaves to catch and conduct water from rain and melting snow to a downspout.

Gypsum Gypsum is a chalk used to make wallboard. It is made into a paste, inserted between two layers of paper, and allowed to dry. This produces a plastered wall with certain fire-resisting characteristics.

Handsaw A handsaw is any saw used to cut wood and operated by manual labor rather than electricity.

Hang a door This term refers to the fact that a door has to be mounted on hinges and aligned with the door frame.

Hangers These are metal supports that hold joists or purlins in place.

Hardware In this case hardware refers to the metal parts of a door. Such things as hinges, locksets, and screws are hardware.

Header A header is a board that fits across the ends of joists.

Head lap This refers to the distance between the top and the bottom shingle and the bottom edge of the shingle covering it.

Hearth A hearth is the part of a fireplace that is in front of the wood rack.

Hex strips This refers to strips of shingles that are six-sided.

Hip rafters A hip rafter is a member which extends diagonally from the corner of the plate to the ridge.

Hip roof A hip roof has four sides, all sloping toward the center of the building.

Hollow-core doors Most interior doors are hollow and have paper or plastic supports for the large surface area between the top and bottom edge and the two faces.

Honeycomb Air bubbles in concrete cause a honeycomb effect and weaken the concrete.

Insert stapling This refers to stapling insulation inside the 2×4 stud. The facing of the insulation has a strip left over. It can be stapled inside the studs or over the studs.

Insulation Insulation is any material that offers resistance to the conduction of heat through its composite materials. Plastic foam and Fiberglas are the two most commonly used types of insulation in homes today.

Insulation batts These are thick precut lengths of insulation designed to fit between studs.

Interlocking Interlocking refers to a type of shingle that overlaps and interlocks with its edges. It is used in high-wind areas.

In the white This term is used to designate cabinets that are assembled but unfinished.

Jack rafters Any rafter that does not extend from the plate to the ridge is called a jack rafter.

Jamb A jamb is the part that surrounds a door or window frame. It is usually made of two vertical pieces and a horizontal piece over the top.

Joist Large dimensional pieces of lumber used to support the flooring platform of a house or building are called joists.

Joist hanger These are metal brackets that hold up the joist. They are nailed to the girder, and the joist fits into the bracket.

Joist header If the joist does not cover the full width of the sill, space is left for the joist header. The header is nailed to the ends of the joists and rests on the sill plate. It is perpendicular to the joists.

Key A key is a depression made in a footing so that the foundation or wall can be poured into the footing and not be able to move with the changes in temperature or settling of the building.

Kicker A kicker is a piece of material installed at the top or side of a drawer to prevent it from falling out of a cabinet when it is opened.

King-post truss This is the type of roof truss used to support low-pitch roofs.

Ladder A ladder is a device used to gain access to higher locations in a building.

Ladder jack Ladder jacks hang from a platform on a ladder. They are most suitable for repair jobs and for light work where only one carpenter is on the job.

Landing A landing is the part of a stairway that is shaped like a platform.

Lap This refers to lap siding. Lap siding fits on

Glossary **501**

the wall at an angle. A small part of the siding is overlapped on the preceding piece of siding.

Laths Laths are small strips of wood or metal designed to hold plastic on the wall till it hardens for a smooth finish.

Ledger A ledger is a strip of lumber nailed along an edge or bottom of a location. It helps support or keep from slipping the girders on which the joists rest.

Left-hand door A left-hand door that has its hinges mounted on the left when viewed from the outside.

Level A level is a tool using bubbles in a glass tube to indicate the level of a wall, stud, or floor. Keeping windows, doors, and frames square and level makes a difference in their fit and operation.

Level-transit This is an optical device that is a combination of a level and a means for checking vertical and horizontal angles.

Load Load refers to the weight of a building.

Load conditions These are the conditions under which a roof must perform. The roof has to support so much wind load and snow. Load conditions vary according to locale.

Lockset The lockset refers to the doorknob and associated locking parts inserted in a door.

Mansard This type of roof is popular in France and is used in the United States also. The second story of the house is covered with the same shingles used on the roof.

Manufactured housing This term is used in reference to houses that are totally or partially made within a factory and then trucked to a building site.

Mastic Mastic is an adhesive used to hold tiles in place. The term also refers to adhesives used to glue many types of materials in the building process.

Military specifications These are specifications that the military writes for the products it buys from the manufacturers. In this case, the term refers to the specifications for a glue used in making trusses and plywood.

Mitre box The mitre box has a backsaw mounted in it. It is adjustable for cutting at angles such as 45 and 90°. Some units can be adjusted by a level to any angle.

Modular houses These houses are made in modules or small units which are nailed or bolted together once they arrive at the foundation or slab on which they will rest.

Moisture barrier A moisture barrier is some type of material used to keep moisture from entering the living space of a building. Moisture barrier, vapor seal, and membrane mean the same thing. It is laid so that it covers the whole subsurface area over sand or gravel.

Moisture control Excess moisture in a well-insulated house may pose problems. A house must be allowed to breathe and change the air occasionally, which in turn helps remove excess moisture. Proper ventilation is needed to control moisture in an insulated house. Elimination of moisture is another method but it requires the reduction of cooking vapors and shower vapors, for example.

Moldings Moldings are trim mounted around windows, floors, and doors as well as closets. Moldings are usually made of wood with designs or particular shapes.

Monolithic slab Mono means "one." This refers to a one-piece slab for a building floor and foundation all in one piece.

Nail creep This is a term used in conjunction with drywall; the nails pop because of wood shrinkage. The nail heads usually show through the panel.

Nailers These are powered hammers that have the ability to drive nails. They may be operated by compressed air or by electricity.

Nailhead stains These occur whenever the iron in the nail head rusts and shows through the paint.

Nail set Finish nails are driven below the surface of the wood by a nail set. The nail set is placed on the head of a nail, and the large end of the nail set is struck with a hammer.

Newels The end posts of a stairway are called newels.

Octagon scale This "eight square" scale is found on the center of the face of the tongue of a steel square. It is used when timber is cut with eight sides.

On-the-job training A person works on a job in order to obtain the needed training.

Open This refers to the type of roofing that allows a joint between a dormer and the main roof. It is an open valley type of roofing. The valley where two roofs interesect is left open and covered with flashing and roofing sealer.

Orbital sander This power sander will vibrate, but in an orbit. This causes the sandpaper to do its job better than it would if used in only one direction. An orbital sander can be used to finish off windows, doors, counters, cabinets, and floors.

Panel This refers to a small section of a door that takes on definite shape, or to the panel in a window made of glass.

Panel door This is a type of door used for the inside of a house. It has panels inserted in the frame to give it strength and design.

Parquet Parquet is a type of flooring made from small strips arranged in patterns. It must be laminated to a base.

Partition A partition is a divider wall or section that separates a building into rooms.

Patio A patio is an area attached to the house that has a flat surface for lounging or enjoying the out-of-doors or backyard.

Peeling This is a term used in regard to paint that will not stay on a building. The paint peels and falls off or leaves ragged edges.

Penny (d) This is the unit of measure of the nails used by carpenters.

Perimeter The perimeter is the outside edges of a plot of land or building. It represents the sum of all the individual sides.

Perimeter insulation Perimeter insulation is placed around the outside edges of a slab.

Pile A pile is a steel or wooden pole driven into the ground sufficiently to support the weight of a wall and building.

Pillar A pillar is a pole or reinforced wall section used to support the floor and consequently the building. It is usually located in the basement, with a footing of its own to spread its load over a wider area than the pole would normally occupy.

Pitch The pitch of a roof is the slant or slope from the ridge to the plate.

Pivot This refers to a point where the bifold door is anchored and allowed to move so that the larger portion of the folded sections can move.

Plane Planes are designed to remove small shavings of wood along a surface. One hand holds the knob in front, and the other hand holds the handle in the back of the plane. A plane is used to shave off door edges to make them fit properly.

Planks This refers to a type of flooring usually made of tongue-and-groove lumber and nailed to the subflooring or directly to the floor joists.

Plaster This refers to plaster of Paris mixed with water and applied to a lath to cover a wall and allow for a finished appearance that will take a painted finish.

Plaster grounds A carpenter applies small strips of wood around windows, doors, and walls to hold plaster. These grounds may also be made of metal.

Plastic laminates These are materials usually employed to make counter tops. Formica is an example of a plastic laminate.

Plate The plate is a roof member which has the rafters fastened to it at their lower ends.

Platform frame This refers to the flooring surface placed over the joists; it serves as a support for further floor finishing.

Plenum A plenum is a large chamber.

Plumb bob This is a very useful tool for checking plumb, or the upright level of a board, stud, or framing member. It is also used to locate points that should be directly under a given location. It hangs free on a string, and its point indicates a specific location for a wall, a light fixture, or the plumb of a wall.

Plumb cut This refers to the cut of the rafter end which rests against the ridge board or against the opposite rafter.

Post and beam Posts are used to support beams, which support the roof decking. Regular rafters are not used. This technique is used in barns and houses to achieve a cathedral-ceiling effect.

Prehung This refers to doors or windows that are already mounted in a frame and ready for installation as a complete unit.

Primer This refers to the first coat of paint or glue when more than one coat will be applied.

Pulls The handle or the part of the door on a cabinet or the handle on a drawer that allows it to be pulled or opened is called a pull.

Purlin Secondary beams used in post-and-beam construction are called purlins.

Radial-arm saw This type of power saw has a motor and blade that moves out over a table which is fixed. The wood is placed on the table, and the blade is pulled through the wood.

Rafter scales This refers to a steel square with the rafter measurements stamped on it. The scales are on the face of the body.

Rail The vertical facing strip on a cabinet is the stile. The horizontal facing strip is the rail.

Rake On a gabled roof, a rake is the inclined edge of the surface to be covered.

Random spacing This refers to spacing that has no regular pattern.

Rebar A rebar is a reinforcement steel rod in a concrete footing.

Reinforcement mesh Reinforcement mesh is made of 10-gauge wires spaced about 4 to 6 inches apart. It is used to reinforce basements or slabs in houses. The mesh is placed so that it becomes a part of the concrete slab or floor.

Remodeling This refers to changing the looks and function of a house.

Residential building A residential building is designed for people to live in.

Resilient flooring This type of flooring is made of plastics rather than wood products. It includes such things as linoleum and asphalt tile.

Ridge board This is a horizontal member that connects the upper ends of the rafters on one side to the rafters on the opposite side.

Right-hand-door A right-hand door has the opening or hinges mounted on the right when viewed from the outside.

Right-hand draw This means that the curtain rod can be operated to open and close the drapes from the right-hand side as one faces it.

Rise In roofing, rise is the vertical distance between the top of the double plate and the center of the ridge board. In stairs, it is the vertical distance from the top of a stair tread to the top of the next tread.

Riser The vertical part at the edge of a stair is called a riser.

Roof brackets These brackets can be clamped onto a ladder used for roofing.

Roof cement A number of preparations are used to make sure that a roof does not leak. Roof cement also can hold down shingle tabs and rolls of felt paper when it is used as a roof covering.

Roofing This term is used to designate anything that is applied to a roof to cover it.

Rough line A rough line is drawn on the ground to indicate the approximate location of a footing.

Rough opening This is a large opening made in a wall frame or roof frame to allow the insertion of a door or window or whatever is to be mounted in the open space. The space is shimmed to fit the object being installed.

Router A router will cut out a groove or cut an edge. It is usually powered and has a number of different shaped tips that will carve its shape into a piece of wood. It can be used to take the edges off counter tops.

Run The run of a roof is the shortest horizontal distance measured from a plumb line through the center of the ridge to the outer edge of the plate.

R values This refers to the unit that measures the effectiveness of insulation. It indicates the relative value of the insulation for the job. The higher the number, the better the insulation qualities of the material.

Saber saw The saber saw has a blade that can be used to cut circles in wood. It can cut around any circle or curve. The blade is inserted in a hole drilled previously, and the saw will follow a curved or straight line to remove the block of wood needed to allow a particular job to be completed.

Saddle A saddle is the inverted V-shaped piece of roof inserted between the vertical side of a chimney and the roof.

Saturated felt Other names for this material are tar paper and builder's felt. It is roll roofing paper and can be used as a moisture barrier and waterproofing material on roofs and under siding.

Scabs Scabs are boards used to join the ends of a girder.

Scaffold A scaffold is a platform erected by carpenters to stand on while they work on a higher level. Scaffolds are supported by tubing or 2×4s. Another name for scaffolding is staging.

Screed A screed may be a board or pipe supported by metal pins. The screed is leveled with the tops of the concrete forms. It is removed after the section of concrete is leveled.

Scribing Scribing means marking.

Sealant A sealant is any type of material that will seal a crack. This usually refers to caulking when carpenters use the term.

Sealer coat The sealer coat ensures that a stain is covered and the wood is sealed against moisture.

Shakes This is a term used for shingles made of hand-split wood, in most cases western cedar.

Sheathing This is a term used for the outside layer of wood applied to studs to close up a house or wall. It is also used to cover the rafters and make a base for the roofing. It is usually made of plywood today. In some cases, sheathing is still used to indicate the 1×6-inch wooden boards used for siding undercoating.

Shed In terms of roofs, this is the flat sloping roof used on some storage sheds. It is the simplest type of roof.

Sheetrock This is another name for panels made of gypsum.

Shim To shim means to add some type of material that will cause a window or door to be level. Usually wood shingles are wedge-shaped and serve this purpose.

Shingles This refers to material used to cover the outside of a roof and take the ravages of weather. Shingles may be made of metal, wood, or composition materials.

Shingle stringers These are nailing boards that can have cedar shingles attached to them. They are spaced to support the length of the shingle that will be exposed to weather.

Side lap The side lap is the distance between adjacent shingles that overlap.

Siding This is a term used to indicate that the studs have been covered with sheathing and the last covering is being placed on it. Siding may be made of many different materials — wood, metal, or plastic.

Sill This is a piece of wood that is anchored to the foundation.

Sinker nail This is a special nail for laying subflooring. The head is sloped toward the shank but is flat on top.

Site This refers to the location or proposed location of a building.

Size Size is a special coating used for walls before the wallpaper is applied. It seals the wall and allows the wallpaper paste to attach itself to the wall and paper without adding undue moisture to the wall.

Skew-back saw This saw is designed to cut wood. It is hand-operated and has a serrated steel blade that is smooth on the noncutting edge of the saw. It is 22 to 26 inches long and can have 5-1/2 to 10 teeth per inch.

Skilled worker A skilled worker is a person who can do a job well each time it is done, or a person who has the ability to do the job a little better each time. Skilled means the person has been at it for some time, usually 4 to 5 years at the least.

Sliding door This is usually a large door made of glass, with one section sliding past the other to create a passageway. A sliding door may be made of wood or glass and can disappear or slide into a wall. Closets sometimes have doors that slide past one another to create an opening.

Sliding window This type of window has the ability to slide in order to open.

Slope Slope refers to how fast the roof rises from the horizontal.

Soffit A soffit is a covering for the underside of the overhang of a roof.

Soil stack A soil stack is the ventilation pipe that comes out of a roof to allow the plumbing to operate properly inside the house. It is usually made of a soil pipe (cast iron). In most modern housing, the soil stack is made of plastic.

Sole plate A sole plate is a 2×4 or 2×6 used to support studs in a horizontal position. It is placed against the flooring and nailed into position onto the subflooring.

Span The span of a roof is the distance over the wall plates.

Spreader Special braces used across the top of concrete forms are called spreaders.

Square This term refers to a shingle-covering area. A square consists of 100 square feet of area covered by shingles.

Square butt strip This refers to shingles for roofing purposes that were made square in shape but produced in strips for ease in application.

Staging This is the planking for ladder jacks. It holds the roofer or shingles.

Stain Stain is a paintlike material that imparts a color to wood. It is usually finished by a clear coating of shellac, varnish or satinlac, or brush lacquer.

Stapler This device is used to place wire staples into a roof's tar paper to hold it in place while the shingles are applied.

Steel square The steel square consists of two parts—the blade and tongue or the body.

Stepped footing This is footing that may be located on a number of levels.

Stile A stile is an upright framing member in a panel door.

Stop This applies to a door. It is the strip on the door frame that stops the door from swinging past the center of the frame.

Storm door A storm door is designed to fit over the outside doors of a house. It may be made of wood, metal, or plastic, and it adds to the insulation qualities of a house. A storm door may be all glass, all screens, or a combination of both. It may be used in summer, winter, or both.

Storm window Older windows have storms fitted on the outside. The storms consist of another window that fits over the existing window. The purpose is to trap air that will become an insulating layer to prevent heat transfer during the winter. Newer windows have thermopanes, or two panes mounted in the same frame.

Stressed-skin panels These are large prebuilt panels used as walls, floors, and roof decks. They are built in a factory and hauled to the building site.

Strike-off After tamping, concrete is leveled with a long board called a strike-off.

Strike plate This is mounted on the door frame. The lock plunger goes into the hole in the strike plate and rests against the metal part of the plate to hold the door secure against the door stop.

Striker This refers to the strike plate. The striker is the movable part of the lock that retracts into the door once it hits the striker plate.

Stringer A carriage is also called a stringer.

Strip flooring Wooden strip flooring is nothing more than the wooden strips that are applied perpendicular to the joists.

Strongback Strongbacks are braces used across ceiling joints. They help align, space, and strengthen joists for drywall installation.

Stucco Stucco is a type of finish used on the outside of a building. It is a masonry finish that can be put on over any type of wall. It is applied over a wire mesh nailed to the wall.

Stud This refers to the vertical boards (usually 2×4 or 2×6) that make up the walls of a building.

Stump A stump is that part of a tree which is left after the top has been cut and removed. The stump remains in the ground.

Subfloor The subfloor is a platform that supports the rest of the structure. It is also referred to as the underlayment.

Sump pump This refers to a pump mounted in a sump or well created to catch water from around the foundation of a house. The pump takes water from the well and lifts it to the grade level or to a storm sewer nearby.

Surveying Surveying means taking in the total scene. In this case, it refers to checking out the plot plan and the relationship of the proposed building with others located within eyesight.

Suspended beams False beams may be used to lower a ceiling like a grid system; these beams are suspended. They use screw eyes attached to the existing ceiling joists.

Sway brace A sway brace is a piece of 2×4 or similar material used to temporarily brace a wall from the wind until it is secured.

Swinging door A swinging door is mounted so that it will swing into or out of either of two rooms.

Table saw A table saw is electrically powered, with a motor-mounted saw blade supported by a table that allows the wood to be pushed over the table into the cutting blade.

Tail The tail is the portion of a rafter that extends beyond the outside edge of the plate.

Tail joist This is a short beam or joist supported in a wall on one end and by a header on the other.

Tamp To tamp means to pack tightly. The term usually refers to making sand tightly packed or making concrete mixed properly in a form to get rid of air pockets that may form with a quick pouring.

Taping and bedding This refers to drywall finishing. Taping is the application of a strip of specially prepared tape to drywall joints; bedding means embedding the tape in the joint to increase its structural strength.

Team A team is a group of people working together.

Terrazzo This refers to two layers of flooring made from concrete and marble chips. The surface is ground to a very smooth finish.

Texture paint This is a very thick paint that will leave a texture or pattern. It can be shaped to cover cracked ceilings or walls or beautify an otherwise dull room.

Thermal ceilings These are ceilings that are insulated with batts of insulation to prevent loss of heat or cooling. They are usually drop ceilings.

Tie A tie is a soft metal wire that is twisted around a rebar or reinforcement rod and chair to hold the rod in place till concrete is poured.

Tin snips This refers to a pair of scissors-type cutters used to cut flashing and some types of shingles.

Tongue and groove Roof decking may have a groove cut in one side and tongue left in the other edge of the piece of wood so that the two adjacent pieces will fit together tightly.

Track This refers to the metal support system that allows the bifold and other hung doors to move from closed to open.

Transit-mix truck In some parts of the country, this is called a Redi-Mix truck. It mixes the concrete on its way from the source of materials to the building site where it is needed.

Traverse rod This is another name for a curtain rod.

Tread The part of a stair on which people step is the tread.

Trestle jack Trestle jacks are used for low platforms both inside and outside. A ledger, made of 2 × 4 lumber, is used to connect two trestle jacks. Platform boards are then placed across the two ledgers.

Trimmer A trimmer is a piece of lumber, usually a 2 × 4, that is shorter than the stud or rafter but is used to fill in where the longer piece would have been normally spaced except for the window or door opening or some other opening in the roof or floor or wall.

Trowel A trowel is a tool used to work with concrete or mortar.

Truss This is a type of support for a building roof that is prefabricated and delivered to the site. The W and King trusses are the most popular.

Try square A try square can be used to mark small pieces for cutting. If one edge is straight and the handle part of the square is placed against this straight edge, the blade can be used to mark the wood perpendicular to the edge.

Underlayment This is also referred to as the subfloor. It is used to support the rest of the building. The term may also refer to the sheathing used to cover rafters and serve as a base for roofing.

Unfaced insulation This type of insulation does not have a facing or plastic membrane over one side of it. It has to be placed on top of existing insulation. If used in a wall, it has to be covered by a plastic film to ensure a vapor barrier.

Union A union is a group of people with the same interests and with proper representation for achieving their objectives.

Utilities Utilities are the things needed to make a house a home. They include electricity, water, gas, and phone service. Sewage is a utility that is usually determined to be part of the water installation.

Utility knife This type of knife is used to cut the underlayment or the shingles to make sure they fit the area assigned to them. It is also used to cut the saturated felt paper over a deck.

Valley This refers to the area of a roof where two sections come together and form a depression.

Valley rafters A valley rafter is a rafter which extends diagonally from the plate to the ridge at the line of intersection of two roof surfaces.

Vapor barrier This is the same as a moisture barrier.

Veneer A veneer is a thin layer or sheet of wood.

Vent A vent is usually a hole in the eaves or soffit to allow the circulation of air over an insulated ceiling. It is usually covered with a piece of metal or screen.

Ventilation Ventilation refers to the exchange of air, or the movement of air through a building. This may be done naturally through doors and windows or mechanically by motor-driven fans.

Vernier This is a fine adjustment on a transit that allows for greater accuracy in the device when it is used for layout or leveling jobs at a construction site.

Vinyl Vinyl is a plastic material. The term usually refers to polyvinyl chloride. It is used in weather stripping and in making floor tile.

Vinyl-asbestos tile This is a floor covering made from vinyl with an asbestos filling.

Water tables This refers to the amount of water that is present in any area. The moisture may be from rain or snow.

Weather stripping This refers to adding insulating material around windows and doors to prevent the heat loss associated with cracks.

Winder This refers to the fan-shaped steps that allow the stairway to change direction without a landing.

Window apron The window apron is the flat part of the interior trim of a window. It is located next to the wall and directly beneath the window stool.

Window stool A window stool is the flat narrow shelf which forms the top member of the interior trim at the bottom of a window.

Wrecking bar This tool has a number of names. It is used to pry boards loose or to extract nails. It is a specially treated steel bar that provides leverage.

Woven This refers to a type of roofing. Woven valley-type shingling allows the two intersecting pieces of shingle to be woven into a pattern as they progress up the roof. The valley is not exposed to the weather but is covered by shingles.

Zoning laws Zoning laws determine what type of structure can be placed in a given area. Most communities now have a master plan which recognizes residential, commerical, and industrial zones for building.

INDEX

Access control during construction, 36–39
Acoustical quality, 335
Acrylic latex sealant, 310
Additions:
 of bathrooms, 442–443
 of drop ceilings, 436, 438
 of space to existing buildings, 452–456
Additives, concrete color, 73
Adhesives, 90, 328, 434
Air conditioning, solar energy and, 484–485
Aluminum siding, 286–288
Aluminum soffit, 161–165
Anchors, 56, 69, 78–79, 145, 157, 403
Andersen Perma-Shield Window, 224
Apartments, premanufactured, 474, 476
Apprenticeship, 463
Aprons (clothing), 176
Asphalt cements, 186–187
Asphalt coatings, 186–187
Asphalt rolls, typical, table, 185
Asphalt roof covering, 177–179
Asphalt shingles:
 repair of, 426
 typical, table, 184
Attics:
 insulating, 296–298
 louvers of, in Tech House, 488
 ventilation, table, 304
 ventilation through, 303
Auxiliary locks, 243–244
Awls, scratch, 8–9
Awning picture windows, 218

Back, framing square, 127
Backsaws, 7
Backsplash, defined, 424
Balloon frames, 77, 83–84
Balloon method of framing, 98
Balusters, defined, 384
Base (see Floors)
Basements:
 and adding space to existing building, 452
 excavating for, 35–36
 footings and, 46
 insulating walls of, 298–299
 preparing, 21–22
 replacing outside doors to, 440–442
 waterproofing walls of, 58
Bathrooms:
 adding, 442–443
 shelves for, 363
 tubs and showers for, 328–329

Batten board, 276
Batter board, 29, 45
Bay windows, 146
Beams:
 in post-and-beam construction, 404–407
 thermal ceiling panels and, 305–306
Bearing plates in post-and-beam construction, 406
Bench mark (surveying), 34
Beveled siding (rabbeted beveled siding; Dolly Varden siding), 276
Bevels, 11
Blades, saw, 16–17
Block (parquet) flooring, 370, 373–374
Block-out (see Buck shape)
Block walls, concrete, 56
Blowouts, 47
Boards:
 batten, 276
 batter, 45
 plain, subfloor, 90–91
 plank-and-beam construction, 401
 plank flooring, 372, 375
 ridge, 131
 for roof decking, 148–149, 152
 for sheathing, 116
 siding of plain, 275–276
 walls of, 331–332
Body, framing square, 127
Body protection, 3
Bow supports for floor insulation, 298
Brace measure, 142–143
Brace scale, 128
Braces:
 buck, 54
 as common rafters, 143
 corner, 108–109
 cradles as, 320
 foundation form, 50
 strongbacks as, 319–320
Bricks:
 concrete and, 73
 as wall cover, 285–286
 windows in brick veneer walls, 223
Bridging, 87–88
Brushing, concrete, 72
Buck keys, 53
Buck shape (block-out):
 buck braces, 54
 buck left as frame, 53
 buck removed, 53–54
 keys, 53
Builder's level, 29–34
Building codes, 468
Building materials:
 new, 465

Building materials: (Continued)
 (See also specific materials)
Built-ins, defined, 351
Built-up corners, defined, 101
Bull floats, 68
Butyloid rubber caulk, 310–311

C clamps, 13
Cabinets, 348–365
 custom, 351–363
 doors of, 356–360
 ready-built, 348–351
Cantilevers, (see Overhangs)
Carpentry in construction industry, 462–481
Carpets:
 base flooring for, 375
 installing, 375–376
Carriage, defined, 383
Casement windows, 216–217
Casing, window, 346–347
Cathedral ceiling, 403
Caulk, 310–311
Caulking guns, 13–14
Cedar-lined closets, 444
Ceiling tiles, installing, 335–339
Ceilings:
 gypsum board, 319–321
 insulating, 294–295
 joists, 146–148
 opening in, for fireplace, 393
 replacing, 435–439
 suspended, 336, 339
 thermal panels in, 303–306
Cements:
 asphalt, 186–187
 roof, 311
Ceramic tiles:
 concrete, 73
 laying, 380–381
Chairs, defined, 45
Chalk and line (roof work), 176
Chicken-wire supports for floor insulation, 298
Chimneys, 403
 flashing for, 192–194
 flues for, 398
Chipboard:
 subfloors of, 89
 (See also Gypsum board)
Chisels, 11, 13
Chord code for roof trusses, table, 125–126
Circle:
 of builder's level, 30
 reading the, 32
Circular saws, 15–16
 portable, for cutting studs to length, 105
Claw hammers, 7

510 Index

Clearing of building site, 34
Closed-cut valleys, roofing, 189, 190
Closets, cedar-lined, 444
Codes:
  building, 468
  chord, for roof trusses, table, 125–126
Cold chisels, 13
Color additives, concrete, 73
Columns:
  footings under, 43
  lolly, 81
Combination blades, 16
Common braces, scale for, 128
Common rafters, 131, 135–138
  braces as, 143
  run of, at dormer, 176
Compasses, 11–12
Concrete, 60–74
  estimating needed, 48
  estimating volume of:
    for foundation, 55
    for slab, 57
  for footing forms, 47–49
  reinforcing foundation of, 54–56
  sidewalks of, 71–72
  slabs of, 60–70
  use chart, table, 48
  waterproofing of concrete block foundations, 58
  waterproofing of walls of, 58
Concrete patch, 311
Condensation:
  insulation and, 303
  (See also Moisture barrier)
Construction procedures, changing, 466
Contraction of concrete, 69
Conversions, 442–458
Coping saws, 7
Corners:
  braces, wall frame, 108–109
  finishing: shingle or shake siding, 284
    for siding, 279–281
  finishing panel, 330
  with gypsum board, 324–325
  nailing studs at, 106
  partition, 112
  wall frame studs at, 101–102
Cornices:
  defined, 267
  enclosed (close), 269, 271
Countertops:
  installing new, 421–422, 424
  making, 360–362
Cracking, concrete, 69–70
Cradles, 320
Crawl space, 43
  energy factors in, 93–95
  insulating, 294, 299

Cripple jacks, 131, 132
Cripple joists, (short), 83
Cripple studs:
  defined, 100
  nailing, 108
  in wall frames, 104
Cripples (window installation), 225
Crisscross wire supports for floor insulation, 298
Cross T's, defined, 336
Crosscutting:
  defined, 6
  power saw, 16–17
Crowbars (wrecking bars), 9
Curtain walls:
  defined, 98
  (See also Walls)
Custom-built storage shed, 457, 460
Custom cabinets, 351–365
Custom work, defined, 348
Cut joints, concrete, 69, 70

Dampers, fireplace, 398
Deck preparation for strip shingles, 195–196
Decking:
  plywood (see Plywood)
  in post-and-beam construction, 406
  roof, 130, 148
  roof covering problem and, 180, 182
Decorative beams, insulation and, 305–306
Deeds to land, 26
Delivery routes for materials, 38–39
Depth:
  finding footing trench, 44
  footing, 43
Design innovations, 466, 468
Diagonal corner braces, 109
Dimple, defined, 317
Dividers, 11–12
Doors:
  in adding to existing building, 455
  adjusting, 411–412
  cabinet, 351–358
  energy factor and, 254
  flush, 226–227
  folding (see Folding doors)
  frames of, 343
  french, 228
  hanging two-door system, 232–234
  hardware for, 344, 360, 362–363
  installing:
    exterior, 228–235
    folding, 236–239
    interior, 342–346

Doors: (Continued)
  lock installing in, 241–247
  masonry or brick-veneer wall installation of gliding, 252
  Perma-Shield gliding, 252–254
  prehung, 225–228
  replacing outside basement, 440–442
  sliding, 228, 247–248
  storm, 245-246
  Tech House, 488
  trimmed interior, 239–240
  wood gliding, 249–251
Dormers, 145–146
  open valley for, 190–191
  valleys, 176
Double-hung windows, 212, 214–216
Double joists, 83
Double-pole scaffolds, 263, 265
Double-ply construction of gypsum board, 325–326
Double trimmers, 83, 86
Downdraft, 398
Drainage:
  foundation wall, 56, 58
  joints for, 274–275
  roof, 168–169, 171
  from sidewalk or driveway, 71
Drains:
  excavating for, 47
  slab, 63
Drapery hardware, 414–416
Drawers:
  guides for, 353, 355
  hardware for, 360, 362–363
  making, 355–356
Drills, 17
Driveways, 71–72
Drop ceilings, 436, 438
Drop siding, 276
Drywall, 315
  (See also Gypsum board)

Easement, defined, 26–27
Eaves:
  close cornices and, 269, 271
  defined, 267
  open, 267–268
Edge spacing, 68, 72, 321
Edges, finishing panel, 330
Elevation:
  establishing building, 34
  porch, 448
End-joined joists, 84
Energy factors:
  concrete and, 74
  of doors and windows, 254
  flood insulation, 93–96
  (See also Insulation)

Index   511

Energy sheathing, 114, 116, 117
Entrance handle locks, 243
Entry vestibules, Tech House, 488
Essex board measure, 128
Excavation:
   for drains and utility lines, 47
   for footing, 44–45
   site, 35–36
   for slabs, 61
Expansion of concrete, 69
Expansion joints, 69
Exposure:
   defined, 171
   wood shingle, 204–205
Extension cords, size of, for portable tools, 14
Extension form, 66
Exterior walls:
   defined, 98
   (See also Walls)

Face, framing square, 127
Face sheilds, 3
Faced insulation, installing, 300
Facing strips, cabinet, 351
Factory-produced housing, 472–480
Factory scaffolds, 265–266
False-bottom beams, insulation and, 305
Farm buildings, roofing, 186
Fascia, 267
Fiberboard:
   subfloors of, 89
   (See also Gypsum board)
Fiberboard sheathing, 114
Fiberglas insulation, cutting, 300
Files, 13
Finishing
   (see specific jobs and materials)
Fire stops, 86
Fireplaces:
   frames for, 392–401
   Tech House, 487
   types of, 394, 396
First jack rafters, 128
Flange, window, 212
Flashing:
   chimney, 192–194
   defined, 171
   eaves, strip shingles and, 196
   roof covering problem, 182
   for siding, 280
   against vertical wall, 191–192
   window (flange), 212
   wood shingle roof, 206
Flat cornices, 271
Flat footings, 41
Floating, 68, 72, 321
Floats, 68

Floors:
   in adding to existing building, 455
   concrete, 70–71
   concrete combined with other materials for, 72–74
   energy savings and concrete, 74
   finishing, 369–375
   frame, 77–96
   insulating, 298
   in job sequence, 22
   kitchen, 449
   minimum plywood flooring standards, table, 88
   replacing, 429–434
   resilient flooring, 377–380
Flues, 398, 401
Flush doors, 226–227
Flush drawers, 355
Folding doors:
   installing, 236–239
   opening sizes, table, 237
Folding rules, using, 5–6
Footings, 41–49
   building forms for, 46–47
   concrete for, 47–49
   laying out, 42–46
   one-piece forms, 52
   sequence for, 42
   slab, 35, 46, 61
   special forms for, 52
   types of, 41–42
Forms:
   building footing, 46–47
   building foundation, 49–56
   for concrete slabs, 62
   footing (see Footings)
   special, 52
Foundations:
   building forms for, 49–56
   concrete block walls and plywood, 56
   drainage and waterproofing for, 56, 58
   layout of, 27–29
   sequence for, 42
Frame floors, 77–96
Frames:
   door, 343
   fireplace, 392–401
   floor, 77–96
   roof (see Roof frames)
   wall (see Wall frames)
Framing square, 10, 126–132
French doors, 228
Frost (freeze) line, 43, 61
Full studs:
   defined, 100
   nailing, 106
   (See also Studs)
Furring strips, 330
   to install ceiling tiles, 336

Gable ends, siding, 271, 273–275
Gable (pitch) roofs, 129
Gable-and-valley (hip-and-valley) roofs, 129
Gambrel-shaped-roof storage sheds, 154
Girders, 77
   joists for metal, 85
   setting, 80–81
Glasses, safety, 2–3
Gliding doors:
   masonry or brick-veneer wall installation of, 252
   Perma-Shield, 252–254
   wood, 249–251
Gloves, 3
Gluing:
   subfloor, 90
   for trusses, 126
Grade line, 34
Grid system, thermal ceiling panels, 304
Grooved boards, subfloor, 91
Grounds (strips):
   defined, 332
   nailing, 332, 334
Guttering, replacing, 427–429
Gypsum board, 315
   installing, 317–328
   repairing walls of, 416, 419–421
Gypsum sheathing, 114

Hammers, 7–8
   in roof work, 176
Hangers, joist, 84
Hardware:
   door, 344, 360, 362–363
   drapery, 414–416
   drawer, 360, 362–363
Hatchets, 8
   shingler's, 205
Head lap, 171
Headers, 81, 82, 86
   nailing, 108
   partition, 112
   size of: table, 104
      in wall frames, 104
   in window installation, 225
Headroom, stair, 387
Heel, framing square, 127
Hinges, door, 345
Hinging doors, 344
Hip, defined, 171
Hip jacks, 131, 132
Hip rafters, 131
Hip and ridge shingles, 198, 200, 203
Hip roofs, 129
   shingles started at, 188
Hip and valley rafters, 128, 138–141

Hips and ridges, wooden shingles
    applied to, 207–208
Hips and valleys, roof covering, 179
Honeycomb, defined, 50
Hundredths scale, 128

Instructions, following, 24
Insulation, 291–312
    installing, 294–302
    perimeter, 64
    for seal, 77–78
    sealants, 310–311
    and solar-heated housing, 491
    storm doors for, 310
    storm windows for, 306–310
        table, 308
    Tech House, 488
    thermal ceilings for, 303–306
    types of, 291–292
    vapor barriers and moisture
        control in, 302–303
    in walls, 315–316
    where to apply, 292–294
    in winterizing a home, 311–312
    and wood over concrete, 74
    (See also Energy factors)
Interior walls (see Walls)

Jack rafters, 131, 141–142
Jack studs, 225
Jambs, defined, 342
Joints:
    in concrete, 69–70
    finishing gypsum board, 326, 328
Joist-girder butts, 84–85
Joists:
    floor, 77
        in job sequence, 22
    floor frame, 81–87
    rafters and ceiling, 146–148
    special, 91–93

Kerosene in roof work, 176
Keyed footings, 44, 45, 47
Keyed notch (slot), footing, 47
Keying, lockset, 244–245
Keys:
    buck, 53
    footing, 44, 45, 47
Kitchens:
    cabinets for (see Cabinets)
    remolding, 447
Knives in roof work, 176

Ladder jacks, 266

Ladders:
    climbing, 5
    in roof work, 176
    scaffold, 266–267
Laminated surfaces, trim for, 362
Landing, defined, 386
Lapped joists, 84
Latex caulk, vinyl, 310
Latex sealant, acrylic, 310
Layout:
    balloon frame, 83–84
    for gambrel-shaped-roof storage
        shed, 154–156
    carriage, 389, 391–392
    footing, 42–46
    footing form, 46
    joist, 81–84
    pattern, for custom cabinets, 351
    rafter, 132–135, 143
    siding, 276, 278
    stair, 160–161
    wall, 98–104
Level-transits, 31–32
Level in window installation, 224
Leveling vial (bubble) of
    builder's level, 30
Levels:
    carpenter's, 12–13
    optical builder's, 29–34
Lip drawers, 355
Load-bearing walls, defined, 98
Local ordinances, 26
Locks:
    adjusting, 412–414
    door, 241–247
Lolly columns, 81
Low-profile building, 91, 93
L-shaped stairs, 384

Maintenance (repairs), 409
    minor, 411–442
Mansard roofs, 157–158
    covering, 198–200
    problems of covering, 168
Manufactured housing, 472–480
Masonry walls:
    finishing, 334–335
    panels over, 330
    windows in, 223
Master stud pattern, 103
Mastic, 74
Materials:
    delivery routes for, 38–39
    storage of, 36–38
Measuring line, rafter, 132
Measuring tools, using, 5–6
    (See also specific measuring
        tools)
Mesh to reinforce concrete
    slab, 64–65

Metal doors, 235
Metal window frames,
    finishing, 347–348
Midfloor supports, 77
Millwork, 348–351
Mitre boxes, 7
Modular housing, 472
Modular standards for spacing, 117
Moisture barrier (vapor seal;
    membrane; vapor barrier), 64
    installing, 316–317
    insulation and, 302–303
    room paneling, 434
    siding and, 259–260
    over subfloor, 94
Moisture control, 63
    insulation and, 302, 303
Molds (see Forms)
Monolithic slabs, 61, 62

Nail sets, 8
Nailers (compressed air nailer;
    nail driver), 2, 18
Nails:
    aluminum siding, 287
    digit system for ordering, table, 20
    for footing forms, 46
    gauge, inch, and penny
        relationships of, 8
    panel, 330
    for plaster grounds, 332
    repair of popping, 420–421
    roll roofing, 202–203
    roofing, 182, 185
        table, 185
    for room paneling, 434
    for shingle siding, 282–283
    for siding, 260, 278–279
    for subfloor, 89
    for subfloor of boards, 91
    for wood shingles, 208
Newels, defined, 384
Nose (nosing), defined, 384
Notched joists, 84
No-wax floors, 430–432

Octagon scale, 128
One-piece forms, 51–52
Open valleys in roofing, 188–190
Openings:
    fireplace, in ceiling and roof, 393
    floor frame, 86
    foundation, 52–54
    in gypsum board, 323–324
    joists for, 83
    partition, 112–114
    roof, 148, 393

Index 513

Openings: (Continued)
  wall frame, 101, 104
  (See also Doors; Windows)
Outdoor work, 5
Overbuilding, 469
Overhangs (cantilevers), 91–93
  and adding space to existing building, 455
  framed, 274

Painting:
  floor, 374–375
  problems of, 409
  wall, 367
Panel doors, 227–228
Panel finish, walls, 330–332
Panel forms, 51
Panels:
  in factory-produced housing, 472, 474
  replacing room, 434–435
  siding, 281–282
  stressed-skin, 406–407
  wall, 330–332
  (See also Gypsum board)
Parquet flooring, 372, 375–376
Partition studs, wall frame, 102
Partitions:
  defined, 98
  (See also Walls)
Patch, concrete, 311
Patching:
  roof, 427
  (See also Maintenance)
Pebble finish, concrete, 72–73
Penny, defined, 8
Perimeter insulation, 64
Perma-Shield gliding doors, 252–254
Permits for additions, 455
Phillips-head screwdrivers, 9
Pile footings, 41, 42
Pillared footings, 41
Pillars (see Columns)
Pipes, insulating around, 302
Pitch:
  defined, 130
  principal roof, 131
  rafter, 134
  roof, 154, 171–172
Planer blades, 16–17
Planes, 11
Plank-and-beam construction, 401
Plank flooring, 370, 373
Planning, community, 468–469
Plaster, wall prepared for, 332–334
Plates:
  bearing, in post-and-beam construction, 406
  defined, 98
  door strike, 345

Plates: (Continued)
  layout of, 98, 100
  nailing studs to, 106
  roof, 131
  solar, 98, 100
Platform method of framing:
  floors, 77
  joists in, 81–82
Plenums, energy, 94–95
Plumb:
  of door frames, 343
  in window installation, 224
Plumb bob, 12
  hanging the, 32
Plywood:
  corner braces of, 109
  foundations of, 56
  for roof decking, 148, 149–152, 160
    problem, 180, 182
    table, 159
  sheathing of, 114
  subfloors of, 88
Plywood blades, 17
Pocket tapes, using, 6
Pole, defined, 82
Porch, enclosing, 447–452
Portable saws, 15, 105
Portland cement, 48
Post-and-beam construction, 401–407
Post-and-beam roofs, 158
Pouring, slab, 67–68
Power, telescope, 32–33
Power tools, 14–18
Precut studs, using, 103–104
Prehung doors, 225–228
Property boundaries, 26–27
Pulls, defined, 362
Purlins, defined, 407
Putty knife in roof work, 176

Radial arm saws, 16, 105
Rafter length, 132, 134–137
Rafter scales, 128
Rafters, 132–142
  in adding to existing building, 455
  ceiling joists and, 146–148
  erecting roof with, 143–145
  hip and valley, 128, 138–141
  jack, 131, 141–142
  layout of, 132–135, 143
  length of main, per foot run, 135
  special, 145–146
  types of, 131–132
  (See also Common rafters)
Rail, defined, 353
Rake:
  closed and solid, 271
  defined, 171

Rake: (Continued)
  exposed nailing parallel to, 202–203
  roof, and asphalt covering, 179
  shingles started at, 187–188
Ready-built cabinets, installing, 348–351
Rebars (reinforcement bars), 47–48
  in concrete block walls, 56
  for concrete slabs, 64–65
  defined, 44
  driveway, 72
  in foundations, 54–55
Reinforcement:
  concrete, 47–48
  of concrete foundation, 54–56
  of concrete slab, 64–65
  for footings, 44
Reliability of carpenter, 23–24
Remodeling, 411
  minor, 411–442
Repairs (see Maintenance)
Resilient flooring, 377–380
Ribbon courses for strip shingles, 197
Ridge, 171
Ridge boards, 131
Rigid foam, 114
Ripping:
  defined, 6
  power saw, 16–17
Rise, 130
  per foot of run, 134
Rods, leveling, 33–34
Roof brackets, 176
Roof cement, 311
Roof frames, 120–165
  aluminum soffit, 161–165
  brace measure in, 142–143
  ceiling joists, 146–147
  decking, 147–154
  erecting roof with rafters, 143–145
  openings in, 147
  rafters, 132–142
  roof load factors, 158–160
  special rafters in, 145–146
  special shapes in, 154–158
  truss, 120–126
Roofs, 167–208
  in adding to existing building, 455
  asphalt, 177–179
  computation of area, 178–179
  finishing edges of, siding and, 267–275
  in job sequence, 23
  leaking, 424–427
  mansard (see Mansard roofs)
  opening in, for fireplace, 393
  problems, 180–182
  quantity estimation, conversion table, 174–175, 179

Roofs: (Continued)
    quantity estimation for covering, 172–173, 176
    roll roofing, 200–203
    shingles (see Shingles)
    steep-slope, 198–200
    terms in roofing, 171
    tools used on, 176–177
    types of, 167
    valleys and, 171, 182, 188–191, 206
Routers, 16
Rubber caulk, butyloid, 310–311
Run:
    stair, defined, 384
    roof, defined, 130
    roof rise per foot of, 134
Runners, defined, 336

Saber saws, 17
Safety, 24
    general rules of, 2–5
    insulation installation, 295
    ladder and scaffold, 267
    roof work, 178
Sanders, 17–18
Saws, 105
    for concrete joints, 69
    hand, 6–7
    power, 15–17
    using, 6–7
Scab, defined, 85
Scaffolds, 262–267
Scales (see specific scales)
Scratch awls, 8–9
Screeds, 68, 69
Screwdrivers, types of, 9
Scribing, room paneling, 434
Sealants, 310–311
Sealers, sill, 77–78
Seams, gypsum board, 326, 328
Sequencing:
    of building layout to start of construction, 26
    correct job, 18, 21–24
    floor frame, 77
    footing and foundation, 42
    interior finishing, 342
    maintenance or remodeling, 410–411
    roof covering, 167
    roof frame, 120
    sheathing, 22
    siding, 23, 258–259
    slab pouring, 60–61
    stair construction, 387
    for wall and ceiling finishing, 315
    wall frame, 22, 98
    window and door installation, 210–211

Setting of joists, 85–87
Shakes, siding of, 283, 284
Sheathing:
    energy, 114, 116, 117
    in job sequence, 22
    plywood roof, with single roofing, 160
    rigid foam, 114
    roof, 148–154
    truss using standard, table, 126
    in wall frame, 98, 110–111, 114–116
Sheds:
    roofs of, 129
    storage, 456–458
Sheetrock, 315
    (See also Gypsum board)
Shelves, 363
Shim in window installation, 225
Shingle stringers, 153, 154
Shingles:
    applying wood roof, 205–208
    coverage of siding wood, table, 262
    interlocking roof, 200
    plywood roof sheathing with roofing of, 160
    repairing, 426
    roof: putting down, 182, 186–194
    strip, 194–198
    siding of, 282–284
    typical asphalt, table, 184
    wood, table, 204
    wood roof, 203–205
Shoes, safety, 3
Showers, preparing wall for, 328–329
Shutters, Tech House, 488
Side lap, 171
Sidewalks, concrete, 71–72
Siding, 257–289
    and adding space to existing building, 452
    aluminum, 286–288
    defined, 257
    finishing roof edges before, 267–275
    installing, 275–281
    in job sequence, 23, 258–259
    needs, table, 261
    panel, 281–282
    preparing for, 259–262
    scaffolds for, 262–267
    shake, 283, 284
    shingle, 282–284
    solid vinyl, 289
    types of, 257–258
Sills:
    nailing, 108
    placement of, floor frame, 77–78
    rough, in window installation, 225

Sills: (Continued)
    wall frame, 104
Single-post scaffolds, 263
Site:
    locating building on, 26–29
    preparing, 18, 21, 34–36
Skew-back saws, standard, 6
Skylights:
    in underground housing, 491–492
    Tech House, 488, 490
Slab footings, 35, 46, 61
Slabs, concrete, 60–70
Sliding doors, 228, 247–248
Sliding windows, horizontal, 212
Slope, roof, 172
Slope cornices, 271
Smoke chamber, 398
Smoke shelf, 397
Sneakers, 3
Soffits, 161–165
    defined, 269
Soil stacks, 194
Soil strength, 42
Solar collectors, 486–487
Solar energy, methods for using, 491–492
Solar-heated housing:
    modifications for, 491
    need for, 490–491
Sole plates, 98, 100
    for partitions, 112, 113
Soundproofing, 113
Spacing:
    room paneling, 434
    standard wall, 117
Span, 130
Special joists, 91–93
Special shapes:
    for concrete slabs, 65–66
    foundation, 52–54
Special walls, 113–114
Spreaders, 47, 50
Square (roofing term), 171
Squares (tool), 10
    framing, 126–132
Stack board (pattern), 56
Staging in roof work, 176
Stains, 368–370
    floor, 376–377
Stairs:
    concrete, 70–71
    design of, 386–387
    inside, 383–391
    layout of, 160–161
Staking out house, 31
Staple-down floor, replacing, 429–430
Staple-down (lock-down) shingles, 426
Staple drivers, 18
Staples:
    faced stapling, insulation, 301

Index   **515**

Staples: (Continued)
   fine wire, table, 19
   insulation, 301
   roof work, 176
   roofing, 186
Stepped footings, 41
Steps, finding number of, in stairs, 386–387
Stile, defined, 353
Stone:
   concrete and, 73
   as wall cover, 285–286
Stool, window, 346
Stops:
   door, 345–346
   fire, 86
Storage:
   additional space for, 444–446
   building for, 456–457, 462
   of building materials, 36–38
   sheds for, 456–458
Storm doors, 245–246, 310
Storm windows, 245–246, 306, 308, 310
   table, 308
Straight-run stairs, 384
Stressed-skin panels, 406–407
Strike-off, defined, 68
Strike plate door, 345
Stringer, defined, 383
Strip flooring, 370
   installing, 371–373
Strip shingles, square-butt, repair, 426
Strongbacks, 319–320
Stucco finish, 284–285
Studs:
   computing lengths of, 103
   cutting lengths of, 104–105
   defined, 98
   full, 100, 106
   insulating narrow spaces between, 302
   jack, in window installation, 225
   layout of, 100–101
   panels over, 330
   partition, 112
   as posts, in post-and-beam construction, 402–404
   regular, in window installation, 225
   (See also Wall frames)
Subfloors (underlayment), laying, 88–91
Subsurface, preparing, for concrete slabs, 63–64
Sunken floors, 91, 93
Surface textures, concrete, 72–74
Surveying instruments, 29
   builder's level as, 29–34
Surveyors, 26

Suspended beams, insulation and, 305
Suspended ceilings, installing, 336, 339

Table saws, 15–16
Tails, rafter, 132
Tamping, 68
Tape measure in roof work, 176
Tech House, 484–490
Telescope:
   of builder's level, 30
   power of, 32–33
Temporary buildings, 36–38
Termite shields, 77, 78
Terrazzo, concrete and, 73
Thermal ceilings, 303–306
Throat, fireplace, 397
Ties, 48
Tiles:
   ceramic (see Ceramic tiles)
   flood, 432–434
   resilient flooring blocks as, 378–380
Tin snips in roof work, 176
Tongue, framing square, 127
Tools:
   hazards of, 5
   (See also specific tools)
Top lap, 171
Top plates, 98, 100
Tread, defined, 384
Tree-stump removal, 35
Trees, cutting, 34–35
Trestle jack, 266
Trim:
   finish, applying, 362–364
   for laminated surfaces, 360
Trimmer studs, 86, 104
   defined, 100
   nailing, 106, 108
   partition, 112
Troweled joints, concrete, 69, 70
Troweling, slab, 68–69
Truss roofs, 120–126
   chord code for, table, 125–126
Try squares, 10
Tubs, preparing walls for, 328–329
Two-door system, hanging, 232–234

Underground housing, 491–492
Underlayment:
   defined, 171
   roof covering, 180
   for strip shingles, 196
   subfloor laying, 88–91
Unfaced insulation, installing, 301
Union activities, 465

U-shaped stairs, 384
Utilities:
   excavating for lines, 47
   slab and, 63
   temporary, 38
Utility knife, 176

Valley jacks, 131, 132
Valley rafters, 131
Valleys in roofing, 171, 182, 188–191, 206
Vapor barrier (see Moisture barrier)
Vernier, reading the, 32
Vinyl latex caulk, 310
Vinyl siding, solid, 289

Wales, defined, 50
Walking area, stair, 387
Wall brackets, 263, 265
Wall frames, 98–118
   corner braces for, 108–109
   erecting, 110–112
   as factors in wall construction, 116–117
   interior walls, 112–114
   in job sequence, 22, 98
   sheathing, 98, 110–111, 114–116
   wall assembly in, 106–108
   wall layout for, 98–104
Wallpaper, 369, 371
Walls:
   and adding to existing building, 455
   applying finish to, 369, 371
   exterior, defined, 98
   gypsum board applied to, 317–328
   insulating (see Insulation)
   joists under, 83
   masonry (see Masonry walls)
   panels (see Panels)
   plaster applied to, 332–334
   preparing, for tubs and showers, 328–329
   siding for exterior (see Siding)
Waste disposal on building site, 38
Water:
   hot, solar collectors for, 487
   slab and lines for, 63
   Tech House conservation of, 490
Waterproofing, foundation, 56, 58
Weatherstripping, window, 306, 308
Western platform method of framing, 98
Width, footing, 43
Wind protection:
   interlocking shingles for, 200
   roll roofing for, 200–203
   shingles and, 198

Winders (stairs), 384–385
Windows:
   in adding to existing building, 455
   bay, 146
   energy factors and, 255
   finishing interiors of, 346–348
   installation terms, 224–225
   opening for, 218, 220–221
   Tech House, 487–488
   trimmed, 240–241

Windows: (Continued)
   types of, 211–218
   wood, installing, 222–225
   (*See also* Storm windows)
Winterizing, 311–312
   (*See also* Insulation)
Wiring, insulating around, 302
Wood beams, solid, and thermal ceiling panels, 305
Wood buck in window installation, 225

Wood floors:
   concrete over, 73–74
   laying, 370–371
Wooden frame windows, finishing, 346–347
Woven valleys, roofing, 189–190
Wrecking bars (crowbars), 9

Zoning, 468–469